*Fundamentals of*

# ELECTRIC MACHINES

*Fundamentals of*

# ELECTRIC MACHINES

**Revised Third Edition**

**B.R. Gupta**
**Vandana Singhal**

PUBLISHING FOR ONE WORLD
**NEW AGE INTERNATIONAL (P) LIMITED, PUBLISHERS**
New Delhi • Bangalore • Chennai • Guwahati • Hyderabad
Jalandhar • Kolkata • Lucknow • Mumbai • Ranchi
Visit us at **www.newagepublishers.com**

Published by New Age International (P) Ltd., Publishers
Third Edition: 2005
**Reprint:** 2015



**BRANCHES**

- **Bangalore** 37/10, 8th Cross (Near Hanuman Temple), Azad Nagar, Chamarajpet, Bangalore-560 018 Tel.: (080) 26756823, Telefax: 26756820, **E-mail: bangalore@newagepublishers.com**
- **Chennai** 26, Damodaran Street, T. Nagar, Chennai-600 017 Tel.: (044) 24353401, Telefax: 24351463, **E-mail: chennai@newagepublishers.com**
- **Cochin** CC-39/1016, Carrier Station Road, Ernakulam South, Cochin-682 016 Tel.: (0484) 2377303, Telefax: 4051303, **E-mail: cochin@newagepublishers.com**
- **Guwahati** Hemsen Complex, Mohd. Shah Road, Paltan Bazar, Near Starline Hotel, Guwahati-781 008 Tel.: (0361) 2513881, Telefax: 2543669, **E-mail: guwahati@newagepublishers.com**
- **Hyderabad** 105, 1st Floor, Madhiray Kaveri Tower, 3-2-19, Azam Jahi Road, Near Kumar Theater Nimboliadda, Kachiguda, Hyderabad-500 027, Tel.: (040) 24652456, Telefax: 24652457 **E-mail: hyderabad@newagepublishers.com**
- **Kolkata** RDB Chambers (Formerly Lotus Cinema) 106A, 1st Floor, S.N. Banerjee Road, Kolkata-700 014 Tel.: (033) 22273773, Telefax: 22275247, **E-mail: kolkata@newagepublishers.com**
- **Lucknow** 16-A, Jopling Road, Lucknow-226 001 Tel.: (0522) 2209578, 4045297, Telefax: 2204098, **E-mail:lucknow@newagepublishers.com**
- **Mumbai** 142C, Victor House, Ground Floor, N.M. Joshi Marg, Lower Parel, Mumbai-400 013 Tel.: (022) 24927869, Telefax: 24915415, **E-mail: mumbai@newagepublishers.com**
- **New Delhi** 22, Golden House, Daryaganj, New Delhi-110 002 Tel.: (011) 23262368, 23262370, Telefax: 43551305, **E-mail: sales@newagepublishers.com**

**ISBN: 978-81-224-1614-5**

**₹ 375.00**

**C-15-02-8287**

Printed in India at Ajit Printing Press, Delhi.
Typeset at Pagitek Graphics, Delhi.

**PUBLISHING GLOBALLY**
**NEW AGE INTERNATIONAL (P) LIMITED, PUBLISHERS**
7/30 A, Daryaganj, New Delhi-110002
Visit us at **www.newagepublishers.com**

*Dedicated to*

***Our Almamater***

**Punjab Engineering College, Chandigarh**

# Preface to the Third Edition

The book has been thoroughly checked and revised. Additional topics have been added in some chapters. More solved examples have been added. The mistakes have been corrected. It is hoped that the readers will find these changes useful.

AUTHORS

# Preface to the First Edition

This book is intended to serve as a text book for the subject of electric machines for the undergraduate engineering students.

There are three main reasons for writing this book. The first is to provide a complete and comprehensive book on the subject. To achieve this goal the syllabi of almost all universities in the country have been kept in view. The second is to make the text as nearly self teaching as possible. To achieve this goal the book has been written in simple and easy to understand language. A large number of solved examples, review questions and problems have also been included. Moreover almost the whole of the material has been class tested. The third is to make the text up to date and as modern as possible.

The book has been presented in such a way that the undergraduate students can grasp the subject matter easily and are motivated to learn more. Several sections of the book will also be useful to post graduate students and practising engineers.

Authors would welcome constructive criticism of the book.

# Contents

*Preface to the Third Edition* vii

*Preface to the First Edition* vii

**1. Basic Principles** **1**

1.1 Faraday's Law of Electromagnetic Induction 1
1.2 Lenz's Law 1
1.3 Methods of Linking Flux 2
1.4 Motional EMF (Dynamically Induced EMF or Speed EMF) 2
1.5 Statically Induced EMF (or Transformer EMF) 4
1.6 General Case of Induction 5
1.7 Coefficient of Self-Inductance 7
1.8 Coefficient of Mutual Inductance 8
1.9 Force on Current Carrying Conductor in a Magnetic Field 10
1.10 Torque on a Current Carrying Coil in a Magnetic Field 11
1.11 Fleming's Right-Hand Rule 12
1.12 Fleming's Left Hand Rule 12
1.13 Generator and Motor Action 12
1.14 Interaction of Magnetic Fields 13
1.15 Hysteresis Loss 15
1.16 Eddy Current Loss 15
1.17 Energy Balance 16
1.18 Losses and Efficiency 17
1.19 Load Types 17
1.20 Considerations Governing Machine Applications 19
1.21 Duty Cycles of Motors 20
1.22 Machine Ratings 21
1.23 Role of Electric Machines in Generation, Transmission and Distribution of Power 22
1.24 Classification of Electric Machines 24
1.24.1 DC Machines 24
1.24.2 AC Machines 24
Summary 25
Job Interview Questions 26

Review Questions 27
Problems 27

**2. Transformers** **29**

2.1 Introduction 29
2.2 Principle of Operation 29
2.3 EMF Equation of Transformer 29
2.4 Types and Construction of Transformers 32
2.5 The Ideal Transformer 33
2.6 Effect of Iron-Losses 33
2.7 Effect of Leakage Flux 34
2.8 Phasor Diagram of Transformer 35
2.9 Equivalent Circuit of Transformer 35
2.10 Approximate Equivalent Circuit 38
2.11 Equivalent Resistance and Reactance 38
2.12 Transformer as a Coupled Circuit 41
2.13 Methods of Reducing Leakage Flux 44
2.14 Per Unit Values 48
2.15 Voltage Regulation 49
2.16 Losses in Transformer 52
2.17 Transformer Efficiency 53
2.18 Condition for Maximum Efficiency 53
2.19 All Day Efficiency (Energy Efficiency) 55
2.20 Polarity of Transformers 57
2.21 Test For Polarity 57
2.22 Open Circuit (or No Load) Test 58
2.23 Short Circuit Test 59
2.24 Back to Back Test (or Regenerative Test) 61
2.25 Separation of Core Losses 62
2.26 Principle of Auto-Transformer 62
2.27 Saving in Copper in Auto-Transformer 64
2.28 Phasor Diagram and Equivalent Circuit of Auto-Transformer 65
2.29 Advantages Disadvantages and Applications of Auto-Transformer 67
2.30 Introduction to Three-Phase Transformers 68
2.31 Star-Star Connection 69
2.32 Delta-Delta Connection 70
2.33 Star-Delta Connection 70
2.34 Delta-Star Connection 71
2.35 Zig Zag or Inter-Connected Star Connection 72
2.36 Phase Groups of Three Phase Transformers 72
2.37 Back-to-Back Test on 3-Phase Transformers 79
2.38 Per Unit System for Three-Phase Transformers 79
2.39 Open Delta (or V-V) Connection 81
2.40 Three-Phase to Two-Phase Conversion 82
2.41 Three Phase to Six-Phase Conversion 86
2.42 Three-Phase to Twelve-Phase Conversion 86

2.43 Conditions for Parallel Operation 87
 2.43.1 Polarity 88
 2.43.2 Voltage Ratio 88
 2.43.3 Impedance 88
 2.43.4 Phase Sequence and Phase Displacement 88
2.44 Load Sharing by Transformers in Parallel 89
 2.44.1 Equal Voltage Ratios 89
 2.44.2 Unequal Voltage Ratios 90
2.45 Parallel-Generator Theorem 90
2.46 Parallel Operation of Three-Phase Transformers 91
2.47 Applications 95
2.48 Equivalent Circuit 96
2.49 Determination of Parameters 96
2.50 Magnetising Current 99
 2.50.1 Wave Shape of Magnetising Current 99
 2.50.2 Inrush of Magnetising Current 100
2.51 Magnetising Current in 3-Phase Transformers 101
 2.51.1 Delta Connected Primaries 101
 2.51.2 Y-Connected Primaries 101
2.52 Transformer Taps 102
 2.52.1 Off-Load Tap Changing 102
 2.52.2 Tap Changing under Load (TCUL) 102
2.53 Induction Regulators 103
 2.53.1 Single-Phase Induction Regulator 103
 2.53.2 Three-Phase Induction Regulator 104
2.54 Moving Coil Voltage Regulator 106
2.55 Instrument Transformers 107
 2.55.1 Potential Transformer 107
 2.55.2 Current Transformer 108
2.56 Transformer Cooling 108
 2.56.1 Oil Immersed Self-cooled Transformer 109
 2.56.2 Oil Immersed Forced Air Cooled Transformer 109
 2.56.3 Oil Immersed Water Cooled Transformer 109
 2.56.4 Oil Immersed Forced Oil Cooled Transformer 109
2.57 Transformer Oil 109
2.58 Conservator and Breather 110
2.59 Transformer Noise 110
2.60 Name Plate and Ratings 110
2.61 Wide-Band Transformer 111
 2.61.1 Classification 111
 2.61.2 Equivalent Circuit 111
 2.61.3 Frequency Response 112
 2.61.4 Special Design Features 112
2.62 Pulse Transformer 113
2.63 Audio-Frequency Transformer 114
2.64 Grounding Transformer 115

2.65 Amorphous Metal Core Transformer 131
2.66 Cast Resin Dry Type Transformers 132
Summary 133
Job Interview Questions 135
Review Questions 136
Problems 138

**3. DC Machines** **142**

3.1 Construction 142
3.1.1 Field Poles 143
3.1.2 Yoke 143
3.1.3 Armature 143
3.1.4 Commutator 144
3.1.5 Brushes 144
3.1.6 Interpoles 145
3.2 Armature Winding 145
3.2.1 Armature Coils 145
3.2.2 Coil Span 146
3.2.3 Single-Layer and Multi-Layer Windings 146
3.2.4 Simplex and Multiplex Windings 146
3.2.5 Back Pitch and Front Pitch 147
3.2.6 Commutator Pitch 147
3.2.7 Progressive and Retrogressive Winding 147
3.2.8 Types of Windings 148
3.3 Lap and Wave Windings 148
3.3.1 Lap Winding 148
3.3.2 Wave Winding 149
3.3.3 Comparison of Lap and Wave Windings 150
3.4 Dummy Coil 150
3.5 Equaliser Rings 151
3.6 Magnetic Circuit and Flux Path 151
3.7 EMF Equation 152
3.8 Commutator Action: Conversion of AC into DC 154
3.8.1 Commutation Process, Reactance Voltage 155
3.9 Armature Reaction in DC Generators 156
3.9.1 Cross-Magnetising Effect 156
3.9.2 Demagnetising Effect 157
3.9.3 Cross Magnetising and Demagnetising Ampere Turns 157
3.10 Solution of Commutation Problems 159
3.11 Flux Density Wave Forms 161
3.12 Classification of DC Generators 162
3.13 Magnetisation Characteristics 162
3.14 Terminal Characteristics of Separately Excited Generator 163
3.15 Voltage Build up in Self-Excited Generator 164
3.16 Terminal Characteristics of Shunt Generator 165
3.17 Series Generator 166

3.17.1 Terminal Characteristics 166
3.17.2 Applications 167
3.18 Compound Generator 167
3.19 Parallel Operation of DC Generators 168
3.19.1 Shunt Generators in Parallel 170
3.19.2 Compound Generators in Parallel 170
3.20 Principle of DC Motor 175
3.21 Electromagnetic Torque 175
3.22 Driving and Retarding Torques, Back EMF 176
3.23 Relation Between Torque and Voltage 176
3.24 Commutation in DC Motors 178
3.25 Armature Reaction in DC Motors 178
3.26 Classification and Characteristics of DC Motors 179
3.26.1 Characteristics of Shunt Motor 179
3.26.2 Series Motor Characteristics 181
3.26.3 Compound Motor Characteristics 182
3.27 Speed Control of Shunt Motors 185
3.27.1 Field Control 186
3.27.2 Armature Control 188
3.28 Speed Control of Series Motors 191
3.28.1 Field Control 191
3.28.2 Armature Control 193
3.29 Ward Leonard Control 195
3.30 DC Motor Starter 196
3.30.1 Problem at Starting 196
3.30.2 Functions of Starter 197
3.30.3 Construction 197
3.30.4 Calculation of Resistance Elements 197
3.31 Modern DC Motor Starters 198
3.31.1 Starter Using Time Delay Relays 199
3.31.2 Starter Using Back-emf Sensing Relays 199
3.32 Electric Braking 201
3.32.1 Plugging 201
3.32.2 Dynamic Braking 201
3.32.3 Regenerative Braking 202
3.33 Permanent Magnet DC Motors 202
3.34 Losses and Power Flow in DC Machines 202
3.34.1 Losses 202
3.35 Efficiency 204
3.36 Brake Test 206
3.37 Swinburne Test 207
3.38 Hopkinson Test 208
3.38.1 Circuit 208
3.38.2 Test Procedure 209
3.38.3 Advantages 210

3.38.4 Disadvantages 210
3.39 Metadyne 211
3.40 Amplidyne 212
3.40.1 Characteristics 212
3.40.2 Voltage Equation 214
3.41 Separately Excited Motor 215
3.42 Separately Excited Generator 216
3.43 Applications of DC Machines 219
3.43.1 DC Generators 219
3.43.2 DC Shunt Motor 219
3.43.3 DC Series Motor 219
3.43.4 Compound Motors 220
Summary 236
Job Interview Questions 239
Review Questions 239
Problems 242

**4. AC Machine Fundamentals 246**

4.1 Introduction 246
4.2 AC Windings 246
4.2.1 Single and Double Layer Winding 247
4.2.2 Phase Spread 247
4.2.3 Multiturn Coil Windings 248
4.2.4 Full-Pitch and Fractional Pitch Windings 248
4.2.5 Integral Slot Windings 248
4.2.6 Fractional Slot Windings 248
4.3 EMF Equation 251
4.3.1 Breadth Factor 253
4.3.2 Pitch Factor 254
4.4 MMF of AC Windings 258
4.4.1 MMF Space Wave of Concentrated Coil 258
4.4.2 MMF of Distributed Single-Phase Winding 259
4.5 MMF of 3-Phase Windings, Rotating Magnetic Field 261
4.6 Reversal of Direction of Rotating Field 265
4.7 Torque in AC Machines 265
4.8 Losses 267
4.8.1 Copper-Losses 268
4.8.2 Core-Losses 268
4.8.3 Mechanical Losses 268
4.8.4 Stray Load Losses 268
4.9 Slot or Tooth Harmonics 269
Summary 277
Job Interview Questions 278
Review Questions 279
Problems 279

**5. Synchronous Generators** **281**

5.1 Introduction 281
5.2 Frequency 282
5.3 Construction 283
5.3.1 Cylindrical Rotor Machines 283
5.3.2 Salient Pole Machines 284
5.4 Cooling 285
5.5 Excitation System 286
5.5.1 Brushless Excitation System 286
5.5.2 Static Excitation System 287
5.6 Phasor Diagram of Cylindrical Rotor Alternator 288
5.6.1 No-Load Condition 288
5.6.2 Effect of Armature Reaction 289
5.6.3 Effect of Armature Resistance and Leakage Reactance 291
5.7 Voltage Regulation 292
5.8 Synchronous Impedance Method 293
5.9 Experimental Determination of Synchronous Impedance 294
5.10 MMF Method 297
5.11 Saturated Synchronous Impedance Method 298
5.12 Zero Power Factor (Potier Reactance) Method 299
5.13 Effect of Power Factor on Characteristics 304
5.14 Short Circuit Ratio (SCR) 305
5.15 Power Angle Characteristics of Cylindrical Rotor Alternator 305
5.15.1 Power Output 305
5.15.2 Power Input 308
5.16 Salient Pole Machines: Two-Reaction Theory 312
5.17 Power Angle Characteristics of Salient Pole Generator 316
5.18 Measurement of $x_d$ and $x_q$: Slip Test 318
5.19 Operating Limits on Alternators 319
5.20 MW-Frequency and MV Ar-Voltage Characteristics of Alternators 322
5.21 Parallel Operation of Alternators 324
5.21.1 Need for Parallel Operation 324
5.21.2 Conditions for Parallel Operation 324
5.21.3 Synchronising 325
5.22 Synchronizing Power, Torque 326
5.23 Two Alternators in Parallel 329
5.23.1 Effect of Change in Excitation 329
5.23.2 Effect of Change in Mechanical Power Input (Steam Supply) 331
5.23.3 Sharing of Load Currents 332
5.24 Operation of Alternator on Infinite Bus Bars 335
5.25 Symmetrical Short Circuit Transient 338
5.26 Equivalent Circuit for Alternator Transient Conditions 341
5.27 Synchronous Machine Reactances 343
5.27.1 Direct Axis Synchronous Reactance $x_d$ 343
5.27.2 Direct Axis Transient Reactance $x_d'$ 343

5.27.3 Direct Axis Sub-Transient Reactance $x_d''$ 344
5.27.4 Potier Reactance 344
5.27.5 Quadrature Axis Synchronous Reactance 344
5.27.6 Quadrature Axis Transient Reactance $x_q'$ and Sub-Transient Reactance $x_q''$ 344
5.27.7 Negative Sequence Reactance $x_2$ 344
5.27.8 Zero Sequence Reactance $x_0$ 345
5.28 Synchronous Machine Time Constants 346
5.28.1 Direct Axis Transient Open Circuit Time Constant $T_{do}$ 346
5.28.2 Direct Axis Transient Short Circuit Time Constant $T_d'$ 346
5.28.3 Armature Short Circuit Time Constant $T_a$ 346
5.28.4 Direct Axis Sub-Transient Time Constants $T''_{do}$ and $T''_d$ 346
5.29 Electro-Mechanical Transients, Hunting 347
5.30 Damper Winding 348
Summary 357
Job Interview Questions 360
Review Questions 360
Problems 361

**6. Synchronous Motors** **365**

6.1 Principle of Operation 365
6.2 Steady State Constant Speed Operation 365
6.3 Construction 366
6.4 Equivalent Circuit and Phasor Diagram: Cylindrical Rotor Theory 367
6.5 Phasor Diagram: Two-Reaction Theory 368
6.6 Power and Torque 369
6.6.1 Cylindrical Rotor Theory 369
6.6.2 Two-Reaction Theory 371
6.7 Change in Load 372
6.8 Speed-Torque Characteristics 373
6.9 Power Factor Control: Effect of Change of Excitation 373
6.10 Armature Current Locus 374
6.10.1 Constant Power 374
6.10.2 Constant Excitation 376
6.11 V Curves 376
6.12 Synchronous Condenser 380
6.12.1 Capacity of Synchronous Condenser Running at No-Load 380
6.12.2 Capacity of Synchronous Condenser Supplying Mechanical Load 381
6.13 Synchronous Reactor 381
6.14 Synchronous Frequency Changer 382
6.15 Starting of Synchronous Motor 384
6.15.1 Auxiliary Motor Starting 384
6.15.2 Induction Motor Starting 385
6.16 Hunting 385
6.17 Applications 386
6.18 Synchronous Induction Motor 386
6.19 Permanent Magnet Synchronous Motor 398

Summary 399
Job Interview Questions 401
Review Questions 401
Problems 402

**7. Polyphase Induction Motors 404**

7.1 Construction 404
7.1.1 Stator 404
7.1.2 Rotor 404
7.1.3 Frame 405
7.2 Principle of Operation 405
7.3 Slip 406
7.4 Production of Torque 406
7.5 EMF and Current 407
7.6 Induction Motor as Transformer 410
7.7 Phasor Diagram 411
7.8 Equivalent Circuit 411
7.8.1 Stator Circuit Model 411
7.8.2 Rotor Circuit Model 412
7.8.3 Exact Equivalent Circuit 412
7.8.4 Approximate Equivalent Circuit 413
7.9 Losses, Power Output 413
7.10 Torque 414
7.11 Torque-Slip Curve 418
7.12 Starting of Induction Motor 423
7.12.1 Wound Rotor Induction Motor 424
7.12.2 Cage Rotor Induction Motor 424
7.13 No-Load Test 428
7.14 Blocked Rotor Test 428
7.15 Brake Test 431
7.16 Computation of Performance by Circle Diagram 432
7.16.1 Theory 432
7.16.2 Construction 432
7.16.3 Computation of Performance 432
7.16.4 Assumptions 434
7.17 Deep Bar Rotor 436
7.18 Double Cage Rotor 437
7.19 Cogging and Crawling 440
7.20 Speed Control 441
7.20.1 Frequency Control 441
7.20.2 Pole Changing 442
7.20.3 Line Voltage Control 445
7.20.4 Rotor Resistance Control 445
7.20.5 Voltage Injection in Rotor Circuit 446
7.20.6 Cascade Connection 447

7.21 Operation with Unbalanced Supply Voltage 449
7.21.1 Symmetrical Components in Three Phase Systems 449
7.21.2 Positive Sequence Equivalent Circuit 451
7.21.3 Negative Sequence Equivalent Circuit 452
7.21.4 Single Phasing 453
7.22 Induction Generator 455
7.22.1 Induction Generator in Parallel with Capacitor Bank 456
7.23 Induction Frequency Changer 458
7.24 Comparison of Induction Motor and Synchronous Motor 459
7.25 Applications of Induction Motors 459
Summary 470
Job Interview Questions 473
Review Question 473
Problem 474

**8. Single-Phase Induction Motor 477**

8.1 Introduction 477
8.2 Construction 477
8.3 Conditions at Starting 478
8.4 Double Revolving Field Theory 478
8.5 Equivalent Circuit (Circuit Model) 480
8.6 Analysis 480
8.7 Determination of Parameters 482
8.7.1 Blocked Rotor Test 482
8.7.2 No-Load Test 483
8.8 Symmetrical Components 486
8.9 Analysis of Single Phase Induction Motor Using Symmetrical Components 487
8.10 Cross-Field Theory 489
8.10.1 Motor at Standstill 489
8.10.2 Motor Running 489
8.11 Starting of Single-Phase Induction Motors 489
8.11.1 Split Phase Motor 490
8.11.2 Capacitor Start Motor 491
8.11.3 Capacitor Start Capacitor Run Motor 491
8.11.4 Permanent Split Capacitor Motor 492
8.11.5 Shaded Pole Motor 493
8.12 Speed Control 497
Summary 498
Job Interview Questions 500
Review Questions 500
Problems 501

**9. AC Commutator Motors 503**

9.1 Schrage Motor 503
9.1.1 Construction 503
9.1.2 Operation 503

9.1.3 Advantages 504
9.1.4 Disadvantages 505
9.1.5 Application 505
9.2 Single-Phase Series Motor 505
9.2.1 Construction 505
9.2.2 Phasor Diagram 506
9.2.3 Analysis 506
9.2.4 Application 508
9.3 Universal Motor 508
9.3.1 Applications 508
9.3.2 Phasor Diagram 508
9.3.3 Analysis 510
9.4 Repulsion Motor 515
9.5 Repulsion Start-Induction Motor 516
9.6 Repulsion-Induction Motor 516
Summary 516
Job Interview Questions 518
Review Questions 518
Problems 518

**10. Special Machines** **520**

10.1 Reluctance Motor 520
10.2 Hysteresis Motor 521
10.3 Two-Phase Servo Motor 522
10.3.1 Construction 522
10.3.2 Analysis Using Symmetrical Components 523
10.3.3 Torque-Speed Curve 526
10.3.4 Transfer Function 526
10.3.5 Drag Cup Rotor 527
10.3.6 Special Features 528
10.4 Stepper Motor 531
10.4.1 Single Stack Variable Reluctance Stepper Motor 533
10.4.2 Multistack Variable Reluctance Stepper Motor 534
10.4.3 Hybrid Stepper Motor 535
10.5 DC Tachometer 535
10.6 AC Tachometer 535
10.7 Synchros and Control Transformers 536
Summary 538
Job Interview Questions 540
Review Questions 540
Problems 540

**11. Solid State Motor Control** **542**

11.1 Introduction 542
11.2 DC Motor Control using Converters 542
11.2.1 Single-Phase Two Quadrant Bridge Converter (Fully Controlled) 542

11.2.2 3-Phase, Six-Pulse Two-Quadrant Bridge Converter (Fully Controlled) 543
11.2.3 Single-Phase, One-Quadrant Bridge Converter 544
11.2.4 3-Phase, One Quadrant Bridge Converter 545
11.3 DC Motor Control Using Choppers 545
11.3.1 Basic Chopper Circuit 545
11.3.2 DC Chopper Commutation 547
11.3.3 Step-Up Chopper 547
11.4 AC Motor Control Using Inverters 548
11.4.1 Single-Phase Bridge Inverter 548
11.4.2 3-Phase Bridge Inverter 549
11.5 Speed Control of Induction Motor Using AC Regulator 550
11.6 Static Kramer Drive (Slip Power Recovery Scheme) 551
11.7 Static Scherbius Drive (Slip Power Recovery Scheme) 552
11.8 Integrated Motors 553
11.9 Speed Control of Single Phase Induction Motor Using Triac (Fan Regulator) 553
Summary 559
Job Interview Questions 561
Review Questions 561
Problems 561
*Appendix A* 563
Objective Questions 563
Answers to Objective Questions 582
*Appendix B* 584
Review Questions 596
Problems 596
*Appendix C* 598
Magnetic Circuits 598
Review Questions 607
Problems 607
*Appendix D* 608
Summary of Important Formulae 608
Chapter 1: Basic Principles 608
Chapter 2: Transformers 609
Chapter 3: DC Machines 611
Chapter 4: AC Machine Fundamentals 612
Chapter 5: Synchronous Generators 613
Chapter 6: Synchronous Motors 614
Chapter 7: Polyphase Induction Motors 615
Chapter 8: Single Phase Induction Motor 617
Chapter 9: Ac Commutator Motors 618
Chapter 10: Special Machines 618
Chapter 11: Solid State Motor Control 618
References 619
*Index* 620

Chapter **1**

# Basic Principles

## 1.1 FARADAY'S LAW OF ELECTROMAGNETIC INDUCTION

Michael Faraday demonstrated through his experiments that an emf is induced in a circuit when the magnetic flux enclosed by the circuit changes with respect to time. In 1831 he proposed the following law known as Faraday's law of electromagnetic induction.

$$e = \frac{d\lambda}{dt} = \frac{Nd\phi}{dt} \quad ...(1.1)$$

where
$e$ = emf induced, volts
$\lambda$ = flux linkage, weber turns
$N$ = number of turns in the winding
$\phi$ = flux, webers
$t$ = time, seconds.

## 1.2 LENZ'S LAW

Every action causes an equal and opposite reaction. The fact that this is true in electromagnetism was discovered by Emil Lenz. The Lenz's law states: the induced current always develops a flux which opposes the motion (or the change producing the induced current). This law refers to induced currents and thus implies that it is applicable to closed circuits only. If the circuit is open, we can find the direction of induced emf by thinking in terms of the response if it were closed.

The motion of a conductor in a field causes an induced emf in the conductor and energy is generated. This is possible if work is done in moving the conductor through the field. If work is to done, a force must oppose the motion of conductor. This opposing force is due to flux set up by induced current. Figure 1.1a illustrates Lenz's law. The motion of conductor causes the deflection of galvanometer to the left. This indicates that direction of induced emf and current are as shown. The current causes a flux in the clockwise direction as shown. This flux strengthens the magnetic field above the conductor and weakens that below it. Thus a force in the downward direction acts on the conductor (Fig. 1.1b). The motion of the conductor is opposed by the magnetic flux due to induced current.

Since induced emf opposes the change in flux, a negative sign is sometimes added in Eq. (1.1). If it is kept in view that direction of induced emf is such as to oppose the change in flux, there is no need of negative sign.

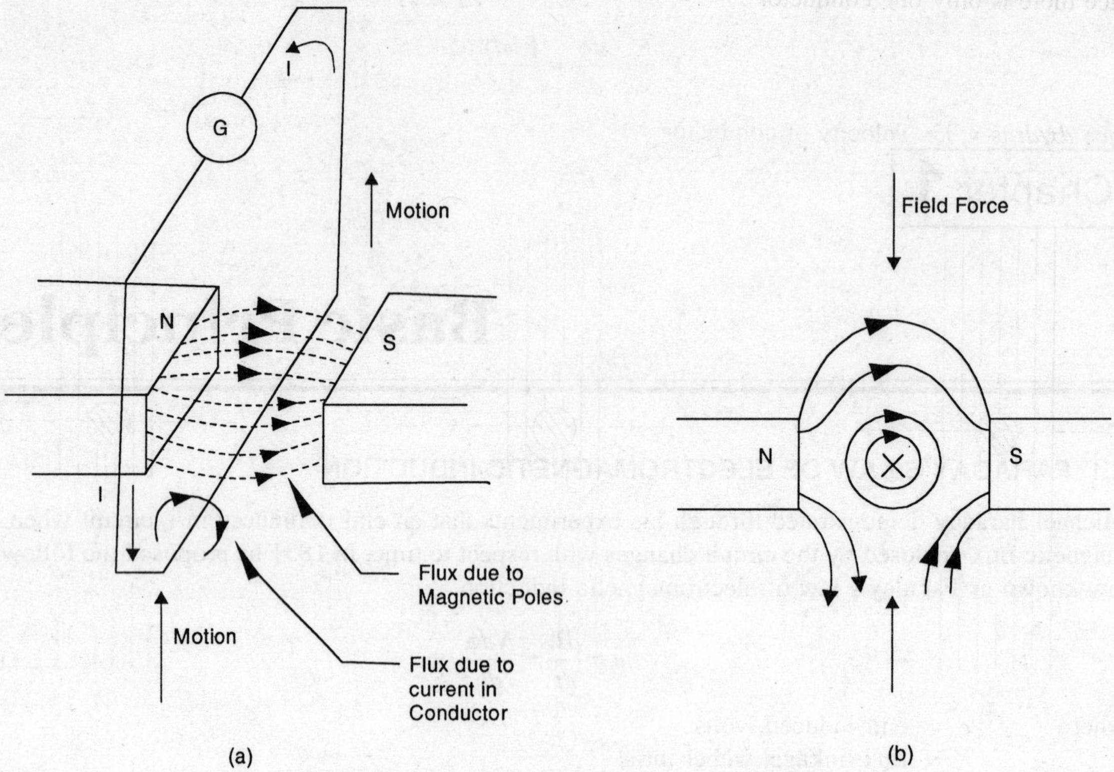


**Fig. 1.1.** Motion of a conductor in a magnetic field.

## 1.3 METHODS OF LINKING FLUX

The induction of emf requires a conductor, a magnetic field and linking or cutting of flux by the conductor. The linking of magnetic field by the conductor can occur in three ways:

(1) Moving conductor and stationary permanent magnet or dc electromagnet. This configuration is used in all dynamos, generators and motors.

(2) Moving dc electromagnet and stationary conductor. This configuration is used in large ac generators and motors.

(3) Stationary conductor, stationary electromagnet and variation of flux by feeding alternating current to the magnet. This configuration is used in transformers.

## 1.4 MOTIONAL EMF (DYNAMICALLY INDUCED EMF OR SPEED EMF)

Figure 1.2 shows three conductors *a, b, c* moving in a magnetic field of flux density *B* in the directions indicated by arrow. Conductor *a* is moving in a direction perpendicular to its length and perpendicular to the flux lines. Therefore it cuts the lines of force and a motional emf is induced in it. Let the conductor move by a distance *dx* in a time *dt*. If the length of conductor is *l*, the area swept by the conductor is *l dx*. Then change in flux linking the coil

$$= d\phi = B \cdot l \cdot dx$$

Since there is only one conductor

$$e = \frac{d\phi}{dt} = \frac{Bl\,dx}{dt}$$

Since $dx/dt$ is $v$, i.e. velocity of conductor

$$e = Blv \text{ volts} \quad ...(1.2)$$

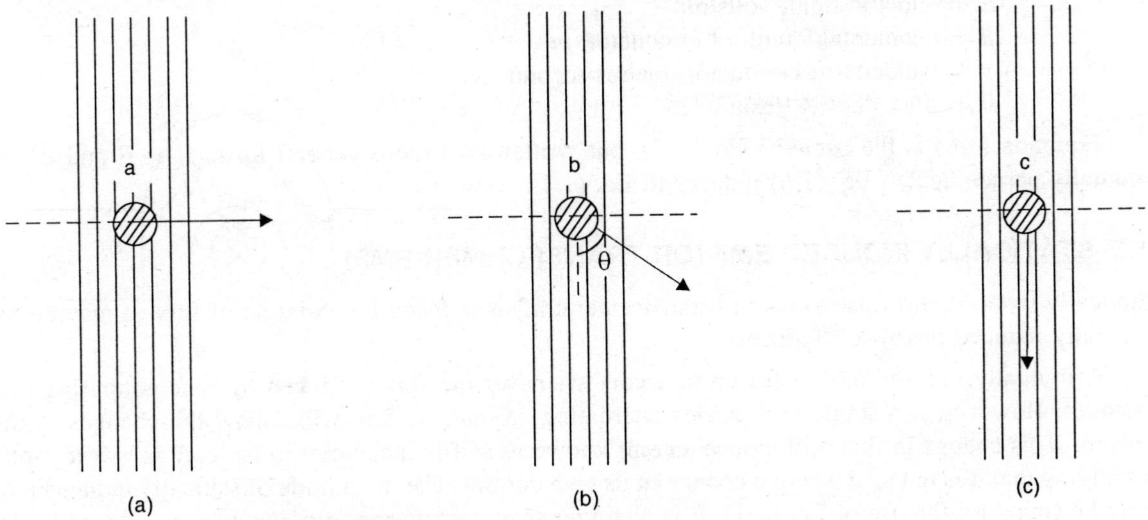


**Fig. 1.2.** Motion of a conductor in a magnetic field.

where $e$ = emf induced, volts
$B$ = flux density, tesla
$v$ = velocity of conductor, metres/second
$l$ = length of conductor, metres.

The motion of conductor $b$ (Fig. 1.2b) is at an angle $\theta$ to the direction of the field. If the conductor moves by a distance $dx$, the component of distance travelled at right angles to the field is $(dx)(\sin\theta)$ and, proceeding as above, the induced emf is

$$e = Blv\sin\theta \text{ volts} \quad ...(1.3)$$

Equation (1.3) includes Eq. (1.2) because when $\theta = 90°$, the two equations become identical. In Fig. 1.2 (c) the motion of conductor $c$ is parallel to the field. Therefore, in this case, no flux is cut, $\theta$ is zero and induced emf is also zero. Dynamically induced emf is also known as speed emf or motional emf or rotational emf.

Equation (1.2) can also be written in a more general vectorial form:

The force $F$ on a particle of charge $Q$ moving with a velocity $v$ in a magnetic field B is

$$F = Q\,(v \times B) \quad N \quad ...(1.4)$$

Dividing $F$ by $Q$ we get the force per unit charge, i.e. electric field $E$, as

$$E = \frac{F}{Q} = v \times B \quad \text{volts/m} \quad ...(1.5)$$

The electric field $E$ is in a direction normal to the plane containing $v$ and $B$. If the charged particle is one of the many electrons in a conductor moving across the magnetic field, the emf $e$ between the end points of conductor is line integral of electric field $E$, or

$$e = \oint E \cdot dl = \oint (v \times B) \cdot dl \qquad \text{...(1.6)}$$

where
$e$ = emf induced, volts
$E$ = electric field, volts/m
$dl$ = elemental length of conductor, $m$
$v$ = velocity of conductor, metres/second
$B$ = flux density, tesla

Equation (1.6) is the same as Eq. (1.3), but written in a more general form. If $v$, $B$ and $dl$ are mutually perpendicular, Eq. (1.6) reduces to Eq. (1.2).

## 1.5 STATICALLY INDUCED EMF (OR TRANSFORMER EMF)

Statically induced emf (also known as transformer emf) is induced by variation of flux. It may be (a) mutually induced or (b) self induced.

A mutually induced emf is set up in a coil whenever the flux produced by a neighbouring coil changes. However, if a single coil carries alternating current, its flux will follow the changes in the current. This change in flux will induce an emf known as self-induced emf in the coil, the word 'self' signifying that it is induced due to a change in its own current. The magnitude of statically induced emf may be found by the use of Eq. (1.1). It is also known as transformer emf, since it is induced in the windings of a transformer.

Equation (1.1) can also be put in a more general form. The total flux linkages $\lambda$ of a coil is equal to the integral of the normal component of flux density $B$ over the surface bounded by the coil, or

$$\lambda = \iint B \cdot ds \qquad \text{...(1.7)}$$

The surface over which the integration is carried out is the surface bounded by the periphery of the coil. Thus, induced emf

$$e = \frac{d\lambda}{dt} = \frac{d}{dt}\iint B \cdot ds$$

or

$$e = \frac{d}{dt}\int_s B \cdot ds \qquad \text{...(1.8)}$$

When the coil is stationary or fixed

$$e = \int_s \frac{\partial B}{\partial t} \cdot ds \qquad \text{...(1.9)}$$

where
$e$ = emf induced, volts
$B$ = flux density, tesla
$ds$ = element of area, $m^2$
$t$ = time, seconds.

## 1.6 GENERAL CASE OF INDUCTION

Equation (1.6) gives the speed emf, while Eq. (1.9) gives the transformer emf. When flux is changing with time and relative motion between coil (or conductor) and flux also exists, both these emfs are induced and the total induced emf $e$ is

$$e = \oint (v \times B) \cdot dl - \int_s \frac{\partial B}{\partial t} \cdot ds \qquad \text{...(1.10)}$$

The first term is Eq. (1.10) is the speed emf and line integral is taken around the coil or conductor. The second term is the transformer emf and the surface integral is taken over the entire surface bounded by the coil. In a particular case, either or both of these emfs may be present. The negative sign in Eq. (1.10) in due to Lenz's law.

*Example 1.1*

A wire 75 cm long moves at right angles to its length at 60 m/s in a uniform field of flux density 1.3 T. Find the emf induced when the motion is (a) perpendicular to the field, (b) parallel to the field and (c) inclined at 60° to the direction of field.

*Solution*

$$e = B\,l\,v \sin\theta$$

(a) $e = (1.3)\,(0.75)\,(60)\,(1) = 58.5$ V

(b) Since motion is parallel to the field, no cutting of flux takes place and $e = 0$.

(c) $e = (1.3)\,(0.75)\,(60)\,(\sin 60°) = 50.66$ V.

*Example 1.2*

A conductor 40 cm long lies along $x$-axis. A magnetic field of flux density 0.04 $T$ is directed along y-axis. What should be the direction of motion of the conductor, if maximum emf is to be induced in it? Velocity = 4 sin $10^3$ t, m/s.

*Solution*

The motion should be perpendicular to the length of conductor as well as perpendicular to the field. Therefore the motion should be along $z$-axis for induced emf to be maximum.

$$e = B\,l\,v = (0.04)(0.4)\left(4\sin 10^3\,\mathrm{t}\right)$$

or

$$e = 0.064 \sin 10^3\,\mathrm{t}\ \text{volts.}$$

*Example 1.3*

An aeroplane having a wing span of 52 m is flying horizontally at 800 km/hr. If the vertical component of earth's magnetic field is $38 \times 10^{-6}$ T, find the emf generated between the wing tips.

*Solution*

$$B = 38 \times 10^{-6}\,\mathrm{T}, \qquad \theta = 90°,\ l = 52\ \mathrm{m}$$

$$v = 800\ \text{km/hr} = 222.2\ \text{m/s}$$

$$e = B\,l\,v\,\sin\theta = \left(38\times10^{-6}\right)(52)(222.2)(1) = 0.44 \text{ volts.}$$

### Example 1.4

A rectangular loop of width $l$ and length $x$ is moving with velocity $v$ in a magnetic field $B = B_0 \cos \omega\, t$. The motion of loop is perpendicular to field and is along the length. Find the emf induced.

### Solution

Because of motion of loop, a speed emf will be induced. Since flux is changing with time, a statically induced emf will also be induced.

The speed emf $e_r$ is

$$e_r = v\,l\,B_0 \cos \omega t$$

The statically induced emf $e_t$ is

$$e_t = -\int_s \frac{\partial B}{\partial t}\cdot ds = \omega x\,l\,B_0 \sin \omega t$$

Total emf

$$= e_r + e_t$$

$$= vB_0 l\cos\omega t + \omega x\,lB_0 \sin\omega t$$

$$= B_0 l\left[v^2 + (\omega x)^2\right]^{0.5} \sin(\omega t + \delta)$$

where $\delta = \tan^{-1}(v/\omega x)$.

### Example 1.5

A circular coil of 200 turns with a mean diameter of 30 cm is rotated about a vertical axis in the earth's field at 32 revolutions per second. Find the instantaneous value of induced emf in the coil, when its plane is (a) parallel, (b) perpendicular and (c) inclined at 30° to the magnetic meridian ($H$ = 14.3 AT/m).

### Solution

When a coil rotates in a magnetic field, the instantaneous value of induced emf is

$$e = N\omega\phi\sin\theta$$

where
$N$ = number of turns
$\omega$ = angular speed, rad/sec
$\phi$ = flux, Wb
$\theta$ = angle between field and direction of rotation

(a) When plane of the coil is parallel to the field, the rotation will be perpendicular to the field, i.e. $\theta = 90°$

$$\phi = B\times\text{area} = \mu_0 H\times\text{area}$$

$$= \left(4\pi\times10^{-7}\right)(14.3)(\pi)(0.15)^2$$

$$= 12.7 \times 10^{-7} \text{ Wb}$$

$$\omega = 32(2\pi) \text{ rad/sec}$$

$$e = (200)(32)(2\pi)\left(12.7 \times 10^{-7}\right) = 0.051 \text{ V}$$

(b) $\theta = 0$ and $e = 0$

(c) $\theta = 60°$

$$e = 0.051 \sin 60 = 0.0044 \text{ V.}$$

*Example 1.6*

The total flux at the end of a long bar magnet is $200 \times 10^{-6}$ Wb. The end of the magnet is withdrawn through a 1200 turn coil in $\frac{1}{150}$ second. Find the generated emf.

*Solution*

$$e = N\frac{d\phi}{dt} = (1200)\frac{\left(200 \times 10^{-6}\right)}{1/150} = 36 \text{ V.}$$

*Example 1.7*

An iron core is circular in shape. The cross-sectional area is 5 $cm^2$ and length of magnetic path is 15 cm. It has two coils *A* and *B*. Coil *A* has 100 turns and coil *B* has 500 turns. The current in coil *A* is changed from zero to 10 A in 0.1 sec. Find the emf induced in Coil *B*. The relative permeability of core material is 300.

*Solution*

$$\text{MMF} = 10 \times 100 = 1000 \text{ AT}$$

$$H = \frac{\text{MMF}}{\text{length}} = \frac{1000}{0.15} = 6666.67 \text{ AT/m}$$

$$B = \mu_0\mu_r H = 4\pi \times 10^{-7} \times 300 \times 6666.67$$

$$= 2.513 \text{ T}$$

$$\phi = B \times A = 2.513 \times 5 \times 10^{-4} = 12.565 \times 10^{-4} \text{ Wb}$$

Therefore, induced emf

$$e = N\left(\frac{d\phi}{dt}\right) = (500)\left(\frac{12.565 \times 10^{-4}}{0.1}\right) = 6.2825 \text{ V.}$$

## 1.7 COEFFICIENT OF SELF-INDUCTANCE

The change in current through a coil causes a change in flux. This change in flux induces an emf in the coil. This self induced emf can be written as

$$e = L\frac{di}{dt} \qquad ...(1.11)$$

where $e$ = emf induced, volts

$\frac{di}{dt}$ = rate of change of current, A/s

$L$ = coefficient of self inductance, H.

$L$, the coefficient of self inductance or simply inductance has the units of henry (symbol H). The inductance of a coil is 1 H if an emf of one volt is induced in it when the current through it changes at the rate of 1 A/sec. By Lenz's law this emf is in a direction so as to oppose the external emf, which is driving current through the coil.

From Eqs. (1.1) and (1.11)

$$e = L\frac{di}{dt} = N\frac{d\phi}{dt}$$

or

$$L = N\frac{d\phi}{di} \qquad ...(1.12)$$

If rate of change of flux is constant, then

$$\frac{d\phi}{di} = \frac{\phi}{i}$$

and

$$L = \frac{N\phi}{i}$$

Since

$$\phi = \frac{\text{MMF}}{\text{Reluctance}} = \frac{Ni}{l/\mu a}$$

Therefore

$$L = N\frac{\left[\frac{Ni}{(l/\mu a)}\right]}{i} = \frac{N^2\mu a}{l} \qquad ...(1.13)$$

## 1.8 COEFFICIENT OF MUTUAL INDUCTANCE

When the flux of one coil links another coil, a mutually induced emf appears across the second coil. By Lenz's law this is also a counter emf. The mutually induced emf can be written as

$$e = M\frac{di}{dt} \qquad ...(1.14)$$

where $e$ is the emf induced in the second coil, $di/dt$ is the rate of change of current in the first coil and $M$ is the coefficient of self-inductance or simply mutual inductance (units henry). The mutual inductance between two coils is 1 H if a current changing at the rate of 1 A/*s* in one coil induces an emf of

1 volt in the second coil.

*Example 1.8*

The current in a coil decreases from 20 A to 5 A in 0.1 sec. If the self-inductance of coil is 3 H, find the induced emf.

*Solution*

$$e = L\frac{di}{dt} = 3\left(\frac{20-5}{0.1}\right) = 450\,\text{V}.$$

*Example 1.9*

A flux of $5 \times 10^{-4}$ Wb is created by a current of 10 A flowing through a 150 turn coil. Find the inductance of the coil corresponding to a complete reversal of current in 0.2 sec. Also find the induced emf.

*Solution*

$$L = N\frac{d\phi}{di} = 150\left[\frac{5 \times 10^{-4} \times 2}{10 \times 2}\right] = 75 \times 10^{-4}\,\text{H}$$

$$e = L\frac{di}{dt} = 75 \times 10^{-4}\left[\frac{10 \times 2}{0.2}\right] = 0.75\,\text{V}.$$

*Example 1.10*

An air cored coil is required to be 3.5 cm long and to have an average cross-sectional area of 3 cm². The coil should have an inductance of 700 μH. Find the number of turns needed.

*Solution*

$$L = \frac{N^2 \mu_r \mu_0 a}{l}$$

or

$$N = \left[\frac{(L)(l)}{\mu_r \mu_0 a}\right]^{0.5} = \left[\frac{700 \times 10^{-6} \times 3.5 \times 10^{-2}}{1 \times 4\pi \times 10^{-7} \times 3 \times 10^{-4}}\right]^{0.5}$$

$$= 255 \text{ turns.}$$

*Example 1.11*

Two identical 1000 turns coils $X$ and $Y$ lie in parallel planes such that 60% of flux produced by one coil links the other. A current of 5 A in coil $X$ produces in it a flux of 0.05 m Wb. If current in $X$ changes from +10 A to −10 A in 0.02 sec, find the emf induced in coil $Y$, self-inductance of each coil and mutual inductance.

*Solution*

$$L = N\frac{\phi}{i} = 1000\left[\frac{0.05 \times 10^{-3}}{5}\right] = 0.01\,\text{H}$$

Since the coils are similar, self-inductance of each coil is 0.01 H

$$k = \text{coefficient of coupling} = 0.6$$

$$M = k(L \times L)^{0.5} = 0.006 \text{ H}$$

$$\text{emf induced in coil } Y = M\frac{di}{dt}$$

$$= 0.006 \times \frac{20}{0.02} = 6 \text{ V}.$$

## 1.9 FORCE ON CURRENT CARRYING CONDUCTOR IN A MAGNETIC FIELD

Figure 1.3 (a) shows a conductor lying in a magnetic field of flux density $B$. The conductor is carrying a current (entering the page). This current sets up a flux in clockwise direction. The external field is in a downward direction. As seen in Fig. 1.3 (a) the field of the conductor assists the external field on the

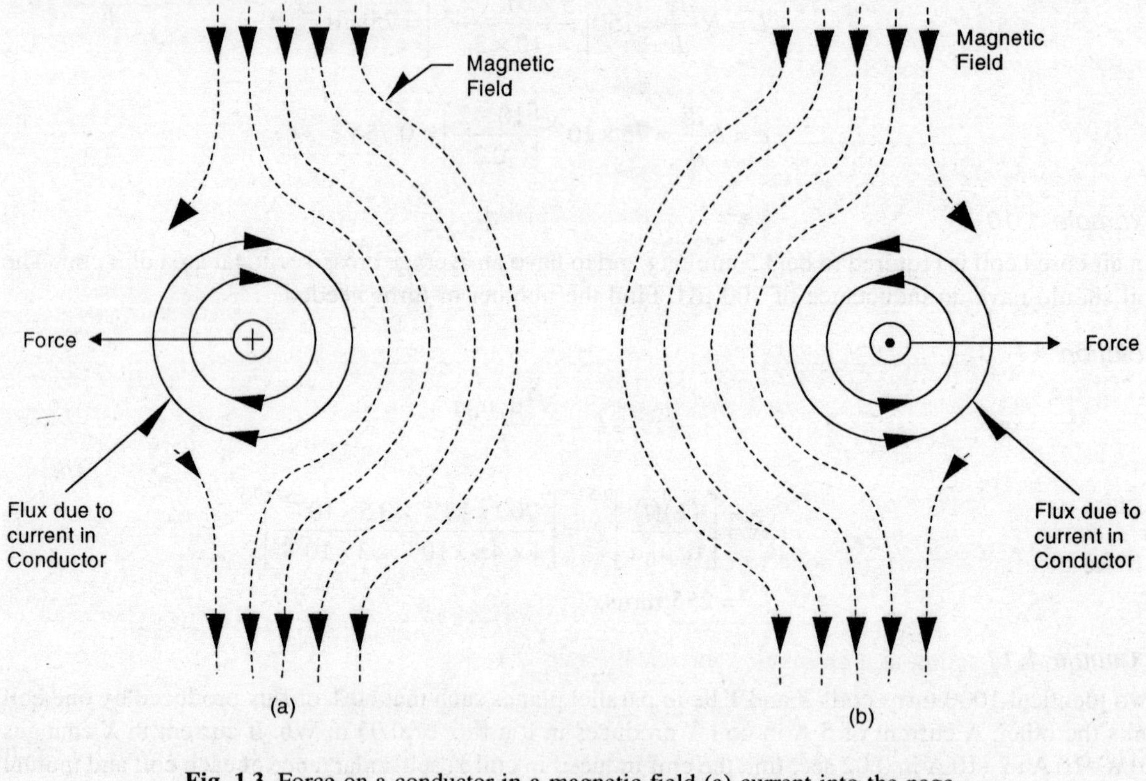


**Fig. 1.3.** Force on a conductor in a magnetic field (a) current into the page, (b) current out of the page

right hand side of the conductor and opposes it on the left hand side. This produces a force on the conductor towards left. If the direction of current is reversed (Fig. 1.3 b), the flux due to this current assumes counter-clockwise direction and the force on the conductor is towards right. In both cases the force is in a direction perpendicular to both the conductor and the field and is maximum if the conductor is at right angles to the field. The magnitude of this force is

$$F = B\,I\,l \text{ newtons} \qquad ...(1.15)$$

where $B$ is flux density in tesla, $I$ is current is amperes and $l$ is the length of conductor in metres.

If the conductor is inclined at an angle $\theta$ to the magnetic field, the force is

$$F = BIl \sin\theta \text{ newtons} \qquad ...(1.16)$$

## 1.10 TORQUE ON A CURRENT CARRYING COIL IN A MAGNETIC FIELD

Figure 1.4 shows a coil carrying $I$ and lying in a magnetic field of flux density $B$. From the discussion in Sec. 1.9, it is seen that an upward force is exerted on the right hand conductor and a downward force on the left hand conductor. Equation (1.15) gives the force on each conductor and the total force is

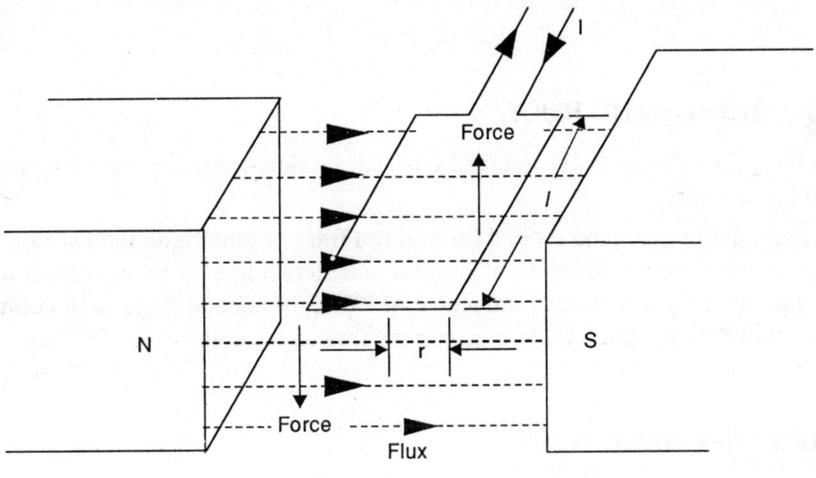


**Fig. 1.4.** Torque on a coil in a magnetic field

$$F = 2BIl \text{ newtons}$$

If the coil has N turns, the total force is

$$F = 2NBIl \text{ newtons}$$

The torque is acting at a radius of $r$ metres and is given by

$$\text{Torque} = 2NBIlr \text{ newtons - metres} \qquad ...(1.17)$$

The configuration of Fig. 1.4 is the basic moving part in an electrical measuring instrument. An electric motor also works on this principle.

*Example 1.12*

A 50 cm long conductor is carrying 5 A current and is situated at right angles to a field having a flux density of 1.1 T. Find the force on the conductor.

*Solution*

$$F = B\,I\,l \text{ newtons}$$

$$= (1.1)(5)(0.5) = 2.75 \text{ N.}$$

*Example 1.13*

A 300-turn coil having an axial length of 8 cm and radius 2 cm is pivoted in a magnetic field of flux density 1.1 T. Find the torque on the coil if $I$ = 2A.

*Solution*

$$\text{Torque} = 2\,N\,B\,I\,l\,r \text{ N-m}$$

$$= 2(300)(1.1)(2)\left(8\times10^{-2}\right)\left(2\times10^{-2}\right)$$

$$= 2.112 \text{ N-m.}$$

## 1.11 FLEMING'S RIGHT-HAND RULE

The direction (polarity) of dynamically induced emf can be determined by the following rule, known as Fleming's right hand rule:

"Hold the thumb, the first and the second (or middle) finger of the right hand at right angles to each other. If the thumb points to the direction of motion and first finger to the direction of the field, the second finger will point in the direction of induced emf" (i.e. the second finger will point to the positive terminal of emf or will indicate the direction of current flow if the ends of the conductor are connected to external circuit).

## 1.12 FLEMING'S LEFT HAND RULE

The direction of force on a current carrying conductor, situated in a magnetic field, can be found from Fleming's left hand rule:

"Hold the thumb, the first and the second (or middle) finger of the left hand at right angles to each other. If the first finger points to the direction of field and the second finger to the direction of current, the thumb will point to the direction of force or motion."

## 1.13 GENERATOR AND MOTOR ACTION

It is seen from $B\,l\,v$ and $B\,l\,I$ equations that generator and motor actions are based on the physical reactions on conductors situated in magnetic fields. When a relative motion between conductor and field exists, an emf is generated in the conductor and when a conductor carries current and is situated in a magnetic field, a force is exerted on the conductor. Both generator and motor actions take place simultaneously in the windings of a rotating machine. Both generators and motors have current carrying conductors in a magnetic field. Thus both torque and speed voltage are produced. Within the winding it is not possible to distinguish between the generator and motor action without finding the direction of power flow. Constructionally a generator and motor of one category are basically identical and differ only in details necessary for its best operation for intended service. Any generator or motor can be used for energy conversion in either direction.

A generator converts mechanical energy into electrical energy. The torque produced, in a generator, is a counter torque opposing rotation. The prime mover must overcome this counter torque. An increase (or decrease) in electrical power output means an increase (or decrease) in counter torque, which finally results in an increase (or decrease) in torque supplied by the prime mover to the generator.

A motor converts electrical energy into mechanical energy. The speed voltage generated in the conductors is a counter or back emf, which opposes the applied voltage. It is through the mechanism of back emf that a motor adjusts its electrical input to meet an increase (or decrease) in mechanical load on the shaft.

## 1.14 INTERACTION OF MAGNETIC FIELDS

The operation of a rotating electric machine can also be regarded as interaction between two magnetic fields. All rotating machines have two basic parts: stator (which remains stationary and is fixed through the frame to the foundations) and rotor, which is free to rotate. Each of them carries a winding and a magnetic field. The machine action is due to the result of these two fields trying to line up, so that the centre line of a north pole on one part is directly opposite the centre line of a south pole on the other part.

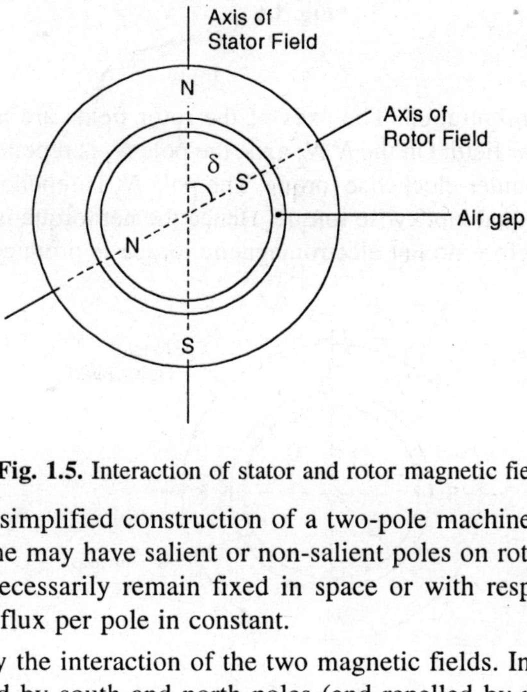


**Fig. 1.5.** Interaction of stator and rotor magnetic fields.

Figure 1.5 shows the simplified construction of a two-pole machine. Both sets of poles are non-salient. A practical machine may have salient or non-salient poles on rotor or stator or both. The axes of the two fields donot necessarily remain fixed in space or with respect to the part producing it. However, in all machines flux per pole in constant.

Torque is produced by the interaction of the two magnetic fields. In Fig. 1.5 the north and south poles of rotor are attracted by south and north poles (and repelled by north and south poles) of the stator, resulting in a torque in the counter-clockwise direction. The same magnitude of torque is exerted on both rotor and stator structures. The torque on the stator is transmitted, through the frame, to the foundations.

The magnitude of the torque in proportional to product of the field strengths of the two fields. The angle $\delta$ between the axes of the two fields is known as torque or power angle. For sinusoidal variation of flux in the air gap (i.e. flux density varying sinusoidally with distance around the gap periphery) the torque is also proportional to sin $\delta$.

*Example 1.14*

Using the concept of interaction of magnetic fields, show the electromagnetic torque cannot be developed, (a) if rotor has 4 poles and stator has 2 poles; (b) if rotor has 2 poles and stator has 4 poles.

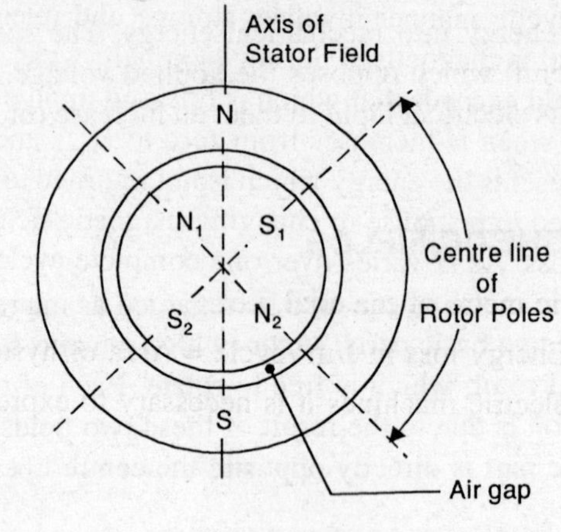


**Fig. 1.6**

*Solution*

(a) Figure 1.6 shows the configuration. The axes of the rotor fields are at an arbitrary angle with respect to the axis of stator field. On the $N_1N_2$ axis, the pole $N_1$ is repelled by pole $N$ and attracted by pole $S$ producing a counter-clockwise torque. The pole $N_2$ is repelled by $N$ pole and attracted by $S$ pole producing an equal clockwise torque. Hence the net torque is zero. A similar situation exists on $S_1S_2$ axis. Therefore no net electromagnetic torque is produced.

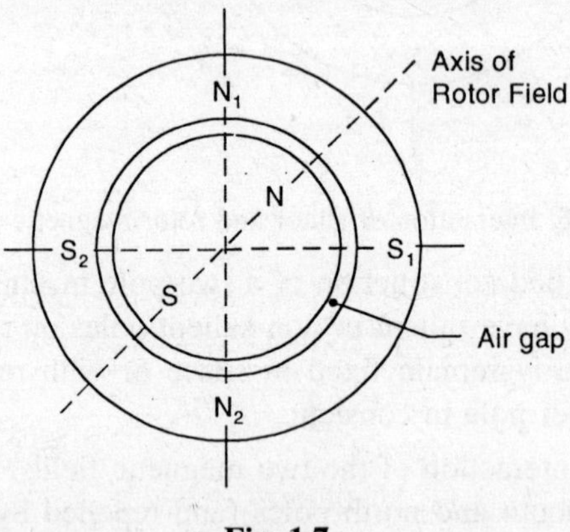


**Fig. 1.7**

(b) Figure 1.7 shows the configuration. The pole $N$ is repelled by $N_1$ and attracted by $S_1$ producing a clockwise torque. The pole $S$ is repelled by $S_2$ and attracted by $N_2$ producing an equal counter-clockwise torque. The net torque is zero.

Whenever the number of poles on rotor and stator are different, net torque is always zero. The conclusion is that all rotating machines (generators and motors) must have the same number of poles on the stator and rotor for steady unidirectional torque.

## 1.15 HYSTERESIS LOSS

When a specimen of ferromagnetic material is taken through a cycle of magnetisation, the hysteresis loop is obtained. The salient feature of hysteresis loop is the delayed re-orientation of the domains in response to a cyclically varying magnetising force. The process of magnetisation and demagnetisation of a ferromagnetic material in a cyclic manner involves storage and release of energy which is not completely reversible. As a material is magnetised during each half-cycle, it is found that the amount of energy stored in the magnetic field exceeds that which is released upon demagnetisation. The difference between the energy absorbed when $H$ increases from zero to $H_{max}$ and the energy released when $H$ decreases from $H_{max}$ to zero represents the energy which is not returned to the source but is dissipated as heat as the domains are re-aligned in response to changing magnetic field intensity. This dissipation of energy is known as hysteresis loss. As $H$ varies over one complete cycle, the area of the hysteresis loop gives the energy loss per cubic metre of material, i.e.

$$\text{Energy loss in J/m}^3\text{/cycle} = \text{Area of hysteresis loop.}$$

In evaluating the performance of electric machines it is necessary to express hysteresis loss in watts. This can be done as under:

$$\text{Area of hysteresis loop} = \frac{\text{Energy loss}}{\text{Volume} \times \text{Cycles}}$$

$$= \frac{\text{Power loss} \times \text{Seconds}}{\text{Volume} \times \text{Cycles}} = \frac{\text{Power loss}}{\text{Volume} \times \text{Cycle/sec}}$$

or $$\text{Hysteresis power loss} = [\text{Area of hysteresis loop}]\ (\text{volume})\ (f)$$

In order to eliminate the need of finding area of hysteresis loop, Steinmitz evolved an empirical formula, based on his experiments, for calculating hysteresis loss. This formula is

$$P_h = k_h(\text{volume})(f)(B_m)^n \qquad \text{...(1.18)}$$

where $P_h$ = Hysteresis loss in watts
$f$ = frequency in Hz
$B_m$ = maximum flux density, T

$n$ varies from 1.5 to 2.5 depending on the material used. Typical value of $n$ for grain oriented silicon sheet steel used for electrical machines is 1.6. The constant $k_h$ also depends on the material. Some typical values are: cast steel 0.025; silicon sheet steel 0.001; permalloy 0.0001. For a particular machine, the volume of material is also constant, so that $P_h$ can be written as

$$P_h = K_h f B_m^n \qquad \text{...(1.19)}$$

where $$K_h = (k_h)\ (\text{volume of material})$$

## 1.16 EDDY CURRENT LOSS

A voltage is induced in a conductor when it is situated in a changing magnetic field. To produce strong magnetic effects, coils of electromagnets are wound on iron cores. When the current in the coil of an electromagnet is changing, the core constitutes a conductor in a changing magnetic field. Voltages are induced in the core and circulating currents are caused to flow. These circulating currents are called eddy currents. The flow of these currents causes losses — a waste of energy and heating of core. Figure 1.8 illustrates flow of eddy currents.

In dc circuits eddy current loss is not serious, since an induced voltage is produced only when the circuit is switched on or off. However, in ac circuits the flux changes continuously and eddy current loss becomes appreciable.

It has been found that eddy currents travel cross-wise in the cores. The magnitude of eddy currents can be reduced to a very small value by increasing the cross-sectional resistance of the iron. The cores are made of thin sheets called laminations. Each lamination is about 0.5 mm thick and is insulated from the others by a coating of oxides or varnish or both, so that it offers a high resistance to the flow of eddy currents. The use of laminated cores reduces the active cross-sectional area by a small amount (about 5-10%) due to thickness of insulation. If a laminated core is taken apart, care must be exercised not to impair the insulations between the laminations. Otherwise when the core is re-assembled and used, large eddy currents may flow and heat the core.

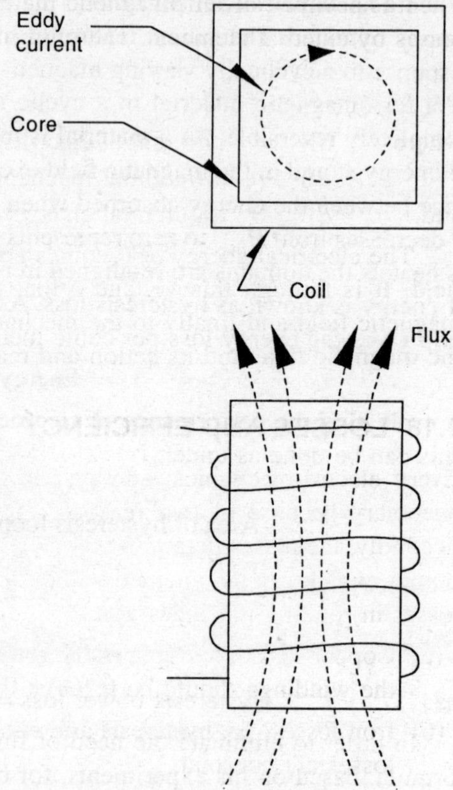


**Fig. 1.8.** Flow of eddy currents in iron core

Eddy current loss is given by

$$P_e = k_e f^2 B_m^2 t^2 \text{ (volume)} \quad ...(1.20)$$

where $P_e$ = eddy current loss, watts
$f$ = frequency, Hz
$B_m$ = maximum flux density, T
$t$ = thickness of laminations
$k_e$ = constant depending on material

For a particular machine $k_e$, $t^2$ and volume can be combined into a single constant $K_e$ so that

$$P_e = K_e B_m^2 f^2 \quad ...(1.21)$$

Taken together the hysteresis and eddy current loss is known as core-loss or iron-loss. Since frequency and maximum flux density are constant, the core-loss in a machine is constant.

## 1.17 ENERGY BALANCE

In all electro-mechanical energy conversion devices, energy in one form is converted into another form. A part of the energy input is converted into losses and dissipated as heat. All these devices have a magnetic field which can store electrical energy. The energy balance equation for a motor can be written as

Elect Energy Input = Mech energy output + increase in energy stored in magnetic field + energy converted into heat ...(1.22)

The above equation is valid for a generator also, if the electrical energy input and mechanical energy output are taken as negative.

It is always convenient to treat the magnetic energy storage elements as lossless and represent the losses by external elements. Therefore the energy balance equation can be written as

$$dW_e = dW_m + dW_f \quad \text{...(1.23)}$$

where $dW_e$ = differential electrical energy input
$dW_m$ = differential mechanical energy output
and $dW_f$ = differential change in energy stored in magnetic field

The electrical energy $dW_e$ equals $e\ i\ dt$, where $e$ is the voltage induced by the changing magnetic field. It is through this voltage $e$ that the external electric system supplies energy to the coupling magnetic field and finally to the mechanical load. All electro-mechanical conversion devices employ the magnetic field and its action and reaction on the electrical and mechanical systems.

## 1.18 LOSSES AND EFFICIENCY

Every electro-mechanical energy conversion is accompanied by losses. A study of these losses is necessary because of two reasons. Firstly, the losses influence the operating cost of the machine. Secondly, the losses determine the heating of the machine and hence fix the machine rating or power output, which can be obtained without deterioration of the insulation due to over-heating. The various losses in rotating machines are:

(a) Copper-loss or $I^2 R$ loss in the rotor and stator windings. To calculate the losses the resistances of the windings should be taken at the operating temperature, which is assumed to be 75°C.

(b) Iron losses, i.e. hysteresis and eddy current losses as discussed in sections 1.15 and 1.16. These losses are constant.

(c) Mechanical or friction and windage losses. These losses are also constant unless speed varies appreciably.

(d) Stray load losses, which mean additional hysteresis and eddy current losses arising from any distortion in flux distribution caused by the load current. These losses are difficult to measure and are usually taken as 1% of the machine output.

$$\text{Efficiency} = \frac{\text{Power Output}}{\text{Power Input}} \quad \text{...(1.24)}$$

For a motor, especially the large ones, it is difficult to measure mechanical power output. Therefore

$$\text{Efficiency of motor} = \frac{\text{Power Input} - \text{losses}}{\text{Power Input}} \quad \text{...(1.25)}$$

For a generator it is almost impossible to measure power input. Therefore

$$\text{Efficiency of generator} = \frac{\text{Power Output}}{\text{Power Output} + \text{losses}} \quad \text{...(1.26)}$$

## 1.19 LOAD TYPES

The characteristics of a motor have to match the characteristics of the load which the motor is driving. These loads can be divided into four main types viz

(a) load torque proportional to square of the shaft speed

(b) load torque proportional to the shaft speed

(c) load torque independent of shaft speed (or constant torque) and

(d) load torque inversely proportional to shaft speed. These four load characteristics are shown in Fig. 1.9. In each case if is important to remember that Power = (torque) (speed). A comparison of these load types is shown is Table 1.1.

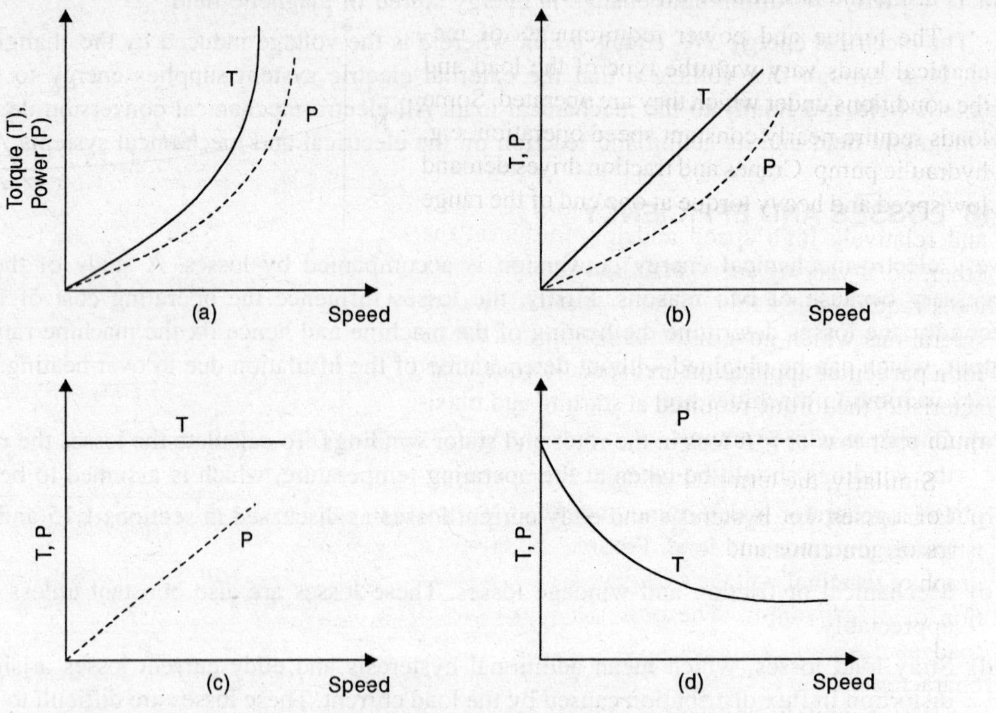


**Fig. 1.9.** Load characteristics (a) Torque ∝ (speed)² (b) Torque ∝ speed (c) Torque constant (d) Torque ∝ 1/speed.

**Table 1.1: Characteristics of loads**

| *Type a* | *Type b* | *Type c* | *Type d* |
|---|---|---|---|
| Torque ∝ (speed)$^2$<br>Power ∝ Torque × speed<br>∝ (speed)$^3$<br>It is very suited for energy conservation<br>Example are:<br>axial and centrifugal pumps, axial and centrifugal ventilators, centrifugal compressors, centrifugal mixers, agitators etc. | Torque ∝ (speed)<br>Power ∝ Torque × speed<br>∝ (speed)$^2$<br>Examples are:<br>mixers, stirrers, etc. | Torque = Constant<br>Power ∝ Torque ∝ speed<br>∝ speed<br>Examples are:<br>conveyor machines.<br>Lift machines, extrusions, draw benches etc. | Torque ∝ 1/speed<br>Power ∝ Torque × Speed<br>or<br>Power = constant<br>Examples are:<br>Lathes, wire drawers, Reciprocating rolling mills, winding machines etc. |

## 1.20 CONSIDERATIONS GOVERNING MACHINE APPLICATIONS

The output and operating point of a motor and generator have to match the load requirement.

The power or torque supplied by a motor depends on the requirement of the mechanical load driven by the motor. The operating speed is fixed by the point, on the characteristic curve, at which the power or torque output of the motor is equal to power or torque required by the load. Figure 1.10 shows the speed-torque characteristic of a motor and that of a fan load coupled to the motor. The operating point *P* is at the intersection of these two curves.

The torque and power requirements of mechanical loads vary with the type of the load and the conditions under which they are operated. Some loads require nearly constant speed operation, e.g. hydraulic pump. Cranes and traction drives demand low speed and heavy torque at one end of the range and relatively high speed and low torque at the other, i.e. a varying speed characteristic. Machine tools require adjustable constant speed. The considerations which govern the selection of a motor for a particular application are: speed-torque characteristic, the torque required at starting and maximum torque while running.

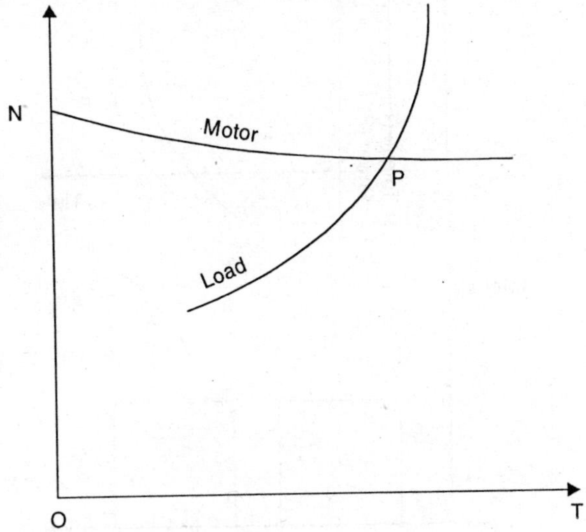


**Fig. 1.10.** Speed torque characteristics of motor and load

Similarly, the terminal voltage and power output of a generator is determined by the characteristics of generator and load. Figure 1.11 shows a graph of terminal voltage of a generator as a function of power output. The power requirement of load at different voltages is also shown by the characteristic curve of load. The operating point *P* is at the intersection of the two curves. Some loads require absolutely constant terminal voltage at all load conditions, while for some other loads the terminal voltage is designed to be varied in a particular fashion.

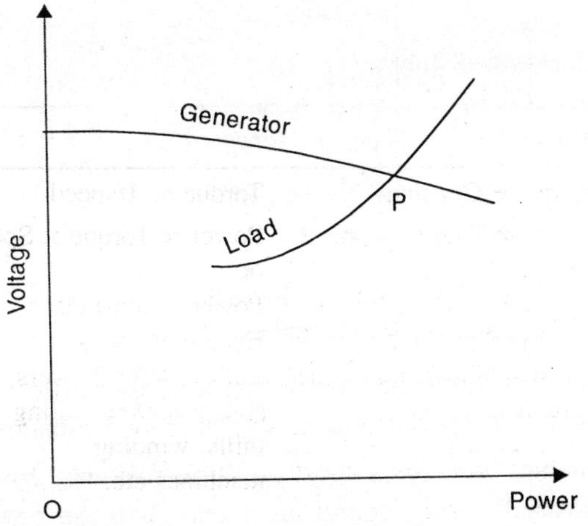


**Fig. 1.11.** Characteristics of generator and electrical load

In addition to above, the factors like efficiency, power factor, comparative costs, etc., also affect the selection of a machine for a particular application.

For most of the applications, the above studies referred to as steady state analyses are sufficient. However many other applications, especially in control systems, the behaviour of machines under transient conditions is also important. Moreover, the various machines are subjected to transient conditions from time to time due to faults and other abnormal conditions in power systems. It is necessary that the machine should give satisfactory operation under both steady state and transient conditions.

## 1.21 DUTY CYCLES OF MOTORS

The size of a motor is generally selected for continuous operation at a near rated output. However, the duty cycle is, many times, not continuous. It is necessary to specify the duty cycles while selecting a motor. Fig. 1.12 shows some duty cycles.

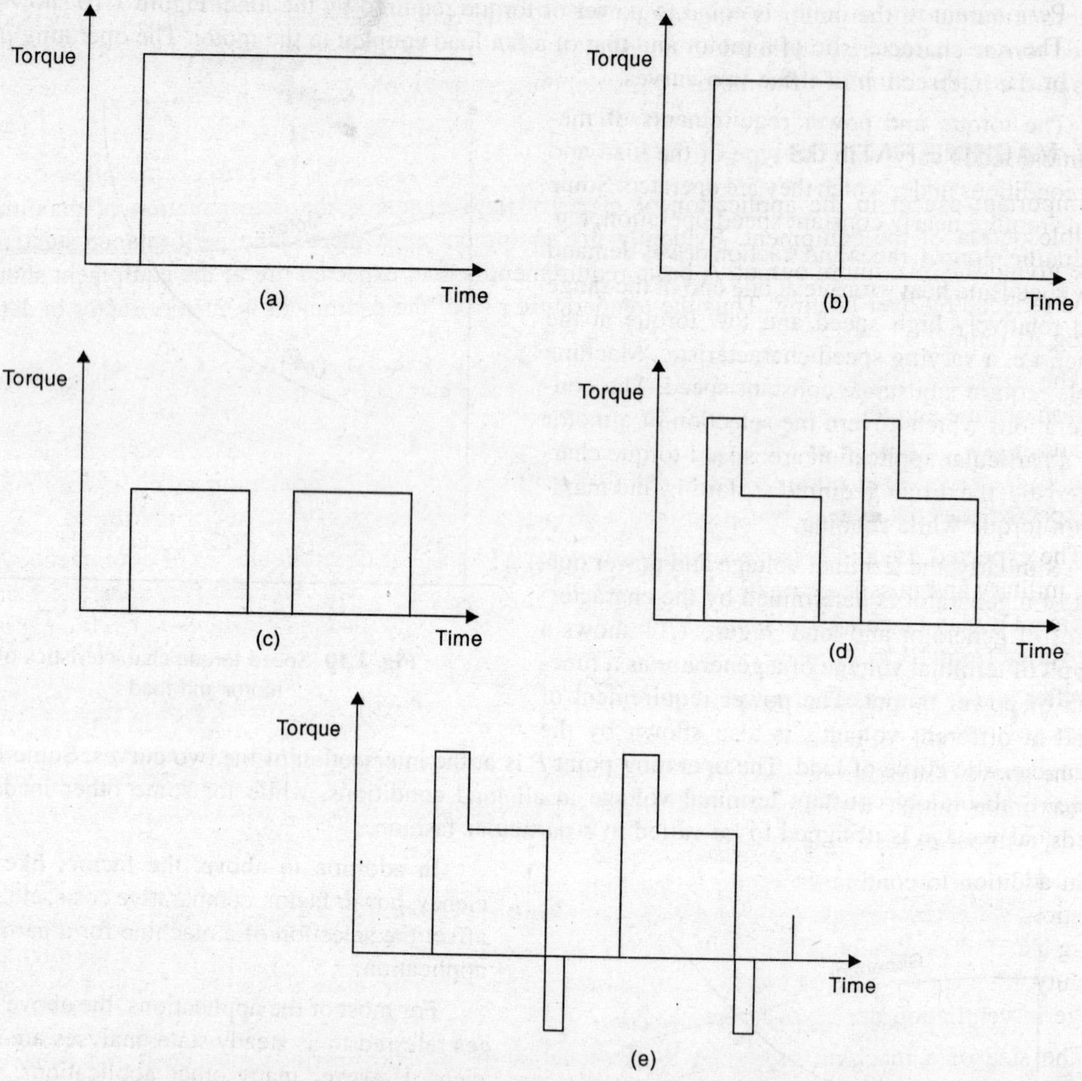


**Fig. 1.12.** Duty Cycles of Motors (a) Continuous (b) Short time (c) Periodic (d) Periodic with high starting torque (e) Periodic with high starting torque and electric braking

(a) Continuous duty (Fig. 1.12 a). This operation means that motor has a constant load for sufficient duration so that thermal equilibrium is reached (hot spot temperature of motor becomes constant)

(b) Short time duty (Fig. 1.12 b). The motor has a constant load for a short duration followed by a long period of rest. The maximum temperature reached during operation is less than the rated maximum temperature.

(c) Periodic duty (Fig. 1.12 c). The motor has a constant load followed by a period of rest and again constant load.

(d) Periodic duty with high starting torque (Fig. 1.12 d). The motor has a periodic duty but the starting torque in each running period is higher than full load rated torque.

(e) Periodic duty with high starting torque and electric braking (Fig. 1.12 e). The duty is periodic. The starting torque is higher than rated full load torque. In each cycle there is a period of electric braking (during which the motor runs in opposite direction).

## 1.22 MACHINE RATINGS

An important aspect in the application of electrical equipments is the determination of maximum possible output of the equipment. Evidently the equipment must meet some performance standards while giving this maximum output. A basic requirement is that, expected life of the equipment should not be reduced by over-heating. Thus the temperature rise of the equipment is a major factor in determining its rating.

The expected life depends on the temperature rise because the deterioration of insulation is a function of time and temperature. This deterioration is a chemical phenomenon involving slow oxidation and brittle hardening and leading to loss of mechanical properties and dielectric strength of insulation. each class of insulation has its maximum permissible operating temperature. The time to failure (or expected life) of organic insulation is reduced to half for each 8 to 10°C temperature rise.

The expected life and service conditions vary very widely for different classes of equipments. For some military and missile equipments, life expectancy may be only a few minutes. For some electronic equipment it may be about 1000 hours, while for electric machines used in power systems, life expectancy varies from 10 to 30 years.

After a machine's expected life is over, its frequency of break down (known as outage rate) increases and it is, generally, more economical to replace it by a new machine.

The above discussion pertains to the continuous rating of the equipment. The continuous rating is the maximum output which can be obtained, if the equipment is used continuously for very long periods, without excessive temperature rise and without deterioration in machine performance.

In addition to continuous rating, a machine has a short time rating for intermittent and periodic operation. Standard periods for short time ratings are 5, 15, 30 and 60 minutes. Evidently a motor having a certain continuous rating has a higher short time rating. When a machine is used for intermittent duty, the average heating must be determined from the motor losses during operating periods and change in ventilation during different periods.

The size of a machine supplying different loads for different periods may be found from the average heating. As in the case of finding rms value of an alternating quantity, the required kW size is given by the equation

$$\text{kW (rms value)} = \left[\frac{\Sigma(\text{kW})^2\,(\text{time})}{\text{operating time} + (\text{standstill time}/k)}\right]^{0.5} \qquad ...(1.27)$$

The constant $k$ accounts for poorer ventilation when the machine is at standstill. For an open motor this constant equals 3. In addition, special consideration should be given to motors which are started frequently, because the current (and losses) during starting are generally more than that under full-load conditions.

*Example 1.15*

A motor operates continuously on the following duty cycle: 50 hp for 20 sec, 100 hp for 20 sec, 150 hp for 10 sec, 120 hp for 20 sec and idling for 15 sec. Find the size of the motor.

*Solution*

Since there is no standstill period, the constant $k$ is not required for calculations.

$$\text{hp (rms)} = \left[\frac{50^2(20)+100^2(20)+150^2(10)+120^2(20)+0^2(15)}{20+20+10+20+15}\right]^{1/2}$$

$$= 94.75 \text{ hp}$$

Choose a 100 hp motor.

## 1.23 ROLE OF ELECTRIC MACHINES IN GENERATION, TRANSMISSION AND DISTRIBUTION OF POWER

Electrical energy occupies the top most grade in the energy hierarchy. It finds innumerable uses in home, industry, agriculture, transport and commercial establishments. The facts that electricity can be transported practically instantaneously, is almost pollution free at the consumer level and its use can be controlled very easily, make it very attractive as compared to other forms of energy. The increasing awareness of environmental problems and anti-pollution measures of government, will compel many users of non-electric energy to go electric in future. The efficient use of electricity has been made possible only due to the role of electric machines in generation, transmission and distribution of power.

Electric energy is generated in large size thermal, hydroelectric and nuclear power stations. The key equipment in these power stations is alternator, which converts mechanical energy of turbine into electrical energy. In addition there are a large number electrical auxillary equipments which help in efficient generation.

The generation voltage in India and many other countries is 11 kV. In some developed countries, it is 33 kV. If electrical energy is transmitted to distant places (load centres) at this voltage, the $I^2R$ losses would be enormous. To decrease the losses, the transmission voltages are much higher 132 kV, 220 kV, 400 kV and 765 kV. The conversion from generation voltage to transmission voltage is done by large size transformers located in step-up sub-stations near the generating station. The efficient transmission of electric energy is possible due to transformers.

At the load centres voltage is stepped down to lower values suitable for consumer use. This is done by step-down transformers located in receiving end sub-stations near the loads.

Electric energy is used very extensively in industrial establishments. The most widely used equipment in these establishments is 3-phase induction motor. For special purposes synchronous motors and dc motors are also used.

Electric energy finds extensive use in agriculture. Tube-wells run by 3-phase induction motors provide a dependable method of irrigation. In addition electricity is also used for harvesting, where the main equipment is, again, 3-phase induction motor.

Electricity finds innumerable uses in domestic and commercial establishments. Single-phase induction motors are sued in fans, coolers, air conditioners, refrigerators, water pumps, mixies, hair dryers, toys, etc.

Electric locomotives use dc and ac series motors.

The per capita consumption of electric energy in any country is an index of the standard of living of the people in that country. The efficient and increasing use of electricity in various fields of daily life is possible only due to the vast number of electric machines used in generation, transmission, distribution and utilisation of electric power.

*Example 1.16*

The armature of a certain motor has 900 conductors, each conductor carrying 24 A. The flux density in the air-gap under the poles is 0.6 T. The armature core is 160 mm long and has a diameter of 250 mm. Assuming that only two-thirds of the conductors are simultaneously in the magnetic field, find (a) torque, (b) mechanical power developed, in kW, if the speed is 700 rpm.

*Solution*

(a) $N$ = number of conductors in magnetic field

$$= 900 \times \frac{2}{3} = 600$$

Force on one conductor = $BIl$

$$= 0.6 \times 24 \times \frac{160}{1000} = 2.304 \text{ N}$$

Total force $= 2.304 \times 600 = 1382.4 \text{ N}$

Torque = force × radius = (1382.4)(125/1000)

$$= 172.8 \text{ N.m}$$

(b) $\omega = \frac{2\pi N}{60} = \frac{2\pi \times 700}{60} = 73.304$ rad/sec

Power = $\omega T$ = 73.304 × 172.8 = 12666.93 W

= 12.667 kW.

*Example 1.17*

A coil with an axial length of 25 cm and diameter of 20 cm has 100 turns. It is placed in a uniform radial flux of 0.002 Wb/m$^2$. (a) If the coil is rotated at 25 rps, find the voltage induced in the coil. (b) What will be the force on each conductor and torque acting on the coil, if it carries a current of 10 A.

*Solution*

(a) $v = \pi D n = \pi \times 0.2 \times 25 = 15.708$ m/sec

Voltage induced in each conductor = $Blv = 0.002 \times 0.25 \times 15.708 = 0.007854$ V

Total number of conductors = 2 × 100 = 200

Voltage induced in coil = 200 × 0.007854

= 1.5708 V

(b) Force on one conductor $= B\,I\,l$

$= 0.002 \times 10 \times 0.25 = 0.005$ N

Total force = 0.005 × 200 = 1 N

$$T = \text{force} \times \text{radius} = 1\left(\frac{10}{100}\right) = 0.1 \text{ N-m}$$

## 1.24 CLASSIFICATION OF ELECTRIC MACHINES

Electric machines can be broadly classified as DC machines and AC machines. Each of these can be further classified into different categories.

### 1.24.1 DC Machines

They can be further classified as dc generators and dc motors. The dc generator converts mechanical energy into electrical energy (dc). The prime mover (i.e. source of mechanical energy) provides rotary motion to the conductor. This relative motion between conductors and the magnetic field causes an emf to be induced in the conductors and dc is generated. A dc motor converts electrical energy (dc) into mechanical energy. The electrical energy to the motor is supplied from a dc source and the mechanical energy produced by the motor is used to drive a mechanical load (e.g. fan, lathe, etc.)

### 1.24.2 AC Machines

They can be classified as transformers, synchronous machines, induction machines, AC commutator machines and special machines.

(a) *Transformers:* A transformer is not an electro-mechanical device. It converts ac electrical energy at one voltage to ac electrical energy at another voltage. Transformers are widely used in electrical power systems, electronic, instrumentation and control circuits.

(b) *Synchronous machines:* In a synchronous machine the rotor moves at a speed which bears a constant relationship to the frequency of ac. This speed is known as synchronous speed. They are all 3-phase machines, because ac systems are all 3-phase systems.

A synchronous generator converts mechanical energy into 3-phase ac energy. It is also known as alternator.

A synchronous motor receives electrical energy from a 3-phase ac supply and converts it into mechanical energy. It produces a continuous positive torque only at its constant synchronous speed.

(c) *Induction machines:* This machine derives its name from the fact that emf in the rotor is induced due to magnetic induction.

A 3-phase induction motor converts 3-phase ac energy into mechanical energy. They are very widely used in small scale, medium and large industries, workshops, etc.

A single-phase induction motor converts single-phase ac into mechanical energy. They are very widely used in household devices, viz. fans, refrigerators, washing machines, etc.

An induction generator can convert mechanical energy into 3-phase ac energy. However, it is not used due to certain limitations.

(d) *AC commutator machines:* An ac commutator motor derives its name from the fact that it has a commutator. These motors have special characteristics and are used for special applications. They can be 3-phase motors or single-phase motors.

(e) *Special machines:* These motors have special constructional features and are used for special applications, e.g. computer peripheral devices, line printers, control circuits, etc.

## SUMMARY

1. When the magnetic flux enclosed by a circuit changes with time, an emf is induced in the circuit. This is called Faraday's law of electromagnetic induction.
2. As per Lenz's law the induced current develops a flux which opposes the change producing the induced current.
3. Induced emf can be dynamically induced emf or statically induced emf.
4. Coefficient of self induction $L = N\dfrac{d\phi}{di}$.
5. When a flux of one coil links another coil a mutually induced emf appears across the second coil.
6. When a conductor of length $l$ is carrying current I and is inclined at an angle $\theta$ to a magnetic field of flux density B, the force F on conductors is
$$F = B\,I\,l \sin\theta$$
7. Direction of dynamically induced emf can be formed by using Fleming's right hand rule.
8. Direction of force on a conductor lying in a magnetic field can be found by using Fleming's left hand rule.
9. A generator converts mechanical energy into electrical energy whereas a motor converts electrical energy into mechanical energy.
10. Both generator and motor action occur in both generators and motors.
11. Torque, in electric machines, is produced by interaction of magnetic fields.
12. Hysteresis loss = $k_h$ (volume) ($f$) $(B_m)^n$
13. Eddy current loss = $k_e\, f^2\, B_m^2\, t^2$ (volume)
14. The various losses in electric machines are copper losses, iron losses, mechanical losses and stray load losses.
15. The output and operating point of a motor and generator have to match the load requirements.
16. Every machine has a continuous rating as well as short time rating.
17. Electric machines are classified as dc machines and ac machines. AC machines are classified as transformers, synchronous machines, induction machines, ac commutator machines and special machines.

## TEST POINT QUESTIONS

A. State whether the following statements are true (T) or false (F)

1.1 A transformer operates on the principle of mutual induction.

1.2 A dynamically induced emf exists in a dc motor.

1.3 Short time rating is always less than continuous rating.

1.4 An induction motor is a singly excited machine.

1.5 Synchronous motor is a constant speed motor.

1.6 Hysteresis loss in proportional to area of hysteresis loop.

1.7 If flux density is increased eddy current loss reduces.

1.8 Torque on a current carrying coil is proportional to number of turns in the coil.

1.9 $e = B\,l\,v \sin\theta$

1.10 Iron losses are constant only if flux density is constant

B. Fill in the blanks with suitable words

1.11 Hysteresis loss is proportional to ........

1.12 If thickness of stampings is increased, eddy current loss .........

1.13 When the number of poles on the stator and rotor of a machine are unequal, torque is ........

1.14 Copper losses are proportional to ........ of current.

1.15 Iron losses are classified as ........ and ........

1.16 If $B_m$ is maximum flux density, eddy current loss is proportional to ........

1.17 Stray load losses =

1.18 Deterioration of insulation is a function of ........ and ........

1.19 Generator converts ........ into ........

1.20 Induction motor is a ........ machine.

## ANSWERS

1.1 T  1.2 T  1.3 F  1.4 T  1.5 T

1.6 T  1.7 F  1.8 T  1.9 T  1.10 T

1.11 frequency  1.12 increases  1.13 zero  1.14 square

1.15 hysteresis losses, eddy current losses

1.16 $B_m^2$  1.17 1% of output  1.18 Time, temperature

1.19 mechanical energy, electrical energy

1.20 singly excited

## JOB INTERVIEW QUESTIONS

1.1 State Faraday's law of electromagnetic induction.

1.2 What is the difference between dynamically induced emf and statically induced emf?

1.3 What is the difference between a generator and a motor?

1.4 State Fleming's left hand rule.
1.5 What are iron losses?
1.6 What is short time rating? Is it more or less than continuous rating?
1.7 What is hysteresis loop?
1.8 What is self inductance?
1.9 What is Fleming's right hand rule.
1.10 What is a synchronous machine?

## REVIEW QUESTIONS

1.1 Since voltage is a form of potential energy, where does the energy represented by an induced voltage come from? Explain.
1.2 Explain the three methods used for linking a conductor with flux. What practical application is made from each of these methods?
1.3 Why does the magnitude of dynamically induced emf depend on the direction of motion of the conductor with respect to the magnetic field? Explain.
1.4 Explain the $B\ l\ v$ and $B\ l\ i$ concept. What practical use is made of these two concepts?
1.5 Derive an equation for dynamically induced emf? On what factors does this emf depend?
1.6 Can an induced emf be produced if the only magnetic field available does not change with time?
1.7 Differentiate between self and mutual inductance. What are their units?
1.8 Derive expressions for (a) force acting on a current carrying conductor situated in a magnetic field, (b) torque on a current carrying coil situated in a magnetic field.
1.9 Explain Fleming's left and right hand rules.
1.10 In every electromagnetic conversion device, both generator and motor action take place simultaneously. Explain.
1.11 Explain the concept of interaction of magnetic fields. Using this concept show that electromagnetic torque cannot be produced, (a) if rotor has 6 poles and stator has 2 poles (b) if rotor has 2 poles and stator has 6 poles.
1.12 Explain the origin of hysteresis and eddy current losses in electric machines. On what factors do these losses depend?
1.13 Discuss the considerations which govern the selection of motor and generator for particular applications.
1.14 Differentiate between continuous and short time ratings of an electric machine.

## PROBLEMS

1.1 A conductor situated at right angles to a magnetic field is carrying a current of 5 A. The flux density is 1.5 T and length of conductor is 12 cm. Find the force on the conductor. [0.9 N]
1.2 A 60-turn coil is pivoted within a magnetic field having a flux density of 1.1 T. The current through the coil is 100 μA, its axial length is 1 cm and its radius is 0.75 cm. Find the torque acting on the coil. [$9.9 \times 10^{-7}$ N-m]
1.3 A 500 turn coil having a radius of 1 cm and a length of 2.5 cm is pivoted between the poles of a magnet at right angles to the magnetic flux. When the coil current is 100 μA, a torque of $0.5 \times 10^{-4}$ N-m acts on the coil. Find flux density. [2 T]

1.4 A straight conductor 100 cm long and carrying a current of 50 A lies perpendicular to a magnetic field of 1 $Wb/m^2$. Find (a) force on the conductor, (b) mechanical power in watts required to move the conductor at a uniform speed of 5 m/sec. [(a) 50 N (b) 250 W]

1.5 A cast iron core (having $\mu_r = 200$) with two coils has a closed magnetic circuit 20 cm long and a cross-sectional area of 5 $cm^2$. One winding has 400 turns and second winding has 250 turns. If the current through the first winding changes from 0 to 1 A in 10 m sec. find the emf induced in the second winding.

[6.28 V]

1.6 A cross bar on the roof rack of a motor car travelling at 60 km/hour is 1.5 m long. Find the emf induced in the bar if the vertical component of earth's magnetic field is $30 \times 10^{-6}$ T. [0.75 mV]

1.7 A conductor 30 cm long on the periphery of an armature of 45 cm diameter rotates at 1000 rpm. If the flux density under the poles is 0.6 T, find the emf induced in the conductor. [4.24 V]

1.8 A square coil of 10 cm side and with 200 turns is rotated at 1000 rpm about an axis at right angles to a uniform magnetic field of flux density 0.5 $Wb/m^2$. Find the instantaneous value of induced emf if the plane of the coil is (a) at right angles to the field, (b) parallel to the field, (c) at 30° to the field.

[(a) 0 (b) 104.8 V (c) 90.6 V]

1.9 A wire of length 50 cm moves at right angles to its length at 40 m per second is a uniform magnetic field of density 1.5 $Wb/m^2$. Find the emf induced in the wire if the motion is (a) inclined at 30° to the direction of field, (b) perpendicular to the field, (c) parallel to field. [(a) 15 V (b) 30 V (c) 0]

1.10 A circuit has 1000 turns enclosing a magnetic circuit 20 $cm^2$ in section. With 4 A current the flux density is 1 $Wb/m^2$ and with 9 A, the flux density is 1.4 $Wb/m^2$. Find the mean value of inductance between these current limits and induced emf if the current fell from 9 A to 4 A in 0.05 seconds. [0.16 H, 16 V]

1.11 A wire of length 65 cm is moved in a field of density 0.8 $Wb/m^2$ at a velocity 35 m/sec. Find the emf induced if the motion is (a) parallel to field, (b) inclined at 45° to the field, (c) perpendicular to the field.

[(a) 0 (b) 12.87 V (c) 18.2 V]

1.12 A square coil of 15 cm side has 150 turns. It is rotated at 800 rpm in a magnetic field of density 0.8 $Wb/m^2$. Find the instantaneous value of induced emf if the plane of the coil is (a) at 90° the field, (b) 45° to the field, (c) parallel to the field. [(a) 0 (b) 159.944 (c) 226.195 V]

1.13 A flat coil has 500 turns and an area of $6 \times 10^{-2}$ $m^2$. It is moving in a field of flux density 80 $mWb/m^2$ at 1600 rpm. The plane of the coil is parallel to the field. Find the peak value of emf induced in the coil.

[402.12 V]

1.14 The flux $\phi$ linked by a coil of 100 turns varies during a period $T$ of one complete cycle as follows:

$0 < t < 0.5T,\ \phi = \phi_m(1 - 4t/T)$

$0.5T < t < T,\ \phi = \phi_m(4t/8 - 3)$

If $T = 0.2$ seconds, plot the flux and emf waves.

1.15 A 75 cms long conductor is carrying a current of 3.6 A and is situated at right angles to a field of flux density 0.95 T. Find the force on the conductor. [2.565 N]

1.16 The coil of a galvanometer is rectangular having a length of 2 cm and radius 1 cm. It has 400 turns, is situated in a field of flux density 0.8T and carries a current of $1.5 \times 10^{-4}$ A. Find the torque on the coil.

[$19.2 \times 10^{-6}$ N-m]

1.17 An air cored solenoid is 1 m long and has a mean diameter of 0.01 m. It has 1000 turns of copper wire 0.5 mm in diameter (a) Find its resistance and inductance, (b) Find the voltage across the coil of the solenoid if it is carrying a dc current of 1 A and increasing at the rate of $10^4$ A/sec.

[(a) 2.77 Ω, 98.7 μH (b) 3.757 V]

1.18 A motor operates continuously on the following duty cycle: 30 kW for 10 sec, 60 kW for 10 sec, 90 kW for 5 sec, 120 kW for 10 secs and idling for 10 sec. Find the size of the motor. [71.41 kW]

Chapter 2

# Transformers

## 2.1 INTRODUCTION

One of the main advantages of ac power transmission and distribution is that the magnitude of ac voltage can be increased or decreased very easily. The general practice is to generate electrical power at about 11 kV (as determined by generator design limitations) and convert it to 132 kV, 220 kV, 400 kV and 765 kV. This conversion of voltage permits transmission of large blocks of power over long distances (from power houses to load centres). The high voltage transmission results in lesser cost of transmission lines and reduced power losses. This conversion of voltage is possible only due to transformers. At distribution points transformers are used to decrease the voltage to a safe level of 400/230 V for use in houses, offices, etc. Without the transformers it would not be possible to use electrical energy in many of the ways it is used now-a-days. Transformers are also widely used in communication circuits, control circuits, radio and television receivers and instrumentation circuits.

## 2.2 PRINCIPLE OF OPERATION

Figure 2.1 (A) shows the general arrangement in a transformer. Its operation is based on the concept of mutual induction. The iron core consists of laminated sheets each about 0.35 mm thick, insulated from each other by thin layers of paper or varnish. The two windings are coupled by magnetic induction. The primary winding is fed by a sinusoidal ac voltage $E_1$. The current in the primary winding sets up an alternating flux in the core. The secondary winding is also linked by this flux and emfs are induced in the two windings. The emf induced in the secondary winding causes a current through the load. Thus energy is transferred from the primary circuit to the secondary circuit through the medium of magnetic field. Figure 2.1(b) shows the transformer symbol.

## 2.3 EMF EQUATION OF TRANSFORMER

Let the flux in the core be written as

$$\phi = \phi_m \sin \omega t \qquad ...(2.1)$$

where $\phi_m$ is the maximum value of flux, $\phi$ is the instantaneous flux and $\omega$ is frequency (radian/sec). The emf induced in a winding having $N$ turns is

$$e = N\frac{d\phi}{dt} = \omega N \phi_m \cos \omega t$$

$$= 2\pi f N \phi_m \sin\left(\omega t + \frac{\pi}{2}\right) \quad ...(2.2)$$

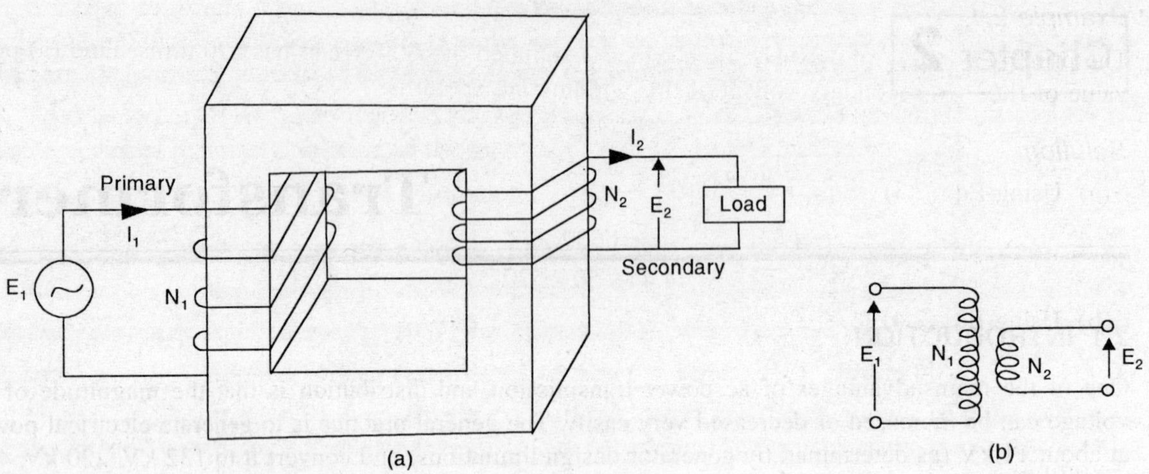


**Fig. 2.1.** (a) Two-winding transformer, (b) transformer symbol

Peak value of induced emf $= 2\pi f N \phi_m$

RMS value of induced emf $= \frac{1}{\sqrt{2}}(2\pi f N \phi_m) = 4.44 f N \phi_m$ volts ...(2.3a)

Therefore

RMS value of induced emf in primary winding $= E_1 = 4.44 f N_1 \phi_m$ volts ...(2.3b)

RMS value of induced emf in secondary windng $= E_2 = 4.44 f N_2 \phi_m$ volts ...(2.3c)

Therefore

$$\frac{E_1}{E_2} = \frac{N_1}{N_2} \quad ...(2.4)$$

The ratio of primary to secondary emf is the same as the ratio of primary and secondary winding turns. This ratio, $N_1/N_2$, is known as transformation ratio. By selecting the values of $N_1$ and $N_2$ properly, we can transform the power from one voltage level to another. If $N_1 > N_2$, the secondary voltage will be less than primary voltage and such a transformer is known as step-down transformer. If $N_1 < N_2$, the secondary voltage will be more than primary voltage and the device is known as step-up transformer.

A transformer may have more than two windings. If it has three windings, it is known as three-winding transformer, the windings being known as primary, secondary and tertiary. If $E_1$, $E_2$, $E_3$ are the induced emfs and $N_1$, $N_2$, $N_3$ are the number of turns in the three windings.

$$E_1 : E_2 : E_3 :: N_1 : N_2 : N_3 \quad ...(2.5)$$

If the losses in the transformer are neglected

$$E_1 I_1 = E_2 I_2$$

or

$$\frac{E_1}{E_2} = \frac{I_2}{I_1} = \frac{N_1}{N_2} \quad ...(2.6)$$

where $I_1$ and $I_2$ are the currents in the primary and secondary windings.

Thus the current, in a transformer, is transformed in the reverse ratio of voltage. If voltage is increased, current is decreased. If voltage is decreased, current is increased.

*Example 2.1*

The primary winding of a 50 Hz transformer is supplied of 6600 V and has 520 turns. Find (a) peak value of flux, (b) secondary voltage if this winding has 25 turns.

*Solution*

(a) Using Eq. (2.3)

$$\phi_m = \frac{6600}{4.44 \times 50 \times 520} = 0.0572 \text{ Wb}$$

(b) Using Eq. (2.4)

$$E_2 = E_1 \times \frac{N_2}{N_1} = 6600 \times \frac{25}{520} = 317.3 \text{ V}.$$

*Example 2.2*

A transformer has three windings: a 230 V primary, a 110 V secondary and a centre tapped 6–0–6 V tertiary. The transformer has a square core and the net cross-sectional area is 12 sq. cm. The flux density should not exceed 0.8 T. Find the number of turns in the windings.

*Solution*

$$\phi_m = 0.8 \times 12 \times 10^{-4} = 9.6 \times 10^{-4}$$

For each half of tertiary winding

$$6 = 4.44 \times 50 \times N_3 \times 9.6 \times 10^{-4}$$

or

$$N_3 = 28$$

Total number of turns in tertiary = 2 × 28

$$\frac{230}{6} = \frac{N_1}{28} \text{ or } N_1 = 1073$$

$$\frac{110}{6} = \frac{N_2}{28} \text{ or } N_2 = 513.$$

*Example 2.3*

The primary winding of a 230 V, 50 Hz transformer has 60 turns. If the flux density is 0.8 T, find active cross-sectional area of the core. Give suitable dimensions of a square core.

*Solution*

$$\phi_m = \frac{E_1}{4.44 \, fN_1} = \frac{230}{4.44 \times 50 \times 60} = 0.0173 \text{ Wb}$$

$$\text{Active cross-sectional area} = \frac{0.0173}{0.8} = 0.0216 \text{ m}^2 = 216 \text{ cm}^2$$

The insulation over the laminations occupies some space. Therefore the active cross-sectional area is less than gross-sectional area. If the ratio of active area to gross area is 0.9, then

$$\text{gross cross-sectional area} = \frac{216}{0.9} = 240 \text{ cm}^2$$

The core size is $\sqrt{240}$ or 15.5 cm square.

## 2.4 TYPES AND CONSTRUCTION OF TRANSFORMERS

Transformers used in power systems are given a variety of names. A transformer used in a big sub-station is known as sub-station or power transformer. The one used to supply loads to consumers, at distribution voltage level, is known as distribution transformer. Special transformers used for testing high voltage equipments are known as testing transformers.

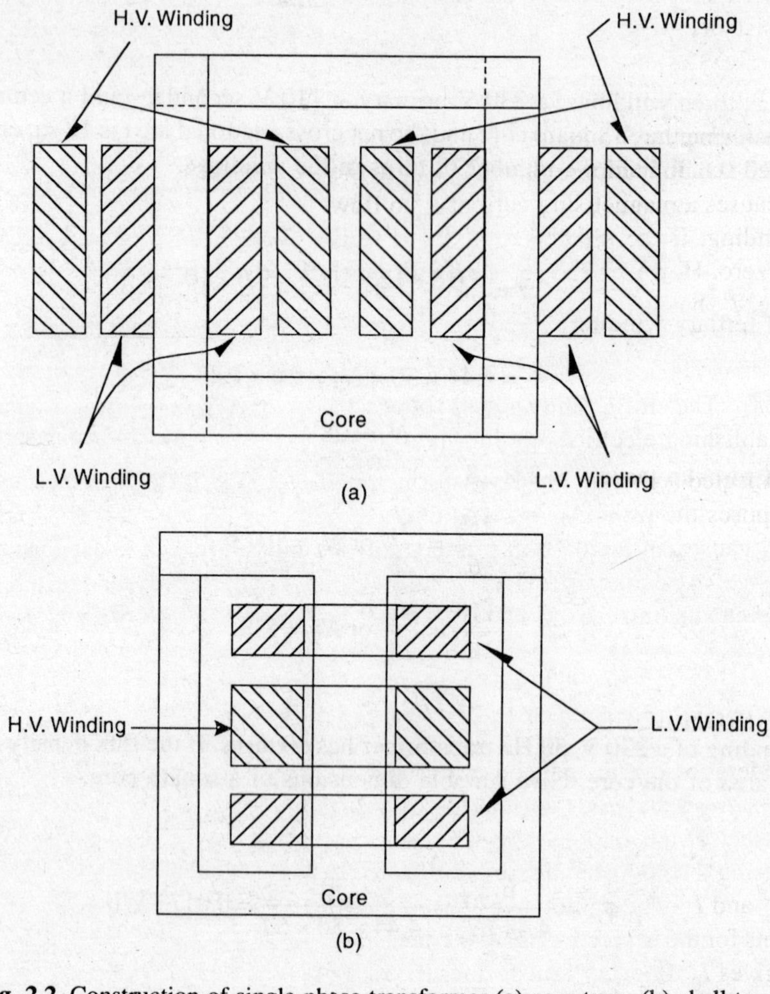


**Fig. 2.2.** Construction of single-phase transformer (a) core type, (b) shell type

A transformer has an iron core and primary and secondary windings. One type of construction has a rectangular core and the windings wrapped around the two sides of the reactangle. The primary and secondary windings are divided into two parts each and wrapped one on the top of the other (Fig. 2.2 a). The low-voltage winding is placed next to the core and the high-voltage winding is on the outside. This construction ensures close magnetic coupling and also simplifies the problem of insulating the high voltage winding from the core. This construction (Fig. 2.2 a) is known as core type transformer and is used in high voltage transformers.

A second type of construction uses a three-legged core and 'sandwich' windings placed on central leg (Fig. 2.2 b). It is seen that in this case the core surrounds the winding (in core type the windings surround the core). This construction (Fig. 2.2 b) is known as shell-type transformer and is more suitable for low voltage transformers.

## 2.5 THE IDEAL TRANSFORMER

The ideal transformer is a hypothetical transformer having the following features: (i) it has no iron losses, (iii) the resistance of its windings is zero, (iii) the magnetic coupling is ideal and there is no magnetic leakage, (iv) the magnetic circuit has infinite permeability and zero reluctance, so that a vanishingly small mmf is sufficient to set up flux in the core.

Figure 2.3 shows the phasor diagram of an ideal transformer on no-load. The application of a voltage $V_1$ to the primary winding causes a magnetising current $I_m$ to flow in the primary winding. If the reluctance of the core is zero, $I_m$ would be zero. However, the reluctance is never zero and, therefore, $I_m$ has a finite value. $I_m$ lags $V_1$ by 90°. The flux $\phi_m$ induces emfs $E_1$ and $E_2$ in the primary and secondary windings. In accordance with Eq. (2.2), $E_1$ and $E_2$ lead $\phi_m$ by 90°. The emf $E_1$ and applied voltage $V_1$ are equal, thus establishing electrical equilibrium.

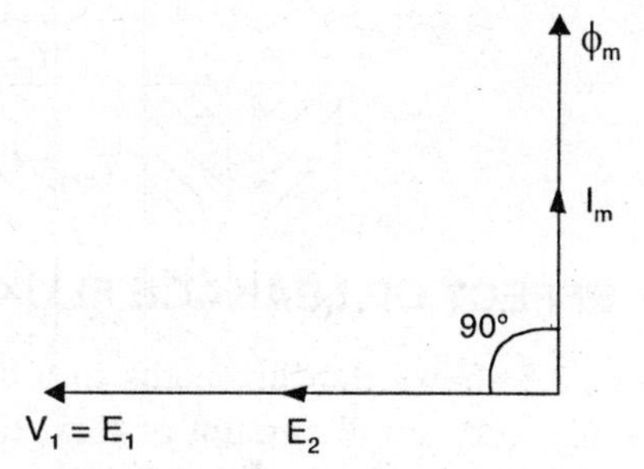


**Fig. 2.3.** Phasor diagram of an ideal transformer under no-load

If a load is connected across an ideal transformer, a load current $I_2$ flows in the secondary winding. The mmf $N_2 I_2$ opposes the primary mmf and tends to reduce $\phi_m$. This reduces $E_1$ and the difference between $V_1$ and $E_1$ causes an increase in primary current and the flux is re-established at its original value. Thus a change in secondary current causes an automatic corresponding change in primary current. If the magnetising current $I_m$ is neglected, the primary mmf $I_1N_1$ is equal and opposite to the secondary mmf $I_2N_2$.

## 2.6 EFFECT OF IRON-LOSSES

When the core undergoes cyclic magnetisation (due to alternating flux) hysteresis and eddy current losses occur in the core. The effect of these losses can be taken into account by considering the no load current $I_0$ to consist of two components $I_m$ and $I_w$. The component $I_w$ is in phase with $E_1$ and accounts for the iron (core) losses. The phasor sum of $I_m$ and $I_w$ gives $I_0$. In every transformer the magnitude of $I_w$ is small as compared to that of $I_m$, so that angle $\theta_0$ (Fig. 2.4) is about 70°.

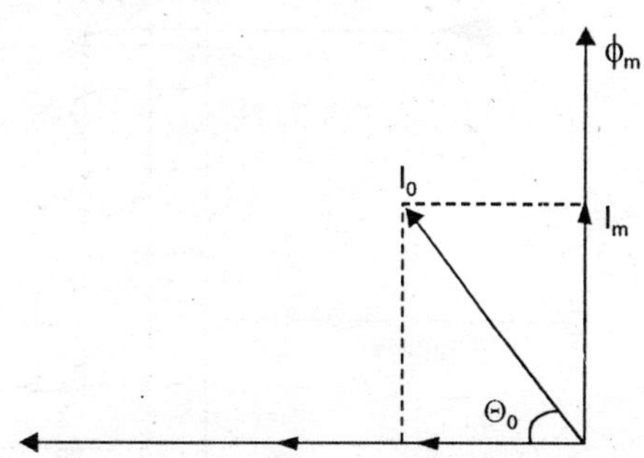


**Fig. 2.4.** Effect of core losses on phasor diagram of a transformer under no load

$$I_0 = \left(I_m^2 + I_w^2\right)^{0.5}$$

*Example 2.4*

A 230/110 V single-phase transformer has a core-loss of 100 W. If the input under no load conditions is 400 V A, find $I_w$, $I_m$ and no load power factor angle $\theta_0$.

*Solution*

$$V_1 I_0 = 400$$

$$V_1 I_0 \cos\theta_0 = 100$$

$$\cos\theta_0 = \frac{100}{400} = 0.25$$

$$\theta_0 = 75.5°$$

$$I_0 = \frac{400}{230} = 1.739 \text{ A}$$

$$I_w = I_0 \cos\theta_0 = 0.435 \text{ A}$$

$$I_m = \left(I_0^2 - I_w^2\right)^{0.5} = 1.68 \text{ A.}$$

## 2.7 EFFECT OF LEAKAGE FLUX

Figure 2.5 shows the flux paths in a transformer. The mutual flux links both the windings. However there is a very small amount of flux, through the air, which links only one winding. Both the primary and secondary windings have leakage fluxes. The path of these leakage fluxes is mostly in air. Since reluctance of air path is constant, the voltages induced by these leakage fluxes vary linearly with currents, i.e. the primary leakage flux varies linearly with primary current and the secondary leakage flux varies linearly with secondary current.

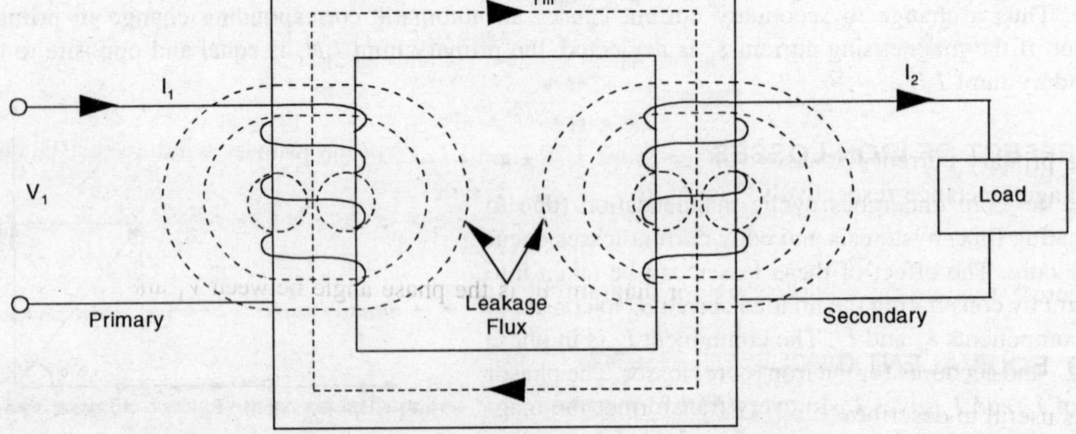


**Fig. 2.5.** Mutual and leakage flux in transformer

The effect of primary leakage flux can be taken into account by a leakage inductance. The magnitude of this inductance is equal to the primary leakage flux per ampere of primary current. The reactance corresponding to this leakage inductance (i.e. $2\pi f \times$ primary leakage inductance) is known as primary leakage reactance $X_1$. Likewise, the effect of secondary leakage flux can be taken into account by a secondary leakage inductance and corresponding secondary leakage reactance $X_2$.

## 2.8 PHASOR DIAGRAM OF TRANSFORMER

An actual transformer has no load current $I_0$ (with components $I_m$ and $I_w$), leakage reactances $X_1$, $X_2$ resistance of primary winding $R_1$ and resistance of secondary winding $R_2$.

The load current $I_2$ lags the secondary terminal voltage $V_2$ by an angle $\theta_2$ determined by the power factor of the load. This current $I_2$ causes voltage drops $I_2 R_2$ and $j I_2 X_2$ in the secondary winding. Thus

$$V_2 = E_2 - I_2 (R_2 + j X_2) \qquad ...(2.7)$$

The load current $I_2$ tends to reduce the mutual flux. This causes an increase in primary current, so that the resultant of the primary and secondary mmfs maintains the mutual flux at a constant value. The primary current is phasor sum of $I_2'$ (to balance the demagnetising effect of $I_2$) and $I_0$.

$$I_2' = \frac{I_2 N_2}{N_1} \qquad ...(2.8)$$

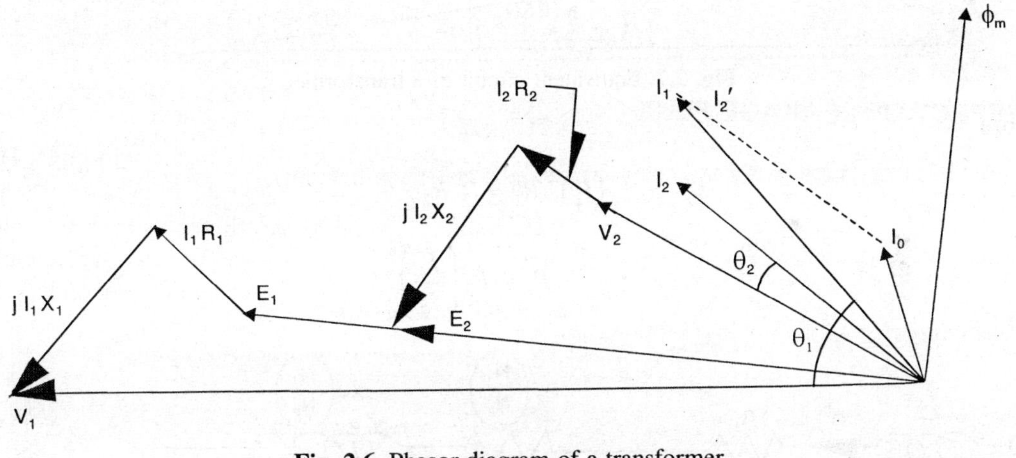


**Fig. 2.6.** Phasor diagram of a transformer

$$I_1 = I_2' + I_0 \qquad ...(2.9)$$

The primary current $I_1$ causes voltage drops $I_1 R_1$ and j $I_1 X_1$ in the primary winding resistance and leakage reactance respectively. Therefore

$$V_1 = E_1 + I_1 (R_1 + j X_1) \qquad ...(2.10)$$

Figure 2.6 shows the complete phasor diagram. $\theta_1$ is the phase angle between $V_1$ and $I_1$.

## 2.9 EQUIVALENT CIRCUIT OF TRANSFORMER

It is useful to describe the behaviour of an electrical equipment in terms of an equivalent circuit. Using the equivalent circuit, the analysis of the equipment can be done by the methods of circuit theory.

The equivalent circtui of a transformer is drawn by assuming the transformer to be ideal, as shown in Fig. 2.7. Resistance $R_0$ represents the effect of core losses. The reactance $X_0$ accounts for magnetising current $I_m$. Winding resistances and leakage reactances are shown separately. It is possible to transfer $R_2$ and $X_2$ to the primary side in such a way that the ratio $E_1/E_2$ is not affected in magnitude or phase. Let $R_2'$ be the resistance added in primary circuit to cause the same copper-loss as caused by resistance $R_2$ in the secondary circuit. Since the current $I_2$ when transferred to primary appears as $I_2'$, this additional resistance causes a copper-loss $(I_2')\,R_2'$.

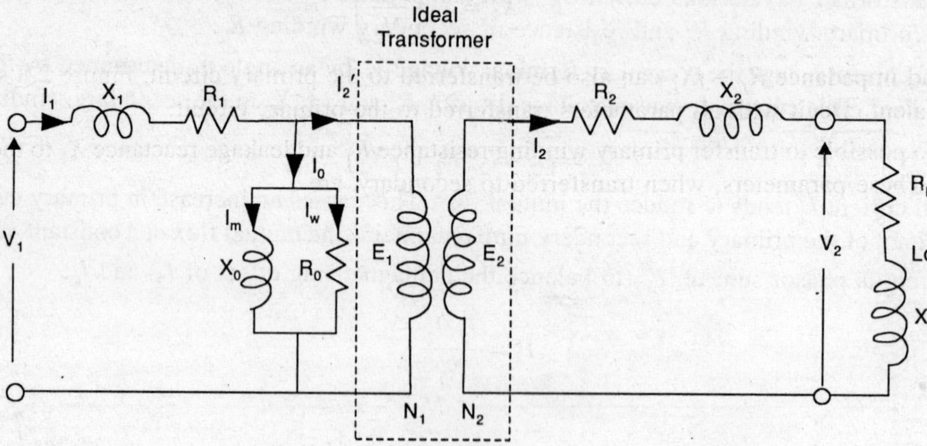


**Fig. 2.7.** Equivalent circuit of a transformer

Therefore

$$(I_2')^2\, R_2' = I_2^2\, R_2$$

or

$$R_2' = R_2\left(\frac{I_2}{I_2'}\right)^2$$

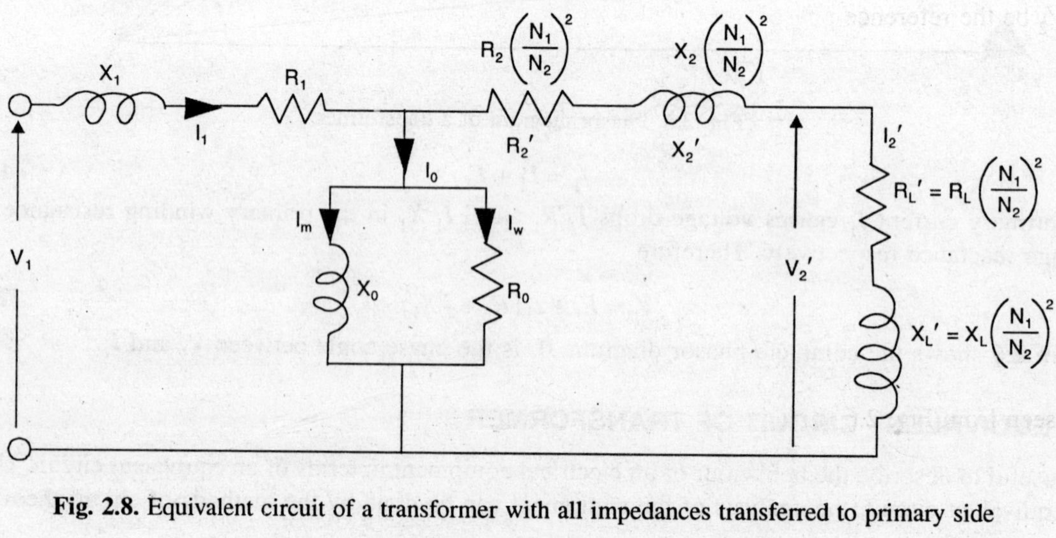


**Fig. 2.8.** Equivalent circuit of a transformer with all impedances transferred to primary side

But $$\frac{I_2}{I_2'} = \frac{N_1}{N_2}$$

Therefore $$R_2' = R_2\left(\frac{N_1}{N_2}\right)^2 \quad ...(2.11)$$

Similarly $$X_2' = X_2\left(\frac{N_1}{N_2}\right)^2 \quad ...(2.12)$$

The load impedance $R_L + jX_L$ can also be transferred to the primary circuit. Figure 2.8 shows the exact equivalent circuit with all parameters transferred to the primary circuit.

It is also possible to transfer primary winding resistance $R_1$ and leakage reactance $X_1$ to the secondary circuit. These parameters, when transferred to secondary, are

$$R_1' = R_1\left(\frac{N_2}{N_1}\right)^2 \quad ...(2.13)$$

$$X_1' = X_1\left(\frac{N_2}{N_1}\right)^2 \quad ...(2.14)$$

where $R_1'$ and $X_1'$ are the primary winding resistance and leakage reactance as transferred to the secondary circuit.

### *Example 2.5*

A step-up transformer supplies a load current of 5A at 200 V and at 0.8 pf lagging. $R_1 = 0.5$ ohm, $X_1 = 1$ ohm, $R_2 = 2$ ohm, $X_2 = 4$ ohm. The ratio $N_2/N_1$ is 2. If $I_w = 0.3$ A, $I_m = 0.5$ A, find $V_1$, $I_1$ and power factor at primary terminals.

### *Solution*

Let $V_2$ be the reference phasor

$$V_2 = 200\ \angle 0\ \text{V}$$

$$\cos\theta_2 = 0.8,\ \theta_2 = 36.87°$$

$$I_2 = 5\angle -36.87°\ \text{A}$$

$$E_2 = V_2 + I_2(R_2 + jX_2)$$

$$= 200\ \angle 0 + (5\angle -36.87°)(2 + j4)$$

$$= 220.2\ \angle 2.6°\ \text{V}$$

$$I_2' = I_2(N_2/N_1) = 10\ \angle -36.87°\ \text{A}$$

It is seen from Fig. 2.6 that $I_w$ is in phase with $E_2$, while $I_m$ lags $E_2$ by 90°.

$$I_w = 0.3\ \angle 2.6°\ \text{A}$$

$$I_m = 0.5(\angle 2.6 - 90°) = 0.5\angle -87.4°\ \text{A}$$

$$\begin{aligned} I_1 &= I_2' + I_w + I_m \\ &= 10\angle -36.87° + 0.3\ \angle 2.6° + 0.5\angle -87.4° \\ &= 8.322 - j6.48 = 10.55\angle -37.9°\text{ A} \\ E_1 &= E_2 (N_1/N_2) = 110.1\ \angle 2.6°\text{ V} \\ V_1 &= E_1 + I_1 (R_1 + jX_1) \\ &= 110.1\ \angle 2.6° + (10.55\angle -37.9°)(0.5 + j1) \\ &= 110.1\ \angle 2.6° + (10.55\angle -37.9°)(1.12\angle 63.4°) \\ &= 120.65 + j10.1 = 121.07\angle 4.8°\text{ V} \\ \theta_1 &= 4.8° + 37.9° = 42.7° \end{aligned}$$

Primary power factor = $\cos\theta_1 = 0.735$ lagging.

## 2.10 APPROXIMATE EQUIVALENT CIRCUIT

The transformer model represented by the equivalent circuit of Fig. 2.7 is rather complex. Since the magnitude of $I_0$ is very small, the equivalent circuit can be simplified by shifting the parallel branch of $R_0$ and $X_0$ to the extreme left position as shown in Fig. 2.9. This approximation leads to simplicity of analysis without any loss of accuracy. Sometimes the parallel branch consisting of $R_0$ and $X_0$ is completely neglected in the analysis.

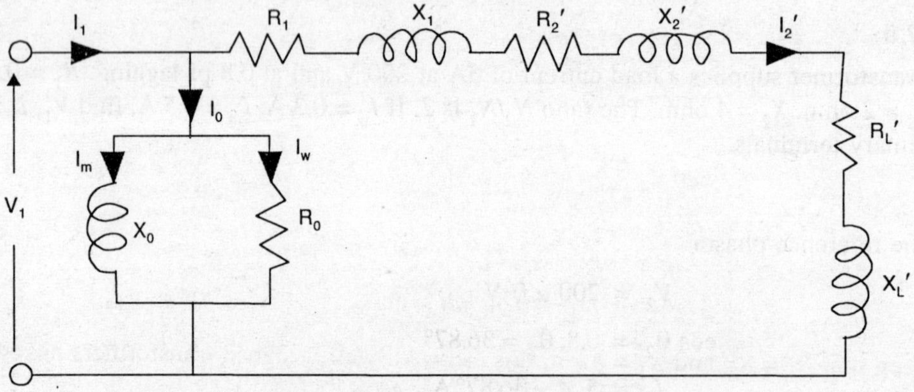


**Fig. 2.9.** Approximate equivalent circuit of a transformer

## 2.11 EQUIVALENT RESISTANCE AND REACTANCE

In the approximate equivalent circuit of Fig. 2.9, the resistance $R_1$ and $R_2'$ can be combined into a single resistance $R_{e1}$, i.e. equivalent resistance as referred to winding 1 (or primary winding). Similarly, $X_1$ and $X'_2$ can be combined into a single reactance $X_{ei}$, i.e. equivalent reactance as referred to winding 1. Thus

$$R_{e1} = R_1 + R_2' = R_1 + R_2 (N_1/N_2)^2 \quad ...(2.15)$$

$$X_{e1} = X_1 + X_2' = X_1 + X_2 (N_1/N_2)^2 \quad ...(2.16)$$

Likewise we can also find the equivalent resistance and reactance as referred to secondary winding (i.e. winding 2). Thus

$$R_{e2} = R_2 + R_1' = R_2 + R_1 (N_2/N_1)^2 \quad ...(2.17)$$

$$X_{e2} = X_2 + X_1' = X_2 + X_1 (N_2/N_1)^2 \quad ...(2.18)$$

*Example 2.6*

A 63 kVA, 1100/220 V single-phase transformer has $R_1 = 0.16$ ohm, $X_1 = 0.5$ ohm, $R_2 = 0.0064$ ohm and $X_2 = 0.02$ ohm. Find (a) equivalent resistance and reactance as referred to primary winding, (b) equivalent resistance and reactance as referred to secondary winding.

*Solution*

(a)

$$\frac{N_1}{N_2} = \frac{1100}{220} = 5$$

$$R_2' = (0.0064)(5)^2 = 0.16 \text{ ohm}$$

$$X_2' = (0.02)(5)^2 = 0.5 \text{ ohm}$$

$$R_{e1} = R_1 + R_2' = 0.16 + 0.16 = 0.32 \text{ ohm}$$

$$X_{e1} = X_1 + X_2' = 0.5 + 0.5 = 1 \text{ ohm}$$

(b)

$$R_1' = 0.16\left(\frac{N_2}{N_1}\right)^2 = \frac{0.16}{25} = 0.0064 \text{ ohm}$$

$$X_1' = \frac{0.5}{25} = 0.02 \text{ ohm}$$

$$R_{e2} = 0.0064 + 0.0064 = 0.0128 \text{ ohm}$$

$$X_{e2} = 0.02 + 0.02 = 0.04 \text{ ohm.}$$

It is seen that $R_1 = R_2'$ and $X_1 = X_2'$ in this case. All well-designed transformers have this feature.

*Example 2.7*

Solve Example 2.5 using approximate equivalent circuit.

*Solution*

$$V_2 = 200\angle 0 \text{ V}$$

$$I_2 = 5\angle -36.87° \text{ A}$$

$$V_2' = V_2 (N_1/N_2) = 100\angle 0 \text{ V}$$

$$I_2' = I_2 (N_1/N_2) = 10\angle -36.87° \text{ A}$$

$$V_1 = V_2' + I_2'[R_1 + R_2' + j(X_1 + X_2')]$$

$$R_2' = R_2(N_1/N_2)^2 = 2(0.5)^2 = 0.5 \text{ ohm}$$

$$X_2' = 4(0.5)^2 = 1 \text{ ohm}$$

$$V_1 = 100\angle 0 + (10\angle -36.87°)[0.5 + 0.5 + j(1+1)]$$

$$= 120.4\angle 4.8° \text{ V}$$

$$I_w = 0.3\angle 4.8° \text{ V}$$

$$I_m = 0.5\angle(4.8 - 90°) = 0.5\angle -85.2° \text{ A}$$

$$I_1 = I_2' + I_w + I_m$$

$$= 10\angle -36.87° + 0.3\angle 4.8° + 0.5\angle -85.2°$$

$$= 8.34 - j6.47 = 10.55\angle -37.8° \text{ A}$$

Primary power factor = cos (4.8° + 37.8°) = 0.736 lagging.

It is seen that solution of Example 2.7 is much easier as compared to that of Example 2.5 and the results are nearly the same.

*Example 2.8*

A 230/2300 V single-phase transformer is excited by a 230 V ac supply. The equivalent resistance and reactanec as referred to primary winding are 0.2 and 0.8 ohm, respectively. $R_0$ = 500 ohm and $X_0$ = 200 ohm. The load impedance is 400 + *j* 600 ohm. Find (a) primary current, (b) secondary terminal voltage, (c) power factor at primary terminals.

*Solution*

With reference to Fig. 2.9

$$R_L' = 400/100 = 4 \text{ ohm}$$

$$X_L' = 600/100 = 6 \text{ ohm}$$

(a) $$I_2' = \frac{V_1}{(R_{e1} + R_L') + j(X_{e1} + X_L')}$$

$$= \frac{230\angle 0}{(0.2 + 4) + j(0.8 + 6)} = 28.78\angle -58.3° \text{ A}$$

$$I_w = \frac{230\angle 0}{500} = 0.46\angle 0 \text{ A}$$

$$I_m = \frac{230\angle 0}{j200} = 1.15\angle -90° \text{ A}$$

$$I_1 = I_2' + I_w + I_m$$

$$= 28.78\angle -58.3° + 0.46 + 1.15\angle -90°$$

$$= 15.58 - j25.64 = 30\angle -58.7° \text{ A}$$

(b) $V_2' = I_2'(R_L' + jX_L')$

$$= (28.78\angle -58.3°)(4 + j6)$$

$$= 207.54\angle -2°$$

$$V_2 = V_2'(N_2/N_1) = 207.54 \times 10 = 2075.4 \text{ V}$$

(c) Power factor at primary terminals = cos (58.7°) = 0.52 lagging.

## 2.12 TRANSFORMER AS A COUPLED CIRCUIT

The transformer is a device with magnetic coupling. Its equivalent circuit can be derived by treating it as a coupled circuit.

The current $i_1$ flowing in the primary winding produces a flux $\phi_1$. This total flux is the sum of primary mutual flux $\phi_{12}$ and the primary leakage flux $\phi_{l1}$. Thus

$$\phi_1 = \phi_{12} + \phi_{l1} \qquad ...(2.19)$$

where $\phi_1$ is the total flux produced by $i_1$. The flux $\phi_{12}$ links both the primary and secondary windings, whereas the primary leakage flux $\phi_{l1}$ links only the primary turns as shown in Fig. 2.10.

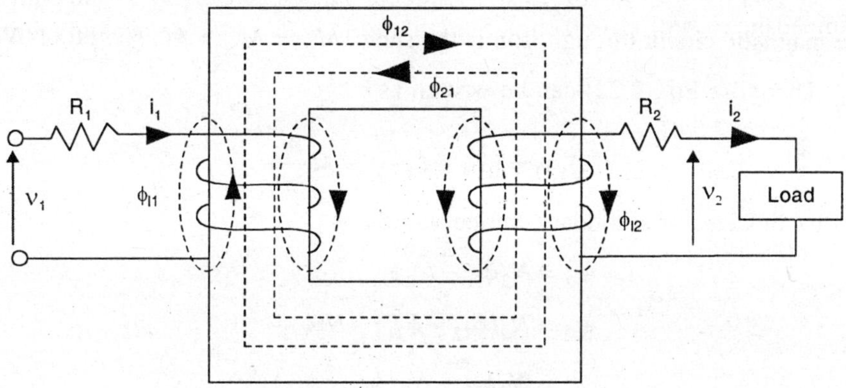


**Fig. 2.10.** Primary and secondary fluxes in a transformer

The current $i_2$ flowing in the secondary winding produces a flux $\phi_2$ which can be written as

$$\phi_2 = \phi_{21} + \phi_{l2} \qquad ...(2.20)$$

where $\phi_{21}$ is the mutual flux (produced by $i_2$) and $\phi_{l2}$ is the secondary leakage flux. The flux $\phi_{21}$ links both the primary and secondary and opposes flux $\phi_{12}$ (as per Lenz's law). The difference $\phi_{12} - \phi_{21}$ is the net mutual flux $\phi_m$, i.e.

$$\phi_m = \phi_{12} - \phi_{21} \qquad ...(2.21)$$

Total flux linkage of primary $\psi_1$ can be written as

$$\begin{aligned}\psi_1 &= N_1\phi_m + N_1\phi_{l1} \\ &= N_1(\phi_{12} - \phi_{21}) + N_1\phi_{l1} \\ &= N_1\phi_1 - N_1\phi_{21}\end{aligned}$$

KVL equation for the primary circuit (Fig. 2.10) is

$$v_1 = i_1R_1 + \frac{d\psi_1}{dt}$$

or

$$v_1 = i_1R_1 + \frac{d}{dt}(N_1\phi_1 - N_1\phi_{21}) \qquad \text{...(2.22)}$$

The self-inductance of a winding is defined as the flux linkage per ampere of current. Therefore $L_1$, the self-inductance of primary winding is

$$L_1 = N_1\frac{d\phi_1}{di_1} = N_1\frac{\phi_1}{i_1} \qquad \text{...(2.23)}$$

$M_{12}$, the mutual inductance between primary and secondary windings is

$$M_{12} = N_1\frac{d\phi_{21}}{di_2} = N_1\frac{\phi_{21}}{i_2} \qquad \text{...(2.24)}$$

Also

$$M_{21} = N_2\frac{d\phi_{12}}{di_1} = N_2\frac{\phi_{12}}{i_1} \qquad \text{...(2.25)}$$

Since the magnetic circuit through iron is the same, $M_{12} = M_{21} = M$. Therefore $N_1\phi_{21} = M i_2$ and $N_2\phi_{12} = M i_1$. Therefore Eq. (2.22) can be written as

$$v_1 = i_1R_1 + L_1\frac{di_1}{dt} - M\frac{di_2}{dt} \qquad \text{...(2.26)}$$

$\psi_2$, i.e. total flux linkages of secondary can be written as

$$\psi_2 = N_2\phi_m - N_2\phi_{l2} \qquad \text{...(2.27)}$$

or

$$\begin{aligned}\psi_2 &= N_2(\phi_{12} - \phi_{21}) - N_2\phi_{l2} \\ &= N_2\phi_{12} - N_2(\phi_{21} + \phi_{l2})\end{aligned}$$

or

$$\psi_2 = N_2\phi_{12} - N_2\phi_2 \qquad \text{...(2.28)}$$

The minus sign in Eq. (2.27) is due to the reason that $\phi_{12}$ is in a direction opposite to that of $\phi_m$. KVL equation for the secondary circuit is

$$\frac{d\psi_2}{dt} = i_2R_2 + v_2$$

or

$$\frac{d}{dt}(N_2\phi_{12} - N_2\phi_2) = i_2R_2 + v_2 \qquad \text{...(2.29)}$$

Since $N_2\phi_{12} = M i_1$ and $N_2\phi_2 = L_2 i_2$ Eq. (2.29) can be written as

$$M\frac{di_1}{dt} = L_2\frac{di_2}{dt} + i_2 R_2 + v_2 \qquad ...(2.30)$$

A magnetically coupled circuit which satisfies Eqs. (2.26) and (2.30) is shown in Fig. 2.11.

Equation (2.30) is for the secondary circuit. To refer it to primary circuit, mutiply it by the turn ratio $a(= N_1/N_2)$. Thus Eq. (2.30) can be written as

$$aM\frac{di_1}{dt} = a^2 L_2\frac{d}{dt}\left(\frac{i_2}{a}\right) + a^2 R_2\left(\frac{i_2}{a}\right) + av_2 \qquad ...(2.31)$$

$i_2/a$ is the secondary current as referred to primary side and is the same as $I_2'$ in Fig. 2.8. Equation (2.31) can be written as

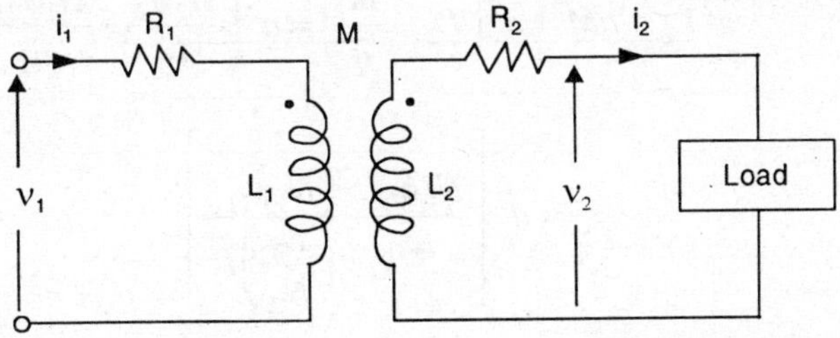


**Fig. 2.11.** Transformer as a coupled circuit

$$aM\frac{d}{dt}\left(i_1 - \frac{i_2}{a}\right) = \left(a^2 L_2 - aM\right)\frac{d}{dt}\left(\frac{i_2}{a}\right) + a^2 R_2\left(\frac{i_2}{a}\right) + av_2 \qquad ...(2.32)$$

Equation (2.26) can be re-written as

$$v_1 = i_1 R_1 + (L_1 - aM)\frac{di_1}{dt} + aM\frac{d}{dt}\left(i_1 - \frac{i_2}{a}\right) \qquad ...(2.33)$$

An equivalent circuit which satisfies Eqs., (2.32) and (2.33) is shown in Fig. 2.12.

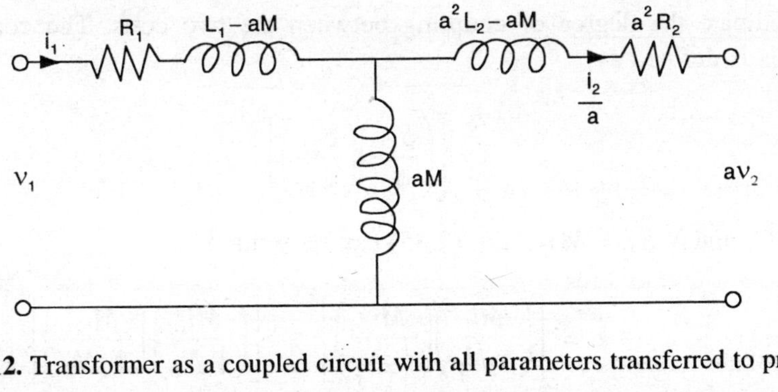


**Fig. 2.12.** Transformer as a coupled circuit with all parameters transferred to primary

The similarity of this circuit with that shown in Fig. 2.8 is apparent. The quantity $aM$ can be written as

$$aM = \left(\frac{N_1}{N_2}\right)\left(\frac{N_2\phi_{12}}{i_1}\right) = \frac{N_1\phi_{12}}{i_1} \qquad \text{...(2.34)}$$

Thus $a\,M$ represents mutual flux linkage of primary per ampere of primary current and is magnetising inductance $L_0$. The corresponding reactance (i.e. $\omega\,L_0$) is $X_0$ as shown in Fig. 2.8. The quantities $L_1 - a\,M$ and $a^2\;L_2 - a\,M$ can be written as

$$L_1 - aM = \frac{N_1\phi_1}{i_1} - \frac{N_1\phi_{12}}{i_1} = \frac{N_1(\phi_1 - \phi_{12})}{i_1}$$

or

$$L_1 - aM = \frac{N_1\phi_{l1}}{i_1} = \text{Primary leakage inductance}$$

$$a^2L_2 - aM = a^2\left(L_2 - \frac{M}{a}\right) = a^2\left(\frac{N_2\phi_2}{i_2} - \frac{N_1\phi_{21}}{ai_2}\right)$$

$$= a^2\left[\frac{N_2\phi_2}{i_2} - \frac{N_1\phi_{21}}{\left(\frac{N_1}{N_2}\right)i_2}\right]$$

$$= a^2\left[\frac{N_2(\phi_2 - \phi_{21})}{i_2}\right] = a^2\,\frac{N_2\phi_{l2}}{i_2}$$

$$= a^2 \text{ (secondary leakage inductance)}$$

The current $i_1 - i_2/a$ is the magnetising current $i_m$. When $i_w$ (i.e. the current representing core-loss) is added to $i_m$, we get no load current $i_0$. Thus it is seen that circuits shown in Figs. 2.8 and 2.12 are identical.

## 2.13 METHODS OF REDUCING LEAKAGE FLUX

The fractional part of $\phi_1$ which links with $N_2$ i.e. $\phi_{12}/\phi_1$ and the fractional part of $\phi_2$ which links with $N_1$, i.e. $\phi_{21}/\phi_2$ indicate the degree of coupling between the two coils. The coefficient of coupling between two coils is defined as

$$k = \left[\left(\frac{\phi_{12}}{\phi_1}\right)\left(\frac{\phi_{21}}{\phi_2}\right)\right]^{0.5} \qquad \text{...(2.35)}$$

Since $N_2\phi_{12} = M\,i_1$ and $N_1\phi_{21} = M\,i_2$, Eq. (2.35) can be written as

$$k = \left[\left(\frac{Mi_1}{\phi_1 N_2}\right)\left(\frac{Mi_2}{\phi_2 N_1}\right)\right]^{0.5} = \left[\left(\frac{Mi_1}{\phi_1 N_1}\right)\left(\frac{Mi_2}{\phi_2 N_2}\right)\right]^{0.5}$$

$$= \frac{M}{\sqrt{L_1 L_2}}$$

or $$M = k\sqrt{L_1 L_2} \qquad ...(2.36)$$

The co-efficient of coupling $k$ may range from about 0.01 (in radio coils) to about 0.98 (in iron cored transformers). This coefficient $k$ can be increased (and leakage flux reduced) by the following methods:

(a) Making the transformer window long and narrow. For the same window area, a greater height of the window means a longer leakage flux path and thus an increase in reluctance offerred to leakage flux. This results is reduction of leakage flux (for the same mmf).

(b) Arranging the primary and secondary windings concentrically. As shown in Fig. 2.2 (a), each limb of the core carries one-half of the primary and one-half of secondary. This ensures a close coupling between the two windings.

(c) Sub-dividing the primary and secondary windings. As shown in Fig. 2.13, the low voltage winding is in four sections. On each limb, the high voltage winding is sandwiched by 2 sections of low voltage windings, thus reducing the leakage flux.

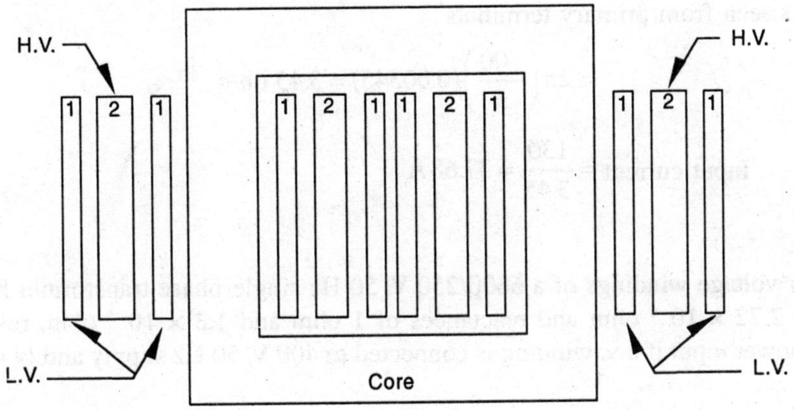


**Fig. 2.13.** Sub-division of low voltage and high voltage windings in a single-phase transformer

(d) Using shell type construction. In shell type construction (Fig. 2.2 b) the low voltage and high voltage windings are sandwiched, thus ensuring close coupling.

*Example 2.9*

The self- and mutual inductances of a two-winding transformer are: $L_1 = 4$ mH, $L_2 = 1.2$ mH, M = 1.8 mH. (a) Find the current which would flow in winding 1, when it is connected to a 130 V, (500/π) Hz source and a load of 0.2 H inductance is connected across winding 2. Neglect power losses. Ratio of turns is 2.

*Solution*

Referring to Fig. 2.12

$$L_1 - aM = 10^{-3}(4 - 2 \times 1.8) = 0.4 \times 10^{-3}\text{ H}$$

$$aM = 2 \times 1.8\text{ mH} = 3.6 \times 10^{-3}\text{ H}$$

$$a^2L_2 - aM = (4 \times 1.2 - 2 \times 1.8)10^{-3} = 1.2 \times 10^{-3}\ \text{H}$$

Load inductance $L$ as transferred to winding 1 is

$$a^2L = 4 \times 0.02 = 0.08\,\text{H}$$

$R_1$, $R_2$, and $R$ are zero.

Total inductance as seen from primary terminals

$$= (L_1 - aM) + \frac{(aM)\left(a^2L_2 - aM + a^2L\right)}{aM + a^2L_2 - aM + a^2L}$$

$$= 0.4 \times 10^{-3} + \frac{\left(3.6 \times 10^{-3}\right)\left(1.2 \times 10^{-3} + 0.08\right)}{3.6 \times 10^{-3} + 1.2 \times 10^{-3} + 0.08}$$

$$= 0.00345\,\text{H}$$

Total reactance as seen from primary terminals

$$= 2\pi\left(\frac{500}{\pi}\right)(0.00345) = 3.45\ \text{ohm}$$

$$\text{Input current} = \frac{130}{3.45} = 37.68\,\text{A}.$$

*Example 2.10*

**The high and low voltage windings of a 6600/250 V, 50 Hz single-phase transformer have resistances of 0.21 ohm and 2.72 × 10⁻⁴ ohm and reactances of 1 ohm and 1.3 × 10⁻³ ohm, respectively. Find current, p.f. and power input if h.v. winding is connected to 400 V, 50 Hz supply and lv winding is short circuited.**

*Solution*

$$R_1 = 0.21\Omega, \quad R_2 = 2.72 \times 10^{-4}\ \Omega$$

$$X_1 = 1\Omega, \quad X_2 = 1.3 \times 10^{-3}\ \Omega$$

Turn ratio = 6600/250 = 26.4

$$R_2' = 2.72 \times 10^{-4}\ (26.4)^2 = 0.1895\ \text{ohm}$$

$$X_2' = 1.3 \times 10^{-3}\ (26.4)^2 = 0.906\ \text{ohm}$$

Total impedance as seen from high voltage terminals when low voltage winding is short circuited $= (R_1 + R_2') + j(X_1 + X_2') = 0.3895 + j1.906\,\Omega$

$$\text{Current} = \frac{400}{0.3895 + j1.906} = 205.66\angle -78.45°\ \text{A}$$

Power factor = cos (78.45°) = 0.2 lagging.

Power Input = (400)(205.66)(0.2) = 16453 W.

*Example 2.11*

A transformer with a 10.1 turn ratio is rated at 100 kVA, 2400/240 V, 50 Hz. The low tension voltage is to be kept at 240 V.

(a) Find the load impedance on 240 V side for full-load conditions.

(b) What is the value of this load impedance referred to high voltage side?

(c) High voltage current for full-load condition. Neglect no load current.

*Solution*

(a) Low voltage current for full-load condition $= \dfrac{100 \times 10^3}{240} = 416.67\,\text{A}$

$$\text{Load impedance} = \frac{240}{416.67} = 0.576\ \Omega$$

(b) Turn ratio = 10

Load impedance as referred to high voltage side $= (0.576)(10)^2 = 57.6\,\Omega$

(c) Primary (high voltage) current for full-load $= \dfrac{416.67}{10} = 41.667\,\text{A}.$

*Example 2.12*

An audio frequency transformer is used to couple a 100 ohm resistive load to a 5 V source having a source resistance of 2000 ohms. (a) Find the turn ratio of the transformer for maximum power transfer, (b) load current, voltage and power. Assume the transformer to be ideal.

*Solution*

(a) For maximum power transfer, load resistance and source resistance should be equal. The turn ratio should be such that 100 ohm load resistance as referred to primary becomes 2000 ohms.

$$2000 = \left(\frac{N_1}{N_2}\right)^2 (100)$$

or

$$\frac{N_1}{N_2} = 4.47$$

(b) Primary currrent $= \dfrac{5}{2000 + 2000} = 1.25 \times 10^{-3}\ \text{A}$

$$I_2 \text{ or secondary current} = \left(1.25 \times 10^{-3}\right)(4.47)$$

$$= 5.59 \times 10^{-3}\ \text{A}$$

$$V_2 = \frac{(5.59)(100)}{1000} = 0.559\,\text{V}$$

$$P_2 = \left(5.59 \times 10^{-3}\right)^2 (100) = 3.1 \times 10^{-3}\ \text{W}$$

## 2.14 PER UNIT VALUES

It is often convinient to carry out calculations relating to electrical machines and other equipments in per unit system. All the quantities are expressed in per unit form, i.e. as fractions of chosen base values.

There are two main advantages in using the per unit system. Firstly the constants of transformers and other equipments fall within very narrow limits when expressed in per unit of their ratings (The actual values of internal impedances, etc., may vary very widely depending on the rating of the equipment). This fact can serve as a useful check in problem solving. Secondly in computations involving transformers the need to refer quantities from one side of transformer to the other side is eliminated. The per unit value is the same on either side of the transformer.

Any quantity may be expressed in per unit by using the equation

$$\text{Per unit value} = \frac{\text{Actual value}}{\text{Base value}} \qquad \text{...(2.37)}$$

The base values of current and voltage may be chosen arbitrarily, but they are usually chosen as equal to rated values. For a single phase system, the base volt amperes are

$$(\text{Volt ampere})_{\text{base}} = (V_{\text{base}})(I_{\text{base}}) \qquad \text{...(2.38)}$$

In practice it is more usual to choose base voltage and base volt amperes and then find $I_{\text{base}}$ using Eq. (2.38). The ratio $V_{\text{base}}/I_{\text{base}}$ is known as base impedance.

The transformer resistance, leakage reactance and impedance are also expressed in per unit. Thus

$$Z_{\text{base}} = \frac{V_{\text{base}}}{I_{\text{base}}} \qquad \text{...(2.39)}$$

$$Y_{\text{base}} = \frac{I_{\text{base}}}{V_{\text{base}}} \qquad \text{...(2.40)}$$

Also

$$Z_{\text{base}} = \left[\frac{V_{\text{base}}}{I_{\text{base}}}\right]\left[\frac{V_{\text{base}}}{V_{\text{base}}}\right] = \frac{(V_{\text{base}})^2}{\text{Volt - Amp}_{\text{base}}} \qquad \text{...(2.41)}$$

$$\text{Per unit resistance} = \frac{\text{Actual resistance}}{Z_{\text{base}}}$$

$$\text{Per unit reactance} = \frac{\text{Actual reactance}}{Z_{\text{base}}}$$

$$\text{Per unit impedance} = \frac{\text{Actual impedance}}{Z_{\text{base}}}$$

In using the per unit system for calculation of regulation, efficiency, etc., for transformers, the transformer rating is taken as the base volt ampere. The respective voltages of high and low voltage sides are taken as the base voltages for the two sides. Then the per unit resistance, reactance and impedance as referred to the high voltage side are the same as referred to the low voltage side.

*Example 2.13*

A 5 kVA, 200/400 V, 50 Hz single-phase transformer has a leakage impedance of 0.12 + $j$ 0.32 ohms as referred to low voltage side. Calculate the per unit values of resistance, reactance and impedance. Take rated quantities as the base values. Check that the per unit values are the same on the two sides.

*Solution*

Base VA = 5000 VA

Base voltage 200 V for low voltage side and 400 V for high voltage side

$$\text{Base current for low voltage side} = \frac{5000}{200} = 25\,\text{A}$$

$$\text{Base impedance for low voltage side} = \frac{200}{25} = 8\,\text{ohm}$$

$$\text{Per unit resistance} = 0.12/8 = 0.015\ \text{p.u.}$$

$$\text{Per unit reactance} = 0.32/8 = 0.04\ \text{p.u.}$$

$$\text{Thus the per unit impedance} = 0.015 + j0.04\ \text{p.u.}$$

For the high voltage side

$$\text{Actual impedance} = (0.12 + j\,0.32)(400/200)^2 = 0.48 + j\,1.28\ \text{ohm}$$

$$\text{Base current} = \frac{5000}{400} = 12.5\,\text{A}$$

$$\text{Base impedance} = \frac{400}{12.5} = 32\ \text{ohm}$$

$$\text{Per unit resistance} = \frac{0.48}{32} = 0.015\,\text{p.u.}$$

$$\text{Per unit reactance} = \frac{1.28}{32} = 0.04\,\text{p.u.}$$

Thus the per unit resistance, reactance and impedance are the same on the two sides of the transformer.

## 2.15 VOLTAGE REGULATION

The transformer windings have resistance and leakage reactances. Voltage drop occurs in these elements. Therefore the secondary terminal voltage under load conditions is different from that under no load conditions. Voltage regulation is defined as the change in secondary terminal voltage, when rated load at a specified power factor is removed. It is expressed as a percentage of the rated secondary voltage.

Figure 2.14 shows the phasor diagram for calculating regulation. $R_{e2}$ and $X_{e2}$ denote the equivalent resistance and reactance as referred to secondary side [Eqs. (2.17) and (2.18)]. $V_1'$ is the primary

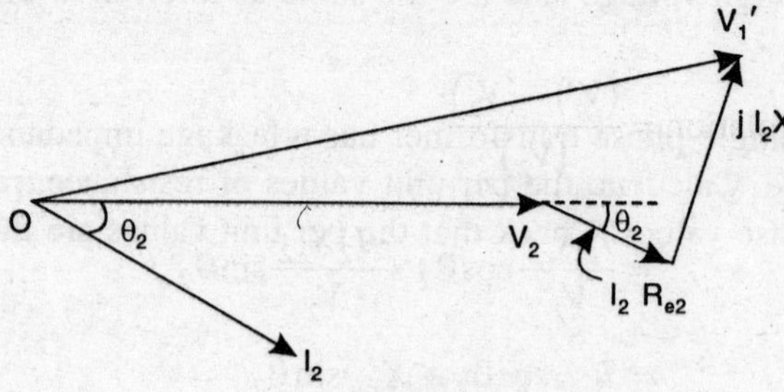


Fig. 2.14

voltage as referred to secondary side and equals $V_1$ $(N_2/N_1)$. At no load, voltage drops are zero and secondary terminal voltage is $V_1'$. From Fig. 2.14

$$(V_1')^2 = (V_2 + I_2R_{e2}\cos\theta_2 + I_2X_{e2}\sin\theta_2)^2 + (I_2X_{e2}\cos\theta_2 - I_2R_{e2}\sin\theta_2)^2 \quad ...(2.42)$$

where $I_2$ is secondary current and $\cos\theta_2$ is secondary power factor.

Equation (2.42) is rather cumbersome. It is sufficient to use the following approximate expression

$$V_1' = V_2 + I_2R_{e2}\cos\theta_2 + I_2X_{e2}\sin\theta_2 \quad ...(2.43)$$

The above expressions are for lagging power factor. Figure 2.15 shows the phasor diagram for leading power factor load and for this condition

$$(V_1')^2 = [V_2 + I_2R_{e2}\cos\theta_2 - I_2X_{e2}\sin\theta_2]^2 + [I_2R_{e2}\sin\theta_2 + I_2X_{e2}\cos\theta_2]^2 \quad ...(2.44)$$

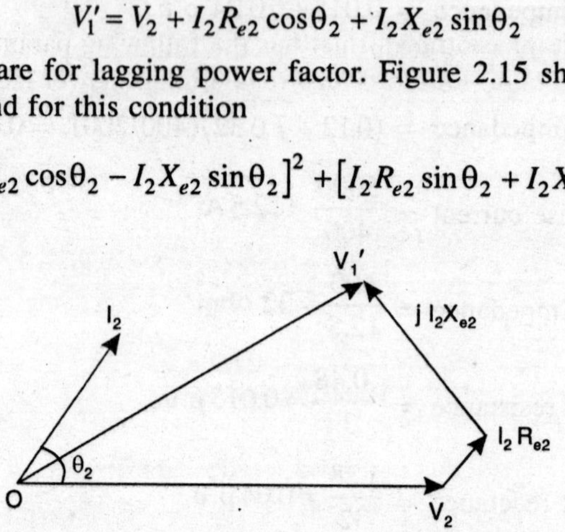


Fig. 2.15

Equation (2.44) can be approximated as

$$V_1' = V_2 + I_2R_{e2}\cos\theta_2 - I_2X_{e2}\sin\theta_2 \quad ...(2.45)$$

$$\text{Voltage regulation} = \frac{(V_1') - (V_2)}{(V_2)} \times 100 \quad ...(2.46)$$

When power factor is lagging or unity, voltage regulation is positive (i.e. output voltage under load conditions is less than that under no load conditions). When power factor is leading, voltage regulation may become negative and output voltage under load condition may become even more than that under no load condition.

For lagging power factor

$$\text{Voltage regulation} = \frac{(V_1') - (V_2)}{(V_2)}$$

$$= \frac{I_2 R_{e2}}{V_2}\cos\theta_2 + \frac{I_2 X_{e2}}{V_2}\sin\theta_2$$

$$= R_{pu}\cos\theta_2 + X_{pu}\sin\theta_2 \quad ...(2.47)$$

For leading power factor

$$\text{Voltage regulation} = R_{pu}\cos\theta_2 - X_{pu}\sin\theta_2 \quad ...(2.48)$$

where $R_{pu}$ and $X_{pu}$ are the total per unit resistance and reactance, respectively. From Eq. (2.48) it is seen that voltage regulation is zero when power factor is leading and such that

$$\tan\theta_2 = \frac{R_{pu}}{X_{pu}} \quad ...(2.49)$$

### *Example 2.14*

A 40 kVA, 6600/250 V single-phase transformer has the following parameters: $R_1$ = 8 ohm, $R_2$ = 0.02 ohm, $X_1$ = 15 ohm, $X_2$ = 0.05 ohm. Find full-load regulation at power factor (a) 0.8 lagging, (b) unity, (c) 0.8 leading.

*Solution*

$$\frac{N_1}{N_2} = \frac{6600}{250} = 26.4$$

$$R_{e2} = \frac{8}{(26.4)^2} + 0.02 = 0.0315$$

$$X_{e2} = \frac{15}{(26.4)^2} + 0.05 = 0.0715$$

At full-load $I_2$ = 40000/250 = 160 A

(a) $\text{Voltage regulation} = \dfrac{I_2 R_{e2}\cos\theta_2 + I_2 X_{e2}\sin\theta_2}{V_2}$

$$= \frac{160 \times 0.0315 \times 0.8 + 160 \times 0.0715 \times 0.6}{250}$$

$$= 0.0436 \text{ or } 4.36\%$$

(b) $\text{Voltage regulation} = \dfrac{160 \times 0.0315 \times 1}{250} = 0.02 \text{ or } 2\%$

(c) $\text{Voltage regulation} = \dfrac{I_2 R_{e2}\cos\theta_2 - I_2 X_{e2}\sin\theta_2}{V_2}$

$$= \frac{160 \times 0.0315 \times 0.8 - 160 \times 0.0715 \times 0.6}{250}$$

$$= -0.011 \text{ or } -1.1\%$$

*Example 2.15*

A 50 kVA, 2000/200 V transformer has impedance drop of 10% and resistance drop of 4%. (a) Find regulation at full-load 0.8 lagging power factor. (b) Find power factor at which regulation is zero.

*Solution*

$$Z_{pu} = \frac{10}{100} = 0.1$$

$$R_{pu} = \frac{4}{100} = 0.04$$

$$X_{pu} = \left(0.1^2 - 0.04^2\right) = 0.0917$$

(a) $$\text{Regulation } = R_{pu} \cos\theta_2 + X_{pu} \sin\theta_2$$

$$= 0.04 \times 0.8 + 0.0917 \times 0.6$$

$$= 0.087 \text{ or } 8.7\%$$

(b) Regulation will be zero when

$$\tan\theta_2 = \frac{R_{pu}}{X_{pu}} = \frac{0.04}{0.0917} = 0.436$$

$$\text{Power factor} = \cos\theta_2 = 0.917 \text{ leading.}$$

## 2.16 LOSSES IN TRANSFORMER

The losses which occur in a transformer under load can be divided into two groups:

(a) iron or core-losses (b) copper-losses

(a) Iron-losses. The iron-losses are the hysteresis loss and eddy current loss and are given by Eqs. (1.19) and (1.21). Since the flux density in a transformer is constant they are also known as constant losses and denoted by $P_i$.

(b) Copper-losses. The copper-losses are due to the resistance of windings.

$$\text{Copper-losses } = I_1^2 R_1 + I_2^2 R_2 \quad \text{...(2.50a)}$$

$$= I_1^2 R_1 + I_1^2 R_2' = I_1^2 R_{e1} \quad \text{...(2.50b)}$$

$$= I_2^2 R_1' + I_2^2 R_2 = I_2^2 R_{e2} \quad \text{...(2.50c)}$$

The resistances $R_1$ and $R_2$ should be determined at the operating temperature of the windings.

## 2.17 TRANSFORMER EFFICIENCY

$$\text{Efficiency } (\eta) = \frac{\text{power output}}{\text{power input}}$$

$$= \frac{\text{Power output}}{\text{Power output + losses}}$$

$$= \frac{V_2 I_2 \cos\theta_2}{V_2 I_2 \cos\theta_2 + I_2^2 R_{e2} + P_i} \qquad ...(2.51)$$

Transformer is a static device and therefore it has no rotational losses. Its efficiency is very high. For power transformer of high ratings used in sub-stations efficiency is around 98%.

## 2.18 CONDITION FOR MAXIMUM EFFICIENCY

Divide the numerator and denominator in Eq. (2.51) by $I_2$. Thus

$$\eta = \frac{V_2 \cos\theta_2}{V_2 \cos\theta_2 + I_2 R_{e2} + P_i / I_2} \qquad ...(2.52)$$

$V_2$ is always constant. Therefore efficiency is maximum when the denominator in Eq. (2.52) is minimum, i.e. when

$$\frac{d}{dI_2}\left( V_2 \cos\theta_2 + I_2 R_{e2} + \frac{P_i}{I_2} \right) = 0$$

or

$$R_{e2} - \frac{P_i}{I_2^2} = 0$$

or

$$I_2^2 R_{e2} = P_i \qquad ...(2.53)$$

Thus for a given value of $V_2$ and cos $\theta_2$, the efficiency of a transformer is maximum when the load current is such that copper-losses are equal to the iron-losses.

### *Example 2.16*

A single-phase transformer is rated at 100 kVA, 5000/250 V. The full-load copper losses are 2000 W and iron losses are 1200 W. Find efficiency at (a) full-load unity p.f., (b) full-load 0.8 p.f. lagging, (c) half-load 0.8 p.f. lagging.

### *Solution*

(a) Output = 100 × 1 = 100 kW

Copper-losses = 2000 W = 2 kW

Iron-losses = 1200 W = 1.2 kW

$$\eta = \frac{100}{100 + 2 + 1.2} = 0.969 \text{ or } 96.9\%$$

(b) Output = 100 × 0.8 = 80 kW

Copper-losses = 2 kW

Iron-losses = 1.2 kW

$$\eta = \frac{80}{80 + 2 + 1.2} = 0.962 \text{ or } 96.2\%$$

(c) Output = 100 × 0.5 × 0.8 = 40 kW

At half load the current is half that at full-load. Since copper-losses are proportional to $I^2$, copper-losses at half-load are one-fourth of those at full-load

Copper-losses = 0.25 × 2 = 0.5 kW

Iron-losses = 1.2 kW

$$\eta = \frac{40}{40 + 0.5 + 1.2} = 0.959 \text{ or } 95.9\%$$

*Example 2.17*

For the transformer of Example 2.16, find the load kVA for maximum efficiency.

*Solution*

Let $n$ be the fraction of full-load kVA at which maximum efficiency occurs. Then

$$n^2 \text{ (copper-losses at full-load)} = P_i$$

or

$$n^2 (2000) = 1200$$

or

$$n = \left(\frac{1200}{2000}\right)^{0.5} = 0.775$$

Load kVA for maximum efficiency = 0.775 × 100 = 77.5 kVA.

*Example 2.18*

A 1000 V, 25 Hz transformer has copper, hysteresis and eddy current losses of 1.5, 0.7 and 0.4% of output. What will be the percentage losses if transformer is used on 2000 V, 50 Hz system, assuming the full-load current to be the same in the two cases. Compare the full-load efficiency at the two frequencies if p.f. is unity.

*Solution*

Let output at 1000 V, 25 Hz be 100 kVA

$$\text{Output} = 100 \times 1 = 100 \text{ kW}$$

$$\text{Copper-losses} = \frac{1.5}{100} \times 100 = 1.5 \text{ kW}$$

$$\text{Hysteresis loss} = 0.7 \text{ kW}$$

$$\text{Eddy current loss} = 0.4 \text{ kW}$$

$$\eta = \frac{100}{100 + 1.5 + 0.7 + 0.4} \times 100 = 97.47\%$$

At 2000 V, 50 Hz, unity power factor

$$\text{Output} = 100 \times \frac{2000}{1000} \times 1 = 200\,\text{kW}$$

Since both the voltage and frequency have been doubled, $B_m$ will remain the same.

Hysteresis loss $\propto f$

$$\text{Therefore hysteresis loss} = 0.7\left(\frac{50}{25}\right) = 1.4 \text{ kW}$$

$$\text{Eddy current loss} \propto f^2$$

$$\text{Therefore eddy current loss} = 0.4\left(\frac{50}{25}\right)^2 = 1.6 \text{ kW}$$

$$\text{Percentage copper-loss} = \frac{1.5}{200} \times 100 = 0.75\%$$

$$\text{Percentage hysteresis loss} = \frac{1.4}{200} \times 100 = 0.7\%$$

$$\text{Percentage eddy current loss} = \frac{1.6}{200} \times 100 = 0.8\%$$

$$\text{Efficiency} = \frac{200}{200 + 1.5 + 1.4 + 1.6} \times 100 = 97.8\%$$

## 2.19 ALL DAY EFFICIENCY (ENERGY EFFICIENCY)

The distribution transformers are installed at the receiving end of feeders and supply the distributors. Their primary windings remain excited all the 24 hours irrespective of whether they are supplying any load or not. Therefore core losses occur continuously. But the copper-losses vary with the load current. A more suitable method of assessing the efficiency of such a transformer is on the basis of energy output and energy input. All day efficiency (or energy efficiency) is defined as

All day efficiency (or Energy efficiency)

$$= \frac{\text{Energy output in 24 hours}}{\text{Energy output in 24 hours} + \text{Energy losses in 24 hours}} \qquad \text{...(2.54)}$$

It is possible that a transformer has a higher efficiency but a lower energy efficiency than the other. Therefore a distribution transformer should be selected on the basis of energy efficiency.

### *Example 2.19*

A 63 kVA distribution transformer has a full-load efficiency of 0.95. The full-load copper-losses are equal to iron-losses. It supplies the following load cycle: full-load 1 p.f. for 6 hours in a day; no-load for 18 hours in a day. Find all day efficiency.

### *Solution*

$$\text{Total losses at full-load} = \frac{5}{100} \times 63 = 3.15 \text{ W}$$

$$\text{Full-load copper-losses} = 0.5 \times 3.15 = 1.575\,\text{kW}$$

$$\text{Iron-losses} = 1.575\,\text{W}$$

$$\text{Energy output} = 63 \times 6 \times 1 = 378\,\text{kWh}$$

$$\text{Energy loss due to iron-losses} = 1.575 \times 24 = 37.8\,\text{kWh}$$

$$\text{Energy loss due to copper-losses} = 1.575 \times 6 = 9.45\,\text{kWh}$$

$$\text{All day efficiency} = \frac{378}{378 + 37.8 + 9.45}$$

$$= 0.89 \text{ or } 89\%.$$

*Example 2.20*

A 100 kVA distribution transformer is supplying the following loads: (a) 80 kW at 0.8 p.f. for 8 hours, (b) 40 kW at unity p.f. for 6 hours and (c) no-load for the remaining 10 hours of the day. Find all day efficiency if full-load copper-losses are 2 kW and iron-losses are 1 kW.

*Solution*

For 80 kW, 0.8 p.f. load

$$\text{Energy output} = 80 \times 8 = 640 \text{ kWh}$$

$$\text{kVA output} = \frac{80}{0.8} = 100 \text{ kVA} = \text{full - load}$$

$$\text{Copper-losses} = 2 \text{ kW}$$

$$\text{Iron-losses} = 1 \text{ kW}$$

$$\text{Total losses} = 2 + 1 = 3 \text{ kW}$$

$$\text{Energy losses} = 3 \times 8 = 24 \text{ kWh}$$

For 40 kW, 1 p.f. load

$$\text{Energy output} = 40 \times 6 = 240 \text{ kWh}$$

$$\text{kVA output} = 40 \text{ kVA}$$

$$\text{Copper-losses} = 2\left(\frac{40}{100}\right)^2 = 0.32\,\text{kW}$$

$$\text{Iron-losses} = 1 \text{ kW}$$

$$\text{Total losses} = 1.32 \text{ kW}$$

$$\text{Energy losses} = 1.32 \times 6 = 7.92 \text{ kWh}$$

For no load condition

$$\text{Energy output} = 0$$

$$\text{Copper-losses} = 0$$

$$\text{Iron-losses} = 1 \text{ kW}$$

$$\text{Total losses} = 1\text{kW}$$

$$\text{Energy losses} = 1 \times 10 = 10 \text{ kWh}$$

$$\text{Total energy output} = 640 + 240 = 880 \text{ kWh}$$

$$\text{Total energy losses} = 24 + 7.92 + 10 = 41.92 \text{ kWh}$$

$$\text{All day efficiency} = \frac{880}{880 + 41.92} = 0.955 \text{ or } 95.5\%$$

## 2.20 POLARITY OF TRANSFORMERS

The two primary terminals of a single-phase transformer are alternately positive and negative. Likewise the secondary terminals are also alternately positive and negative. When two transformers are to be connected in parallel on the primary and secondary sides, it is essential that terminals which have identical instantaneous polarity are connected together. The h.v. terminals are marked as $A_1$, $A_2$ and l.v. terminals as $a_1$, $a_2$. These markings indicate that when instantaneous voltage is directed from $A_1$ to $A_2$ in the h.v. winding, it is directed from $a_1$ to $a_2$ in the l.v. winding; i.e. when $A_1$ is positive with respect to $A_2$, $a_1$ will be positive with respect to $a_2$ at that instant. The arrangement in Fig. 2.16(a) shows subtractive polarity, while that in Fig. 2.16(b) shows additive polarity.

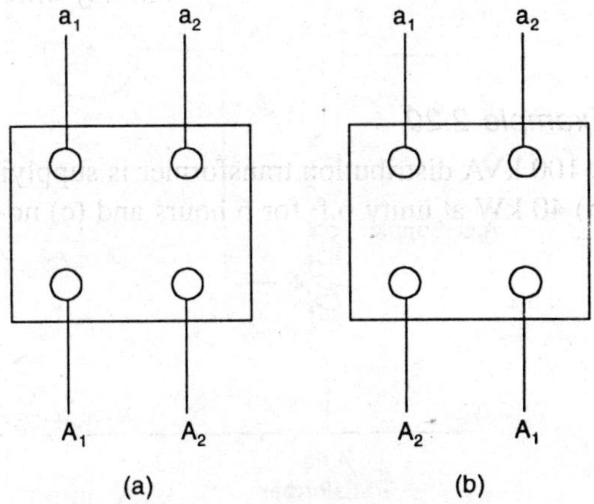


**Fig. 2.16.** (a) Subtractive polarity, (b) Additive polarity

## 2.21 TEST FOR POLARITY

Figure 2.17 shows the connection diagram to identify the terminals having the same instantaneous polarity. One HT and one LT terminals are connected together and an ac voltmeter is connected between the remaining two terminals. The HT winding is excited by a moderate value of ac voltage. If $V' > V$, the polarity is additive. If $V' < V$, the polarity is substrative. The terminals can then be marked accordingly.

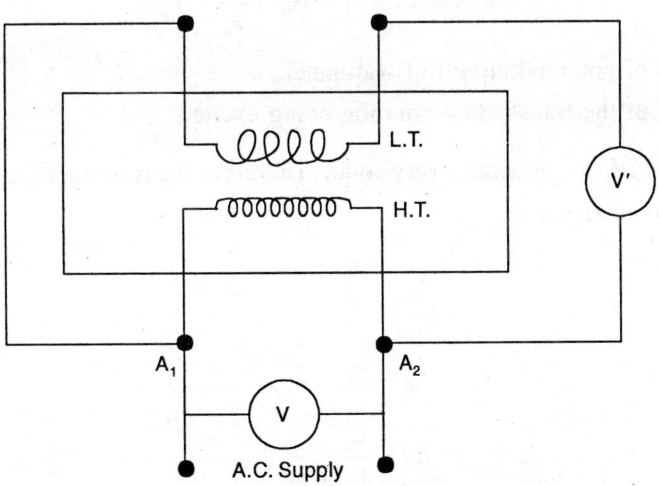


**Fig. 2.17.** Connections for polarity test

## 2.22 OPEN CIRCUIT (OR NO LOAD) TEST

This test is conducted to find core losses and currents $I_w$ and $I_m$ (and parameters $R_0$ and $X_0$).

Figure 2.18 shows the circuit for this test. One of the windings is excited by the rated voltage at rated frequency and the other winding is kept open circuited. Usually the low voltage winding is excited. The core losses are the same, measured on either side, provided the rated voltage of that winding is applied. Nevertheless if the measurements were made on the high voltage side, the required voltage would be rather high and the current $I_0$ would be inconveniently small.

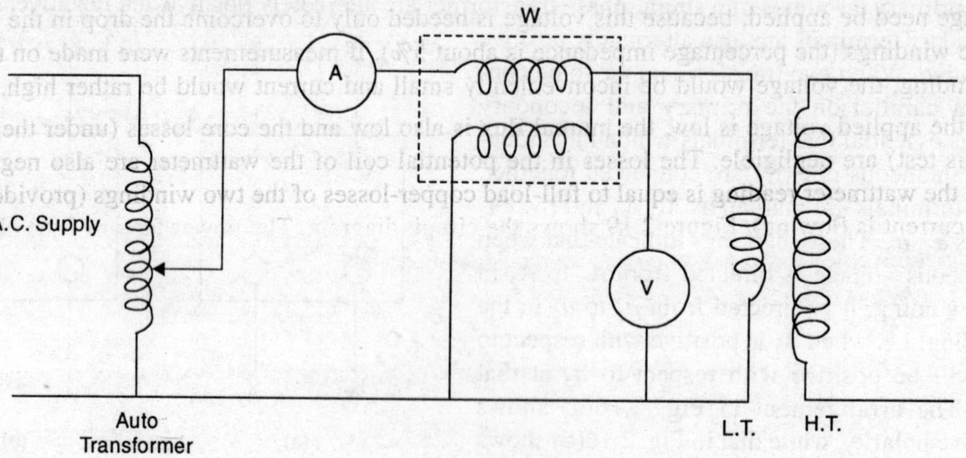


**Fig. 2.18.** Connections for no-load test on a single-phase transformer

An auto-transformer is used to adjust the input voltage to the rated value. It is pertinent to add that the power factor under no-load conditions is very low (around 0.3 or so). Therefore it is advisable to use a low power factor wattmeter to ensure accurate value of no-load losses.

Let $I_0$, $V_0$ and $W_0$ be the readings of current, voltage and power. Then

$$\text{Core losses} = P_i = W_0 - \frac{V_0^2}{r_p} - I_0^2 R_1 \quad \text{...(2.55)}$$

where $r_p$ = resistance of potential circuit of wattmeter

$R_1$ = resistance of the transformer winding being excited

The values $V_0^2/r_p$ and $I_0^2 R_1$ are generally very small. Therefore the reading $W_0$ can be taken as the core losses. $I_0$ is the no-load current.

$$I_w = \frac{W_0}{V_0} \quad \text{...(2.56)}$$

$$I_m = \left(I_0^2 - I_w^2\right)^{0.5} \quad \text{...(2.57)}$$

$$R_0 = \frac{V_0}{I_w} \quad \text{...(2.58)}$$

$$X_0 = \frac{V_0}{I_m} \quad \text{...(2.59)}$$

## 2.23 SHORT CIRCUIT TEST

This test is conducted to find out the equivalent resistance and leakage reactance of the windings. In this test one of the windings (usually low voltage one) is short circuited and the other winding is excited by a suitable voltage, so that rated full-load current may flow in the windings. Only a few per cent of the rated voltage need be applied, because this voltage is needed only to overcome the drop in the impedance of the windings (the percentage impedance is about 5%). If measurements were made on the low voltage winding, the voltage would be inconveniently small and current would be rather high.

Since the applied voltage is low, the mutual flux is also low and the core losses (under the conditions of this test) are negligible. The losses in the potential coil of the wattmeter are also negligible. Therefore, the wattmeter reading is equal to full-load copper-losses of the two windings (provided full-load rated current is flowing). Figure 2.19 shows the circuit diagram. The power factor of the circuit is low and it is advisable to use a low power factor wattmeter.

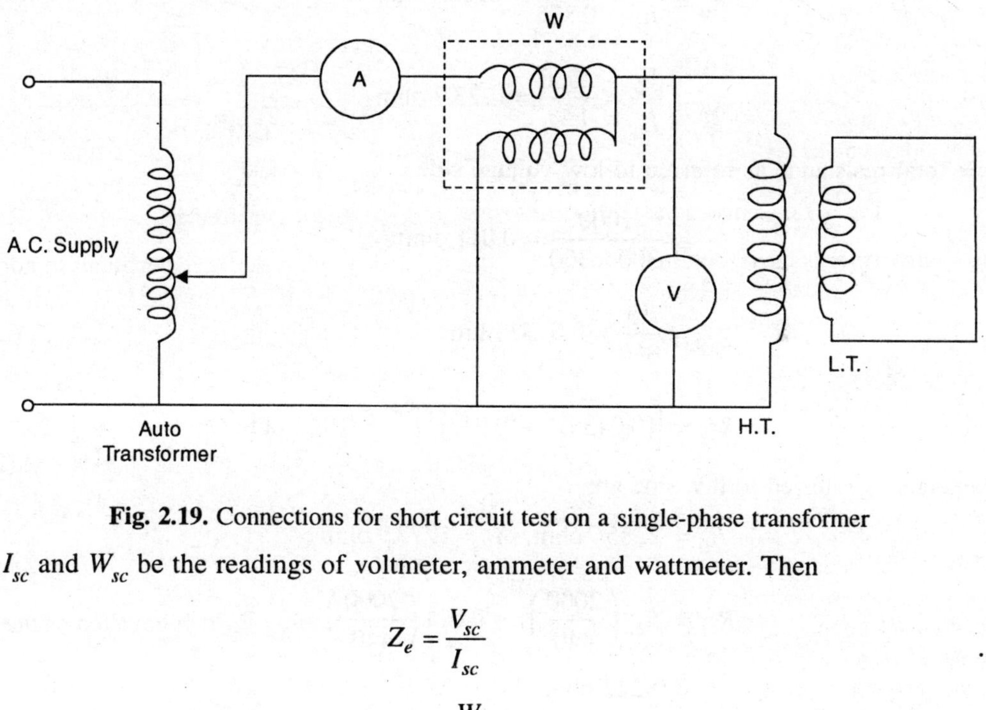


**Fig. 2.19.** Connections for short circuit test on a single-phase transformer

Let $V_{sc}$, $I_{sc}$ and $W_{sc}$ be the readings of voltmeter, ammeter and wattmeter. Then

$$Z_e = \frac{V_{sc}}{I_{sc}} \quad \text{...(2.60)}$$

$$R_e = \frac{W_{sc}}{I_{sc}^2} \quad \text{...(2.61)}$$

$$X_e = \left(Z_e^2 - R_e^2\right)^{0.5} \quad \text{...(2.62)}$$

where $Z_e$, $R_e$ and $X_e$ denote the equivalent impedance, resistance and leakage reactance, as referred to the winding being excited.

*Example 2.21*

A 200 kVA, 2000/440 V, 50 Hz single-phase transformer gave the following test results:

| | | | |
|---|---|---|---|
| oc test (hv) | 2 kV, | 1.75 kW, | 1.8 A |
| sc test (lv) | 13 V, | 1 kW, | 300 A |

Find (a) parameters of equivalent circuit as referred to h.v. side, (b) regulation and efficiency at full-load, 0.8 p.f. lagging.

*Solution*

(a)

$$I_w = \frac{1.75 \times 1000}{2000} = 0.875\,\text{A}$$

$$I_m = \left(I_0^2 - I_w^2\right)^{0.5} = \left(1.8^2 - 0.875^2\right)^{0.5}$$

$$= 1.57\text{ A}$$

$$R_0 = \frac{V_0}{I_w} = \frac{2000}{0.875} = 2285.7\text{ ohm}$$

$$X_0 = \frac{V_0}{I_m} = \frac{2000}{1.57} = 1273.9\text{ ohm}$$

$R_{e2}$ = Total resistance as referred to low voltage side

$$= \frac{1000}{300 \times 300} = 0.011\text{ ohm}$$

$$Z_{e2} = \frac{13}{300} = 0.0433\text{ ohm}$$

$$X_{e2} = \left[(0.0433)^2 - (0.011)^2\right]^{0.5} = 0.042\text{ ohm}$$

Parameters as referred to h.v. side are

$$R_0 = 2285.7\,\text{ohm},\ \ X_0 = 1273.9\text{ ohm}$$

$$R_{e1} = R_{e2}\left(\frac{2000}{440}\right)^2 = 0.011\left(\frac{2000}{440}\right)^2$$

$$= 0.227\text{ ohm}$$

$$X_{e1} = X_{e2}\left(\frac{2000}{440}\right)^2 = 0.042\left(\frac{2000}{400}\right)^2$$

$$= 0.868\text{ ohm}$$

(b) Full load current on low voltage side

$$I_2 = \frac{200 \times 1000}{440} = 454.54\,\text{A}$$

$$\text{Regulation} = \frac{I_2 R_{e2} \cos\theta_2 + I_2 X_{e2} \sin\theta_2}{V_2} \times 100$$

$$= \frac{454.54(0.011 \times 0.8 + 0.042 \times 0.6)}{440} \times 100$$

$$= 3.5\%$$

$$\text{Output} = 200 \times 0.8 = 160 \text{ kW}$$

$$\text{Iron-losses} = 1.75 \text{ kW}$$

$$\text{Full-load copper-losses} = (454.54)^2\,(0.011)$$

$$= 2272.67 \text{ W}$$

$$\text{Efficiency} = \frac{\text{output}}{\text{output + losses}}$$

$$= \frac{160 \times 1000}{160 \times 1000 + 1.75 \times 1000 + 2272.67}$$

$$= 0.975 \text{ or } 97.5\%$$

## 2.24 BACK TO BACK TEST (OR REGENERATIVE TEST)

Sometimes it is necessary to conduct a full-load test on a transformer to determine the maximum temperature rise. If this test is conducted by directly loading a large transformer, a considerable expenditure of energy occurs. An artificial load to absorb the large output of the transformer under test is also necessary. A regenerative test (or back-to-back) is very convenient for this purpose. It is also known as Sumpner's test and requires two similar transformers.

Figure 2.20 shows the circuit for this test. The primary windings of the two transformers are connected in parallel to ac mains of rated voltage and frequency. The two secondaries are connected in opposition. For this purpose any two terminals (say *B* and *C*) are joined together and a high range voltmeter (of twice the rated secondary voltage) is connected to remaining terminals *A* and *D*. If the voltmeter reads zero, the two secondaries are in opposition and terminals *A* and *D* are used for test. If voltmeter reads a large value (equal to approximately twice the rated secondary voltage), the secondaries are acting in the same direction. Then terminals *A* and *C* are joined and the terminals *B* and *D* used for the test.

When the two secondaries are connected in opposition, the two secondary emfs oppose each other and no current can flow in the secondary circuit. A regulating transformer, excited by ac mains supply, is used to inject a voltage in the secondary circuit. The magnitude of injected voltage is adjusted till the ammeter $A_2$ reads full-load secondary current. This secondary current causes full-load current to flow through the primary windings, but the primary current remains confined to the dotted path shown.

It is seen that wattmeter $W_1$ indicates total core losses, $W_2$ indicates total copper-losses and ammeter $A_1$ indicaets total no-load current of the two transformers. Thus by this indirect method we have loaded the two transformers to full-load and the expenditure of energy is just equal to that required for the losses only. The test can be continued for a long time for determination of maximum temperature rise.

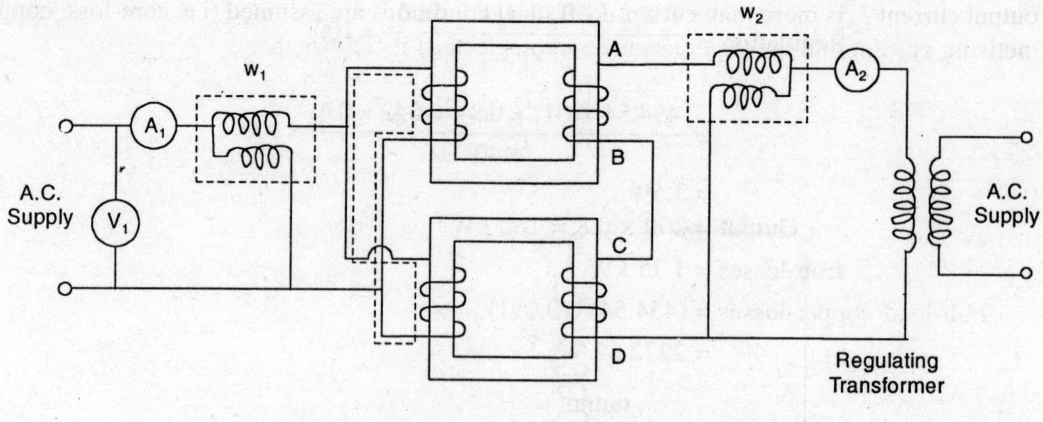


**Fig. 2.20.** Connections for back-to-back test

## 2.25 SEPARATION OF CORE LOSSES

It is possible to determine the two components of the core losses, i.e. hysteresis and eddy current loss. The circuit is the same as for no-load test (Fig. 2.18). However, a source with variable voltage and frequency is necessary.

The voltage and frequency are varied simultaneously, so that the ratio $V_0/f$ remains constant. From Eq. (2.3) it is seen that if this ratio is constant, $\phi_m$ and hence $B_m$ will remain constant. From Eqs. (1.19) and (1.21) we can write the total iron losses are

$$W_0 = K_h' f + K_e' f^2$$

or

$$\frac{W_0}{f} = K_h' + K_e' f \qquad \text{...(2.63)}$$

If iron-losses $W_0$ are determined at a number of different values of $f$ (and the ratio $V_0/f$ is kept constant), we can plot Fig. 2.21. The intercept of the line on Y-axis gives $K_h'$ and the slope of the line is equal to $K_e'$. At a frequency $f$, the hysteresis loss is $K_h' f$ and eddy current loss is $K_e' f^2$.

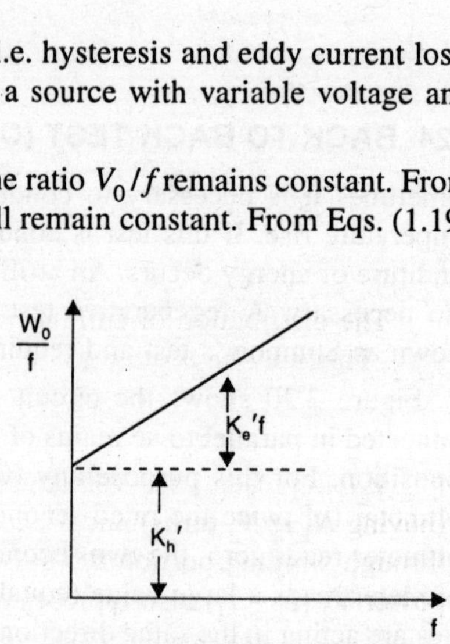


**Fig. 2.21.** Separation of core losses

## 2.26 PRINCIPLE OF AUTO-TRANSFORMER

Figure 2.22 shows a winding AC having $N_1$ turns and wound on an iron core. The winding is connected to an ac voltage $V_1$. An emf $E_1$ will be induced in the winding.

Let the number of turns from end $C$ to point $B$ be $N_2$. The emf between points $C$ and $B$ is $E_2$. The emfs $E_1$ and $E_2$ are related by the expression $E_1/E_2 = N_1/N_2$. A load current $I_2$ is supplied to the load. The mmfs $N_1I_1$ and $N_2I_2$ will be equal and opposite. This arrangement is known as auto-transformer. If terminal $B$ is a sliding contact, the output voltage $V_2$ can be varied.

### (a) Step down auto-transformer

Figure 2.22 shows a step down auto-transformer. The output voltage $V_2$ is less than input voltage $V_1$ and the output current $I_2$ is more than current $I_1$. If ideal conditions are assumed (i.e. core-loss, copper-loss, magnetising current and leakage reactance are neglected) it follows that

$$\frac{V_1}{V_2} = \frac{I_2}{I_1} = \frac{N_1}{N_2} = a \qquad \text{...(2.64)}$$

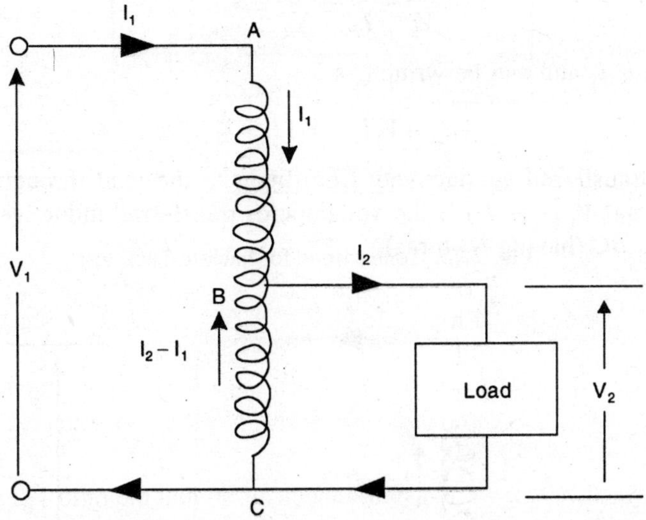


**Fig. 2.22.** Step down auto-transformer

The distribution of currents in the windings is as shown. The voltampere delivered to the load is $V_2I_2$. This may be written as

$$V_2I_2 = V_2I_1 + V_2(I_2 - I_1) \qquad \text{...(2.65)}$$

$V_2I_1$ represents the voltamperes transferred conductively to the load through winding portion *AB* (having $N_1 - N_2$ turns) and $V_2\,(I_2 - I_1)$ represents the voltamperes transferred inductively to the load through winding portion *BC* having $N_2$ turns (i.e. power transferred through transformer action). This power $V_2\,(I_2 - I_1)$ also represents the rating of the equivalent two winding transformer

$$\frac{\text{Output VA of auto-transformer}}{\text{Transformed VA}} = \frac{\text{Output VA of auto-transformer}}{\text{Output VA of equivalent two-winding transformer}}$$

$$= \frac{V_2I_2}{V_2(I_2 - I_1)} = \frac{a}{a-1} \qquad \text{...(2.66)}$$

Thus a two-winding transformer connected as a step down auto-transformer will have a voltampere rating $a/(a-1)$ times its rating as a two-winding transformer. This increase in rating becomes larger as $a$ approaches unity. In the limiting case when $a = 1$, the power transferred conductively is equal to the entire input and the power transferred inductively becomes zero, which means that there is no need for auto-transformer. On the other hand, if an auto-transformer is to be designed for a rating *P*, it will consist of a two-winding transformer having a rating $[P\,(a - 1)/a]$, which becomes smaller as $a$ approaches unity. When designed in this way the material in the auto-transformer will be utilised to the maximum extent.

### (b) Step up auto-transformer

Figure 2.23 shows a step up auto-transformer. The number of turns in portions $AB$ and $BC$ are $(N_1 - N_2)$ and $N_2$ respectively. $V_2$ is more than $V_1$ and $I_2$ is less than $I_1$. The distribution of currents in the winding is as shown. Assuming ideal conditions

$$\frac{V_2}{V_1} = \frac{I_1}{I_2} = \frac{N_1}{N_2} = a \qquad ...(2.67)$$

The voltampere input is $V_1 I_1$ and can be written as

$$V_1 I_1 = V_2 I_2 + V_1 (I_1 - I_2) \qquad ...(2.68)$$

$V_1 I_2$ is the voltampere transferred conductively from input to the load through winding portion $AB$ (having $N_1 - N_2$ turns) and $V_1 (I_1 - I_2)$ is the voltamperes transferred inductively from input to load through winding portion $BC$ (having $N_2$ turns).

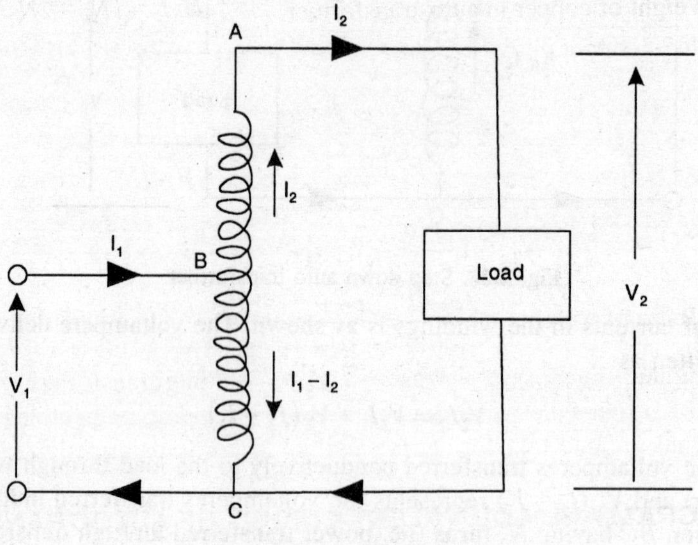


**Fig. 2.23.** Step up auto-transformer

$$\frac{\text{Output VA of auto-transformer}}{\text{Transformed VA}} = \frac{V_1 I_1}{V_1 (I_1 - I_2)} = \frac{a}{a-1} \qquad ...(2.69)$$

Thus the advantage of increased power rating in this case is also the same as in the case of step down auto-transformer.

## 2.27 SAVING IN COPPER IN AUTO-TRANSFORMER

For the same capacity and voltage ratio, an auto-transformer requires less winding material than a two-winding transformer. The cross-section of a conductor is proportional to the current carried by the conductor and the length of conductor is proportional to the number of turns. Therefore the weight of copper (or aluminium) is proportional to the product of number of turns and current. For a two-winding transformer,

$$\text{Weight of copper in primary} \propto N_1 I_1$$
$$\text{Weight of copper in secondary} \propto N_2 I_2$$
$$\text{Total weight of copper} \propto N_1 I_1 + N_2 I_2$$

For an auto-transformer (Fig. 2.22), the winding section $AB$ carries a current $I_1$ and has $(N_1 - N_2)$ turns. Thus the weight of copper in section $AB \propto (N_1 - N_2) I_1$. The winding section $BC$ carries a current $(I_2 - I_1)$ and has $N_2$ turns. Thus the weight of copper in section $BC \propto N_2 (I_2 - I_1)$.

Total weight of copper in auto-transformer

$$\propto (N_1 - N_2) I_1 + N_2 (I_2 - I_1)$$
$$\propto N_2 I_2 + (N_1 - 2N_2) I_1$$

Therefore

$$\frac{\text{Weight of copper in auto-transformer}}{\text{Weight of copper in two-winding transformer}} = \frac{N_2 I_2 + (N_1 - 2N_2) I_1}{N_1 I_1 + N_2 I_2}$$

$$= \frac{(N_1/N_2) - 2 + (I_2/I_1)}{(N_1/N_2) + (I_2/I_1)}$$

But

$$\frac{N_1}{N_2} = \frac{I_2}{I_1} = a$$

$$\text{Ratio of weights} = \frac{2a-2}{2a} = \frac{a-1}{a} = 1 - \frac{1}{a} \qquad \text{...(2.70)}$$

Thus the use of an auto-transformer, in place of a two-winding transformer, results in saving in copper to the extent of $1/a$ times the weight of copper in two-winding transformer. In $a$ is close to 1, saving is very large. Thus an auto-transformer is more useful for transformation ratio close to 1.

## 2.28 PHASOR DIAGRAM AND EQUIVALENT CIRCUIT OF AUTO-TRANSFORMER

The phasor diagram and equivalent circuit of an auto-transformer can be drawn in a similar way as that of a two-winding transformer. Figure 2.22 shows a step down auto-transformer. The winding portion $AB$ has $N_1 - N_2$ turns and portion $BC$ has $N_2$ turns. $R_1$ and $X_1$ are the resistance and leakage reactance of $AB$, while $R_2$ and $X_2$ are the resistance and leakage reactance of $BC$. In the analysis of auto-transformers, the magnetising current is neglected. The applied voltage $V_1$ creates a mutual flux, which induces emfs $E_1$ and $E_2$ in windings $AC$ and $BC$, to give

$$\frac{E_1}{E_2} = \frac{N_1}{N_2} = a = \frac{I_2}{I_1} \qquad \text{...(2.71)}$$

Neglecting the magnetising current, the total mmf must be zero, so that

$$N_2 (I_2 - I_1) - (N_1 - N_2) I_1 = 0$$

or

$$I_2 - I_1 = \left( \frac{N_1}{N_2} - 1 \right) I_1 = (a - 1) I_1 \qquad \text{...(2.72)}$$

Let $\cos\theta_2$ be the power factor of load. Then

$$V_2 = E_2 - (I_2 - I_1)(R_2 + jX_2) \qquad ...(2.73)$$

The applied voltage $V_1$ balances the induced emf $E_1$, voltage drop from $C$ to $B$ and voltage rise from $B$ to $A$. Thus

$$V_1 = E_1 - (I_2 - I_1)(R_2 + jX_2) + I_1(R_1 + jX_1) \qquad ...(2.74a)$$

or

$$V_1 = aE_2 - I_1(a-1)(R_2 + jX_2) + I_1(R_1 + jX_1) \qquad ...(2.74b)$$

Substituting the value of $E_2$ from Eq. (2.73) into Eq. (2.74) and re-arranging the terms

$$V_1 = aV_2 + I_1\left[(a-1)^2 R_2 + R_1 + jX_1 + j(a-1)^2 X_2\right] \qquad ...(2.75)$$

Figure 2.24 shows the phasor diagram. The phasor sum of $V_2$ and $(I_2 - I_1)(R_2 + jX_2)$ gives the emf $E_2$ [Eq. (2.73)]. The voltages $E_1$, $E_2$ and currents $I_1$, $I_2$ are related by Eq. (2.71). On the primary side $V_1$ and $E_1$ are related by Eq. (2.74b) and the phasor diagram of Fig. 2.24 shows this relationship.

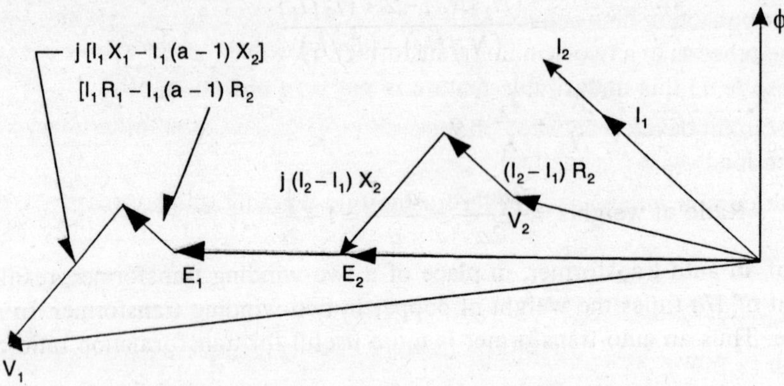


**Fig. 2.24.** Phasor diagram of auto-transformer

Equation (2.75) is the basis for the equivalent circuit shown in Fig. 2.25. The equivalent internal impedance as seen from primary terminals is

$$Z_1 = R_1 + (a-1)^2 R_2 + j\left[X_1 + (a-1)^2 X_2\right]$$

The parallel branch consisting of $R_0$ and $X_0$ accounts for the core-loss and magnetising current.

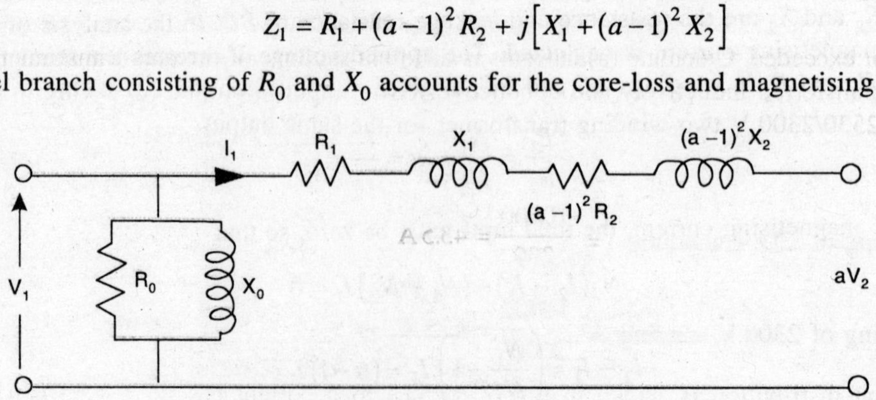


**Fig. 2.25.** Equivalent circuit for auto-transformer

## 2.29 ADVANTAGES DISADVANTAGES AND APPLICATIONS OF AUTO-TRANSFORMER

### (a) Advantages

An economically designed auto-transformer has some or all of the following advantages over a two-winding transformer of the same rating:

1. Smaller size
2. Lower cost
3. Greater efficiency
4. Smaller exciting current
5. Better voltage regulation

All the above advantages depend on the transformation ratio $N_1/N_2$ and little is gained by the use of an auto-transformer when $a$ is large. As such auto-transformers are used only when $N_1/N_2$ is less than 3.

### (b) Disadvantages

(1) There is a direct connection between the high voltage and low voltage sides. One side is no longer isolated from the other as in a two-winding transformer. However, when $V_1$ and $V_2$ are of the same order (i.e. $a$ close to 1) this undesirable feature is not very objectionable.

(2) Should an open circuit develop between points $B$ and $C$ (Fig. 2.22) the full primary voltage would be applied to the load on the secondary side.

(3) The short circuit current is larger than that for the two-winding transformer.

### (c) Applications

1. Continuously variable ac voltage.
2. Inter connection of power systems of different voltage levels, e.g. 132 kV and 220 kV.
3. Starting of ac machines.
4. Boosting or bucking of mains ac voltage by a small amount.

*Example 2.22*

A 5 kVA, 2300/230 V two-winding transformer is reconnected as an auto-transfomer as shown in Fig. 2.26 and excited by a 2530 V source. The transformer is loaded so that the rated currents of the windings are not exceeded. Calculate (a) currents is different sections of the auto-transfomer, (b) kVA output, (c) VA transferred inductively and conductively from input to output, (d) saving in copper as compared to a 2530/2300 V two-winding transformer for the same output.

*Solution*

$$\text{Current rating of 230 V winding} = \frac{10000}{230} = 43.5\text{ A}$$

$$\text{Current rating of 2300 V winding} = \frac{10000}{2300} = 4.35\text{ A}$$

(a) The current distribution is shown in Fig. 2.26. The load current is 43.5 + 4.35 = 47.85 A at 2300 V

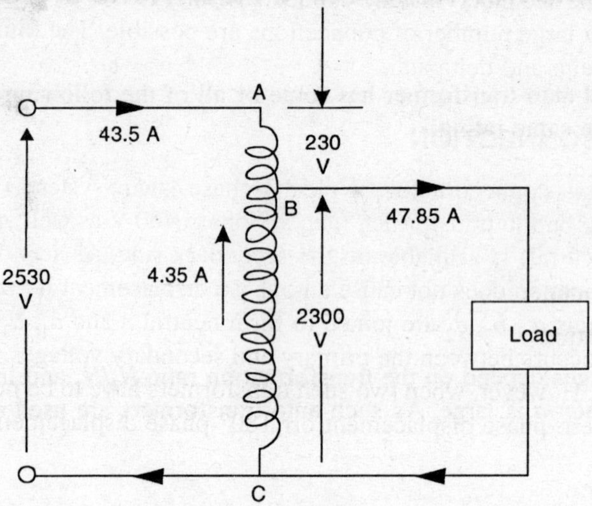


**Fig. 2.26**

(b) kVA rating $= \dfrac{2300 \times 47.85}{1000} = 110$ kV

(c) VA transferred inductively $= 2300 \times 4.35 = 10000$ VA

VA transferred conductively $= 2300 \times 43.5 = 100000$ VA

(d) $a = \dfrac{N_1}{N_2} = \dfrac{2530}{2300} = 1.1$

Saving in copper $= \dfrac{1}{a} = 0.909$ or 90.9%

## 2.30 INTRODUCTION TO THREE-PHASE TRANSFORMERS

The present day power systems are all 3-phase ac systems. The change of voltage in a 3-phase system can be done either through a single three-phase transformer or by a bank of 3 single-phase transformers. A single unit three-phase transformer has a three limbed core, one limb for each phase winding. On each limb the low voltage winding is placed over the core and the high voltage winding is placed over the low voltage winding with suitable insulation between the core and low voltage winding as well as between the two windings. A single unit 3-phase transformer is lighter, cheaper and more efficient than a bank of 3 single-phase units.

The high voltage terminals are brought out on one side and marked as *A*, *B*, *C*; similarly the low voltage terminals are brought out on the other side and denoted by small letters *a*, *b*, *c*. The ends of phase *A* are labelled as $A_1A_2$ on h.v. side and $a_1$, $a_2$ on the l.v. side. For windings having tappings or sections, the markings are $A_1$, $A_2$, $A_3$, etc. In such cases the terminal $A_1$ and the last numbered one are the two ends. The polarity is such that if $A_1$ is positive with respect to $A_2$ at a particular instant, then $a_1$ is positive with respect to $a_2$ at the same instant.

The primary and the secondary can be connected either in star or delta. Moreover each phase can be reversed too. Thus a large number of connections are possible. The four main connections are star-star, delta-delta, star-delta and delta-star.

## 2.31 STAR-STAR CONNECTION

Figure 2.27 (a) shows this connection for a typical 3-phase 4-wire system. When this connection is used for a 11000/400 V distribution transformer, then a 3-phase 400 V as well as 230 V single-phase supply (between phase and neutral) is available on the secondary side. As seen from the phasor diagram in Fig. 2.27 (b), this connection does not cause any phase displacement between primary and secondary. However, if the terminals $a_2$, $b_2$, $c_2$ are joined to form neutral $n$ and $a_1$, $b_1$, $c_1$ are brought out a phase displacement of 180° occurs between the primary and secondary voltages. Both the arrangements give the same performance. However, when two such transformers have to be put in parallel, both should be connected either for zero-phase displacement orr 180° phase displacement.

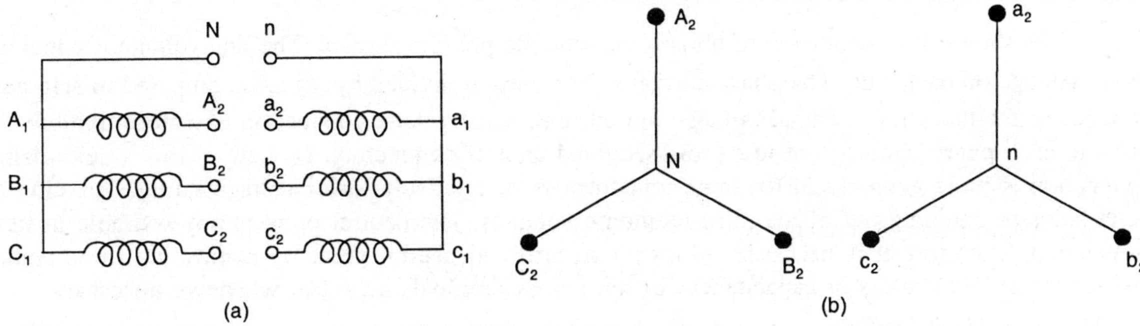


**Fig. 2.27.** Three phase star-star transformer (a) connection, (b) primary and secondary emfs.

The use of star-star connnection on a 3-phase 3-wire system is undesirable. Even if a sine wave voltage is impressed on the primary winding, the exciting current is not a pure sine wave, but contains a third harmonic component. Consequently, the three exciting currents do not add upto zero and neutral is needed to carry the resultant current. If the neutral is not present, the three currents are forced to add upto zero. As a result of this, the flux is not purely sinusoidal and therefore the secondary voltages are distorted from the sine wave. The resultant line-to-neutral voltage may exceed the rated value, so that harmful insulation stresses may develop. Moreover, if the loads are unbalanced, the transformer line-to-neutral voltages can become unbalanced, again causing insulation stresses. Another serious disadvantage of a star-star connection without a primary neutral is that, it is practically impossible to supply power to a load connected between line and neutral on the secondary side. This is because of the reason that source current has to flow through two unloaded primary phase windings. These two-phase windings have high impedance (equal to magnetising reactance). This also leads to unbalanced line-to-neutral voltages.

The problems of unbalanced voltages and third harmonic components can be solved by either of the following two methods.

### (a) Solid grounding of neutrals, especially the primary neutral

The additive third harmonic components will cause a current flow in the neutral and the unbalanced line-to-neutral voltages will be avoided. The neutral also provides a return path for any unbalanced load currents.

### (b) Addition of a delta connected tertiary winding

If a third delta connected winding is added to the transformer, the third harmonic components of voltage will cause a circulating current in this winding. The third harmonic voltage will thus be suppressed. The tertiary winding may, in addition, be used to supply lights and auxillary loads in the sub-station.

The star-star connection is the most economical connection for small high voltage transformers, as the number of turns per phase and the amount of insulation is a minimum. However it is very rarely used because of problems mentioned above. If at all it is to be used, either of the above two correction techniques must be used.

## 2.32 DELTA-DELTA CONNECTION

Figure 2.28 shows this connection of transformer and the phasor diagram. The line voltage is equal to phase voltage on each side. The phase current is line current divided by $\sqrt{3}$ . As compared to star-star connection for the same terminal voltage and current, a delta-delta connection has more numbre of turns in each phase winding but less cross-sectional area of conductors. In view of this a delta-delta connection is more economical for large transformers of relatively lower voltage rating. The closed delta helps in damping out of the third harmonic voltages. The neutral point is not available in this connection. Therefore it is necessary to create an artificial neutral point by connecting three equal resistances or inductances or capacitances in star across terminals $a_2$, $b_2$, $c_2$ whenever necessary.

The connections in Fig. 2.28 (a) cause zero phase displacement between primary and secondary voltages. If the secondary windings are reversed, a displacement of 180° is obtained. When two delta-delta transformers are to be operated in parallel, they should be connected for the same phase displacement.

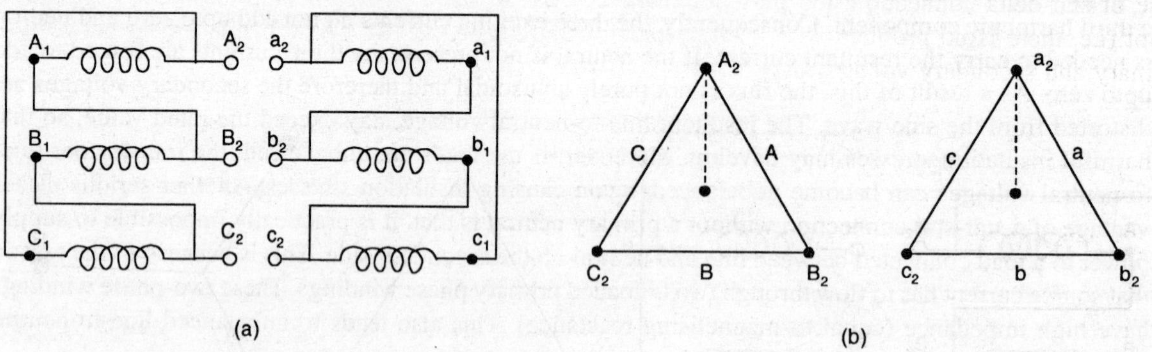


**Fig. 2.28.** Three phase delta-delta transformer (a) connection, (b) primary and secondary emfs

## 2.33 STAR-DELTA CONNECTION

Figure 2.29 shows the connections and phasor diagram for this arrangement. On the primary side the line voltage is $\sqrt{3}$ times the phase voltage, while the line and phase voltages are equal on the secondary side.

This connection has no problem with third-harmonic components, because they are consumed in a circulating current in the delta winding. Moreover this connection is more stable with respect to unbalanced loads, because the delta partially redistributes any unbalance which may occur.

However, this connection causes a 30° phase displacement between primary and secondary voltages (Fig. 2.29b). This has to be taken into account when connecting two such transformers in parallel (both the transformers must be wound for the same phase displacement). Generally the h.v. winding is star connected to result in saving in the cost of insulation. This connection is very commonly used for step-down transformers at receiving end sub-stations.

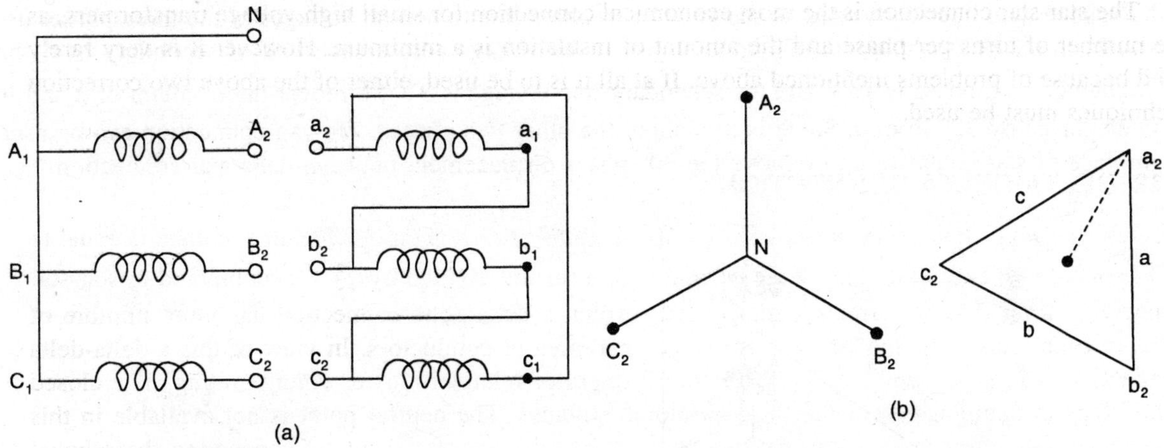


**Fig. 2.29.** Three-phase star-delta transformer (a) connection, (b) primary and secondary emfs

## 2.34 DELTA-STAR CONNECTION

Figure 2.30 shows this connection and the phasor diagram. On the primary side the line and phase voltages are equal, but on the secondary side the line voltage is $\sqrt{3}$ times the phase voltage. As in the case of star-delta connection, the presence of delta winding helps in stabilising the potential of star point (i.e. there is no problem with third harmonic components). A 30° phase displacement between primary and secondary voltages occurs.

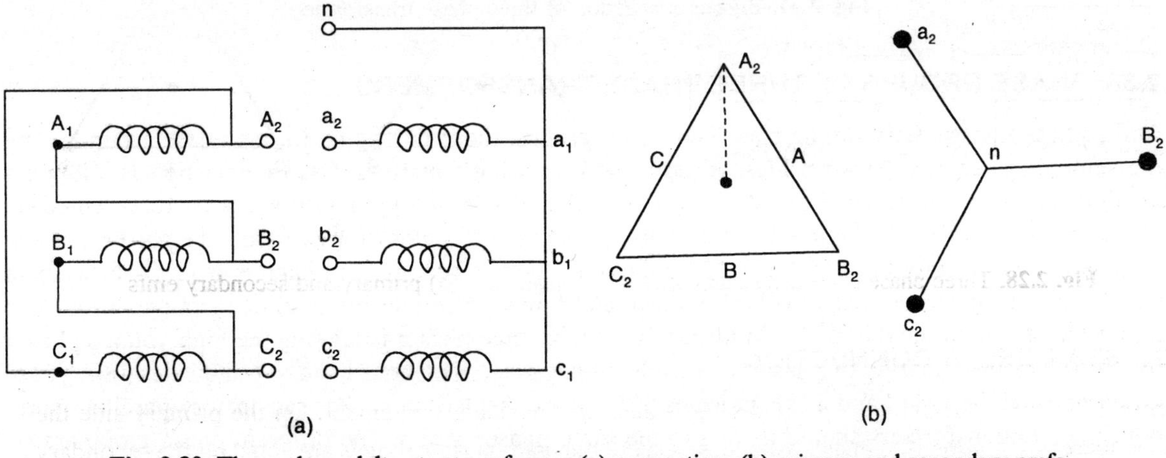


**Fig. 2.30.** Three-phase delta-star transformer (a) connection, (b) primary and secondary emfs

The delta-star connection is very popular for transformers at sending end and receiving end sub-stations in power systems. At power stations the generator feeds the delta winding and the star winding is connected to h.v. transmission lines. In distribution transformers, feeders are connected to delta winding and the star winding supplies 3-phase 4-wire distributors.

## 2.35 ZIG ZAG OR INTER-CONNECTED STAR CONNECTION

Sometimes the secondary windings of a 3 phase transformer are connected in zig-zag or inter-connected star shown in Fig. 2.31 (a). Each transformer secondary is made into two halves and the six halves are then inter-connected as shown. The resultant line to neutral voltages of the three phases are shown in the phasor diagram of Fig. 2.31 (b). If the voltage across each half of secondary is V, the voltage of each phase is $\sqrt{3}$ V and the secondary line voltage is 3V. Moreover the resultant of *a* and *b′* is displaced by 30° from *a*. Same is true about the other two phases. Zig-zag connection can be used in place of star connection to remove the 30° phase displacement between delta star connection.

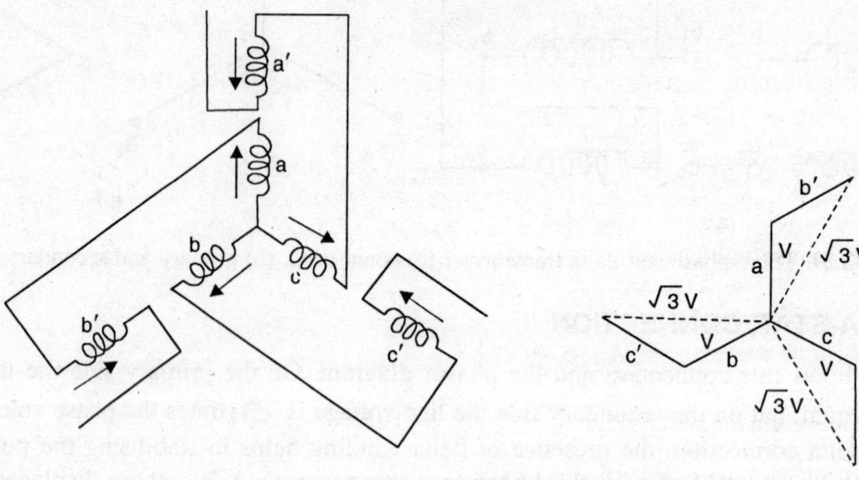


**Fig. 2.31.** Zigzag connection of three phase transformer

## 2.36 PHASE GROUPS OF THREE PHASE TRANSFORMERS

Three phase transformers can be classified into 4 groups, those belonging to one phase group having the same displacement between high voltage and low voltage phasors. Each connection is allotted a symbol. The phase difference between high voltage and low voltage phasors can be represented by clockface hour figure. The high voltage phasor is given the 12'O clock (or zero) and the position of low voltage phasor is represented by the hour hand of the clock. When the hour hand is at 12'O clock, there is no phase displacement between high voltage and low voltage phasors (e.g. connections Yyo, Ddo, etc). When the hour hand is at 1, it indicates that low voltage phasor is lagging the high voltage phasor by 30° (clockwise rotation of phasor is considered negative). The hour hand at 6 indicates 180° phase difference and the hour hand at 11 indicates +30° phase displacement. For satisfactory parallel operation, the two transformers should belong to the same phasor group. The different phase group are:

Group 1: Zero phase displacement. The connections which belong to this group are Yyo, Ddo and Dzo (i.e. star-star, delta-delta and delta-zigzag). The connections and their phasor diagrams are shown in Fig. 2.32.

Group 2: 180° phase displacement. The connections in this group are Yy 6, Dd 6 and Dz 6 and are shown in Fig. 2.35.

Group 3: 30° lag (or –30°) phase displacement. The connections in this phase group are Dy 1, Yd 1 and Yz 1 and are shown in Fig. 2.34.

Group 4: 30° lead (or +30°) phase displacement. The connections in this group are Dy 11, Yd 11 and Yz 11 and are shown in Fig. 2.33.

### *Example 2.23*

A 3-phase step down transformer is connected to 6600 V on the primary side. The ratio of turns per phase is 12 and the line current drawn from mains is 20 A. Find secondary line voltage, line current and output if the transformer is (a) star-star, (b) delta-delta, (c) star-delta, (d) delta star. Neglect losses.

*Solution*

$$\text{Input} = \sqrt{3} \times 6600 \times 20/1000 = 228.62 \text{ kVA}$$

Since losses have been neglected, output in each case is 228.62 k VA

(a) Primary phase voltage $= \dfrac{6600}{\sqrt{3}}$

Secondary phase voltage $= \dfrac{6600}{\sqrt{3} \times 12}$

Secondary line voltage $= \dfrac{6600}{\sqrt{3} \times 12} \times \sqrt{3} = 550 \text{ V}$

Primary phase current $= 20 \text{ A}$

Secondary phase current $= 20 \times 12 = 240 \text{ A}$

Secondary line current $= 240 \text{ A}$

(b) Primary phase voltage $= 6600 \text{ V}$

Secondary phase voltage $= \dfrac{6600}{12} = 550 \text{ V}$

Secondary line voltage $= 550 \text{ V}$

Primary phase current $= \dfrac{20}{\sqrt{3}} \text{ A}$

Secondary phase current $= \dfrac{20}{\sqrt{3}} \times 12 \text{ A}$

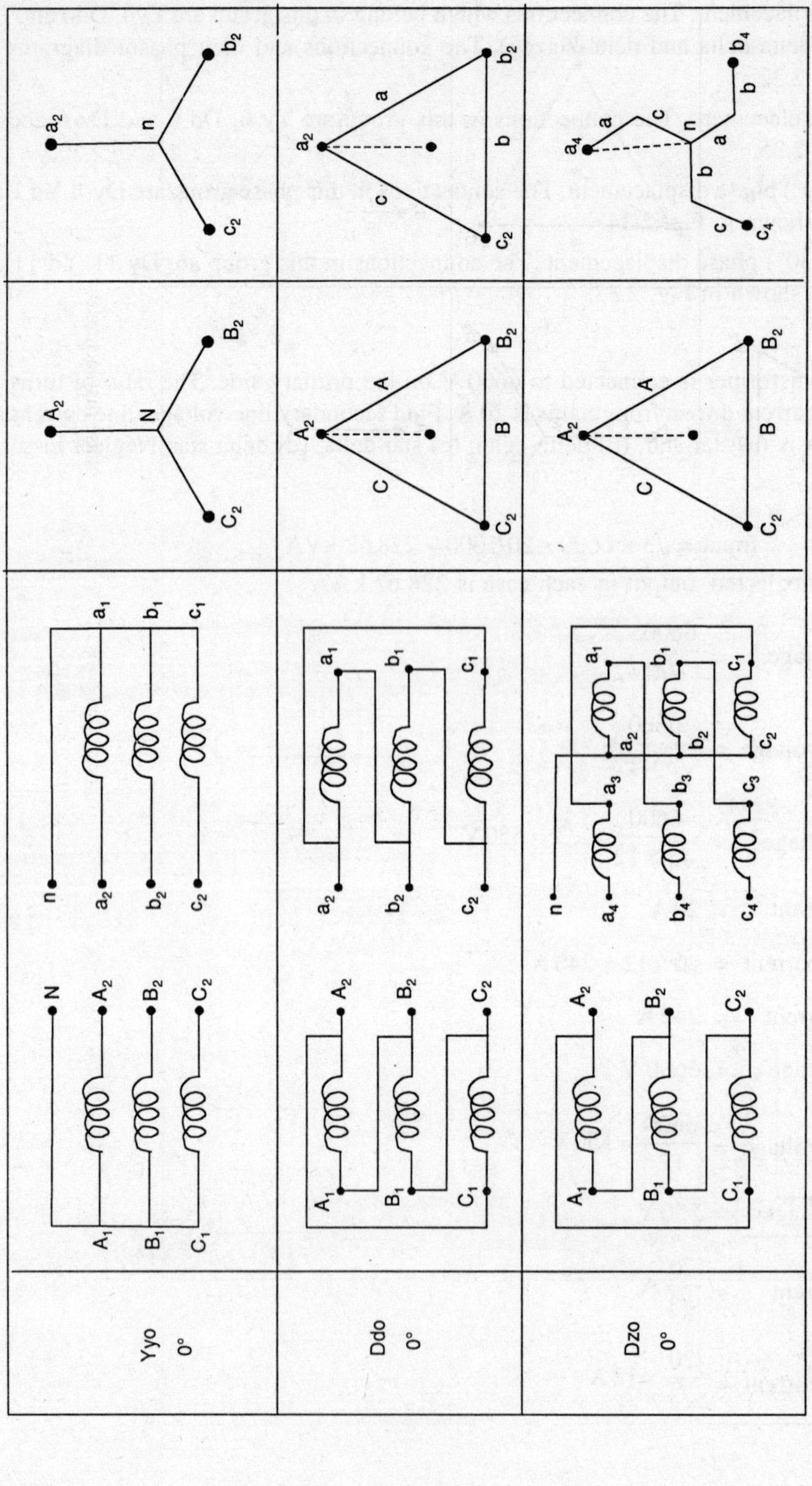


**Fig. 2.32.** Three phase transformer connections for zero phase displacement between primary and secondary voltages

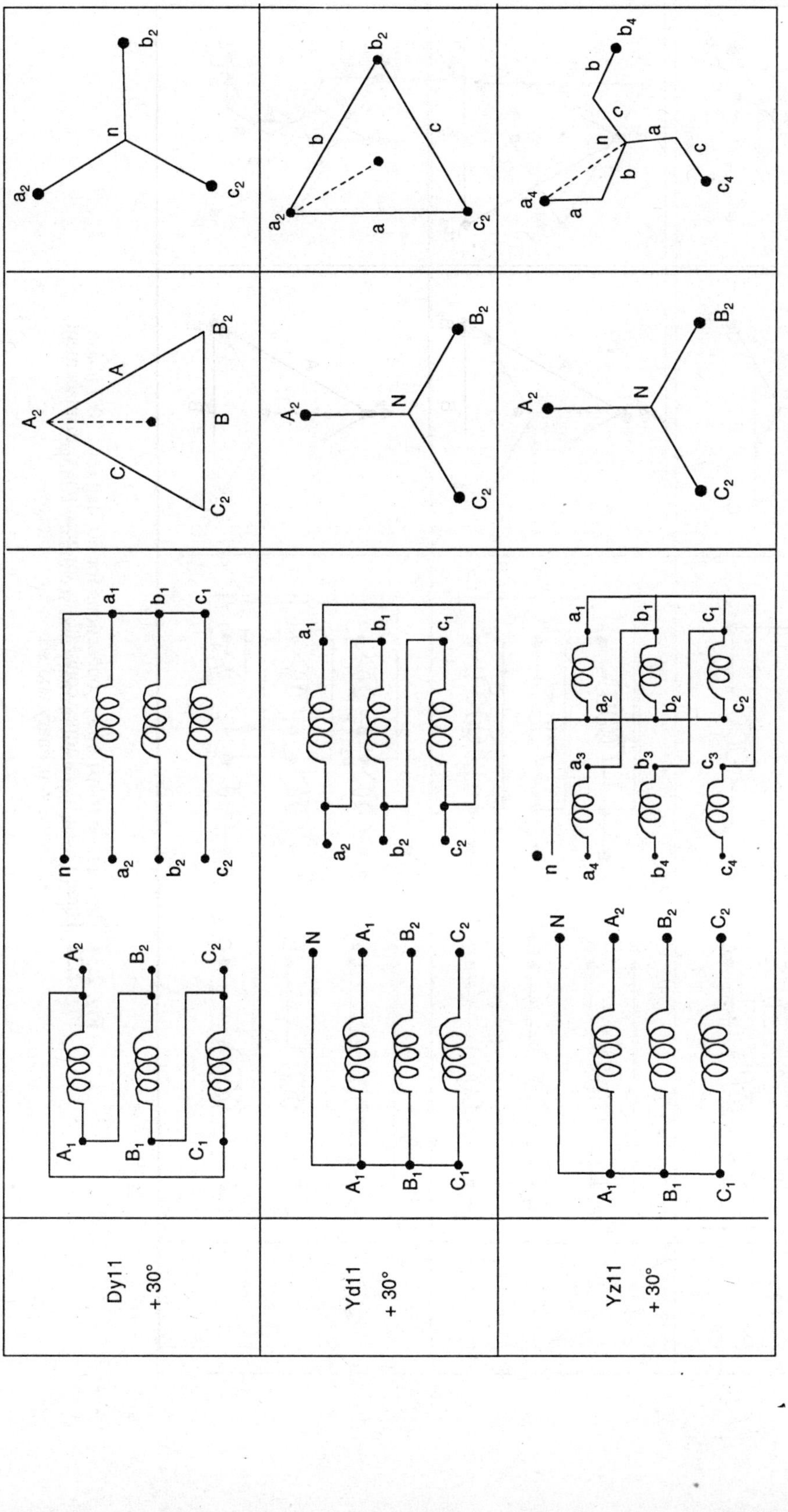


**Fig. 2.33.** Three phase transformer connections for 30° lead (+30°) phase displacement between primary and secondary voltages

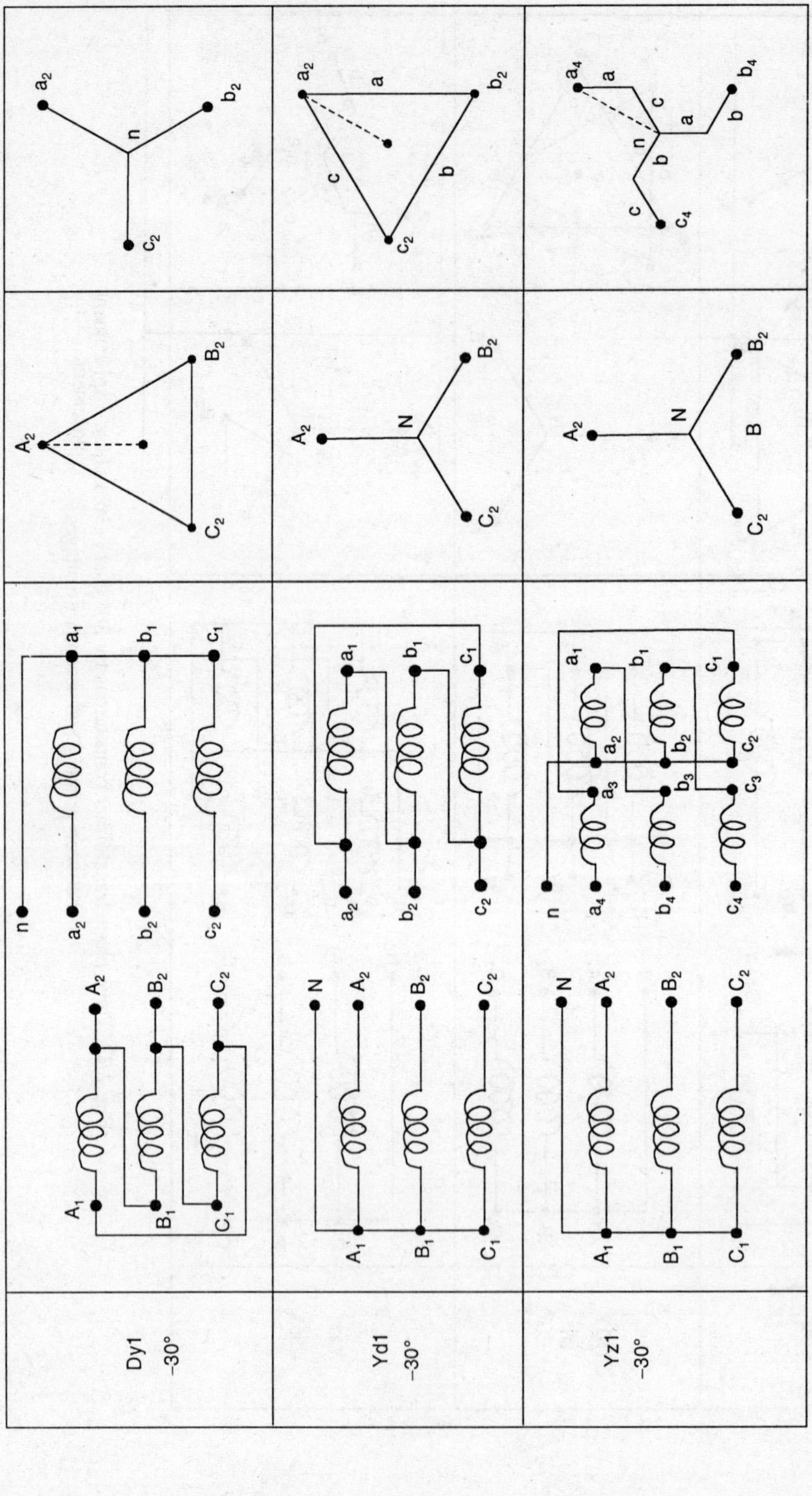


**Fig. 2.34.** Three phase transformer connections for 30° lag (or –30°) phase displacement between primary and secondary voltages

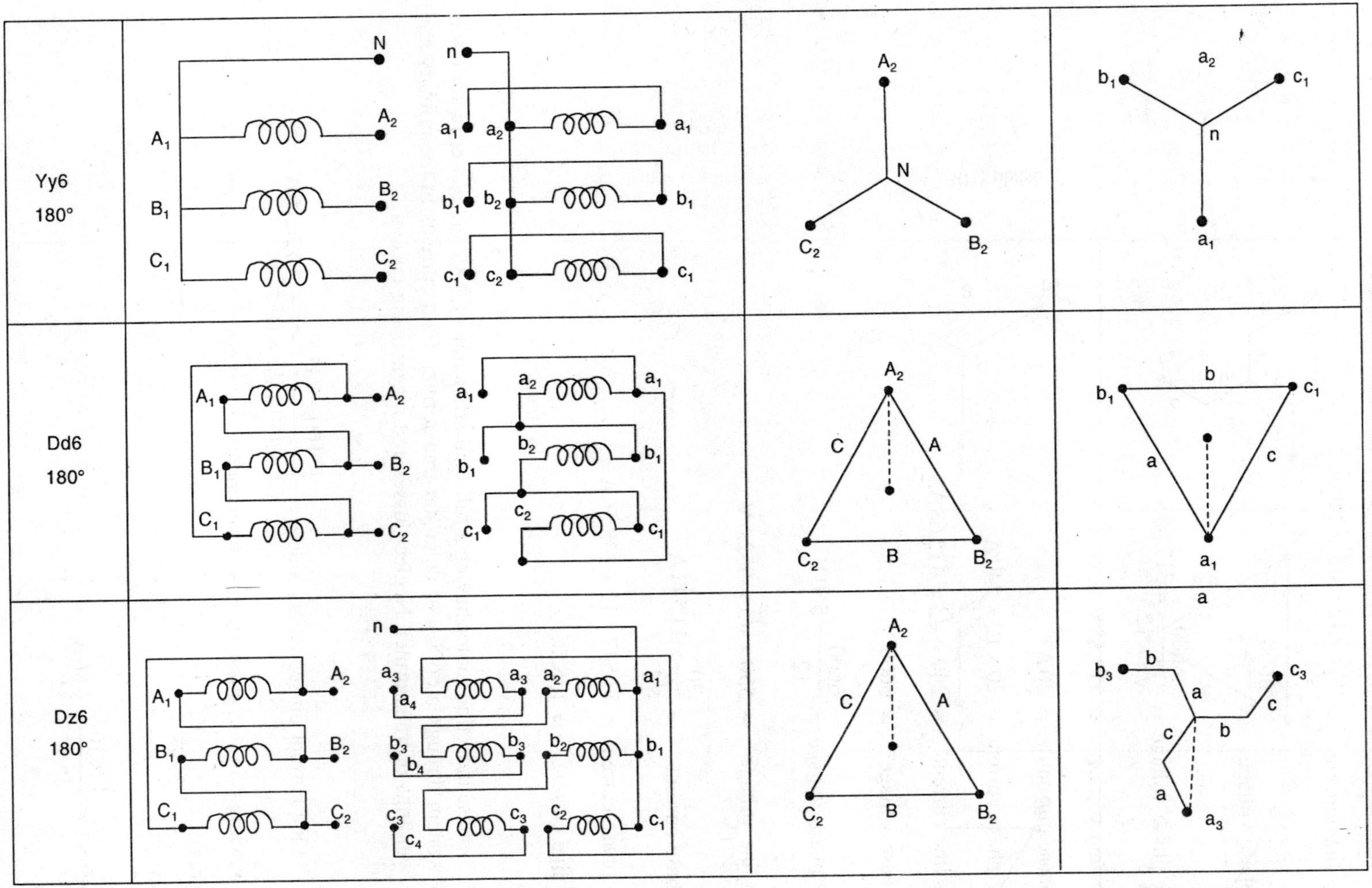


**Fig. 2.35.** Three phase transformer connections for 180° phase displacement between primary and secondary voltages

Secondary line current $= \frac{20}{\sqrt{3}} \times 12 \times \sqrt{3} = 240\text{ A}$

(c) Primary phase voltage $= \frac{6600}{\sqrt{3}}$

Secondary phase voltage $= \frac{6600}{\sqrt{3} \times 12} = 317.55\text{ V}$

Secondary line voltage $= 317.55\text{ V}$

Primary phase current $= 20\text{ A}$

Secondary phase current $= 20 \times 12 = 240\text{ A}$

Secondary line current $= 240 \times \sqrt{3} = 415.68\text{ A}$

(d) Primary phase voltage $= 6600\text{ V}$

Secondary phase voltage $= \frac{6600}{12} = 550\text{ V}$

Secondary line voltage $= 550 \times \sqrt{3} = 952.6\text{ V}$

Primary phase current $= \frac{20}{\sqrt{3}} = 11.547\text{ A}$

Secondary phase current $= 11.547 \times 12 = 138.56\text{ A}$

Secondary line current $= 138.56\text{ A}.$

*Example 2.24*

**A 100 hp, 440 V, 3-phase induction motor having a full-load efficiency of 0.85 and power factor 0.9 is fed from 6600 V system through a 6600/440 delta-star transformer. Find currents in the primary and secondary phase windings of transformer. Neglect losses and magnetising current.**

*Solution*

$$\text{Line current on secondary side} = \frac{100 \times 735.5}{\sqrt{3} \times 440 \times 0.9 \times 0.85} = 126.16\text{ A}$$

Since secondary winding is in star, phase current on secondary side is equal to line current, i.e. 126.16 A

$$\text{Turn ratio} = \frac{6600}{\left(440/\sqrt{3}\right)} = 25.98$$

$$\text{Primary phase current} = \frac{126.16}{25.98} = 4.86\text{ A}.$$

## 2.37 BACK-TO-BACK TEST ON 3-PHASE TRANSFORMERS

A variety of connections are available for conducting a back-to-back test (or temperature rise test) on 3-phase transformers. One circuit is shown in Fig. 2.36. The test requires only one transformer. The low voltage windings are connected in delta and fed from a 3-phase supply at rated voltage and frequency. The power input is measured by two-wattmeter method. The sum of $W_1$ and $W_2$ gives the core loss. The secondaries are connected in open-delta and a low single-phase rated frequency voltage is applied to it as shown in Fig. 2.36. The voltage applied to open-delta is adjusted, so that full-load current circulates. This voltage, evidently equals 3 × full-load current × impedance per phase as referred to h.v. side. Wattmeter $W_3$ gives the total copper-loss.

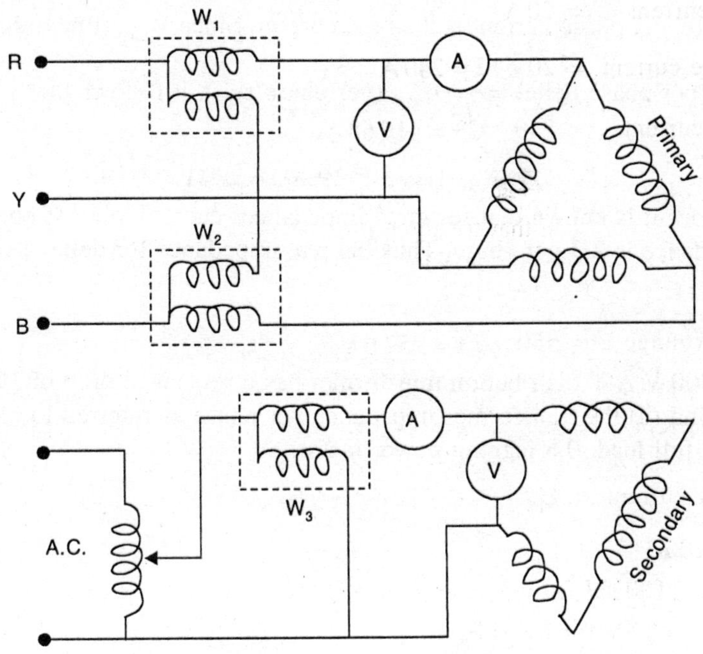


**Fig. 2.36.** Back-to-back test on 3-phase transformer

## 2.38 PER UNIT SYSTEM FOR THREE-PHASE TRANSFORMERS

When applied to 3-phase systems, the base values are so selected as to satisfy the 3-phase balanced circuit relations between line and phase quantities. The 3-phase kVA (or MVA) is taken as base kVA (or MVA) and the line-to-line voltage (in volts or kV) is taken as base voltage.

Assuming star connection (a delta connection can always be converted into equivalent star)

$$V_{phase} = \frac{V_{line}}{\sqrt{3}} \text{ and } I_{phase} = \frac{\text{Volt - amp}}{\sqrt{3}\, V_{line}} \qquad ...(2.76)$$

and

$$Z_{base} = \frac{V_{phase}}{I_{phase}} \qquad ...(2.77)$$

Combining Eqs. (2,76) and (2.77)

$$Z_{base} = \frac{(kV)^2}{MVA} \quad ...(2.78)$$

In Eq. (2.78), kV is line-to-line and MVA is 3-phase MVA. Hence

$$Z_{pu} = \frac{[Z(\Omega)][MVA]}{(kV)^2} \quad ...(2.79)$$

For delta connections, the value of $Z_{base}$ is again given by Eq. (2.78), but in this case line-to-line voltage is base voltage and phase current is the base current. Since $V_{base}$ (line to line) is $\sqrt{3}\ V_{base}$ (line to neutral) and $I_{base}$ (per phase delta) $= \frac{1}{\sqrt{3}} I_{base}$ (per phase star), it follows that

$$Z_{base,\text{ per phase }\Delta} = 3Z_{base,\text{ per phase }Y} \quad ...(2.80)$$

From circuit theory it is known that for three impedances each of value $Z$ connected in delta, the equivalent star impedance is $Z/3$ per phase. Thus the p.u. impedance for delta or equivalent star is the same.

*Example 2.25*

A 100 kVA, 11000/400 V Δ-Y distribution transformer has a resistance drop of 1% and leakage reactance drop of 6%. Find (a) the transformer impedance per phase as referred to high voltage side, (b) voltage regulation at full-load, 0.8 lagging power factor.

*Solution*

(a) Base kVA = 100

Base voltage 11000 V for h.v. side

400 V for l.v. side

Since the primary is delta connected,

Base VA per phase = 100 × 1000/3

Base voltage = 11000 V

$$\text{Base current} = \frac{\text{Base VA per phase}}{\text{Base voltage per phase}}$$

$$= \frac{100 \times 1000}{3 \times 11000} = 3.03\,\text{A}$$

Base impedance for h.v side

i.e.

$$Z_b = \frac{11000}{3.03} = 3630 \text{ ohm}$$

$$R_{pu} = 0.01$$

$$Z_{pu} = 0.06$$

$$X_{pu} = \left(Z^2 - R^2\right)^{0.5} = 0.0592 \text{ pu}$$

$$Z_{pu} = 0.01 + j\,0.0592$$

$$Z = \left(Z_{pu}\right)\left(Z_b\right) = (0.01 + j0.0592)(3630)$$

$$= 36.3 + j214.9 \text{ ohm}$$

(b) Voltage regulation $= R_{pu}\cos\theta + X_{pu}\sin\theta$

$$= (0.01)(0.8) + (0.0592)(0.6)$$

$$= 0.0435 \text{ or } 4.35\%.$$

## 2.39 OPEN DELTA (OR V-V) CONNECTION

A bank of three separate transformers connected to Δ-Δ will continue to operate, if one of the transformers gets damaged and is removed from the circuit. The resulting connection is known as open-delta or V connection. However, the load that can be supplied is less than the capacity of the remaining two transformers.

Figure 2.37 shows the transformer bank connected in Δ-Δ feeding a resistive load. If the rated voltage of transformer in each phase is $V_p$ and rated current is $I_P$, the power rating of the bank is

$$P_{3\text{-phase}} = 3V_pI_p \qquad ...(2.81)$$

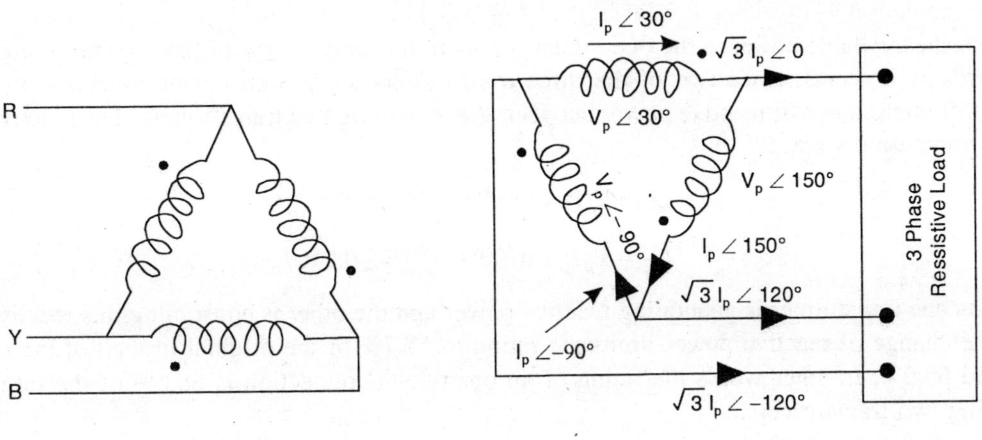


**Fig. 2.37**

Figure 2.38 shows the open delta connection. The voltages and currents in the remaining two transformers are also shown. Since one of the transformers is missing, the line current supplied to the load is equal to phase current in each transformer. For transformer in phase 1, the voltage has a phase angle of 150° and the current has a phase angle of 120°.

Therefore $$P_1 = V_pI_p\cos(150°-120°) = \frac{\sqrt{3}}{2}V_pI_p.$$

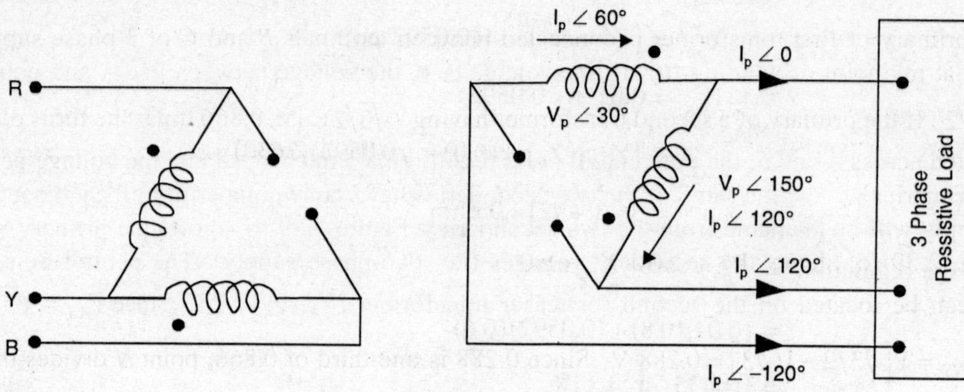


**Fig. 2.38.** Open delta connection

For transformer 2, voltage has a phase angle of 30° and the current has a phase angle of 60°. Therefore

$$P_2 = V_p I_p \cos(30°–60°) = \frac{\sqrt{3}}{2} V_p I_p$$

Therefore $$P_{\text{open-}\Delta} = P_1 + P_2 = \sqrt{3}\, V_p I_p \qquad ...(2.82)$$

and $$\frac{P_{\text{open-}\Delta}}{P_{3\text{–phase}}} = \frac{\sqrt{3}\, V_p I_p}{3 V_p I_p} = \frac{1}{\sqrt{3}} = 0.577 \qquad ...(2.83)$$

Thus the available power in the open-delta connection is only 57.7% of the original rating (and not two thirds as it seems, since two of the three transformers are present). This reduction in rating is because of interchange of reactive power between the remaining two transformers. The reactive powers of two transformers are

$$Q_1 = V_p I_p \sin(150°–120°) = 0.5\, V_p I_p$$

$$Q_2 = V_p I_p \sin(30°–60°) = –0.5\, V_p I_p$$

Thus one transformer is generating reactive power and the other is consuming this reactive power. This interchange of reactive power limits the rating to 57.7% of the original instead of the otherwise expected 66.6%. In other words the rating of an open-delta connection is 86.6% of the rating of the remaining two transformers.

Open-delta connection is, sometimes, used to supply a system which requires a small amount of 3-phase power along with a large amount of single-phase power. In such a case, on of the transformers (across which the single-phase load is connected) is bigger than the other.

## 2.40 THREE-PHASE TO TWO-PHASE CONVERSION

This arrangement is needed to supply two-phase furnaces from a 3-phase source. The most commonly used connection is known as Scott connection shown in Fig. 2.39 (a). It requires two transformers of different ratings, but for interchangeability and provision of spares, the transformers may be identical, both being provided with required tappings.

The primary of first transformer is connected between terminals $B$ and $C$ of 3-phase supply. The point $S$ is at midpoint of winding $BC$. If line voltage is $V$, the voltage between line $A$ and point $S$ will be $(\sqrt{3}V)/2$. If the primary of a second transformer having $(\sqrt{3}/2)$, i.e. 0.866 times the turns of the first is connected across $A$ and $S$, the potential of point $S$ will remain unchanged and the voltage per turn in the two primaries will be the same. If the two secondaries have same number of turns, a balanced two-phase supply will be available from the two secondaries. Figure 2.39 (b) shows the primary voltages, while Fig. 2.39 (c) shows the secondary voltages (i.e. two-phase supply). The neutral point on the primary can be located on the second (or teaser transformer). $V_{AN} = V/\sqrt{3}$. Since $V_{AS} = \sqrt{3}V/2$, the voltage $V_{NS} = V\left[\sqrt{3}/2 - 1/\sqrt{3}\right] = 0.288$ V. Since 0.288 is one-third of 0.866, point $N$ divides the teaser winding $AS$ in the ratio of 2:1.

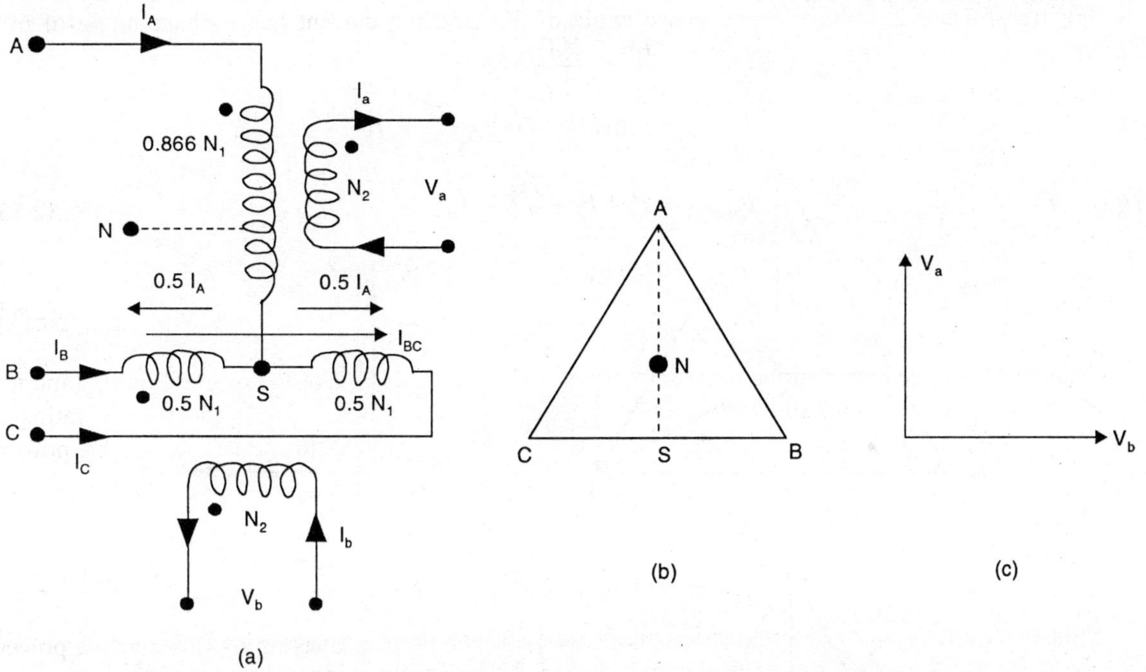


**Fig. 2.39.** Scott connection (a) circuit, (b) primary emfs, (c) secondary emfs

If the secondary currents are $I_a$ and $I_b$, the primary currents can be found from the phasor diagrams of Fig. 2.39.

$$I_A = \frac{2I_a N_2}{\sqrt{3}\,N_1} = \frac{2I_a}{\sqrt{3}}, \text{ if } N_1 = N_2 \qquad ...(2.84a)$$

$$I_{BC} = \frac{I_b N_2}{N_1} = I_b \quad \text{if } N_1 = N_2 \qquad ...(2.84b)$$

$$I_B = I_{BC} - 0.5 I_A \qquad ...(2.85)$$

$$I_C = -I_{BC} - 0.5 I_A \qquad \text{...(2.86)}$$

Figure 2.40 (a) shows the phasor diagram for balanced two-phase load at unity power factor. It is seen that primary currents are also balanced. Figure 2.40(b) shows the phasor diagram for unbalanced load currents.

### *Example 2.26*

Two single-phase furnaces are supplied at 120 V through a Scott connection from a 3-phase 6600 V system. The voltage of furnace *A* is leading. The loads are (a) 400 kW and 800 kW both at unity power factor, (b) 400 kW at 0.8 lagging power factor (i.e. furnace *A*) and 800 kW (i.e. furnace *B*) at unity power factor. Find line currents on 3-phase side.

### *Solution*

With reference to Fig. 2.39 (a)

$$\frac{N_1}{N_2} = \frac{6600}{120} = 55$$

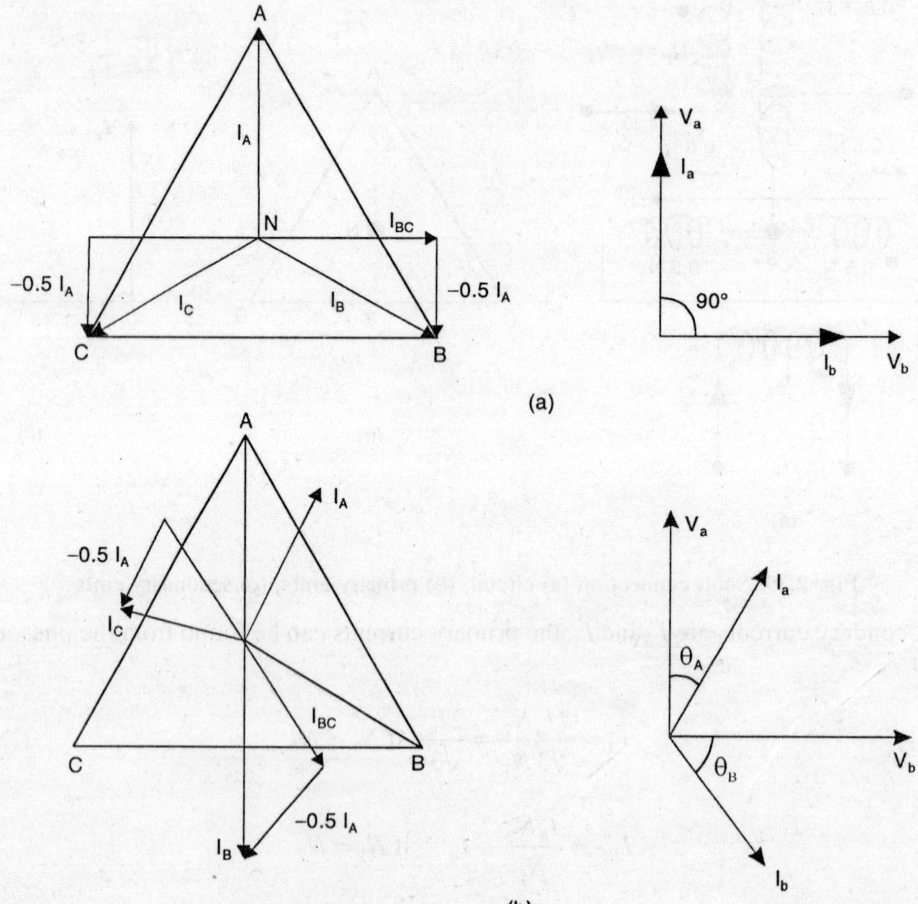


**Fig. 2.40.** Phasor diagram for Scott connection (a) balanced load currents, (b) unbalanced load currents

(a) With reference to phasor diagram of Fig. 2.40 (a)

$$I_a = \frac{400 \times 1000}{120} = 3333.33 \angle 90° \text{ A}$$

$$I_b = \frac{800 \times 1000}{120} = 6666.66 \angle 0 \text{ A}$$

The phase angle of $I_a$ has been written as 90°, because $I_a$ is in phase with $V_a$ and $V_a$ is leading $V_b$ by 90° as seen in phasor diagram of Fig. 2.40 (a)

$$I_A = \frac{2I_a N_2}{\sqrt{3}\, N_1} = \frac{2(3333.33 \angle 90°)}{\sqrt{3} \times 55} = 69.98 \angle 90° \text{ A}$$

$$I_{BC} = \frac{I_b N_2}{N_1} = \frac{6666.66}{55} = 121.21 \angle 0 \text{ A}$$

$$I_B = I_{BC} - 0.5 I_A = 121.21 - 0.5(69.98 \angle 90°)$$

$$= 121.21 - j34.99 = 126.16 \text{ A}$$

$$I_C = -I_{BC} - 0.5 I_A = -121.21 - 0.5(69.98 \angle 90°)$$

$$= -121.21 - j34.99 = 126.16 \text{ A}$$

(b)

$$I_a = \frac{400 \times 1000}{120 \times 0.8} = 4166.66 \text{ A}$$

$$I_b = \frac{800 \times 1000}{120} = 6666.66 \text{ A}$$

Taking $V_b$ as reference

$$I_b = 6666.66 \angle 0 \text{ A}$$

$$I_a = 4166.66 \angle \left(90 - \cos^{-1} 0.8\right)$$

$$= 4166.66 \angle 53.13° \text{ A}$$

$$I_A = \frac{2I_a N_2}{\sqrt{3} N_1} = \frac{2(4166.66 \angle 53.13°)}{\sqrt{3} \times 55}$$

$$= 87.48 \angle 53.13° \text{ A}$$

$$= 52.5 + j70 \text{ A}$$

$$I_{BC} = I_b \frac{N_2}{N_1} = \frac{6666.66}{55} = 121.21 \angle 0 \text{ A}$$

$$I_B = I_{BC} - 0.5 I_A = 121.21 - 0.5(52.5 + j70)$$

$$= 94.96 - j35 = 101.2\,\text{A}$$

$$I_C = -I_{BC} - 0.5I_A = -121.21 - 0.5(52.5 + j70)$$

$$= -147.46 - j35 = 151.56\,\text{A}.$$

## 2.41 THREE PHASE TO SIX-PHASE CONVERSION

A six-phase supply is often needed for rectifier and thyristor circuits. A 3-phase to six-phase transformation is possible by using three single-phase transformers, each transformer having two independent but identical secondaries. A commonly used connection known as double star connection is shown in Fig. 2.41. The three primaries have been connected in delta, but may also be connected in star subject to what has been mentioned about third harmonics in 3-phase star-star connection. One set of three secondaries is connected in star and the other in reversed star, thereby giving six voltages displaced by 60° from each other. The two neutral points may be connected together and this neutral will serve as the neutral wire of dc supply from the rectifier.

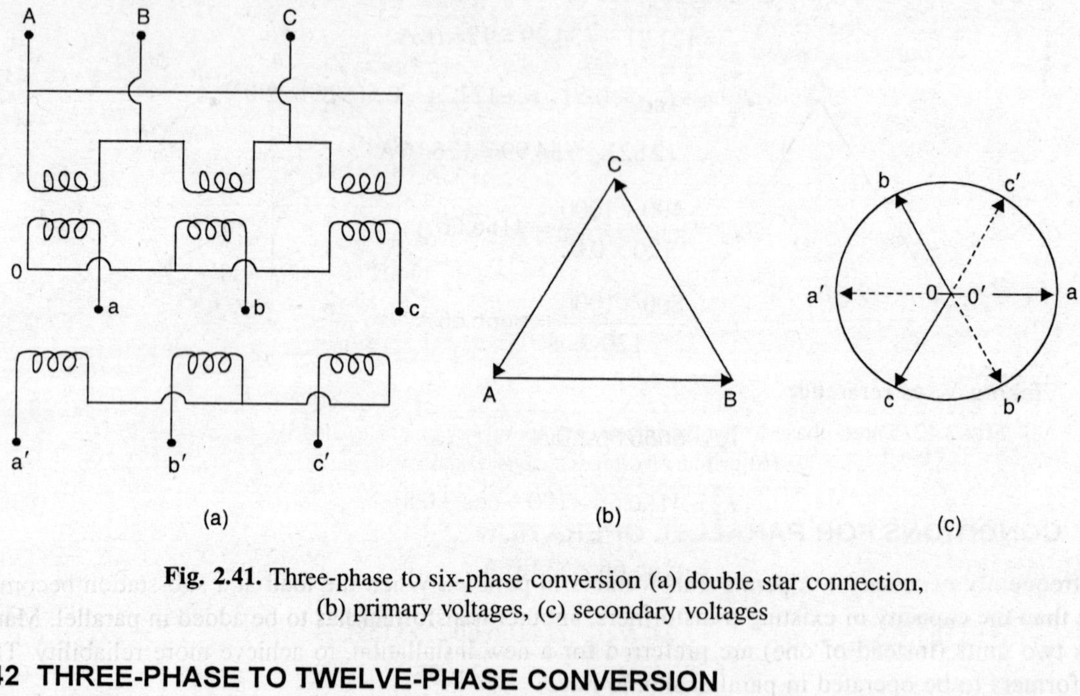


**Fig. 2.41.** Three-phase to six-phase conversion (a) double star connection, (b) primary voltages, (c) secondary voltages

## 2.42 THREE-PHASE TO TWELVE-PHASE CONVERSION

The efficiency of conversion from ac to dc by rectifier and thyristor circuits increases with increase in number of phases and this may necessitate a 12-phase supply. Figure 2.42 shows a connection known as Y – Δ, double star connection. It requires two banks of three transformers. Each bank has two secondaries arranged for double-star connection. The primaries of one bank are connected in star, while those of second are connected in delta (Fig. 2.42) thus creating a 30° phase angle between the secondary star voltages of the two systems. If $V_1$ is the input line voltage, each phase winding of star connected primary is designed for $V_1/\sqrt{3}$ volts, while that of delta connected primary is designed for $V_1$ volts. If $V_2$ is the line-to-line voltage of the twelve-phase circuit, each secondary winding should be designed for $V_2/(2\sin 15°)$ or 1.932 $V_2$ volts.

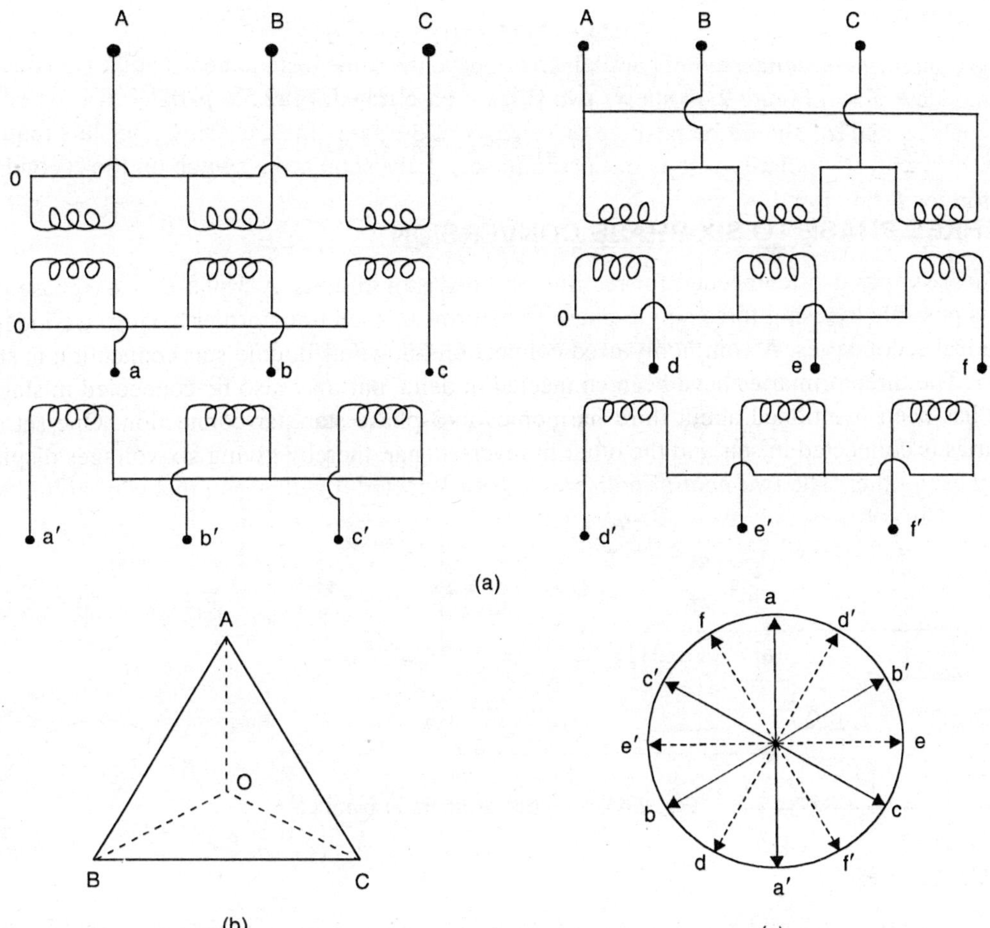


**Fig. 2.42.** Three-phase to twelve-phase conversion (a) star-delta, double star connection, (b) primary voltages, (c) secondary voltages

## 2.43 CONDITIONS FOR PARALLEL OPERATION

It is frequently necessary to operate transformers in parallel. When the load at a sub-station becomes more than the capacity of existing transformers, another transformer has to be added in parallel. Many times two units (instead of one) are preferred for a new installation, to achieve more reliability. The transformers to be operated in parallel should have

(a) the same polarity
(b) the same voltage ratio
(c) equal per unit impedance in magnitude and phase angle (i.e. equal $X/R$ ratios)
(d) the same phase sequence and zero relative phase displacement (in case of 3-phase transformers).

The conditions (a) and (d) are essential and must be satisfied. Otherwise parallel operation is not possible. The condition (b) must also be satisfied to a very close degree. If condition (c) is satisfied, the load sharing will be proper. Otherwise one of the transformers may be over loaded and the other may remain under loaded.

### 2.43.1 Polarity

If the two transformers, connected in parallel, do not have the same instantaneous polarity, a dead short-circuit may take place. Figure 2.43 shows two single-phase transformers in parallel. It is essential that the terminals $a_1$ and $a_1'$ should be positive or negative at the same instant. Only after this requirement has been met with (by polarity test as discussed in sec. 2.20) should the switch on the secondary side of transformer $B$ be closed.

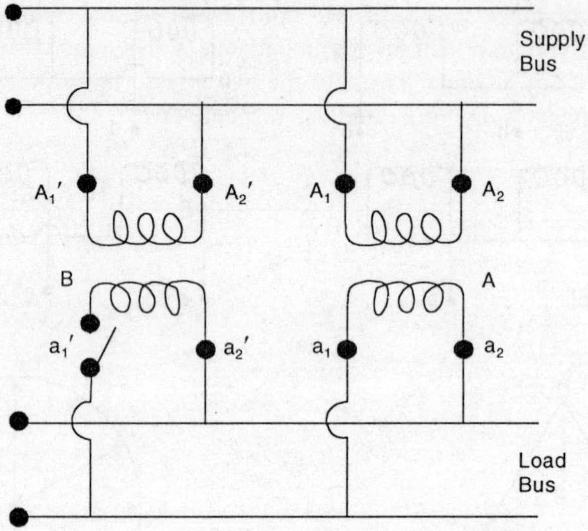


**Fig. 2.43.** Two transformers in parallel

### 2.43.2 Voltage Ratio

The two transformers should have the same voltage ratio. Otherwise, circulating current would flow. Even a small difference in voltage ratio can cause large circulating current, because the internal impedance of a transformer is very low. It is pertinent to note that in case of transformers having high values of $N_1/N_2$, the voltage ratio $V_2/V_1$ may not be the same, even if the ratio $N_2/N_1$ is the same. This may happen if the per unit impedances are not the same.

### 2.43.3 Impedance

Transformers connected in parallel may not have the same kVA rating. However, it is essential that they share the load in the ratio of their ratings. Since they are in parallel on the primary as well as secondary side, it is necessary that they have the same voltage regulation. This would happen if they have the same per unit impedance.

In case the $X/R$ ratios of the two transformers are not equal, they would not operate at the same power factor.

### 2.43.4 Phase Sequence and Phase Displacement

The three-phase transformers connected in parallel must have the same phase sequence and phase displacement (between primary and secondary voltages). Otherwise a dead short may occur. To ensure correct phase sequence, the terminals on h.v. and l.v. sides are labelled for conventional positive phase sequence. The phase difference between primary and secondary voltages depends on the connections (i.e. star or delta) of the primary and secondary winding. Only the transformers which have the same phase displacement can be put in parallel.

## 2.44 LOAD SHARING BY TRANSFORMERS IN PARALLEL

Figure 2.44 shows the equivalent circuit for two transformers $A$ and $B$ in parallel. They supply load currents $I_A$ and $I_B$, respectively to a load having impedance $Z$. The load current is $I$ and voltage across the load in $V$. The open-circuit secondary emfs of the two transformers are $E_A$ and $E_B$ and their impedances as referred to secondary side are $Z_A$ and $Z_B$, respectively.

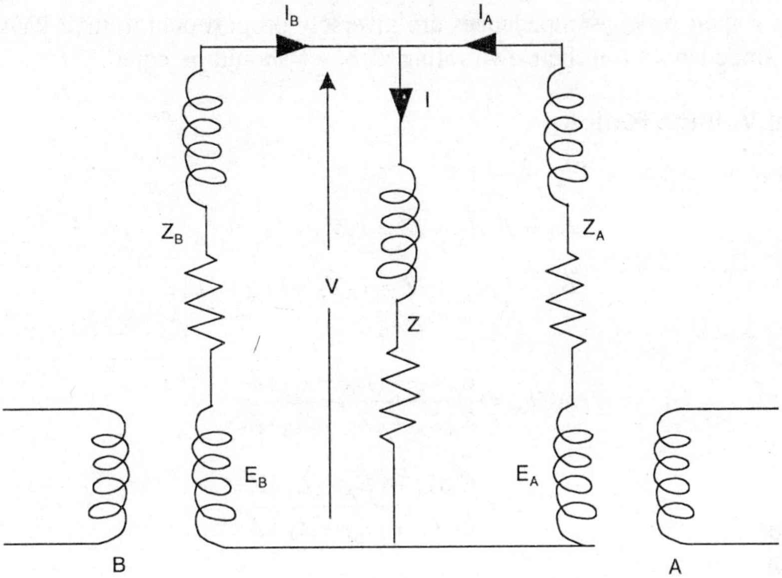


**Fig. 2.44.** Equivalent circuit of two transformers in parallel

### 2.44.1 Equal Voltage Ratios

If $E_A$ and $E_B$ are equal in magnitude and are in phase, there will not be any circulating current between the transformers. Then

$$I_A Z_A = I_B Z_B \qquad ...(2.87)$$

$$I = I_A + I_B \qquad ...(2.88)$$

From Eqs. (2.87) and (2.88)

$$I_A = I \frac{Z_B}{Z_A + Z_B} \qquad ...(2.89)$$

$$I_B = I \frac{Z_A}{Z_A + Z_B} \qquad ...(2.90)$$

If both sides of Eqs. (2.89) and (2.90) are multiplied by $V$, we get the equations for VA delivered by each transformer. $VI_A$ is $S_A$, the $VA$ delivered by transformer $A$, $VI_B$ is $S_B$ and $VI$ is the total volt-amp. Thus

$$S_A = S\frac{Z_B}{Z_A + Z_B} \qquad ...(2.91)$$

$$S_B = S\frac{Z_A}{Z_A + Z_B} \qquad ...(2.92)$$

In above equations $Z_A$ and $Z_B$ can be expressed in ohms or in p.u. If p.u. values are to be used, they should be with respect to a common base. $S_A$, $S_B$ and $S$ can be in VA, kVA, MVA or in p.u.

From above equations, it is evident that the transformers will share the load in proportion to their kVA ratings only if their leakage impedances are inversely proportional to their kVA ratings. In other words, their p.u. impedances (on their own rating as base) should be equal.

### 2.44.2 Unequal Voltage Ratios

KVL equations for the circuit in Fig. 2.44 are

$$E_A = I_A Z_A + IZ = I_A Z_A + (I_A + I_B)Z \qquad ...(2.93)$$

$$E_B = I_B Z_B + IZ = I_B Z_B + (I_A + I_B)Z \qquad ...(2.94)$$

The solution is

$$I_A = \frac{E_A Z_B + (E_A - E_B)Z}{Z_A Z_B + (Z_A + Z_B)Z} \qquad ...(2.95)$$

$$I_B = \frac{E_B Z_A - (E_A - E_B)Z}{Z_A Z_B + (Z_A + Z_B)Z} \qquad ...(2.96)$$

$$I = \frac{E_A Z_B + E_B Z_A}{Z_A Z_B + (Z_A + Z_B)Z} \qquad ...(2.97)$$

$$V = IZ = \frac{E_A Z_B + E_B Z_A}{\frac{Z_A Z_B}{Z} + Z_A + Z_B} \qquad ...(2.98)$$

It is clear from the above equations that if $E_A$ and $E_B$ are not equal, there is a circulating current between the transformers. This circulating current is given by

$$I = \frac{(E_A - E_B)Z}{Z_A Z_B + (Z_A + Z_B)Z} \qquad ...(2.99)$$

## 2.45 PARALLEL-GENERATOR THEOREM

A solution for any number of transformers (or generators) in parallel is possible by the use of this theorem. Referring to Fig. 2.44

$$V = (I_A + I_B)Z = \left[\frac{E_A - V}{Z_A} + \frac{E_B - V}{Z_B}\right]Z$$

$$= \left[\frac{E_A}{Z_A} + \frac{E_B}{Z_B}\right]Z - V\left[\frac{1}{Z_A} + \frac{1}{Z_B}\right]Z$$

or

$$V\left[\frac{1}{Z} + \frac{1}{Z_A} + \frac{1}{Z_B}\right] = \frac{E_A}{Z_A} + \frac{E_B}{Z_B} \qquad ...(2.100)$$

The left hand side of Eq. (2.100) contains the sum of reciprocals of impedances of transformers and load. The terms $E_A/Z_A$, $E_B/Z_B$ are the short-circuit currents of the transformers taken separately. Equation (2.100) is applicable to any number of transformers (or generators).

## 2.46 PARALLEL OPERATION OF THREE-PHASE TRANSFORMERS

Three-phase transformers can be operated in parallel if they have the same ratio of primary to secondary line-to-line voltages and the secondary voltages of the same phase of the two transformers are in phase. If two star-star transformers or delta-delta transformers are to be operated in parallel, both should be connected for zero phase displacement or 180° phase displacement. A delta-delta transformer can also be operated in parallel with star-star transformer. A Δ – Y transformer can be operated in parallel with another Δ –Y one. Similarly, two Y – Δ transformers can be operated in parallel. However, a Δ – Δ or Y – Y transformer cannot be put in parallel with a Δ – Y or Y – Δ one, because this would result in 30° phase displacement between primary and secondary line voltages and consequently a large circulating current. As regards the parallel operation of Y – Δ and Δ – Y transformers, it is seen from Fig. 2.29 and 2.30 that the secondary line voltages of Δ – Y transformer lead the corresponding secondary voltages of Y – Δ transformer by 60°. This phase difference can be nullified by a simultaneous cyclic interchange of primary leads and reversal of primary terminals of Y – Δ transformer.

### *Example 2.27*

Two 100 kVA single-phase transformers are to be operated in parallel. Transformer A has an internal impedance of 0.005 + *j* 0.08 pu while transformer B has an internal impedance of 0.0075 + *j* 0.04 pu. How will they share the following loads and what will their power factor be: (a) 180 kW at 0.9 lagging power factor, (b) 200 kW at unity power factor.

### *Solution*

$$Z_A = 0.005 + j0.08\,\text{pu}$$

$$Z_B = 0.0075 + j0.04\,\text{pu}$$

(a) $$S = \frac{180}{0.9}\angle -\cos^{-1} 0.9 = 200\angle -25.84°\ \text{kVA}$$

Using Eqs. (2.91) and (2.92)

$$S_A = (200\angle -25.84°)\left[\frac{0.0075 + j0.04}{0.005 + j0.08 + 0.0075 + j0.04}\right]$$

$$= (200\angle -25.84°)\left(\frac{0.0407\angle 79.39°}{0.1206\angle 84.05°}\right)$$

$$= 67.5\angle -30.5° \text{ kVA}$$

$$= 58.16 \text{ kW at } 0.86 \text{ lagging p.f.}$$

$$S_B = (200\angle -25.84°)\left[\frac{0.005 + j0.08}{0.005 + j0.08 + 0.0075 + j0.04}\right]$$

$$= (200\angle -25.84°)\left(\frac{0.0801\angle 86.42°}{0.1206\angle 84.05°}\right)$$

$$= 132.8\angle -23.47°$$

$$= 121.85 \text{ kW at } 0.92 \text{ lagging power factor}$$

(b) $$S = 200\angle 0 \text{ kVA}$$

$$S_A = (200\angle 0)\left[\frac{0.0407\angle 79.39°}{0.1206\angle 84.05°}\right]$$

$$= 67.5\angle -4.66°$$

$$= 67.3 \text{ kW at } 0.997 \text{ lagging p.f.}$$

$$S_B = (200\angle 0)\left[\frac{0.0801\angle 86.42°}{0.1206\angle 84.05°}\right]$$

$$= 132.8\angle 2.37°$$

$$= 132.7 \text{ kW at } 0.999 \text{ leading p.f.}$$

*Example 2.28*

A 500 kVA transformer with $0.01 + j\,0.05$ pu impedance and a 250 kVA transformer with $0.015 + j\,0.04$ pu impedance are operating in parallel. The no-load secondary voltages of the two are equal. How will they share a 750 kVA, 0.8 lagging power factor load?

*Solution*

Let base kVA = 500 kVA

$$Z_A = 0.01 + j0.05 \text{ p.u.}$$

$$Z_B = (0.015 + j0.04)\left(\frac{500}{250}\right) = 0.03 + j0.08 \text{ p.u.}$$

$$S = 750\angle -\cos^{-1} 0.8 = 750\angle -36.9° \text{ kVA}$$

$$S_A = (750\angle -36.9°)\left[\frac{0.03 + j0.08}{0.01 + j0.05 + 0.03 + j0.08}\right]$$

$$= (750\angle -36.9°)\left(\frac{0.085\angle 69.4°}{0.136\angle 72.9°}\right)$$

$$= 471\angle -40.4°\,\text{kVA}$$

$$= 471\ \text{kVA at } 0.762 \text{ lagging p.f.}$$

$$S_B = (750\angle -36.9°)\left(\frac{0.01 + j0.05}{0.136\angle 72.9°}\right)$$

$$= (750\angle -36.9°)\left(\frac{0.051 + \angle 78.7°}{0.136\angle 72.9°}\right)$$

$$= 281\angle -31.1°\ \text{kVA}$$

$$= 281\ \text{kVA at } 0.856 \text{ lagging p.f.}$$

*Example 2.29*

Two equal ratio, parallel connected, 3-phase transformers supply 1000 kVA at 0.8 p.f. lagging at 11 kV. Their internal impedances are $Z_A = 1 + j5$ and $Z_B = 1 + j3$ ohm.

Find kVA, kW, current and power factor of each transformer.

*Solution*

For transformer $A$

$$S_A = \left[1000\angle -\cos^{-1} 0.8\right]\left[\frac{1 + j3}{2 + j8}\right]$$

$$= (1000\angle -36.87°)\left(\frac{3.16\angle 71.56°}{8.25\angle 75.96°}\right)$$

$$= 383.03\angle -41.27°\,\text{kVA}$$

$$P_A = 383.03(\cos 41.27°) = 287.27\,\text{kW}$$

Power factor = cos 41.27° = 0.75 lagging

$$I_A = \frac{383.03}{\sqrt{3} \times 11} = 20.1\ \text{A}$$

For transformer $B$

$$S_B = \left[1000\angle -\cos^{-1} 0.8\left(\frac{1 + j5}{2 + j8}\right)\right]$$

$$= (1000\angle -36.87°)\left(\frac{5.1\angle 78.69°}{8.25\angle 75.96°}\right)$$

$$= 618.18\angle -34.14°\,\text{kVA}$$

$$P_B = 618.18(\cos 34.14°) = 511.65$$

$$\text{Power factor} = 0.83 \text{ lagging}$$

$$I_B = \frac{618.18}{\sqrt{3}\times 11} = 32.45\,\text{A}.$$

*Example 2.30*

A single-phase transformer A, 500 kVA, 0.01 pu resistance, 0.05 pu reactance and another single-phase transformer *B*, 250 kVA, having 0.015 pu resistance, 0.04 pu reactance are supplying a 750 kVA load at 400 V and at 0.8 p.f. lagging. If $E_A = 405$ V and $E_B = 415$ V find (a) circulating current at no-load, (b) current supplied by each transformer, (c) kVA and p.f of each transformer.

*Solution*

$$\text{Rated current of } A = \frac{500000}{400} = 1250\,\text{A}$$

$$\text{Rated current of } B = \frac{250000}{400} = 625\,\text{A}$$

$$Z_A = 0.01 + j0.05\,\text{pu}$$

$$= (0.01 + j0.05)\frac{400}{1250}$$

$$= 0.0032 + j0.016 \text{ ohm} = 0.0163\angle 78.7° \text{ ohm}$$

$$Z_B = 0.015 + j0.04\,\text{pu}$$

$$= (0.015 + j0.04)\frac{400}{625}$$

$$= 0.0096 + j0.0256 \text{ ohm}$$

$$= 0.0274\angle 69.4° \text{ ohm}$$

$$\text{Total load current } I = \frac{750000}{400} = 1875\,\text{A}$$

$$\text{Load Impedance } Z = \frac{V}{I} = \frac{400}{1875} = 0.2133\angle 36.8° \text{ ohm}$$

(a) Circulating current at no-load

$$I_B = -I_A = \frac{E_B - E_A}{Z_B + Z_A} = \frac{415 - 405}{0.0274\angle 69.4° + 0.0163\angle 78.7°}$$

$$= \frac{10}{0.0436\angle 72.9°}$$

$$= 230\,\text{A at p.f. of } \cos(72.9°) \text{ or } 0.294 \text{ lag.}$$

(b) Using Eqs. (2.95) and (2.96)

$$I_A = \frac{(405)\,(0.0274\angle 69.4°) - (10)(0.2133\angle 36.9°)}{(0.0163\angle 78.7°)(0.0274\angle 69.4°) + (0.2133\angle 36.9°)(0.0436\angle 72.9°)}$$

$$= 970\angle -35°\,\text{A}$$

$$I_B = \frac{(415)\,(0.0163\angle 78.7°) + 10(0.2133\angle 36.9°)}{(0.0163\angle 78.7°)(0.0274\angle 69.4°) + (0.2133\angle 36.9°)(0.0436\angle 72.9°)}$$

$$= 875\angle -42.6°\,\text{A}$$

$$\text{kVA of transformer } A = \frac{(400)(970)}{1000} = 388\text{ kVA}$$

$$\text{kVA of transformer } B = \frac{(400)(875)}{1000} = 350\text{kVA}$$

Power factor of $A = \cos(-35°) = 0.82$ lagging

Power factor of $B = \cos(-42.6°) = 0.736$ lagging

It is seen that slight differences in open-circuit secondary voltages and per unit impedances has resulted in improper load sharing. Transformer A is under loaded while transformer B is over loaded.

## THREE WINDING TRANSFORMERS

### 2.47 APPLICATIONS

A two-winding transformer inter-connects two circuits of different voltage ratings. A multi-circuit transformer inter-connects three or more circuits. A 3-circuit transformer has primary, secondary and tertiary windings and is also known as three-winding transformer. Some situations which require the use of these transformers are as follows:

(i) To feed a distribution system from two transmission circuits of different voltages.

(ii) A generating station feeding two outgoing transmission systems of different voltages.

(iii) Supplying an auxiliary load (e.g. a synchronous condenser or static capacitors or sub-station load) at a voltage different from that of secondary.

(iv) Transformers fed from split winding generators. Such transformers require two primary windings of the same voltage.

(v) To sub-divide the secondary load into two separate secondary windings for reducing short-circuit current.

(vi) Testing transformers with potential coils for direct voltmeter connections.

(vii) A delta connected tertiary winding is provided in 3-phase star-star transformers, so that third harmonic currents may flow in it, thereby reducing the third harmonic components in phase to neutral voltages.

(viii) Load ratio control circuits require multi-circuit transformers.

## 2.48 EQUIVALENT CIRCUIT

A 3-winding transformer may be represented by a mesh or star equivalent circuit. A star equivalent circuit is more convenient and is shown in Fig. 2.45. The terminals 1, 2 and 3 indicate the primary, secondary and tertiary terminals. $R_1$, $R_2$, $R_3$ are the resistances and $X_1$, $X_2$, $X_3$ the leakage reactances of primary, secondary and tertiary windings. $V_1$, $V_2$, $V_3$ are voltages and $I_1$, $I_2$, $I_3$ currents of primary, secondary and tertiary windings. All the impedances are to a common kVA and respective voltage base. The external circuits are connected between terminals 1, 2, 3 and common terminal 0. The magnetising branch is usually omitted, but can be added at the beginning of the circuit (i.e. between terminals 1 and 0).

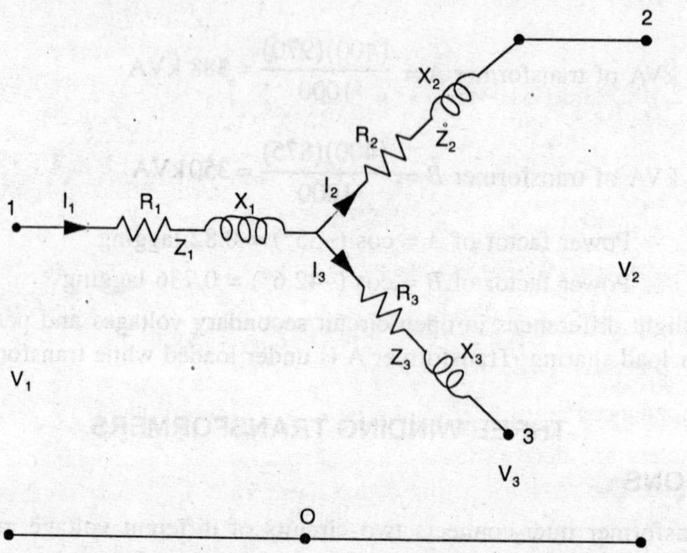


**Fig. 2.45.** Equivalent circuit of three-winding transformer

## 2.49 DETERMINATION OF PARAMETERS

The parameters of equivalent circuit can be determined from three short-circuit tests. In the first test (Fig. 2.46a) winding 2 is short circuited, winding 3 is left open circuited and a low voltage (so as to circulate full rated current in winding 2) is applied to winding 1. The voltage, current and power input to winding 1 are measured. The impedance $Z_{12}$ thus found is (Fig. 2.46b)

$$Z_{12} = Z_1 + Z_2 \qquad \text{...(2.101)}$$

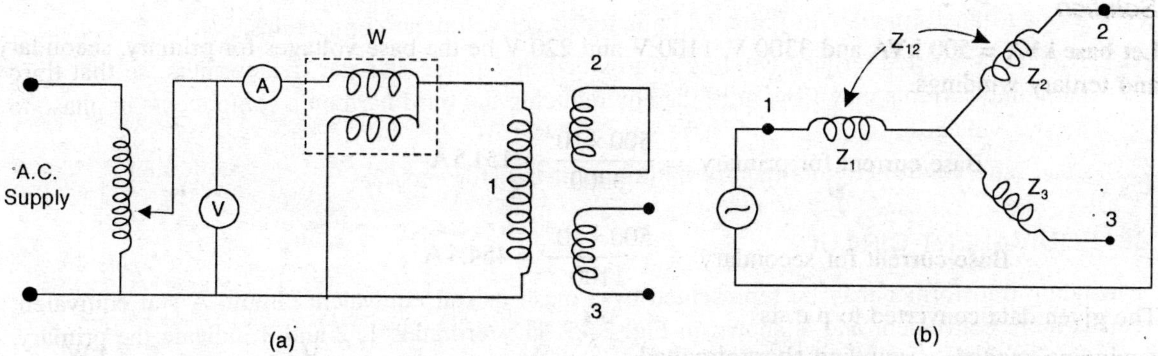


**Fig. 2.46.** Short circuit test on a three-winding transformer (a) connection diagram, (b) equivalent circuit

In the second test, winding 1 is excited, winding 3 is short circuited and winding 2 is left open circuited. The impedance measured is $Z_{13}$. In the third test winding 2 is excited, winding 3 is short circuited and winding 1 is open circuited. The impedance thus measured is $Z_{23}$. Thus

$$Z_{13} = Z_1 + Z_3 \quad \text{...(2.102)}$$

$$Z_{23} = Z_2 + Z_3 \quad \text{...(2.103)}$$

From Eqs. (2.101), (2.102) and (2.103)

$$Z_1 = 0.5(Z_{12} + Z_{13} - Z_{23}) \quad \text{...(2.104)}$$

$$Z_2 = 0.5(Z_{12} + Z_{23} - Z_{13}) \quad \text{...(2.105)}$$

$$Z_3 = 0.5(Z_{23} + Z_{13} - Z_{12}) \quad \text{...(2.106)}$$

where $Z_1 = R_1 + jX_1, \; Z_2 = R_2 + jX_2$

and $Z_3 = R_3 + jX_3$

All the above impedances are to a common base.

When the parameters are determined as above, some value of resistance and/or reactance may come out to be negative. The reason for this is that these values are the parameters of the equivalent circuit and not the actual resistances and leakage reactances of the windings. These negative values can be regarded as mathematical fictions. They reproduce the terminal characteristics faithfully, but are not the actual values for any winding. The resistance and reactance between any pair of terminals will always be positive. These negative values should be taken as negative for calculations of efficiency, voltage regulation, etc.

*Example 2.31*

A 3-winding transformer has the following ratings: primary 500 kVA, 3300 V, secondary 500 kVA, 1100 V, tertiary 250 kVA, 220V. The results of short-circuit tests are as follows:

| winding excited | winding short circuited | A | V | kW |
|---|---|---|---|---|
| 1 | 2 | 151.5 | 363 | 4 |
| 1 | 3 | 75.5 | 332 | 2.75 |
| 2 | 3 | 227.27 | 78.5 | 3.3 |

Find the parameters of equivalent circuit.

*Solution*

Let base kVA = 500 kVA and 3300 V, 1100 V and 220 V be the base voltages for primary, secondary and tertiary windings.

$$\text{Base current for primary} = \frac{500 \times 10^3}{3300} = 151.5\,\text{A}$$

$$\text{Base current for secondary} = \frac{500 \times 10^3}{1100} = 454.4\,\text{A}$$

The given data converted to p.u. is

| winding excited | winding short circuited | A | V | kW |
|---|---|---|---|---|
| 1 | 2 | 1 | 0.11 | 0.008 |
| 1 | 3 | 0.5 | 0.1 | 0.0055 |
| 2 | 3 | 0.5 | 0.0714 | 0.0066 |

From the above data

$$Z_{12} = 0.11 \text{ pu}, R_{12} = 0.008 \text{ pu}, X_{12} = \left(0.11^2 - 0.008^2\right)^{0.5} = 0.11 \text{ pu}$$

or
$$Z_{12} = 0.008 + j0.11\,\text{pu}$$

$$Z_{13} = \frac{0.1}{0.5} = 0.2 \text{ pu}, \; R_{13} = \frac{0.055}{0.5^2} = 0.022\,\text{pu}$$

$$X_{13} = \left(0.2^2 - 0.022^2\right)^{0.5} = 0.2\,\text{pu}$$

or
$$Z_{13} = 0.022 + j0.2 \text{ pu}$$

$$Z_{23} = \frac{0.0714}{0.5} = 0.1428 \text{ pu}, \; R_{23} = \frac{0.0066}{0.5^2} = 0.0264$$

$$X_{23} = \left(0.1428^2 - 0.0264^2\right)^{0.5} = 0.14\,\text{pu}$$

or
$$Z_{23} = 0.0264 + j0.14\,\text{pu}$$

Using Eqs. (2.104) to (2.106)

$$Z_1 = 0.5(Z_{12} + Z_{13} - Z_{23})$$

$$= 0.5\,(0.008 + j0.11 + 0.022 + j0.2 - 0.0264 - j0.14)$$

$$= 0.0018 + j0.085\,\text{pu}$$

$$Z_2 = 0.5(Z_{12} + Z_{23} - Z_{13})$$

$$= 0.5\,(0.008 + j0.11 + 0.0264 + j0.14 - 0.022 - j0.2)$$

$$= 0.0062 + j0.025 \text{ pu}$$

$$Z_3 = 0.5(Z_{23} + Z_{13} - Z_{12})$$

$$= 0.5\,(0.0264 + j0.14 + 0.022 + j0.2 - 0.008 - j0.11)$$

$$= 0.0202 + j0.115 \text{ pu}$$

The equivalent circuit is shown in Fig. 2.45.

## MISCELLANEOUS TOPICS

## 2.50 MAGNETISING CURRENT

### 2.50.1 Wave shape of Magnetising Current

If the flux is a transformer core is sinusoidal, the induced emf is also sinusoidal and leads the flux by 90°. However, the magnetising current is not sinusoidal but contains the fundamental component and third, fifth, seventh and other higher order harmonics. This is due to the non-linear *B–H* curve for core material. A graphical procedure for determining the wave shape of magnetising current is shown in Fig. 2.47. At instant 0 in Fig. 2.47 (b) the flux is zero and the corresponding current as seen from hysteresis loop in Fig. 2.47 (a) is 0 A. At instant *B*, the flux is *OB* and corresponding magnetising current is *BC*. When the flux is a maximum (*OF*) the current is also maximum (*FG*). Thus by finding the values of current for different values of flux, the wave shape of magnetising current can be drawn. The wave form, thus obtained, includes magnetising current and hysteresis loss component, but not the component corresponding to eddy current loss. The wave form is non-sinusoidal and contains odd harmonics. The amplitude of harmonics decreases with increase in the order of harmonic. The strongest is the third harmonic, which may have an amplitude of about 20% of fundamental. The amplitudes of harmonics increases (and the wave shape becomes more non-sinusoidal) with increase in maximum flux density in the core.

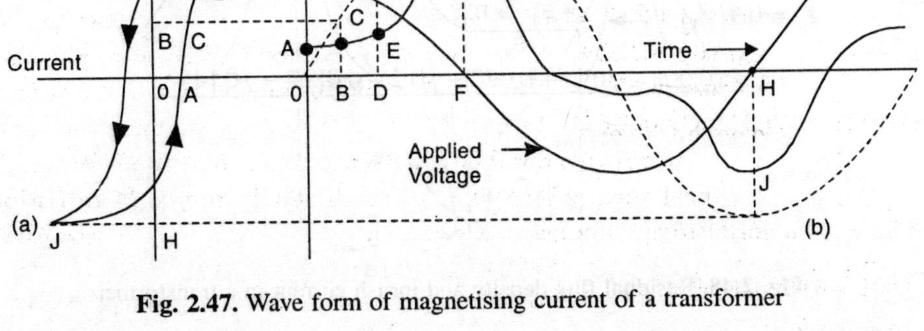


**Fig. 2.47.** Wave form of magnetising current of a transformer

The total current drawn by primary under load conditions is the phasor sum of load current and no-load current. Since the magnitude of load current is very large as compared to that of no-load current, the current drawn from source, under load conditions, is almost sinusoidal.

### 2.50.2 Inrush of Magnetising Current

When a transformer is switched on, a high inrush of magnetising current takes place. This transient current has a completely offset wave shape and negative half-cycles may get wiped out. Though this transient current lasts only a few cycles, it may cause a severe dip in system voltage and may cause tripping of over load and differential relays.

If a transformer could be switched on at that instant of voltage wave, which normally corresponds to the actual magnetic density in the core at that moment, the switching would be smooth without giving rise to any transient. However the instant of switching is not under control and a magnetic transient appears.

When a transformer is switched off at $t = t_1$ the magnetising current becomes zero, but the flux density will have some residual value $B_r$. Figure 2.48 shows this residual flux density. If the transformer were not switched off, $I_1$ and $B_1$ would have followed the dotted curves. However after switching off, the current $I_2$ is zero and flux density $B_2$ is equal to residual density $B_r$. If the transformer is switched on again at instant marked $t_2$, the flux density would normally be at its negative maximum value $-B_{max}$. Since flux cannot be instantly created or destroyed, the flux density wave instead of starting at ($-B_{max}$) starts at $B_2$ and traces the curve $B_3$, which is a displaced sinusoid (because with sinusoidal applied voltage, the induced emf and flux have to be sinusoidal). The magnetising current wave, corresponding to flux density $B_3$, is shown as $I_3$ in Fig. 2.48. The maximum theoretical value of $B_3$ is ($2B_{max} + B_r$). Since the transformer is designed for flux density $B_{max}$, the crest of $B_3$ will cause super saturation of the magnetic circuit and result in a very high value of magnetising current.

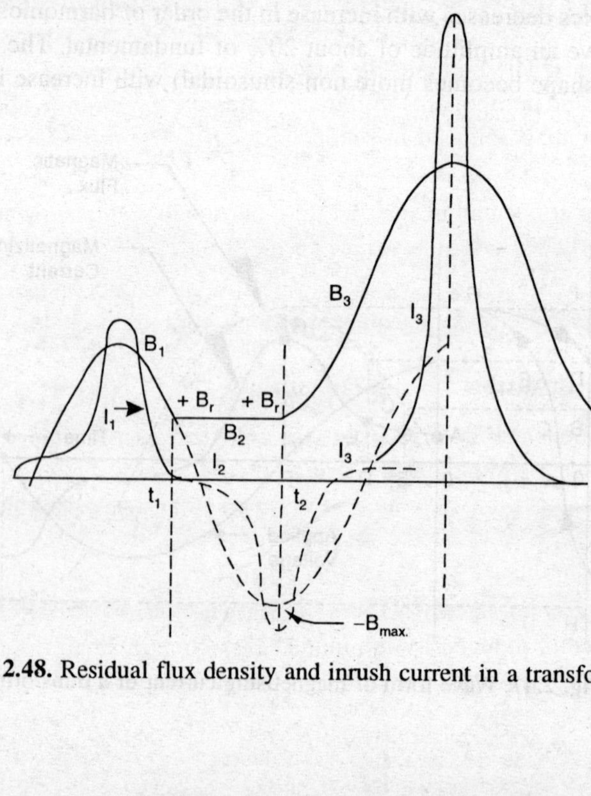


**Fig. 2.48.** Residual flux density and inrush current in a transformer

The above mentioned crest value of $B_3$ (= $2B_{max} + B_r$) is the theoretical maximum value. In actual practice the instant of switching on (i.e. $t_2$) is not likely to occur when flux density is at negative peak value. It is more likely that circuit will be switched on at peak value of voltage wave (which means zero value of flux density) by the bridging of the contacts by a spark. The actual value of $B_3$ and therefore the magnetising current would always be much less than the theoretical maximum shown in Fig. 2.48. The normal value of magnetising current is very small (about 0.05 pu), but, at the time of switching on the transformer, the transient magnetising current may be about 2 pu.

As seen from above discussion, the maximum value of inrush current depends on instant of switching on and residual flux density. The residual flux density in transformer cores may be about 50-60% of the maximum flux density and is responsible for making the peak of inrush current about 2-5 times as high as it would be otherwise. The residual flux density is, in general, not affected by the instant of switching off.

The decay of inrush current starts with a high average initial rate and progressively becomes slower, lasting many seconds.

## 2.51 MAGNETISING CURRENT IN 3-PHASE TRANSFORMERS

It is not possible to switch in the three phases simultaneously without producing an inrush current. If the instant of switching is most favourable for one of the phases, it will be unfavourable for the other two. The inrush currents in 3-phase transformers occur more often, but are less in intensity than in single-phase transformers. The transformers connections have important effect on magnetising current harmonics.

### 2.51.1 Delta Connected Primaries

The third harmonic components of magnetising currents in Δ-connected primaries cause third harmonic leakage impedance voltage drops. Since the applied voltages are balanced, the induced emfs contain third harmonic components. The mutual flux must adjust itself to generate these non-sinusoidal emfs. However the third harmonic components in induced emfs are small, only about 0.1% of rated voltage. A small third harmonic emf is induced in secondaries also. The third harmonic components in the 3 phases are in time phase.

If the secondaries are connected in star, the third harmonic components are present only in line-to-neutral voltages. However they are not present in line-to-line voltages.

If the secondaries are connected in Δ, the third harmonic components of the three line-to-neutral secondary voltages produce a circulating current around the closed delta.

### 2.51.2 Y-Connected Primaries

A third harmonic component will be present in the phase current of primary. Since the third harmonic components in the 3 phases are in phase, the current in neutral wire is sum of the three components and is thus 3 times the third harmonic component of exciting current in each phase. The magnitude of this neutral current is about 6 per cent of the rated current. This may cause inductive interference with parallel communication circuits.

The third harmonic components of exciting current produce third harmonic leakage impedance drops and therefore the emfs induced by the mutual flux contain small third harmonic components.

If secondaries are connected in star, third harmonic components are present in each line-to-neutral secondary voltage, but are not present in secondary line-to-line voltage.

If the secondaries are connected in delta, the third harmonic components of secondary emfs cause a circulating current in the closed delta.

## 2.52 TRANSFORMER TAPS

Electric utilities are required to keep the voltage at consumer terminals within statutory limits. This can be achieved by altering the ratio of transformation by tapping the winding to change the number of turns. A typical installation may have four taps, in addition to the normal settings, with a spacing of 2.5 per cent of rated voltage. Thus the output voltage available at various taps would be 1.05, 1.025, 1, 0.975 and 0.95 pu.

The taps may be provided either on low voltage side or high voltage side. It is generally more difficult to tap low voltage winding, since it is placed next to the core. Moreover tap changing gear, if placed on low voltage side, would have to handle larger magnitude of current. In the low voltage winding (having only a few turns) the voltage per turn is a large percentage of total voltage and if taps are placed on low voltage winding, the choice of available output voltages would be very much limited. An 11 kV/440 V 3-phase transformer (Δ–Y) having an emf of 13 V per turn would have only 20 turns per phase on low voltage side and it would not be possible to have a tapping for less than 5 per cent. Because of above considerations, taps are usually provided on the high voltage side.

The tapped part of the winding may be at phase end, at neutral end or in the middle of the winding. The insulation between several phase parts, when taps are near the neutral end, is small. This has economic advantage in case of high voltage transformers. When a comparatively large voltage ratio change is required, the taps should be at the centre of the phase windings, so that the magnetic dissymmetry is the least.

The tap changing may be done under off-load conditions or under load conditions.

### 2.52.1 Off-load Tap Changing

The leads from the different taps are brought out to a vertical switch or to a dial switch, which can be operated from outside the transformer case. A simple arrangement, shown in Fig. 2.49 uses four taps, in the winding, placed at 2.5 per cent intervals. The contacts are arranged on a circle, so that they can be bridged by a rotatable arm A. When A bridges studs 1 and 2, the output voltage is 1 pu, when it bridges studs 2 and 3, the output voltage is 0.975 pu, for studs 3 and 4, it is 0.95 pu, for studs 4 and 5, it is 0.925 pu and finally for studs 5 and 6, the output voltage is 0.9 pu. This tap changer is used for seasonal variations of voltage.

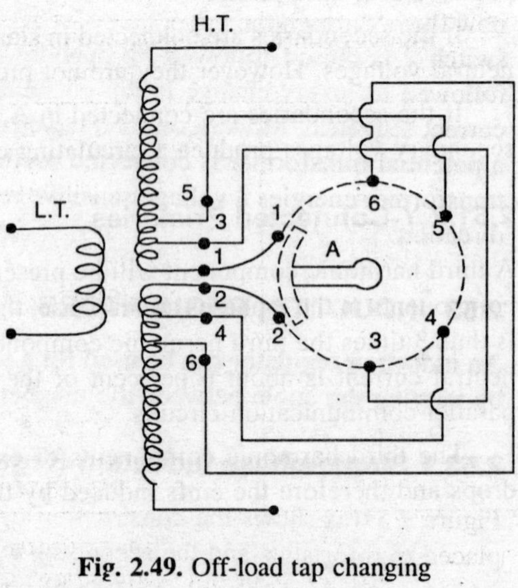

Fig. 2.49. Off-load tap changing

### 2.52.2 Tap Changing under Load (TCUL)

Every on-load tap changing circuit possesses two fundamental features.

(i) Some form of impedance (resistor or reactor) which is introduced in the circuit to prevent short circuiting of tapping section.

(ii) A duplicate circuit is provided, so that the load current can be carried by one circuit while tap selection is being undertaken in the other.

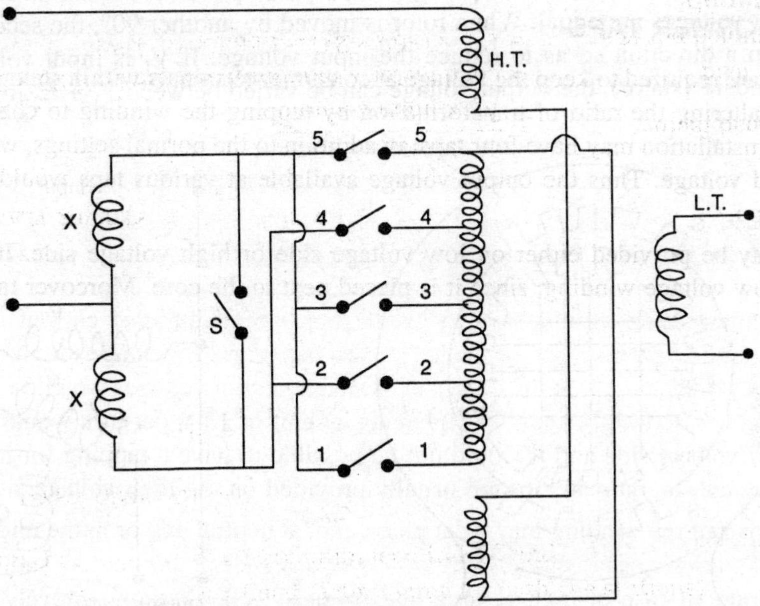


**Fig. 2.50.** Tap changing under load

Figure 2.50 shows one simple form of on-load tap changer. The reactor *X* prevents the short circuiting of tapping section. The switch *S* is closed during normal operation. Let the switch 1 be closed and 2, 3, 4, 5 open initially. In this position the complete winding is in circuit. The two halves of reactor carry equal currents in opposite directions. Thus the mmfs of the two halves cancel each other and voltage drop in the reactor is negligible. To change to tap 2, the switch *S* is opened. The total current now flows through the upper half of reactor and there may be considerable voltage drop across it. Then switch 2 is closed followed by opening of switch I and closing of switch *S*. A similar sequence is followed for every change in tap position. It is necessary that the switches are opened and closed in correct sequence. Modern on-load tap changers are designed for automatic operation. In such a circuit a potential transformer is connected on the output side of the transformer. The output of the potential transformer energies a voltage sensitive relay with contacts, which initiate a tap change in the required direction.

## 2.53 INDUCTION REGULATORS

An induction regulator can be used for a smooth variation of ac voltage. The output voltage is changed by varying the angle between the magnetic fluxes of primary and secondary windings.

### 2.53.1 Single-phase Induction Regulator

Figure 2.51(a) shows the construction of a single-phase induction regulator. The primary winding is placed in rotor slots and the secondary winding in stator slots. The sliprings are not needed, since the movement of rotor is limited to 180° and only flexible leads are sufficient. The secondary is connected in series with the output circuit Fig. (2.51b). The emf induced in the secondary is added to (or subtracted from) the primary voltage. When the axes of primary and secondary windings coincide, the emf induced in the secondary will be maximum and the output voltage will be sum of input voltage and secondary emf. When the axes of two windings are perpendicular, no emf is induced in secondary and output and input voltages are equal. When rotor is moved by another 90°, the secondary emf is again maximum but in a direction so as to reduce the input voltage. If $V_1$ is input voltage and $E_2$ is emf induced in secondary (stator), the output voltage can be varied between $V_1 + E_2$ to $V_1 - E_2$ by moving the rotor through 180°.

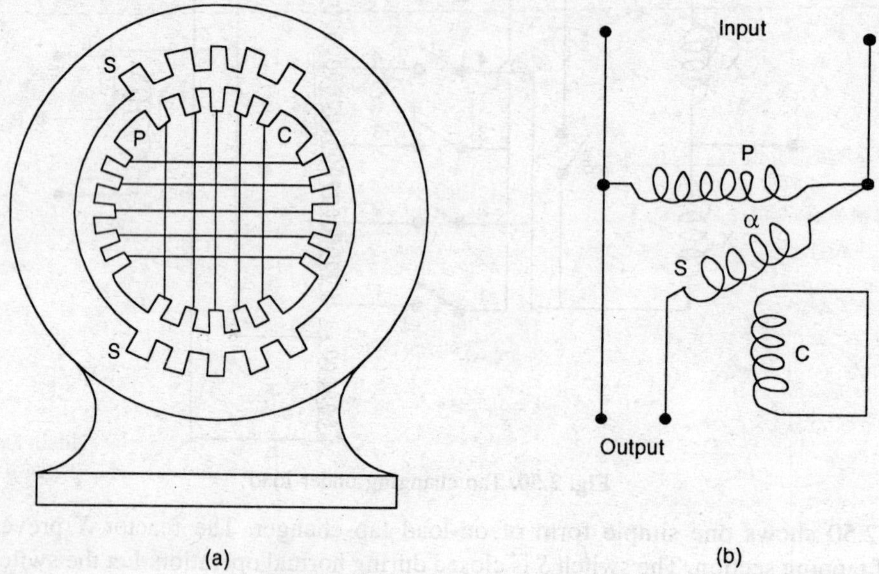


**Fig. 2.51.** Single-phase induction regulator (a) construction, (b) connection diagram

A single-phase regulator is provided with a compensating winding, placed on rotor and short circuited, on an axis in quadrature with primary. If this winding were not provided, the leakage reactance of secondary winding would be large when the rotor is the position of zero boost (i.e. $E_2 = 0$). In this position the short circuited compensating winding has maximum mutual inductance with the secondary and the effective reactance is reduced.

For intermediate positions of rotor, the degree of boost or buck is proportional to the sine of the angle of displacement between rotor and stator.

### 2.53.2 Three-phase Induction Regulator

A three-phase induction regulator is very similar to a three-phase induction motor. Since the motion of rotor is limited, sliprings are not required and only flexible connections are sufficient.

When the primary windings are connected to 3-phase ac supply, the primary currents produce a constant amplitude rotating magnetic field as in a 3-phase induction motor. The voltages induced in the three secondary windings are constant in magnitude (i.e. independent of rotor position) but the phase of the secondary voltage changes, relative to that of the primary, as the rotor is moved. Thus a variation in output voltage is obtained by altering the angular position of rotor causing a phase shift (but not a change in magnitude) of the secondary emf. Figure 2.52 (a) shows the connection diagram. As seen in Fig. 2.52 (b) the locus of output voltage $V_2$ is a circle with centre at extremity of $V_1$ and of radius $E_2$. When $\theta = 0$, $V_2 = V_1 + E_2$ and when $\theta = 180°$, $V_2 = V_1 - E_2$. Figure 2.52 (c) shows the phasor diagram for all the three phases.

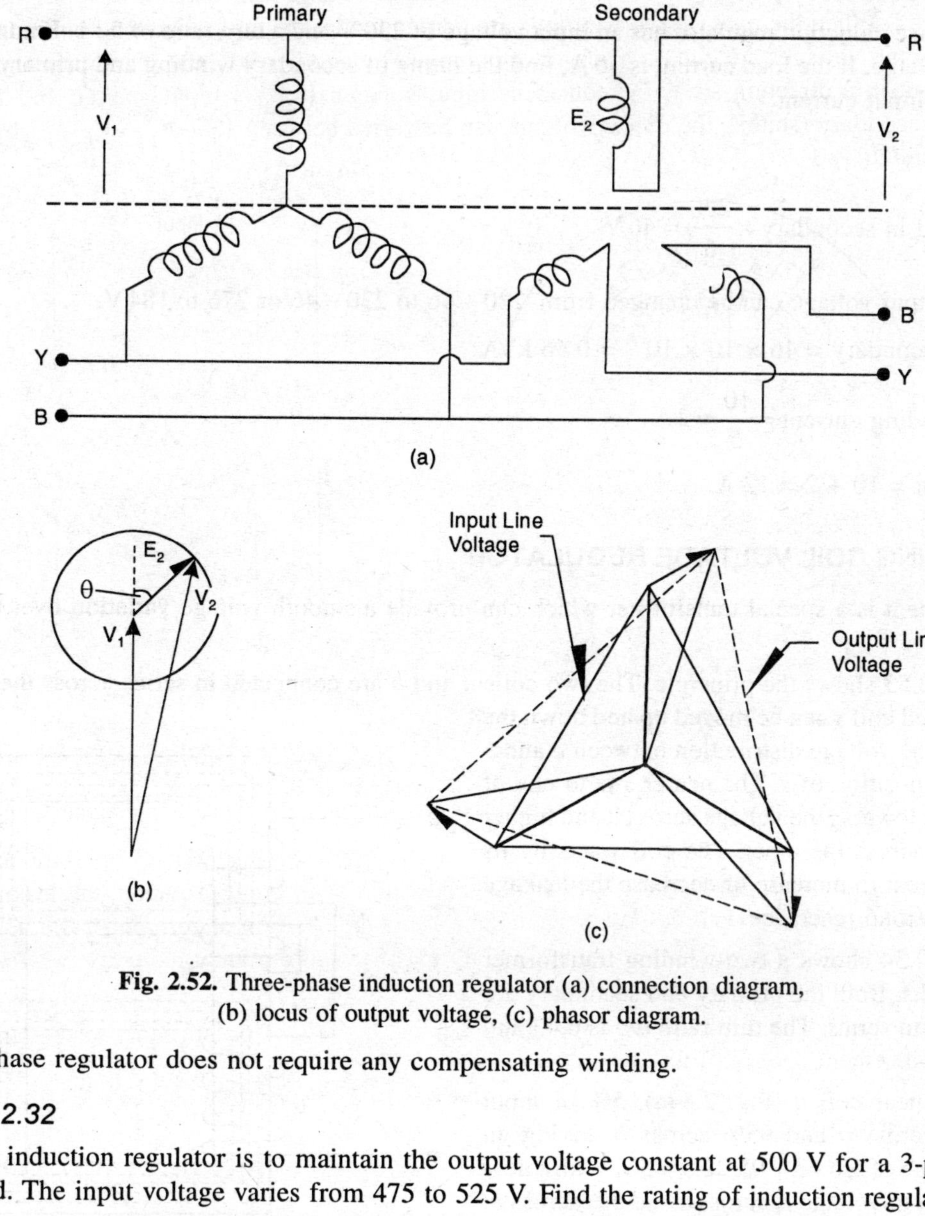


**Fig. 2.52.** Three-phase induction regulator (a) connection diagram, (b) locus of output voltage, (c) phasor diagram.

A 3-phase regulator does not require any compensating winding.

### *Example 2.32*

A 3-phase induction regulator is to maintain the output voltage constant at 500 V for a 3-phase 100 k V A load. The input voltage varies from 475 to 525 V. Find the rating of induction regulator.

### *Solution*

$$\text{Line current} = \frac{100 \times 1000}{\sqrt{3} \times 500} = 115.47\,\text{A}$$

Line voltage rating of regulator = $0.5(525 - 475) = 25\,\text{V}$

kVA rating $= \sqrt{3} \times 25 \times 115.47 \times 10^{-3} = 5\,\text{kVA}$..

*Example 2.33*

**A single-phase induction regulator has an input voltage of 230 V and a turn ratio of 5 : 1. Find the range of output voltage. If the load current is 10 A, find the rating of secondary winding and primary winding current and input current.**

*Solution*

Emf induced in secondary $= \dfrac{230}{5} = 46\,\text{V}$

Thus the output voltage can be changed from $230 + 46$ to $230 - 46$ or 276 to 184 V.

Rating of secondary = $46 \times 10 \times 10^{-3} = 0.46$ kVA

Primary winding current $= \dfrac{10}{5} = 2\,\text{A}$

Input current = 10 + 2 = 12 A.

## 2.54 MOVING COIL VOLTAGE REGULATOR

This equipment is a special transformer which can provide a smooth voltage variation over a desired range.

Figure 2.53 shows the principle. The two coils *a* and *b* are connected in series across the input. A short circuited coil *s* can be moved up and down the core limb. The voltage distribution between *a* and *b* depends on position of *s*. The nearer *s* is to one of the coils, the lower is the voltage across it and higher the voltage across the other. The coil *s* acts by its induced current to increase or decrease the leakage flux and the total reactance.

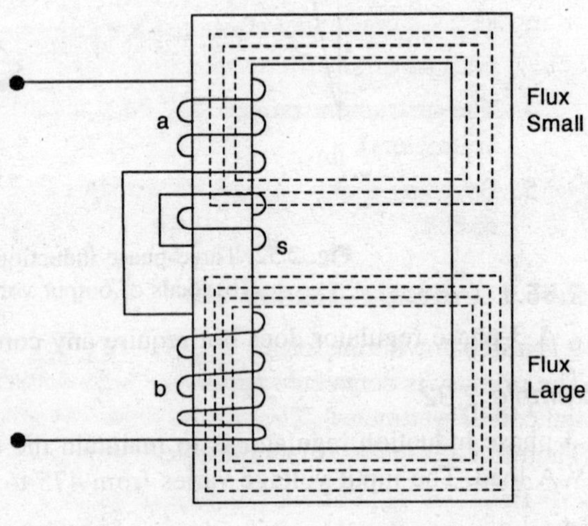


**Fig. 2.53.** Moving coil voltage regulator

Figure 2.54 shows a two-winding transformer with the coil *s*. Both the primary and secondary are in two parts in series. The turn ratio *a/p* is 6.43 and *b/q* is 22.5.

With *s* near coil *a* (Fig. 2.54a), 5% of input voltage is across *a* and 95% across *b*, giving an output voltage of 5/6.43 + 95/22.5 = 5V. With *s* near *b*, 95% of input voltage is across *a* and 5% across *b*, thereby giving an output voltage of 95/6.43 + 5/22.5 = 15 V (Fig. 2.54b). A variety of such arrangements are possible.

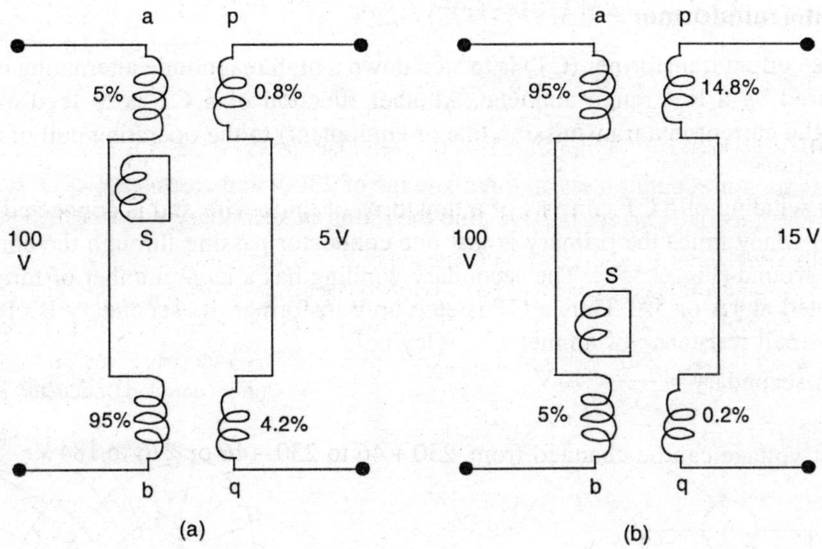


Fig. 2.54

## 2.55 INSTRUMENT TRANSFORMERS

Current and potential transformers are used to extend the range of ac instruments in preferences of shunts and series resistors for the following reasons:

1. Errors due to stray inductance and capacitance in shunts, multipliers and their leads are eliminated.
2. The measuring circuit is isolated from the mains by the transformer and may be earthed.
3. The length of leads from the transformer to the instrument is of lesser importance and the leads may be of small cross-section.
4. The instrument ranges may be standardised (usually 1A and 5A for ammeters and 110 V for voltmeters).
5. By using a clip on type of transformer core, the current can be measured without breaking the circuit.

### 2.55.1 Potential Transformer

It is a step down transformer and is used along with a low range voltmeter for measuring a high voltage. The primary is connected across the high voltage supply and the secondary to the voltmeter (or potential coil of wattmeter). The current drawn by a voltmeter (and potential coil of wattmeter) is very small. Therefore the potential transformer is like a two-winding transformer under no-load.

Potential transformers are fabricated from good quality iron core and operated at low flux densities. Therefore magnetising current is very small. It is necessary that ratio error (i.e. change of voltage ratio with load) and phase angle error (i.e. phase angle between input and output) are low. The secondary of these transformers is usually designed for 110 V.

### 2.55.2 Current Transformer

The function of a current transformer (CT) is to step down a high magnitude alternating current, so that it can be measured by a low range ammeter. Another function of a CT is to feed a small current (proportional to the current in a transmission line or equipment) to the operating coil of a relay for the purpose of protection.

The primary winding of a CT consists of a few turns of thick wire and is connected in series with high current line. Many times the primary is just one conductor passing through the core, over which the secondary is wound (Fig. 2.55b). The secondary winding has a large number of turns of thin wire and is usually rated at 1A or 5A. Thus a CT is step up transformer. Its secondary is effectively short circuited by the small resistance of ammeter or relay coil.

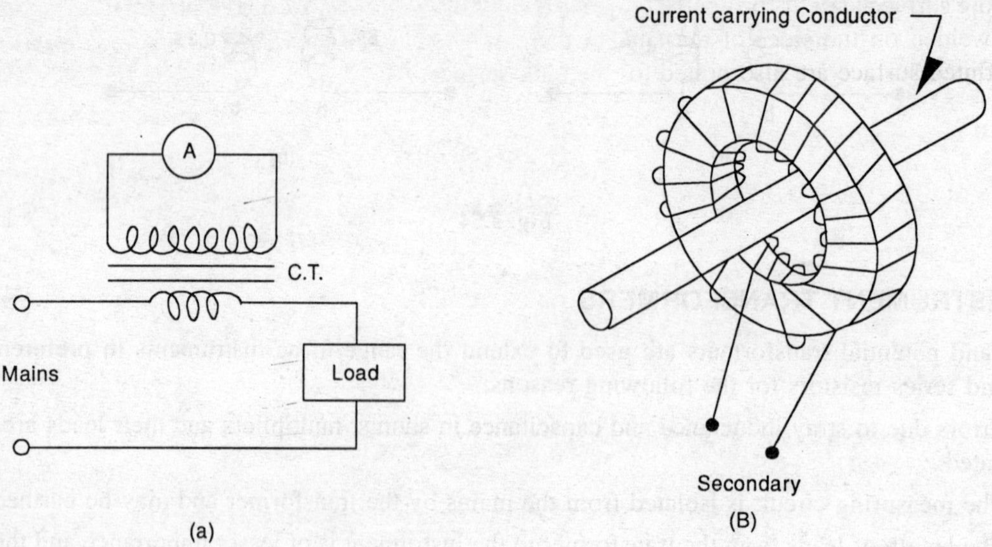


**Fig. 2.55.** Current transformer (a) with wound primary, (b) with line conductor acting as primary

A current transformer draws a magnetising current from the mains (like any other transformer). This current causes ratio error (i.e. actual ratio of primary-to-secondary current is not equal to the ratio of secondary-to-primary number of turns) and phase angle error (i.e. primary and secondary currents are not in phase). These errors should be as low as possible.

The secondary of a CT should never be open circuited, otherwise there will be no secondary mmf. Since the primary is carrying a large current, the magnetic flux would become very high resulting in high eddy current and hysteresis losses. This would increase the temperature of the core and the insulation and core may get damaged. Moreover a high voltage would be induced in the open circuited secondary and this high voltage may be dangerous to the equipment and personnel.

## 2.56 TRANSFORMER COOLING

The core and copper losses cause heating of transformer. It is necessary to ensure that the maximum temperature rise is less than that which may cause damage to the insulation. In rotating machines the heat dissipation is easier due to the turbulent air flow. Transformers require an elaborate cooling system (inspite of the fact that losses in transformers are less than in rotating machines).

Small transformers (upto 25 kVA rating) are of dry type. The air convection current and radiation from transformer tank are sufficient to keep the temperature within limits.

A majority of transformers are oil immersed. The core and the windings are completely immersed in mineral oil. Oil is a good insulating material and provides better heat dissipation than air. These transformers can be further classified as under:

### 2.56.1 Oil Immersed Self-cooled Transformer

The dissipation of heat by the transformer tank to the atmosphere is by natural process. Hot oil flows upwards through ducts (in the core and windings) and then flows down along the inner walls of the tank. The tank surface dissipates heat to the atmosphere. Sometimes the tank surface is corrugated to provide more surface area (a corrugated tank has about 50% more surface area than plain tank). Vertical tubes are welded on the sides of the tank to increase the surface area. Sometimes radiators having finned or fluted surface are also bolted to the tank surface.

### 2.56.2 Oil Immersed Forced Air Cooled Transformer

This transformer is also immersed in oil. The cooling is improved further by an air blast over the outside surface of the tank.

### 2.56.3 Oil Immersed Water Cooled Transformer

Metallic tubes are situated inside the tank, below the oil level. Water is circulated through these tubes to extract heat from oil.

### 2.56.4 Oil Immersed Forced Oil Cooled Transformer

This method uses forced circulation of oil by a pump. Oil is pumped through the ducts and then through external radiators, which are cooled by fans. The oil pumping equipment adds to the cost, but cooling is more effective. The temperature difference between oil at the top and bottom is less.

## 2.57 TRANSFORMER OIL

Oil serves the double purpose of cooling and insulating. Minerals oils alone are suitable for electrical purposes. The main characteristics of transformer oil are:

**(a) Viscosity**

This determines the rate of cooling. A high viscosity is an obvious disadvantage, because of sluggish flow through ducts and tubes.

**(b) Insulating Property**

The dielectric strength of transformer oil is generally good. However moisture and dust tend to lower the dielectric strength appreciably. A dielectric strength of 30 kV is considered a good value.

**(c) Flash Point**

It is the temperature at which the vapours above an oil surface ignite spontaneously. Flash point of transformer oil should not be less than 140°C.

**(d) Fire Point**

The temperature at which an oil will ignite and continue burning is known as fire point. It should be above 200°C.

**(e) Purity**

Oil should not contain impurities like sulphur and its compounds, which may cause corrosion of metal parts and accelerate formation of sludge.

**(f) Sludging**

Sludging means slow formations for semi-solid hydrocarbons, which may get deposited on windings and tank walls. The formation of sludge is due to heat and oxidation. The maximum permissible value of sludge is 0.1%.

The deterioration of oil during use can be retarded by the use of antioxidants, particularly oxidation 'inhibitors'. They are well defined chemical products, which help in retarding the formation of acid and sludge.

## 2.58 CONSERVATOR AND BREATHER

In oil cooled transformers the expansion of oil forces out the air above it, so that on cooling fresh air is again drawn in thus bringing is oxygen and moisture. A conservator consists of an airtight cylindrical metal drum supported on transformer lid. It takes up the expansion and contraction of oil with changes in temperature. The tank is filled with oil when cold and the expansion is taken up in the conservator. In the absence of conservator condensation of water soluble acids on the undersurface of tank lid may occur and acid droplets may fall back into the oil. The displacement of air due to change of oil volume takes place through a breather containing calcium chloride or silica gel, which extracts the moisture from air. Silica gel should be checked regularly and dried or replaced when necessary.

## 2.59 TRANSFORMER NOISE

Most of the transformers are located near residential areas and its noise may be troublesome. The 'hum' originates in the core where the laminations tend to vibrate by magnetic forces. This noise is transmitted through oil to tank and thence to surroundings. The essential factors in noise production are the following:

(i) Magnetostriction, i.e. very small extension with corresponding reduction in cross-section of sheet steel strips when magnetised.

(ii) The degree of mechanical vibration developed by laminations depending on the tightness of clamping, size, gauge, etc.

(iii) Mechanical vibration of the tank.

Though magnetostriction is the first cause of hum, much of the noise depends on the natural frequency of vibration of mechanical parts. Any construction changes which remove this frequency outside the audible range will, evidently, mean an improvement.

## 2.60 NAME PLATE AND RATINGS

The specifications of transformers used in power industry are contained in BIS 2026. As per this standard, every transformer must be provided with a name plate giving the following information:

Type (power, auto, booster, etc.), year of manufacture, number of phases, rated kVA (for multiwinding transformers rated kVA of each winding), rated frequency, rated voltage of each winding, connection symbol, per cent impedance voltage at rated current (measured value corrected to 75°C), type of cooling, total mass, mass and volume of insulating oil.

As per BIS 2026, the preferred kVA ratings are 6.3, 10, 16, 25, 40, 63, 100, 160, 250, 400, 630, 1000 kVA, etc.

## 2.61 WIDE-BAND TRANSFORMER

### 2.61.1 Classification

Transformers used in electric power systems serve the basic function of stepping up or stepping down the voltage level for efficient transmission of electric power. On the other hand, another class of transformers, used in electronics and telecommunication circuits, are concerned with transfer of information. The information is in the form of a signal, e.g. voice, music, a train of pulses, etc. In transformers used in electric power systems the two important requirements are good efficiency (i.e. ratio of power output to power input) and good voltage regulation. In the transformers used in electronics and telecommunication system the focus is on the undistorted or accurate transmission of information or the signal.

A wide band transformer is designed to pass a band of frequencies generally exceeding a few decades. They can be classified on the basis of frequency spectrum, i.e. subsonic, audio, ultrasonic, high, very high and ultra high frequency transformers. Another classification is on the basis of system in which they are used, e.g. servo, carrier, video, data-transmission, etc. Still another classification is on the basis of specific function, e.g. line, coupling, driver, output, modulation, bridge, hybrid, etc. A hybrid transformer has a very specialised application, viz to couple two or more signal sources and provide isolation between input and output ports.

### 2.61.2 Equivalent Circuit

A transformer has stray capacitance between turns, between layers, between windings, between terminals, between core and terminals, etc. These stray capacitances influence the frequency response and affect the degree of distortion and attenuation of signal. These stray capacitances can be combined into input capacitance $C_1$, output capacitance $C_2$ and bridge capacitance $C_b$. The equivalent circuit after addition of these capacitances is shown in Fig. 2.56.

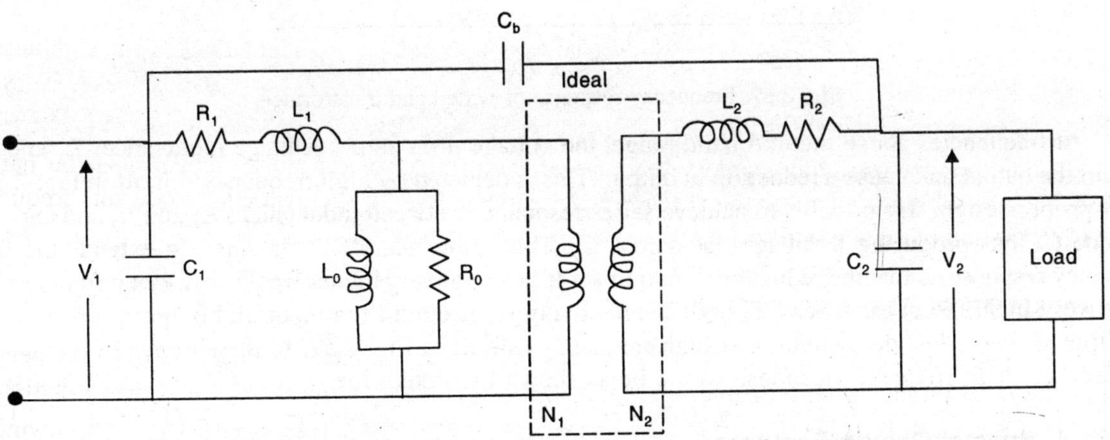


**Fig. 2.56.** Equivalent circuit for wide band transformer

### 2.61.3 Frequency Response

The complete frequency range can be classified into low, intermediate and high frequencies.

In the intermediate or mid band range, the effect of inductances and capacitances can be neglected and the transformer behaves as a simple resistance attenuator with zero phase shift. It is usual to define a mid band frequency $\omega_m = (\omega_1 \omega_2)^{0.5}$ where $\omega_1$ and $\omega_2$ are lower cut off and higher cut off frequencies. The mid band frequency $\omega_m$ is the geometric mean of $\omega_1$ and $\omega_2$. At frequencies $\omega_1$ and $\omega_2$, the response is 3 dB lower than that at $\omega_m$. The circuit behaviour is simple at $\omega_m$ and the transformer characteristics are specified at $\omega_m$. These characteristics are: rated output power, desired input (determined by turn ratio), insertion loss, i.e. loss in signal due to insertion of transformer between input and load. The power level and insertion loss are, generally, specified in decibels. The response is flat in the mid band range.

At very low frequencies the capacitances behave as open circuits and are not effective. However the reactance of magnetising inductance $L_0$ is small and shunts the input of the transformer. When the frequency is very very low, inductance $L_0$ is virtually a short circuit and output of transformer is zero. As frequency increases the shunting effect of $L_0$ decreases and the output rises. At $\omega = \omega_1$, the output is 3 dB lower as compared to that at $\omega_m$. As frequency increases the response flattens and approaches the mid band response. In addition to attenuation, low frequency distortion is also important. Ferromagnetic distortion due to hysteresis phenomenon occurs because harmonic currents or voltages result from the non-linear relationship between flux and core.

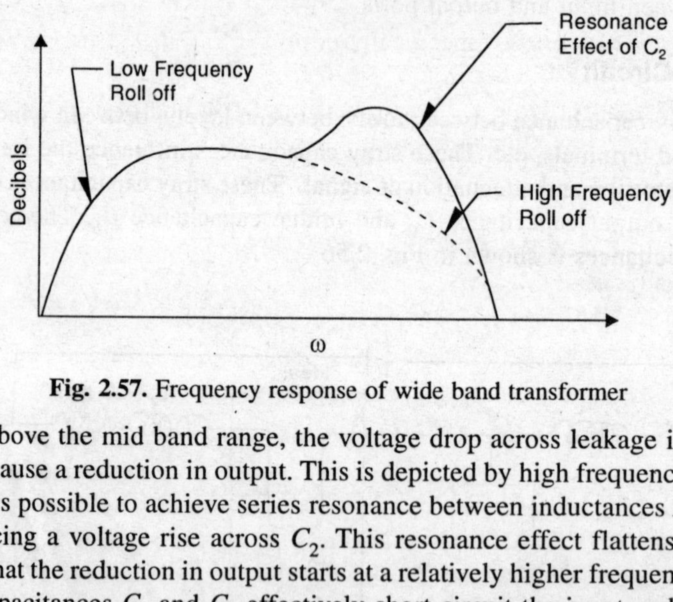


**Fig. 2.57.** Frequency response of wide band transformer

At frequencies above the mid band range, the voltage drop across leakage inductances $L_1$ and $L_2$ limit the output and cause a reduction in output. This is depicted by high frequency roll off in Fig. 2.57. By proper design, it is possible to achieve series resonance between inductances $L_1$ and $L_2$ and capacitance $C_2$ thus producing a voltage rise across $C_2$. This resonance effect flattens and extends the frequency response so that the reduction in output starts at a relatively higher frequency. When frequencies are very high, the capacitances $C_1$ and $C_2$ effectively short circuit the input and output ports and the output shows a fast decay shown as high frequency roll off in Fig. 2.57. In the absence of resonance effect, the high frequency response would be as shown by dotted line.

### 2.61.4 Special Design Features

Wide band transformer design and fabrication include some special features which improve the frequency response. These are:

(a) Use of thicker conductors for primary and secondary winding to reduce winding resistances.

(b) Use of small number of turns $N_1$ and $N_2$ to reduce leakage inductances.

(c) Use of special core materials and core geometry. Sometimes composite core structure is also used. A composite core consists of two cores combined into a single assembly and encircled by common windings, e.g. one core is made of permalloy (which has high value of permeability × quality factor, i.e. $\mu Q$ in the kilohertz range) and the other is made of ferrite or powdered iron (which has high $\mu Q$ in the megahertz range).

(d) Use of high permeability cores of large cross-sectional area.

(e) Use of small length of magnetic circuit so as to increase magnetising inductance $L_0$. Large cross-sectional area of core and high permeability also increase $L_0$ because $L_0 = (\mu N^2 \times \text{area})/\text{length}$ as depicted by Eq. (1.13).

(f) Use of proper coil geometry to reduce leakage inductance and stray capacitance.

(g) Use of thin, long and similar primary and secondary coils.

It is seen that some of the requirements are contrary to each other. Low leakage reactances require small values of $N_1$ and $N_2$ but high value of $L_0$ requires large number of turns. Therefore a proper compromise is necessary in selection of number of turns.

## 2.62 PULSE TRANSFORMER

Transformer used in digital circuits are specifically designed to transmit pulses of voltage or current with some reasonable degree of fidelity. Such transformers are known as pulse transformers. They perform functions like isolation (i.e. to isolate dc between input and output), inversion (i.e. to invert polarity of a pulse) and impedance transformation.

Pulse transformers handle a variety of pulse waveforms, the most popular being a rectangular pulse (Fig. 2.58). The pulse width may vary from about 0.5 micro seconds to about 25 micro-seconds. A large time may elapse between the end of one pulse and the start of the next one. A transient analysis using the equivalent circuit can be used to get the waveform of the output. An ordinary two winding transformer is designed for a limited frequency range. A periodic train of pulses can be represented by a very large number of sine waves of different frequencies (Fourier analysis). A pulse transformer designed for faithful transmission should therefore, be a wide band transformer.

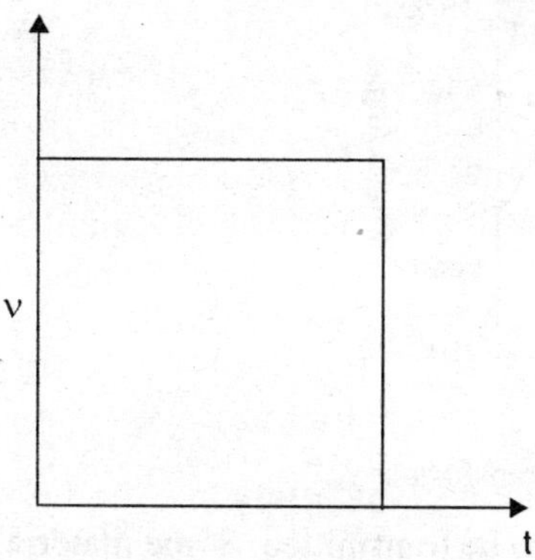


**Fig. 2.58.** Rectangular pulse

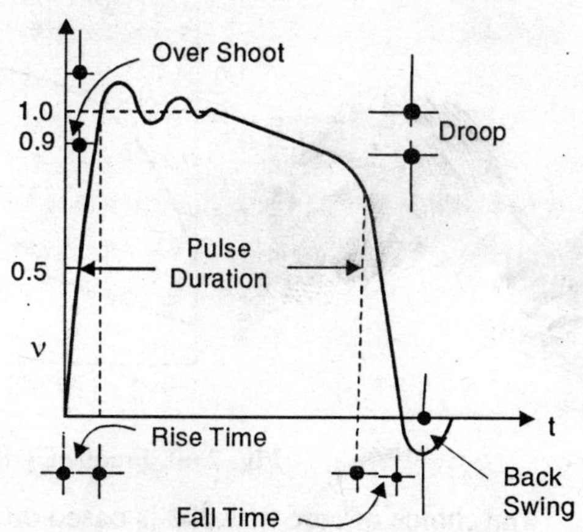


**Fig. 2.59.** Pulse transformer output wave shape

The output wave of a pulse transformer is shown in Fig. 2.59. A comparison with Fig. 2.58 shows the deformation. The main alterations in the shape are caused by (1) a finite rise time ($t_r$), (2) a droop or tilt, (3) a finite fall time ($t_f$), (4) a backswing as shown in Fig. 2.59. These are caused by leakage inductances and stray capacitances. Pulse transformer has only a few turns to reduce leakage inductance.

The permeability of core material depends on the type of waveform. The permeability of the core excitation by pulses is known as pulse permeability. Ferrites and high permeability alloys (e.g. nickel-iron alloy, cobalt-iron alloy, metallic glass) have high pulse permeability (about 1000 or so) even when pulse repetition frequency (expressed in pulses per second) is high and are therefore used for cores of pulses transformers.

## 2.63 AUDIO-FREQUENCY TRANSFORMER

It is a transformer used in audio-frequency range (20 to 20000 Hz). They are used in electronic circuits (control, communication, measurement, etc.) for stepping up the voltage or impedance matching. They are generally small in size and have iron cores.

It is necessary that distortion should be as low as possible. The input voltages should be amplified equally and phase shift should be zero for all frequencies. Two characteristics curves, viz. amplitude-frequency characteristics (i.e. ratio of output voltage to input voltage versus (frequency) and phase characteristics (i.e. phase angle between output voltage and input voltage versus frequency) are used for frequency response. A flat frequency response over the audio frequency range is desirable. The response can be determined using the exact equivalent circuit of the transformer. At low frequencies the leakage inductances are not important, but the shunting effect of magnetising branch is very important. Because of high magnetising current, the ratio $V_2/V_1$ decreases with decrease in frequency. In the intermediate frequency range, none of the inductances is effective and the equivalent circuit reduces to a purely resistive circuit. The ratio $V_2/V_1$ is constant during this range. At high frequencies leakage reactances become increasingly important. Moreover the inter-turn, turn-to-ground and stray capacitances are also important. Due to the effect of these inductances and capacitances, the ratio $V_2/V_1$ decreases with increasing frequency. A typical response is shown in Fig. 2.60. The frequency range between lower 3 db and upper 3 db frequencies is known as bandwidth.

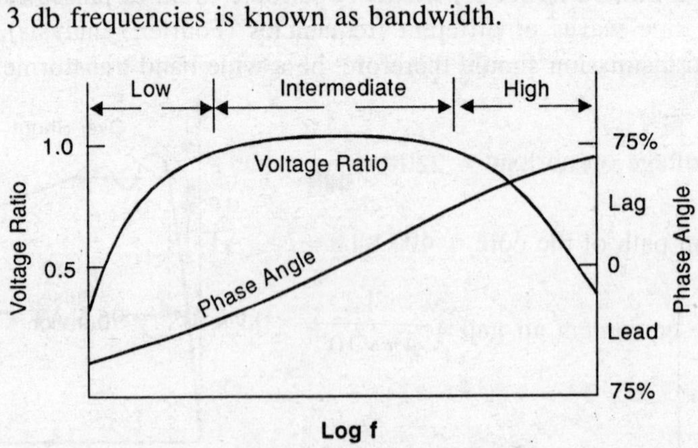


**Fig. 2.60.** Frequency response of audio-frequency transformer

The choice of core material is based on the minimum frequency to be transmitted. Some materials used are supermalloy, silectron, deltamax, ferrite, etc. Sometimes composite cores (i.e. two cores into a single assembly, each core with one of the desired characteristics) are also used. The desired high frequency response generally determines the shape of the core, which may be *EI* core, toroid or pot core. The size of the core is generally governed by the allowable distortion.

## 2.64 GROUNDING TRANSFORMER

Sometimes the neutral of the power transformer is not available for grounding (e.g. when a Δ–Δ transformer is used). In such situations a special Y–Δ transformer is used solely for neutral grounding. This transformer is known as a grounding transformer and it is a step down transformer. The star connected primaries are connected to the system and its neutral is grounded. The secondaries are in delta and generally donot supply any load but provide a closed path for triplen harmonic currents to circulate in them. Under balanced conditions the current in a grounding transformer is only its own exciting current. However, large currents may flow in it in the event of single line-to-ground fault conditions. Therefore a grounding transformer should be of sufficient rating to withstand the effects of line-to-ground faults. Transformers with 'zigzag' connection are also used as grounding transformers.

*Example 2.34*

A single-phase transformer has 480 turns on the primary and 120 turns on the secondary. The mean length of the flux path in the core in 1.8 m and joints are equivalent to an air gap of 0.1 mm. The primary voltage is 2200 V at 50 Hz. The maximum value of flux density should be 1.2 T. Find (a) gross area of the core, taking an iron factor of 0.9, (b) secondary voltage on no-load, (c) primary current and power factor on no-load. Assume that for a flux density of 1.2 T, *H* is 400 A/m, core-loss is 1.8 W/kg and density of core material is 7800 kg/m³. [*P.U. 2002*]

*Solution*

(a) $$2200 = 4.44 \times 480 \times 50 \times \phi_m$$

or $$\phi_m = 0.0206 \text{ Wb}$$

$$\text{Active area of core} = \frac{0.0206}{1.2} = 0.01717\,\text{m}^2$$

$$\text{Gross area of core} = \frac{0.01717}{0.9} = 0.019\,\text{m}^2$$

(b) $$\text{Secondary voltage on no-load} = 2200 \times \frac{120}{480} = 550\text{ V}$$

(c) $$\text{MMF for iron path of the core} = 400 \times 1.8 = 720 \text{ AT}$$

$$\text{MMF for the equivalent air gap} = \frac{1.2}{4\pi \times 10^{-7}} \times 0.1 \times 10^{-3} = 95.5 \text{ AT}$$

$$\text{Total MMF} = 720 + 95.5 = 815.5 \text{ AT}$$

$$\text{Maximum value of } I_m = \frac{815.5}{480} = 1.7\text{ A}$$

Assuming $I_m$ to be sinusoidal

rms value of $I_m = 0.707 \times 1.7 = 1.2$ A

Volume of core = $1.8 \times 0.01717 = 0.031 \text{ m}^3$

Mass of core = $0.031 \times 7800 = 241.8$ kg

Iron-loss = $(241.8)(1.8) = 434.24$ W

$$I_w = \frac{435.24}{2200} = 0.198 \text{ A}$$

$$\text{No-load current} = I_0 = \left(I_w^2 + I_m^2\right)^{0.5}$$

$$\text{or } I_0 = \left[(0.198)^2 + (1.2^2)\right]^{0.5} = 1.216 \text{ A}$$

$$\text{Power factor on no-load} = \frac{I_w}{I_0} = \frac{0.198}{1.216} = 0.163 \text{ lagging.}$$

*Example 2.35*

A single-phase transformer is rated as 240/120 V, 50 Hz. Find voltage and frequency at secondary terminals at no-load if (a) $V_p = 120$ V, 25 Hz, (b) $V_p = 240$ V, dc.

*Solution*

$$\frac{N_1}{N_2} = \frac{240}{120} = 2$$

(a) Secondary voltage $= \frac{120}{2} = 60$ V

The frequency at secondary terminals will be the same as that at primary terminals, i.e. 25 Hz.

(b) Since a dc voltage is being applied, no emf would be induced (An emf is induced only due to change of flux). As such secondary voltage will be zero. The current flowing in the primary would be very large (equal to $V_p/R_1$ where $R_1$ is the resistance of primary winding). This may damage the winding. Secondary voltage = 0.

*Example 2.36*

A 20 kVA, 250/2500 V, 50 Hz, single-phase transformer gave the following test results:

| | | | |
|---|---|---|---|
| O.C. (l.v. side) | 250 V, | 1.4 A, | 105 W |
| S.C. (h.v. side) | 120 V, | 8 A, | 320 W |

Find the parameters and draw equivalent circuit as referred to low voltage side. *[B.U. 2003]*

*Solution*

$$I_w = \frac{105}{250} = 0.42 \text{ A}$$

$$I_m = \left(I_0^2 - I_w^2\right)^{0.5} = \left(1.4^2 - 0.42^2\right)^{0.5} = 1.34\,\text{A}$$

$$R_0 = \frac{V_0}{I_w} = \frac{250}{0.42} = 592.2\,\text{ohm}$$

$$X_0 = \frac{V_0}{I_m} = \frac{250}{1.34} = 186.6\text{ ohm}$$

Let high voltage winding be designated as winding 2 and low voltage winding as 1. Then

$$R_{e2} = \frac{320}{8^2} = 5\text{ ohm}$$

$$Z_{e2} = \frac{120}{8} = 15\text{ ohm}$$

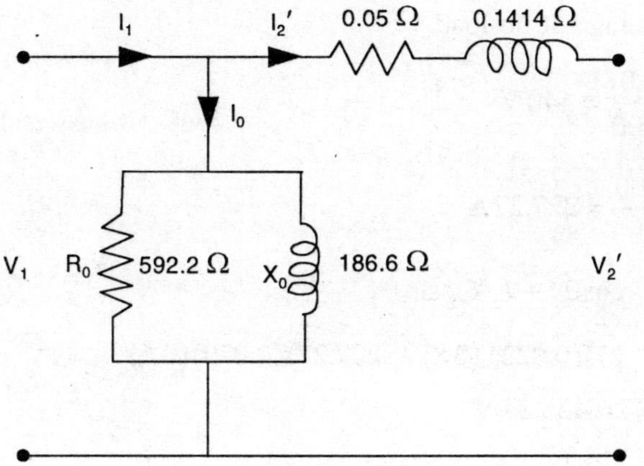


**Fig. 2.61**

$$X_{e2} = \left(15^2 - 5^2\right)^{0.5} = 14.14\text{ ohm}$$

The equivalent resistance and reactance as referred to low voltage side are

$$R_{e1} = \frac{5}{10^2} = 0.05\text{ ohm}$$

$$X_{e1} = \frac{14.14}{10^2} = 0.1414\text{ ohm}$$

The equivalent circuit is shown in Fig. 2.61.

*Example 2.37*

A 100 kVA single-phase transformer has $N_1$ = 400, $N_2$ = 80, $R_1$ = 0.3 ohm, $X_1$ = 1.1 ohm, $R_2$ = 0.01 ohm, $X_2$ = 0.035 ohm. The applied voltage is 2200 V. Find (a) total impedance as referred to primary, (b) voltage regulation and output-voltage at 0.8 lagging power factor full-load. [*N.U. 2002*]

*Solution*

(a) $\dfrac{N_1}{N_2} = 5$

The secondary resistance $R_2$ as referred to primary $= R_2' = (0.01)(5)^2 = 0.25$ ohm

Total resistance as referred to primary $= 0.3 + 0.25 = 0.55$ ohm

The secondary reactance $X_2$ as referred to primary $= X_2' = (0.035)(5)^2 = 0.875$ ohm

Total reactance as referred to primary $= 1.1 + 0.875 = 1.975$ ohm

Hence the impedance as referred to primary $= 0.55 + j1.975$ ohm

(b) $R_{e2}$ = total resistance as referred to secondary $= 0.55/(5)^2 = 0.022$ ohm

$X_{e2}$ = total reactance as referred to secondary $= 1.975/(5)^2 = 0.079$ ohm

$V'_1$ = output voltage at no-load

$$= 2200 \times \frac{80}{400} = 440 \text{ V}$$

$$I_2 = \frac{100 \times 1000}{440} = 227.27 \text{ A}$$

$$V_1' = V_2 + I_2 R_{e2} \cos\theta_2 + I_2 X_{e2} \sin\theta_2$$

or $$V_2 = 440 - (227.27)(0.022)\,(0.8) - (227.27)(0.079)(0.6)$$

$$= 440 - 14.77 = 425.23 \text{ V}$$

$$\text{Voltage regulation} = \frac{14.77}{425.23} \times 100 = 3.47\%.$$

*Example 2.38*

For the transformer of example 2.37, find output voltage and voltage regulation for full-load 0.8 leading power factor.

*Solution*

$$V_1' = V_2 + I_2 R_{e2} \cos\theta_2 - I_2 X_{e2} \sin\theta_2$$

or $$V_2 = 440 - (227.27)(0.022)(0.8) + (227.27)(0.079)(0.6)$$

$$= 440 + 6.77 = 446.77 \text{ V}$$

$$\text{Voltage regulation} = \frac{V_1' - V_2}{V_2} \times 100$$

$$= \frac{440 - 446.77}{446.77} \times 100 = -1.5\%.$$

*Example 2.39*

A 100 kVA transformer has iron losses of 1.2 kW and full-load copper losses of 1.5 kW. (a) Find the kVA for maximum efficiency, (b) efficiency at the kVA found in part (a) for power factors of 1, 0.8 lagging and 0.8 leading. [*K.U. 2002*]

*Solution*

(a) Let $x$ be the fraction of full-load kVA at which maximum efficiency occurs. Then

$$x^2(1.5) = 1.2 \text{ or } x = \sqrt{\frac{1.2}{1.5}} = 0.894$$

Load kVA for maximum efficiency = 0.894 (100) = 89.4 kVA

(b) Unity power factor

$$\text{Output} = 89.4 \times 1 = 89.4 \text{ kW}$$

$$\text{Iron losses} = 1.2 \text{ kW}$$

$$\text{Copper losses} = 1.2 \text{ kW}$$

$$\text{Efficiency} = \frac{89.4}{89.4 + 1.2 + 1.2} \times 100 = 97.39\%$$

0.8 lagging power factor

$$\text{Output} = 89.4 \times 0.8 = 71.52 \text{ kW}$$

$$\text{Iron losses} = 1.2 \text{ kW}$$

$$\text{Copper losses} = 1.2 \text{ kW}$$

$$\text{Efficiency} = \frac{71.52}{71.52 + 1.2 + 1.2} \times 100 = 96.75\%$$

0.8 leading power factor

The kW output and losses will be the same as at 0.8 lagging power factor.

Hence efficiency = 96.75%.

*Example 2.40*

The maximum efficiency of a transformer is 0.97 and occurs at three-fourth full-load, unity power factor. The p.u. impedance is 0.08. Find regulation at full-load, 0.8 lagging power factor.

*Solution*

Maximum efficiency occurs when copper losses are equal to iron losses.

Total losses at 3/4 full-load = 0.03 pu

Copper losses at 3/4 full-load = 0.5 × 0.03 = 0.015 pu

$$\text{Copper losses at full-load} = 0.015\left(\frac{4}{3}\right)^2 = 0.027 \text{ pu}$$

$$R_{pu} = 0.027,\ Z_{pu} = 0.08$$

$$X_{pu} = \left(0.08^2 - 0.027^2\right)^{0.5} = 0.0753 \text{ pu}$$

$$\text{Full-load regulation} = R_{pu}\cos\theta_2 + X_{pu}\sin\theta_2$$

$$= (0.027)\,(0.8) + (0.0753)\,(0.6)$$

$$= 0.0666 \text{ or } 6.66\%$$

*Example 2.41*

A 100 kVA, 10000/200 V, 50 Hz single-phase transformer needs 300 V on h.v. side to circulate full-load current with l.v. winding short circuited. The power input is 1000 W. Find maximum possible regulation and power factor at which it occurs. What is $V_2$ under this condition?

*Solution*

$$\text{Full-load current on h.v side} = \frac{100 \times 1000}{10000} = 10 \text{ A}$$

Taking 100 kVA and 10000 V as base values for high voltage side

$$\text{Base Impedance} = \frac{10000}{10} = 1000 \text{ ohm}$$

$$\text{Resistance as referred to h.v. side} = \frac{1000}{10 \times 10} = 10 \text{ ohm}$$

$$\text{Impedance as referred to h.v. side} = \frac{300}{10} = 30 \text{ ohm}$$

$$\text{Reactance as referred to h.v. side} = \left(30^2 - 10^2\right)^{0.5} = 28.28 \text{ ohm}$$

$$R_{pu} = \frac{10}{1000} = 0.01 \text{ pu}$$

$$X_{pu} = \frac{28.28}{1000} = 0.02828 \text{ pu}$$

$$\text{Regulation} = R_{pu}\cos\theta_2 + X_{pu}\sin\theta_2$$

Regulation will be maximum when differential of above expression with respect to $\theta_2$ is zero.

$$\frac{d(\text{Regulation})}{d\theta_2} = -R_{pu}\sin\theta_2 + X_{pu}\cos\theta_2 = 0$$

or

$$\tan\theta_2 = \frac{X_{pu}}{R_{pu}} = \frac{0.02828}{0.01} = 2.828$$

$$\text{Power factor} = \cos\theta_2 = 0.333 \text{ lagging}$$

$$\sin\theta_2 = 0.943$$

$$\text{Regulation} = (0.01)(0.333) + (0.02828)(0.943)$$

$$= 0.03 \text{ pu or } 3\%$$

$$V_2 = \frac{200}{1.03} = 194.17 \text{ V.}$$

*Example 2.42*

A 200/400 V single-phase transformer has core loss of 100 W. It has a resistance of 0.45 ohm and leakage reactance of 1.2 ohm, both referred to h.v. side. Find the p.f. at which regulation is zero. Full-load primary current is 20 A.

*Solution*

$$\text{Full-load secondary current} = 20 \times \frac{200}{400} = 10\,\text{A}$$

$$R_{pu} = \frac{10 \times 0.45}{400} = 0.01125 \text{ pu}$$

$$X_{pu} = \frac{10 \times 1.2}{400} = 0.03 \text{ pu}$$

Regulation can be zero only when p.f. is leading

$$\text{Regulation} = R_{pu}\cos\theta_2 - X_{pu}\sin\theta_2 = 0$$

or

$$\tan\theta_2 = \frac{R_{pu}}{X_{pu}} = \frac{0.01125}{0.03} = 0.375$$

Power factor = $\cos\theta_2$ = 0.936 leading.

*Example 2.43*

In a sumpner's test on two identical single-phase transformers rated 500 kVA, 11/0.4 kV, 50 Hz the wattmeter reading on h.v. side is 6000 W and on l.v. side is 15000 W. Find the efficiency of each transformer on half full-load and 0.8 power factor. What will be its maximum efficiency.

*Solution*

$$\text{Iron losses of one transformer} = \frac{15000}{2} = 7500\,\text{W} = 7.5\,\text{kW}$$

$$\text{Copper losses of one transformer} = \frac{6000}{2} = 3000\,\text{W}$$

$$\text{Copper losses at half full-load} = (0.5)^2(3000) = 750\,\text{W} = 0.75\,\text{kW}$$

$$\text{Efficiency} = \frac{0.5 \times 500 \times 0.8 \times 100}{0.5 \times 500 \times 0.8 + 7.5 + 0.75} = 96.04\%$$

Let $x$ be the fraction of full-load at which maximum efficiency occurs

$x^2$ (copper loss at full-load) = Iron losses

or
$$x = \sqrt{\frac{7500}{3000}} = 1.58$$

kVA for maximum efficiency = 1.58 × 500 = 790 kVA

Maximum efficiency will always occur at unity power factor

$$\text{Maximum Efficiency} = \frac{790 \times 1 \times 100}{790 \times 1 + 7.5 + 7.5} = 98.14\%.$$

*Example 2.44*

The core losses in a 2200 V transformer at 60 Hz are: eddy current loss 80 W, hysteresis loss 190 W. Find the total core-loss when it is operated on 2200 V, 50 Hz supply.

*Solution*

$$E = 4.44\, fN\phi_m$$

Since voltage is same but frequency is 5/6 of original value, $\phi_m$ will be 6/5 of original value. Hence $B_m$, at 50 Hz, will 6/5 times the value of 60 Hz.

$$\text{Hysteresis loss at 50 Hz} = 190 \times \frac{50}{60} \times \left(\frac{6}{5}\right)^{1.6} = 211.96\,\text{W}$$

$$\text{Eddy current loss} = K_e f^2 B_m^2$$

$$\text{Eddy current loss at 50 Hz} = 80\left(\frac{50}{60}\right)^2\left(\frac{6}{5}\right)^2 = 80\,\text{W}$$

$$\text{Total core-loss} = 211.96 + 80 = 291.96\ \text{W}.$$

*Example 2.45*

Two transformers have the following characteristics.

| | | | | |
|---|---|---|---|---|
| *A* | 150 kVA, | 6600/225 V, | IR = 1.5%, | IX = 4% |
| *B* | 300 kVA, | 6600/215 V, | IR = 1%, | IX = 5% |

They are connected in parallel. Find the no-load terminal voltage and current when 6600 V is applied at h.v. bus bars. [*K.U.* 2002]

*Solution*

$$\text{Full-load secondary current of transformer } A = \frac{150 \times 1000}{225} = 666.67\,\text{A}$$

$$Z_A = (0.015 + j0.04)\frac{225}{666.67} = 0.005 + j0.0135\,\Omega$$

Full-load secondary current of transformer $B = \dfrac{300 \times 1000}{215} = 1395.35\,\text{A}$

$$Z_B = (0.01 + j0.04)\left(\frac{215}{1395.35}\right) = 0.0015 + j0.006\,\Omega$$

Circulating current at no-load $= \dfrac{E_A - E_B}{Z_A + Z_B}$

$$= \frac{225 - 215}{0.005 + j0.0135 + 0.0015 + j0.006}$$

$$= \frac{10}{0.0065 + j0.0195} = 486.5\angle -71.56^\circ\,\text{A}$$

Therefore $I_A = 486.5$ A, $I_B = -486.5$ A

$$V = E_A - I_A Z_A$$

$$= 225 - (486.5\angle -71.56^\circ)(.005 + j0.0135)$$

$$= 225 - (486.5\angle -71.56^\circ)(0.0144\angle 69.68^\circ)$$

$$= 225 - 7\angle -1.88^\circ = 218\,\text{V}.$$

### *Example 2.46*

An auto-transformer is used to step down 250 V to 200 V. The load on 200 V side is 100 A. Find the power transferred conductively and ratio of kVA rating of auto-transformer to that of 2-winding transformer for the same load conditions. State assumptions made. [*M.D.U. 2001*]

### *Solution*

Referring to Fig. 2.22

$$V_2 = 200\,\text{V},\; I_2 = 100\,\text{A},\; V_1 = 250\,\text{V}$$

$$I_1 = \frac{V_2 I_2}{V_1} = \frac{200 \times 100}{250} = 80\,\text{A}$$

Power transferred conductively $= V_2 I_2 = 200 \times 80 = 16000\,\text{VA} = 16\,\text{kVA}$

Power transferred inductively $= V_2(I_2 - I_1) = 200(100 - 80) = 4000\,\text{VA} = 4\,\text{kVA}$

Rating of auto-transformer $= 4$ kVA

Rating of two-winding transformer for same load conditions

$$= 200 \times 100 \times 10^{-3} = 20\,\text{kVA}$$

$$\frac{\text{Rating of auto-transformer}}{\text{Rating of two-winding transformer}} = \frac{4}{20} = 0.2$$

The assumptions are (a) iron losses have been neglected, (b) copper losses have been neglected, (c) there is no magnetic leakage, (d) the magnetising current has been assumed to be zero.

*Example 2.47*

Show that the per unit regulation of a single-phase two-winding transformer is given by $V_{sc}\cos(\theta_{sc}-\theta_2)/V_{\text{rated}}$ where $V_{sc}$ is the voltage required to circulate full-load current on short circuit and $\theta_{sc}$ and $\theta_2$ are power factor angles on short circuit and load conditions. [*P.U. 2001*]

*Solution*

Let $R_{e2}$, $X_{e2}$, $Z_{e2}$ be the total resistance, reactance and impedance as referred to secondary. Then

$$Z_{e2} = \frac{V_{sc}}{I_{sc}}$$

$$R_{e2} = Z_{e2}\cos\theta_{sc}$$

$$X_{e2} = Z_{e2}\sin\theta_{sc}$$

$$R_{pu} = \frac{I_2 R_{e2}}{V_{\text{rated}}} = \frac{I_2 Z_{e2}\cos\theta_{sc}}{V_{\text{rated}}}$$

Since $V_{sc}$ is the voltage required to circulate full-load current, $I_{sc} = I_2$

Therefore
$$R_{pu} = \frac{I_{sc} Z_{e2}\cos\theta_{sc}}{V_{\text{rated}}}$$

and
$$X_{pu} = \frac{I_{sc} Z_{e2}\sin\theta_{sc}}{V_{\text{rated}}}$$

$$\begin{aligned}\text{Regulation} &= R_{pu}\cos\theta_2 + X_{pu}\sin\theta_2 \\ &= \frac{I_{sc}Z_{e2}}{V_{\text{rated}}}\left[\cos\theta_{sc}\cos\theta_2 + \sin\theta_{sc}\sin\theta_2\right] \\ &= \frac{V_{sc}}{V_{\text{rated}}}\cos(\theta_{sc}-\theta_2).\end{aligned}$$

*Example 2.48*

An auto-transformer has an applied voltage of 9.5 V at its input terminals *AC*. The output terminals *BC* are short circuited. The ratio of transformation is 2:1. The resistance and leakage reactance of section *AB* are 0.06 ohm and 0.2 ohm, while those for section *BC* are 0.04 ohm and 0.1 ohm, respectively. Find the input current and power.

*Solution*

Referring to Sec. 2.28, Fig. 2.24 and 2.25

$a = 2$

$R_1 = 0.06$ ohm, $\quad X_1 = 0.2$ ohm

$R_2 = 0.04$ ohm, $\quad X_2 = 0.1$ ohm

Since output terminals are short circuited, the impedance as seen from input terminals is

$$Z_1 = 0.06 + (2-1)^2(0.04) + j\left[0.2 + (2-1)^2(0.1)\right]$$

$$= 0.1 + j0.3 \text{ ohm}$$

$$\text{Input current} = \frac{9.5}{0.1 + j0.3} = 30.04\angle -71.57° \text{ A}$$

$$\text{Power Input} = (9.5)(30.04)(\cos 71.57°) = 90.22 \text{ W.}$$

### *Example 2.49*

A 6600/400/110 V, star/star/mesh, three-phase transformer has a magnetising current of 5.5 A and balanced 3-phase loads of 1000 kVA at 0.8 lagging power factor on the secondary and 200 kVA at 0.5 leading power factor on tertiary. Neglect losses. Find primary current, kVA and power factor.

### *Solution*

Let Base kVA = 1000 and base line-to-line voltages for primary, secondary and tertiary be 6600 V, 400 V and 110 V, respectively.

Then base currents are

$$\text{Primary (Y)} = \frac{1000 \times 1000}{\sqrt{3} \times 6600} = 87.48 \text{ A}$$

$$\text{Secondary (Y)} = \frac{1000 \times 1000}{\sqrt{3} \times 400} = 1443.42 \text{ A}$$

$$\text{Tertiary } (\Delta) = \frac{1000 \times 1000}{3 \times 110} = 3030.3 \text{ A}$$

$$\text{Secondary load current per phase} = \frac{1000 \times 1000}{\sqrt{3} \times 400} = 1443.42 \text{ A}$$

$$\text{Secondary load current per unit} = \frac{1443.42}{1443.42}\angle -\cos^{-1}.8$$

$$= 1(0.8 - j0.6) \text{ pu}$$

$$\text{Tertiary load current per phase} = \frac{200 \times 1000}{3 \times 110} = 606.06 \text{ A}$$

$$\text{Tertiary load current per unit} = \frac{606.06}{3030.3}\angle \cos^{-1} 0.5$$

$$= 0.2(0.5 + j0.866)\text{ pu}$$

$$\text{Magnetising current per unit} = \frac{-j5.5}{87.48} = -j0.063\text{ pu}$$

$$\text{Total current} = 1(.8 - j.6) + .2(.5 + j0.866) - j0.063$$

$$= 0.9 - j0.5 = 1.03\angle -29^\circ\text{ pu}$$

$$\text{Primary current} = 1.03 \times 87.48 = 90.1\text{ A}$$

$$\text{Primary kVA} = 1.03 \times 1000 = 1030\text{ kVA}$$

$$\text{Primary power factor} = \cos(29^\circ) = 0.875\text{ lagging.}$$

*Example 2.50*

For the three-winding transformer of example 2.49 find the regulation (a) from primary to secondary, (b) from primary to tertiary for the given load. Assume $R_{pu}$ for the three windings as 0.0053, 0.0063 and 0.0083 and $X_{pu}$ as 0.0265, 0.025 and 0.0332 pu (c). Also find the regulation from secondary to tertiary.

*Solution*

Regulation contributed by primary $= 1.03(0.0053 \times 0.875 + 0.0265 \times 0.48) = 0.0178\text{ pu}$

Regulation contributed by secondary $= 1\,(0.0063 \times 0.8 + 0.025 \times 0.6) = 0.02\text{ pu}$

Regulation contributed by tertiary $= 0.2(0.0083 \times 0.5 - 0.0332 \times 0.866) = -0.0049\text{ pu}$

(a) Regulation from primary to secondary $= 0.0178 + 0.02 = 0.0378$ pu or 3.78%

(b) Regulation from primary to tertiary $= 0.0178 - 0.0049 = 0.0129$ pu or 1.29%

(c) Regulation from secondary to tertiary $= 0.02 - (-0.0049) = 0.0249$ pu or 2.49%.

*Example 2.51*

A 3-winding transformer has the following ratings. Primary 300 kVA, 3.3 kV, secondary 240 kVA, 1.1 kV and tertiary 60 kVA, 220 V. The O.C. and S.C. test data are as under:

$$Y_0 = 0.03 - j0.06, \quad Z_{12} = 0.01 + j0.06\text{ pu}$$

$$Z_{13} = 0.016 + j0.1, \quad Z_{23} = 0.02 + j0.06\text{ pu}$$

The base kVA is 300 kVA and respective winding voltages are the base voltages. The secondary is supplying its rated kVA at 0.8 p.f. lagging at rated voltage while tertiary is open circuited. Find (a) voltage at tertiary terminals, (b) voltage at primary terminals, (c) primary current, (d) primary power factor. [*P.U. 2001*]

*Solution*

Using Eqs. (2.104 to 2.106), the internal impedances of the three winding are

$$Z_1 = 0.5(Z_{12} + Z_{13} - Z_{23}) = 0.003 + j0.05\text{ pu}$$

$$Z_2 = 0.5(Z_{12} + Z_{23} - Z_{13}) = 0.007 + j0.01\,\text{pu}$$

$$Z_3 = 0.5(Z_{23} + Z_{13} - Z_{12}) = 0.013 + j0.05\,\text{pu}$$

$$I_2 = \frac{240}{300}(0.8 - j0.6) = 0.64 - j0.48\,\text{pu}$$

$$V_2 = 1\angle 0\,\text{pu}$$

Referring to the equivalent circuit of Fig. 2.45, $V_0$, the voltage of star point is

$$V_0 = V_2 + I_2 Z_2 = 1\angle 0 + (0.64 - j0.48)(0.007 + j0.01)$$

$$= 1.00928 + j0.00304\,\text{pu}$$

(a) Since tertiary current is zero, terminal voltage of tertiary is equal to $V_0$

$$V_3 = 1.00928 + j0.00304\,\text{pu}$$

$$= (1.00928 + j0.00304)(220) = 222\,\text{V}$$

(b) $V_1 = V_0 + I_1 Z_1$

Since $I_3 = 0$, $I_1 = I_2$

$$V_1 = 1.00928 + j0.00304 + (0.64 - j.48)(0.003 + j0.05)$$

$$= 1.0352 + j0.0336\,\text{pu}$$

Primary voltage $= (1.0352 + j0.0336)3.3$

$$= 3.42\angle 1.86°\,\text{kV}$$

(c) No-load current $= V_1 Y_0 = I_0$

or $I_0 = (1.0352 + j0.0336)(0.03 - j0.06)\,\text{pu}$

or $I_0 = 0.033 - j0.061\,\text{pu}$

Primary current $= I_2 + I_0 = 0.673 - j0.541\,\text{pu}$

Base current for primary $= \dfrac{300 \times 1000}{3.3 \times 1000} = 90.91\,\text{A}$

Primary current $= (90.91)(0.673 - j\,0.541)$

$$= 61.18 - j49.18 = 78.5\angle -38.8°\,\text{A}$$

(d) Primary power factor $= \cos(1.86° + 38.8°) = 0.758$ lagging.

*Example 2.52*

A 200 VA, 120/12 V two-winding transformer is to be used as an auto-transformer. The input voltage is 120 V.

(a) What is the secondary voltage?

(b) What is the maximum VA rating of auto-transformer?

*Solution*

Referring to Fig. 2.23

(a) The winding portion $BC$ is the 120 V winding and $AB$ is the 12 V winding

$$V_2 = 120 + 12 = 132\,\text{V}$$

(b) The current ratings of the windings of the two-winding transformer are

$$\text{Primary winding (portion } BC) = \frac{200}{120} = \frac{5}{3}\,\text{A}$$

$$\text{Secondary winding (portion AB)} = \frac{200}{12} = \frac{50}{3}\,\text{A}$$

Rating when used as auto-transformer

$$= V_2 I_2 = 132 \times \frac{50}{3} = 2200\,\text{VA}$$

The rating can also be calculated using Eq. (2.69)

$$a = \frac{132}{120} = 1.1$$

$$\text{Output VA of auto-transformer} = \left(\frac{a}{a-1}\right)(200) = 2200\,\text{VA.}$$

*Example 2.53*

For the two windings transformer of Example 2.52, the resistances of primary and secondary windings are 1.5 ohm and 0.015 ohm, respectively. Find the efficiency at full-load unity power factor when it is used (a) as a two-winding, 120/12 V step down transformer, (b) as a 120/132 V auto-transformer. Iron losses = 5 W.

*Solution*

(a) Primary current $= \dfrac{200}{120} = \dfrac{5}{3}\,\text{A}$

Secondary current $= \dfrac{200}{12} = \dfrac{50}{3}\,\text{A}$

$$\text{Total copper losses} = \left(\frac{5}{3}\right)^2 (1.5) + \left(\frac{50}{3}\right)^3 (0.015) = 8.33\,\text{W}$$

Iron losses $= 5$ W

$$\text{Efficiency} = \frac{\text{output}}{\text{output + losses}} = \frac{200 \times 1}{200 \times 1 + 8.33 + 5} \times 100 = 93.75\%$$

(b) Referring to Fig. 2.23

Output = 2200 VA (as found in Example 2.52)

$$I_2 = \frac{2200}{132} = \frac{50}{3}\,\text{A}$$

$$I_1 = \frac{2200}{120} = \frac{55}{3}\,\text{A}$$

$$\text{Current in winding } AB = \frac{50}{3}\,\text{A}$$

$$\text{Current is winding } BC = I_1 - I_2 = \frac{5}{3}\,\text{A}$$

$$\text{Copper-loss in winding } AB = \left(\frac{50}{3}\right)^2 (0.015)\,\text{W}$$

$$\text{Copper-loss in winding } BC = \left(\frac{5}{3}\right)^2 (1.5)\,\text{W}$$

$$\text{Total copper-loss} = \left(\frac{50}{3}\right)^2 (0.015) + \left(\frac{5}{3}\right)^2 (1.5) = 8.33\,\text{W}$$

$$\text{Iron-losses} = 5\ \text{W}$$

$$\text{Efficiency} = \frac{\text{output}}{\text{output + losses}} = \frac{2200 \times 1 \times 100}{2200 \times 1 + 8.33 + 5}$$

$$= 99.4\%.$$

*Example 2.54*

A 3-phase star connected auto-transformer is used to supply a 350 V load from a 440 V 3-phase system. The load draws 50 kW at 0.8 lagging power factor. (a) Draw a diagram of connections. (b) Calculate the currents in different windings of auto-transformer and show them on the diagram. Neglect losses.

*Solution*

(a) The connection diagram is shown in Fig. 2.62.

(b) $\text{Load current} = \dfrac{50 \times 1000}{\sqrt{3}(350)(0.8)} = 103.1\ \text{A}$

$$\text{Line current drawn from source} = \frac{50 \times 1000}{\sqrt{3}(440)(0.8)} = 82\,\text{A}$$

The currents in winding portions *na, nb* and *nc* = 103.1 – 82 = 21.1 A.

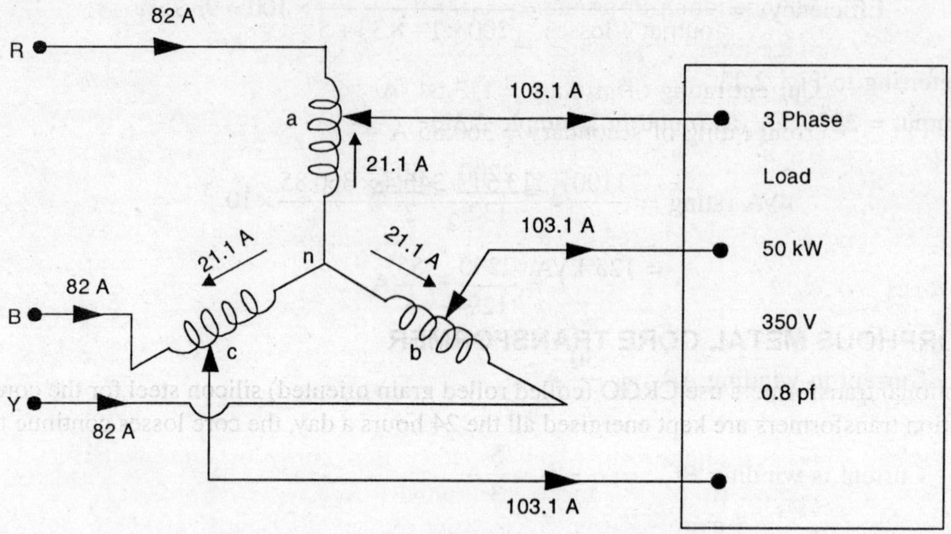


**Fig. 2.62**

*Example 2.55*

A balanced 3-phase, 200 kW load at 400 V and 0.8 power factor lagging is to be supplied from a two-phase 1100 V supply. Determine voltage and current rating of each winding of scott connected transformers and kVA rating of each unit. [*K.U. 2002*]

*Solution*

Referring to Fig. 2.39 (a), the 3-phase side is secondary and two phase side is primary.

$$\text{Line current on 3-phase side} = \frac{200 \times 1000}{\sqrt{3} \times 440 \times 0.8} = 360.85\,\text{A}$$

A balanced load on one side results in balanced load on the other side. Therefore input of each transformer is 100 kW, 1100 V, 0.8 power factor lagging.

$$\text{Current in primary of each transformer} = \frac{100 \times 1000}{1100 \times 0.8} = 113.64\,\text{A}$$

Main transformer

Voltage rating of primary = 1100 V

Voltage rating of secondary = 400 V

Current rating of primary = 113.64 A

Current rating of secondary = 360.85 A

$$\text{kVA rating} = \frac{1100 \times 113.64 + 400 \times 360.85}{2} \times 10^{-3} = 134.67\text{ k VA}$$

Teaser transformer

$$\text{Voltage rating of primary} = 1100 \text{ V}$$

$$\text{Voltage rating of secondary} = 0.866 \times 400 = 346.4 \text{ V}$$

$$\text{Current rating of primary} = 113.64 \text{ A}$$

$$\text{Current rating of secondary} = 360.85 \text{ A}$$

$$\text{kVA rating} = \frac{1100 \times 113.64 + 346.4 \times 360.85}{2} \times 10^{-3}$$

$$= 125 \text{ kVA.}$$

## 2.65 AMORPHOUS METAL CORE TRANSFORMER

The conventional transformers use CRGO (colled rolled grain oriented) silicon steel for the core. Since the distribution transformers are kept energised all the 24 hours a day, the core losses continue to occur throughout the day. This causes huge energy loss. To reduce this energy loss amorphous metal core can be used in distribution transformers. The use of amorphous metal core reduces the core losses by as much as 75%. Table 2.1 shows a comparison of properties of amorphous metal and CRGO silicon steel.

**Table 2.2:** Comparison of properties of Amorphous Metal and CRGO Silicon Steel

| *Property* | *Amorphous Metal* | *CRGO Silicon Steel* |
|---|---|---|
| Density gm/cm$^3$ | 7.15 | 7.65 |
| Resistivity | 130 | 45 |
| Saturation flux density Wb/m$^2$ | 1.56 | 2.03 |
| Core loss W/kg at 50 Hz and 1.4 Wb/m$^2$ | 0.2 | 0.9 |
| Thickness | 0.025 mm | 0.27 |
| Space factor | 0.86 | 0.97 |
| Available form | Ribbon/foil | Sheet/Roll |
| Annealing Temperature | 360°C | 810°C |
| Annealing atmosphere | inert gas | inert gas |
| Special annealing requirement | magnetic field annealing | – |

Amorphous metals are alloys and donot have any atomic order. As such they lack systematic structure or form. Because of amorphous nature, they are also called metallic glasses. Rapid cooling of molten metal prevents crystallisation and results in a vitrified solid in the form of strips.

Some properties of amorphous metals are

1. Amorphous metals are in the form of thin strips
2. Amorphous metals are hard, strong, corrosion resistant and have long life.
3. Amorphous metals need magnetic field annealing.
4. After annealing these metals become less ductile.

Advantages of amorphous metals over CRGO silicon steel are

1. Less core losses. In amorphous metal core transformers core losses are about 25% of the core losses in transformers using CRGO silicon steel core. In a 100 kVA CRGO silicon steel core distribution transformer the core losses are about 250 W. However in a 100 kVA amorphous metal core transformer, the core losses are only about 60 W. In one year this means an energy saving of (250–60) 8760 Wh or about 1665 kWh.
2. Amorphous metal core distribution transformers have less losses due to harmonic distortion.
3. Amorphous metal core transformers have high capability to withstand short circuit forces.
4. Amorphous metal core transformers depict lower temperature rise thus resulting is lesser insulation stresses.
5. Amorphous metal core transformers are less noisy than CRGO silicon steel core transformers.

   Amorphous metal core transformers were developed by General Elect. Company (USA) in 1978.

## 2.66 CAST RESIN DRY TYPE TRANSFORMERS

Oil immersed transformers have been in use for the past many decades. These transformers use mineral oil for cooling and insulation. However these transformers are fire prone and can cause huge loss of life and property due to explosion. To guard against these risks dry type transformers should be used for densely populated areas and multi-storeyed buildings.

A dry type transformers may be an encapsulated or non-encapsulated. In encapsulated transformer the windings are encapsulated in solid insulation. However in non-encapsulated transformer the windings are not encapsulated with solid insulation.

Dry type cast resin transformer uses a metal mould for manufacture of winding. The mould holds the epoxy in position during filling and cure cycles. In the absence of metal mould the epoxy will draw in during the cure cycle. A vacuum filling process is necessary to eliminate the formation of voids and trapped gases in the casting. The windings are permanently embedded between layers of fibre glass mat.

Two methods viz filled system and unfilled system have been developed. In filled system a filler (e.g. quartz) is used along with resin. However in unfilled system it is not necessary to mix resin with any filler. The resin filling is by capillary action. In the unfilled system the coils are cast under high vacuum, have high mechanical strength and casting is void free. Therefore unfilled system is superior and preferable.

The resin system, in unfilled casting method has epoxy resin, hardner, accelerator and platiciser as its components. The epoxy resin is a molecule containing more than one alpha epoxy group capable of being converted into a thermoset form. The liquid epoxy resin, used in dry type cast resin transformers, has low viscosity, low shrinkage, high adhesive strengths, good electrical properties, good mechanical strength and good chemical resistance. Hardner is an acid anhydride having anhydride content of more than 90%. Accelerator is free tertiary amine and accelerates the reaction process. Plasticiser is a solvent free polyglycol and provides elastic properties.

The assembled cores are treated with anti corrosive paint for protection against weather. The winding inter-layer insulation consists of epoxy compatible glass fibre. The gap between HT and LT windings consists of epoxy resin and air. Resistance elements are used to sense the winding temperature. Windings are first assembled into steel moulds and preheated before transferring them to casting vessel. Vacuum is applied to the casting vessel. The resin is introduced at the bottom of the mould when the casting vessel is at desired pressure and temperature. The resin filling is by capillary action. When resin filling is complete the mould is transferred to the curing oven. Controlled curing ensures optimum galling rate and controls exhotherm and thus prevents shrinkage, cracks and voids.

Advantages of cast resin dry type transformers (over oil immersed transformers) include non-hygroscopic insulation, fire safety, high mechanical strength, freedom from leakage and pilferage of oil, low maintenance cost, freedom from partial discharges, high impulse strength, high short circuit withstand capability, high short time overload capacity, lesser weight, lower losses, higher efficiency and high reliability.

## SUMMARY

1. A transformer works on the principle of mutual induction. It steps up or steps down ac voltage.
2. RMS emf induced in a coil = 4.44 fN $\phi_m$.
3. Transformers are classified as core type and shell type.
4. An ideal transformer has no magnetic leakage, zero resistance of windings and its core has infinite permeability.
5. Phasor diagram of a transformer shows the phasor relationship between primary and secondary voltages, currents and voltage drops.
6. An equivalent circuit of transformer helps in analysis.
7. Resistances and reactances of transformers can be transferred from primary to secondary and vice versa for ease of calculations.
8. Per unit method of calculations is very convenient.
9. Voltage regulation is the change in secondary terminal voltage when rated load at a specified power factor is removed. It is expressed as a percentage of rated secondary voltage.
10. Losses in a transformer are copper losses and iron losses
11. Efficiency $= \dfrac{V_2 I_2 \cos\theta_2}{V_2 I_2 \cos\theta_2 + I_2^2 R_{e2} + P_i}$
12. Condition for maximum efficiency of transformer
    Copper losses = Iron losses
13. All day efficiency $= \dfrac{\text{Energy output in 24 hours}}{\text{Energy output in 24 hours} + \text{Energy losses in 24 hours}}$
14. Parameters of equivalent circuit can be determined from open circuit and short circuit tests.
15. Sumpner's test is a full load test on transformer. It requires two similar transformers.
16. An auto transformer has only one winding. It is useful when transformation ratio is nearly 1.
17. 3 phase transformer can be star-star, delta-delta, delta-star and star-delta. Delta winding is useful in eliminating triplen harmonics.
18. Scott connection is used for 3-phase to two phase conversion.
19. Transformers for 3-phase to 6-phase conversion and 3-phase to 12 phase conversion are used in rectifier circuits.

20. Conditions for parallel operation for two transformers are: same polarity, same voltage ratio, same phase sequence and zero relative phase displacement and equal per unit resistance and reactance.
21. A three winding transformer has 3 windings primary, secondary and tertiary.
22. The magnetising current of transformer contains harmonics.
23. When a transformer is switched on an inrush of magnetising current occurs.
24. Taps are provided on high voltage side of transformer. Tap changing helps in keeping secondary terminal voltage constant.
25. Induction regulator and moving coil voltage regulator are used for smooth variation of voltage.
26. Potential and current transformers are used in instrumentation and protection circuits.
27. Large size transformers are oil cooled.
28. Wide band transformer and pulse transformers are used in electronic circuits.
29. Amorphous metal core transformer has very low iron losses.
30. Cast resin dry type transformers are preferable for distribution systems in densely populated areas.

## TEST POINT QUESTIONS

### A. State whether the following statements are true (T) or false (F)

2.1 Transformers are used only in power industry.

2.2 Transformer operates on the principle of mutual induction.

2.3 In a step down transformer secondary voltage and current are less than primary voltage and current.

2.4 Core type transformers are used in high voltage circuits.

2.5 In an ideal transformer iron losses are zero.

2.6 Leakage flux exists only in primary circuit of a transformer.

2.7 If the load power factor is 0.8 lagging, the power factor at primary terminals of the transformer will be worse than 0. 8 lagging.

2.8 In a step up transformer the resistances of primary and secondary windings are nearly equal.

2.9 If the transformer window is long and narrow, the leakage flux is less.

2.10 Voltage regulation, at rated load and 0.8 lagging power factor, of a large size transformer is about 30%.

2.11 Iron losses, in a transformer, are very very small.

2.12 All day efficiency in important for transformers used in big power houses.

2.13 In open circuit test on a transformer, the secondary winding supplies a small load.

2.14 In short circuit test, the iron losses are negligible.

2.15 Auto transformer requires less winding material than a two winding transformer of the same size.

### B. Fill in the blanks with suitable words

2.16 In an auto transformer with transformation ratio 1.1 the saving in copper will be about ......... as compared to that in similar two winding transformer.

2.17 Delta winding helps in damping out .........

2.18 Back to back test on transformers is a ......... test.

2.19 Three phase to twelve phase conversion is often needed in ......... circuits.

2.20 Two transformers connected in parallel must have the same .........

2.21 In a three winding transformer the windings are called, ........., ........., ......... .

2.22 Magnetising current of a transformer is .........

2.23 The maximum value of inrush magnetising current depends on .........

2.24 Transformer taps are generally on ......... of transformers.

2.25 A 3-phase induction regulator is similar to .........

2.26 The secondary of a CT must not be .........

2.27 Dielectric strength of transformer oil is about .........

2.28 Transformer noise is due to .........

2.29 Moisture ......... the dielectric strength of oil.

2.30 Amorphous core transformer has very low .........

## ANSWERS

| | | | | |
|---|---|---|---|---|
| 2.1 F | 2.2 T | 2.3 F | 2.4 T | 2.5 T |
| 2.6 F | 2.7 T | 2.8 F | 2.9 T | 2.10 F |
| 2.11 F | 2.12 F | 2.13 F | 2.14 T | 2.15 T |

| | | |
|---|---|---|
| 2.16 90% | 2.17 third harmonic voltage | 2.18 regenerative |
| 2.19 rectifier | 2.20 polarity | 2.21 primary, secondary, tertiary |
| 2.22 non-sinusoidal | 2.23 residual flux density | 2.24 high voltage winding |
| 2.25 3 phase inductor motor | 2.26 open circuited | 2.27 30 kV |
| 2.28 magnetostriction | 2.29 decreases | 2.30 iron losses |

## JOB INTERVIEW QUESTIONS

2.1 Name some applications of transformers.

2.2 What is an air cored transformer? Where is it used?

2.3 What is leakage flux?

2.4 What is all day efficiency?

2.5 What is the function of oil in a transformer?

2.6 What is meant by voltage regulation?

2.7 What is the need of load test on a transformer?

2.8 What are iron losses due to?

2.9 What is an auto transformer?

2.10 Name some uses of auto transformer.

2.11 Name some applications of three winding transformers.

2.12 Why are copper losses negligible during open circuit test on transformer.

2.13 What is the reason for transformer noise?

2.14 Why is the core of a transformer made from thin stampings?

2.15 What is a grounding transformer?

2.16 What is amorphous metal core?

2.17 Where is a pulse transformer used?

2.18 What is a potential transformer?

2.19 Where are instrument transformers used?

2.20 Why is magnetising current of a transformer non-sinusoidal?

## REVIEW QUESTIONS

2.1 (a) Why does a transformer have iron core?
(b) How does flux leakage occur in a transformer?
(c) What is the effect of leakage flux in a transformer?
(d) What is meant by magnetising current?

2.2 (a) Why is a transformer core laminated?
(b) What is meant by step up and step down transformers? Which quantity is being stepped up or stepped down?
(c) What is turn ratio? What is its significance?
(d) How is emf induced in transformer windings?

2.3 In a core type transformer, copper surrounds iron and in a shell type iron surrounds copper. Explain the statement.

2.4 When the load current of a transformer increases how does the input current increase to meet the new condition?

2.5 What is meant by 'equivalent resistance referred to primary' and 'equivalent resistance referred to secondary'? How are they determined?

2.6 In what ways is the approximate equivalent circuit of a transformer different from the exact equivalent circuit? How does the approximate circuit simplify the calculations?

2.7 What is all-day efficiency? For which type of transformers is all-day efficiency important and why?

2.8 Discuss the principle of operation and construction of a single-phase transformer.

2.9 Starting from first principles develop the equation of emf induced in transformer winding. Show that the primary and secondary voltages are in the ratio of primary and secondary turns and that primary and secondary currents are in the inverse ratio of turns.

2.10 Define an ideal transformer. How is the concept of ideal transformer useful in understanding the behaviour of an actual transformer?

2.11 Draw and explain the phasor diagram of a transformer.

2.12 Discuss the effect of (a) core loss (b) magnetic leakage on the performance of a transformers.

2.13 Starting from first principles develop the equivalent circuit of a transformer.

2.14 What is voltage regulation? What is its importance? Develop an expression for calculating the voltage regulation of a two-winding transformer.

2.15 What losses are present in a transformer? Write an expression for efficiency and develop the condition for maximum efficiency.

2.16 On what factors do (a) hysteresis loss, (b) eddy current loss depend. Why can these losses be assumed constant?

2.17 Is the voltage regulation at 0.6 lagging power factor better or worse than that at 0.8 lagging power factor for the same kVA output? Give reason for your answer.

2.18 Explain why

(a) The open circuit test on a transformer is conducted at rated voltage;

(b) Only a low voltage is applied to a transformer during short circuit test;

(c) Usually the low voltage winding is excited and the high voltage winding open circuited for the open circuit test; and

(d) Usually the low voltage winding is short circuited and high voltage winding excited for the short circuit test.

2.19 Give reasons for the following:

(a) An auto-transformer is generally used for transformation ratio of less than 3.

(b) During the open circuit test, the copper losses are negligible and during short circuit test, the iron losses are negligible.

(c) It is very difficult to perform a load test, on a large size transformer, by direct loading.

(d) The secondary of a current transformer should not be open circuited.

2.20 Why is the condition of correct polarity the most important and essential condition for parallel operation of transformers?

2.21 Give circuit diagrams and explain the procedures of open circuit and short circuit tests on a two-winding transformer. How can the parameters of equivalent circuit be determined from these tests?

2.22 What is the effect of

(a) unequal voltage ratio.

(b) unequal p.u. impedances.

(c) unequal *X/R* ratio on the parallel operation of two transformers?

2.23 Is it necessary that the two transformers to be operated in parallel should be of the same capacity? Discuss.

2.24 In an auto-transformer the power transfer from primary circuit to secondary circuit is partly by conduction and partly by induction. Discuss.

2.25 (a) Draw the phasor diagram and equivalent circuit of an auto-transformer.

(b) An auto-transformer has a transformation ratio *a*. Derive the equation giving its rating as a fraction of the rating of a two-winding transformer.

2.26 Discuss the advantages, disadvantages and applications of auto-transformers. Compare it with a two-winding transformer.

2.27 (a) What are the distinguishing features of $Y$–$Y$, $Y$–$\Delta$, $\Delta$–$Y$ and $\Delta$–$\Delta$, 3-phase connections?

Compare their advantages and disadvantages.

(b) How can the problems of unbalanced voltages and third harmonic current be overcome in $Y$–$Y$ connection?

2.28 Explain the following connections giving suitable diagrams:

(a) Scott connection,

(b) Open-delta connection,

(c) 3-phase to 6-phase conversion.

(d) 3-phase to 12-phase conversion.

2.29 What are the applications of 3-winding transformer? How can this transformer be represented by an equivalent circuit. Give the details of tests to find out the parameters of this equivalent circuit.

2.30 (a) Why is the waveshape of magnetising current of a transformer non-sinusoidal?

(b) Discuss the phenomenon of inrush of magnetising current. What factors contribute to the magnitude of inrush current?

## PROBLEMS

2.1 Estimate the active cross-sectional area of the core of a 20-turn, 1-phase inductor for a terminal voltage of 100 V at 50 Hz. The flux density should be about 1 Wb/m$^2$. [225 cm$^2$]

2.2 A single-phase transformer has 400 primary and 1000 secondary turns. The net cross-sectional area of the core is 60 cm$^2$. The primary winding is connected to a 500 V supply. Find (a)peak value of core flux density, (b) emf induced in secondary winding. [(a) 0.94 T; (b) 1250 V]

2.3 A transformer core has a square cross-section of 2 cm side. The flux density should not exceed 1 tesla. The primary winding is to be designed for 230 V, secondary for 110 V. Another centre tapped 6-0-6 V tertiary winding is also provided. Find the number of turns in each winding. [2607, 1247, 2 × 68]

2.4 A transformer with an output voltage of 75 V supplies a load consisting of 33.5 ohm resistance and 22 ohm inductive reactance. Find input voltage, current and power factor, if the transformer has following parameters: $N_1/N_2 = 3/2$, $R_1 = 2\ \Omega$, $X_1 = 5\ \Omega$, $R_0 = 7500\ \Omega$, $X_0 = 3000\ \Omega$, $R_2 = 0.25\ \Omega$, $X_2 = 1.2\ \Omega$. [121 V, 1.27 A, 0.682 lagging]

2.5 A single-phase transformer has 90 and 180 turns in the primary and secondary windings. The winding resistances are 0.067 Ω and 0.233 Ω respectively. Find (a) equivalent resistance of secondary as referred to primary, (b) equivalent resistance of primary as referred to secondary, (c) total winding resistance as referred to secondary. [(a) 0.058 Ω (b) 0.268 Ω (c) 0.501 Ω]

2.6 The parameters as referred to low voltage side for a 4 kVA, 50 Hz, 200/400 V, single-phase transformer are $R_0 = 600\ \Omega$, $X_0 = 300\ \Omega$, $R_1 = 0.15\ \Omega$, $X_1 = 0.31\ \Omega$. For a high voltage current of 10 A at a power factor of 0.8 lagging, find (a) low voltage input current, (b) efficiency, (c) output voltage. Take input voltage as 200 V. Use approximate equivalent circuit. [(a) 20.65 A (b) 96.08% (c) 386 V]

2.7 The efficiency at unity power factor of a 6600/384 V, 200 kVA 1-phase transformer is 98% for both full-load and half-load conditions. The power factor at no-load is 0.2 and full-load regulation at a power factor of 0.8 lagging is 4%. Draw approximate equivalent circuit as referred to low voltage side and insert the parameters. [$R_0 = 108\ \Omega$, $X_0 = 22.1\ \Omega$, $R_2 = 0.01\ \Omega$, $X_2 = 0.036\ \Omega$]

2.8 A 30 kVA, 6000 V/230 V single-phase transformer has primary and secondary winding resistances of 10.2 and 0.016 ohm, respectively. The reactance as referred to primary is 34 ohms. Find (a) primary voltage to circulate full-load current with secondary short circuited (b) percentage regulation of the transformer for a load of 30 kVA at 0.8 lagging power factor. [(a) 199.5 V (b) 3.1%]

2.9 Compute the regulation of a transformer in which the ohmic loss in 1% of output and reactance drop is 5% of voltage, when the power factor is, (a) 0.8 lagging, (b) unity, (c) 0.8 leading.

[(a) 3.8% (b) 1% (c)–2.2%]

2.10 The maximum efficiency of a 500 kVA, 3300/500 V, single-phase transformer occurs at 75% full-load unity power factor and is 0.97. If the impedance is 10%, find regulation at full-load 0.8 lagging p.f. [7.51%]

2.11 A 50 Hz, 2000/400 V, 125 kVA single-phase transformer gave the following test results:

Open circuit test (l.v. winding open circuited) 2000 V, 1A, 1000 W

Short circuit test (h.v. winding short circuited) 13 V, 200 A, 750 W

Find (a) $I_m$ and $I_w$, (b) full-load efficiency at power factors of 1, 0.8 lag, 0.8 lead, (c) secondary terminal voltage at full-load, at p.f. of 1, 0.8 lag and 0.8 lead. The primary is excited at 2000 V.

[(a) 0.866 A, 0.5 A (b) 0.978, 0.972, 0.972 (c) 394 V, 384 V, 406 V]

2.12 The following results pertain to a 200/400 V, 5 kA, 50 Hz single-phase transformer. O.C. test (l.v. data) 200 V, 60 W, 0.7 A. S.C. test (h.v. data) 22 V, 120 W, 16 A. Find (a) voltage regulation at rated kVA, 0.9 lagging p.f., (b) kVA for maximum efficiency and the value of this efficiency. [(a) 3.08% (b) 4.52 kVA, 0.974]

2.13 A no-load test on a single-phase transformer gave the following test results:

| Applied Voltage V | 200 | 150 | 100 | 50 |
|---|---|---|---|---|
| Frequency Hz | 50 | 37.5 | 25 | 12.5 |
| Power Input W | 55 | 38 | 23 | 10.25 |

Find hysteresis and eddy current loss at frequency of (a) 60 (b) 50 (c) 25.

[(a) 44.4, 25.9 W (b) 37, 18 W, (c) 18.5, 4.5 W]

2.14 Find the parameters of a transformer as referred to l.v. side. The transformer is rated at 4 kVA, 200/400 V, 50 Hz and test results are: no-load test (l.v. data) 200 V, 70 W, 0.7A, S.C. test (h.v. data) 15 V, 80 W, 10 A.

[571.4 ohm, 330 ohm, 0.2 ohm, 0.317 ohm]

2.15 A no-load test on a transformer gave this data.

| Applied Voltage ($V$) | 250 | 200 | 150 | 100 |
|---|---|---|---|---|
| Frequency ($f$) | 54.4 | 43.5 | 32.6 | 21.7 |
| Input ($W$) | 130 | 93.5 | 62.4 | 36.2 |

Find hysteresis and eddy current loss at 230 V, 50 Hz. [59.5 W, 55.2 W]

2.16 A transformer gave the following test results:

No-load test: $V_1 = 120\ V$, $V_2 = 35\ V$, $W_0 = 5W$, $I_0 = 0.125$ A

S.C. test: $V_1 = 4V$, $W_1 = 3W$, $I_1 = 0.8$ A.

Find turn ratio and parameters of equivalent circuit. [3.43, 2880 Ω, 1020 Ω, 4.69 Ω, 1.73 Ω]

2.17 A single-phase 110 kVA transformer has a voltage ratio of 11/0.44 kV. Iron losses are 1100 W. The short circuit test with full-load current and 500 V applied to primary yielded a wattmeter reading of 1000 W. Find secondary terminal voltage and efficiency, if 250 A, 0.8 p.f. load is connected to secondary and primary is excited by 11 kV. [425 V, 0.976]

2.18 Find the efficiency of a 150 kVA transformer at 25%, 33% and 100% full-load at (a) unity power factor, (b) power factor 0.8 lagging if copper loss is 1600 W at full-load and iron-loss is 1400 W.

[(a) 96.15%, 96.94%, 98.04% (b) 95.28%, 96.21%, 97.56%]

2.19 In a 25 kVA, 2000/200 V transformer the iron and full-load copper losses are 350 W and 400 W respectively. Find the efficiency at unit power factor at (a) full-load, (b) half-load, (c) Determine the load for maximum efficiency. [(a) 97.1% (b) 96.5% (c) 23.4 kVA]

2.20 Calculate the efficiency at half-load, full-load and 25% overload for a 100 kVA transformer at power factor of (a) unity, (b) 0.8. The core-loss is 1000 W and full-load copper loss is 1000 W.

[(a) 0.976, 0.98, 0.98 (b) 0.97, 0.976, 0.975]

2.21 An 800 kVA transformer has a core-loss of 7.5 kW and full-load copper loss of 14.2 kW. Find all day efficiency for the following duty cycle: 500 kW at 0.8 p.f. for 6 hours, 700 kW at 0.9 p.f. for 4 hours, 300 kW at 0.95 p.f. for 4 hours and no-load for 10 hours. [95.86%]

2.22 A transformer has its maximum efficiency of 0.98 at 15 kVA, unity power factor load. Find all day efficiency for the following loading cycle: 2 kW at 0.5 p.f. for 12 hours, 12 kW at 0.8 p.f. for 6 hours and 18 kW at 0.9 p.f. for 6 hours. [0.97]

2.23 Two transformers operate in parallel to supply a load of impedance $2 + j\,1.5$ ohm. The transformer impedances as referred to secondary windings are $0.15 + j\,0.5\ \Omega$ and $0.1 + j\,0.6\ \Omega$ respectively. The open circuit secondary emfs are 207 V and 205 V. Find (a) voltage across load, (b) power supplied to load, (c) power output of each transformer. [(a) 189 V (b) 11.5 kW (c) 6.55, 4.95 kW]

2.24 A 400 kVA transformer of impedance $0.01 + j\,0.05$ p.u. operates in parallel with a 200 kVA transformer of $0.012 + j\,0.04$ p.u. impedance. Find how will they share a load of 600 kVA at 0.8 lagging p.f.

[373 $\angle -39°$ kVA, 227 $\angle -33.6°$ kVA]

2.25 A 11500/230 V transformer is rated at 100 kVA as a two-winding transformer. If the two windings are connected in series to form an auto-transformer, what will be the voltage ratio and output.

[13.8/11.5 kV, 600 kVA, 13.8/2.3 kV, 120 kVA]

2.26 A 100 kVA single phase two-winding transformer has iron losses of 1300 W and full load copper losses of 1200 W. Find (a) efficiency at full load and half load unity p.f., (b) kVA for maximum efficiency.

[(a) 97.56%, 96.9% (b) 104.08 kVA]

2.27 A single phase transformer has 500 turns in one winding and 100 turns in the second. Find (a) transformation ratio when used as step up and step down transformer, (b) primary current if used as a step down transformer to supply 24A, (c) primary current if used as a step up transformer to supply 24 A. Neglect no-load current.

[(a) 0.2, 5 (b) 4.8 A (c) 120 A]

2.28 A transformer has 2000 turns in high voltage windings. It is it used to step down from 12 kV to 120 V find (a) number of turns in low voltage winding, (b) peak value of flux for frequencies of 50 and 25 Hz.

[(a) 20 (b) 0.027 Wb, 0.054 Wb]

2.29 A 100 kVA single phase transformer gave the following test results:

O.C. test (secondary winding open): primary volts 1000, secondary volts 100, power input 975 W.

S.C. test (secondary winding short circuited): primary volts 22, full-load current (100 A), power input 1050 W.

Find (a) percentage impedance, resistance and leakage reactance, (b) efficiency at 75% load 0.8 lagging p.f. (c) full load regulation at 0.9 lagging p.f.

[(a) 2.2%, 1.05%, 1.933%, (b) 97.46%, (c) 1.788%]

2.30 A single phase 50 Hz 1200/120 V transformer gave the following results of open circuit test with high voltage winding open circuited: Applied voltage 120 V, current 16 A, power input 400 W. Find (a) core losses, (b) no-load power factor, (c) magnetising and core loss components of no-load current.

[(a) 400 W (b) 0.2083 lagging (c) 15.65 A, 3.333 A]

2.31 A transformer is used to couple an audio amplifier to an 8 ohm load. The primary has 500 turns and the secondary has 25 turns. Find (a) load impedance as seen by amplifier, (b) primary current if output of amplifier is 15 V. Neglect no-load current. [(a) 3200 Ω (b) 4.6875 mA]

2.32 A single phase transformer has the following data: $N_1/N_2 = 15$,

Primary winding, resistance 384 Ω, leakage reactance 200 Ω

Secondary winding, resistance 1.3 Ω, leakage reactance 0.9 Ω.

If transformer is supplying 300 mA at 2.3 V at p.f. 0.8 lagging find input voltage. Draw phasor diagram.

[50.2 V]

2.33 A 20 kVA single phase, 2400/240 V, 50 Hz transformer gave the following short circuit test data with low voltage winding short circuited: input voltage for rated current = 72 V, input power 275 W. Find (a) rated currents for primary and secondary, (b) percentage impedance, resistance and leakage reactance, (c) p.f. on short circuit. [(a) 8.333 A, 83.33 A (b) 3%, 1.375%, 2.666% (c) 0.458 lagging]

2.34 For the transformer of problem 2.33 find winding resistance, leakage reactance and impedance as referred to high voltage side. [3.96 Ω, 7.68 Ω, 8.64 Ω]

2.35 For the transformer of problem 2.33 find winding resistance, leakage reactance and impedance as referred to low voltage side. [0.0396 Ω, 0.0768 Ω, 0.0864 Ω]

2.36 For the transformer of problem 2.33 find full load regulation at p.f. of (a) unity, (b) 0.8 lagging, (c) 0.8 leading. [(a) 1.375% (b) 2.7% (c) –0.5%]

2.37 A three-winding, single phase, 50 Hz transformer has $N_1 = 600$, $N_2 = 150$ and $N_3 = 300$ turns. The primary is excited by a 16 V, 50 Hz source. A resistance of 30 ohm is connected across secondary and a resistance of 15 ohm is connected across tertiary. Find (a) total impedance as seen by primary, (b) current drawn from supply, (c) secondary and tertiary voltages, (d) secondary and tertiary power dissipated. Neglect transformer resistance and reactances. [(a) 53.333 Ω (b) 0.3 A (c) 4V, 8 V (d) .533 W, 4.266 W]

2.38 Two single phase transformers each 250 kVA are operating in parallel and supplying 500 kVA at 0.707 lagging p.f. Their turn ratios are equal. Their impedance drops are (1.5 + $j$ 3.33)% and (0.9 + j 4)% respectively. Find their kVA loadings and power factors. [263 kVA, 237 kVA, 0.77, 0.63 lagging]

2.39 A single phase transformer, 2400/240 V, 10 kVA, 50 Hz has core loss of 153 W and full load copper loss of 224 W. Find all day efficiency for the following loading cycle: 25% over-load for 2-hours, full-load for 6 hours, half-load for 8 hours, quarter-load for 4 hours and no-load for 4 hours. All loads are at unity p.f. [95.6%]

2.40 A power amplifier has a no-load output voltage of 20 V and an internal resistance of 18 ohms. It is desired to feed its output to an 8 ohm speaker. Find (a) power delivered to speaker when it connected directly to amplifier, (b) turn ratio of an intervening transformer to maximize power transfer, (c) power delivered when transformer of part (b) is used. [(a) 4.73 W (b) 1.5 (c) 5.555 W]

2.41 A 3-phase, two-winding, 50 Hz, 11/3.3 kV, Y/Δ transformer has the primary and secondary resistances of 0.375 Ω and 0.095 Ω per phase respectively. The leakage reactances per phase are 9.5 ohms and 2 ohms respectively. Find the voltage which must be applied to primary so that full load rated current may flow if secondary terminals are short circuited. Also find the power input. The kVA rating of transformer is 1000 kVA. [1540 V, 6 kW]

2.42 A single phase, 2400/240 V, 50 Hz transformer has a primary winding impedance of (0.1 + $j$ 0.3) ohm and a secondary winding impedance of 0.001 + $j$ 0.003 ohm. Find (a) total impedance as referred to primary, (b) total impedance as referred to secondary, (c) primary current if primary voltage is 2400 V and a resistive load of 0.1058 ohm is connected across secondary. Neglect no-load current.

[(a) 0.2 + $j$ 0.6 Ω, (b) 0.002 + $j$ 0.006 Ω (c) 222.29 A]

2.43 A 5 kVA, 440/110 V, single phase, two-winding, 50 Hz transformer is used as an auto-transformer by connecting the two windings in series to supply 440 V from a 550 V source. Each coil operates at its rated voltage and current. Find (a) kVA rating of auto-transformer, (b) kVA transferred conductively, (c) kVA transferred inductively. [(a) 25 kVA, (b) 20 kVA (c) 5 kVA]

2.44 Two transformers connected in open delta supply a 375 kVA, 0.9 lagging p.f. load. The load voltage is 480 V. Find (a) kVA supplied by each transformer, (b) kW supplied by each transformer.

[(a) 216.5 kVA each (b) 121.6 kW, 215.9 kW]

Chapter **3**

# DC Machines

## 3.1 CONSTRUCTION

A dc machine may be a dc generator or a dc motor. The dc generator converts mechanical energy into dc, electrical energy, while a dc motor converts de electrical energy into mechanical energy.

A dc generator works on the principle that the motion of a conductor through a magnetic field causes as emf to be induced in the conductor. A dc motor works on the principle that a mechanical force acts on a conductor carrying current and lying in a magnetic field.

A dc machine consists of two parts: the stator, i.e. stationary part and the rotor i.e. the rotating part. Figure 3.1 shows a sectional view of a 4-pole dc machine and Fig. 3.2 shows a longitudinal view of armature (i.e. rotor) along with two coils of the winding. The stator consists of field poles and the outer frame (called yoke) and the rotor consists of armature, commutator and the shaft. The radial distance between the pole face and armature is known as air gap.

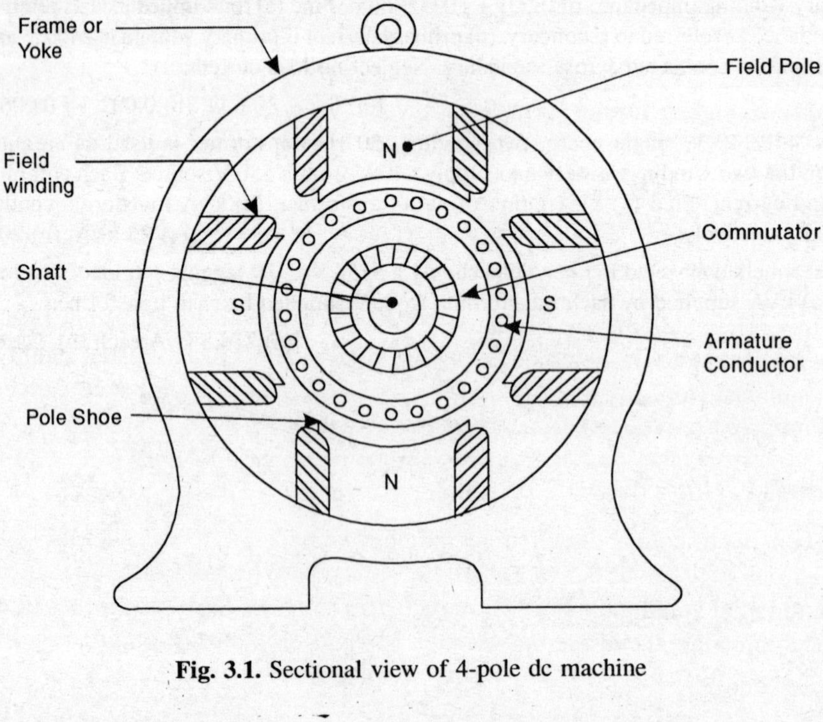


**Fig. 3.1.** Sectional view of 4-pole dc machine

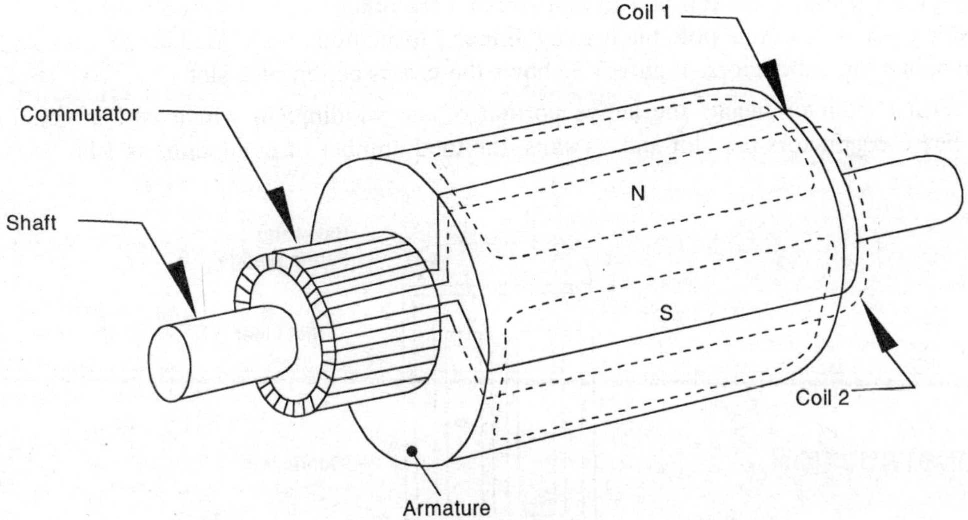


**Fig. 3.2.** Longitudinal view of armature and commutator

### 3.1.1 Field Poles

Small sized dc machines (usually known as magnetos) use permanent magnets as field poles. All other machines use electromagnets. The number of poles, may be 4 or more. A pole consists of pole core, pole shoe and field winding. The present day dc machines have associated solid state equipment and there is a high ac content in the power supplied to such dc motors. Therefore the pole core is made of laminated material (i.e. built-up thin steel sheets). DC machines have salient poles, i.e. the poles stick out from the surface of the stator. The pole core is welded or bolted to the yoke. The field winding consists of concentrated coil placed around the pole core. The pole shoe is the portion of the pole nearest to the air gap. It is curved and is wider than the pole core. It supports the field winding and helps in spreading the flux more uniformly. The surface of the pole shoe is chemferred, i.e. the outer tips of the shoe are spaced slightly farther from the armature surface than the centre of the pole shoe. This is done to increase the reluctance at the tips of the shoe, so that the flux bunching effect of armature reaction is reduced.

### 3.1.2 Yoke

Yoke is the outer frame of the machine. As seen in Fig. 3.12 (page 152) yoke also serves as a path for the magnetic flux. The field poles are fixed to the yoke. In small machines the yoke is made of cast iron, which is cheaper than steel. In large machines the yoke is made of fabricated steel, which has higher permeability than cast iron. In some modern dc motors, driven by solid state converter supplies, the yoke is also laminated to reduce eddy current losses.

### 3.1.3 Armature

The armature consists of core and armature winding. As discussed in Chapter 1 the core is always made of thin laminations, each about 0.5 mm thick, to reduce eddy current loss. The core is provided with longitudinal and radial ventilating ducts for circulation of cooling air. The surface of the armature core is slotted to accommodate the armature winding. Modern machines invariably use double layer winding, one layer of a coil in the top half of one slot and the other layer of the coil in the bottom half of another slot spaced about one pole pitch away. Enough insulation is provided in the slots and over the coils to insulate the conductors. Figure 3.3 shows the cross-section of a slot.

The term conductor means the active portion of the winding, in which emf is generated. If a machine has 6 conductors per slot and 24 slots, the total number of conductors is 144.

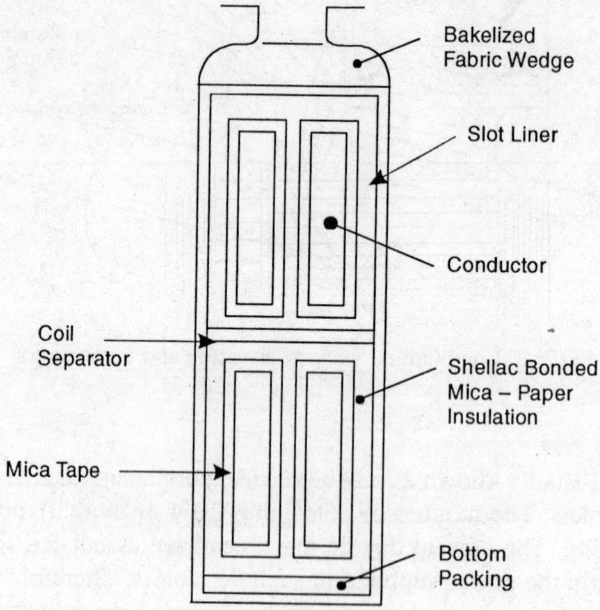


**Fig. 3.3.** Cross-section of armature slot

### 3.1.4 Commutator

The emf generated in the conductors of a dc machine is an alternating emf. The commutator helps in converting this alternating emf into a direct one. It is made from copper bars or segments assembled side-by-side to form a ring. the segments are insulated from each other by mica sheets. The ends of armature coils are connected to the commutator segments. It is cylindrical in shape and is mounted on the shaft alongside the armature. Figure 3.4 shows the construction of a commutator.

### 3.1.5 Brushes

The brushes collect the armature current from the commutator segments and supply it to the load (in case of generator) or feed the current into the commutator segments (in case of motor). They are made of graphite or a mixture of carbon and graphite. They have good conductivity to reduce electrical losses and a low co-efficient of friction to reduce excessive wear. The wear on the commutator surface is a result of the fact that brushes must rub over the commutator surface to convert ac emf in armature coils into a direct one. The brush pressure on the commutator surface needs very careful adjustment. If brush pressure is too small, excessive arcing may occur at the brush-commutator segment interface. If brush pressure is too high, excessive wear of brushes and commutator surface will take place. The brushes are held in position by brush holders.

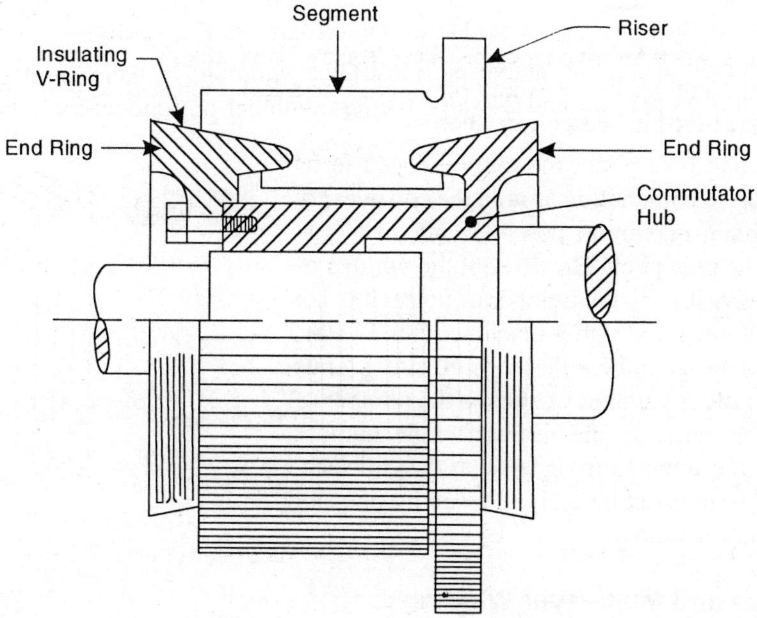


**Fig. 3.4.** Construction of commutator

### 3.1.6 Interpoles

The interpoles help in reducing the arcing due to commutation of current from ac to dc. They are also fixed to the yoke and are located in between the main poles. The winding of these poles has a few turns of thick wire and is connected in series with the armature circuit. Therefore the mmf of an interpole is proportional to armature current.

## 3.2 ARMATURE WINDING

All rotating machines use distributed armature windings, i.e. the windings are distributed in slots around the periphery of armature. A distributed winding may be a closed winding or an open winding. A closed winding is one in which if we start from any point on the winding and traverse it, we will return to the starting point. All dc and ac commutator machines use closed windings. On the other hand open windings terminate at suitable number of sliprings and are used in ac synchronous machines and induction machines. Some of the terms commonly used in closed windings are as under:

### 3.2.1 Armature Coils

The winding consists of diamond shaped preformed coils which are inserted into the slots as a unit. Each coil consists of two coil sides. The number of turns in a coil may be one or more. Usually multi-turn coils are used. Figure 3.5 shows a 4-turn coil. The total number of conductors in a machine is given by

$$Z = 2CN_c \qquad ...(3.1)$$

where $Z$ = total number of conductors

$C$ = number of coils

$N_c$ = number of turns per coil

### 3.2.2 Coil Span

This is the distance between the two sides of a coil. It may be expressed in terms of slots, coil sides or electrical degrees. If a coil spans an exact pole pitch, it is a full pitch coil. If it spans less than a pole pitch, it is known as short pitched coil or chorded coil.

If a coil spans one pole pitch, the emfs induced in its two coil sides will act in conjunction around the coil at every instant. Thus, to obtain maximum possible emf, coil span should be equal to the pole pitch. However, if the number of slots is not exactly divisible by the number of poles, it is not possible to have full pitch coils (in a machine with 39 slots and 4 poles, the coil span can be either 9 or 10 slots). Thus is a case where the ratio of number of slots to the number of poles is not a whole number, the nearest whole number (greater or less than the actual ratio) is taken as the coil span. Sometimes the coil span is deliberately reduced by one slot pitch to facilitate commutation.

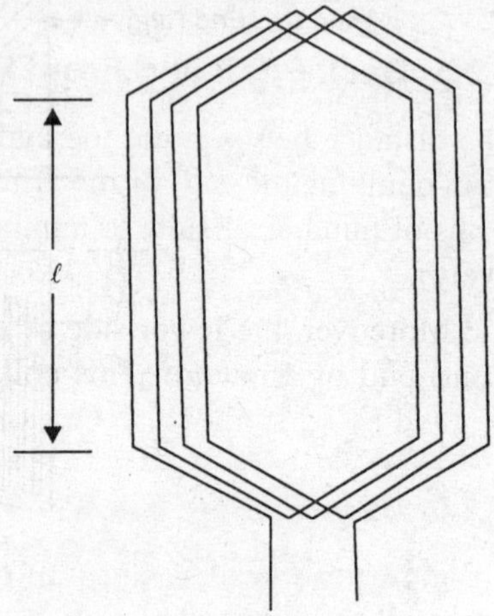


**Fig. 3.5.** Armature coil having 4 turns

### 3.2.3 Single-layer and Multi-layer Windings

A winding having only one coil side in each slot is known as single-layer winding. Multi-layer windings have 2, 4, 6, 8 coil sides in each slot. All modern machines have multi-layer windings and most machines have two layers and have, thus, double-layer windings.

In multi-layer windings the coil sides in top half of each slot are assigned odd numbers, while those in the bottom half of each slot are assigned even numbers, as shown in Fig. 3.6. In a double-layer winding, the upper side of a coil in placed at the top of one slot and the lower side of this coil is situated at the bottom of another slot, about one pole pitch away from the upper side. Thus the upper conductors in a slot do not form the part of the same coil as the lower conductors. Generally the complete coil is made out of one length of conductor, so that there are no joints at the back end of the machine, the change over from the upper to the lower portion of the coil being effected by a bend introduced at the back end of the coil.

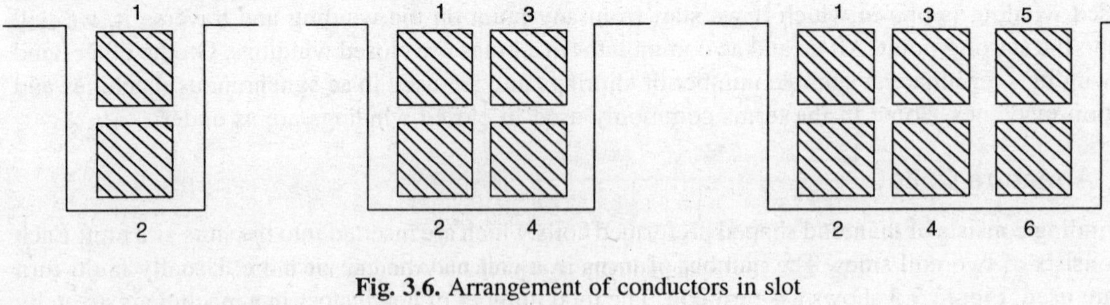


**Fig. 3.6.** Arrangement of conductors in slot

### 3.2.4 Simplex and Multiplex Windings

A simplex winding is a single complete closed winding on the rotor. A duplex winding is a rotor with two complete and independent sets of rotor windings. If a rotor has duplex winding, then each of the windings will be associated with every alternate commutator segment. One winding will be connected to segments 1, 3, 5, etc. and the other winding will be connected to segments 2, 4, 6, etc. Similarly a triplex winding will have three complete and independent sets of windings, each winding connected to every third commutator segment on the rotor. All armatures with more than one set of windings are said to have multiplex windings.

### 3.2.5 Back Pitch and Front Pitch

The distance between the top and bottom sides of a coil measured at the back end (i.e. the end away from commutator end) of the armature is known as back pitch (denoted by $y_b$). It may be expressed in terms of number of slots or number of coil side (provided the coil sides have been numbered consecutively).

Moreover, the lower side of a coil is connected to the start of adjacent coil at the commutator (or front) end of armature. This coil side displacement at the front end, is known as front pitch and is denoted by $y_f$. It may also be expressed in terms of number of slots or coil sides.

### 3.2.6 Commutator Pitch

For most machines the finish of each coil is connected to the start of the neighbouring coil. However, for small machines the finish of one coil is connected to the start of another coil well away from first coil as shown in Fig. 3.7. The displacement between coils is generally called total pitch or winding pitch. Similarly, the displacement between the commutator segments to which the two ends of a coil are connected is called commutator pitch (symbol $y_c$). Figure 3.7 shows coil span, winding pitch and commutator pitch.

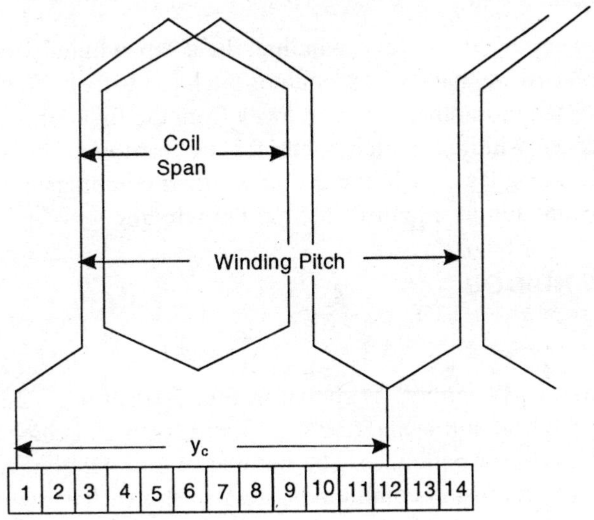


**Fig. 3.7.** Coil span, commutator pitch and winding pitch

### 3.2.7 Progressive and Retrogressive Winding

In a progressive winding the end of a coil (or a set number of coils for wave winding) is connected to a commutator segment ahead of the one to which its beginning is connected. If the end of a coil is connected to a commutator segment behind the one to which its beginning is connected, the winding is retrogressive. If everything is identical, the direction of rotation of a progressive winding machine will be opposite to that with a retrogressive winding. The commutator pitch is +1 for a progressive lap winding and is –1 for a retrogressive lap winding. Figures 3.8 and 3.9 show the coils of progressive and retrogressive lap windings.

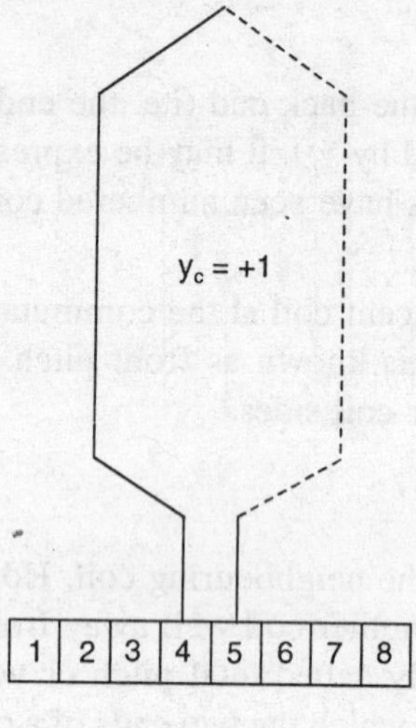


**Fig. 3.8.** Coil of progressive lap winding

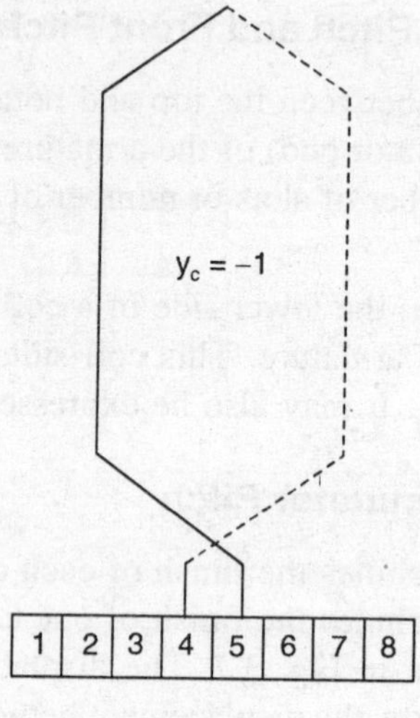


**Fig. 3.9.** Coil of retrogressive lap winding

### 3.2.8 Types of Windings

A winding may be a lap winding or a wave winding. In a lap winding the finish of each coil is connected to the start of next coil, so that the commutator pitch is 1 (or –1). In a wave winding the finish of a coil is connected to the start of another coil well away from the first coil. In addition there is a third type of winding called frog leg winding, which combines the features of lap and wave windings on a single rotor. In this winding a frog leg or self equalising winding combines a lap and wave winding, so that the wave winding can function as equaliser for the lap winding.

## 3.3 LAP AND WAVE WINDINGS

### 3.3.1 Lap Winding

A developed view of a simplex lap winding is shown in Fig. 3.10. It is a 12-slot 4-pole winding, the number of coil sides being 24. The coil span is 12/4 or 3 slots or 6 coil sides. 1, 3, 5,... denote upper coil sides and 2, 4, 6,... denote lower coil sides. The number of commutator segments is 12, i.e. equal to the number of slots, and connections to commutator are also shown. The winding is progressive. The directions of induced emfs are also shown. If emfs in the conductors under north pole are assumed downward, the emfs in the conductors under south pole would be upward. The conductors (or coil sides) in the interpolar region would not have any induced emf (e.g. conductors 1, 7, 8, etc.) At the positions of meeting points of two emfs, positive brushes are placed. The positive brushes are connected together and connected to the positive terminal of the output. Similarly the negative brushes are connected at the separating points of emfs. The negative brushes are connected together to the negative terminal of the output. It is seen from Fig. 3.10 that there are 4 parallel paths in the winding. In general the number of parallel paths in a lap winding is equal to the number of poles. The total induced emf is the emf induced in one path and the current delivered to the external circuit is equal to the current in each armature coil multiplied by the number of parallel paths.

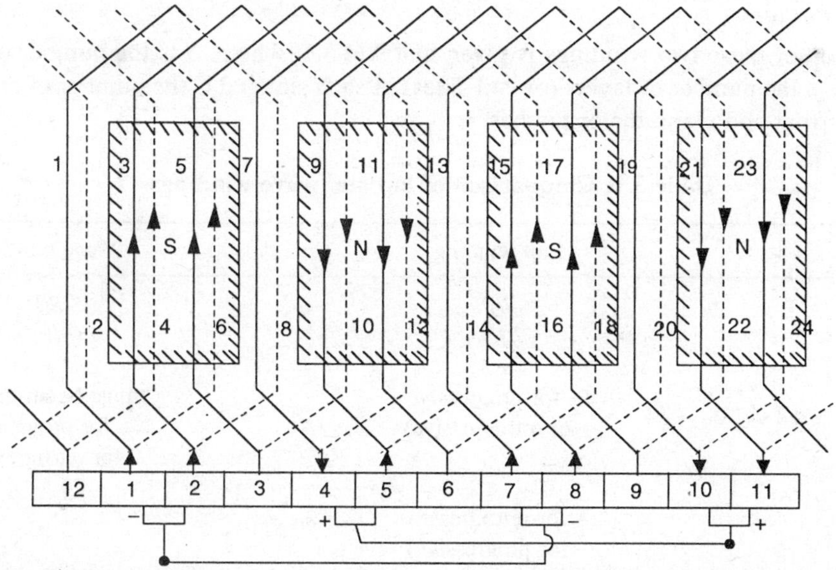


**Fig. 3.10.** Developed view of simplex lap winding (4 poles, 12 slot double layer)

### 3.3.2 Wave Winding

In a lap winding all the coil sides which have emf induced in the same direction at any instant are connected in parallel by the brushes. In a wave winding they are connected in series. Figure 3.11 shows the developed view of a 4-pole, 16-slot simplex wave winding. 1, 3, 5,... denote upper coil sides and 2, 4, 6,... denote lower coil sides. The directions of emfs and positions of brushes are also shown. The coil span is 4 slots. The back pitch is 9 coil sides and front pitch is 7 coil sides. The commutator pitch is 8. Irrespective of the number of poles, a wave winding has only 2 parallel paths.

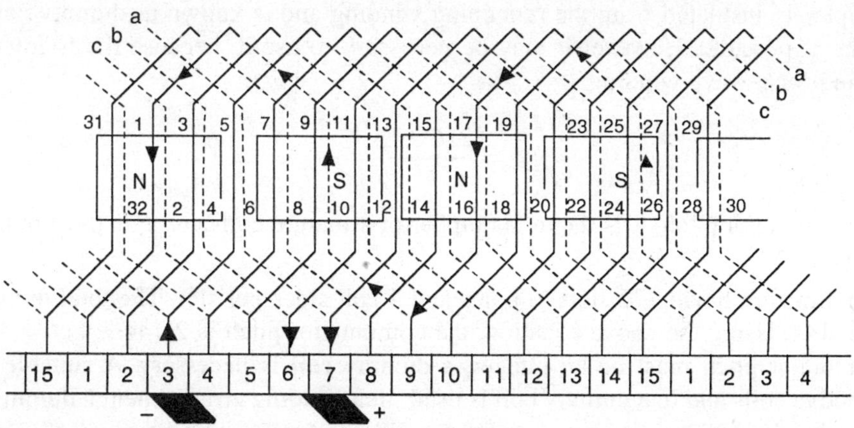


**Fig. 3.11.** Developed view of simplex wave winding (4-pole, 16-slot, double layer)

### 3.3.3 Comparison of Lap and Wave Windings

Wave winding is well suited for high voltage dc machines, because in this winding more number of armature coils are in series and thus a high voltage can be induced more easily than in a lap winding. Since a lap winding has more parallel paths (than a wave winding), it is more suited for low voltage high current dc machines.

A comparison of these two windings is given in Table 3.1, where, $S$ is the number of slots, $P$ the number of poles, $u$ the number of layers (or coil sides) in each slot and $C$ the number of coils; $y_b$, $y_f$ and $y_c$ are the back, front and commutator pitches.

**Table 3.1: Comparison of lap and wave windings**

| | Lap winding | Wave winding |
|---|---|---|
| 1. Coil span | $S/P$ | $S/P$ |
| 2. $y_b$ | $u(S/P) + 1$ | $u(S/P) + 1$ |
| 3. $y_c$ | $\pm 1$<br>(+ for progressive<br>−for retrogressive) | $2(C \pm 1)/P$<br>(must be an integer,<br>+ for progressive<br>− for retrogressive) |
| 4. $y_f$ | $y_b \pm 2$<br>(+ for retrogressive<br>− for progressive) | $2y_c - y_b$ |
| 5. Parallel paths | $P$ | 2 |
| 6. Armature current | (conductor current) × $P$ | (conductor current) × 2 |
| 7. Dummy coils | not needed | may be needed |
| 8. Equaliser rings | needed | not needed |
| 9. Application | low voltage, high current machines | High voltage machines |

## 3.4 DUMMY COIL

It is desirable to use standard armature laminations from economy considerations. When a machine has a wave winding, it is necessary to use one extra coil to maintain the mechanical balance of armature. This coil is completely insulated from the remaining winding and is known as dummy coil. It should be avoided as far as possible. However it may be necessary to use it, because the following relation should be satisfied for a wave winding.

$$y_c = \frac{2(C \pm 1)}{P} \qquad ...(3.2)$$

where $y_c$ is the commutator pitch, $C$ is the total number of armature coils and $P$ is the number of poles. $y_c$ must be an integer.

Consider an armature having 72 slots, 6 poles and 4 coil sides per slot. The total number of coils is (72 × 4/2), i.e. 144. Using the above equation, the commutator pitch is 2(144 ± 1)/6 = 47 2/3 or 48 1/3. Since commutator pitch must be an integer, a dummy coil is necessary. A suitable winding is possible if 143 active coils and one dummy coil is used. In a winding arrangement a dummy coil is not assigned any number.

## 3.5 EQUALISER RINGS

The existence of many parallel paths, in a lap winding, can lead to the serious problem of circulating currents. The fluxes from all the poles are not exactly equal. Because of wear on the bearings, the air gap does not remain uniform around the whole periphery. As the armature conductors rotate, the voltage induced in some conductors may be slightly more than that in the others. Since all the parallel paths are in parallel, a resultant emf acting around a closed path may cause circulating current in the winding. The resistance of winding being very small, even a small imbalance in the emfs can give rise to a large circulating current. Evidently these circulating currents cause energy loss and heating. Therefore the points which should be at the same potential, in different parallel paths, are connected together by a ring made of copper. Each ring is insulated from the other. These rings, known as equaliser rings or bars, help in keeping the circulating currents inside the small sections shorted together, so that these circulating currents may not flow through the brushes.

## 3.6 MAGNETIC CIRCUIT AND FLUX PATH

The main or useful flux in a dc machine, starts from a north pole, crosses the air gap and passes down to the armature core. It then divides into two parts, each half going to each adjacent south pole and completes its path through the yoke as seen in Fig. 3.12. Each flux line crosses the air gap twice. There are as many main magnetic paths as the number of poles. Some flux lines do not enter the armature and go directly from north pole to south pole. This is leakage flux and does not contribute to generation of emf.

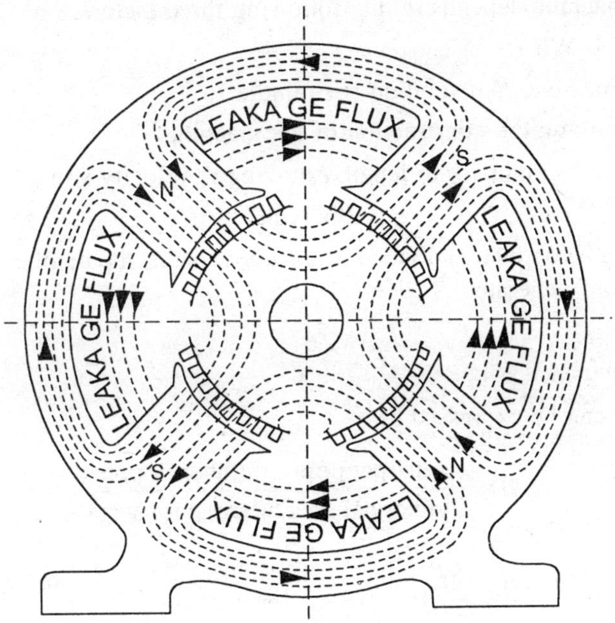


**Fig. 3.12.** Magnetic circuit of a 4-pole dc machine

Figure 3.13 shows as electric circuit analogous to the magnetic circuit of Fig. 3.12. The mmf is represented by battery. $R_b$, $R_g$, $R_t$, $R_a$ and $R_y$ represent the reluctance of pole, air gap, armature teeth, armature core and yoke, respectively.

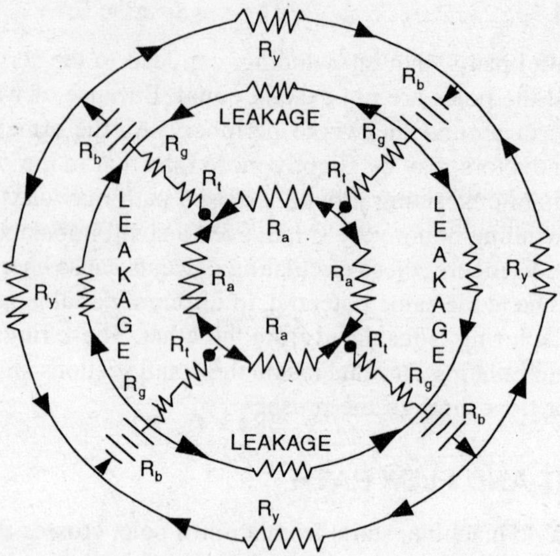


**Fig. 3.13.** Electric circuit analogous to magnetic circuit of a 4-pole dc machine

## 3.7 EMF EQUATION

The emf generated in a machine depends on the following three factors:

1. The magnetic flux $\phi$, Wb
2. The speed of the machine, $N$ revolution per minute
3. A constant depending on the construction of the machine.

The emf generated in a conductor of length $l$ moving at velocity $v$ in a field of flux density $B$ is

$$e = Blv \text{ volts}$$

In a machine having $Z$ conductors divided in $A$ parallel paths, the number of conductors in series is $Z/A$. Therefore emf generated is

$$E = \frac{ZBlv}{A} \qquad \text{...(3.3)}$$

The flux density $B$ can be expressed as

$$B = \frac{\text{Flux per pole} \times \text{number of poles}}{\text{Cylindrical area of armature surface}}$$

or

$$B = \frac{\phi P}{2\pi r l}$$

where $P$ is the number of poles, $r$ is the radius of armature and $l$ is the length of armature core (i.e. active length of each conductor).

The velocity $v$ is equal to $2\pi rN/60$. Substituting the values of $B$ and $v$ in Eq. (3.3), we get

$$E = \frac{ZP\phi N}{60\,A} \qquad \text{...(3.4)}$$

or $$E = k_e \phi N \qquad ...(3.5)$$

where $k_e = \dfrac{ZP}{60\,A}$ is constant for a machine.

*Example 3.1*

A four-pole dc generator has 48 slots and 8 conductors per slot. The useful flux per pole is 30 mWb and speed is 800 rpm. Find generated emf, if the machine is (a) lap wound, (b) wave wound.

*Solution*

(a) $$Z = 48 \times 8 = 384$$

For lap winding $A = P = 4$

$$k_e = \frac{ZP}{60\,A} = \frac{384 \times 4}{60 \times 4} = \frac{32}{5}$$

$$E = k_e \phi N = \frac{32}{5} \times 30 \times 10^{-3} \times 800 = 153.6\text{ V}$$

(b) for wave winding $A = 2$

$$k_e = \frac{ZP}{60\,A} = \frac{384 \times 4}{60 \times 2} = \frac{64}{5}$$

$$E = k_e \phi N = \frac{64}{5} \times 30 \times 10^{-3} \times 800 = 307.2\text{ V.}$$

*Example 3.2*

A 6-pole dc generator has a simplex wave winding with 144 coils of 10 turns each. The speed is 2300 rpm and resistance of each turn is 0.01 ohm. The armature is 50 cm long, 50 cm diameter. The flux density in the air gap is 0.6 Wb/m$^2$ and each pole subtends an angle of 120°. (a) How may current paths are there in the winding? (b) Find the generated emf. (c) Find the resistance of armature winding.

*Solution*

(a) The machine has simplex wave winding. The number of parallel paths or current paths = 2

(b) Area of air gap under one pole $= \dfrac{2\pi r l}{\text{number of poles}} \times \dfrac{120}{180}$

$$= \frac{2\pi \times 0.5 \times 0.5}{2 \times 6} \times \frac{120}{180} = 0.08727\text{ m}^2$$

$$\phi = 0.08727 \times 0.6 = 0.052\text{ Wb}$$

$$Z = 144 \times 10 \times 2 = 1440 \times 2 = 2880$$

$$K_e = \frac{ZP}{60\,A} = \frac{2880 \times 6}{60 \times 2} = 144$$

$$E = k_e \phi N = 144 \times 0.052 \times 200 = 1497.6 \text{ V}$$

(c) Total number of turns $= 144 \times 10 = 1440$

$$\text{Number of turns per path} = \frac{1440}{2} = 720$$

$$\text{Resistance per path} = 720 \times 0.01 = 7.2 \text{ ohm}$$

Since two paths are in parallel, we have

$$\text{Resistance of armature} = \frac{7.2}{2} = 3.6 \text{ ohm}$$

## 3.8 COMMUTATOR ACTION: CONVERSION OF AC INTO DC

The emf generated in the conductor of a dc machine is an alternating emf. The current in the conductor is in one direction when the conductor is moving under the north pole and in the reverse direction, when it is moving under the south pole. One method to convert ac to dc is shown in Fig. 3.14(a). Two semi-circular conducting rings are attached to the ends of the loop. The two rings are insulated from each other. Two fixed contacts are arranged, such that at the instant when voltage induced in the loop is zero, the contacts short circuit the two segments. These fixed contacts are called brushes.

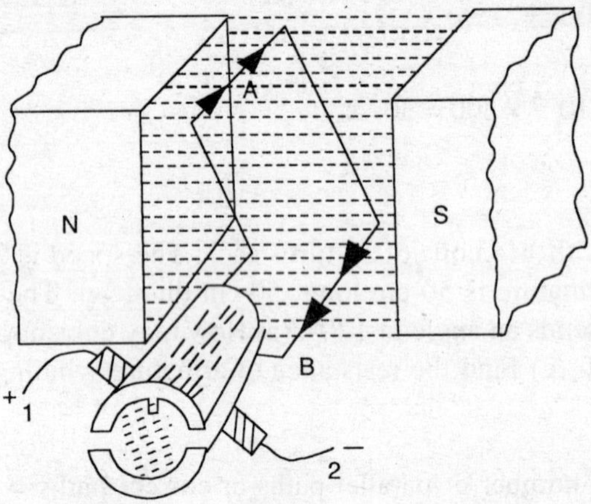


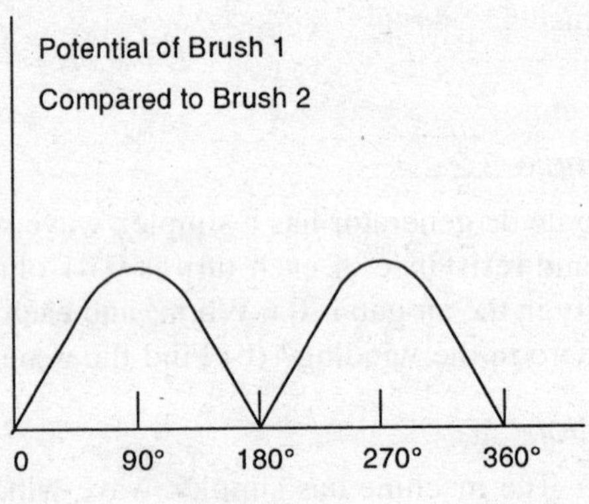


**Fig. 3.14.** Two segment commutator (a) connections, (b) output voltage

Initially let conductor *A* move down and conductor *B* move up. Brush 1 acquires positive charge and brush 2 acquires negative charge. In the first half revolution (i.e. 0 to 180°) the output voltage will be positive sine wave (Fig. 3.14b). After one-half revolution conductor *A* comes to the bottom position and begins to move up, while conductor *B* goes to top position and begins to move down. The directions of induced emf in both the conductors reverse. However, conductor *A* is now making contact with brush 2, while conductor *B* is in contact with brush 1. The emf in conductor *B* is now positive, so that brush 1 is also at positive potential. The emf in conductor *A* is negative and brush 2 continues to be at negative potential. The output voltage is shown in Fig. 3.14(b). The direction of voltage and current in the two half cycles is the same, but the magnitude is not constant.

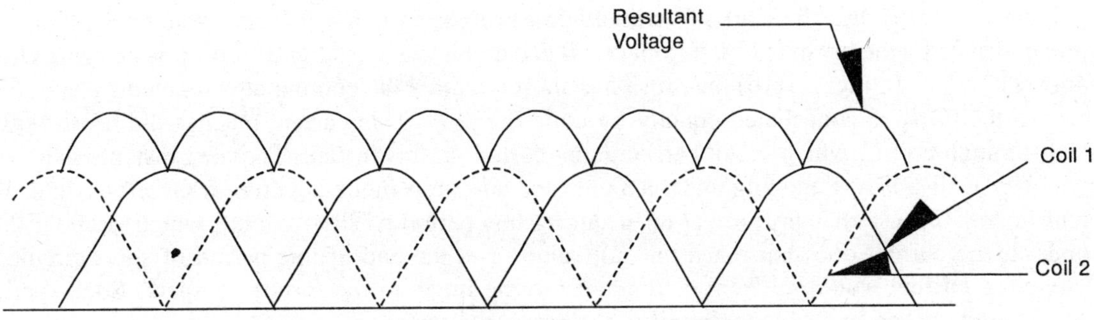


**Fig. 3.15.** Voltage output from two coils at right angle

An actual machine has a number of coils and many such rings (known as commutator segments). Figure 3.15 shows the output voltage for a machine having two coils placed at 90° to each other. The voltages of the two coils will be out of phase by 90°. Its resultant voltage has lesser variation in magnitude as compared to that in Fig. 3.14(b). For a machine with a large number of coils distributed over the armature periphery, the output voltage and current are nearly constant.

The process of converting the ac voltages and currents in the armature coils to dc voltages and currents at the terminals is known as commutation. This process, in an actual machine, is influenced by two problems, viz reactance voltage and armature reaction.

### 3.8.1 Commutation Process, Reactance Voltage

A brush has to make contact with two adjacent commutator segments during the commutation process. In this process the current in a coil has to change from $+I$ to $-I$. The time taken for this change is known as commutation period.

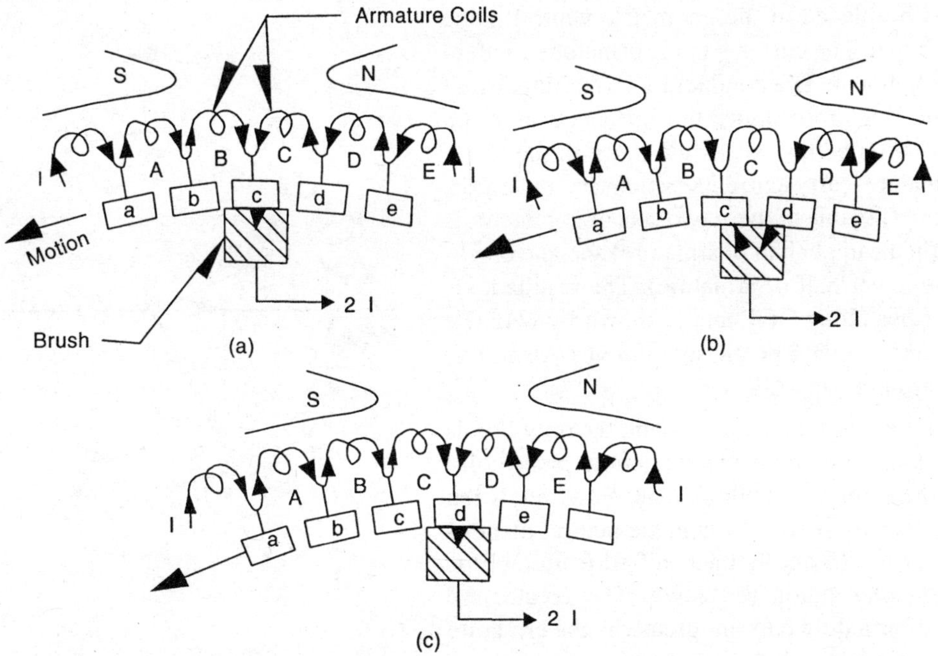


**Fig. 3.16.** Commutation

Figure 3.16(a) shows the start of commutation process in coil *C* of the armature winding. The current collected by the brush is *I* each from coil *B* and coil *C*, i.e. a total of $2I$. Coil *C* is carrying current *I* from right to left. In Fig. 3.16(b) the brush has short circuited the commutator segments *c* and *d*. The current to the brush is contributed equally by coils *B* and *D*. This current reaches the brush without passing through coil *C*, which is not carrying any current at this instant. The next instant is shown in Fig. 3.16(c). Coil *C* is now moving under a *S* pole and current is in the direction from left to right. Thus current in coil *C* must change from $+I$ to $-I$ during this period of short circuit, which is about 0.002 seconds. If the current does not reach the full value $-I$ at the end of this period of short circuit, the difference of current would go from commutator to the brush in the form of a spark. Such sparking causes heating, pitting and roughening of the commutator surface.

Let the inductance of coil be *L*. If a change in current from $+I$ to $-I$ takes place in time $T_c$ (known as period of commutation), a self-induced emf $L\,di/dt$, known as reactance voltage appears.

$$\text{Reactance voltage} = \frac{2LI}{T_c} \qquad ...(3.6)$$

The direction of this induced emf is so as to oppose the change in current.

## 3.9 ARMATURE REACTION IN DC GENERATORS

### 3.9.1 Cross-magnetising Effect

When a current flows through armature conductors a magnetic field is produced. This magnetic field, due to armature current, distorts the magnetic field produced by the field poles. This effect is known as armature reaction.

Figure 3.17 shows a two-pole dc generator. Let the brushes be placed in the geometric neutral plane (GNP) as shown. The currents in the armature conductors are also shown. The conductors in the upper half of armature (i.e. those under the influence of north pole) carry currents inwards, while those in the lower half of armature carry currents which are outwards. The armature flux due to these two sets of conductor is clockwise (in the upper half of armature) and anti clockwise (in the lower half of armature). The resultant armature flux lies along GNP and is shown by *OA*. The main field flux is *OD*. The phasor sum of *OA* and *OD* gives the resultant flux *OE*.

When armature flux is not present, the only flux is the field flux *OD* and the magnetic neutral plane coincides with the geometric neutral plane. However, when armature flux is present, the magnetic neutral plane is perpendicular to *OE* and is thus shifted from GNP is the direction of rotation as shown. The greater the magnitude of armature current, greater is the armature flux and greater is the shift of magnetic neutral plane.

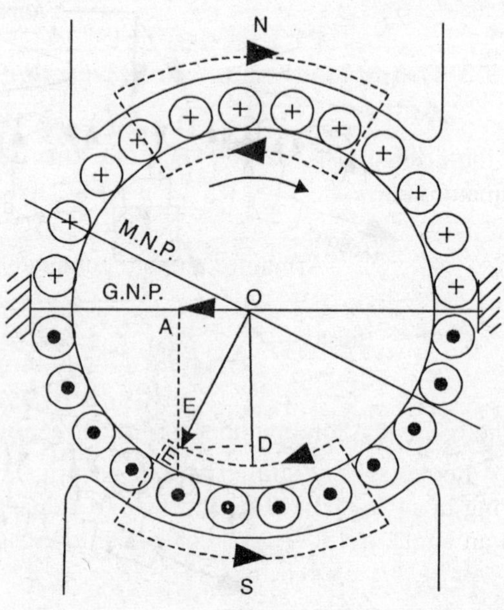


**Fig. 3.17.** Armature reaction in a dc generator

Figure 3.17 shows two flux lines due to armature mmf. The field flux is downwards (i.e. from *N* pole to pole). It is seen that direction of field and armature flux are opposite to each other at leading pole tip at each pole, but the directions of these two fluxes are the same at trailing pole tips. This is known as cross-magnetising effect of armature reaction. The field at trailing pole tips is strengthened and that at leading pole tips is weakened. If the armature teeth were not saturated, the total flux would remain constant. Due to saturation of armature teeth, it is possible that strengthening of trailing pole tips is less than the weakening of leading pole tips. Therefore a small amount of demagnetisation may also occur.

### 3.9.2 Demagnetising Effect

In Fig. 3.17 the brushes are not at geometric neutral plane. To improve commutation it is necessary to move the brushes forward to the magnetic neutral plane. Figure 3.18 shows this condition. It is seen that armature mmf *OA* can be decomposed into two components *OC* and *OB*. The component *OC* is in opposition to the main field mmf *OD*. Therefore *OC* has a demagnetising effect. The component *OB* is the cross-magnetising component. The resultant mmf is *OE*. The armature ampere turns between the horizontals $PQ$ and $T$ are the demagnetising ampere turns. The remaining armature ampere turns, i.e. those above $PQ$ and below $T$ are cross-magnetising ampere turns.

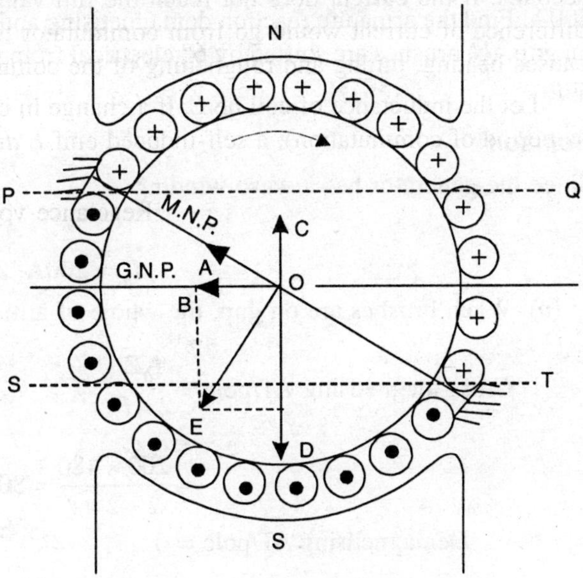


**Fig. 3.18.** Demagnetising and cross-magnetising effects of armature reaction

### 3.9.3 Cross Magnetising and Demagnetising Ampere Turns

If the armature has *A* parallel paths, the armature current per parallel path is $I_a/A$, where $I_a$ is the armature current.

$$\text{Armature ampere conductors per pole} = \frac{I_a}{A} \cdot \frac{Z}{P}$$

$$\text{Armature ampere turns per pole} = \frac{I_a Z}{2AP}$$

where Z is the total number of armature conductors and *P* is the number of poles.

Let brushes be shifted by $\angle\theta°$ electrical from gnp. Then in every inter-polar region, the conductors lying in an angle of $2\theta°$ contribute to demagnetising effect, while the remaining conductors (i.e. lying in an angle of $180 - 2\theta°$) contribute to cross-magnetising effect.

$$\text{Demagnetising AT/pole} = \frac{2\theta}{180°}\left(\frac{I_a Z}{2AP}\right) = \frac{I_a Z}{AP}\left(\frac{\theta}{180°}\right) \qquad \text{...(3.7)}$$

$$\text{Cross-magnetising AT/pole} = \left(\frac{180 - 2\theta}{180}\right)\left(\frac{I_a Z}{2AP}\right) \quad ...(3.8)$$

Evidently if $\theta = 0$, i.e. brushes are on gnp, demagnetising ampere turns are zero and the entire armature ampere turns are cross-magnetising.

*Example 3.3*

An 8-pole dc generator has 480 armature conductors and a wave winding. The armature current is 200A. Find the armature reaction demagnetising and cross-magnetising ampere turns, if (a) brushes are on gnp, (b) brushes are shifted by 6° electrical from gnp, (c) brushes are shifted by 6° mechanical from gnp.

*Solution*

Since the generator has a wave winding

$$A = 2$$

$$I_a = 200\text{A},\ Z = 480,\ P = 8$$

(a) When brushes are on gnp, the whole of armature reaction is cross-magnetising.

$$\text{Cross-magnetising AT/pole} = \frac{I_a Z}{2\,AP}$$

$$= \frac{200 \times 480}{2 \times 2 \times 8} = 3000 \text{ AT/pole}$$

Demagnetising AT/pole = 0

(b) $$\text{Demagnetising AT/pole} = \frac{I_a Z}{AP}\left(\frac{\theta}{180°}\right)$$

$$= \frac{200 \times 480}{2 \times 8}\left(\frac{6}{180}\right) = 200 \text{ AT/pole}$$

Cross-magnetising AT/pole = 3000 – 200 = 2800 AT/pole

(c) The angle of shift is 6° mechanical. Since it is an 8-pole machine, 1° mechanical equal *P*/2, i.e. 4° electrical.

Therefore $\theta = 4 \times 6 = 24°$ electrical

$$\text{Demagnetising AT/pole} = \frac{I_a Z}{AP}\left(\frac{\theta}{180°}\right)$$

$$= \frac{200 \times 480}{2 \times 8}\left(\frac{24}{180}\right) = 800 \text{ AT/pole}$$

Cross-magnetising AT/pole = 3000 – 800

= 2200 AT/pole.

## 3.10 SOLUTION OF COMMUTATION PROBLEMS

The commutation process is adversely affected by armature reaction and reactance voltage. Four methods have been suggested to overcome these problems: (a) Brush shifting, (b) Interpoles, (c) High resistance brushes, (d) Compensating winding.

(a) *Brush shifting:* The armature reaction causes a shift in the magnetic neutral plane. One method to improve commutation is to shift the brushes from the geometric neutral plane to a little ahead of new position of magnetic neutral plane, so that the brushes lie in the fringe of next pole. Because of this an emf is induced in the coil (undergoing commutation) due to the influence of next pole and this emf cancels the reactance voltage. However, this method involves some practical difficulties. The extent of brush shift depends on magnitude of armature current (which keeps on varying). Moreover the direction of brush shift has to be changed when the machine goes from generator operation to motor operation. The brush shifting also increases the demagnetising ampere terms. In view of these difficulties, brush shifting is used only in small sized machines.

(b) *Interpoles:* The interpoles (also known as commutating poles or compoles) are small auxiliary poles situated between the main poles as shown in Fig. 3.19. The presence of these poles introduces a commutating field in the air gap, which induces an emf in the coil undergoing commutation. If the interpole is of correct polarity, this emf would oppose and cancel the reactance voltage. In a generator an interpole should have a polarity which is the same as that of next main pole. As seen from Eq. (3.6), the reactance voltage is directly proportional to armature current. Therefore an interpole must be connected in series with armature, so that commutating field is also proportional to armature current.

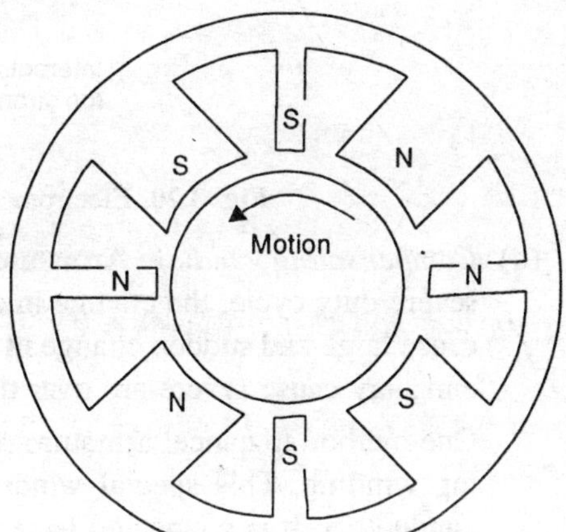


**Fig. 3.19.** Interpoles in a dc generator

The commutating field should be of correct magnitude, neither too weak nor too strong. If the commutating field is too weak, the reactance voltage may not be cancelled fully. In this situation (known as under commutation) the current in the coil is not able to reach its full value in reverse direction at the end of period of commutation and consequent arcing occurs. If the commutating field is too strong, the current may reach a higher value, in the reverse direction, at the end of commutation period. This results in arcing in reversed direction. Figure 3.20 illustrates this effect.

The methods of brush shifting and interpoles are together classified as emf commutation, because in both these methods an additional emf is induced in the coil undergoing commutation.

(c) *High resistance brushes:* The use of brushes having high contact resistance improves the commutation. In Fig. 3.16(b) the brush is making contact with segments *c* and *d*. When the current from coil *B* reaches segment *c*, it flows partly direct to the brush and partly through coil *C* and segment *d*. If the contact resistance of brush is high, more and more current tends to take the second path (via coil *C*). This is due to the fact that the area of segment *c* (in contact with brush) is decreasing and the area of segment *d* (in contact with brush) is increasing. This process helps commutation. The use of carbon and graphite for the brush material provides high contact resistance and is used in all dc machines.

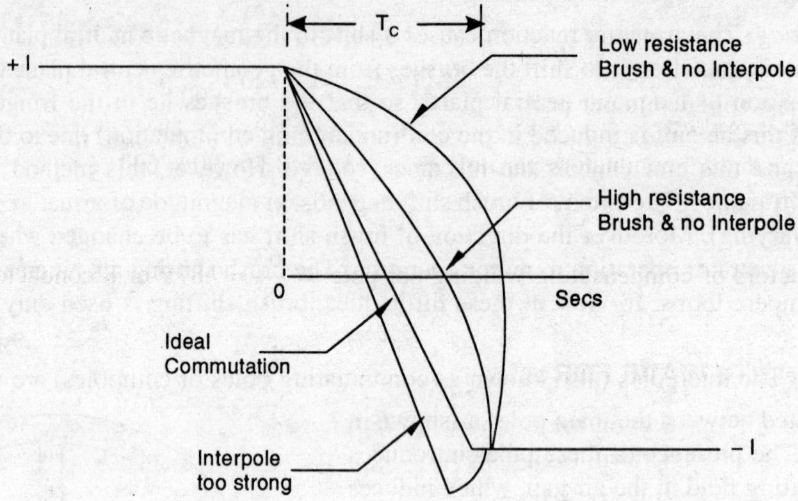


**Fig. 3.20.** Electromotive force (emf) and resistance commutation

(d) *Compensating winding:* Armature reaction alters the flux distribution. If the machine has a very severe duty cycle, the change in current may be very large. This large change in current would cause large and sudden change in flux, which would induce a high emf in the armature coil. This emf may cause severe arc over the commutator surface.

One method to cancel armature reaction, neutral shift and flux weakening is to use compensating winding. This special winding is placed in slots cut in pole faces, parallel to the rotor conductors. It is connected in series with rotor (i.e. armature) in such a way that ampere turns of compensating winding are equal and opposite to those of armature conductors, which are opposite to pole face. Thus a compensating winding cancels the effect of armature reaction. Evidently the ampere conductors of compensating winding should be equal to the active ampere conductor of armature per pole, or

$$\frac{I_a}{A} \cdot Z_a = I_a Z_p$$

or
$$Z_p = \frac{Z_A}{A} \qquad ...(3.9)$$

where $Z_p$ is the number of conductors of compensating winding in each pole face, $Z_a$ is the number of active armature conductors per pole and $A$ is the number of parallel paths.

The use of compensating winding increases the cost of the machine and also increases the copper losses in the machine. It is used (i) if the motor is likely to be subjected to very severe duty cycle as in rolling mill, (ii) if the generator has to supply high current at a low voltage as in Ward-Leonard set in steel mills.

*Example 3.4*

An 8-pole lap wound dc machine has 720 armature conductors. the pole covers 70% of pole pitch. Find the number of conductors of compensating winding in each pole face.

*Solution*

Number of armature conductors per pole $= \dfrac{720}{8} = 90$

Number of active armature conductors per pole $= 0.7 \times 90 = 63$

Number of parallel paths $= 8$

Number of conductors of compensating winding per pole $= \dfrac{63}{8} = 7.875$ or 8 conductors/pole.

## 3.11 FLUX DENSITY WAVE FORMS

The effect of armature reaction on the flux density distribution can be understood very clearly if the wave form of flux density is drawn. Figure 3.21(a) shows the developed diagram of one pole-pair with

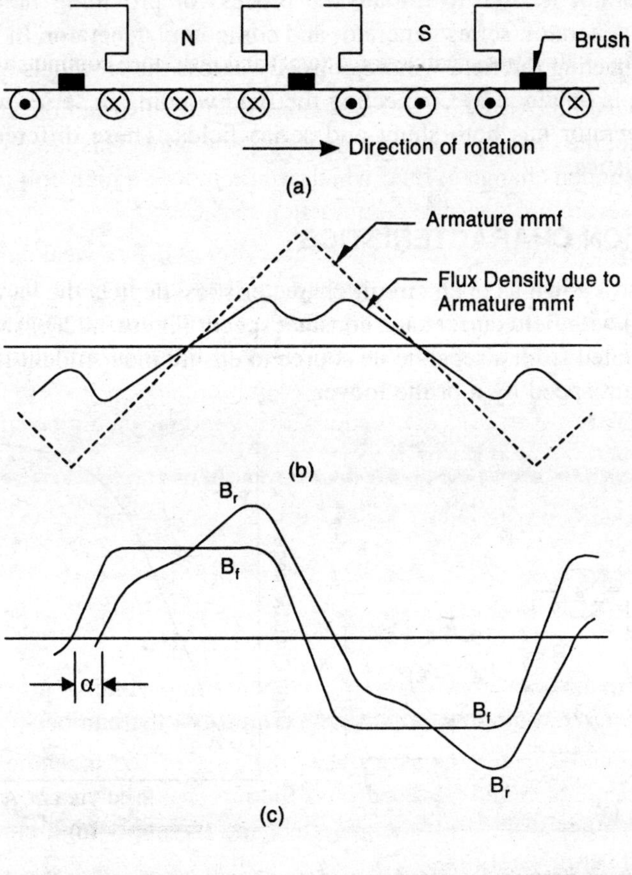


**Fig. 3.21.** Flux density wave forms in a dc machine (a) one pole pair and conductors, (b) armature mmf and flux density distribution, (c) resultant flux density distribution

brushes in the geometric neutral plane. Figure 3.21(b) shows the armature mmf and flux density distribution. The armature mmf is a stepped wave. However, it can be approximated by a triangular wave. The axis of armature mmf is fixed at 90° electrical from the main field axis by the brush position. The flux density wave produced by this armature mmf is saddle shaped because of the high reluctance of the interpolar region. Figure 3.21(c) shows the main field ($B_f$). When this flux density $B_f$ is combined with the flux density due to armature mmf (shown in Fig. 3.21b), we get the resultant flux density distribution ($B_r$). It is seen that the net effect is to increase the flux in the trailing half of pole and to decrease the flux in the leading half of the pole (cross-magnetisation effect). If the increase in trailing half is less than the decrease in the leading half (due to saturation), demagnetisation occurs.

When armature mmf is absent, the magnetic neutral plane coincides with the geometric neutral plane. As seen in Fig. 3.21(c), the effect of armature mmf is to shift the magnetic neutral plane by $\alpha°$ from the geometric neutral plane.

## 3.12 CLASSIFICATION OF DC GENERATORS

DC generators are classified according to the method of producing the main field flux. In a separately excited generator, the field flux is provided by a separate auxiliary source. This method of excitation has the obvious disadvantage of the need of a separate dc source. In self-excited generators the residual magnetism in the generator is used to initiate the process of providing field flux. The self-excited generator can be shunt generator, series generator and compound generator. In shunt generator the field flux is obtained by connecting the field winding directly across the terminals of the generator. In series generator the field flux is obtained by connecting the field winding in series with the generator terminals. A compound generator has both shunt and series fields. These different generators have very different *V–I* characteristics.

## 3.13 MAGNETISATION CHARACTERISTICS

This characteristics, also known as open circuit characteristics, depicts the variation of generated voltage (or no-load voltage) with field current at a constant speed. Figure 3.22(a) shows the circuit diagram. The field winding is excited from a separate dc source to ensure independent flux control. The generator is driven at a constant speed by a prime mover.

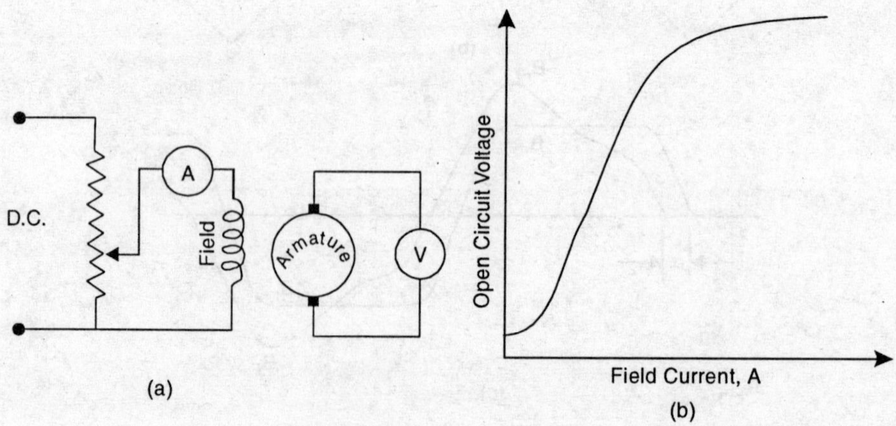


**Fig. 3.22.** Magnetisation characteristics of dc generator (a) circuit, (b) characteristics

Figure 3.22(b) shows the magnetization characteristics. Since speed is constant, the generated voltage is directly proportional to field flux. When the field current is zero, a small residual flux is present in the machine and, therefore, a small voltage is generated even when field current is zero. As field current is increased, the flux increases proportionately and voltage increases linearly with $I_f$. At some value of $I_f$, the magnetic circuit starts getting saturated. Therefore a further increase in $I_f$ does not increase the flux in the same proportion. Therefore voltage increases only slightly in this region.

The generated voltage is directly proportional to speed. Therefore the characteristics at any speed can be obtained from the test curve at one speed.

## 3.14 TERMINAL CHARACTERISTICS OF SEPARATELY EXCITED GENERATOR

The terminal characteristics, also known as external characteristics, is a plot of the terminal voltage versus line current (which is the same as armature current in this case), when field current and speed are kept constant. Figure 3.23(a) depicts a test set up. The generator is driven by a prime mover at a constant speed. The field winding is excited from a separate dc source and $I_f$ is kept constant at a value so that the terminal voltage, when $I_L = 0$, is equal to the rated value. Then line current is increased in steps and values of $I_L$ and $V$ are noted. Figure 3.23(b) shows the terminal characteristics.

It is seen that as $I_L$ increases, the terminal voltage drops. This is due to two reasons:

1. Armature reaction tends to reduce the field flux, so that the actual generated voltage $E_G$ is less than that under no-load condition (i.e. $E_0$). The graph of $E_G$ versus $I_L$ is usually known as internal characteristics.

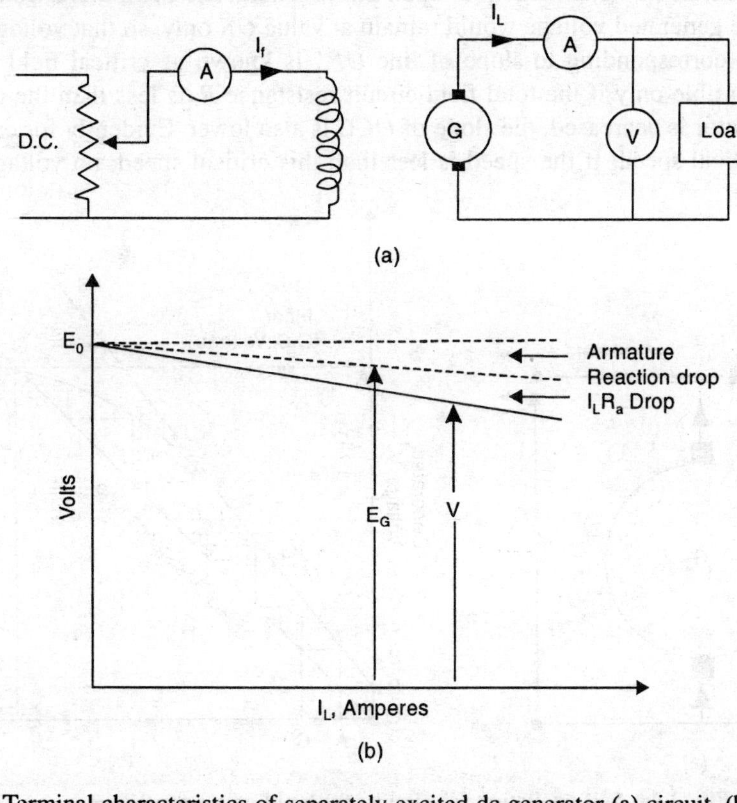


**Fig. 3.23.** Terminal characteristics of separately excited dc generator (a) circuit, (b) characteristics

2. There is a voltage drop across the resistance of armature winding ($= I_L R_a$). Voltage regulation is defined by the equation

$$\text{Voltage regulation} = \frac{E_0 - V}{V} \times 100 \qquad ...(3.10)$$

## 3.15 VOLTAGE BUILD UP IN SELF-EXCITED GENERATOR

Figure 3.24(a) shows a self-excited shunt generator on no-load. The field winding (having resistance $r_f$) is connected across the armature terminals. A variable resistance $r$ (to control the current $I_f$) is connected in series with field winding. The total resistance in the field circuit is $R_f (= r_f + r)$. Figure 3.24(b) shows the process of voltage build up. The line *OP* has a slope equal to $R_f$. Due to residual magnetism a small voltage *OR* is generated. When field switch *S* is closed, this voltage causes field current *OA* to flow. This current increases the field flux and causes generated voltage *OB* which in turn results in field current *OC* which will generate a still higher voltage. This process continues and the generated voltage continues to increase. At every point the excess of generated voltage $E_0$ over $I_f R_f$ causes an increase in field current, which strengthens the field flux resulting in increased generated voltage.

This process of voltage build-up continues till point *Q* on the no-load characteristics in Fig. 3.24(b). At point *Q*, $E_0$ is equal to $I_f R_f$ and no further increase in $E_0$ is possible. If resistance $r$ is decreased, $R_f$ will decrease, slope of line *OP* will be less and the final generated voltage would be higher. If $r$ is increased, the final generated voltage would be lower. If the resistance $r$ is such that the resistance $R_f$ is equal to the slope of line *OP′* (i.e. tangent to open-circuit characteristics), there would be no excess of $E_0$ over $I_f R_f$ and the generated voltage would remain at value *OR* only, so that voltage will not build up. This value of $R_f$, corresponding to slope of line *OP′*, is known as critical field resistance. The voltage build up is possible only if the total field circuit resistance $R_f$ is less than the critical value. If the speed of the generator is decreased, the slope of *OCC* is also lower. Evidently for each value of $R_f$, there is a value of critical speed. If the speed is less than this critical speed, no voltage build up will occur.

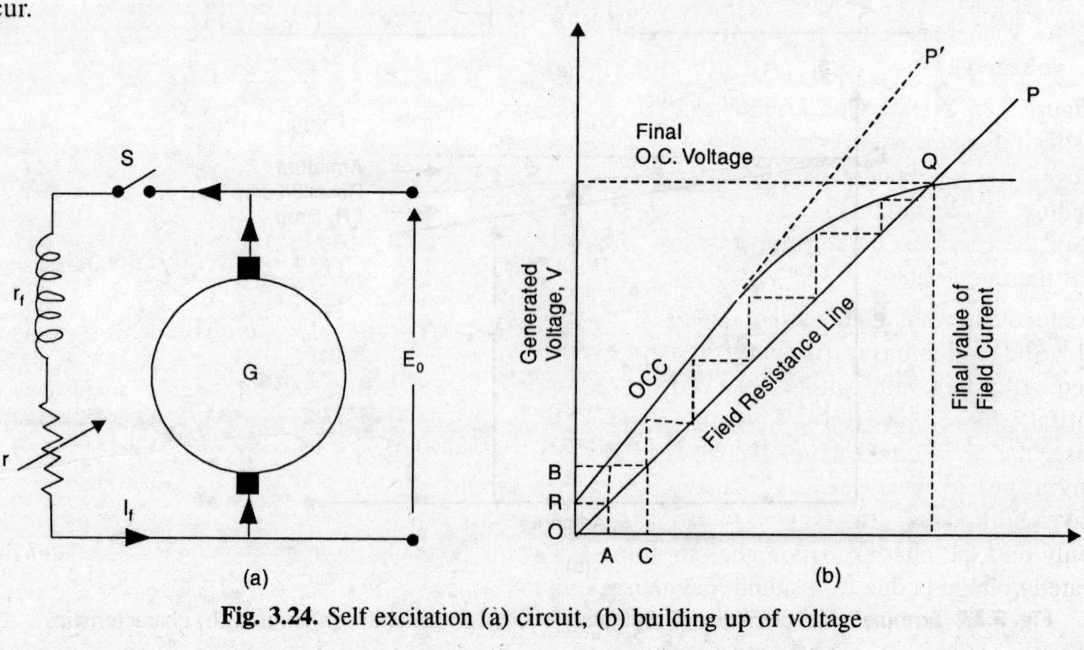


**Fig. 3.24.** Self excitation (a) circuit, (b) building up of voltage

In the above discussion it has been assumed that connections of the field circuit are such that the field current strengthens the residual flux. If connections are reverse, the field current would decrease the residual magnetism and voltage build up would not take place. If, due to some reason, the generator does not have residual magnetism, no voltage build up is possible.

In a series generator the field winding is connected in series with the armature and load. The current can flow through the field winding only if the load current exists. The resistance of load should be less than the critical resistance.

Thus the necessary conditions for voltage build up due to self-excitation are:

1. Residual magnetism must be present.
2. The connections of field winding should be such that the field current strengthens the residual magnetism.
3. The resistance of the field circuit should be less than the critical resistance.
4. The speed of the machine should be higher than critical speed.
5. In a series generator, the load should be connected and load current should flow. The resistance of load should be less than critical resistance and speed should be higher than critical speed.

## 3.16 TERMINAL CHARACTERISTICS OF SHUNT GENERATOR

Figure 3.25(a) shows the circuit of a dc shunt generator supplying a load current $I_L$. The armature current $I_a$ is given by

$$I_a = I_L + I_f \qquad ...(3.11)$$

The field current $I_f$ depends on the terminal voltage $V$. As load current $I_L$ increases, the terminal voltage $V$ becomes less than no-load voltage $E_0$ due to the following three reasons:

1. The armature reaction causes a reduction in the flux.
2. A voltage drop $I_a R_a$ occurs across armature resistance.
3. As $V$ decreases, the field current $I_f$ (= $V/R_f$) decreases, causing a further decrease in terminal voltage $V$.

Figure 3.25(b) shows the terminal characteristics (also known as external characteristics). It can be obtained experimentally by loading the machine (Fig. 3.25a). The speed of the machine should be kept constant at the rated value.

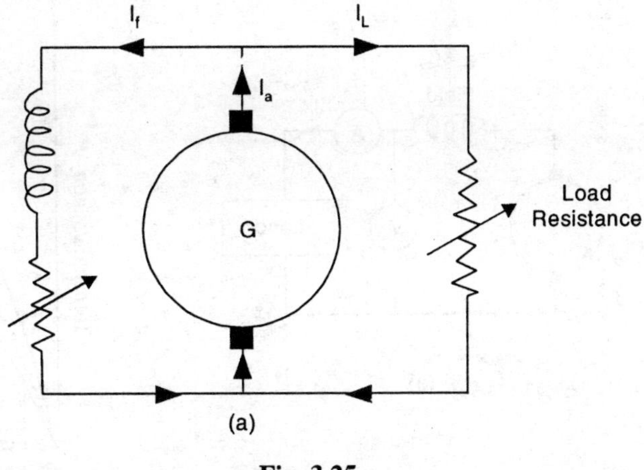

(a)

Fig. 3.25

As load current $I_L$ is increased, the terminal voltage decreases till point $B$ is reached. After this point, an increase of $I_L$ cannot take place, because as $I_L$ tends to increase, the armature reaction increases, causing a further decreases in $V$. This cumulative action causes the voltage to drop suddenly and the characteristics goes to point $A$. At this point the terminal voltage is zero and the generated voltage is due to residual magnetism only.

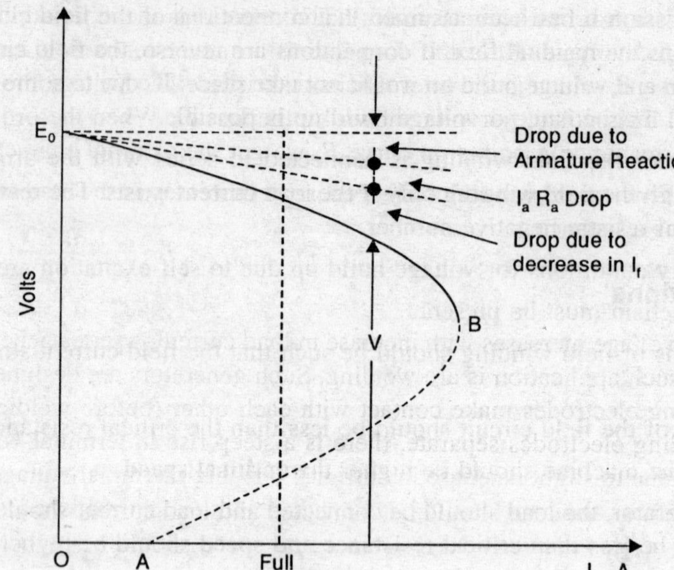


(b)

**Fig. 3.25.** Terminal characteristics of dc shunt generator (a) circuit, (b) characteristics

## 3.17 SERIES GENERATOR

### 3.17.1 Terminal Characteristics

Figure 3.26(a) shows the connections for a dc series generator. The field winding is connected in series with the armature, so that armature current, field current and load current are the same. Figure 3.26(b)

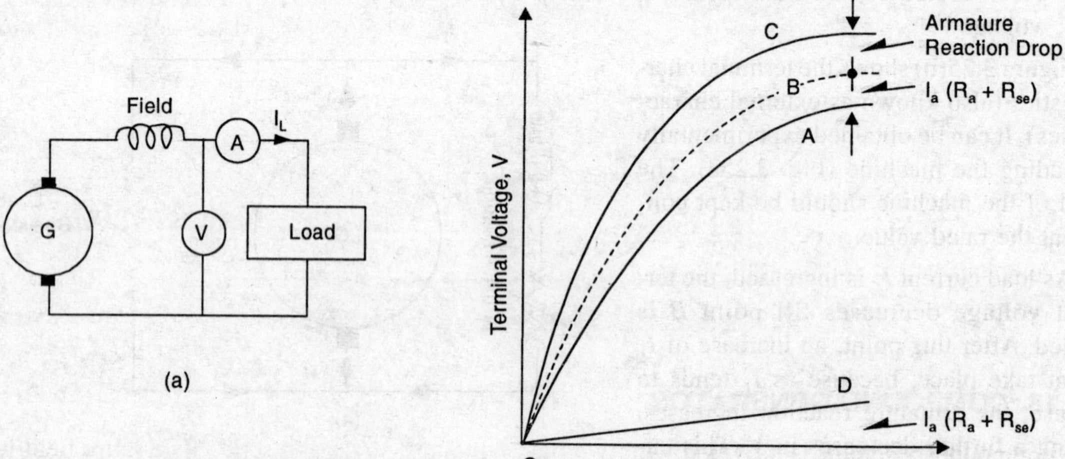


(b)

**Fig. 3.26.** Terminal characteristics of dc series generator (a) circuit, (b) characteristics

shows the terminal (or external characteristics). The curve *A* is the external characteristics. The curve *C* gives the generated voltage. After the drop due to armature reaction is subtracted from generated voltage, we get curve *B*, which is the internal characteristics. The voltage drop in armature and series field is $I_a(R_a+R_{se})$. This voltage drop is shown by the line *D*. When the ordinates of this line *D* are subtracted from corresponding points on curve *B*, we get the external characteristics (curve *A*).

It is seen that the terminal voltage of this generator increases as the load current increases. As such voltage regulation is a large negative number.

### 3.17.2 Applications

Since the terminal voltage increases with increase in load current, a series generator is suited for special applications. One such application is arc welding. Such generators are designed to have high armature reaction. As welding electrodes make contact with each other (before welding starts), a high current flows. As the welding electrodes separate, there is a steep rise in terminal voltage (because armature reaction decreases due to slight decrease in current). The high terminal voltage ensures that arc will be maintained in air across the contacts, so that welding will be accomplished.

Another application of dc series generator is as series booster. A booster is a series generator connected in series with the line as shown in Fig. 3.27. Its function is to add a voltage to the line so as to neutralise the voltage drop in the line. Since the voltage drop in the line is proportional to line current, and terminal voltage of a series generator also increases with increase in current, a series generator is particularly suited for this application. The series field winding is provided with a variable resistance connected in parallel into it. The output voltage of booster can be adjusted with the help of this resistance.

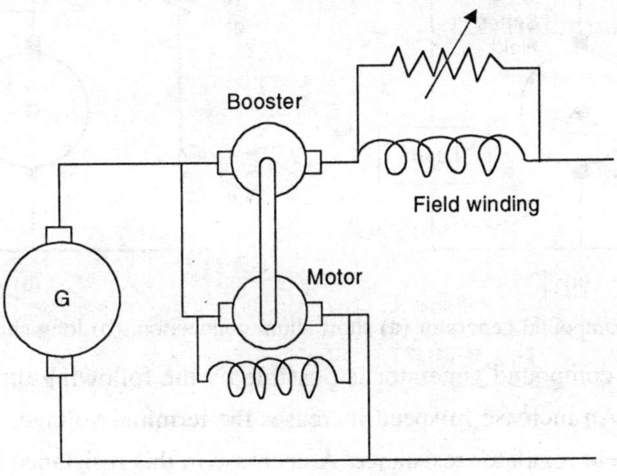


**Fig. 3.27.** DC series generator used as booster

## 3.18 COMPOUND GENERATOR

A compound generator is provided with two field windings, shunt and series. The shunt field winding consists of a large number of turns of thin wire. It has a high resistance and is connected in parallel with armature terminals. The series field winding consists of a few turns of thick wire. It has a low resistance and is connected in series with the armature.

The shunt and series fields can be connected in two ways as shown in Fig. 3.28a and b. In short shunt connection, the shunt winding is directly across the armature. The current through the series field winding is the line current $I_L$ (which is equal to armature current minus shunt field current). In long shunt connection (Fig. 3.28b), the series field is directly in series with armature and carries the armature current. The shunt field is across the series connection of armature and series field. The voltage across shunt field is the armature voltage less the voltage drop in series field. In both cases a diverter resistance is connected across the series field to adjust the terminal voltage.

Figure 3.29 shows the terminal characteristics of a compound generator. The shunt and series characteristics are also shown separately.

When the mmf of series field opposes the mmf of shunt field, the generator is differentially compounded. The terminal voltage decreases sharply with increase in load current. Evidently this connection is not suitable and is not used.

In cumulative compounding, the connections of the two fields are such that their mmfs add and help each other. If the series field is very strong, the terminal voltage may increase as the load current increases and it is called over compounding. When terminal voltages on full-load and no-load are equal, it is known as flat compounded generator. If the series field is not strong, the terminal voltage will decrease with increase in load current (under compounding).

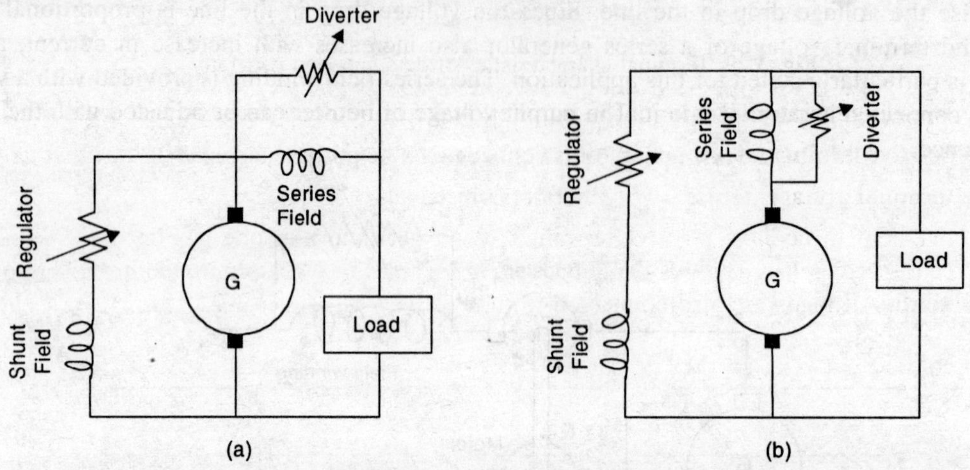


**Fig. 3.28.** Dc compound generator (a) short shunt connection, (b) long shunt connection

The voltage control of a compound generator is possible by the following three ways:

(1) Change is speed. An increase in speed increases the terminal voltage.

(2) change in shunt field regulator resistance. A decrease in this resistance increases the shunt field current, which increase terminal voltage.

(3) Change in diverter resistance. An increase in diverter resistance decreases the current through this resistance, and therefore, increases series field current. This action increases the terminal voltage.

## 3.19 PARALLEL OPERATION OF DC GENERATORS

Many times it is necessary to operate two or more generators in parallel. There are a number of good reasons for this, viz.

1. The load is more than the capacity of one generator.
2. It is necessary to take a generator off line for maintenance and repairs, without interrupting the power supply to the loads.
3. In the event of a fault on one generator, the remaining ones may continue to supply the loads.

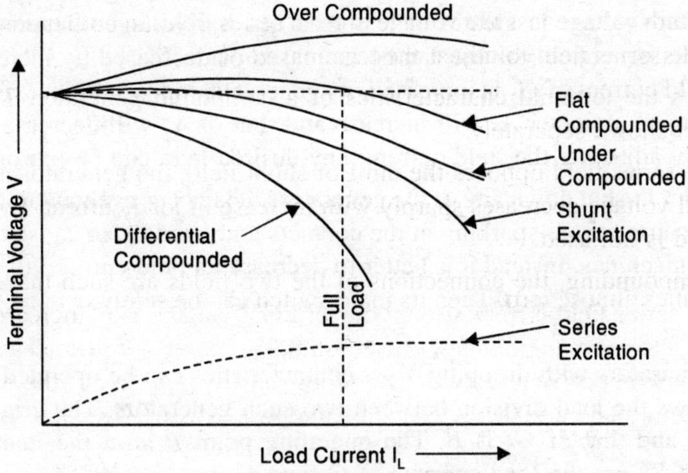


**Fig. 3.29.** Terminal characteristics of dc compound generator

Before two dc generators can be put in parallel, two requirements must be met with, viz.

1. The positive terminal of one generator is connected to the positive terminal of the other generator.
2. The terminal voltages of the two generators are equal.

It is convenient to connect positive terminals of all generators to one bus bar and the negative terminals to another bus bar. The bus bars are made of copper (or aluminium) and are located behind the switch board. All loads are also connected to the bus bars as seen in Fig. 3.30.

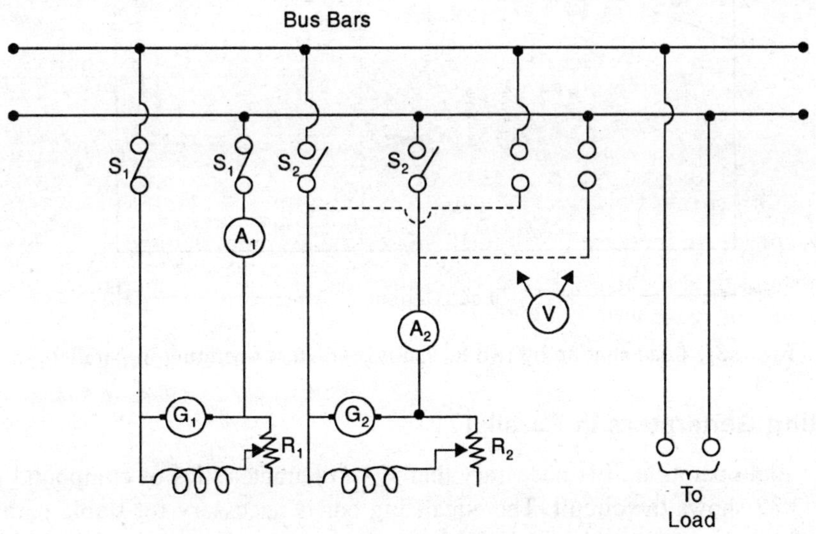


**Fig. 3.30.** Two dc shunt generators operating in parallel

### 3.19.1 Shunt Generators in Parallel

Figure 3.30 shows two shunt generators operating in parallel. Generator $G_1$ is already working and generator $G_2$ is to be put in parallel. The polarity of $G_2$ can be checked by the moving coil voltmeter, which can be connected to either bus bar or terminals of $G_2$ through two-way key. After polarity check, main switch $S_2$ of generator $G_2$ is closed. The generator $G_2$ will work as generator or motor, depending on whether its generated voltage is greater or less than bus bar voltage. Generator $G_2$ is running at constant rated speed. Its generated voltage can be increased or decreased by increasing or decreasing its field current. As field current of $G_2$ is increased, its generated emf would increase and it will supply more load. The reading of ammeter $A_2$ will increase and that of $A_1$ will decrease (because total load current is constant). By adjusting the field current, any desired load can be put on $G_2$.

When it is necessary to shut down any of the generators, it is not convenient to open its main switch suddenly. This would cause severe sparking at the contacts and would also cause a sudden increase in the loads on the other machines. Instead it is better to decrease its field current till the current supplied by this generator becomes almost zero. Then its main switch can be safely opened and its prime mover can be shut down.

Generally only generators with drooping $V - I$ characteristics can be operated in parallel successfully. Figure. 3.31 shows the load division between two such generators. The origin of terminal characteristics of $G_1$ is $O$ and that of $G_2$ is $B$. The operating point $P$ is at the intersection of the two characteristics. $OA$ and $BA$ are the load currents of $G_1$ and $G_2$.

The load division can be changed by changing the field current or speed of one of the machines. If the field current or speed in increased, the $V - I$ characteristics of this generator is moved up and its load current would increase.

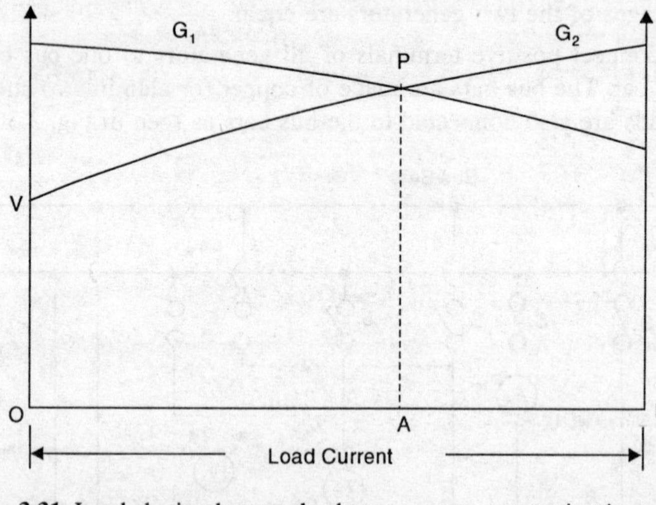


**Fig. 3.31.** Load sharing by two dc shunt generators operating in parallel

### 3.19.2 Compound Generators in Parallel

For successful parallel operation, it is necessary that $V - I$ characteristics of compound generators are drooping. Figure 3.32 shows the circuit. The equalising bar is necessary for stable parallel operation. In the absence of this bar, an accidental increase in generated voltage of one machine causes it to supply more load current, which in turn strengthens it series field. This increases its generated emf further, with still further increase in its load current.

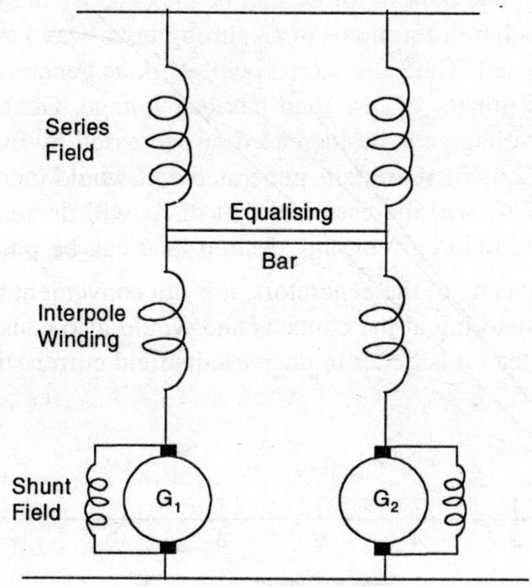


**Fig. 3.32.** Connection of equalising bar for two compound generators operating in parallel

When the equalising bar has been connected, this cumulative action cannot take place. An increase in generated emf of one machine increases it armature current, but the series field is not strengthened, so that successful parallel operation is possible. The equalising bar should be connected at the junction of interpole and series field windings. This is necessary, so that the interpole winding carries the armature current.

### *Example 3.5*

The magnetisation curve for a dc generator, running at 800 rpm, is as under:

| Field current (A) | 1.6 | 3.2 | 4.8 | 6.4 | 8 | 9.6 | 11.2 |
|---|---|---|---|---|---|---|---|
| $E_0$ (V) | 148 | 285 | 390 | 460 | 520 | 560 | 590 |

(a) Find the generated emf when the machine is used as a shunt generator with a resistance of 60 ohm in the field circuit.

(b) Find the critical resistance at this speed.

### *Solution*

(a) The magnetisation curve is shown in Fig. 3.33. Line *OP* has a slope of 60 ohm and is the line representing field resistance. Point *A* gives the operating point. The voltage corresponding to point *A* is the generated emf and is 550 V.

(b) Line *OQ* is tangent to the magnetisation curve and represents the critical resistance line. The slope of this line is the critical resistance and is 91 ohm.

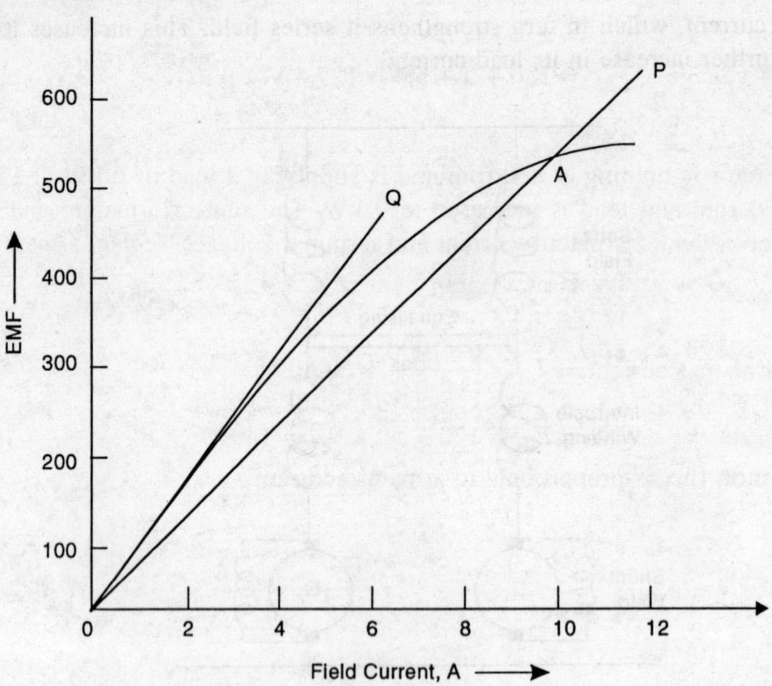


**Fig. 3.33**

*Example 3.6*

A 4-pole dc shunt generator has a wave winding. The armature and field resistance are 0.2 ohm and 60 ohms. The brush contact drop is 1 V per brush. The generator is supplying 50 lamps, each of 60 W at 120 V. Find

(a) current in each armature conductor,

(b) generated voltage. Neglect armature reaction. [*P.U. 2000*]

*Solution*

(a) Current of each lamp $= \dfrac{60}{120} = 0.5\text{ A}$

Total load current $= I_L = 0.5 \times 50 = 25\text{ A}$

Field current $I_f = \dfrac{120}{60} = 2\text{ A}$

$$I_a = I_L + I_f = 25 + 2 = 27\text{ A}$$

A wave winding has two parallel paths.

Current in each armature conductor $= \dfrac{27}{2} = 13.5\text{ A}$

Armature resistance drop $= I_a R_a = 27 \times 0.2 = 5.4\text{ V}$

Brush drop $= 2 \times 1 = 2$ V

Generated voltage $= 120 + 2 + 5.4 = 127.4$ V.

*Example 3.7*

A dc series generator is running at 800 rpm and is supplying a load of 6 kW at 120 V. The speed is increased to 1200 rpm and load is increased to 9 kW. The sum of armature and field resistance is 0.4 ohm. Find new value of armature current and terminal voltage. Neglect armature reaction.

*Solution*

Initial value of armature current $= I_{a1} = \dfrac{6 \times 1000}{120} = 50\text{ A}$

Initial generated emf $= E_{01} = 120 + 50 \times 0.4 = 140$ V

In a series generator, flux is proportional to armature current

$$E_{01} \propto (N_1)(I_{a1}) = kN_1 I_{a1}$$

or
$$k = \frac{140}{800 \times 50} = \frac{7}{2000}$$

Let the new armature current be $I_{a2}$ and terminal voltage $V$

$$I_{a2} = \frac{9000}{V}$$

$$\text{Generated emf } E_{02} = V + \frac{9000}{V}(0.4)$$

Also the generated emf
$$E_{02} = kN_2 I_{a2}$$

$$= \frac{7}{2000}(1200)\left(\frac{9000}{V}\right)$$

$$V + \frac{9000}{V}(0.4) = \frac{7}{2000}(1200)\left(\frac{9000}{V}\right)$$

which gives
$$V = 184.93\text{ V}$$

$$I_{a2} = \frac{9000}{184.93} = 48.67\text{ A}$$

*Example 3.8*

A compound generator is supplying a load of 120 A at 120 V. The shunt field, series field and armature resistances are 30 ohm, 0.05 ohm and 0.1 ohm, respectively. Find the generated voltage in:

(a) long shunt connection,

(b) short shunt connection,

(c) a diverter of 0.08 ohm resistance is connected in parallel with the series field in long shunt connection. Find the change in series field mmf. Neglect armature reaction.

*Solution*

(a) Fig. 3.28(b) shows the long shunt connection

$$\text{Shunt field current } I_f = \frac{120}{30} = 4\text{ A}$$

$$I_a = I_L + I_f = 120 + 4 = 124\text{ A}$$

$$E_0 = V + I_a(R_a + R_{se})$$

$$= 120 + 124(0.1 + 0.05) = 138.6\text{ V}$$

(b) Fig. 3.28(a) shows the short shunt connection

$$\text{Voltage across shunt field} = V + I_L(R_{se})$$

$$= 120 + 120 \times 0.05 = 126\text{ V}$$

$$I_f = \frac{126}{30} = 4.2\text{ A}$$

$$I_a = I_L + I_f = 120 + 4.2 = 124.2\text{ A}$$

$$E_0 = 126 + I_a R_a$$

$$= 126 + 124.2 \times 0.1 = 138.42\text{ V}$$

(c) Initial value of series field current (without diverter) = 124 A

$$\text{Current in series field with diverter in parallel} = 124 \times \frac{0.08}{0.08 + 0.05} = 73.85\text{ A}$$

Series field mmf changes in the same proportion as the series field current.

$$\%\text{ reduction in series field mmf} = \frac{124 - 73.85}{124} \times 100$$

$$= 40.4\%.$$

*Example 3.9*

Two shunt generators are operating in parallel. Generator 1 has a no-load emf of 405 V and a terminal voltage of 375 V when supplying 80 A. Generator 2 has a no-load emf of 405 V and a terminal voltage of 370 V when supplying 80 A. If total load is 120 A, find the currents supplied by the two generators and the terminal voltage. [*P.U. 2001*]

*Solution*

Let current supplied by generator 1 be $I_1$, A. Then current supplied generator 2 will be $(120 - I_a)$ A. Then

$$V = 405 - I_1 \frac{(405 - 375)}{80}$$

$$V = 405 - (120 - I_1)\frac{(405 - 370)}{80}$$

From above equations

$$I_1 = 64.6 \text{ A}$$

Current supplied by generator 2 $\quad = 120 - 64.6 = 55.4 \text{ A}$

Terminal voltage $\quad V = 405 - (64.6)\dfrac{(405 - 375)}{80} = 280.77 \text{ V}.$

## 3.20 PRINCIPLE OF DC MOTOR

A dc motor is a dc machine working as a motor to convert dc electrical energy into mechanical energy (Dc) motors are very commonly used in cars, trucks, aircraft, etc. They are also used in large ratings, where wide range of speed control is necessary.

When a conductor is carrying current and is lying in a magnetic field, a force acts on the conductor. This is the principle of a dc motor. The conductors are housed in armature slots and current is supplied from a dc source through brushes and commutator. The magnetic field is provided by the field poles.

The construction of a dc motor and generator are similar. However their external appearances may be somewhat different. A motor has a rather closed frame, because it may be operating in dusty environment and unskilled workers may be working in its vicinity.

## 3.21 ELECTROMAGNETIC TORQUE

The force on a conductor of length $l$, carrying a current $I$ and lying in a field of flux density $B$ is given by

$$F = B\,I\,l \text{ newtons} \qquad \text{...(3.12)}$$

In a motor the flux density $B$ is not constant for all conductors. When Eq. (3.12) is used for a motor, $B$ is taken as the average flux density at air-gap radius $r$ and $l$ is taken as length of the armature core. If the motor has $Z$ conductors, the torque acting at radius $r$ is

$$T_e = Z\,B\,I\,l\,r \text{ newton-m} \qquad \text{...(3.13)}$$

The conductor current $I$ equals $I_a/A$ where $I_a$ is the armature current and $A$ is the number of parallel paths. Therefore

$$T_e = \frac{Z\,B\,I_a\,l\,r}{A} \text{ newton - m} \qquad \text{...(3.14)}$$

Flux density $B$ equals $\phi/a$ where $a$ is the cross-sectional area of flux path per pole at radius $r$ and $\phi$ is flux per pole.

Therefore
$$B = \frac{\phi}{2\pi r l/P} = \frac{\phi P}{2\pi r l} \qquad \text{...(3.15)}$$

Then
$$T_e = \frac{Z\phi P I_a l r}{2\pi r l A} = \frac{Z\phi P I_a}{2\pi A} \qquad \text{...(3.16)}$$

$$= k_t \phi I_a \text{ newton-m} \qquad \text{...(3.17)}$$

where $k_t = \dfrac{ZP}{2\pi A}$ is a constant for any machine.

## 3.22 DRIVING AND RETARDING TORQUES, BACK EMF

Figure 3.34(a) shows a dc machine working as a generator. The prime mover drives the armature in anticlockwise direction. An emf is generated in the conductors and a current is supplied to the load. The current in armature conductors is as shown. Since the armature conductors are carrying current and lying in a magnetic field, an electromagnetic torque is developed in clockwise direction (as per Fleming's left hand rule). This torque is in a direction opposite to the torque of the prime mover. The prime mover must supply enough driving torque to overcome this opposing (retarding) torque and also to overcome friction at bearing, brushes, windage forces and magnetic drag caused by hysteresis and eddy currents in the armature coils.

Figure 3.34(b) shows a dc machine working as a motor. The motor develops a torque in the anticlockwise direction and supplies this torque to the mechanical load coupled to its shaft. As the armature conductors rotate in the magnetic field, an emf is induced in them. This emf opposes the flow of current in armature conductors and thus opposes the applied voltage. Therefore this emf is known as back emf (or counter emf). The applied voltage is equal to the back emf plus the voltage drop in the armature.

It is seen that both the motor and generator action take place simultaneously in a generator as well as in a motor. In a generator the generated emf is greater than the terminal voltage, whereas in a motor the terminal voltage is greater than the generated emf (or back emf). In a generator the torque supplied by the prime mover is greater than the electromagnetic torque developed by the machine. In a motor the torque supplied to the load is less than the electromagnetic torque developed by the motor.

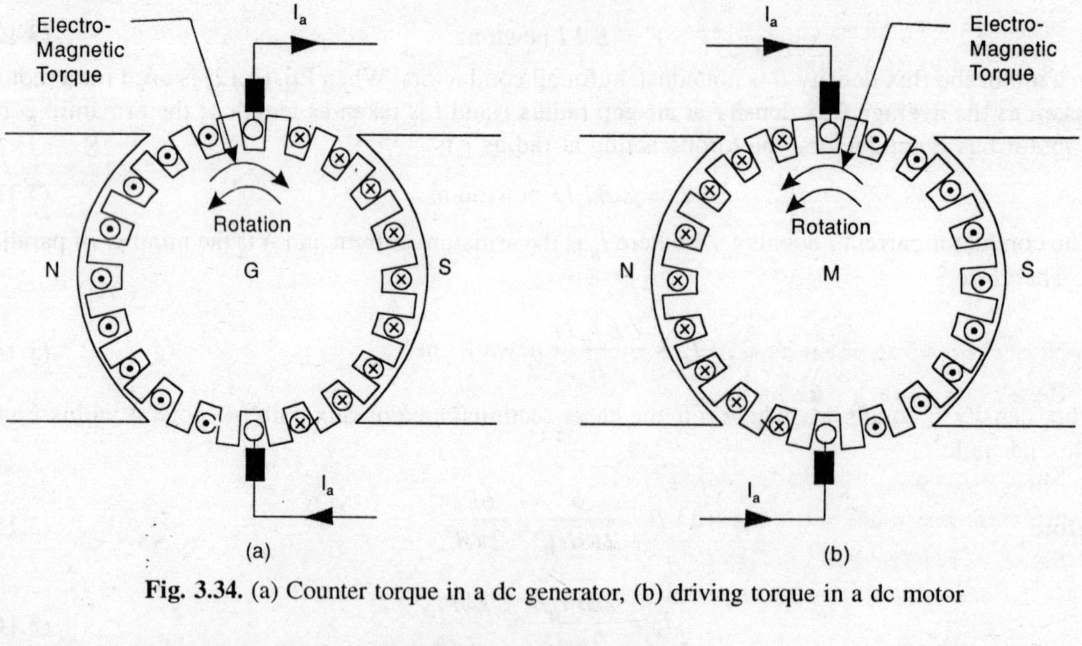


**Fig. 3.34.** (a) Counter torque in a dc generator, (b) driving torque in a dc motor

## 3.23 RELATION BETWEEN TORQUE AND VOLTAGE

From Eq. (3.4)

$$\phi = \frac{60AE}{PNZ} \quad ...(3.18)$$

Substituting Eq. (3.18) in Eq. (3.16)

$$T_e = \frac{E\,I_a}{2\pi N/60} = \frac{E\,I_a}{\omega_m} \qquad ...(3.19)$$

where $2\pi N/60$ is the angular speed $\omega_m$ in radians per second.

Equation (3.19) is the power balance relationship that underlines the operation of electromechanical energy conversion devices. Neglecting losses, the power input must be equal to power output.

We have derived the torque equation [Eq. (3.16)] on the basis of force acting on a conductor carrying current. However, it should be remembered that in an actual machine the mechanism of torque production is different. Due to the large reluctance of the air path of slots, the flux density in the slots is very low, so that there is practically no force on the conductors. The force is produced by the interaction of the main flux (due to field) and the flux produced by armature conductors carrying current. This force is exerted on the teeth. It is really fortunate that there is no force on conductors. If all the forces were to act on conductors, the slot insulation would be damaged.

### *Example 3.10*

A 12-pole dc generator has a simplex wave wound armature having 144 coils of 10 turns each. The resistance of each turn is 0.011 ohm, the flux per pole is 0.05 Wb and speed is 200 rpm.

(a) What is the induced emf?

(b) What is the armature resistance?

(c) What is the induced counter torque on the shaft of the machine, if a 1000 ohm resistance is connected to its terminals? Neglect the effect of armature resistance.

### *Solution*

(a) Since the machine has wave winding, the number of parallel paths is 2. Using Eq. (3.4)

$$E = \frac{\phi PNZ}{60A} = \frac{(0.05)(12)(200)(2 \times 144 \times 10)}{60 \times 2}$$

$$= 2880 \text{ V}$$

(b) Number of turns in each parallel path = 144 × 10/2 = 720

Resistance of each parallel path = 720 × 0.011

= 7.92 ohm

Since there are two parallel paths,

armature resistance = 0.5 × 7.92 = 3.96 ohm

(c) Since $R_a$ is neglected

$$I_L = I_a = \frac{2880}{1000} = 2.88\,\text{A}$$

$$T_e = \frac{Z\phi I_a P}{2\pi A} = \frac{(2880)(0.05)(2.88)(12)}{2\pi(2)}$$

$$= 396.03 \text{ N-m}$$

Also $$T_e = \frac{EI_a}{\omega_m} = \frac{(2880)(2.88)}{2\pi(200)/60} = 396.03\,\text{N-m.}$$

*Example 3.11*

**A dc motor takes an armature current of 110 A at 480 V. The resistance of armature is 0.2 ohm. The motor has 6 poles and the armature has a lap winding with 864 conductors. The flux per pole is 0.05 Wb. Find (a) speed, (b) torque.**

*Solution*

(a) Back emf = 480–(110 × 0.2) = 458 V

Using Eq. (3.4)

$$458 = \frac{(0.05)(6)(N)(864)}{60 \times 6}$$

or $$N = 636.1 \text{ rpm}$$

(b) $$T_e = \frac{EI_a}{2\pi N/60} = \frac{(458)(110)}{2\pi\,(636.1)/60} = 756.3\,\text{N-m.}$$

## 3.24 COMMUTATION IN DC MOTORS

The direction of force on a conductor depends on the direction of current and magnetic field. As a conductor moves from the region of influence of a north pole to the region of influence of a south pole, the direction of current in the conductor must be reversed, in order that the direction of force (and torque) on the conductor may remain the same. The function of commutator and brush gear is to cause this reversal of current.

In a motor, the armature current is flowing in the opposite direction to the generated emf (called back emf). Therefore, in a motor, the current to be reversed in a coil, when the coil is short circuited by brush, is in the opposite direction to that of the current in the generator. Therefore the emf required for good commutation (to balance the reactance voltage) is also in opposite direction to that in a generator. If a motor does not have interpoles, the brushes must be retarded to give good commutation (in a generator brushes need a forward lead). If the motor has interpoles, the polarity of an interpole is the same as the polarity of the main pole immediately preceding the interpole. When the operation of a dc machine changes from generator to motor, the direction of armature current reverses. since the interpoles are excited by the armature current, the polarity of interpole reverses automatically to the correct polarity.

## 3.25 ARMATURE REACTION IN DC MOTORS

Figure 3.35(a) shows the effect of armature mmf on a motor with brushes on geometric neutral plane. The direction of rotation is clockwise and direction of currents in the conductors must be as shown. *OD* is the mmf due to field poles, *OA* is armature reaction mmf and *OE* is the resultant mmf. It is seen that a cross-magnetising effect is produced. However, the armature reaction strengthens the field under leading pole tips and weakens it under trailing pole tips. The magnetic neutral plane is perpendicular to *OE* and is shifted backwards.

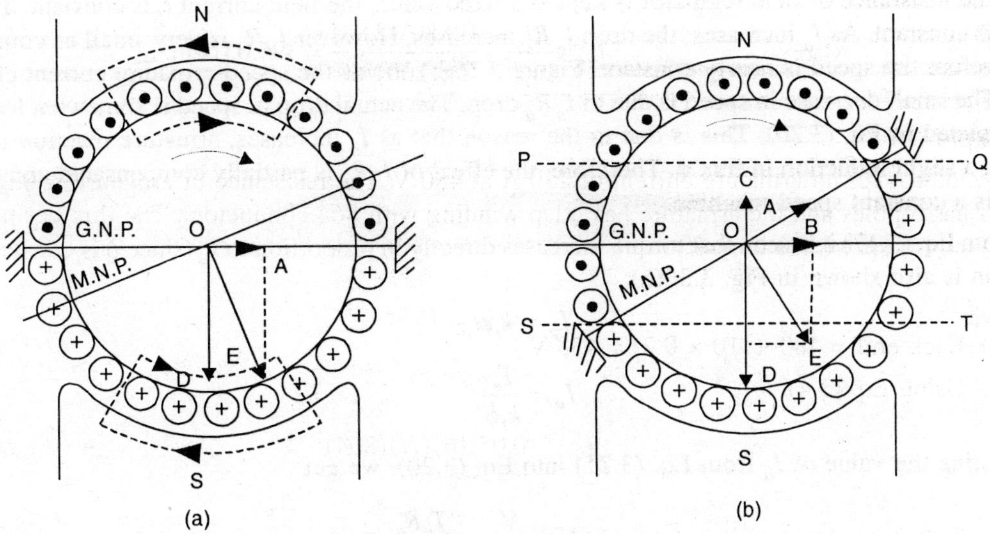


**Fig. 3.35.** Armature reaction in a dc motor (a) brushes on GNP, (b) brushes shifted to MNP

Figure 3.35(b) shows the position after the brushes have been shifted to lie along mnp. The armature conductors above the line *PQ* and below the line *ST* cause cross-magnetising effect, while those between *PQ* and *ST* cause demagnetising effect. The components *OB* and *OC* are the cross-magnetising and demagnetising components, respectively. *OE* is the resultant mmf.

If a dc motor is designed to rotate in both the directions, the brushes should be on gnp. In such machines the air-gap is slightly larger. Since each line of force of cross-flux has to cross the air-gap twice, the magnitude of cross-flux is reduced. In such motors the main field system should also have a higher mmf, because the reluctance of the air-gap is more.

## 3.26 CLASSIFICATION AND CHARACTERISTICS OF DC MOTORS

Dc motors are also classified according to the connections of the field winding. In a shunt motor the field winding consists of a large number of turns of thin wire and is connected across the armature terminals. Such a field winding has high resistance and carries a small current. In a series motor, the field winding consists of a few turns of thick wire and is connected in series with the armature. This field winding has a low resistance and carries the full armature current. A compound motor has both series and shunt field windings.

### 3.26.1 Characteristics of Shunt Motor

Figure 3.36(a) shows the connections of a shunt motor. A field regulator resistance is connected in the field circuit to control the speed of the motor.

The back emf $E_b$ is

$$E_b = V - I_a R_a = k_e N\phi$$

or

$$N = \frac{V - I_a R_a}{k_e \phi} \qquad \text{...(3.20)}$$

If the resistance of field regulator is kept at a fixed value, the field current $I_f$ is constant. Therefore flux $\phi$ is constant. As $I_a$ increases, the drop $I_a R_a$ increases. However $I_a R_a$ is very small as compared to $V$. Therefore the speed is nearly constant. Figure 3.36(b) shows the speed-armature current characteristics. The small decrease in speed is due to $I_a R_a$ drop. The actual drop in speed is somewhat lesser than that depicted in Eq. (3.20). This is due to the reason that as $I_a$ increases, armature reaction increases causing a slight reduction in flux $\phi$. Therefore, the effect of $I_a R_a$ is partially compensated and dc shunt motor is a constant speed machine.

From Eq. (3.17) it is seen that torque increases directly in proportion to $I_a$ (since $\phi$ is constant). This variation is also shown in Fig. 3.36(b).

Moreover

$$T_e = k_t \phi I_a$$

or

$$I_a = \frac{T_e}{k_t \phi} \qquad \text{...(3.21)}$$

Substituting the value of $I_a$ from Eq. (3.21) into Eq. (3.20), we get

$$N = \frac{V}{k_e \phi} - \frac{T_e R_a}{k_t k_e \phi^2} \qquad \text{...(3.22)}$$

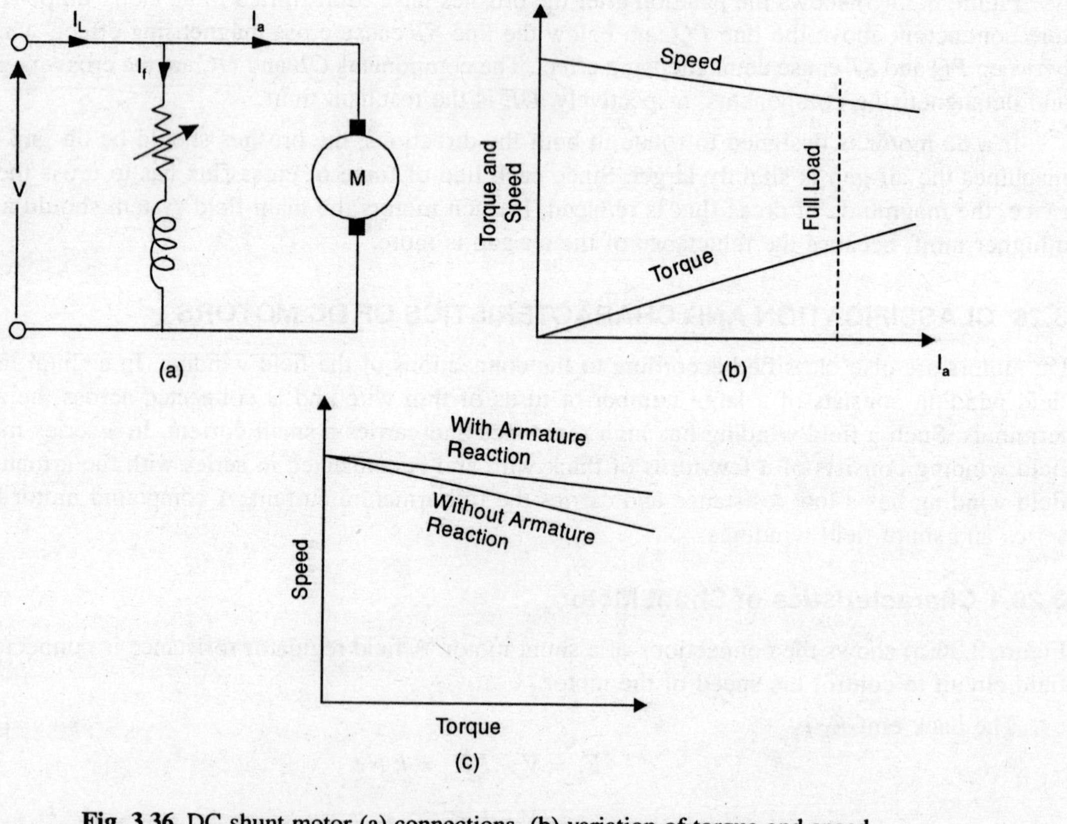


**Fig. 3.36.** DC shunt motor (a) connections, (b) variation of torque and speed with armature current, (c) speed torque characteristics

Equation (3.22) gives a relation between torque and speed. As torque increases speed decreases. However, the decrease in speed is small, because $R_a$ is small.

The armature reaction tends to reduce the flux $\phi$ to some extent. The effect of reduction in flux is to increase the speed. Figure 3.36(c) shows the variation of speed with torque with and without the armature reaction. If the motor has compensating windings, there would be no flux weakening due to armature reaction and the characteristics would be the same as that without armature reaction.

The total input current $I_L$ is sum of $I_a$ and $I_f$.

### 3.26.2 Series Motor Characteristics

Figure 3.37(a) shows the connections of a series motor. The function of diverter resistance (connected in parallel with field winding) is to control the speed.

$$E_b = V - I_a(R_a + R_{se}) = k_e N\phi$$

or

$$N = \frac{V - I_a(R_a + R_{se})}{k_e\phi} \quad ..(3.23)$$

where $I_a$ is armature current, $R_a$ is armature resistance, $R_{se}$ is series field resistance, $\phi$ is flux per pole and $N$ is speed in rpm. Moreover $T = k_t\phi I_a$.

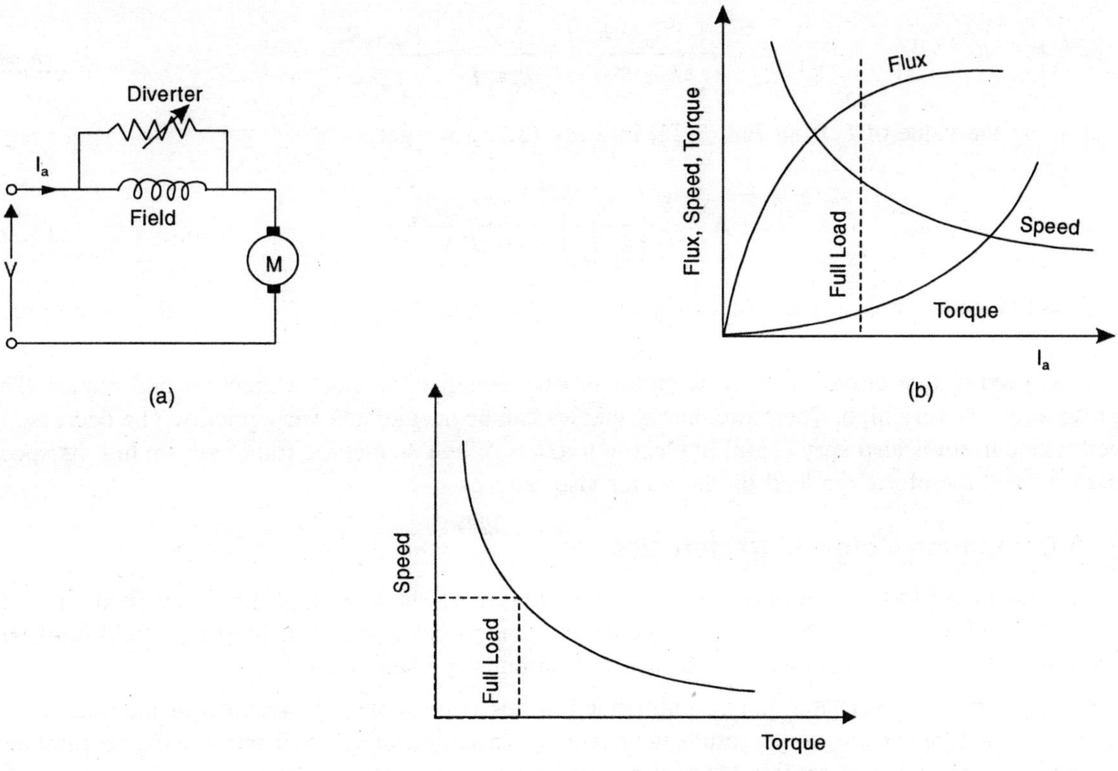


**Fig. 3.37.** DC series motor (a) connections, (b) variation of flux, speed and torque with armature current, (c) speed torque characteristics

Figure 3.37(b) shows the variation of torque, flux and speed with armature current. The current in the field winding is the same as the armature current. When this current is small, flux varies directly with current. However, for high values of $I_a$, flux does not increase in the same proportion as the current due to the saturation of the magnetic circuit. Therefore torque increases as $(I_a)^2$ for small values of $I_a$. However for high values of $I_a$, torque is proportional to $I_a$.

Since $\phi \propto I_a$, speed is inversely proportional to current. However, at high values of current, speed tends to remain constant, because flux becomes nearly constant. When $I_a$ is very small, speed can be very high. Therefore a series motor should not be run at no-load. Moreover a series motor should be coupled to the load either directly or through gears and not through a belt (because if the belt slips or breaks, the speed would become very high).

The speed-torque characteristics can be derived by assuming a linear variation of flux with $I_a$, i.e.

$$\phi = k_f I_a$$

$$T_e = k_f \phi I_a = k_t k_f I_a^2$$

or

$$I_a = \left[\frac{T_e}{k_t k_f}\right]^{0.5} \qquad \text{...(3.24)}$$

$$N = \frac{V - I_a(R_a + R_{se})}{k_e \phi} = \frac{V}{k_e k_f I_a} - \frac{R_a + R_{se}}{k_e k_f} \qquad \text{...(3.25)}$$

Substituting the value of $I_a$ from Eq. (3.24) into Eq. (3.25), we get

$$N = \frac{V}{k_e}\left(\frac{k_t}{k_f T_e}\right)^{0.5} - \frac{R_a + R_{se}}{k_e k_f} \qquad \text{...(3.26)}$$

Thus the speed varies as reciprocal of square root of torque. Figure 3.37(c) shows the speed-torque curve.

The speed-torque curve of a series motor is very suitable for electric traction and cranes. The starting torque is very high. Therefore, heavy masses can be accelerated very quickly. The decrease in speed with current is also very useful in electric traction. When an electric train goes up hill, its speed decreases, and therefore, the load on the motor also reduces.

### 3.26.3 Compound Motor Characteristics

A compound motor has a shunt field winding as well as series field winding. If the series field mmf and shunt field mmf help each other, it is a cumulatively compounded motor. If the series field mmf and shunt field mmf subtract from each other, it is a differentially compounded motor.

The operation of a differentially compounded motor is unstable. An increase in load causes an increase in armature current, which results in an increase in series field mmf. Therefore the net mmf and flux decrease and speed increases. Moreover, a decrease in flux causes a decrease in back emf, and therefore, an increase in armature current which causes a further increase in speed. Due to this cumulative action the motor starts running at a very high speed and draws a very heavy current. The high speed causes high centrifugal stresses and high current causes over heating. In view of this a differentially compounded motor is never used.

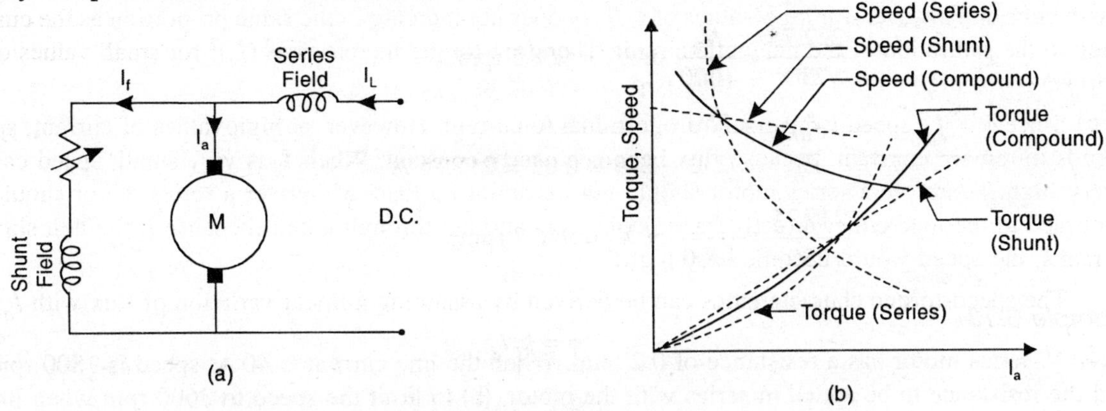


**Fig. 3.38.** DC compound motor (a) connections, (b) characteristics

In a cumulatively compounded motor the shunt flux is constant, but the series field flux increases with increase in armature current. Therefore its characteristics are in between the characteristics of the shunt and series motors. At light loads the series field has a very small effect, so that the motor behaves as a shunt motor. At high loads the effect of series field is very pronounced and the torque speed curve nearly resembles that of series motor. By varying the relative proportion of shunt and series field, any desired characteristics in between those of shunt and series motor can be obtained. Figure 3.38(a) shows the circuit diagram and Fig. 3.38(b) shows the variation of speed and torque with armature current.

*Example 3.12*

A 60 hp, 230 V dc shunt motor has armature resistance of 0.05 ohm and field circuit resistance $R_f$ of 46 ohm. The no-load speed is 1000 rpm. Find the speed when line current is (a) 75A, (b) 150A, (c) 250 A. Assume that the motor has compensating winding.

*Solution*

At no load, $I_{a0} = 0$ and $E_{b0} = V = 230$ V

$N_0 = 1000$ rpm. Since the motor has compensating winding, the armature reaction does not cause any demagnetisation and flux is constant.

Therefore, $E_b \propto N$

$I_f = 230/46 = 5\text{A}$

(a)

$$I_{a1} = 75 - 5 = 70 \text{ A}$$

$$E_{b1} = V - I_{a1}R_a = 230 - 70 \times 0.05 = 226.5 \text{ } V$$

$$\frac{E_{b1}}{E_{b0}} = \frac{N_1}{N_0}$$

or

$$N_1 = \frac{226.5}{230} \times 1000 = 984.8 \text{ rpm}$$

(b)
$$I_{a2} = 150 - 5 = 145 \text{ A}$$
$$E_{b2} = 230 - 145 \times 0.05 = 217.75 \text{ A}$$
$$\frac{222.75}{230} = \frac{N_2}{1000} \text{ or } N_2 = 968.5 \text{ rpm}$$

(c)
$$I_{a3} = 250 - 5 = 245 \text{ A}$$
$$E_{b3} = 230 - 245 \times 0.05 = 217.75 \text{ A}$$
$$\frac{217.75}{230} = \frac{N_3}{1000} \text{ or } N_3 = 946.7 \text{ rpm.}$$

*Example 3.13*

**A 240 V, series motor has a resistance of 0.2 ohm. When the line current is 40 A, speed is 1800 rpm. Find the resistance to be added in series with the motor, (a) to limit the speed to 3600 rpm when line current is 10A [Assume that between line currents of 10 A and 40 A, flux is proportional to current]; (b) to make the motor run at 900 rpm when line current is 60 A [Assume that flux at 60 A is 1.18 times the flux at 40 A]. (c) Find the speed of motor when it is connected directly to mains and line current is 60 A.**

*Solution*

$$E_{b1} = 240 - 40 \times 0.2 = 232 \text{ V}$$
$$N_1 = 1800 \text{ rpm}$$

(a) Let the resistance connected in series be $R_2$

$$\phi_2 = \frac{10}{40}\phi_1 = 0.25\phi_1$$
$$E_{b1} = k_e\phi_1 N_1$$
$$E_{b2} = k_e\phi_2 N_2$$
$$E_{b2} = \frac{E_{b1}\phi_2 N_2}{\phi_1 N_1} = \frac{232 \times 0.25 \times 3600}{1800}$$
$$= 116 \text{ V}$$
$$E_{b2} = V - I_a(R_2 + 0.2)$$

or $$116 = 240 - 10(R_2 + 0.2)$$

or $$R_2 = 12.2 \text{ ohm}$$

(b) Let the resistance connected in series with motor be $R_3$

$$\phi_3 = 1.18\ \phi_1$$
$$E_{b3} = \frac{E_{b1}\phi_3 N_3}{\phi_1 N_1} = \frac{232 \times 1.18 \times 900}{1800}$$
$$= 136.88 \text{ V}$$
$$E_{b3} = V - 60(R_3 + 0.2)$$

or $$136.88 = 240 - 60(R_3 + 0.2)$$

or $$R_3 = 1.519 \text{ ohm}$$

(c) $$E_{b4} = 240 - 60 \times 0.2 = 228 \text{ V}$$

$$\phi_4 = \phi_3 = 1.18\ \phi_1$$

$$228 = E_{b4} = \frac{E_{b1}\phi_4 N_4}{\phi_1 N_1} = \frac{232 \times 1.18 \times N_4}{1800}$$

or $$N_4 = 1499 \text{ rpm}$$

*Example 3.14*

A 35 hp, 230 V dc series motor has an armature resistance of 0.07 ohm, series field resistance of 0.08 ohm and brush voltage drop of 2V. When line current is 95 A, speed is 700 rpm. Find (a) speed when line current is 130 A, (b) speed when line current is 130 A and a diverter having a resistance of 0.08 ohm is connected across the field. Assume that flux is proportional to line current.

*Solution*

$$I_{se} = I_{a1} = 95 \text{ A},\ N_1 = 700 \text{ rpm}$$

$$E_{b1} = V - I_{a1}(R_a + R_{se}) - \text{brush drop}$$

$$= 230 - 95\ (0.07 + 0.08) - 2 = 213.75 \text{ V}$$

$$E_b \propto \phi N \propto I_{se} N$$

(a) $$E_{b2} = 230 - 130\ (0.07 + 0.08) - 2 = 208.5$$

The series field current = 130 A

Thus $$\frac{208.5}{213.75} = \frac{130 \times N_2}{95 \times 700} \text{ or } N_2 = 499 \text{ rpm}$$

(b) The resistance of parallel combination of series field and diverter

$$= \frac{0.08 \times 0.08}{0.16} = 0.04 \text{ ohm}$$

$$E_{b3} = 230 - 130(0.07 + 0.04) - 2 = 213.7 \text{ V}$$

$$\text{Series field current} = \frac{130 \times 0.08}{0.08 + 0.08} = 65 \text{ A}$$

$$\frac{213.7}{213.75} = \frac{65 \times N_3}{95 \times 700} \text{ or } N_3 = 1022.8 \text{ rpm.}$$

## 3.27 SPEED CONTROL OF SHUNT MOTORS

The speed of a shunt motor is nearly constant. However its speed is adjustable. When equipped with suitable control resistances, it can provide a fine speed control over a wide range.

Equation (3.20) indicates the philosophy of speed control. If $\phi$ is reduced, by reducing the field current, speed increases. If the applied voltage is reduced the speed decreases. These two methods are known as field control and armature control, respectively.

### 3.27.1 Field Control

The speed of a shunt motor is inversely proportional to flux. If flux is reduced the speed will increase. This is achieved by connecting a resistance (i.e. field regulator) in the field circuit (Fig. 3.39a). An increase of field regulator resistance decreases the field current with a consequent decrease in flux and an increase in speed. This method is very simple and economical, but obviously it can only give speeds higher than the normal speed. The necessary value of regulator resistance, to provide a required increase in speed, can be determined from the magnetisation characteristics of the machine. If $E_{b1}$ and $\phi_1$ are the values of back emf and flux at speed $N_1$ and $E_{b2}$ and $\phi_2$ at speed $N_2$,

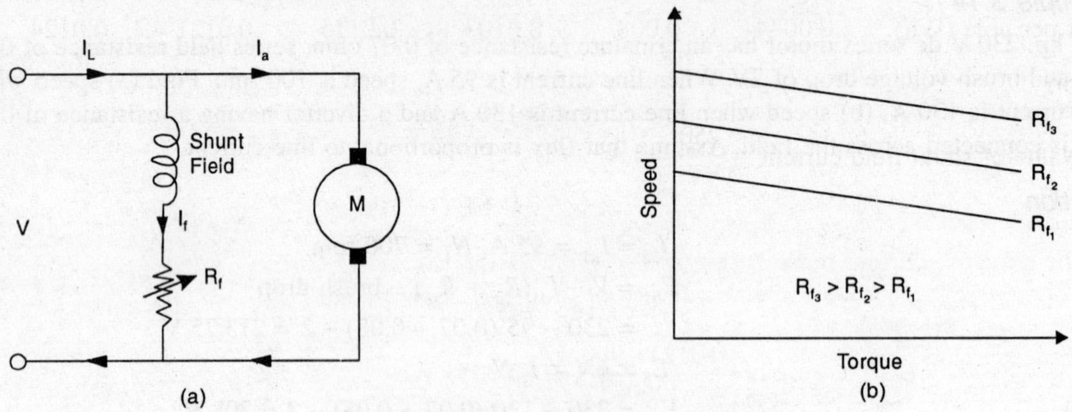


**Fig. 3.39.** Speed control of dc shunt motor by controlling field current (a) connections, (b) characteristics

$$E_{b1} = k_e \phi_1 N_1 \quad \text{...(3.27)}$$

$$E_{b2} = k_e \phi_2 N_2 \quad \text{...(3.28)}$$

or

$$\phi_2 = \phi_1 \left[\frac{E_{b2}}{E_{b1}}\right]\left[\frac{N_1}{N_2}\right] \quad \text{...(3.29)}$$

From the values of armature currents, $E_{b1}$ and $E_{b2}$ can be calculated and $\phi_2$ determined. The magnetisation curve gives the value of field current corresponding to the value of $\phi_2$. From the value of field current, the shunt regulator resistance can be calculated.

Figure 3.39(b) shows the speed torque characteristic of a motor using field control for three values of field regulator resistance $R_f$. The limitations of this method are:

(1) speeds lower than rated speed cannot be obtained.

(2) The speed is inversely proportional to flux, while the torque is directly proportional to it for a given armature current. Therefore, this method is suitable only for loads where torque decreases with increases in speed (i.e. constant kW or constant power loads).

(3) The motor operation at high speeds may become unstable. This is because of the reason that a high speed requires a weak field. However, the required armature current may be high (for a constant torque drive). The demagnetising armature reaction mmf may weaken the field still further, giving rise to unstable operation or poor commutation. Interpoles may be used to improve commutation.

Field control is used for about 2 to 1 speed range for large motors and 4 to 1 for medium and small motors.

### *Example 3.15*

A 250 V, dc shunt motor has a no-load speed of 900 rpm. The resistance of field winding is 50 ohm. Find the additional resistance which should be connected in the field circuit to increase the no-load speed to 1200 rpm. Assume that the armature current remains the same. The magnetisation characteristics of the motor is as under.

| Field current (A) | 1 | 2 | 3 | 4 | 5 | 6 |
|---|---|---|---|---|---|---|
| Flux per pole (Wb) | 0.0044 | 0.008 | 0.0103 | 0.0115 | 0.0121 | 0.0124 |

### *Solution*

Initial value of shunt field current $= \dfrac{250}{50} = 5$ A

$$\phi_1 = 0.0121 \text{ Wb}$$

Since the armature current remains the same, $E_{b1} = E_{b2}$

or $$k_e \phi_1 N_1 = k_e \phi_2 N_2$$

or $$\phi_2 = \frac{0.0121 \times 900}{1200} = 0.009075 \text{ Wb}$$

From the magnetisation curve, the field current for a flux of 0.009075 Wb is 2.46 A.

Total resistance in field circuit $= \dfrac{250}{2.46} = 101.6$ ohm

Field regulator resistance $= 101.6 - 50 = 51.6$ ohm.

### *Example 3.16*

A 250 V dc shunt motor has a field resistance of 125 ohm and armature resistance of 1 ohm. It takes an armature current of 25 A at a speed of 900 rpm. It is required to increase the speed to 1100 rpm, keeping the torque constant. Find the additional resistance of field regulator to achieve this speed. Assume that the magnetic circuit is unsaturated.

### *Solution*

Let $I_{f1}, I_{a1}, N_1, \phi_1, T_1, E_{b1}$ denote initial values of field current, armature current, speed, flux, torque and back emf (i.e. at a speed of 900 rpm) and $I_{f2}, I_{a2}, N_2, \phi_2, T_2$ and $E_{b2}$ denote the final values (i.e. at a speed of 1100 rpm) of these quantities.

$$E_{b1} = 250 - 25 \times 1 = 225 \text{ V}$$

$$I_{f1} = \frac{250}{125} = 2 \text{ A}$$

$$T_1 = T_2$$

$$\text{Torque} \propto \text{flux} \times \text{armature current}$$

$$\propto \text{field current} \times \text{armature current}$$

or $$I_{f1}I_{a1} = I_{f2}I_{a2}$$

or $$\frac{I_{a1}}{I_{a2}} = \frac{I_{f2}}{I_{f1}}$$

Back emf ∝ flux × speed ∝ field current × speed

or $$\frac{E_{b1}}{E_{b2}} = \frac{I_{f1}N_1}{I_{f2}N_2} = \frac{I_{a2}N_1}{I_{a1}N_2} = \frac{I_{a2} \times 900}{25 \times 1100}$$

$$E_{b2} = 250 - I_{a2} \times 1$$

Therefore $$\frac{225}{250 - I_{a2}} = \frac{I_{a2} \times 900}{25 \times 1100}$$

or $$I_{a2}^2 - 250\,I_{a2} - 6875 = 0$$

or $$I_{a2} = 218.54\text{ A or } 31.46\text{ A}$$

The armature current cannot be allowed to vary over a very wide range. Therefore armature current cannot be 218.54 A

Hence $I_{a2}$ = 31.46 A

$$I_{f2} = \frac{I_{a1}}{I_{a2}} \times I_{f1} = \frac{25}{31.46} \times 2 = 1.59\text{ A}$$

If additional regulator resistance is $R_f$,

then $$I_{f2} = \frac{250}{125 + R_f}$$

or $$1.59 = \frac{250}{125 + R_f} \text{ or } R_f = 32.23\text{ ohm.}$$

### 3.27.2 Armature Control

(a) *Armature resistance control*: If an additional adjustable resistance $R$ is connected in series with armature, the speed of the motor can be reduced, (Fig. 3.40a). This method can give speeds lower than the normal speed. By a combination of field control and armature control, we can obtain speeds above or below the normal speed.

Let $E_{b1}$ be the back emf at a speed $N_1$, with armature current $I_{a1}$ and no extra resistance in the circuit. Insertion of a resistance $R$ in the armature circuit causes the back emf to change to $E_{b2}$, speed to $N_2$ and armature current to $I_{a2}$. The flux $\phi$ is the same in the two cases, because the field current is the same.

$$E_{b1} = k_e N_1 \phi = V - I_{a1}R_a \quad \text{...(3.30)}$$

$$E_{b2} = k_e N_2 \phi = V - I_{a2}(R_a + R) \quad \text{...(3.31)}$$

From above equations

$$\frac{N_2}{N_1} = \frac{V - I_{a2}(R_a + R)}{V - I_{a1}R_a} \quad ...(3.32)$$

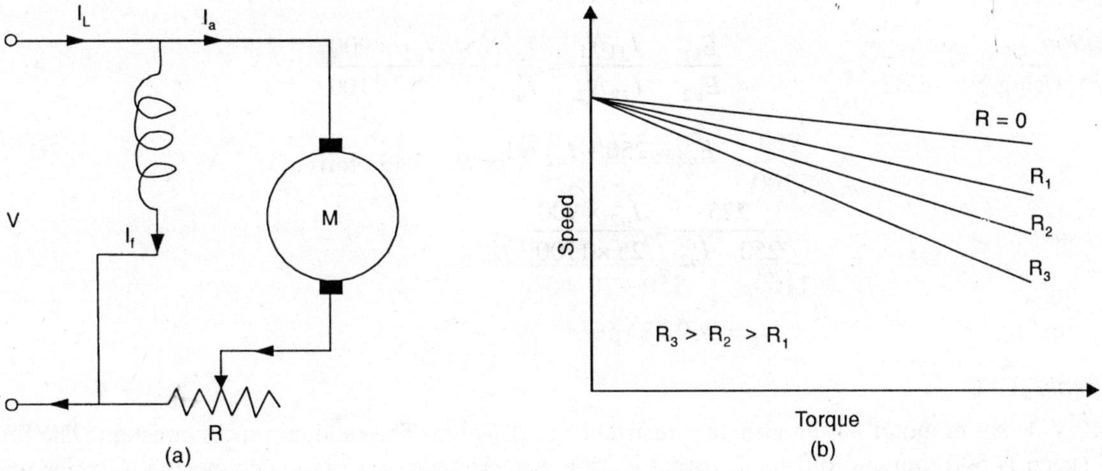


**Fig. 3.40.** Speed control of dc shunt motor by changing applied voltage (a) connections, (b) characteristics

From the values of $V$, $I_{a1}$, $I_{a2}$, $R_a$, $N_1$, $N_2$ the additional resistance $R$ to be connected in series with the armature, to obtain speed $N_2$, can be determined. The speed torque curve for different values of $R$ is shown in Fig. 3.40(b).

This method has the disadvantage that controller resistance $R$ is a function of speed as well as load current. Another disadvantage is the loss of energy in the controller resistance. Since the resistance carries full armature current, the loss of energy becomes appreciable especially at low speeds.

(b) *Shunted armature control*: In this method the motor is connected to dc supply through a potential divider circuit as shown in Fig. 3.41. This method gives a better speed regulation in the low speed range than the armature resistance control. For a constant supply voltage $V$, the voltage applied to the armature depends on resistance $R_1$ and $R_2$. The field current does not depend on these resistances, because the field winding is connected directly across the mains.

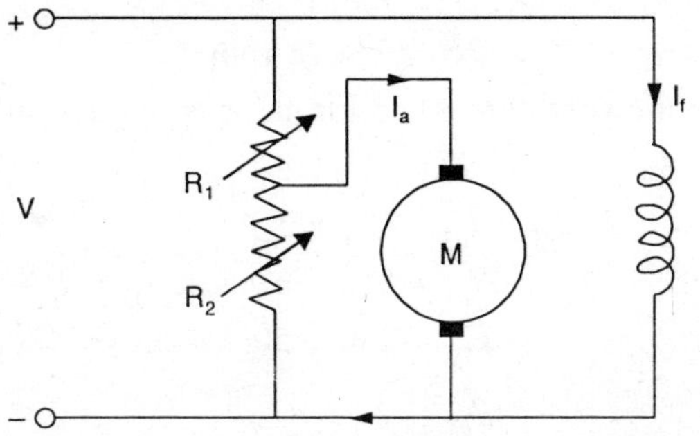


**Fig. 3.41.** Shunted armature control of dc shunt motor

### Example 3.17

A 250 V dc shunt motor has an armature resistance of 0.5 ohm. When the armature current is 30 A, the speed is 1100 rpm. (a) It is required to reduce the speed to 750 rpm. The armature current remains the same. Find additional resistance $R$ to be connected in series with armature. (b) If the current decreases to 15A (with resistance $R$ connected) find the speed of the motor.

### Solution

(a) Using Eq. (3.32)

$$\frac{750}{1100} = \frac{250 - 30(R + 0.5)}{250 - 30 \times 0.5} \quad \text{or } R = 2.49 \text{ ohm}$$

(b)

$$\frac{N_2}{1100} = \frac{250 - 15(2.49 + 0.5)}{250 - 30 \times 0.5}$$

or

$$N_2 = 960.28 \text{ rpm.}$$

### Example 3.18

A 240 V dc shunt motor has an armature resistance of 0.4 ohm. The field current is constant. The full-load speed is 600 rpm and full-load current is 25A. A resistance of 1 ohm is connected in series with the armature. Find (a) speed at full-load torque, (b) speed at twice the full-load torque, (c) stalling torque.

### Solution

(a) Let the subscripts 1 and 2 denote the initial and final conditions

$$T_1 = T_2$$

or

$$I_{f1} \times I_{a1} = I_{f2} \times I_{a2}$$

or

$$I_{a1} = I_{a2} = 25\text{A (because field current is constant)}$$

$$E_{b1} = 240 - 25 \times 0.4 = 230 \text{ V}$$

$$E_{b2} = 240 - 25(1 + 0.4) = 205 \text{ V}$$

Since flux is constant

$$\frac{E_{b1}}{E_{b2}} = \frac{N_1}{N_2} \quad \text{or} \quad \frac{230}{205} = \frac{600}{N_2}$$

or

$$N_2 = 534.78 \text{ rpm}$$

(b) Let subscript 3 denote the condition when torque is twice the full-load torque

$$T_3 = 2T_1$$

$$I_{f3} \times I_{a3} = 2I_{f1} \times I_{a1}$$

or

$$I_{a3} = 2I_{a1} = 50 \text{ A (because } I_{f3} = I_{f1})$$

$$E_{b3} = 240 - 50(1 + 0.4) = 170 \text{ V}$$

$$\frac{E_{b1}}{E_{b3}} = \frac{N_1}{N_3}$$

or $$N_3 = \frac{E_{b3} N_1}{E_{b1}} = \frac{170 \times 600}{230} = 443.48 \text{ rpm}$$

(c) Let subscript 4 denote the new conditions. When motor stalls, speed is zero

Therefore $E_{b4} = 0$

or $$E_{b4} = 240 - I_{a4}(1 + 0.4) = 0$$

or $$I_{a4} = 171.43 \text{ A}$$

$$\frac{T_4}{T_1} = \frac{I_{f4} I_{a4}}{I_{f1} I_{a1}} = \frac{I_{a4}}{I_{a1}}$$

$$= \frac{171.43}{25} = 6.857$$

Thus the stalling torque is 6.857 times full-load torque.

## 3.28 SPEED CONTROL OF SERIES MOTORS

The two methods, viz. field control and armature control can be used for series motors also.

### 3.28.1 Field Control

(a) *Diverter field control:* A variable resistance (known as diverter) is connected across the field winding as shown in Fig. 3.42 (a). If the diverter resistance is $R_d$, then

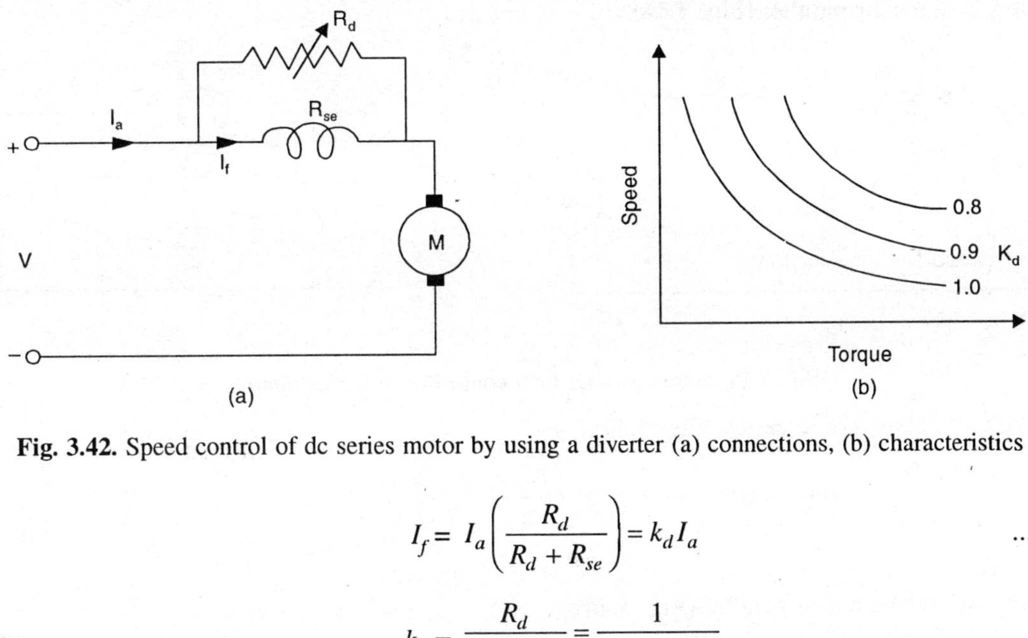


**Fig. 3.42.** Speed control of dc series motor by using a diverter (a) connections, (b) characteristics

$$I_f = I_a \left( \frac{R_d}{R_d + R_{se}} \right) = k_d I_a \quad ...(3.33)$$

where $$k_d = \frac{R_d}{R_d + R_{se}} = \frac{1}{1 + R_{se}/R_d}$$

If saturation is neglected

$$\phi_{se} = k_f k_d I_a \quad ...(3.34)$$

Therefore Eq. (3.26) can be written as

$$N = \frac{V\sqrt{k_t}}{k_e\sqrt{T_e k_f k_d}} - \frac{R_a + R_{se}}{k_e k_d k_f} \quad \text{...(3.35)}$$

Equation (3.35) gives the speed torque curve, which is plotted in Fig. 3.42(b) for 3 different values of $k_d$.

(b) *Tapped field control:* As shown in Fig. 3.43, the field winding is tapped. Thus the number of field winding turns can be changed. This changes the field mmf and hence the speed. The speed torque curve is similar to that in Fig. 3.42(b).

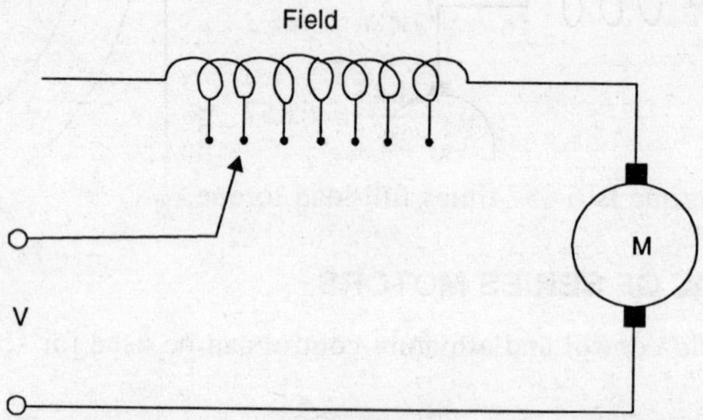


**Fig. 3.43.** Tapped field control of dc series motor

(c) *Series-parallel field control:* The series field winding is divided into two parts which can be connected in series or parallel (Fig. 3.44).

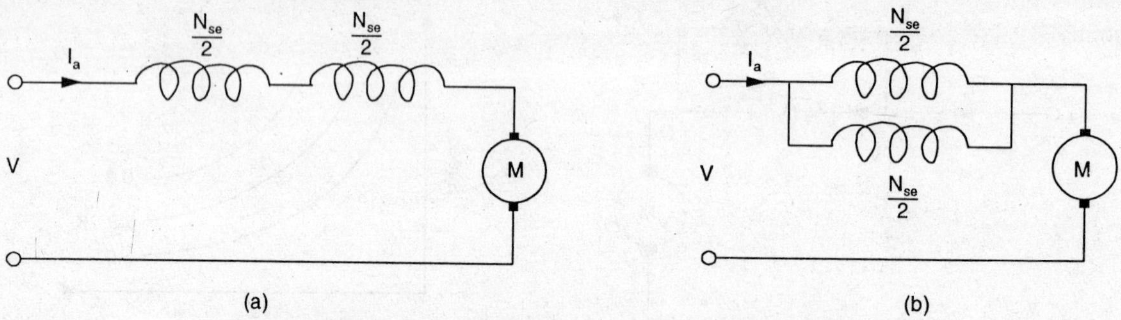


**Fig. 3.44.** Series parallel field control of dc series motor

When the two halves are in series (Fig. 3.44a),

$$\text{total ampere turns} = I_a\left(\frac{N_{se}}{2} + \frac{N_{se}}{2}\right)$$
$$= I_a N_{se}$$

When the two halves are in parallel (Fig. 3.44b),

$$\text{total ampere turns} = \left(\frac{I_a}{2}\right)\left(\frac{N_{se}}{2}\right)2$$
$$= 0.5\, I_a N_{se}$$

Since the ampere turns for parallel connection are half of those for series connection, the speed of the machine with parallel fields (Fig. 3.44b) is double that in the case of series fields (Fig. 3.44a).

### 3.28.2 Armature Control

(a) *Armature resistance control:* A variable resistance $R$ is connected in series with armature (Fig. 3.45a). As the value of $R$ is increased, the voltage applied to the armature decrease and the speed decreases. Figure 3.45(b) shows the speed-torque curve.

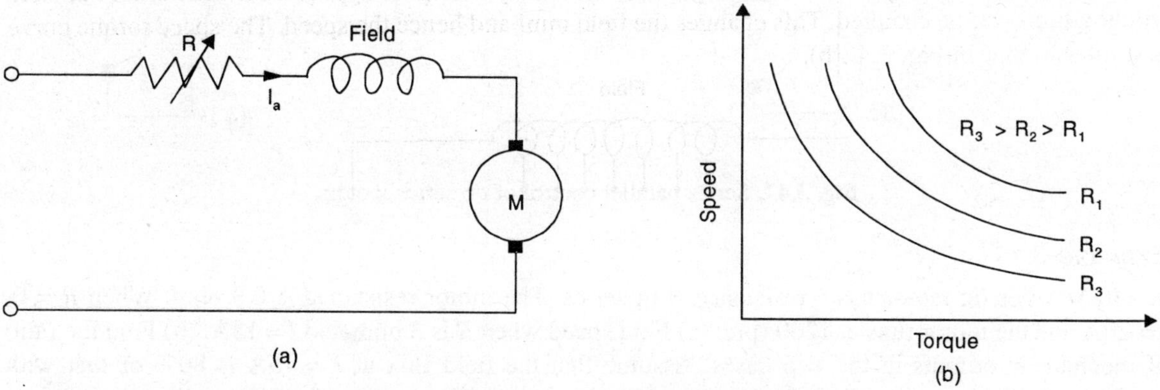


**Fig. 3.45.** Speed control of dc series motor by using a resistance in series with armature (a) connections, (b) characteristics

(b) *Shunted armature control:* The motor is connected to the dc supply through a potential divider consisting of resistance $R_1$ and $R_2$ as shown in Fig. 3.46. By adjusting $R_1$, $R_2$, the voltage across armature is changed and this changes the motor speed. The no-load speed can be adjusted to a finite reasonable value and good speed regulation in low-speed range can be obtained.

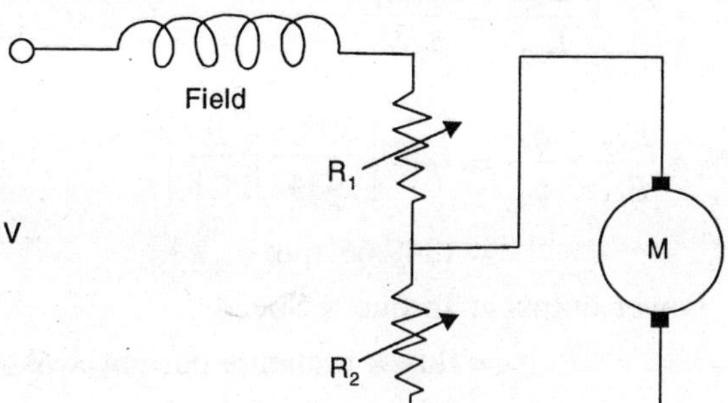


**Fig. 3.46.** Shunted armature control of dc series motor

(c) *Series-parallel control:* As shown in Fig. 3.47, two identical motors mechanically coupled to a common load are used in this case. By using a controller, the motors can be connected in series (Fig. 3.47b) or in parallel (Fig. 3.47a). When in series, the voltage across each motor is 0.5 V and when in parallel, the voltage across each is $V$. Thus two speeds can be obtained. This method is very commonly used for speed control of traction motors.

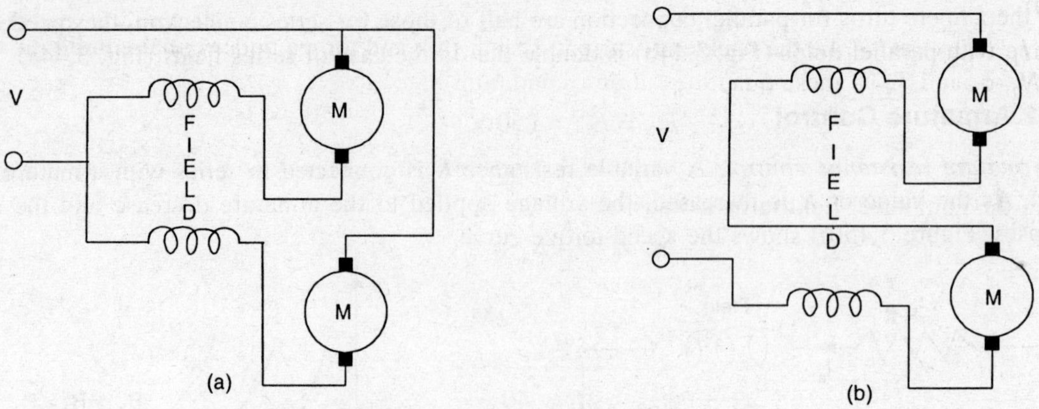


**Fig. 3.47.** Series parallel control of dc series motors

*Example 3.19*

A 440 V series dc motor has a resistance $R$ in series. The motor resistance is 0.3 ohm. When $R = 0$, $I = 20$A and the motor runs at 1200 rpm. (a) Find speed when $R$ is 3 ohm and $I = 15$A. (b) Find the ratio of mechanical outputs in the two cases. Assume that the field flux at $I = 15$A is 80% of that with $I = 20$A.

*Solution*

$$E_{b1} = 440 - 20 \times 0.3 = 434 \text{ V}$$

$$E_{b2} = 440 - 15(3 + 0.3) = 390.5 \text{ V}$$

$$\phi_2 = 0.8\ \phi_1$$

(a)
$$\frac{E_{b1}}{E_{b2}} = \frac{\phi_1 N_1}{\phi_2 N_2}$$

or
$$N_2 = N_1 \times \frac{E_{b2}}{E_{b1}} \times \frac{\phi_1}{\phi_2} = 1200\left[\frac{390.5}{434}\right]\left[\frac{1}{0.8}\right]$$

$$= 1349.65 \text{ rpm}$$

(b)
$$\text{Power output} \propto \text{Torque} \times \text{Speed}$$

$$\propto \text{flux} \times \text{armature current} \times N$$

$$\frac{\text{Mech. power output in first case}}{\text{Mech. power output in second case}} = \frac{\phi_1 \times 20 \times 1200}{\phi_2 \times 15 \times 1349.5} = \frac{20 \times 1200}{0.8 \times 15 \times 1349.5} = 1.48.$$

*Example 3.20*

A 4-pole series wound fan motor has a speed of 600 rpm, when fed from 250 V dc mains and input current is 20A. The field coils are all in series. The load torque increases as the square of speed. The field coils are now connected in two parallel groups of two each. Find the speed and input current. Ignore losses. Assume that flux is proportional to current.

*Solution*

Let $E_{b1}$, $I_1$, $N_1$, $\phi_1$ and $T_1$ be the back emf, current, speed, flux and torque under initial conditions and $E_{b2}$, $I_2$, $N_2$, $\phi_2$ and $T_2$ be these quantities in new condition. Since losses are neglected,

$$E_{b1} = E_{b2} = 250 \text{ V}$$

$$I_1 = 20 \text{ A}, N_1 = 600 \text{ rpm}$$

If $N_{se}$ are the number of field turns in each coil

$$\phi_1 \propto 4(I_1)(N_{se})$$

$$\phi_2 \propto 2\left(\frac{I_2}{2}\right)(2N_{se})$$

or

$$\frac{\phi_1}{\phi_2} = \frac{2I_1}{I_2} = \frac{40}{I_2}$$

$$\frac{E_{b1}}{E_{b2}} = 1 = \frac{\phi_1 N_1}{\phi_2 N_2} = \frac{40}{I_2} \times \frac{600}{N_2} \qquad \text{...(i)}$$

$$\frac{T_1}{T_2} = \frac{600^2}{N_2^2} \qquad \text{...(ii)}$$

Moreover

$$\frac{T_1}{T_2} = \frac{\phi_1 I_1}{\phi_2 I_2} = \frac{40 \times 20}{I_2^2} = \frac{800}{I_2^2} \qquad \text{...(iii)}$$

From Eqs. (ii) and (iii)

$$\frac{600^2}{N_2^2} = \frac{800}{I_2^2} \qquad \text{...(iv)}$$

From Eqs. (i) and (iv)

$$I_2 = 33.6 \text{ A}, \; N_2 = 714.3 \text{ rpm}.$$

## 3.29 WARD LEONARD CONTROL

The speed of a shunt motor or separately excited motor can be controlled very easily by field control and armature control methods. Ward Leonard control is an armature control method and was very commonly used in steel mills, seamless tube mills and shears, mine hoists, elevators, paper mill drives, etc. It permits wide range of speed control as well as reversal of direction of rotation. Nowadays this method has been superseded by solid state control.

Figure 3.48 shows the circuit diagram. *M* is the dc motor, whose speed is required to be controlled. Its armature is fed by a separately excited generator *G*. The field windings of motor *M* and generator *G* are fed by an exciter *E*, which is mounted on the same shaft as generator *G*. Both generator *G* and exciter *E* are driven by a motor *DM*, which may be a 3-phase induction motor (if the system is to work on ac). *DM* may also be a dc shunt motor. If a separator dc supply is available, this can be used for feeding field windings of *M* and *G* and the exciter *E* is not necessary. The field current of generator *G* can be adjusted by the help of potential divider $R$. The double throw switch $S$ helps to reverse the direction of field current of generator $G$.

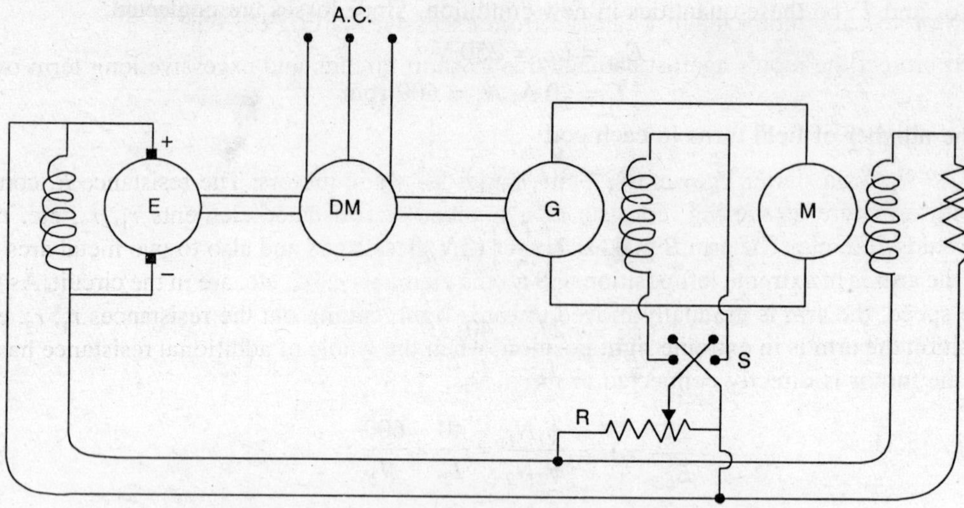


**Fig. 3.48.** Connections for Ward Leonard method of speed control

The field current of generator $G$ can be adjusted to any value from zero to maximum by adjustment of potential divider $R$. Thus the output voltage of generator $G$ can be controlled. This voltage is applied to motor $M$. Thus the voltage applied to the armature of motor $M$ can be controlled. Hence the speed of motor $M$ can be controlled. The reversal of excitation of generator $G$ (by the help of switch $S$) changes the polarity of output voltage of generator $G$. Thus the direction of rotation of $M$ can be reversed.

A special advantage of Ward Leonard Systems is that, it can regenerate or return and motor's energy of motion back to the mains supply. When a heavy load is first raised and then lowered by the motor $M$, this motor acts as a generator (when the load is lowered) returning electrical energy to the mains. This reduces the operating cost of the system.

The main disadvantage of this system is the high initial cost. The overall efficiency is also low. Because of these disadvantages, it has been replaced by solid state control circuits in all modern installations.

## 3.30 DC MOTOR STARTER

### 3.30.1 Problem at Starting

When a motor is at standstill, its back emf is zero. If the motor is switched on to the supply directly, the current would be $V/R_a$. This may be about 20 times the rated current. In a motor equipped with a fuse, this current would blow off the fuse. In case a fuse is not used, this large current may damage the motor.

The only way to limit this current is to insert an additional resistance in series with the armature. As the motor picks up speed, the back emf increases and it is necessary to cut out this resistance gradually or in steps, so that when the motor has come to its rated speed, the whole of this additional resistance has been removed.

### 3.30.2 Functions of Starter

The functions of a dc motor starter are:

(1) To protect the motor against damage from excessive starting current.

(2) To protect the motor against damage due to short circuits and excessive long term overloads.

### 3.30.3 Construction

Figure 3.49 shows a starter known as 3-point starter for shunt motors. The resistance $R$, connected in series with armature at starting, consists of a number of resistance elements $r_1$, $r_2$, etc. connected between studs 1, 2, etc. The arm $B$ makes contact with these studs and also to two metal arcs $M$ and $N$. Initially the arm is in extreme left position and all the elements $r_1$, $r_2$, etc. are in the circuit. As the motor picks up speed, the arm is gradually moved towards right, cutting out the resistances $r_1$, $r_2$, etc. In the final position the arm is in extreme right position, when the whole of additional resistance has been cut out and the motor is directly connected to the mains.

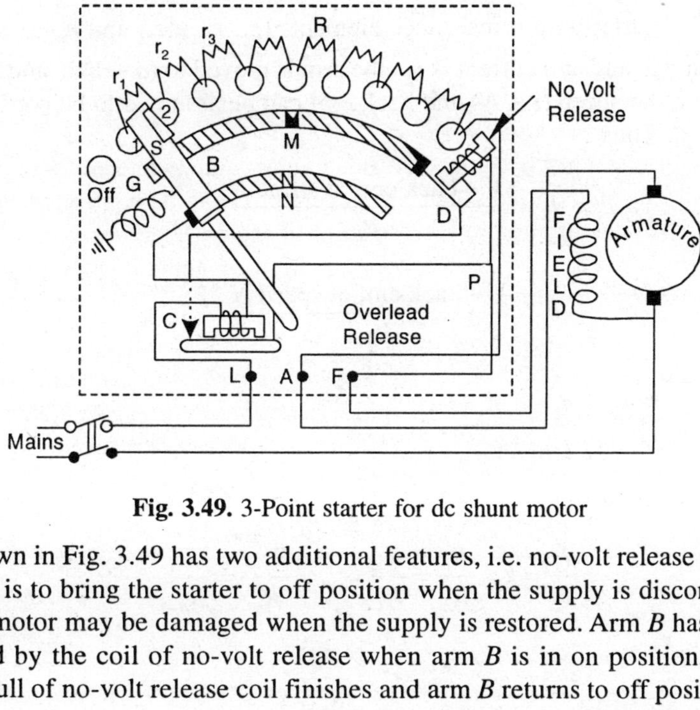


**Fig. 3.49.** 3-Point starter for dc shunt motor

The starter shown in Fig. 3.49 has two additional features, i.e. no-volt release and overload release. The no-volt release is to bring the starter to off position when the supply is disconnected. If it remains in on position, the motor may be damaged when the supply is restored. Arm $B$ has a soft iron armature $S$ which is attracted by the coil of no-volt release when arm $B$ is in on position. When the supply is disconnected, the pull of no-volt release coil finishes and arm $B$ returns to off position due to the action of spring $G$. When the line current becomes more than about 1.3 times full-load current, the overload release coil pulls up its armature to terminal $C$. This action short circuits the coil $DP$ of no-volt release and arm $B$ returns to off position. Thus the motor is protected from overloads. Such an overload release is used for only small and medium size motors. Large size motors are provided with automatic circuit breakers.

### 3.30.4 Calculation of Resistance Elements

The various resistance elements $r_1$, $r_2$, etc. should be so selected that current during starting remains between the maximum value $I_1$ and minimum value $I_1$. When one resistance element is cut out, the current increases to $I_1$. As speed increases, back emf increases and current decreases. When current reaches the value $I_2$, the next section is cut out, increasing the current to $I_1$ and so on, as shown in Fig. 3.50.

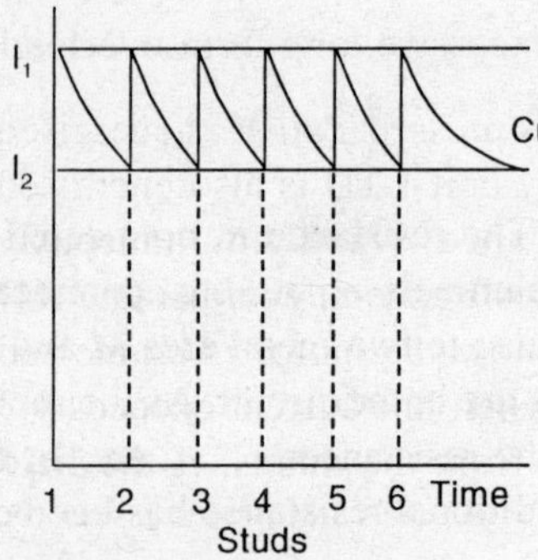


**Fig. 3.50.** Variation of current during starting

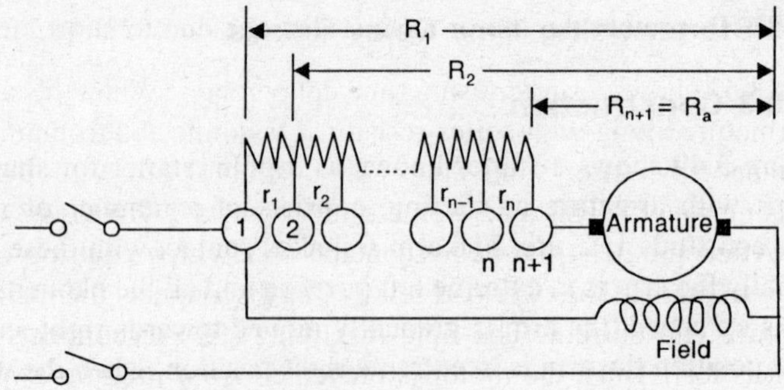


**Fig. 3.51.** Resistance elements of dc shunt motor starter

Figure 3.51 shows a starter with $n$ resistance elements ($r_1$, $r_2$, etc.) and $n + 1$ studs.

When the arm is on $q$th stud and current is $I_2$, the arm is moved to $(q + 1)$th stud, raising the current to $I_1$. During this instant, the speed (i.e. $N$) and back emf can be assumed to be constant and resistance changes from $R_q$ to $R_{q+1}$. Thus

$$I_2 = \frac{V - \text{back emf at speed } N}{R_q}$$

$$I_1 = \frac{V - \text{back emf at speed } N}{R_{q+1}}$$

or

$$\frac{I_1}{I_2} = \frac{R_q}{R_{q+1}}$$

Substituting $q = 1, 2, 3$, etc., we get

$$\frac{I_1}{I_2} = k = \frac{R_1}{R_2} = \frac{R_2}{R_3} = \ldots = \frac{R_{n-1}}{R_n} = \frac{R_n}{R_{n+1}}$$

or

$$k^n = \frac{R_1}{R_{n+1}} \qquad \text{...(3.36)}$$

When all additional resistances have been cut out, $R_{n+1}$ equals armature resistance $R_a$. Therefore

$$k^n = \frac{R_1}{R_a} \qquad \text{...(3.37)}$$

Using Eq. (3.37) we can find $n$ and hence the resistance $r_1$, $r_2$, etc.

## 3.31 MODERN DC MOTOR STARTERS

The starter shown in Fig. 3.49 is subject to the problem of human error. If the handle is moved too quickly the line current may be large and may burn the motor. If the handle is moved too slowly, the resistor $R$ may be burnt (this resistor $R$ is designed for short term rating only). Modern starters have automatic circuits. These starters either use time delay relays or back emf sensing relays.

### 3.31.1 Starter Using Time Delay Relays

Figure 3.52 shows a starter using time delay relays. When the start button is pressed, coil $M$ is energised and armature along with series resistor $R$ is connected to mains. The relay coil 1 TD is also energised simultaneously. After some time contacts 1 TD close cutting out resistance $r_1$ and simultaneously energising relay 2 TD. After some time contacts 2 TD close cutting out resistance $r_2$ energising 3 TD relay. After another time delay contacts 3 TD close cutting out $r_3$ and now the motor is connected directly to mains. The overload relay (*OL*) disconnects the mains supply when line current exceeds 1.3 times full-load current. The field-loss relay (FL) disconnects the motor from the mains, if the field current is lost. Thus, the motor is prevented from running away.

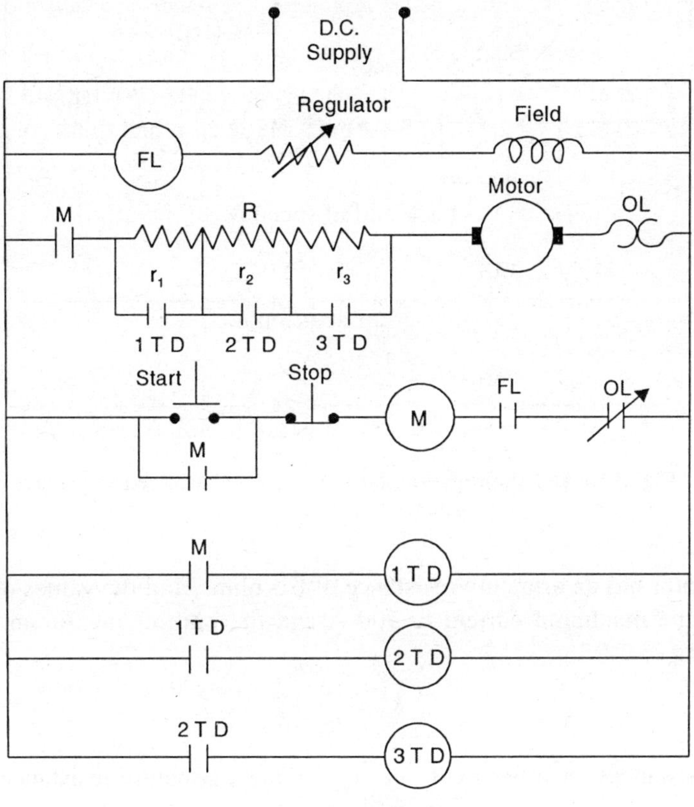


**Fig. 3.52.** DC shunt motor starter using time delay relays

### 3.31.2 Starter Using Back-emf Sensing Relays

Figure 3.53 shows a dc motor starter, which use back-emf sensing relays 1AR and 2AR. When the start button is pressed, coil $M$ is energised and the armature along with series resistor $R$ is connected to mains. When back-emf exceeds 50% line voltage, the relay 1AR closes contacts 1AR. This closes contacts 1A and resistance $r_1$ is cut out. When back-emf exceeds 80% of line voltage, the relay 2AR closes the contacts 2AR and the second resistance $r_2$ is cut out. The overload relay (OL) disconnects the motor from the mains supply when line current exceeds 1.3 times full-load current. The field loss relay (FL) disconnects the motor if field current is lost.

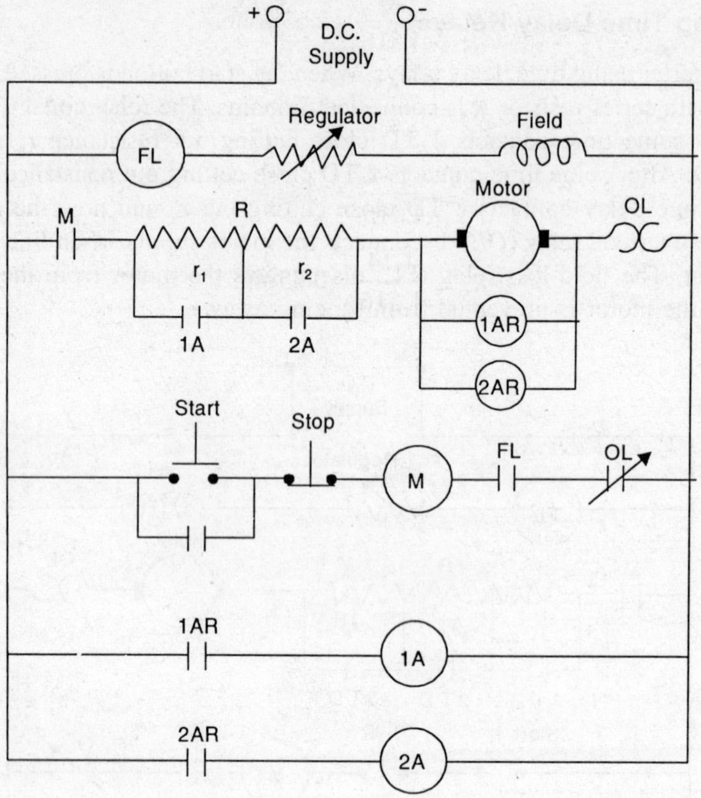


**Fig. 3.53.** DC shunt motor starter using back-emf sensing relays

*Example 3.21*

A 400 V dc shunt motor has an armature resistance of 0.5 ohm. Find the values of resistance elements of a starter to limit the maximum current to 100 A and the ratio of maximum current to minimum current during starting to 1.5.

*Solution*

Referring to Fig. 3.51

$$R_1 = \frac{400}{100} = 4 \text{ ohm}, \; k = 1.5$$

$$k = \left(\frac{R_1}{R_a}\right)^{1/n}$$

or
$$1.5 = \left(\frac{4}{0.5}\right)^{1/n} \quad \text{or } n = 5.13$$

Take $n = 5$

$$k = (8)^{1/5} = 1.516$$

$$R_2 = \frac{R_1}{1.516} = 2.638\,\text{ohm}$$

$$R_3 = \frac{2.638}{1.516} = 1.74\,\text{ohm}$$

$$R_4 = \frac{1.74}{1.516} = 1.14\,\text{ohm}$$

$$R_5 = \frac{1.14}{1.516} = 0.75\,\text{ohm}$$

$$R_6 = \frac{0.75}{1.516} = 0.5\,\text{ohm}$$

$$r_1 = R_1 - R_2 = 4 - 2.638 = 1.362\ \Omega$$

$$r_2 = R_2 - R_3 = 2.638 - 1.74 = 0.898\ \Omega$$

$$r_3 = R_3 - R_4 = 1.74 - 1.14 = 0.6\ \Omega$$

$$r_4 = R_4 - R_5 = 1.14 - 0.75 = 0.39\ \Omega$$

$$r_5 = R_5 - R_6 = 0.75 - 0.5 = 0.25\ \Omega$$

## 3.32 ELECTRIC BRAKING

There is no basic difference between a dc generator and a dc motor. If the induced emf is greater than the terminal voltage, the machine works as a generator and if the induced emf is less than the terminal voltage, it works as a motor. This reversibility of dc machine provides a convenient method of electric braking. The three types are: (a) plugging, (b) dynamic braking, (c) regenerative braking.

### 3.32.1 Plugging

In this method the connections to the armature are reversed suddenly (Fig. 3.54). A high braking torque is produced. The effective voltage is $E_b + V$ initially and a braking resistance is required in the circuit, so that the maximum permissible current may not be exceeded. The supply must be switched off as soon as the speed is near zero, otherwise the motor will reverse its rotation. The kinetic energy is dissipated in the armature and braking resistance.

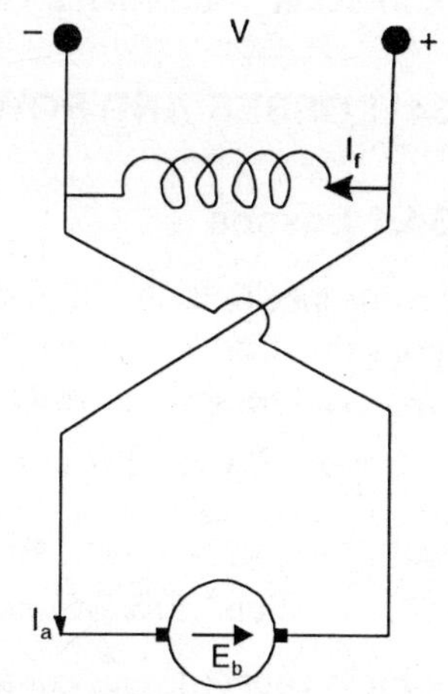

**Fig. 3.54.** Plugging

### 3.32.2 Dynamic Braking

In this method the motor is disconnected from the supply and a braking resistance is connected to its terminals. The motor works as a generator, driven by the stored mechanical energy and the power generated is dissipated in the braking resistance. The field circuit is kept connected to the mains. If the field is left connected to the machine, the braking torque will start falling sharply with speed and time required for braking will be more. If this method is used for two or more series machines operating in parallel, an equalising bar must be used, so that the parallel operation may be stable.

### 3.32.3 Regenerative Braking

In this method the field current is increased, so that the induced emf becomes more than the supply voltage. The armature current and torque are reversed. The kinetic energy of the machine is converted into electric energy and returned to supply. Thus a braking torque is provided upto the speed at which emf and supply voltage are equal. Thus braking is possible only for a limited range of speed, because as speed is decreasing, it is not possible to maintain emf at a higher value than supply voltage. The method can be easily used for a shunt or separately excited motor. A series motor needs separate excitation for regeneration. This method is commonly used in electric traction, where the duty cycle requires frequent braking. A part of the energy is recovered, thus leading to economy and energy conservation.

## 3.33 PERMANENT MAGNET DC MOTORS

It is a motor whose poles are made of permanent magnets. It is essentially a shunt motor with the field circuit replaced by permanent magnets. A major application of these motors is in motor vehicles as fan motors, wind shield wiper driving motor, air conditioners, etc. A dc motor is a natural choice, because of the available dc supply from battery. The motor size is pretty small and permanent magnet offers the advantages of low cost, simple construction and high efficiency. In some situations permanent magnet motors are also used in large ratings.

The material for magnets of these motors is either alnico or ceramic (ferrite). These materials have high residual flux density and high coercive force. Alnico has a residual flux density of about 1T and is used for motors of large ratings. In these motors speed control is possible below the rated speed (by armature voltage control). Ceramic materials have a residual flux density of about 0.5T and are used for motors is the fractional horse power range.

## 3.34 LOSSES AND POWER FLOW IN DC MACHINES

### 3.34.1 Losses

In every machine only a part of the power input is converted into useful power out-put, the remainder is consumed in the losses. The losses in dc machines can be classified as copper losses, iron losses, mechanical losses and stray load losses.

(a) *Copper losses:* These losses are the $I^2R$ losses due to flow of current.

Armature copper losses $= I_a^2 R_a$, where $I_a$ is armature current and $R_a$ is armature resistance. They constitute about 30% of total full-load losses.

Field copper losses equal $I_f^2 R_f$ in the shunt machine (where $I_f$ is shunt field current and $R_f$ is field-resistance) and $I_a^2 R_{se}$ in the series machine (where $R_{se}$ is series field resistance). In a compound machine, both shunt and series field losses occur. They are about 20% of total full-load losses.

Brush contact loss occurs due to voltage drop across the brushes. It is equal to $2v_d I_a$ where $v_d$ is the voltage drop across one brush.

Loss in interpole winding = $I_a^2$ × Resistance of interpole winding.

Loss in compensating winding = $I_a^2$ × resistance of compensating winding.

(b) *Iron losses:* The iron (or core) losses are the hysteresis loss and eddy current loss. They have been discussed in Chap. 1. Since machines are usually operated at constant flux density and constant speed, these losses are almost constant. They constitute about 20% of total full-load losses.

(c) *Mechanical losses:* These are the losses associated with mechanical effects, i.e. friction and windage. Friction losses are caused by the friction at the bearings of the machine. The windage losses are caused by the friction between the moving parts of the machine and the air inside the machine casing. These losses vary as cube of the speed of rotation. Since machines are usually operated at constant speed, they are also assumed to be constant.

(d) *Stray load losses:* In addition to above losses, there are some losses which cannot be included in any of the above categories. One such component is that, due to increased flux density at some pole tips, because of armature reaction. All such losses are included in this category. In most machines stray losses are assumed to be 1% of full-load.

In a generator the input mechanical power is equal to

$$P_{in} = T_{app}\omega_m \quad \text{...(3.38)}$$

where $T_{app}$ = Applied torque N-m

$\omega_m$ = speed, rad./sec.

$$P_{\text{con}} = P_i - \text{stray losses} - \text{mechanical losses} - \text{core losses} \quad \text{...(3.39)}$$

where $P_{\text{con}}$ = Power converted into electrical power = $T_e\omega_m$

$T_e$ = induced torque, N-m

$$P_0 = P_{\text{con}} - \text{copper losses} \quad \text{...(3.40)}$$

where $P_0$ = Output electrical power = $V I_L$

In a motor the input electrical power $P_i$ is

$$P_i = V I_L$$

The power converted is $P_{\text{con}}$ and is equal to

$$P_{\text{con}} = P_i - \text{copper losses} \quad \text{...(3.41)}$$

$P_0$, the output power is given by

$$P_0 = T_L\omega_m \quad \text{...(3.42)}$$

where $T_L$ = load torque N-m

$\omega_m$ = speed, rad./sec.

$$P_0 = P_{\text{con}} - \text{core losses} - \text{mechanical losses} - \text{stray losses} \quad \text{...(3.42)}$$

## 3.35 EFFICIENCY

Efficiency is the ratio of power output to power input. In case of generators it is very difficult to measure power input. In case of motors it is difficult to measure power output. Therefore efficiency of generator and motor is calculated as:

$$\text{Efficiency of generator} = \frac{\text{Power output}}{\text{Power output} + \text{losses}} \quad \text{...(3.43)}$$

$$\text{Efficiency of motor} = \frac{\text{Power input} - \text{losses}}{\text{Power input}} \quad \text{...(3.44)}$$

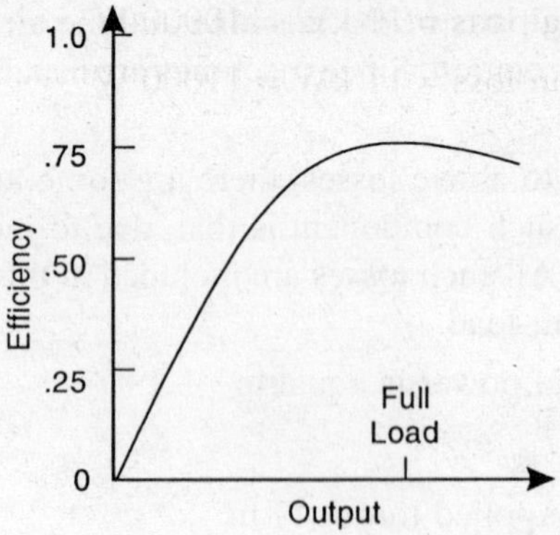


**Fig. 3.55.** Efficiency curve of a dc machine

Figure 3.55 shows a typical efficiency curve. The efficiency increases with increase in output, is the maximum at or near the rated full-load and then again shows a decreasing trend.

*Example 3.22*

An 800 kW, 500 V dc shunt generator has the following data:

$$\text{Armature resistant } R_a = 0.005 \text{ ohm}$$
$$\text{Mechanical losses} = 10 \text{ kW}$$
$$\text{Iron losses} = 11 \text{ kW}$$
$$\text{Shunt field resistance} = 50 \text{ ohm}$$
$$\text{Brush contact drop} = 1 \text{ volt per brush}$$
$$\text{Stray losses} = 1\% \text{ of output}$$

Find (a) efficiency at full-load, (b) efficiency at half-load, (c) input torque and induced torque at full-load, if speed is 1500 rpm.

*Solution*

(a)
$$\text{Output} = 800 \text{ kW} = 800000 \text{ W}$$

$$\text{Load current } I_L = \frac{800000}{500} = 1600 \text{ A}$$

$$\text{Field current } I_f = \frac{500}{50} = 10 \text{ A}$$

$$\text{Armature current } I_a = I_L + I_f = 1610 \text{ A}$$

$$\text{Armature copper-loss} = I_a^2 R_a$$

$$= (1610)^2\,(0.005) = 12960.5 \text{ W}$$

$$\text{Field copper-loss} = I_f^2 R_f$$

$$= (10)^2\,(50) = 5000 \text{ W}$$

$$\text{Brush contact loss} = 2 \times 1610 = 3220 \text{ W}$$

$$\text{Mechanical loss} = 10 \text{ kW} = 10000 \text{ W}$$

$$\text{Iron-loss} = 11 \text{ kW} = 11000 \text{ W}$$

$$\text{Stray losses} = 1\% \text{ of } 800000 = 8000 \text{ W}$$

$$\text{Efficiency} = \frac{\text{ouput}}{\text{output + losses}}$$

$$= \frac{800000}{[800000 + 12960.5 + 5000 + 3220 + 10000 + 11000 + 8000]}$$

$$= 0.94 \text{ or } 94\%$$

(b)

$$\text{Output} = 400 \text{ kW} = 400000 \text{ W}$$

$$I_L = \frac{400000}{500} = 800 \text{ A}$$

$$I_f = 10 \text{ A}$$

$$I_a = 800 + 10 = 810 \text{ A}$$

$$\text{Armature copper-losses} = (810)^2\,(0.005) = 3280.5 \text{ W}$$

$$\text{Field copper-losses} = 5000 \text{ W}$$

$$\text{Brush contact losses} = 2 \times 810 = 1620 \text{ W}$$

$$\text{Mechanical losses} = 10000 \text{ W}$$

$$\text{Iron-losses} = 11000 \text{ W}$$

$$\text{Stray losses} = 1\% \text{ of } 400000 = 4000 \text{ W}$$

$$\text{Efficiency} = \frac{400000}{[400000 + 3280.5 + 5000 + 1620 + 10000 + 11000 + 4000]}$$

$$= 0.9197 \text{ or } 91.97\%$$

(c)

$$\omega_m = \frac{1500 \times 2\pi}{60} = 157.08 \text{ rad/sec.}$$

$$\text{Input torque} = \frac{\text{Power Input}}{157.08}$$

$$= \frac{[800000 + 12960.5 + 5000 + 3220 + 10000 + 11000 + 8000]}{157.08}$$

$$= 5412.4 \text{ N-m}$$

$$P_{con} = 800000 + 12960.5 + 5000 + 3220 = 821180.5 \text{ W}$$

$$\text{Induced torque} = \frac{821180.5}{157.08} = 5227.78 \text{ N-m}$$

## 3.36 BRAKE TEST

The efficiency of a small motor can be measured directly by a brake test. The motor is loaded directly by a belt or rope brake placed on a water cooled pulley mounted on the shaft of the motor. The desired load is put by tightening the rope or belt. An eddy current brake or calibrated fan can also be used. The terminal voltage $V$ and line current $I_L$ are measured.

Let $W_1$ = tension on tight side of brake, kg

$W_2$ = tension on slack side of brake, kg

$N$ = speed, rpm

$r$ = radius of pulley, m

$$\text{Output torque} = (W_1 - W_2)\, r \times 9.81 \text{ N-m}$$

$$\text{Power output} = \frac{(W_1 - W_2) r \times 9.81 \times 2\pi N}{60} \text{ W} \qquad ...(3.45)$$

$$= \frac{(2\pi N)(9.81 r)(W_1 - W_2)}{60 \times 735.5} \text{ hp} \qquad ...(3.46)$$

$$\text{Input} = VI_L$$

$$\text{Efficiency} = \frac{(2\pi N)(9.81 r)(W_1 - W_2)}{60\, VI_L} \qquad ...(3.47)$$

*Example 3.23*

The brake test on a dc shunt motor gave the following results: Tensions 3.2 kg and 0.3 kg, radius of pulley 10 cm, speed 1500 rpm, $V$ = 230 V, $I_L$ = 2.8 A. Find output torque, output power, horse power and efficiency.

*Solution*

$$\text{Torque} = (3.2 - 0.3)\left(\frac{10}{100}\right)(9.81) = 2.845 \text{ N-m}$$

$$\text{Power output} = \frac{2\pi(1500)(2.845)}{60} = 446.9 \text{ W}$$

$$\text{Horse power} = \frac{446.9}{735.5} = 0.608 \text{ hp}$$

$$\text{Efficiency} = \frac{446.9}{230 \times 2.8} = 0.694 \text{ or } 69.4\%.$$

## 3.37 SWINBURNE TEST

This is a no-load test to determine the losses of the machine. The machine is run as a motor on no-load. The input voltage and current are measured. The armature resistance and field resistance are measured using a battery, voltmeter and ammeter. These resistances obvioüsly, pertain to the ambient temperature. The probable values of resistances, under operating conditions are determined by assuming a suitable value of temperature rise (say 40° C). The stray load losses are assumed as 1% of full-load output.

Let $V$ = supply voltage

$I_0$ = Input line current at no-load

$R'_a$, $R'_f$ = measure values of armature and field resistances, at ambient temperature

$R_a$, $R_f$ = armature and field resistances at operating temperature

Input at no-load = $VI_0$

$$\text{Field copper loss at no-load} = \frac{V^2}{R'_f}$$

$$\text{Armature copper-loss at no-load} = \left(I_0 - \frac{V}{R'_f}\right)^2 R'_a$$

$$\text{Mechanical and Iron losses} = VI_0 - \frac{V^2}{R'_f} - \left(I_0 - \frac{V}{R'_f}\right)^2 R'_a$$

Armature copper-loss at load for armature current of $I_a = I_a^2 R_a$

$$\text{Field copper-loss at load} = \frac{V^2}{R_f}$$

Total losses at load = Mechanical and Iron-losses + Armature copper-loss + Field copper-loss + Stray-losses

It is not desirable to run a series motor on no-load. Therefore this test should not be conducted on a series machine.

*Example 3.24*

A dc shunt motor gave the following results of Swinburne test:

Voltage 600 V, current 8 A,

armature resistance 0.1 ohm,

field resistance 300 ohm

Find efficiency of the machine, (a) as a generator delivering 100 A at 600 V, (b) as a motor having a line current of 100 A at 600 V. Neglect increase in resistance due to temperature rise. Assume stray-losses as 1% of output.

*Solution*

$$\text{Input at no-load} = 600 \times 8 = 4800 \text{ W}$$

$$I_f = \frac{600}{300} = 2 \text{ A}$$

$$\text{Field copper-loss } I^2R = (2)^2(300) = 1200 \text{ W}$$

$$\text{Armature copper-loss at no-load} = (8 - 2)^2 (0.1) = 3.6 \text{ W}$$

(a)

$$\text{Output} = 600 \times 100 = 60000 \text{ W}$$

$$\text{Armature current} = 100 + 2 = 102 \text{ A}$$

$$\text{Armature copper-loss} = (102)^2 (0.1) = 1040.4 \text{ W}$$

$$\text{Stay-losses} = \frac{1}{100} \times 60000 = 600 \text{ W}$$

$$\begin{aligned}\text{Total losses} &= 1040.4 + 600 + 3596.4 + 1200 \\ &= 6436.8 \text{ W}\end{aligned}$$

$$\text{Efficiency} = \frac{60000}{60000 + 6436.8} = 0.903 \text{ or } 90.3\%$$

(b)

$$\text{Input} = 600 \times 100 = 60000 \text{ W}$$

$$\text{Armature current} = 100 - 2 = 98 \text{ A}$$

$$\text{Armature copper-loss} = (98)^2 (0.1) = 960.4 \text{ W}$$

$$\begin{aligned}\text{Output} &= 60000 - 960.4 - 1200 - 3596.4 \\ &= 54243.2 \text{ W}\end{aligned}$$

$$\text{Stray losses} = \frac{1}{100} \times 54243.2 = 542.4 \text{ W}$$

$$\text{Net output} = 54243.2 - 542.4 = 53700.8 \text{ W}$$

$$\text{Efficiency} = \frac{53700.8}{60000} = 0.895 \text{ or } 89.5\%$$

## 3.38 HOPKINSON TEST

### 3.38.1 Circuit

Figure 3.56 shows the circuit diagram of this test. It is a regenerative test and requires two identical dc shunt machines. As seen in Fig. 3.56 the two machines are mechanically coupled. The field currents of the two machines are so adjusted that one machine works as a motor and drives the other machine, which works as a generator. The power input from the mains is only that needed for supplying the losses of the two machines. The two machines can be tested under full-load conditions (for determining efficiency and maximum temperature rise).

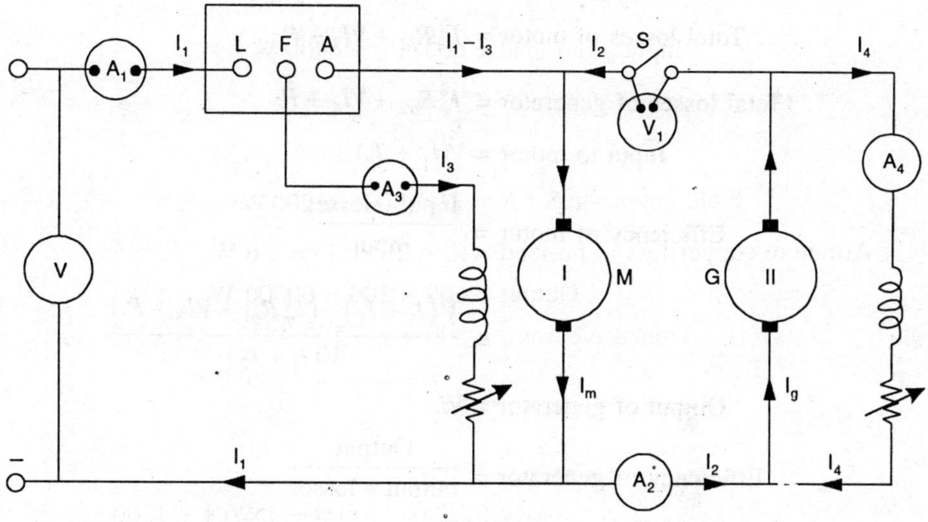


**Fig. 3.56.** Connections for Hopkinson test on dc shunt machines

### 3.38.2 Test Procedure

The machine I is started as a motor with the help of starter and its field regulator is adjusted, so that its speed is equal to the rated value. Machine I will drive machine II. The field current of machine II is adjusted till the voltmeter $V_1$ reads zero. Then switch $S$ is closed. At this position machine II is floating (i.e. it is neither taking any input from mains nor delivering any output to the mains). Any desired load can be put on the set by adjustment of field regulators. The machine with lower field current will act as motor and the other will work as generator. The power input from the mains is only the power required to supply the losses of the two machines. At the end of the test, the resistance $R_{a1}$ and $R_{a2}$ of the armature circuit (i.e. armature winding + brush contact + interpole winding) are measured by passing full-load current through them (with the machines at standstill) and measuring the voltage drops with field circuits open. The efficiency can be calculated as under:

$$\text{Total input} = VI_1$$

$$\text{Armature current of generator (machine II)} = I_g = I_2 + I_4$$

$$\text{Armature current of motor (machine I)} = I_m = I_1 - I_3 + I_2$$

$$\text{Armature copper-loss of machine I} = I_m^2 R_{a1}$$

$$\text{Armature copper loss of machine II} = I_g^2 R_{a2}$$

$$\text{Field copper-loss of machine I} = VI_3$$

$$\text{Field copper-loss of machine II} = VI_4$$

Total iron and mechanical losses of the two machines

$$= VI_1 - I_m^2 R_{a1} - I_g^2 R_{a2} - VI_3 - VI_4$$

$$\text{Iron and mechanical losses of each machine} = P_c$$

$$= 0.5\left(VI_1 - I_m^2 R_{a1} - I_g^2 R_{a2} - VI_3 - VI_4\right)$$

$$\text{Total losses of motor} = I_m^2 R_{a1} + VI_3 + P_c$$

$$\text{Total losses of generator} = I_g^2 R_{a2} + VI_4 + P_c$$

$$\text{Input to motor} = V(I_1 + I_2)$$

$$\text{Efficiency of motor} = \frac{\text{Input} - \text{losses}}{\text{Input}}$$

$$= \frac{V(I_1 + I_2) - \left(I_m^2 R_{a1} + VI_3 + P_c\right)}{V(I_1 + I_2)} \quad \text{...(3.48)}$$

$$\text{Output of generator} = VI_2$$

$$\text{Efficiency of generator} = \frac{\text{Output}}{\text{output} + \text{losses}}$$

$$= \frac{VI_2}{VI_2 + I_g^2 R_{a2} + VI_4 + P_c} \quad \text{...(3.49)}$$

### 3.38.3 Advantages

(1) The two machines are tested under full-load conditions.
(2) The total power input is only that required to supply the losses.
(3) It can be used to determine efficiency as well as temperature rise.
(4) There is no need to arrange for the load. This is especially important for large machines.

### 3.38.4 Disadvantages

(1) The iron and mechanical losses are assumed equal for the two machines. This may not be correct, especially for small machines.
(2) The test requires two identical machines.

*Example 3.25*

The Hopkinson test on two identical shunt machines gave the following results:

Input voltage 500V
Input current 15A
Output current of generator 120A
Field current of generator 4A
Field current of motor 3A
Armature resistance of each machine 0.06 ohm
Find efficiency of motor and generator.

*Solution*

Total input = 500 × 15 = 7500 W

Armature current of generator = 120 + 4 = 124 A

Armature copper-loss of generator = $124^2 \times 0.06 = 922.56$ W

Armature current of motor = 120 + 15 − 3 = 132 A

Armature copper-loss of motor = $132^2 \times 0.06 = 1045.44$ W

Field copper-loss of generator = 500 × 4 = 2000 W

Field copper-loss of motor = 500 × 3 = 1500 W

Total iron- and mechanical-losses = 7500 − 922.56 − 1045.44 − 2000 − 1500
= 2032 W

$P_c$ = Iron- and mechanical-losses of each machine = 2032/2 = 1016 W

Generator output = 500 × 120 = 60000 W

Generator losses = 1016 + 922.56 + 2000
= 3938.56 W

Generator efficiency $= \dfrac{60000}{60000 + 3938.56} = 0.938$ or 93.8%

Motor Input = 500(120 + 15) = 67500 W

Motor losses = 1016 + 1045.44 + 1500 = 3561.44 W

Motor efficiency $= \dfrac{67500 - 3561.44}{67500}$
= 0.947 or 94.7%.

## CROSS-FIELD MACHINES

### 3.39 METADYNE

**The dc machines considered so far have brushes only on the quadrature axis. Cross field machines are special dc machines having an additional set of brushes on the *d*-axis. By this arrangement, the armature mmf can be used to provide most of the excitation and high gain can be obtained. One such machine is metadyne.**

**A dc generator can be converted into a metadyne by providing an additional pair of brushes on *d*-axis as shown in Fig. 3.57. The quadrature axis brushes are short circuited and the output is obtained from *d*-axis brushes. The stator has a control field winding, through which a current $i_f$ flows. When the machine is rotating at a constant speed, the flow of $i_f$ generates an emf $e_{aq}$ between the quadrature axis brushes *qq*′. This emf is**

$$e_{aq} = k_{qf} i_f \qquad ...(3.50)$$

**where $k_{qf}$ is a constant.**

**Since brushes *qq*′ are short circuited, a quadrature axis armature current flows. Because of the low resistance in the short circuited path, only a small field current $i_f$ is required for a much larger quadrature**

axis armature current and the associated flux density wave centered on the $q$-axis. Due to commutator action this armature field ($F_q$) is stationary in space.

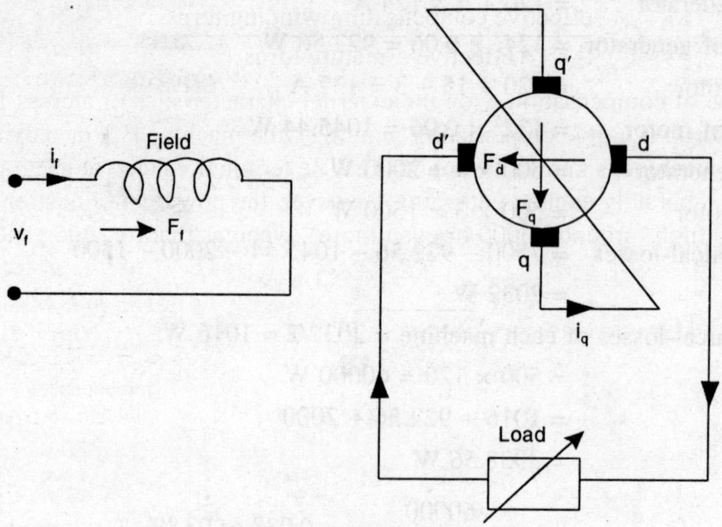


**Fig. 3.57.** Metadyne

The rotation of machine in the $q$-axis flux generates an emf, which appears across direct axis brushes $dd'$. This emf $e_{ad}$ is

$$e_{ad} = k_{dq} i_q \qquad \text{...(3.51)}$$

where $i_q$ is quadrature axis current and $k_{dq}$ is a constant if saturation is neglected and constant speed is assumed. If a load is connected across brushes $dd'$, a current $i_d$ flows. This current produces an mmf ($F_d$), which opposes the control field mmf. Each stage of voltage generation produces a current whose magnetic field is 90° ahead of the flux wave producing the voltage. Because of the two stages of generation, the mmf of $i_d$ is shifted by 90° twice and thus opposes the control field mmf. Therefore the voltage $e_{aq}$ now becomes

$$e_{aq} = k_{qf} i_f - k_{qd} i_d \qquad \text{...(3.52)}$$

where $k_{qd}$ is also a constant if magnetic saturation is neglected and speed is constant.

The above analysis shows that a metadyne is a two-stage amplifier with a negative feedback from the final output to input. Any increase in $i_d$ causes a reduction in $e_{aq}$ [Eq. (3.52)], which results in decrease in current $i_q$. This means a reduction in voltage $e_{ad}$ [(Eq. (3.51)] and a decrease in current $i_d$. Thus for a fixed value of control field current $i_f$, the output current $i_d$ remains substantially constant and metadyne behaves as a constant current generator.

## 3.40 AMPLIDYNE

### 3.40.1 Characteristics

The $d$-axis mmf which opposes the control field mmf in a metadyne can be cancelled by a $d$-axis compensating winding as shown in Fig. 3.58. This compensating winding carries the current $i_d$ and produces a flux which opposes the flux produced by $d$-axis armature current. The degree of compensation $C$ is defined as

$$C = \frac{\text{effective compensating winding turns}}{\text{effective armature turns}} \qquad ...(3.53)$$

The effect of degree of compensation $C$ on the external characteristics of a cross-field machine is shown in Fig. 3.59. If $C = 0$ (i.e. no compensating winding), the machine is a metadyne. If $C = 1$ (i.e. full compensation), the machine is known as amplidyne. The terminal voltage of an amplidyne remains almost constant as in a separately excited generator. However, the power amplification (i.e. $e_{ad}\, i_d / v_f\, i_f$) of an amplidyne is very high (around 20000 or even more), whereas that for a dc generator it may be only about 100.

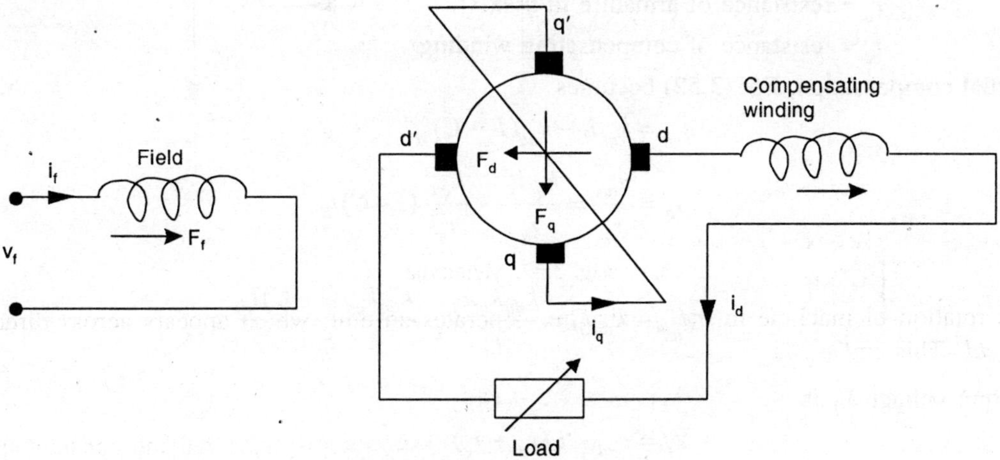


**Fig. 3.58.** Amplidyne

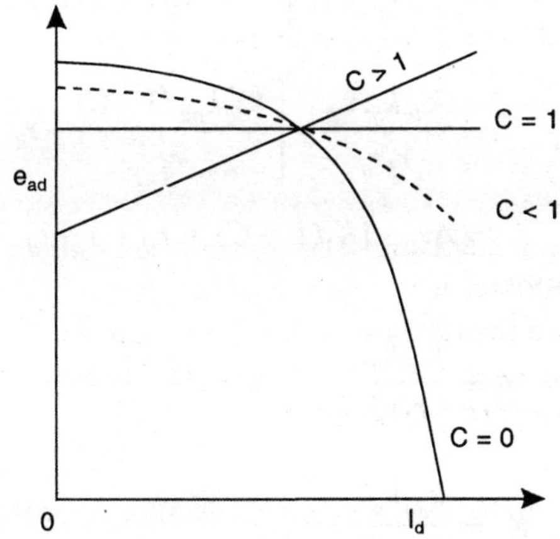


**Fig. 3.59.** Effect of degree of compensation on characteristics of dc cross-field machine

To improve the performance of an amplidyne, a series field winding is added on the $q$-axis and connected in series with $q$-axis current. This winding improves $q$-axis commutation.

An amplidyne can be used for power amplification in feedback control systems, e.g. as a voltage regulator in the excitation system of a large alternator to add a boost or buck voltage in series with the field winding of the main exciter. It can also be used as the generator in Ward-Leonard speed control system and position control system. However, in modern control systems thyristors are preferred.

### 3.40.2 Voltage Equation

Let $R_f$ = resistance of field winding

$v_f$ = voltage applied to field winding

$C$ = degree of compensation

$r_d$ = resistance of armature in $d$-axis

$r_q$ = resistance of armature in $q$-axis

$r_c$ = resistance of compensating winding

For partial compensation, Eq. (3.52) becomes

$$e_{aq} = k_{qf} i_f - k_{qd}(1 - C) i_d \quad ...(3.54)$$

$$i_q = \frac{e_{aq}}{r_q} = \frac{k_{qf} i_f}{r_q} - \frac{k_{qd}}{r_q}(1 - C) i_d \quad ...(3.55)$$

$$e_{ad} = k_{dq} i_q = \frac{k_{dq} k_{qf} i_f}{r_q} - \frac{k_{dq} k_{qd}(1 - C) i_d}{r_q} \quad ...(3.56)$$

The output voltage $V_L$ is

$$V_L = e_{ad} - i_d(r_d + r_c)$$

$$= \frac{k_{dq} k_{qf} i_f}{r_q} - \left[\frac{k_{dq} k_{qd}(1 - C)}{r_q} + r_d + r_c\right] i_d$$

$$= \frac{k_{dq} k_{qf} v_f}{r_f r_q} - \left[\frac{k_{dq} k_{qd}(1 - C)}{r_q} + r_d + r_c\right] i_d$$

$$= A v_f - [K_1(1 - C) + r_d + r_c]\, i_d \quad ...(3.57)$$

where $A$ is the voltage amplification

$$A = \frac{k_{dq} k_{qf}}{r_f r_q}$$

and constant

$$K_1 = \frac{k_{dq} k_{qd}}{r_q}$$

When output terminals are short circuited, $V_L$ is zero. The ratio $i_d/i_f$ under this condition is known as current amplification. From Eq. (3.57), when $V_L = 0$

$$A v_f = [K_1(1 - C) + r_d + r_c]\, i_d$$

or $$Ai_f r_f = [K_1(1 - C) + r_d + r_c]\, i_d$$

or $$\frac{i_d}{i_f} = \frac{Ar_f}{K_1(1-C)+r_d+r_c} \qquad ...(3.58)$$

## DC MACHINE DYNAMICS

### 3.41 SEPARATELY EXCITED MOTOR

Figure 3.60 shows a separately excited dc motor. The dynamic behaviour can be predicted if a transfer function relating angular speed $\omega_m$ with applied voltage $V$ is found. The analysis is done under the following assumptions:

(1) The axis of armature mmf is fixed in space along the $q$-axis.

(2) The demagnetisation due to armature reaction is neglected.

(3) The magnetic circuit is unsaturated. Thus the armature and field inductances can be assumed constant.

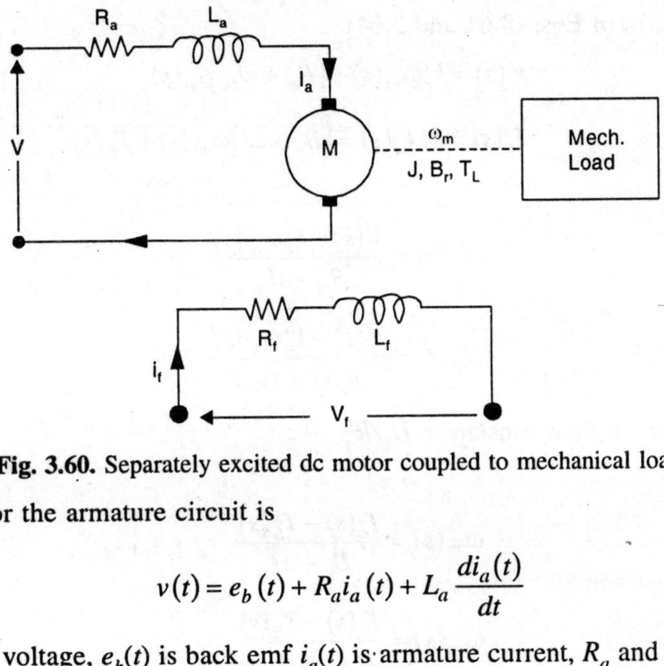


**Fig. 3.60.** Separately excited dc motor coupled to mechanical load

KVL equation for the armature circuit is

$$v(t) = e_b(t) + R_a i_a(t) + L_a \frac{di_a(t)}{dt} \qquad ...(3.59)$$

where $v(t)$ is applied voltage, $e_b(t)$ is back emf $i_a(t)$ is armature current, $R_a$ and $L_a$ are armature resistance and inductance. Since we are interested in dynamic behaviour, $v$, $e_b$ and $i_a$ are functions of time.

The back emf is

$$e_b(t) = k_e \phi N = k_e' \omega_m(t) \qquad ...(3.60)$$

Since $i_f$ is constant, $\phi$ is constant. Therefore the back emf is directly proportional to angular speed $\omega_m$. Combining Eqs. (3.59) and (3.60), we get

$$v(t) = k_e' \omega_m(t) + R_a i_a(t) + L_a \frac{di_a(t)}{dt} \qquad ...(3.61)$$

At the motor shaft the electromagnetic torque $T_e$ must be equal to the sum of opposing torques. Thus

$$T_e(t) = J\frac{d}{dt}\omega_m(t) + B_r\omega_m(t) + T_L(t) \quad ...(3.62)$$

where $T_e(t)$ = electromagnetic torque, N-m
$J$ = moment of inertia of load and motor
$T_L$ = Load torque, N-m
$B_r$ = rotational viscous friction coefficient (or mechanical rotational resistance) of motor and load, N-m/rad/sec.

Since flux is constant, electromagnetic torque is proportional to armature current $i_a$

Thus
$$T_e(t) = k_t' i_a(t) \quad ...(3.63)$$

From Eqs. (3.62) and (3.63)

$$T_e(t) = k_t' i_a(t) = J\frac{d}{dt}\omega_m(t) + B_r\omega_m(t) + T_L(t) \quad ...(3.64)$$

Taking Laplace transform of Eqs. (3.61 and 3.64)

$$V(s) = k_e'\omega_m(s) + [R_a + sL_a]I_a(s) \quad ...(3.65)$$

$$T_e(s) = k_t' I_a(s) = [B_r + sJ]\omega_m(s) + T_L(s) \quad ...(3.66)$$

Eq. (3.65) gives

$$I_a(s) = \frac{V(s) - k_e'\omega_m(s)}{R_a + sL_a}$$

or

$$I_a(s) = \frac{V(s) - k_e'\omega_m(s)}{R_a(1 + s\tau_a)} \quad ...(3.67)$$

where $\tau_a$ = armature circuit time constant = $L_a/R_a$.

Equation (3.66) gives

$$\omega_m(s) = \frac{T_e(s) - T_L(s)}{B_r + sJ}$$

or

$$\omega_m(s) = \frac{T_e(s) - T_L(s)}{B_r(1 + s\tau_m)} \quad ...(3.68)$$

where $\tau_m = J/B_r$ = mechanical time constant.

Equations (3.63), (3.67) and (3.68) describe the dynamic behaviour of the motor. The block diagram corresponding to these equations is shown in Fig. 3.61.

## 3.42 SEPARATELY EXCITED GENERATOR

A dc generator is a power amplifier. A small amount of power in the field circuit (= $v_f i_f$) can control a large power in the armature circuit (= $E_a I_a$). The magnitude of this amplification is about 100 or so.

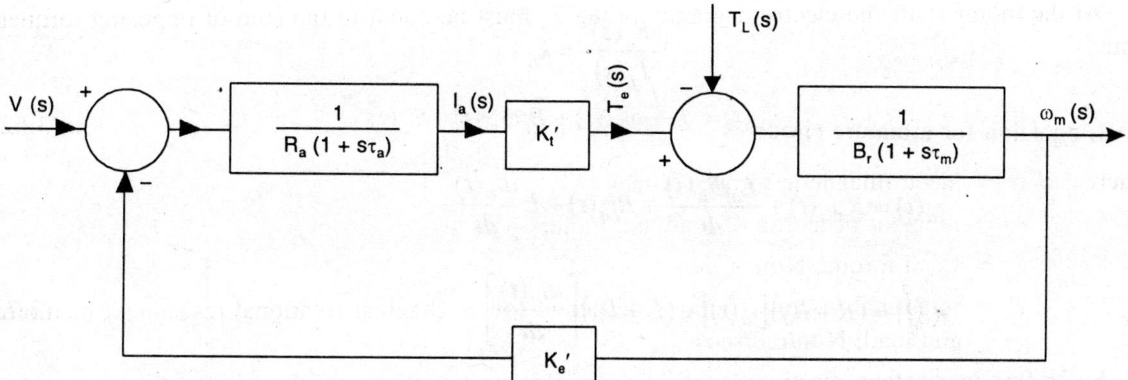


**Fig. 3.61.** Block diagram of a dc separately excited motor

Figure 3.62 shows a separately excited generator. $R_f$ and $L_f$ are the resistance and inductance of field circuit, $R_a$ and $L_a$ of the armature circuit and $R$ and $L$ of the load. $i_f$ is the field current and $i_a$ the armature current. The field circuit equation is

$$v_f(t) = i_f(t)R_f + L_f \frac{di_f(t)}{dt} \qquad \text{...(3.69)}$$

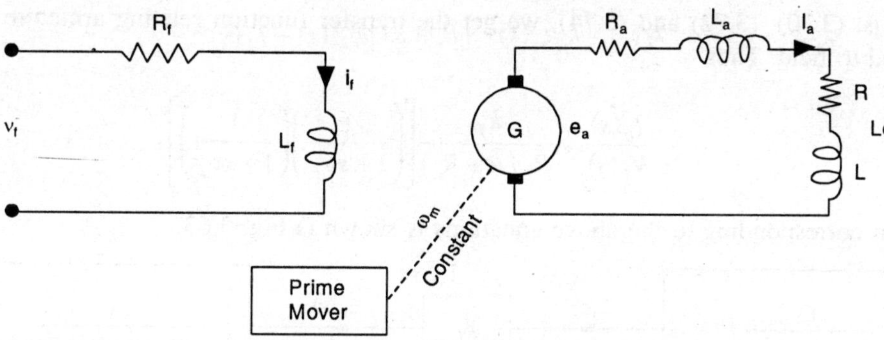


**Fig. 3.62.** Separately excited dc generator

Taking Laplace transform

$$V_f(s) = I_f(s)[R_f + sL_f]$$

or

$$I_f s = \frac{V_f(s)}{R_f(1 + s\tau_f)} \qquad \text{...(3.70)}$$

where $\tau_f = L_f/R_f$ = field circuit time constant.

The induced emf is $E_a$ and can be calculated as follows:

$$e_a(t) = k_f i_f(t) \qquad \text{...(3.71)}$$

Taking Laplace transform

$$E_a(s) = k_f I_f(s)$$

or
$$\frac{E_a(s)}{I_f(s)} = k_f \qquad \text{...(3.72)}$$

KVL equation for armature circuit is

$$e_a(t) = R_a i_a(t) + \frac{L_a di_a(t)}{dt} + Ri_a(t) + L\frac{di_a(t)}{dt}$$

or
$$e_a(t) = (R + R_a)[i_a(t)] + (L + L_a)\left[\frac{di_a(t)}{dt}\right] \qquad \text{...(3.73)}$$

Taking Laplace transform

$$E_a(s) = I_a(s)[R + R_a] + sI_a(s)[L + L_a]$$

or
$$\frac{I_a(s)}{E_a(s)} = \frac{1}{(R + R_a)(1 + s\tau_a)} \qquad \text{...(3.74)}$$

where $\tau_a = \dfrac{L + L_a}{R + R_a}$ = armature circuit time constant.

Combining Eqs. (3.70), (3.72) and (3.74), we get the transfer function relating armature current to voltage applied to field. Thus

$$\frac{I_a(s)}{V_f(s)} = \frac{k_f}{R_f(R + R_a)}\left[\left(\frac{1}{1 + s\tau_f}\right)\left(\frac{1}{1 + s\tau_a}\right)\right] \qquad \text{...(3.75)}$$

Block diagram corresponding to the above equations is shown is Fig. 3.63.

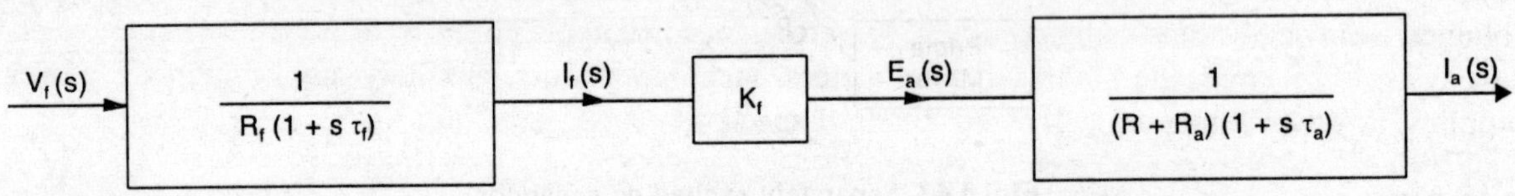


**Fig. 3.63.** Block diagram of separately excited dc generator

*Example 3.26*

A separately excited dc generator has field winding resistance and inductance of 25 ohm and 5H. The induced emf is 100 V per field ampere. The armature winding resistance and inductance are 0.2 ohm and 0.4 H and those of load, 4.8 ohm and 2.1 H. The machine is running at constant speed. Find an expression for armature current when a 100 V supply is connected to field.

*Solution*

In Eq. (3.70) we have

$$k_f = 100, \quad V_f(s) = \frac{100}{s}$$

$$R_f = 25 \text{ ohm}, \quad R + R_a = 5 \text{ ohm}$$

$$\tau_f = \frac{L_f}{R_f} = \frac{5}{25} = 0.2 \text{ sec}$$

$$\tau_a = \frac{(L + L_a)}{(R + R_a)} = 0.5 \text{sec}$$

Substituting the values in Eq. (3.75)

$$I_a(s) = \left(\frac{100}{s}\right)\left(\frac{100}{25 \times 5}\right)\left[\frac{1}{(1 + 0.2s)(1 + 0.5s)}\right]$$

$$= \frac{800}{s(s+5)(s+2)} = \frac{80}{s} + \frac{53.33}{s+5} - \frac{133.33}{s+2}$$

Taking Laplace transform

$$i_a(t) = 80 + 53.33e^{-5t} - 133.33e^{-2t} \text{ A.}$$

## 3.43 APPLICATIONS OF DC MACHINES

### 3.43.1 DC Generators

Whenever mechanical energy is to be converted into electric energy in dc form, a dc generator is needed. The initial application of electricity started with the use of direct current. The first power station in the world was a 110 V dc station at New York (USA), set up in 1882. However after the invention of transformer and 3-phase induction motors, all power systems became ac systems. Nowadays whenever dc is required (for HV DC transmission, variable speed drives, electrolysis, etc.) it is obtained from ac systems using rectifiers. Nevertheless some applications of dc generators are: excitation systems of large alternators, dynamometers, tacho-generators, auxiliary and emergency power supplies, as control generator in Ward-Leonard speed control system, etc.

### 3.43.2 DC Shunt Motor

This motor has medium starting torque, a speed regulation 5-10%, and excellent adaptability to adjustable speed service. It can be used for constant speed drives like pumps, fans, blowers, etc. However for all these applications, a 3-phase induction motor is preferable, because it is cheaper, more rugged and requires lesser maintenance than a dc motor. However a dc shunt motor is cheaper for low speed drives. Moreover, whenever a wide range of speed control is required, dc shunt motor is used. The main application of dc shunt motors is in steel and aluminium rolling mills and Ward-Leonard speed control system.

### 3.43.3 DC Series Motor

This motor has a very high starting torque. Moreover its speed decreases with increase in load torque, so that its power input remains more or less constant. It is best suited for hoists, cranes and traction.

Speed control is generally obtained through armature control. They are also used for battery powered vehicles.

### 3.43.4 Compound Motors

A differential compound motor is rarely used as discussed in sec. 3.26. The characteristics of a cumulative compound motor depend on the relative strengths of series and shunt fields. It has good starting torque and drooping speed-load characteristics. They can be used for pulsating loads needing flywheel action, viz. shears, conveyors, crushers, hoists, plunger pumps, etc. Whenever the supply voltage is likely to vary considerably, a compound motor is preferred. Because of its inductance and contribution to back emf, the series field helps in reducing the fluctuations of current drawn from mains.

*Example 3.27*

The magnetisation curve for a dc shunt generator is as under at a speed of 1500 rpm:

| $I_f$ | 0 | 0.4 | 0.8 | 1.2 | 1.6 | 2.0 | 2.4 | 2.8 | 3.0 |
|---|---|---|---|---|---|---|---|---|---|
| $E_0$ | 6 | 60 | 120 | 175.5 | 202.5 | 221 | 231 | 237 | 240 |

(a) If $R_f$ is 100 Ω, find open circuit voltage.

(b) Find critical field resistance for this speed.

(c) Plot a magnetisation curve for $N = 1200$ rpm and find open circuit voltage if $R_f = 100$ Ω.

*Solution*

(a) The magnetisation curve at 1500 rpm is shown in Fig. 3.64 (curve 1). *OP* represents field resistance of 100 ohm. The intersection of magnetisation curve and line OP is at point *A*, which gives the open circuit voltage of 235 V.

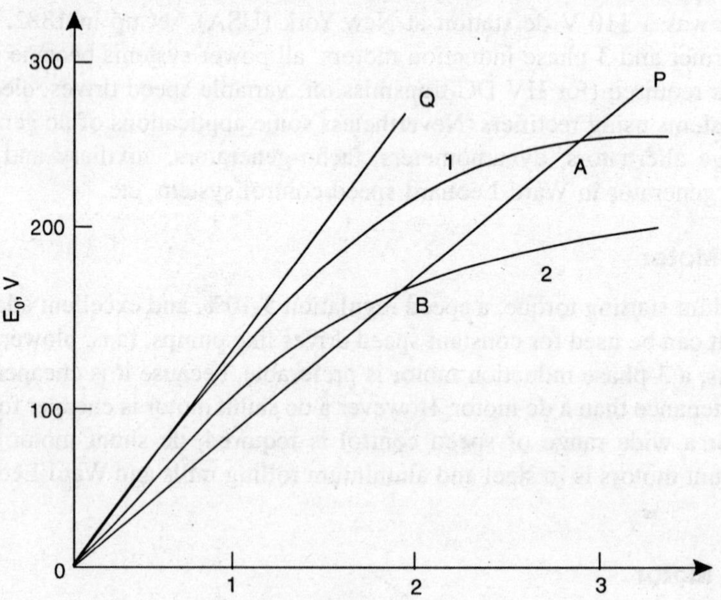


Fig. 3.64

(b) Line $OQ$ is drawn tangent to curve 1. The slope of this line gives critical resistance for $N = 1500$ rpm. The value of critical resistance is 160 ohm.

(c) For 1200 rpm, the values of $E_0$, the different values of $I_f$, are 12/15 times the values at 1500 rpm. Thus magnetisation curve at 1200 rpm is

| $I_f$ | 0 | 0.4 | 0.8 | 1.2 | 1.6 | 2 | 2.4 | 2.8 | 3.0 |
|---|---|---|---|---|---|---|---|---|---|
| $E_0$ | 4.8 | 48 | 96 | 138 | 162 | 176.8 | 184.8 | 189.6 | 192 |

This is shown as curve 2 in Fig. 3.64. Line OP (100 ohm resistance line) intersects curve 2 at point $B$, which gives open circuit voltage of 178 V.

### Example 3.28

A 220 V shunt motor has $r_a = 0.2\ \Omega$, $R_f = 110\ \Omega$, no-load current is 5A and no-load speed is 1500 rpm. If full-load current is 52A, find $N$ and $T$ at full-load. Assume that mechanical losses are the same at no-load and full-load. [*K.U. 2001*]

#### Solution

Field current $= 220/110 = 2\text{A}$

Armature current at no-load $= 5 - 2 = 3\text{A}$

Back emf at no-load $= 220 - 3 \times 0.2 = 219.4$ V

Armature current at full-load $= 52 - 2 = 50$ A

Back emf at full-load $= 220 - 50 \times 0.2 = 210$ V

$$\frac{\text{Back emf at full-load}}{\text{Back emf at no-load}} = \frac{\text{Speed at full-load}}{\text{Speed at no-load}}$$

Therefore speed at full-load $= \dfrac{210}{219.4} \times 1500 = 1435.73$ rpm

Mechanical losses $= 220 \times 52 - 2^2 \times 110 - 3^2 \times 0.2 = 658.2$ W

Motor output $= 220 \times 52 - 2^2 \times 110 - 50^2 \times 0.2 - 658.2$

$= 9841.8$ W

$$\frac{2\pi \times 1435.73 \times T}{60} = 9841.8$$

or $$T = 65.46 \text{ N-m.}$$

### Example 3.29

A shunt generator is to be converted into level compounded generator. Shunt field current is 3.1 A, to give 400 V at no-load and 4.8 A to give 400 V when output current is 200 A. Shunt winding has 1200 turns per pole. Find number of series field turns.

#### Solution

Additional ampere turns needed to maintain constant voltage $= (4.8 - 3.1)\ 1200 = 2040$ AT

Assuming short shunt connection, current through series field $= 200$ A

$$\text{Number of series field turns} = \frac{2040}{200} = 10.2$$
$$= 10 \text{ turns (say).}$$

*Example 3.30*

**A 400 V dc shunt motor takes 3.5 A on no-load and 59.5 A on full-load. $R_f$ = 267 ohm, $R_a$ = 0.2 ohm. Brush drop is 2V. If armature reaction weakens the field by 2%, find percentage change in speed from no-load to full-load.** [*P.U. 2002*]

*Solution*

$$\text{Field current} = \frac{400}{267} = 1.5\,\text{A}$$
$$I_a \text{ at no-load} = 3.5 - 1.5 = 2 \text{ A}$$
$$I_a \text{ at full-load} = 59.5 - 1.5 = 58 \text{ A}$$
$$\text{Back emf at no-load} = 400 - 2 \times 0.2 - 2 = 397.6 \text{ V}$$
$$\text{Back emf at full-load} = 400 - 58 \times 0.2 - 2 = 386.4 \text{ V}$$
$$\text{Flux at full-load} = 1 - \frac{2}{100} = 0.98 \text{ of flux at no-load}$$
$$\text{Back emf} \propto \text{flux} \times \text{speed}$$

or
$$\frac{\text{Back emf at full-load}}{\text{Back emf at no-load}} = \frac{\text{Flux at full-load}}{\text{Full at no-load}} \times \frac{\text{Speed at full-load}}{\text{Speed at no-load}}$$

$\therefore$
$$\frac{\text{Speed at full-load}}{\text{Speed at no-load}} = \frac{386.4}{397.6} \times \frac{1}{0.98} = 0.9917$$

Percentage change in speed from no-load to full-load = (1 – 0.9917) 100 = 0.83%.

*Example 3.31*

**What resistance must be placed in series with armature of the machine in example 3.30 to reduce full-load speed to 50% with gross torque remaining the same. Assume no change in flux.**

*Solution*

Torque $\propto$ flux × armature current

Since torque is the same, armature current is also the same, i.e. 58 A

Since flux is constant,

Back emf $\propto$ speed

$$\frac{400 - 58(0.2 + R) - 2}{400 - 58 \times 0.2 - 2} = 0.5$$

or
$$R = 3.33 \text{ ohm.}$$

*Example 3.32*

The OCC of a shunt generator running at 1000 rpm is as under:

| $I_f$ | 0 | 2 | 4 | 6 | 8 | A |
|---|---|---|---|---|---|---|
| $E_0$ | 10 | 90 | 170 | 214 | 240 | V |

(a) Find critical field resistance when machine is operated at 1000 rpm.

(b) Machine is run at 1500 rpm with the above value of field resistance. Find $E_0$.

(c) Find critical speed if field resistance is 60 Ω.

(d) The machine is run at 1500 rpm. Field resistance is as calculated in part (a). The armature current is 400 A, armature resistance is 0.063 ohm and brush drop is 2 V. Find output voltage and load current.

*Solution*

(a) The OCC at 1000 rpm is drawn in Fig. 3.65 (curve 1). Draw line *OP* through origin *O* and touching maximum portion of curve 1. This line represents critical field resistance. The slope of this line is 46.5 ohm, which is critical field resistance at 1000 rpm.

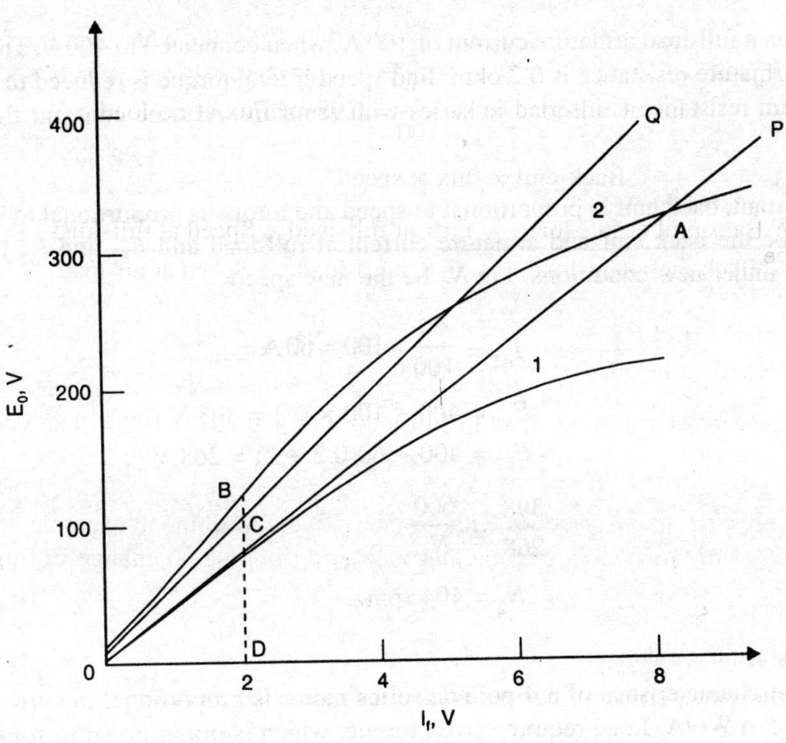


**Fig. 3.65**

(b) OCC at 1500 rpm can be found by multiplying different given values of $E_0$ by 1500/1000. Thus OCC at 1500 rpm is

| $I_f$ | 0 | 2 | 4 | 6 | 8 | A |
|---|---|---|---|---|---|---|
| $E_0$ | 15 | 135 | 255 | 321 | 360 | V |

This is shown as curve 2. The intersection of curve 2 and $OP$ at point $A$ gives $E_0$, which 355 V.

(c) Line $OQ$ is the 60 ohm resistance line. Draw vertical from any point $B$ on curve 2

$$\text{Critical speed} = \frac{\text{Intercept } CD}{\text{Intercept } BD} \times 1500$$

$$= \frac{120}{140} \times 1500 = 1285.7 \text{ rpm}$$

(d) $E_0 = 355$ V

$$\text{Output voltage} = 355 - I_a R_a - 2$$

$$= 355 - 400 \times 0.063 - 2 = 327.8 \text{ V}$$

$$I_f \text{ or field current} = \frac{327.8}{46.5} = 7.05 \text{ A}$$

$$\text{Load current} = I_a - I_f = 400 - 7.05 = 392.95 \text{ A.}$$

*Example 3.33*

A shunt motor has a full-load armature current of 100 A, when connected to 400 V. The full-load speed is 600 rpm and armature resistance is 0.2 ohm. find speed if total torque is reduced to 60% of full-load value and a 2 ohm resistance is inserted in series with armature. Assume constant flux.

*Solution*

Since flux is constant, back emf is proportional to speed and torque is proportional to armature current. Let $E_{b1}$ and $I_{a1}$ be the back emf and armature current at full-load and $E_{b2}$ and $I_{a2}$ be back emf and armature current under new conditions. Let $N_2$ be the new speed.

Then

$$I_{a2} = \frac{60}{100} \times 100 = 60 \text{ A}$$

$$E_{b1} = 400 - 100 \times 0.2 = 398 \text{ V}$$

$$E_{b2} = 400 - 60(0.2 + 2) = 268 \text{ V}$$

Therefore

$$\frac{398}{268} = \frac{600}{N_2}$$

$$N_2 = 404 \text{ rpm.}$$

*Example 3.34*

The magnetisation characteristics of a 4-pole dc series motor is proportional to current. On this basis flux per pole is 4.5 mWb/A. Load requires gross torque, which is proportional to square of speed and is 30 N-m at 1000 rpm. Armature is wave wound and has 492 active conductors. Find speed and armature current, if the motor is fed from 220 V supply and motor resistance is 2 ohm.

*Solution*

For a current $I$

$$\text{Flux } \phi = 4.5 \times 10^{-3} I$$

At a speed $N$, the torque is $\dfrac{30\,N^2}{(1000)^2}$

Back emf = 220 – 2 $I$

$$\frac{30N^2}{(1000)^2} = \frac{Z\phi IP}{2\pi A} = \frac{492 \times 4.5 \times 10^{-3} I^2 \times 4}{2\pi \times 2}$$

$$220 - 2I = \frac{P\phi ZN}{60A} = \frac{4 \times 4.5 \times 10^{-3} I \times 492N}{60 \times 2}$$

From the above equations, $I = 4.32$ A

$N = 662$ rmp.

*Example 3.35*

A dc series generator is running at 1150 rpm and delivering 10 kW at 125 V. The flux per pole $\propto$ $I/(60 + I)$, $r_f = 0.015\ \Omega$, $r_a = 0.024\ \Omega$. The armature demagnetising ampere-turns/pole = 8% of field amp-turns/pole. Find terminal voltage $V$ when $N = 1200$ rpm and $I = 100$ A. Brush drop = 2V.

*Solution*

Let $I_1$, $\phi_1$, $N_1$ and $E_{b1}$ be the current, flux, speed and emf at original conditions and $I_2$, $\phi_2$, $N_2$ and $E_{b2}$ be these quantities under new conditions.

Then $I_1 = 10 \times \dfrac{1000}{125} = 80\,\text{A},\ N_1 = 1150\ \text{rpm},\ I_2 = 100\,\text{A}$

$$E_1 = 125 + I_1(r_f + r_a) + 2$$

$$= 125 + 80(0.015 + 0.024) + 2 = 130.12\ \text{V}$$

The ratio of net fluxes is

$$\frac{\phi_1}{\phi_2} = \frac{[I_1/(60+I_1)] \times 0.92}{[I_2/(60+I_2)] \times 0.92} = \frac{80}{60+80} \times \frac{160}{100}$$

$$= 0.914$$

$$\frac{E_1}{E_2} = \frac{\phi_1}{\phi_2} \times \frac{N_1}{N_2} \quad \text{or} \quad \frac{130.12}{E_2} = \frac{0.914 \times 1150}{1200}$$

or $E_2 = 148.55$ V

Terminal voltage $= E_2 - I_2(r_f + r_a) - 2$

$= 148.55 - 100(0.015 + 0.024) - 2$

$= 142.56$ V.

*Example 3.36*

A dc shunt generator has $R_f = 100\ \Omega$. The terminal voltage is 50 V at 1000 rpm, 226 V at 2000 rpm and

400 V at 3000 rpm. Draw magnetisation curve at 2000 rpm and find terminal voltage if $R_f$ is reduced to 80 ohm. Neglect losses. [*B.U. 2002*]

*Solution*

Since losses are neglected, terminal voltage is equal to generated emf, which is given by the intersection of OCC and field resistance line.

At 1000 rpm, V = 50 V, $I_f = \dfrac{50}{100} = 0.5\,\text{A}$

Therefore for $I_f = 0.5$ A and $N = 2000$ rpm,

$$V = \frac{(50 \times 2000)}{1000} = 100\ \text{V}$$

At 2000 rpm, $V = 226$ V, $I_f = \dfrac{226}{100} = 2.26\,\text{A}$

At 3000 rpm, $V = 400$ V, $I_f = \dfrac{400}{100} = 4\,\text{A}$

Therefore for $I_f = 4$ A and $N = 2000$ rpm,

$$\text{V} = \frac{(400 \times 2000)}{3000} = 266.67\ \text{V.}$$

Using above three values, magnetisation curve is plotted is Fig. 3.66. Line *OP* is the 80 ohm field resistance line. The intersection of OCC and line *OP* is at *A*, which gives a voltage of 245 V.

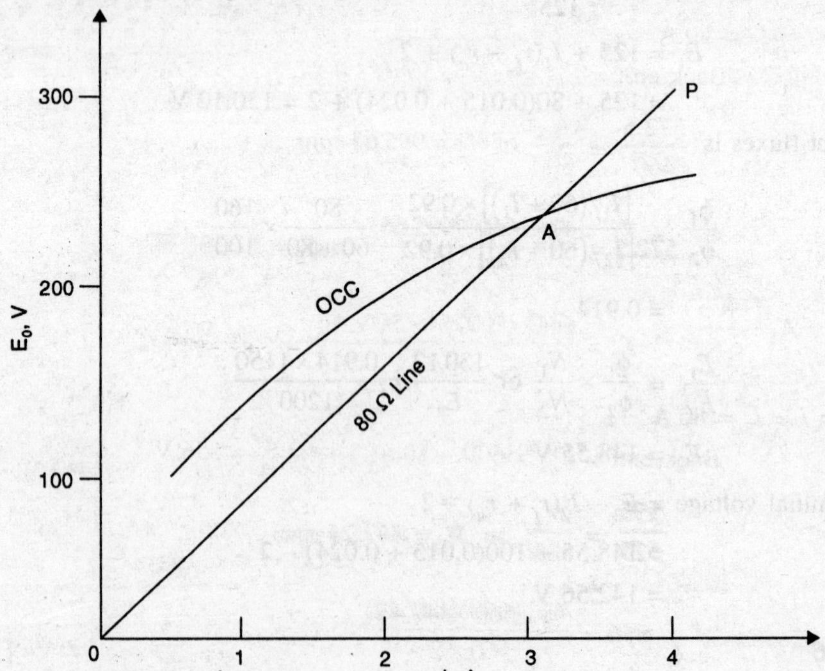


**Fig. 3.66**

*Example 3.37*

A 4-pole dc series motor with 2-circuit winding has the following OCC at 800 rpm and separate excitation.

| $I$ | 10 | 20 | 30 | 40 | 50 | 60 | 70 | 80 |
|---|---|---|---|---|---|---|---|---|
| $E$ | 160 | 295 | 375 | 425 | 460 | 485 | 505 | 525 |

Find speed and torque for currents of (a) 10 A (b) 50 A (c) 80 A. The supply voltage is 600 V, number of armature conductors is 660, $r_a = 0.3\ \Omega$, $r_f = 0.25\ \Omega$.

*Solution*

(a) When $I = I_1 = 10$ A

Back emf $= E_{b1} = 600 - 10(0.3 + 0.25) = 594.5$ V

Since back emf $\propto$ speed

$$\frac{594.5}{160} = \frac{N_1}{800} \text{ or } N_1 = 2972.5 \text{ rpm}$$

$$E_b = \frac{P\phi ZN}{60\,A}$$

or $$594.5 = \frac{(4\phi_1)(660)(2972.5)}{60 \times 2} \text{ or } \phi_1 = 0.0091 \text{ Wb}$$

$$T_1 = \frac{Z\phi_1 I_1 P}{2\pi A} = \frac{600 \times 0.091 \times 10 \times 4}{2\pi(2)} = 19.11 \text{ N-m}$$

(b) When $I = I_2 = 50$ A

Back emf $= E_{b2} = 600 - 50\ (0.3 + 0.25) = 572.5$ V

$$\frac{572.5}{460} = \frac{N_2}{800} \text{ or } N_2 = 995.65 \text{ rpm}$$

$$572.5 = \frac{4\phi_2(660)(995.65)}{60 \times 2} \text{ or } \phi_2 = 0.0261 \text{ Wb}$$

$$T_2 = \frac{660 \times 0.0261 \times 50 \times 4}{2\pi(2)} = 274.16 \text{ N-m}$$

(c) When $I = I_3 = 80$ A

Back emf $= E_{b3} = 600 - 80\ (0.3 + 0.25) = 556$ V

$$\frac{556}{525} = \frac{N_3}{800} \text{ or } N_3 = 847.24 \text{ rpm}$$

$$556 = \frac{4\phi_3(660)(847.42)}{60 \times 2}$$

or $$\phi_3 = 0.0298 \text{ Wb}$$

$$T_3 = \frac{660 \times 0.0298 \times 80 \times 4}{2\pi(2)} = 500.8 \text{ N-m}$$

*Example 3.38*

A dc generator has two field windings, (i) separately excited winding having 1000 turns/pole, a resistance of 80 ohm and fed from 110 V supply, (ii) shunt winding having 500 turns/pole and a resistance of 200 ohm. At 500 rpm OCC is as under.

| AT | 500 | 1000 | 1500 | 2000 | 2500 | 3000 |
|---|---|---|---|---|---|---|
| $E$ | 154 | 302 | 396 | 458 | 505 | 538 |

Find no-load voltage at (a) 500 rpm (b) 600 rpm, with both windings connected.

*Solution*

(a) The OCC at 500 rpm is drawn in Fig. 3.67 (curve 1)

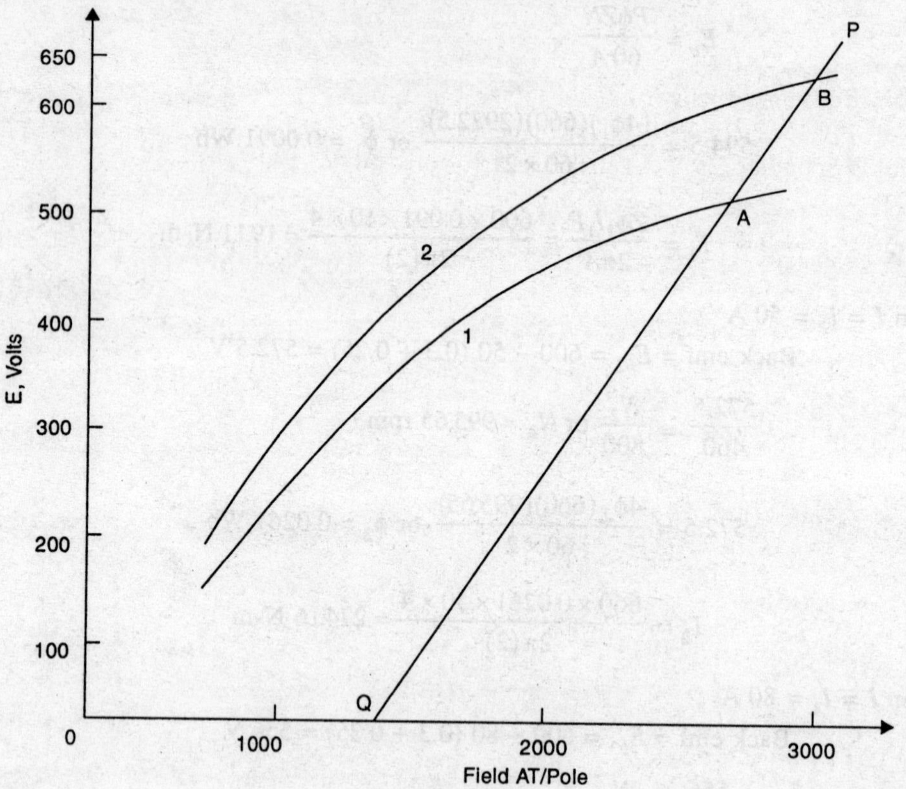


**Fig. 3.67**

$$\text{AT of shunt winding} = \frac{E}{200} \times 500 = 2.5\, E \text{ AT/pole}$$

$$\text{AT of separately excited winding} = \frac{110}{80} \times 1000 = 1375\,\text{AT}$$

Total field AT = 1375 + 2.5 $E$

when $E$ = 0, AT = 1375

when $E$ = 100, AT = 1375 + 250 = 1625

Line $QP$ passing through above two points is as shown. It is clear that this line has a slope of 80 ohm and is shifted from origin by the AT of separately excited winding, i.e. 1375 AT.

Line $QP$ intersects curve 1 at point $A$, which gives $E$ = 520 V.

(b) The OCC at 600 rpm can be obtained by multiplying the given values of $E$ by 600/500. The OCC is

| AT | 500 | 1000 | 1500 | 2000 | 2500 | 3000 |
|---|---|---|---|---|---|---|
| $E$ | 184.8 | 362.4 | 475.2 | 549.6 | 606 | 645.6 |

This is shown as curve 2 in Fig. 3.67. Line $QP$ intersects curve 2 at point $B$, which gives $E$ = 645 V.

### *Example 3.39*

A 45 kW, 225 V dc shunt generator runs at 500 rpm at full-load. The field and armature resistance are 45 ohm and 0.03 ohm, respectively. Calculate the speed of the machine, when running as a shunt motor and taking 45 kW input at 225 V. Assume brush contact drop of 1 V per brush.

### *Solution*

When working as generator

$$I_L = \frac{45 \times 1000}{225} = 200\text{ A}$$

$$I_f = \frac{225}{45} = 5\text{ A}$$

$$I_a = I_L + I_f = 205\text{ A}$$

$$E = V + I_a R_a + \text{Brush drop}$$

$$= 225 + 205 \times 0.03 + 2 = 233.15\text{ V}$$

$$E = k_e \phi N_1 = 233.15$$

or

$$k_e \phi = \frac{233.15}{500} = 0.4663$$

When working as motor

$$\text{Input current} = \frac{45 \times 1000}{225} = 200\text{ A}$$

$$I_f = \frac{225}{45} = 5\text{ A}$$

$$I_a = \text{Input current} - I_f = 195 \text{ A}$$

$$\text{Back emf} = E_b = V - I_a R_a - \text{brush drop}$$

$$= 225 - 195 \times 0.03 - 2 = 217.15 \text{ V}$$

Since field current is the same in the two cases, the flux is also the same. Hence $k_e \phi$ is constant. Let $N_2$ be the speed of the motor

$$k_e \phi N_2 = 217.15$$

$$N_2 = \frac{217.15}{0.4663} = 465.69 \text{ rpm.}$$

*Example 3.40*

A dc series motor is driving a fan load, whose torque varies as cube of the speed. The total armature and series field resistance is 1 ohm. It takes 10 A from 200 V mains and runs at 1000 rpm. Find the resistance to be connected in series with the motor to make it run at 800 rpm.

*Solution*

$$T_e \propto \phi I_a \propto I_a^2$$

$$\text{Load torque } T_L \propto N^3$$

$$T_e = T_L$$

or

$$I_a^2 \propto N^3$$

Let $I_{a1}$, $E_{b1}$ and $N_1$ denote the armature current, back emf and speed under initial conditions and $I_{a2}$, $E_{b2}$, and $N_2$ be these quantities under new conditions.

Then $$I_{a1} = 10 \text{ A}, \ N_1 = 1000 \text{ rpm}, \ N_2 = 800 \text{ rpm}$$

Now $$\frac{10^2}{I_{a2}^2} = \frac{1000^3}{800^3} \text{ or } I_{a2} = 7.155 \text{ A}$$

$$E_{b1} = 200 - 10 \times 1 = 190 \text{ V}$$

$$E_b \propto \phi N \propto I_a N$$

$$\frac{E_{b1}}{E_{b2}} = \frac{I_{a1} N_1}{I_{a2} N_2}$$

or $$\frac{190}{E_{b2}} = \frac{10 \times 1000}{7.155 \times 800}$$

or $$E_{b2} = 108.756 \text{ V}$$

$$E_{b2} = 200 - I_{a2}(1 + R)$$

or $$108.756 = 200 - 7.155(1 + R)$$

$$\text{Additional resistance } R = 11.75 \text{ ohm.}$$

### *Example 3.41*

A dc shunt motor runs at 1500 rpm and takes an input of 880 W at 220 V under no-load conditions. The shunt field current is 2 A and armature resistance is 0.1 ohm. Find the efficiency when the machine is used as a generator supplying 60 A at 220 V. [*P.U. 2000*]

*Solution*

Input at no-load = 880 W

Shunt field losses = 220 × 2 = 440 W

Armature current at no-load $= \dfrac{880}{220} - 2 = 2$ A

Armature copper-loss at no-load = $2^2$ (0.1) = 0.4 W

Mechanical- and iron-losses = 880 – 440 – 0.4 = 439.6 W

Efficiency when generator supplies 60 A

Output = 60 × 220 = 13200 W

Armature current = 60 + 2 = 62 A

Armature copper-loss = $62^2$ (0.1) = 384.4 W

Shunt field loss = 440 W

Mechanical- and iron-losses = 439.6 W

Total losses = 384.4 + 440 + 439.6 = 1264 W

$$\text{Efficiency} = \frac{\text{output}}{\text{output + losses}} = \frac{13200}{13200 + 1264} \times 100$$

$$= 91.26\%.$$

### *Example 3.42*

A dc shunt machine runs at no-load at 1000 rpm taking 480 W at 120 V. The shunt field current is 2A. The armature resistance is 0.1 ohm. Compute the line current, both as motor and generator, at which maximum efficiency occurs. Find the value of maximum efficiency.

*Solution*

At no-load,

$$\text{Armature current} = \frac{480}{120} - 2 = 2\text{ A}$$

Armature copper-loss = $2^2 \times 0.1$ = 0.4 W

Constant losses = 480 – 0.4 = 479.6 W

Maximum efficiency occurs when armature copper-loss is equal to constant losses

$$I_a^2 R_a = 479.6$$

or

$$I_a = \sqrt{\frac{479.6}{0.1}} = 69.25\text{ A}$$

when working as generator

$$\text{Line current} = I_a - I_f = 69.25 - 2$$
$$= 67.25 \text{ A}$$
$$\text{Output} = 67.25 \times 120 \text{ W} = 8070 \text{ W}$$
$$\text{Losses} = 2 \times 479.6 = 959.2 \text{ W}$$
$$\text{Efficiency} = \frac{8070 \times 100}{8070 + 959.2} = 89.38\%$$

When working as motor

$$\text{Line current} = I_a + I_f = 69.25 + 2 = 71.25 \text{ A}$$
$$\text{Input} = 71.25 \times 120 = 8550 \text{ W}$$
$$\text{Losses} = 2 \times 479.6 \text{ W}$$
$$\text{Efficiency} = \frac{\text{Input} - \text{losses}}{\text{Input}} \times 100$$
$$= \frac{8550 - 2 \times 479.6}{8550} \times 100 = 88.78\%.$$

*Example 3.43*

A cross field machine is rated at 3 kW and has a short circuit current amplification of 2 with zero compensation. The other data are as follows:

Voltage amplification = 50,
Resistance of *d*-axis winding = 5 ohm
Resistance of compensating winding = 1 ohm
Resistance of field winding = 100 ohm
If field current is 60 mA, find output voltage expression.

*Solution*

Using Eq. (3.58) with $C = 0$, $r_c = 0$

$$\frac{i_d}{i_f} = 2 = \frac{50 \times 100}{K_1 + 5 + 0} \text{ or } K_1 = 2495$$

The output voltage is given by Eq. (3.57)

$$V_L = Av_f - [K_1(1 - C) + r_d + r_c]i_d$$
$$= 50 \times 60 \times 10^{-3} \times 100 - [2495(1 - C) + 6]\, i_d$$

or $$V_L = 300 - i_d[2495(1 - C) + 6].$$

*Example 3.44*

Find the number of turns on each interpole of a 6-pole, 240 kW, 250 V dc generator having lap winding and 540 armature conductors. The flux density under the inter poles is 0.3 Wb/m² and effective gap length is 10 mm. Neglect mmf for iron parts and magnetic leakage.

*Solution*

The ampere-turns of interpole must cancel the armature mmf and also create the necessary flux density

$$\text{Armature mmf} = \frac{I_a Z}{2AP}$$

$$\text{MMF to create flux density} = \frac{B}{\mu_0}(\text{length of gap})$$

$$\text{Total AT of interpole} = \frac{I_a Z}{2AP} + \frac{B}{\mu_0}(\text{length of gap})$$

$$I_a = \frac{240 \times 1000}{250} = 960\text{ A}$$

$$A = P = 6, \; Z = 540$$

$$AT = \frac{960 \times 540}{2 \times 6 \times 6} + \frac{0.3}{4\pi \times 10^{-7}}\left(\frac{10}{1000}\right) = 5587.32$$

Since interpole winding carries armature current,

$$\text{Number of turns on interpole} = \frac{5587.32}{960} = 5.82 \text{ (say 6) turns/pole.}$$

*Example 3.45*

A dc generator has 12000 AT/pole as armature mmf. The ratio of pole arc to pole pitch is 0.7, length of interpolar gap is 1.25 cm and flux density in the interpolar air gap is 0.3 T. The machine is provided with compensating winding and interpoles. Find mmf for compensating winding and interpoles.

*Solution*

Total mmf to be neutralised

$$= \text{Armature mmf} + \frac{B}{\mu_0}(\text{length of air gap})$$

$$= 1200 + \frac{0.3}{4\pi \times 10^{-7}} \times \frac{1.25}{100} = 14984\text{ AT/pole}$$

AT of compensating winding

$$= \text{Armature mmf} \times \frac{\text{Pole arc}}{\text{Pole pitch}}$$

$$= 12000 \times 0.7 = 8400 \text{ AT/pole}$$

The remaining mmf must be neutralised by interpoles

Interpole mmf = 14984 − 8400 = 6584 AT/pole.

### *Example 3.46*

A commutator has a diameter of 74 cm and rotates at 900 rpm. If brush width is 1.5 cm, find time of commutation.

*Solution*

Speed = 900 rpm = 15 rps

$$\text{Time of commutation} = \frac{1.5}{\pi \times 74 \times 15} = 0.43 \times 10^{-3}\ s.$$

### *Example 3.47*

Find the back and front pitch of a simplex wave winding for a 4-pole, 24-slot, dc machine having 2 coil sides per slot. Give the winding table.

*Solution*

For a winding with a coil sides per slot, $C$ or the number of coils = number of slots = 24.

Using Eq. (3.2)

$$y_c = \frac{2(24 \pm 1)}{4} = 12\frac{1}{2} \text{ or } 11\frac{1}{2}$$

Since commutator pitch must be an integer, a wave winding with 24 active coils in not possible. If one coil is made dummy, the number of active coils is 23. Then $y_c$ = 12 or 11. Take $y_c$ = 12.

Coil span = 24/4 = 6 slots

Back pitch = 2 (6) + 1 = 13 coil sides

Front pitch = 2 (12) – 13 = 11 coil sides

Therefore this wave winding should have 23 active coils or 46 active coil sides (numbered 1–46) and one dummy coil. $y_b$ = 13, $y_f$ = 11 and $y_c$ = 12. The winding table is as under:

$$1 \overset{+y_b}{—} 14 \underset{+y_f}{—} 25 \overset{+y_b}{—} 38 \underset{+y_f}{—} 49 — 46 = 3 \overset{+y_b}{—} 16 \underset{+y_f}{—} 27$$

$$\overset{+y_b}{—} 40 \underset{+y_f}{—} 51 - 46 = 5 \overset{+y_b}{—} 18 \underset{+y_f}{—} 29 \overset{+y_b}{—} 42 \underset{+y_f}{—}$$

$$53 - 46 = 7 \overset{+y_b}{—} 20 \underset{+y_f}{—} 31 \overset{+y_b}{—} 44 \underset{+y_f}{—} 55 - 46 = 9 \overset{+y_b}{—}$$

$$22 \underset{+y_f}{—} 33 \overset{+y_b}{—} 46 \underset{+y_f}{—} 57 - 46 = 11 \overset{+y_b}{—} 24 \underset{+y_f}{—} 35$$

$$\overset{+y_b}{—} 48 — 46 = 2 \underset{+y_f}{—} 13 \overset{+y_b}{—} 26 \underset{+y_f}{—} 37 \overset{+y_b}{—} 50 - 46 = 4$$

$$\underset{+y_f}{—} 15 \overset{+y_b}{—} 28 \underset{+y_f}{—} 39 \overset{+y_b}{—} 52 - 46 = 6 \underset{+y_f}{—} 17 \overset{+y_b}{—} 30 \underset{+y_f}{—}$$

$$41 \overset{+y_b}{—} 54 - 46 = 8 \underset{+y_f}{—} 19 \overset{+y_b}{—} 32 \underset{+y_f}{—} 43 \overset{+y_b}{—} 56 - 46 = 10 \underset{+y_f}{—}$$

$$21 \overset{+y_b}{—} 34 \underset{+y_f}{—} 45 \overset{+y_b}{—} 58 - 46 = 12 \underset{+y_f}{—} 23 \overset{+y_b}{—} 36 \underset{+y_f}{—} 47 \text{ or } 1.$$

### *Example 3.48*

**A two-pole 8-slot lap wound armature has 2 coil sides per slot. (a) Find back pitch and front pitch. (b) Draw a winding table. (c) Draw armature winding in radial form. The machine has a lap winding.**

### *Solution*

(a) Number of coil sides = 8 × 2 = 16

Coil span = 8/2 = 4 slots

$y_b = 2\,(8/2) + 1 = 9$ coil sides

$y_f = y_b - 2 = 7$ coil sides for progressive winding

(b)

$$1 \overset{+y_b}{—} 10 \underset{-y_f}{—} 3 \overset{+y_b}{—} 12 \underset{-y_f}{—} 5 \overset{+y_b}{—} 14 \underset{-y_f}{—} 7 \overset{+y_b}{—} 16$$

$$\underset{-y_f}{—} 9 \overset{+y_b}{—} 18 \text{ or } 2 \underset{-y_f}{—} 11 \overset{+y_b}{—} 20 \text{ or } 4 \underset{-y_f}{—} 13$$

$$\overset{+y_b}{—} 22 \text{ or } 6 \underset{-y_f}{—} 15 \overset{+y_b}{—} 24 \text{ or } 8 \underset{-y_f}{—} 1$$

(c) The winding is shown in radial form in Fig. 3.68. The connections to commutator segments are also shown.

Fig. 3.68

*Example 3.49*

**Find the back pitch and front pitch for a 6-pole 18-slot, lap wound machine having 4 coil sides per slot. Prepare a winding table.**

*Solution*

Coil span = 18/6 = 3 slots

$y_b = 4(3) + 1$ = 13 coil sides

$y_f = 13 - 2$ = 11 coil sides for progressive winding

The coil sides are numbered as shown in Fig. 3.6. The winding table is as under:

$$1 \overset{+y_b}{\text{—}} 14 \underset{-y_f}{\text{—}} 3 \overset{+y_b}{\text{—}} 16 \underset{-y_f}{\text{—}} 5 \overset{+y_b}{\text{—}} 18 \underset{-y_f}{\text{—}} 7 \overset{+y_b}{\text{—}} 20 \underset{-y_f}{\text{—}} 9 \overset{+y_b}{\text{—}} 22 \underset{-y_f}{\text{—}}$$

$$11 \overset{+y_b}{\text{—}} 24 \underset{-y_f}{\text{—}} 13 \overset{+y_b}{\text{—}} 26 \underset{-y_f}{\text{—}} 15 \overset{+y_b}{\text{—}} 28 \underset{-y_f}{\text{—}} 17 \overset{+y_b}{\text{—}} 30 \underset{-y_f}{\text{—}} 19 \overset{+y_b}{\text{—}}$$

$$32 \underset{-y_f}{\text{—}} 21 \overset{+y_b}{\text{—}} 34 \underset{-y_f}{\text{—}} 23 \overset{+y_b}{\text{—}} 36 \underset{-y_f}{\text{—}} 25 \overset{+y_b}{\text{—}} 38 \underset{-y_f}{\text{—}} 27 \overset{+y_b}{\text{—}}$$

$$40 \underset{-y_f}{\text{—}} 29 \overset{+y_b}{\text{—}} 42 \underset{-y_f}{\text{—}} 31 \overset{+y_b}{\text{—}} 44 \underset{-y_f}{\text{—}} 33 \overset{+y_b}{\text{—}} 46 \underset{-y_f}{\text{—}} 35 \overset{+y_b}{\text{—}}$$

$$48 \underset{-y_f}{\text{—}} 37 \overset{+y_b}{\text{—}} 50 \underset{-y_f}{\text{—}} 39 \overset{+y_b}{\text{—}} 52 \underset{-y_f}{\text{—}} 41 \overset{+y_b}{\text{—}} 54 \underset{-y_f}{\text{—}} 43 \overset{+y_b}{\text{—}}$$

$$56 \underset{-y_f}{\text{—}} 45 \overset{+y_b}{\text{—}} 58 \underset{-y_f}{\text{—}} 47 \overset{+y_b}{\text{—}} 60 \underset{-y_f}{\text{—}} 49 \overset{+y_b}{\text{—}} 62 \underset{-y_f}{\text{—}} 51 \overset{+y_b}{\text{—}}$$

$$64 \underset{-y_f}{\text{—}} 53 \overset{+y_b}{\text{—}} 66 \underset{-y_f}{\text{—}} 55 \overset{+y_b}{\text{—}} 68 \underset{-y_f}{\text{—}} 57 \overset{+y_b}{\text{—}} 70 \underset{-y_f}{\text{—}} 59 \overset{+y_b}{\text{—}}$$

$$72 \underset{-y_f}{\text{—}} 61 \overset{+y_b}{\text{—}} 74 \text{ or } 2 \underset{-y_f}{\text{—}} 63 \overset{+y_b}{\text{—}} 76 \text{ or } 4 \underset{-y_f}{\text{—}} 65 \overset{+y_b}{\text{—}}$$

$$78 \text{ or } 6 \underset{-y_f}{\text{—}} 67 \overset{+y_b}{\text{—}} 80 \text{ or } 8 \underset{-y_f}{\text{—}} 69 \overset{+y_b}{\text{—}} 82 \text{ or } 10 \underset{-y_f}{\text{—}}$$

$$71 \overset{+y_b}{\text{—}} 84 \text{ or } 12 \underset{-y_f}{\text{—}} 1.$$

## SUMMARY

1. A dc machine may be a dc generator or dc motor. DC generator converts mechanical energy into dc electrical energy whereas a dc motor converts dc electrical energy into mechanical energy.

2. The main parts of a dc machine are field poles, yoke, armature, commutator, brushes and interpoles.
3. Armature winding of dc machine may be lap winding or wave winding. In a lap winding the number of parallel paths is equal to the number of poles but in a wave winding the number of parallel paths is two.
4. EMF induced $= \dfrac{ZP\phi N}{60\,A}$
5. Commutator converts ac to dc and vice versa.
6. Armature reaction is the effect of armature flux on field flux. It may be demagneting or cross magnetising.
7. The use of high resistance carbon brushes and interpoles helps the commutation process and reduces arcing due to commutation.
8. DC generators may be separately excited or self excited. The self excited generator may be shunt, series or compound generator.
9. Magnetisation characteristics shows the variation of open circuit armature voltage with field current when the generator is running at a constant speed.
10. Self excitation is due to residual magnetism.
11. Shunt generator has the field winding in parallel with the armature. A series generator has its field winding in series with the armature. A compound generator has both shunt and series field windings.
12. Generators are often connected in parallel to meet the load requirement.
13. In a dc motor the induced emf is the back emf and opposes the applied voltage.
14. DC motor may be shunt motor, series motor or compound motor.
15. A dc shunt motor has its field winding connected in parallel with the armature. Its field flux is constant and speed in also nearly constant.
16. A dc series motor has its field winding connected in series with the armature. Therefore its field flux depend on armature current. It speed is inversely proportional to torque.
17. A compound motor has both series and shunt fields. Its characteristics are in between the characteristics of series and shunt motor.
18. The speed of dc shunt motor can be controlled by field control or armature control. These two methods can also be used for a dc series motor.
19. A starter is used with a dc motor to protect the motor against damage due to excessive starting current.
20. Three types of electric braking are: plugging, dynamic braking and regenerative braking.
21. Losses in a dc machine are: copper losses in armature winding and field winding, iron losses, mechanical losses and stray load losses. Efficiency is the ratio of power output to power input.
22. The various tests on dc machines are: brake test, Swinburne test, Hopkinson test.
23. Cross field machines have an additional set of brushes on the d-axis. Metadyne and Amplidyne are such machines.
24. DC generators are used for excitation of ac generators dynamometers, tachogenerator etc. DC

shunt motor is used where a wide range of speed control is required. DC series motor is used in lifts, cranes, electric traction etc.

## TEST POINT QUESTIONS

### A. State whether the following statements are true (T) or false (F)

3.1 The field system of dc machine is on rotor.

3.2 Armature core of a dc machine is laminated.

3.3 Brushes of a dc machine are made of copper.

3.4 Modern dc machines use single layer armature winding.

3.5 Wave winding is used for high voltage dc machines.

3.6 Wave winding of a dc machine has two parallel paths.

3.7 The induced emf of dc machine is directly proportional to speed.

3.8 The induced emf in a dc motor is called back emf and is always less than terminal voltage.

3.9 Reactance voltage of a dc machine is inversely proportional to time of commutation.

3.10 If the brushes are on gnp, demagnetising ampere turns are zero.

3.11 The commutating field, due to interpoles, should be as strong as possible.

3.12 The magnetisation characteristics of dc generator is linear.

3.13 In a separately excited dc generator there is no voltage drop due to armature reaction.

3.14 For self excitation of a dc generator the resistance in field circuit should be less than critical resistance.

3.15 The terminal voltage of a dc series generator increases with load current.

### B. Fill in the blanks with suitable words:

3.16 A dc series generator can be used as a ...........

3.17 In a differentially compounded dc generator, the terminal voltage ............. with increase in load current.

3.18 compound dc generators need ........... for successful parallel operation.

3.19 Torque of a dc motor is directly proportional to ........... current.

3.20 If a dc motor is designed to run in both directions, brushes should be on ...........

3.21 In a dc shunt motor the decrease in speed with increasing load current is about ........... %.

3.22 A diverter is connected in ........... with the field winding.

3.23 The operation of a differentially compounded motor is ...........

3.24 In a dc shunt motor using field control, speeds ........... Than rated speed cannot be obtained.

3.25 Three point starter is used for dc ........... motor.

3.26 Stray load losses are assumed to be equal to ........... of full load.

3.27 Swinburne test is a ........... Test.

3.28 Hopkinson test is a ........... Test.

3.29 In a dc shunt motor very ........... speed control can be achieved.

3.30 DC series motor is very suitable for ...........

## ANSWERS

| | | | | |
|---|---|---|---|---|
| 3.1 T | 3.2 T | 3.3 F | 3.4 F | 3.5 T |
| 3.6 T | 3.7 T | 3.8 T | 3.9 T | 3.10 T |
| 3.11 F | 3.12 F | 3.13 F | 3.14 T | 3.15 T |
| 3.16 booster | 3.17 decreases sharply | | 3.18 equalising bar | |
| 3.19 armature | 3.20 g n p | 3.21 5 | 3.22 parallel | |
| 3.23 unstable | 3.24 lower | 3.25 shunt | 3.26 1% | 3.27 no load |
| 3.28 regenerative | | 3.29 wide | 3.30 electric traction | |

## JOB INTERVIEW QUESTIONS

3.1 Write emf equation of a dc machine.

3.2 The terminal voltage of dc shunt generator is less than induced emf but the terminal voltage of a dc shunt motor is more than induced emf. Why?

3.3 Why are brushes of dc machine made of carbon?

3.4 What is the difference between lap and wave winding?

3.5 Draw magnetisation characteristics of dc generator.

3.6 What is a compound dc generator?

3.7 Draw torque speed curve of dc shunt motor

3.8 Name the methods for speed control of dc shunt motor.

3.9 Name some applications of dc generators

3.10 Name some application of dc motors

3.11 What is the function of commutator in a dc machine?

3.12 Why is starter used in dc shunt motor?

3.13 What is the difference between short shunt and long shunt?

3.14 What is differential compounding?

3.15 What is back emf?

3.16 What is a cross field machine?

3.17 Name the various losses in a dc machine?

3.18 Write torque equation of a dc motor.

3.19 What is reactance voltage?

3.20 What is the use of interpoles in dc machines?

## REVIEW QUESTIONS

3.1 List the different parts of a dc machine along with their functions. Name the materials used in the construction of each of these parts.

3.2 Describe the construction of armature, field magnet and commutator of a dc machine.

3.3 Discuss the main features of dc machine windings. Differentiate between lap and wave winding and name their field of application.

3.4 Explain the terms: pole pitch, coil span, winding pitch, commutator pitch, progressive winding, retrogressive winding, back pitch, front pitch, dummy coil.

3.5 Draw developed diagrams for (a) 4-pole 13-slot wave winding, (b) 4-pole 18-slot lap winding. Show the position of brushes in each case.

3.6 Derive the equation for induced emf of a dc machine. Which quantities in the expression are constant?

3.7 Draw a diagram showing magnetic circuit of a dc machine. Indicate the paths of main flux and leakage flux. Draw an electric circuit analogous to the magnetic circuit.

3.8 Why are some parts of a dc machine laminated? Which parts are laminated in this machine?

3.9 (a) What is the function of interpoles in a dc machine?

(b) What would happen if the interpole winding is disconnected?

(c) If the direction of rotation of a generator provided with interpoles is reversed, the polarity of main poles remaining the same, will it be necessary to reverse the interpole winding connections? Explain.

(d) What can be the harm if the interpoles produce a very strong commutating field?

3.10 What is commutation? What causes sparking on the commutator surface? How can it be avoided?

3.11 Define reactance voltage. What is its effect? Why is it necessary to neutralise it? How can it be done?

3.12 Distinguish between the following:

(a) emf and resistance commutation,

(b) over and under commutation,

(c) cross magnetisation and demagnetisation effect of armature reaction.

3.13 What is a compensating winding? What is its function? Where is it housed? How is it connected? In which machine is it very necessary? How can the number of conductors of this winding be estimated?

3.14 When both compensating winding and interpoles are provided in a dc generator, is it necessary to shift the brushes from the geometric neutral plane? Explain.

3.15 With reference to OCC of a dc generator, explain the following:

(a) Why is the emf not zero when field current is zero?

(b) Will the residual flux induce a voltage in the machine, if speed of machine is zero?

(c) Why does the slope of the curve change, after a certain value of field current?

3.16 Discuss the process of self-excitation in a dc machine. What conditions must be fulfilled for self-excitation?

3.17 A dc machine fails to excite. List the various causes for this phenomenon.

3.18 Define the terms 'critical resistance' and 'critical speed' and bring out their role in the process of self-excitation of a dc machine.

3.19 Explain the term 'voltage regulation'. Can it be negative for a dc generator? Discuss.

3.20 List the reasons which cause the terminal voltage under load conditions to be different from the terminal voltage under no-load condition of a dc.

(a) separately excited generator,

(b) shunt generator,

(c) series generator.

3.21 The shunt field winding has a high resistance, while the series field winding has a low resistance. Why?

3.22 Distinguish between external and internal characteristics of a dc generator. How can internal characteristics be derived from external characteristics of:

(a) separately excited generator,

(b) series generator.

3.23 How is the output voltage controlled in a:

(a) series generator,

(b) shunt generator?

Explain with the help of connection diagram.

3.24 Draw a circuit diagram of two shunt generators in parallel. Explain the process of:

(a) putting a generator in parallel with a bus bar and loading it,

(b) taking a generator off the bus bars and shutting it down.

3.25 How can the load sharing between two shunt generators operating in parallel be determined? Explain.

3.26 Is the polarity of a shunt generator changed by:

(a) reversing the direction of rotation,

(b) connections of shunt field coils? Explain.

3.27 In every dc generator, a motor action occurs and in every dc motor a generator action occurs? Explain.

3.28 What is meant by torque of a motor? On what factors does the torque of a dc motor depend? Derive an expression for the torque.

3.29 Give reasons for the following:

(a) A shunt motor operates at almost constant speed irrespective of load.

(b) A series motor develops a high staring torque.

(c) A differential compound motor is rarely used.

(d) A series motor should not be connected to load through a belt.

3.30 How can speed of dc shunt motor and dc series motor be controlled? Explain the different methods with connection diagram and characteristics curves.

3.31 Explain the speed current and torque current characteristics of dc:

(a) shunt motor,

(b) series motor.

3.32 Drive the speed-torque characteristic of:

(a) dc shunt motor,

(b) dc series motor,

(c) dc compound motor.

Discuss the applications of these motors in the light of these characteristics.

3.33 How does a dc motor adjust its input automatically to match the load on the shaft? Explain.

3.34 (a) What is back emf?

(b) Does the back emf ever equal the applied voltage? Explain.

(c) What is counter torque? How is it produced and in which dc machine?

3.35 (a) Explain why a small change in speed and back emf will produce a larger change in armature current of a dc motor.

(b) If speed is decreased, what happens to back emf and armature current and why?

3.36 Derive a relation between emf and torque in a dc machine. Is this relation valid for generator or motor? Explain.

3.37 Draw a neat diagram of dc shunt motor starter. How can the values of resistance elements be calculated? Explain the function of overload release and no-volt release coils?

3.38 List the different losses in a dc machine. Which of these losses are constant? Drive a condition for maximum efficiency.

3.39 How is brake test conducted on a dc motor? How can the efficiency be calculated from the results of brake test? Why is this test not suitable for large machines?

3.40 How is Swinburne test conducted on a dc machine? How can the efficiency be determined from the results of this test when the machine works as:

(a) motor, (b) generator?

Why can't this test be applied to series machines?

3.41 What is Hopkinson test? Draw a diagram and explain the procedure of this test. What are the advantages of this method of testing?

3.42 What is a cross-field machine? Explain its working with the help of a diagram.

3.43 What is an amplidyne? Draw its connection diagram and explain its working. Derive an expression for its output voltage.

3.44 Derive an equation for the short circuit current amplification of a compensated cross-field machine.

3.45 Derive an equation to describe dynamic behaviour of a separately excited motor. Draw a block diagram and explain.

3.46 Draw a block diagram showing the dynamic behaviour of a separately excited dc generator. Derive the necessary equations on which this diagram is based.

3.47 Discuss the applications of dc:

(a) generator,

(b) shunt, series and compound motor.

## PROBLEMS

3.1 A 4-pole lap wound dc generator has 740 armature conductors, useful flux of 40 mWb per pole and runs at 500 rpm. Find open circuit emf. [246.66 V]

3.2 A 4-pole lap wound machine has 124 coils each having 2 turns. The flux per pole is 0.015 Wb. If speed is 1500 rpm, find induced emf. [186 V]

3.3 The armature of a 2-pole 400 V dc generator has 800 conductors and runs at 3000 rpm. Find the useful flux per pole. If the number of turns in each field coil is 600, find the average value of emf induced in the field coil on breaking the field, if the flux decays to zero in 0.2 sec. [0.1 Wb, 300 V]

3.4 A 4-pole dc generator has 51 slots and each slot contains 20 conductors. The machine has a useful flux of 0.007 Wb and runs at 1500 rpm. Find the induced emf, if the machine is wave wound. [357 V]

3.5 A lap wound dc generator has 100 armature conductors and 10 poles. The rated armature current is 800 A. Find the number of conductors of compensating winding per pole to give full armature reaction compensation, if the pole face covers two-third of pole pitch. [7]

3.6 For the machine of problem 3.5, Calculate the following:

(a) Cross magnetising AT/pole, if brushes are set on geometric neutral plane.

(b) Cross magnetising AT/pole if brushes are shifted by 5° electrical from gnp.

(c) Demagnetising AT/pole if brushes are shifted by 5° electrical from gnp.

[(a) 4000 AT/pole (b) 3778 AT/pole (c) 222 AT/pole]

3.7 A 4-pole generator has 492 armature conductors and an armature current of 144A. The brushes are given a lead of 10° actual. Find demagnetising AT/pole if the machine is:

(a) lap connected, (b) wave connected. [(a) 492 AT/pole (b) 984 AT/pole]

3.8 A 110 kW short shunt compound generator has a no-load voltage of 500 V and is over compounded by 10%, producing a linear external characteristic. If the armature resistance is 0.09 ohm, series field resistance is 0.04 ohm and shunt field resistance is 100 ohm. Plot the internal characteristic.

[500 V at 0 A rising linearly to 590 V at 250 A]

3.9 Two shunt generators together supply 2000 A. The machines have armature resistances of 0.025 ohm and 0.04 ohm, field resistances of 20 ohm and 25 ohm and induced emfs of 420 V and 440, respectively. Calculate the power output of each machine. [370 kW, 430 kW]

3.10 The power input to a 230 V dc shunt motor is 8.477 kW. The field resistance is 230 ohm and armature resistance is 0.28 ohm. find armature current and back emf. [35.8 A, 220 V]

3.11 The OCC of a shunt generator running at 850 rpm is an under:

| $I_f$ (A) | 0.8 | 1.6 | 2.4 | 3.2 | 4 | 4.8 | 5.6 |
|---|---|---|---|---|---|---|---|
| $E$ (V) | 28 | 55 | 75 | 82 | 100 | 108 | 115 |

(a) Find the open circuit induced emfs for field resistances of 22 ohm and 30 ohm.

(b) Find the field resistance, so that the open circuit induced emf at 850 rpm may be 100 V.

(c) Find the critical speed for the field resistance found in (b).

(d) Find critical field resistance at a speed of 850 rpm.

[(a) 110 V, 81 V (b) 25 ohm (c) 575 rpm (d) 37 ohm]

3.12 The external characteristics of a series generator is a straight line. When $I_a = 0$, $V = 0$ and when $I_a = 300$ A, $V = 75$ V. This generator is used as a booster between a station bus bar and a feeder of resistance 0.4 ohm. Find the net voltage drop between the station bus bar and the far end of the feeder, if the feeder carries a current of (a) 200 A, (b) 100 A. [(a) 30 V (b) 15 V]

3.13 Estimate the reduction in speed of a separately excited generator feeding a 500 V bus bar to decrease the load on the generator from 500 kW to 250 kW. The resistance between the terminals is 0.015 ohm. [1.456%]

3.14 A long shunt compound generator is supplying 60 A at 400 V. The resistances of shunt field, armature and series field are 200 ohm, 0.06 ohm and 0.04 ohm, respectively. The brush contact drop is 1 volt per brush. Find armature current and induced emf. [62A, 408.2 V]

3.15 A shunt generator has the following external characteristics:

| $I_L$ | 0 | 5 | 10 | 15 | 20 | 25 | A |
|---|---|---|---|---|---|---|---|
| $V$ | 100 | 99 | 97.5 | 95.5 | 93 | 90 | V |

The shunt field and armature resistances are 50 ohm and 0.2 ohm. Plot internal characteristics.

3.16 A shunt generator has an induced emf of 510 V and a terminal voltage of 500 V. The resistances of shunt field and armature are 50 ohm and 0.05 ohm, respectively. Find the load current. Neglect armature reaction. [190 A]

3.17 A series dc generator has the following external characteristics:

| $I_L$ | 10 | 20 | 30 | 40 | 50 | 60 | 70 | A |
|---|---|---|---|---|---|---|---|---|
| $V$ | 200 | 380 | 520 | 620 | 680 | 720 | 725 | V |

The total armature and series field resistance is 0.6 ohm. Plot internal characteristics.

3.18 Two identical 20 hp, 1000 rpm, 250 V series motors are connected in series across a 250 V supply. They drive the same shaft through reduction gearing 5:1 and 4:1, respectively. The load torque is 810 N-m. Find speed of shaft, voltage across each motor and line current. Neglect losses.

[140 rpm, 111 V, 139 V, 47.5 V]

3.19 Two 220 V generators operate in parallel. One machine has a terminal voltage of 270 V on on-load and 220 V when supplying 35 A. The second machine has a voltage of 280 V on no-load and 220 V when supplying 50 A. Find the output voltage and current of each machine when total current is 60 A. Assume the external characteristics to be linear. [236.3 V, 23.6 A, 36.4 A]

3.20 A station has 6 generators in parallel. Each generator has an armature current of 1000 A, induced emf of 250 V and terminal voltage of 240 V. The load resistance is constant. Find the new terminal voltage and current supplied by each, if the emf of machine 1 is increased by 5%, the emfs of other machines remaining the same. [$V = 242$ V, $I_1 = 2050$ A, $I_2 = I_3 = I_4 = I_5 = I_6 = 800$ A]

3.21 A series motor has an armature resistance of 0.7 ohm and field resistance of 0.3 ohm. It takes a current of 15A from a 200 V supply and runs at 800 rpm. Find the speed at which it will run, when connected in series with a 5 ohm resistance and taking the same current at the same supply voltage. [476 rpm]

3.22 A 240 V series motor has a resistance of 0.2 ohm. At a speed of 1800 rpm it takes a current of of 40 A. Find the resistance to be added,

(a) to limit the speed to 3600 rpm, when current is 10 A, assuming that flux is proportional to current between 10 A and 40 A.

(b) to make the speed 900 rpm when current is 60 A, given that flux at 60 A is 18% greater than flux at 40 A.

(c) At what speed will it run when connected to mains, taking 60 A.

[(a) 12.2 Ω (b) 1.52 Ω (c) 1500 rpm]

3.23 Find the no-load and full-load speeds of a 220 V 4-pole shunt motor having following data: flux 0.04 mWb, armature resistance 0.04 Ω, 160 armature conductors, wave connection, full-load line current 95 A, no-load line current 9 A, field resistance 44 Ω. [1030.5 rpm, 1014.4 rpm]

3.24 A shunt generator delivers 50 kW a 250 V and 400 rpm. The armature and field resistances are 0.02 ohm and 50 ohm, respectively. Find the speed when the machine works as a motor, taking an input of 50 kW at 250 V. Allow 1 V per brush as contact drop. [381.25 rpm]

3.25 A 4-pole wave wound dc armature has 294 conductors. Find:

(a) flux per pole to generate 230 V at 1500 rpm,

(b) electromagnetic torque at this flux when armature current is 120 A. [(a) 0.0156 Wb (b) 175.2 N-m]

3.26 A 220 V series motor runs at 700 rpm when operating at full-load current of 20 A. The motor resistance is 0.5 Ω and magnetic path may be assumed to be unsaturated. What will be the speed if:

(a) load torque is increased by 44%.

(b) motor current is 10 A. [(a) 578 rpm (b) 1433 rpm]

3.27 Find the number of steps and resistance elements of a starter for a 250 V shunt motor having an armature resistance of 0.5 ohm. The current should not exceed 67 A. The minimum current should be 75% of maximum current. [7 sections, 0.94, 0.7, 0.53, 0.39, 0.3, 0.21, 0.17 Ω]

3.28 The maximum current during starting for a 500 V shunt motor is to be limited to 125 A. The resistance of armature is 0.25 ohm. Find the resistance elements for 12-element starter.

[0.825, 0.655, 0.52, 0.413, 0.327, 0.26, 0.206, 0.164, 0.13, 0.103, 0.082, 0.065 Ω]

3.29 A simplex dc machine has 800 conductors, 10 poles and is lap wound. The armature current is 1000 A. The pole face covers 70% of pole pitch. Find the number of pole face conductors to give full compensation for armature reaction. [6 conductors/Pole]

3.30 A 350 kW, 500 V, 8-pole dc generator has a core length of 0.35 m and an armature diameter of 1.3 m. It has a 4-circuit wave winding accommodated in 114 slots with 6 coil sides per slot. Assume specific permeance = 6, interpole axial length = 0.2 m and a gap of 1 cm. Find the mmf and number of turns for each interpole.

[9980 AT/pole, 14]

3.31 A 4-pole dc machine has 60 conductors per parallel path and a simplex lap winding. There are 3 turns per coil and two coil sides per slot. Find (a) number of coils, (b) number of slots, (c) number of commutator segments, (d) back pitch, (e) front pitch, (f) commutator pitch. Assume progressive winding.[(a) 40 (b) 40 (c) 40 (d) 10 slots (e) 9 slots (f) 1 segment]

3.32 A 100 kW, 500 V, long shunt compound dc generator has armature resistance of 0.03 ohm, shunt field resistance of 125 ohm and series field resistance of 0.01 ohm. The current in diverter is 54A. Find (a) diverter resistance at full load, (b) generated emf at full load. [(a) 0.0278 Ω (b) 507.62 V]

3.33 In a dc machine the hysteresis and eddy current losses are 250 W and 100 W respectively when the speed is 1000 rpm. At what speed will the total iron losses be 175 W, if flux remains constant. [570 rpm]

3.34 A motor is developing a torque of 130 N-m. Find the torque if the field flux is decreased by 25% and the armature current is increased by 10%. [107.25 N-m]

3.35 A shunt motor takes a current $I$ at speed $n$. Find the current at speed $2n$ (a) with field control, (b) with armature voltage control when (i) output is constant, (ii) torque is constant. Neglect losses.
[(a) (i) I (ii) 2I (b) (i) I/2 (ii) I]

3.36 Two dc shunt generators are operating in parallel. Their no-load voltages are 250 V and characteristics are linear. At 220 V, generator $A$ can deliver 300 kW and generator $B$ can deliver 600 kW. Find the total load $P$ and load supplied by each at (a) 250 V, (b) 240 V, (c) 230 V.
[(a) $P = 0$, $P_A = 0$, $P_B = 0$ (b) $P = 300$ kW, $P_A = 100$ kW, $P_B = 200$ kW
(c) $P = 600$ kW, $P_A = 200$ kW, $P_B = 400$ kW]

3.37 Find the efficiency of a long shunt compound generator rated at 250 kW, 230 V when supplying 75% of rated load at rated voltage. The resistance of armature and series field are 0.009 ohm and 0.003 ohm. The shunt field current is 12 A. When the machine is run as a motor at no-load the armature current is 25 A at rated voltage. [90.8%]

3.38 A 50 kW belt driven shunt generator is running at 300 rpm and delivering rated load to 250 V bus bar. Armature resistance is 0.025 ohm, shunt field resistance is 50 ohm and brush drop is 1V per brush. Suddenly the belt breaks. The machine continues to run as motor taking 5 kW from bus bars. Find the motor speed. Neglect armature reaction. [288.92 rpm]

Chapter **4**

# AC Machine Fundamentals

## 4.1 INTRODUCTION

AC machines can be classified into two major categories: synchronous machines and induction machines. Synchronous machines have a fixed relation between speed, frequency and number of poles. Its field winding is excited from a separate dc source. A synchronous generator (also known as alternator) converts mechanical energy to ac electric energy. All generators in present day small, medium and big power plants are synchronous generators. A synchronous motor converts ac electric energy to mechanical energy. In induction machines (or asynchronous machines) the field current is supplied by magnetic induction (transformer action).

AC machines are mostly 3-phase machines (except small motors used in domestic appliances). There have armature winding on stator and field winding on rotor. The rotating magnetic field from the rotor field windings of an ac machine induces a 3-phase set of ac voltages in the stator winding. Conversely, the flow of 3-phase ac currents in stator windings produces a rotating magnetic field, which interacts with rotor magnetic field, thus producing a torque. These two effects are ac generator and motor action.

## 4.2 AC WINDINGS

AC windings differ from dc windings in many respects. No commutation is needed. Therefore the winding need not be a closed one. All ac windings are open windings. Since most ac machines are of 3-phase type, the three windings of the three phases are identical, but spaced 120 electrical degrees apart. The windings may be connected in star or delta. However, if the three windings are connected in delta, it forms a closed winding.

The ac windings are characterised by the following:

(1) The number of phases are usually three.

(2) The number of circuits in parallel per phase, may be one or more.

(3) The armature windings of synchronous generators and motors are generally connected in star. In the case of three-phase induction motors both star and delta connections are used.

(4) The number of coil layers per slot may be one or two, but the two layer windings are more common.

(5) The angular spread of the consecutive conductors belonging to a given phase belt.

(6) The pitch of the individual coils of the winding.
(7) The arrangement of end connections.

### 4.2.1 Single and Double Layer Winding

In a single layer winding one coil side occupies one slot completely. Therefore, the number of coils is equal to half the number of slots. In a double layer winding, one coil side lies in the upper half of one slot, while the other coil side lies in the lower half of another slot spaced about one pole pitch from the first one. In a single layer winding the coils are arranged in groups and the overhang of each group of coils is made to cross the overhang of other groups by adjusting the size and shape of different coil groups. However, in double layer windings all the coils are identical in shape and size. Thus a double layer winding results in a cheaper machine. All synchronous machines and most induction machines (except below a few kW rating) use double layer windings.

### 4.2.2 Phase Spread

A group of adjacent slots belonging to one phase under one pole pair is known as 'phase belt'. The angle subtended by a phase belt is known as phase spread.

Consider an arrangement of 2 slots per pole per phase (i.e. 12 slots per pole pair). The slot angle is 30°. One possible arrangement is to have winding of phase *a* in slots 1–4, that of phase *b* in slots 5–8 and that of phase *c* in slots 9–12. Each phase belt has a spread of 120° [Fig. 4.1(a)]. The emfs of conductors in adjacent slots have a phase difference equal to slot angle. Therefore the resultant phase emf is less than the sum of individual conductor emfs. The phasor sum of conductor emfs in slots 1-4 gives the phase emf of phase *a*. This is as shown in the phasor diagram. Another possible arrangement is shown in Fig. 4.1(b). The complete winding has been divided into six groups, each having a spread of 60°. The conductors in slots 7 and 8 serve as return conductors for those in slots 1 and 2. As such the conductor emfs for slots 7 and 8 have been marked negative. The phasor sum of conductor emfs again gives the phase emf for phase *a*. It is seen that magnitude of phase emf in Fig. 4.1(b) is more

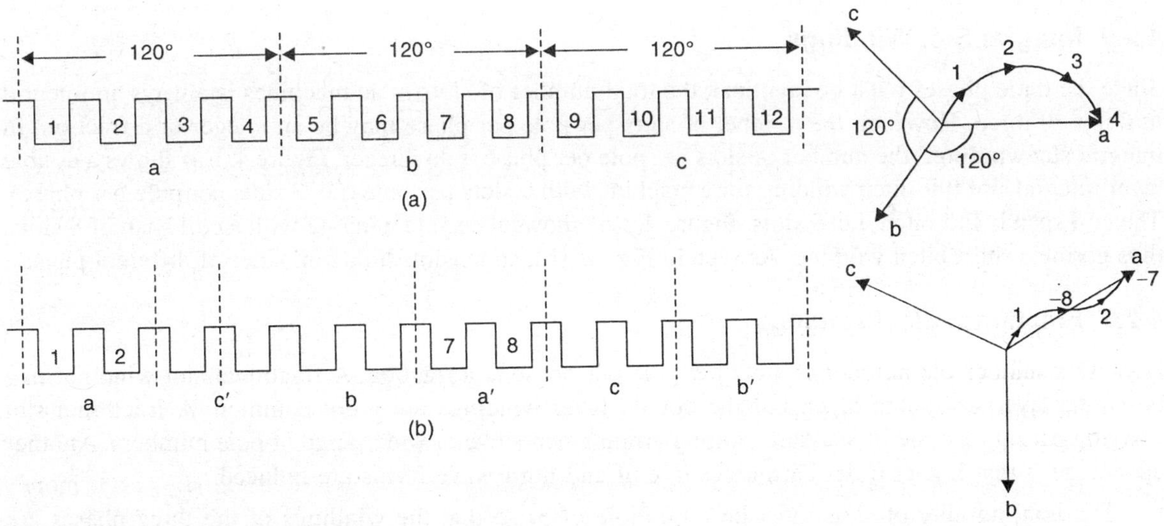


**Fig. 4.1.** Effect of phase spread on generated emf (a) 120° phase spread, (b) 60° phase spread

than that in Fig. 4.1(a). Therefore 3-phase windings are always designed for 60° phase spread. Moreover the arrangement shown in Fig. 4.1(a) cannot be used with a single layer winding, because there are no conductors one pole pitch apart to form return conductors.

### 4.2.3 Multiturn Coil Windings

Machines with small number of poles and low value of flux per pole require a large number of conductors. Same is true for high voltage machines. Such machines have multiturn coils.

### 4.2.4 Full-pitch and Fractional Pitch Windings

When the span from centre-to-centre of the coil sides comprising a phase belt is equal to pole pitch, the winding is full pitch (Fig. 4.2a). When this span is less than the pole pitch, the winding is fractional pitch or short pitch. A fractional pitch winding is also known as chorded winding (Fig. 4.2b).

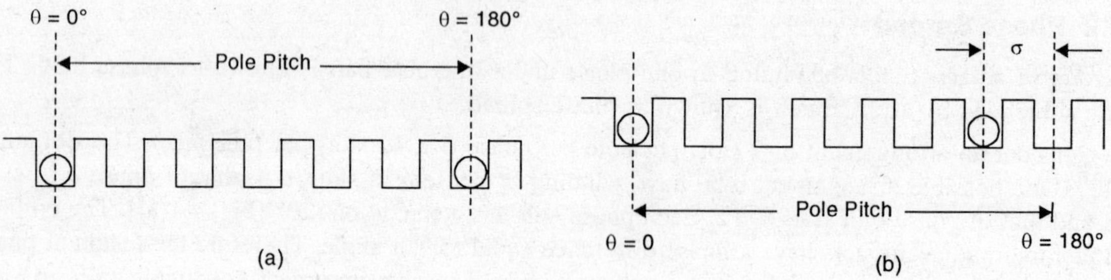


**Fig. 4.2.** (a) Full pitch coil, (b) short pitch coil

Fractional pitch windings are extensively used, because the resulting voltage wave form is more nearly sinusoidal than with full-pitch winding and also because of saving in copper and greater stiffness of coils due to shorter end connections. The only disadvantage of fractional pitch windings is that, terminal voltage is a bit less than that with full-pitch coils.

### 4.2.5 Integral Slot Windings

Since the three phases must be identical, the total number of slots in ac machines is always an integral multiple of three. However, the number of slots per pole per phase may be an integer or a fraction. In integral slot windings, the number of slots per pole per phase is an integer. Figure 4.3(a) shows a double layer integral slot full-pitch winding for a machine with 9 slots per pole (i.e. 3 slots per pole per phase). The coil span is full pitch, i.e. 9 slots. Figure 4.3(b) shows the same winding with a coil span of 8 slots, thus giving a short pitch winding. As seen in Fig. 4.3(b), some slots hold coil sides of different phases.

### 4.2.6 Fractional Slot Windings

In these windings the number of slots per pole per phase is a fraction. A fractional slot winding may be single layer or double layer, but the double layer windings are more common. A fractional slot winding permits the use of standard slotting arrangements over a wide range of pole numbers. Another advantage is that higher order harmonics in emf and mmf wave forms are reduced.

The total number of slots must be a multiple of 3, so that the windings of the three phases are symmetrical.

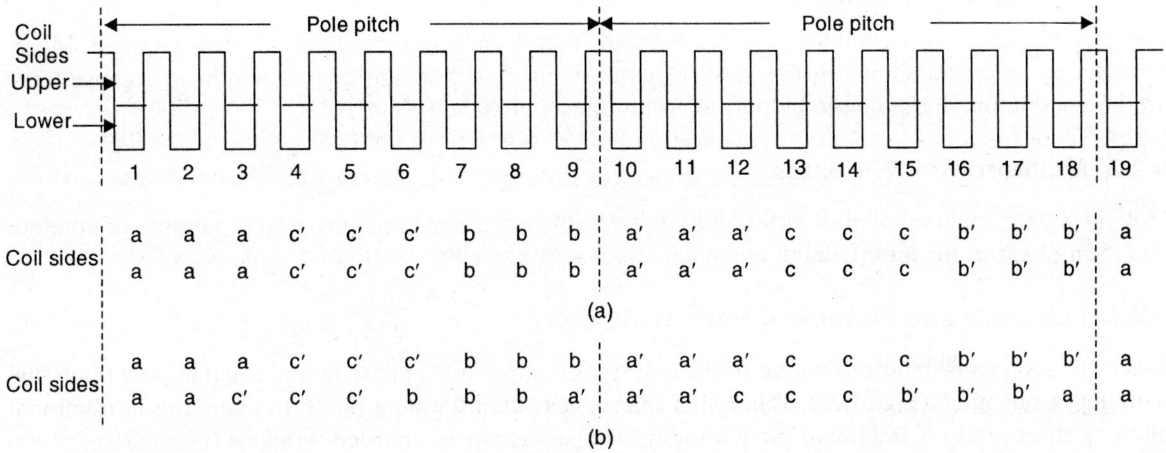


**Fig. 4.3.** Double layer integral slot winding (a) full pitch coils, (b) short pitch coils

For a 3-phase winding with $S$ slots the number of slots per pole per phase is $(S/3)/P$. If there is a common factor $k$ between $S/3$ and $P$, the characteristic ratio is $s_k/p_k$,

where
$$s_k = \frac{S/3}{k}$$

and
$$p_k = \frac{P}{k}.$$

The common factor $k$ expresses the number of times the slot arrangement repeats itself in one complete traverse around the armature periphery. For example in a machine having 30 slots and 4 poles, the ratio $\dfrac{S/3}{P}$ is 10/4 or 5/2. The common factor $k$ is 2, so that the slot arrangement is repeated twice. In the number 5/2, the denominator 2 is the number of poles required for a complete pattern and the numerator 5 is the number of slots for each phase in each pole pair. The two groups of each phase will occupy 3 and 2 slots.

In a double layer fractional slot winding, the arrangement of coil sides is repeated for the bottom layer, with corresponding coil sides located one coil span away. The coil span is always slightly less than one pole pitch. The start of phase $b$ is displaced 120° from start of phase $a$ and that of phase $c$ is further displaced by 120°.

### *Example 4.1*

A synchronous machine stator has 90 slots and 8 poles. Prepare a table showing the various slot positions for one repeatable basic section of the winding. Write the phases to which the coils in various slots belong. Choose a suitable coil span. Show the inter-connection of coils.

### *Solution*

$$\text{Slot angle} = \frac{8 \times 180}{90} = 16$$

$$\text{Slots per pole per phase} = \frac{90}{3 \times 8} = \frac{15}{4}$$

Thus one basic repeatable section of winding will have 4 poles and 15 slots per phase or 45 total slots. The arrangement of coils in this section will be repeated in the next section of 4 poles.

Beginning with zero, calculate the angles for different slots. Every time the total angle exceeds 180°, subtract 180° (i.e. angle of one pole pitch).

The phase groups are located as under

| | |
|---|---|
| Phase *a* | $0 \le \text{angle} < 60°$ |
| Phase *c* | $60° \le \text{angle} < 120°$ |
| Phase *b* | $120° \le \text{angle} < 10°$ |

The order of phases is *a c′ b a′ c b′*. Table 4.1 shows the results.

**Table 4.1**

| Slot number | 1 | 2 | 3 | 4 | 5 | 6 | 7 | 8 | 9 | 10 | 11 | 12 |
|---|---|---|---|---|---|---|---|---|---|---|---|---|
| Angle | 0° | 16° | 32° | 48° | 64° | 80° | 96° | 112° | 128° | 144° | 160° | 176° |
| Phase | a | a | a | a | c′ | c′ | c′ | c′ | b | b | b | b |
| Pole Pitch 1 | | | | | | | | | | | | |
| Slot number | 13 | 14 | 15 | 16 | 17 | 18 | 19 | 20 | 21 | 22 | 23 | |
| Angle | 12° | 28° | 44° | 60° | 76° | 92° | 108° | 124° | 140° | 156° | 172° | |
| Phase | a′ | a′ | a′ | c | c | c | c | b′ | b′ | b′ | b′ | |
| Pole Pitch 2 | | | | | | | | | | | | |
| Slot number | 24 | 25 | 26 | 27 | 28 | 29 | 30 | 31 | 32 | 33 | 34 | |
| Angle | 8° | 24° | 40° | 56° | 72° | 88° | 104° | 120° | 136° | 152° | 168° | |
| Phase | a | a | a | a | c′ | c′ | c′ | b | b | b | b | |
| Pole Pitch 3 | | | | | | | | | | | | |
| Slot number | 35 | 36 | 37 | 38 | 39 | 40 | 41 | 42 | 43 | 44 | 45 | |
| Angle | 4° | 20° | 36° | 52° | 68° | 84° | 100° | 116° | 132° | 148° | 164° | |
| Phase | a′ | a′ | a′ | a′ | c | c | c | c | b′ | b′ | b′ | |
| Pole Pitch 4 | | | | | | | | | | | | |

The arrangement shown in Table 4.1 will be repeated for the next 4 poles. Evidently this arrangement can also be used for 180 slot 16 pole, 270 slot 24 pole windings.

The pole pitch is $11\frac{1}{4}$ slots. The coil span can be 11 slots or 10 slots. If it is made of 11 slots, there will not be any decrease in third harmonic content in emf. However, a coil span of 10 slots will offer reasonable reduction in harmonic content. Hence coil span should be 10 slots.

The bottom layer of slots will have an arrangement similar to that shown in Table 4.1, except that corresponding coil sides will be located 10 slots away.

The layout of the winding, and connections of coil for the three phases for the first 4 poles is shown in Fig. 4.4(a). The layout of the remaining 4 poles would be exactly identical.

The complete layout showing interconnection of all the coils is shown in Fig. 4.4(b).

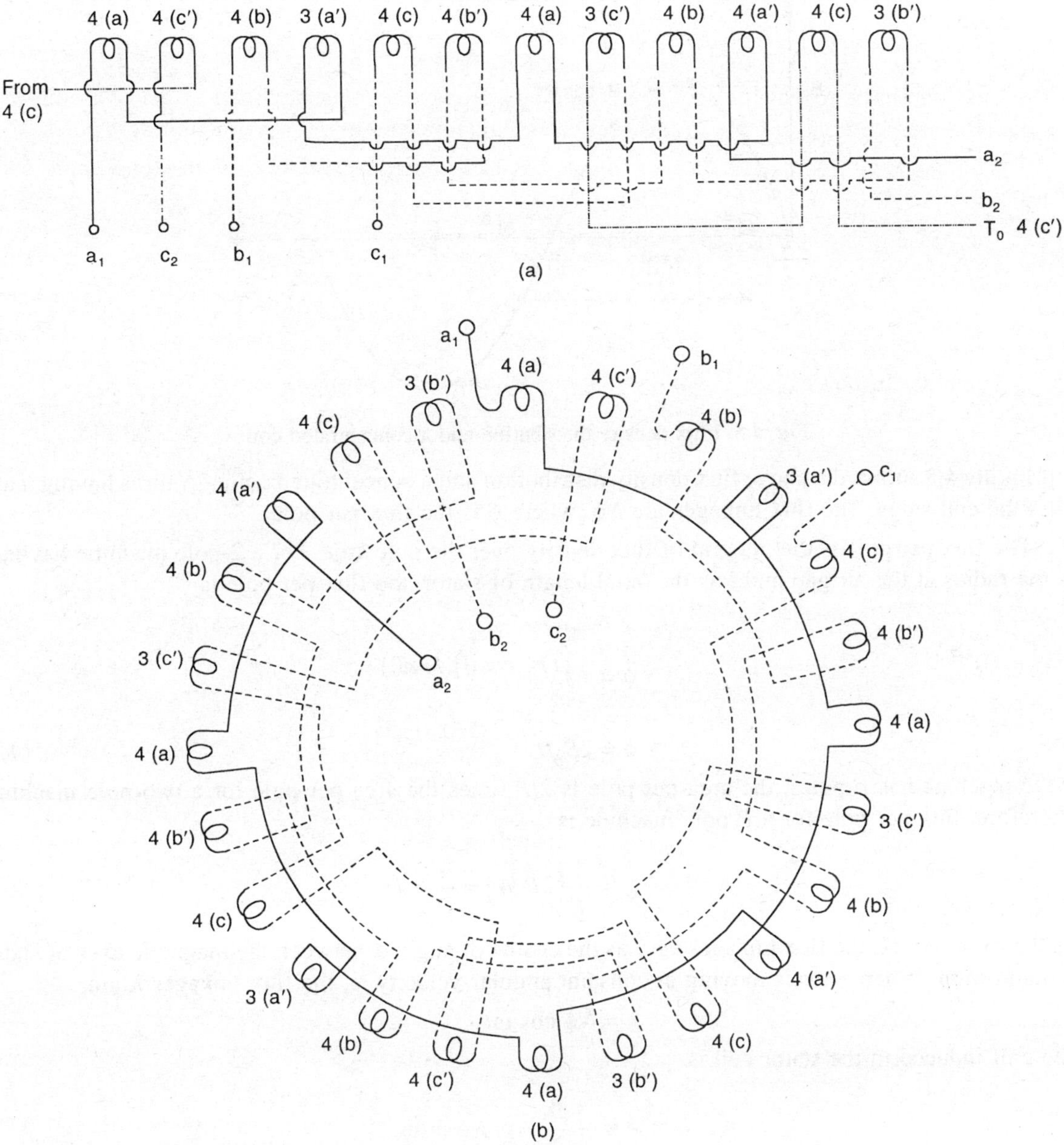


**Fig. 4.4.** Fractional slot armature winding for ac machine 8 poles 90 slots (a) layout for first 4 poles, (b) complete interconnection of coils

## 4.3 EMF EQUATION

The flux density wave in the air gap is sinusoidally distributed, i.e.

$$B = B_p \cos \theta \qquad ...(4.1)$$

where $B_p$ is the peak value of flux density (at the centre of rotor pole) and $\theta$ is the angle is radians measured from the rotor pole axis.

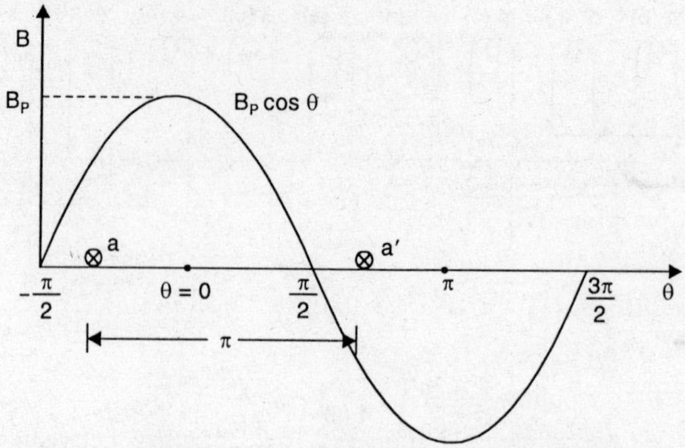


**Fig. 4.5.** Flux density distribution and a concentrated coil

Figure 4.5 shows the above flux density distribution and a concentrated coil of $N$ turns having $a$ and $a'$ as the coil sides. The flux linkages are $N\phi$, where $\phi$ is the flux per pole.

The flux per pole is the integral of flux density over the pole area. For a 2-pole machine having $r$ as the radius at the air gap and $l$ as the axial length of stator, the flux per pole is

$$\phi = \int_{-\pi/2}^{\pi/2} (B_p \cos\theta)(l\, r\, d\theta)$$

or

$$\phi = 2B_p lr \qquad \text{...(4.2)}$$

If the machine has $p$ poles, the area per pole is $2/P$ times the area per pole for a two-pole machine. Therefore, flux per pole for a $P$ pole machine is

$$\phi = \frac{2}{P}(2B_p lr) = \frac{4}{P}B_p lr \qquad \text{...(4.3)}$$

As the rotor moves, the flux linkages vary as the cosine of angle $\alpha$ between the magnetic axes of stator coil and rotor. When rotor is moving at constant angular velocity $\omega$, the flux linkages $\lambda$ are

$$\lambda = N\phi \cos \omega t$$

The emf induced in the stator coil is

$$e = -\frac{d\lambda}{dt} = \omega N\phi \sin \omega t$$

Thus the maximum value of induced emf is $\omega N \phi$. The rms value $E$ is

$$E = \frac{\omega N\phi}{\sqrt{2}} = \frac{2\pi f N\phi}{\sqrt{2}}$$

or

$$E = 4.44\, fN\, \phi \qquad \text{...(4.4)}$$

When the winding is distributed in slots, the above expression must be multiplied by a factor $k_b$ (known as breadth factor or distribution factor). Since the windings are short pitched, the expression must be multiplied by another factor $k_p$ (known as pitch factor). Thus the rms value of induced emf is

$$E = 4.44\ fN\phi\ k_b\ k_p \qquad ...(4.5a)$$

or

$$E = 4.44\ fN\phi k_\omega \qquad ...(4.5b)$$

where $E$ = rms emf per phase, volts

$f$ = frequency, Hz

$\phi$ = flux per pole, Wb

$N$ = number of turns in series

$k_b$ = breadth factor

$k_p$ = pitch factor

$k_\omega = k_b \times k_p$ = winding factor

Equation (4.5) is the emf for fundamental frequency. To distinguish between the emfs of various harmonics, it is better to write this emf as $E_1$, flux per pole as $\phi_1$ and winding factor as $k_{\omega 1}$. Thus

$$E_1 = 4.44 fN\phi_1 k_{\omega 1} \qquad ...(4.6)$$

For $n$ harmonic, emf per phase is $E_n$, flux per pole is $\phi_n$, frequency is $nf$ and windinig factor is $k_{\omega n}$. Thus

$$E_n = 4.44\, nfNf\phi_n k_{\omega n} \qquad ...(4.7)$$

For $n$th harmonic, pole pitch is $1/n$ th of the pole pitch for fundamental frequency. If $B_1$ and $B_n$ are the peak flux densities for fundamental and $n$th harmonic respectively,

$$\phi_n = \frac{1}{n}\frac{B_n}{B_1}\phi_1 \qquad ...(4.8)$$

From Eqs. (4.18), (4.19) and (4.20)

$$\frac{E_n}{E_1} = \frac{k_{\omega n} B_n}{k_{\omega 1} B_1} \qquad ...(4.9)$$

If the flux density wave contains harmonics, the phase voltage will contain all the harmonics present in the flux density wave. However, the triplen harmonics (3rd, 9th, etc.) will not be present in the line-to-line voltage, because the triplen harmonic emfs in the 3 phases are equal and in phase.

### 4.3.1 Breadth Factor

When the coils comprising a phase of the winding are distributed in two or more slots per pole, the emfs in the adjacent slots will be out-of-phase with respect to one another and their resultant will be less than their algebraic sum. Figure 4.6 shows the component emfs of coils ($E_1$, $E_2$ and $E_3$) and the resultant emf $E$ due to a phase group for a winding having 3 slots per pole per phase. Each component phasor is équal to rms value of coil emf. It is displaced from the component emf of adjacent slot by slot angle $\alpha$ electrical degrees. The distribution (or breadth factor) is defined as

$$k_b = \frac{\text{phasor sum of component emfs}}{\text{arithmetic sum of component emfs}} \qquad ...(4.10)$$

From the geometry of Fig. 4.6

$$k_b = \frac{2R\sin(q\alpha/2)}{q \times 2R\sin(\alpha/2)} = \frac{\sin(q\alpha/2)}{q\sin(\alpha/2)} \qquad ...(4.11)$$

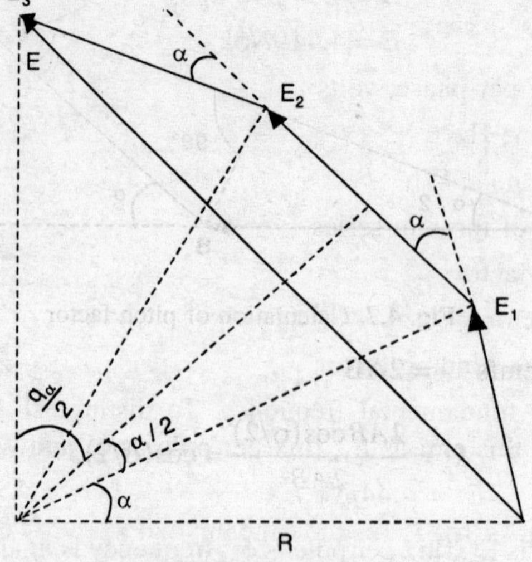


**Fig. 4.6.** Calculation of breadth factor

where $q$ = number of slots per pole per phase or the number of slots in a group.

When $q$ is very large, α becomes small and $k_b$ approaches the ratio chord/arc

$$k_b = \frac{\sin(q\alpha/2)}{q\alpha/2} \quad ...(4.12)$$

$q\alpha$ is also known as phase spread and is expressed in electrical radians.

The flux density wave may contain harmonics. Since the positive and negative halves of flux density wave are identical, only odd harmonics can be present. The harmonic poles of the $n$th space harmonic have a pitch equal to $1/n$ of the fundamental pole pitch. Therefore the angle between adjacent slots is $n\alpha$ for the $n$th harmonic and breadth factor for $n$th harmonic is

$$k_b = \frac{\sin(nq\alpha/2)}{q\sin(n\alpha/2)} \quad ...(4.13)$$

### 4.3.2 Pitch Factor

In a full pitch coil, the emfs in the two coil sides are in phase and therefore the coil emf is twice the emf of each coil side. In a short pitch coil, the emfs of the two coil sides are not in phase, but must be added vectorially to give the coil emf. The factor by which the emf per coil is reduced, because of the pitch being less than full pitch, is known as pitch factor (or coil span factor) $k_p$, given by

$$k_p = \frac{\text{phasor sum of coil side emfs}}{\text{arithmetic sum of coil side emfs}} \quad ...(4.14)$$

Figure 4.7 shows the coil side emfs $AB$ and $BC$ and the resultant coil emf $AC$ when the coil pitch is short of full pitch by electrical angle σ.

Phasor sum of coil side emfs = $AC$ = $2AB\cos(\sigma/2)$

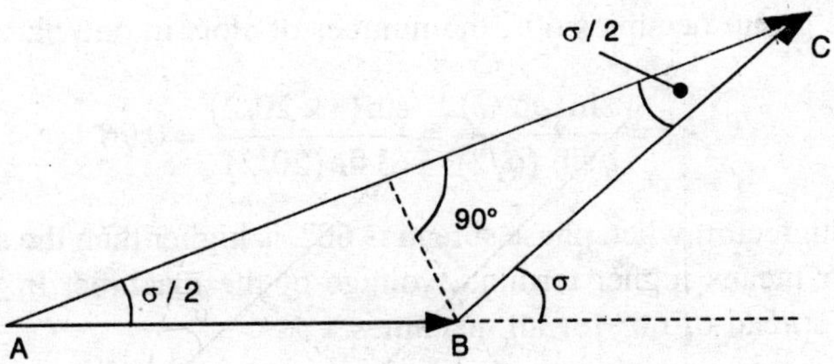


**Fig. 4.7.** Calculation of pitch factor

Arithmetic sum of coil side emfs = 2AB

$$k_p = \frac{2AB\cos(\sigma/2)}{2AB} = \cos(\sigma/2) \qquad ...(4.15)$$

The pitch factor given by Eq. (4.15) is for the fundamental frequency. If the flux density distribution contains space harmonics, the pitch factor for *n*th harmonic is

$$k_{pn} = \cos(n\sigma/2) \qquad ...(4.16)$$

The *n*th harmonic emf is reduced to zero if the angle σ is such that

$$\cos(n\sigma/2) = 0 \text{ or } \frac{n\sigma}{2} = 90° \qquad ...(4.17)$$

The flux density distribution contains some odd harmonics (because of symmetry, even harmonics are absent). Therefore phase voltage may contain third, fifth, seventh and higher order harmonics. The components of triplen (i.e. multiples of 3) harmonics are identical in the three phases and do not appear in the line-to-line voltage. Since the strength of harmonic components of voltage decreases with increasing frequency; only fifth and seventh harmonics are important. These are known as belt harmonics.

Equation (4.16) shows that pitch factor is different for different harmonics. By a proper selection of coil span, the pitch factor for 5th and 7th harmonics can be made small and thus these harmonics can be nearly eliminated from voltage wave. If coil span is 5/6 of pole pitch, the pitch factor is 0.259 for both fifth and seventh harmonics.

*Example 4.2*

Calculate the breadth factor for a machine having 9 slots per pole for the following cases and comment on the results: (a) a 3-phase winding with 120° phase groups, (b) a 3-phase winding with 60° phase groups.

*Solution*

$$\text{Slot angle} = \frac{180}{9} = 20°$$

(a) Since one phase occupies 120°, the number of slots in one phase group = $q = 6$

$$k_b = \frac{\sin(q\alpha/2)}{q\sin(\alpha/2)} = \frac{\sin(6\times 20/2)}{6\sin(20/2)} = 0.831$$

(b) Since one phase group occupies 60°, the number of slots in one phase group = $q = 3$

$$k_b = \frac{\sin(q\alpha/2)}{q\sin(\alpha/2)} = \frac{\sin(3\times 20/2)}{3\sin(20/2)} = 0.96$$

It is seen that breadth factor, when phase spread is 60°, is higher than the same for a spread of 120°. A higher breadth factor means higher terminal voltage of the machine. In view of this, 3-phase, ac windings have a phase spread of 60° for all machines.

*Example 4.3*

A 50 Hz, 600 rpm, salient pole synchronous generator has a sinusoidal flux density having a maximum value of 1 tesla. The generator has 180 slots wound with 2-layer 3-turn coils. The coil span is 15 slots and phase spread is 60°. The armature diameter is 1.25 m and core length 0.45 m. Find (a) peak value of emf per conductor, (b) peak value of emf per coil, (c) rms phase and line voltage, if the machine is star connected. [*P.U. 2001*]

*Solution*

$$P = \frac{(2f)}{n_s} = \frac{(2\times 50)}{(600/60)} = 10$$

$$\text{Area under one pole pitch} = \frac{\pi dl}{10} = \frac{\pi\times 1.25\times 0.45}{10} = 0.1767\,\text{m}^2$$

$$\text{Average flux density} = 1\left(\frac{2}{\pi}\right) = \frac{2}{\pi}\ \text{tesla}$$

$$\text{Flux per pole} = \left(\frac{2}{\pi}\right)(0.1767) = 0.1125\,\text{Wb}$$

(a) $\text{RMS emf per conductor} = 2.22\, f\,\phi$

$$= 2.22\,(0.1125)\,(50) = 12.49\ \text{V}$$

$$\text{Peak value of emf per conductor} = \sqrt{2}\,(12.49) = 17.66\ \text{V}$$

(b) RMS emf per coil = RMS emf per conductor × number of conductors in the coil × pitch factor

$$\text{Number of slots per pole} = \frac{180}{10} = 18$$

$$\text{Slot angle} = \frac{180}{10} = 10°\ \text{electrical}$$

The full-pitch coil will have a span of 18 slots. Since the actual coil span is 15 slots, the winding is short pitched by 3 slots.

$$\sigma = 3\times 10 = 30°$$

$$\text{Pitch factor} = k_p = \cos\left(\frac{\sigma}{2}\right) = \cos 15° = 0.966$$

Each coil has 3 turns or 6 conductors

$$\text{RMS emf per coil} = 12.49 \times 6 \times 0.966 = 72.39 \text{ V}$$

$$\text{Peak value of emf per coil} = \sqrt{2}(72.39) = 102.37 \text{ V}$$

(c) Number of slots per pole = 18

Since each phase group occupies 60°, the number of slots per pole per phase = $q = 6$

$$k_b = \frac{\sin(6 \times 10/2)}{6 \sin(10/2)} = 0.956$$

$$\text{Total number of coils per phase} = 180/3 = 60$$

$$\text{RMS value of phase voltage} = 72.39 \times 60 \times 0.956 = 4152 \text{ V}$$

$$\text{Line voltage} = \sqrt{3}(4152) = 7191.5 \text{ V.}$$

*Example 4.4*

Find the rms value of phase voltage for a 3-phase, 50 Hz, 20 pole, 180 slot synchronous generator having a single layer winding with full-pitch coils, the coils being connected in 60° phase groups, each coil having 6 turns. All the coils of a phase are is series. Flux per pole = 0.025 Wb.

*Solution*

$$\text{Number of slots/pole} = \frac{180}{20} = 9$$

$$\text{Slot angle} = \alpha = \frac{180}{9} = 20°$$

$$q = \text{number of slots per pole per phase} = \frac{180}{20 \times 3} = 3$$

$$k_b = \frac{\sin(q\alpha/2)}{q \sin(\alpha/2)} = \frac{\sin(3 \times 20/2)}{3 \sin(20/2)} = 0.96$$

Since it is a single layer winding, total number of conductors = 180 × 6 = 1080

$$\text{Number of conductors per phase} = \frac{1080}{3} = 360$$

$$N = \text{number of coils/phase} = \frac{360}{2} = 180$$

$$E = 4.44\, k_b k_p f \phi N$$

$$= 4.44(0.96)(1)(50)(0.025)(180) = 959 \text{ V.}$$

*Example 4.5*

Find the rms. value of different harmonic components and total emf per phase for a 50 Hz, 3-phase, synchronous generator having the following parameters: Number of poles = 10; slots/pole/phase = 2;

conductors per slot (2 layers) = 4; coil span = 150°; flux per pole (fundamental) = 0.12 Wb. The analysis of gap flux density shows a 20% third harmonic. All coils of a phase are in series.

*Solution*

For the fundamental

$$k_{b1} = \frac{\sin(q\alpha/2)}{q\sin(\alpha/2)} = \frac{\sin(2\times 30/2)}{2\sin(30/2)} = 0.966$$

$$k_{p1} = \cos\left[\frac{(180-150)}{2}\right] = 0.966$$

$$\text{Turns per phase} = 2\times 10\times\frac{4}{2} = 40$$

$$E_1 = 4.44\, k_{b1}k_{p1}f\phi_1 N = 4.44 \times 0.966 \times 0.966 \times 50 \times 0.12 \times 40$$

$$= 994.37 \text{ V}$$

For the third harmonic

$$\phi_n = \frac{1}{n}\frac{B_n}{B_1}\phi_1$$

or

$$\phi_3 = \frac{1}{3}\times 0.2\times 0.12 = 0.008\,\text{Wb}$$

$$k_{b3} = \frac{\sin(3\times 2\times 30/2)}{2\sin(3\times 30/2)} = 0.707$$

$$k_{p3} = \cos\left[\frac{3(180-150)}{2}\right] = 0.707$$

$$E_3 = 4.44k_{b3}k_{p3}3f\phi_3 N$$

$$= 4.44 \times 0.707 \times 0.707 \times 150 \times 0.008 \times 40 = 106.53 \text{ V}$$

$$\text{Induced emf per phase } E = \left[E_1^2 + E_3^2\right]^{0.5}$$

$$= \left[994.37^2 + 106.53^2\right]^{0.5} = 1000.06\,\text{V}.$$

## 4.4 MMF OF AC WINDINGS

### 4.4.1 MMF Space Wave of Concentrated Coil

Consider a single $N$ turn concentrated full-pitch coil carrying current $i$ and located in two slots of the stator of a 2-pole ac machine as shown in Fig. 4.8 (a). The magnetic field produced by this current is shown. As per right-hand rule the filed is clockwise for current entering the page (indicated by ⊗) and is anticlockwise for current coming out of page (indicated by ⊙). The mmf $F$ for any closed path is net ampere turns $Ni$ enclosed by that path. Assuming infinite permeability for iron, this mmf is consumed in the air gap only. Since each flux line crosses the air gap twice, the mmf for each gap is 0.5 $Ni$. The air gap mmf on the opposite sides of rotor are equal in magnitude and opposite in direction. Figure 4.8(b) shows the air gap mmf along with rotor and stator surfaces. The rectangular mmf wave can be decomposed into fundamental component, third, fifth and higher order harmonics. The fundamental ($F_{a1}$) and third harmonica ($\mathcal{F}_{a3}$) are shown in Fig. 4.8(b). By Fourier analyses the fundamental component $F_{a1}$ is

$$F_{a1} = \frac{4}{\pi}(0.5Ni)\cos\theta \qquad \text{...(4.18)}$$

where θ is the electrical angle measured from the magnetic axis of the coil which coincides with the positive peak of fundamental component as shown in Fig. 4.8(b).

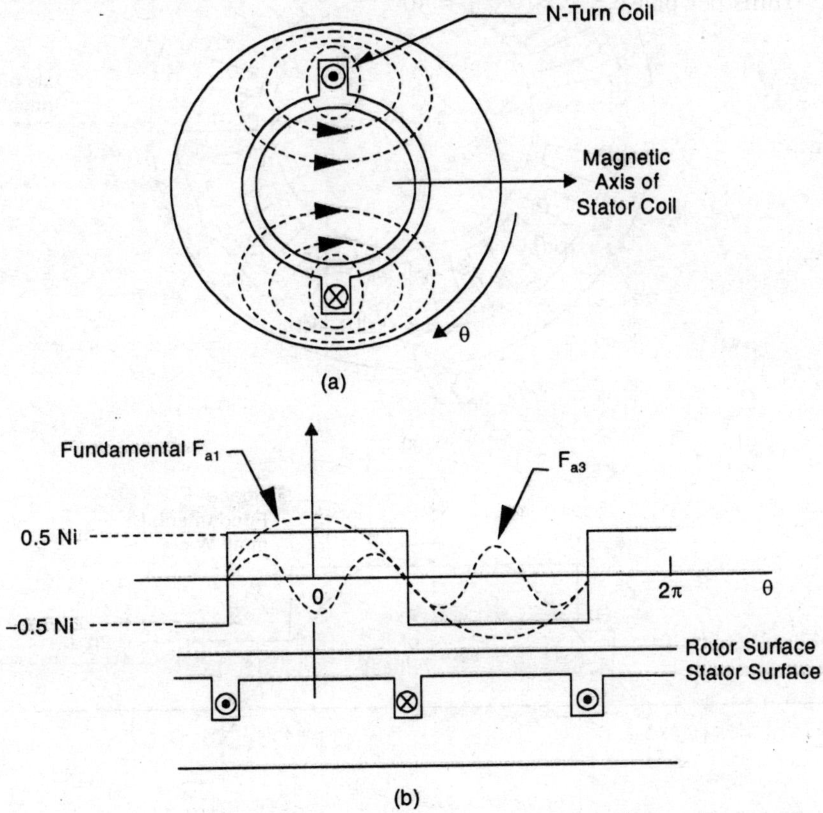


**Fig. 4.8.** MMF of concentrated full pitch coil (a) configuration, (b) mmf wave

### 4.4.2 MMF of Distributed Single-phase Winding

Figure 4.9(a) shows the distributed winding of phase $a$ of a 2-pole 3-phase ac machine. The empty slots house phases $b$ and $c$. It is a two-layer winding, each coil of $n_c$ turns having one side in the top half of one slot and the other side in the bottom half of another slot, nearly one pole pitch away. Figure 4.9(b) shows the winding laid out flat and mmf wave, which is a series of steps each of height 2 $n_c\, i_c$, i.e.

ampere conductors in one slot, when $i_c$ is the coil current. The distribution of winding produces a closer approximation to a sinusoid as compared to a concentrated coil (Fig. 4.8). The mmf wave can be decomposed into fundamental component and higher order harmonics. The mmf of a slot is displaced from that of adjacent slot by slot angle $\alpha$. The resultant mmf wave can be found by phasor addition of slot mmfs, the phase difference between adjacent phasors being $\alpha°$ electrical. (The effect is similar to that considered in Sec. 4.3 for finding out emf). Thus the effect of winding distribution on mmf wave can be accounted for by using a multiplication factor $k_b$ (same as that for finding emf). The effect of

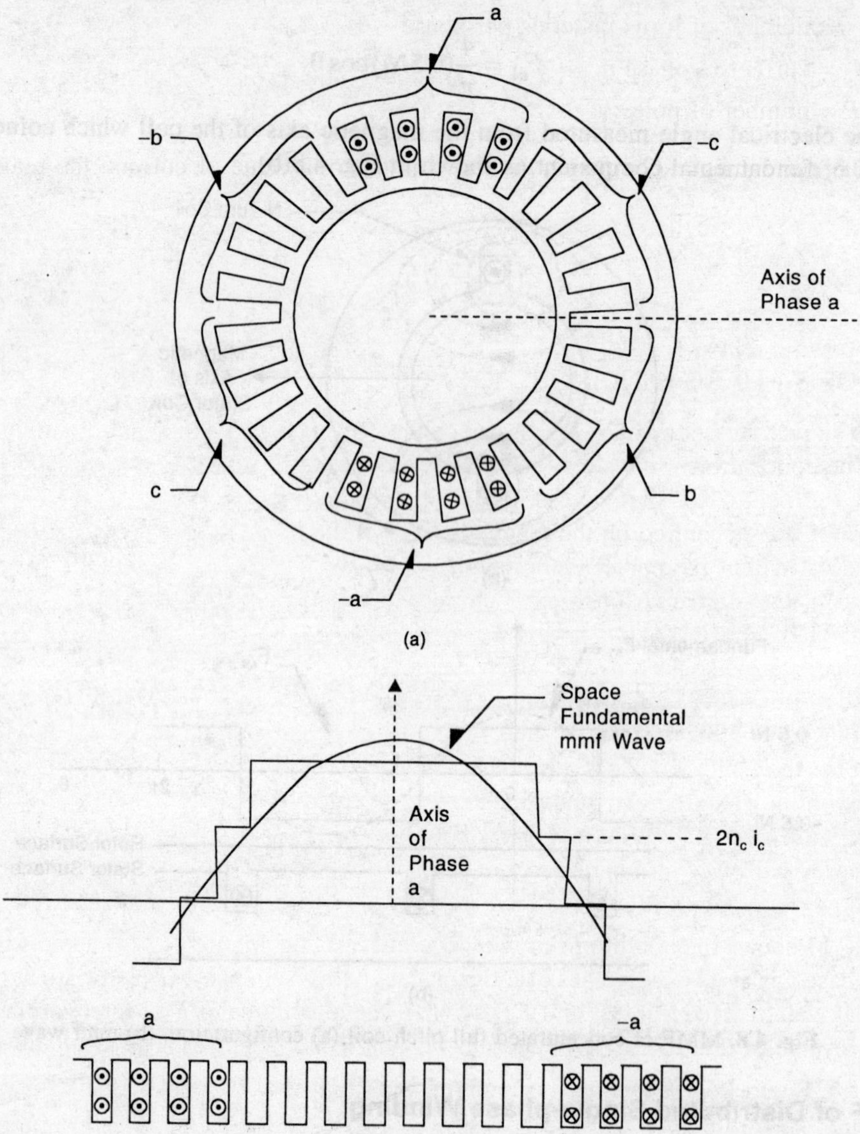


**Fig. 4.9.** MMF of one phase of a 2-pole 3-phase winding having full pitch coils (a) configuration, (b) mmf wave

pitch being less than full-pitch can be accounted for by using the multiplication factor $k_p$ (as in the case of emf). Thus the fundamental component of mmf of phase $a$ is $F_{a1}$ and equals

$$F_{a1} = \frac{4}{\pi} k_b k_p \frac{N_{ph}}{P} i_a \cos\theta \qquad ...(4.19)$$

where $k_b$ = breadth factor
$k_p$ = pitch factor
$N_{ph}$ = number of turns in series per phase
$i_a$ = current in phase $a$
$P$ = number of pole..

Since the current $i_a = \sqrt{2}\, I \cos\omega t$, , where $I$ is the rms value of current, the maximum value of $F_{a1}$ is

$$F_{max} = \frac{4}{\pi} k_b k_p \frac{N_{ph}}{P} \left(\sqrt{2} I\right) \qquad ...(4.20)$$

## 4.5 MMF OF 3-PHASE WINDINGS, ROTATING MAGNETIC FIELD

Figure 4.10 shows the stator of a 3-phase machine. To make the explanation simple the winding is assumed to be concentrated in 6 slots. These concentrated full-pitch coils represent the distributed windings producing sinusoidal mmf waves centred on the magnetic axes of the respective phases. $a$ and $a'$ represent the start and finish of a phase, $b$ and $b'$ the start and finish of $b$ phase and $c$ and $c'$ represent the start and finish of $c$ phase. The windings of the individual phases are displaced by 120 electrical degrees in the space around the air gap. Therefore the mmfs of these windings are also displaced by 120° in space. $F_a$ and $F_b$ and $F_c$ represent the mmfs of the three phases, when currents in them are at theii positive maximum values. Each phase current is alternating sinusoidally with time. As the currents alternate, the mmfs of the three phases will vary in magnitude and change direction, so as to follow the changes in currents.

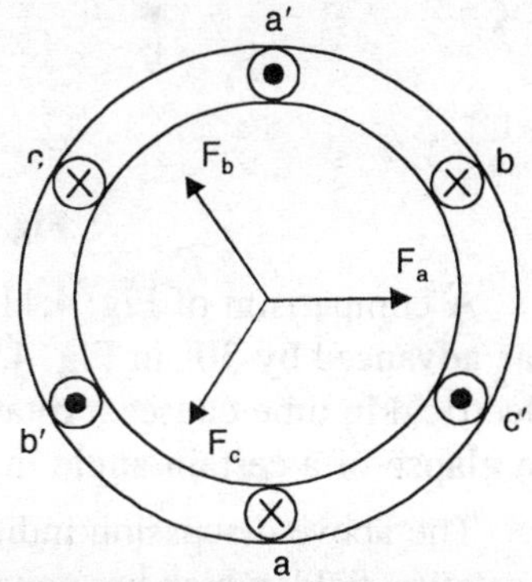


**Fig. 4.10.** Simplified configuration of a 2-pole 3-phase stator winding

Figure 4.11 shows the conditions at two different instants. Figure 4.11(a) depicts the instant when current in phase a is at its positive peak (at this instant currents in phases $b$ and $c$ are at one-half of their negative peak value). Therefore $F_b$ and $F_c$ have half the magnitude of $F_a$. Moreover the direction of $F_b$ and $F_c$ are reverse of that in Fig. 4.10 (because $I_b$ and $I_c$ are negative). The resultant of $F_a$, $F_b$ and $F_c$ is total mmf $F$. Figure 4.11(b) shows the conditions at an instant 30° later. $I_a$ is still positive but $\sqrt{3}/2$ of its peak value. Accordingly $F_a$ has the same direction but 0.866 times the magnitude of that in Fig. 4.11(a). $I_b$ and hence $F_b$ are zero. $I_c$ is negative and equal to $\sqrt{3}/2$ of its negative maximum value. Therefore $F_c$ has a direction opposite to that in Fig. 4.11(a). Magnitude of $F_a$ and $F_c$ are equal (since $I_a$ and $I_c$ are equal in magnitude). The resultant mmf $F$ is also shown.

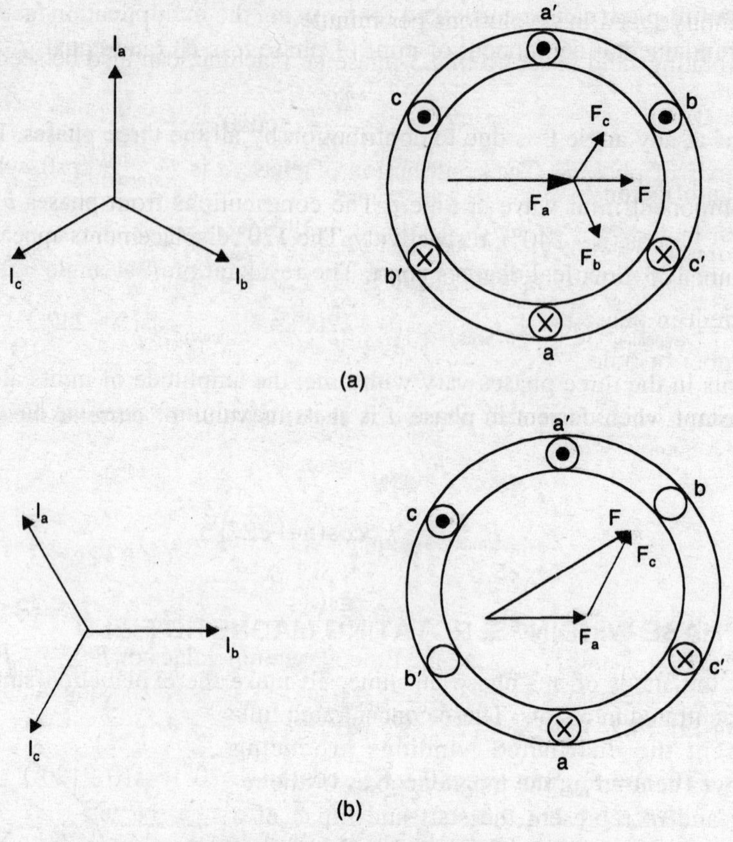


**Fig. 4.11.** Production of rotating field in ac machines

A comparison of Fig. 4.11 (a and b) shows that the resultant MMF has a constant magnitude, but has advanced by 30° in Fig. 4.11(b) as compared to its position is Fig. 4.11(a). Thus an elapse of 30° electrical in time causes a rotation of mmf by 30°. If we consider more instants of time, it is seen that an elapse of a certain angle in time results in rotation of mmf by the same angle.

The above discussion indicates that application of 3-phase currents to 3-phase windings produces a rotating field, which has constant amplitude and speed. The fact that its speed is constant is borne out by the fact that field travels by 30° in space for every 30 electrical degrees variation in time for the current phasors. In a two-pole machine (where electrical and mechanical angles are identical), each cycle of variation of current causes one complete revolution of mmf. Thus the speed of rotating field has a fixed relationship with the frequency of current and the number of poles of the machine. A winding designed for 4 poles requires two cycles of variation of current for one cycle of rotation of mmf. The speed of rotation of the field is known as synchronous speed and is given by

$$f = \left(\frac{P}{2}\right)(n_s) \qquad ...(4.21)$$

where $f$ is frequency (Hz), $P$ is the number of poles and $n_s$ is synchronous speed in revolutions per second. From Eq. (4.21)

$$N_s = \frac{120f}{P} \qquad ...(4.22)$$

where $N_s$ is synchronous speed in revolutions per minute.

The fact that a rotating field is set up in a 3-phase ac machine, can also be seen by the following analysis.

The air gap mmf at any angle θ is due to contribution by all the three phases. Let the angle θ be measured from the axis of phase *a*. The contribution of phase *a* is $F_{a(peak)}$ cosθ, where $F_{a(peak)}$ is the amplitude of the component mmf wave at time *t*. The contributions from phases *b* and *c* are $F_{b(peak)}$ cos(θ – 120°) and $F_{c(peak)}$ cos (θ – 240°), respectively. The 120° displacements appear because the axes of the three phases are 120 electrical degrees apart. The resultant mmf at angle θ is

$$F_\theta = F_{a(\text{peak})} \cos\theta + F_{b(\text{peak})} \cos(\theta - 120°) + F_{c(\text{peak})} \cos(\theta - 240°) \quad ...(4.23)$$

Since the currents in the three phases vary with time, the amplitude of mmfs also vary with time. Starting from the instant when current in phase *a* is at its maximum value, we have

$$F_{a(\text{peak})} = F_{a(\text{max})} \cos\omega t \quad ...(4.24a)$$

$$F_{b(\text{peak})} = F_{b(\text{max})} \cos(\omega t - 120°) \quad ...(4.24b)$$

$$F_{c(\text{peak})} = F_{c(\text{max})} \cos(\omega t - 240°) \quad ...(4.24c)$$

Evidently $F_{a(max)}$, $F_{b(max)}$ and $F_{c(max)}$ are the time maximum values of $F_{a(peak)}$, $F_{b(peak)}$ and $F_{c(peak)}$. Since the current $I_a$, $I_b$, $I_c$ are balanced, $F_{a(max)}$, $F_{b(max)}$ and $F_{c(max)}$ are all equal and can be denoted by $F_{max}$. By substituting Eq. (4.24) in Eq. (4.23), we have

$$F(\theta, t) = F_{\text{max}} \cos\omega t \cos\theta + F_{\text{max}} \cos(\omega t - 120°)\cos(\theta - 120°)$$

$$+ F_{\text{max}} \cos(\omega t - 240°)\cos(\theta - 240°) \quad ...(4.25)$$

Each of the three components of *F*(θ, *t*) is a pulsating standing wave. In each term the trigonometric function of *t* indicates the variation of amplitude with time, while the trigonometric function of θ defines the space distribution of a stationary sinusoid.

Using the trigonometric identity

$$\cos A \cos B = 0.5[\cos(A + B) + 0.5\cos(A - B)] \quad ...(4.26)$$

We can write Eq. (4.25) as

$$F(\theta, t) = 0.5[F_{\text{max}} \cos(\theta - \omega t) + F_{\text{max}} \cos(\theta + \omega t)$$

$$+F_{\text{max}} \cos(\theta - \omega t) + F_{\text{max}} \cos(\theta + \omega t - 240°)$$

$$+ F_{\text{max}} \cos(\theta - \omega t) + F_{\text{max}} \cos(\theta + \omega t - 480°)] \quad ...(4.27)$$

The terms $F_{max}$ cos (θ + ωt), $F_{max}$ cos (θ + ωt – 240°) and $F_{max}$ cos (θ + ωt – 480°) represent three sinusoids having the same magnitude $F_{max}$ and displaced by 120°. Their phasor sum is zero. Therefore

$$F(\theta, t) = 1.5 F_{\text{max}} \cos(\theta - \omega t) \quad ...(4.28)$$

The resultant mmf wave given by Eq. (4.28) has a constant magnitude. The term ω*t* provides rotation of mmf around the gap at a constant angular velocity ω. At a time $t_1$, the mmf wave is a sinusoid in space with its positive peak displaced $\omega t_1$ electrical radians from the fixed point (on the winding), which is origin for $\theta$. At a later instant $t_2$ the positive peak is displaced $\omega t_2$ electrical radians from the origin. This means that during the time $(t_2 - t_1)$, the wave has moved $\omega (t_2 - t_1)$ electrical radians around the air gap. At $t = 0$, $I_a$ is maximum and the resultant wave is directed along the axis of phase $a$. One-third cycle (i.e. 120°) later the current in phase $b$ is maximum and the resultant wave is directed along the axis of phase $b$ and so on. The angular velocity of the wave is $\omega = 2\pi f$ electrical radians per second. For a machine with $P$ pole, the velocity of rotating mmf is

$$\omega_s = \omega\left(\frac{2}{P}\right) \text{rad/sec}$$

and the synchronous speed is

$$N_s = \frac{(120 f)}{P} \text{ rpm}$$

It is important to mention that the speed of rotation of the field is always relative to the speed of the windings carrying 3-phase current. If the winding is stationary, $N_s$ is the absolute speed of the field. If the winding is itself revolving (e.g. rotor winding of 3-phase induction motor) the speed of rotation of the field relative to inertial space is the algebraic sum of the speed of rotating field and the speed of the winding.

The maximum value of mmf for a single-phase winding is given by Eq. (4.20). The 3-phase mmf has a constant amplitude equal to 1.5 times the maximum value of single phase mmf. Therefore, combining Eqs. (4.20 and 4.28), 3-phase mmf is

$$F(\theta, t) = 1.5\left[\frac{4}{\pi} k_b k_p \frac{N_{ph}}{P} \sqrt{2} I\right] \cos(\theta - \omega t)$$

$$= \left[\frac{2.7}{P} k_b k_p N_{ph} I\right] \cos(\theta - \omega t) \text{ AT/Pole} \qquad \text{...(4.29)}$$

*Example 4.6*

The rms value of armature current per phase of the synchronous machine of example 4.5 is 200 A. Find (a) synchronous speed, (b) magnitude of fundamental component of armature mmf.

*Solution*

(a) $N_s = \dfrac{120 f}{p} = \dfrac{120 \times 50}{10} = 600$ rpm

(b) Using Eq. (4.29)

$$\text{Magnitude of mmf} = \frac{2.7}{P} k_b k_p N_{ph} I$$

$$= \frac{2.7}{10} \times 0.966 \times 0.966 \times 40 \times 200$$

$$= 2015.62 \text{ AT/pole.}$$

## 4.6 REVERSAL OF DIRECTION OF ROTATING FIELD

Let the supply terminal $b$ and $c$ be interchanged. Then Eq. (4.24) takes the form

$$F_{a(\text{peak})} = F_{a(\max)} \cos \omega t \qquad \text{...(4.30}a\text{)}$$

$$F_{b(\text{peak})} = F_{b(\max)} \cos(\omega t - 240°) \qquad \text{...(4.30}b\text{)}$$

$$F_{c(\text{peak})} = F_{c(\max)} \cos(\omega t - 120°) \qquad \text{...(4.30}c\text{)}$$

Writing $F_{a(\max)}$, $F_{b(\max)}$ and $F_{c(\max)}$ as $F_{(\max)}$ and substituting Eq. (4.30) in Eq. (4.23)

$$F(\theta, t) = F_{\max} \cos \omega t \cos \theta + F_{\max} \cos(\theta - 120°) \cos(\omega t - 240°)$$

$$+ F_{\max} \cos(\theta - 240°) \cos(\omega t - 120°) \qquad \text{...(4.31)}$$

Using Eq. (4.26), we can write Eq. (4.31) as

$$F(\theta, t) = 0.5 F_{\max} [\cos(\theta - \omega t) + \cos(\theta + \omega t) + \cos(\theta - \omega t + 120°)$$

$$+ \cos(\theta - \omega t - 360°) + \cos(\theta + \omega t - 120°) + \cos(\theta + \omega t - 360°)] \qquad \text{...(4.32)}$$

The terms cos $(\theta - \omega t)$, cos $(\theta - \omega t + 120°)$ and cos $(\theta - \omega t - 120°)$ add upto zero. Moreover, cos $(\theta + \omega t)$ and cos $(\theta + \omega t - 360°)$ are equal. Therefore Eq. (4.32) reduces to

$$F(\theta, t) = 1.5 F_{\max} \cos(\theta + \omega t) \qquad \text{...(4.33)}$$

A comparison of Eqs. (4.28) and 4.33) shows that direction of rotation of field has reversed. Thus an interchange of any two supply terminals causes a reversal of direction of rotation of the magnetic field produced by 3-phase currents.

The rotating mmf wave creates a coincident flux density wave in the air gap. If the reluctance of iron path is negligible, the peak value of flux density is

$$B_{\text{peak}} = \frac{(1.5 F_{\max}) \mu_0}{\text{length of air gap}} \qquad \text{...(4.34)}$$

## 4.7 TORQUE IN AC MACHINES

Figure 4.12(a) shows a round rotor (also known as cylindrical rotor) ac machine. The flow of current in stator winding creates a magnetic flux, which completes its path through air gap, stator iron and rotor iron. This stator field causes the appearance of north and south poles on the stator surface as shown in Fig. 4.12(a). Similarly the flow of current in rotor winding also creates a magnetic flux which also causes appearance of north and south poles on rotor surface. These two magnetic field tend to align themselves and electromagnetic torque is developed. The magnitude of this torque is proportional to the product of stator and rotor mmf and sine of angle between their magnetic axes. The assumptions for derivation of torque equation are as follows:

(1) The stator and rotor iron has infinite permeability. MMF is required to set up flux in the air gap only. Evidently saturation and hysteresis are absent.

(2) Rotor is cylindrical, so that the air gap is uniform around the periphery.

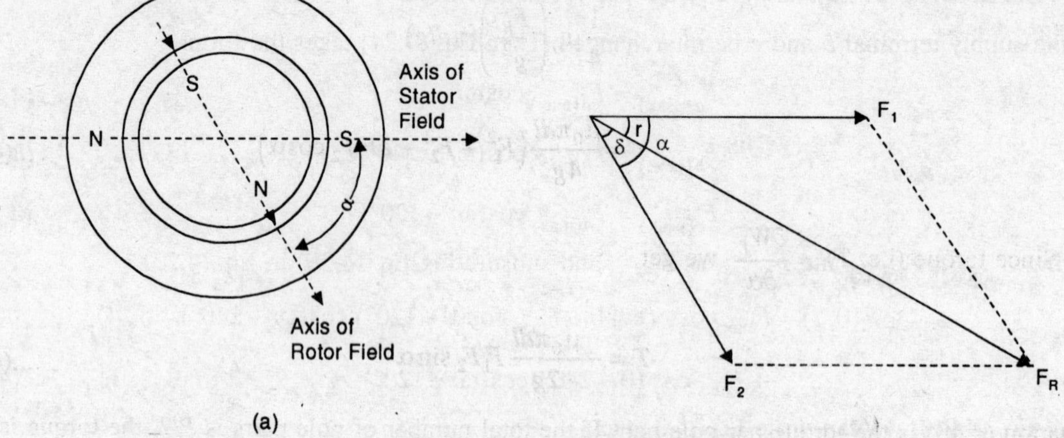


**Fig. 4.12.** 2-pole machine (a) simplified configuration, (b) phasor diagram showing stator and rotor mmfs

(3) Only the fundamental components of stator and rotor mmfs are considered. The harmonics are neglected.

(4) The leakage flux is negligible and whole of the flux is mutual flux.

(5) The path of mutual flux, in the air gap, is radial and the flux density is constant throughout the air gap.

For production of torque it is necessary that the two mmfs should be stationary with respect to each other and have the same number of poles. If they have a relative velocity, the torque would be alternating in nature and average torque will be zero.

Let $F_1$ and $F_2$ be the fields of stator and rotor, respectively. Since they are spatial sine waves, they can be represented by phasors (Fig. 4.12b). $F_R$ is the resultant field and is given by

$$F_R^2 = F_1^2 + F_2^2 + 2F_1F_2\cos\alpha \qquad ...(4.35)$$

The resultant field intensity is

$$H_R = \frac{F_R}{g} \qquad ...(4.36)$$

where $g$ is the length of air gap.

$$\text{Total energy in an air gap} = \frac{1}{2}\mu_0 H^2 \text{ (volume of gap)} \qquad ...(4.37)$$

Since $H_R$ is sinusoidally distributed, average value of $H^2 = 0.5\,H_R^2$

$$\text{The volume of air gap} = \pi dlg \qquad ...(4.38)$$

where $d$ and $l$ are average diameter of core (at the air gap) and axial length of core, respectively.

Therefore the total energy in the air gap, i.e. $W_f$ is

$$W_f = \frac{1}{2}\mu_0\left(0.5H_R^2\right)(\pi dlg)$$

$$= \frac{1}{4}\mu_0\left(\frac{F_R^2}{g^2}\right)(\pi dlg)$$

$$= \frac{\mu_0 \pi dl}{4g}\left(F_1^2 + F_2^2 + 2F_1F_2\cos\alpha\right) \qquad \text{...(4.39)}$$

Since torque (i.e. $T$) $= \dfrac{\partial W_f}{\partial \alpha}$, we get,

$$T = -\frac{\mu_0 \pi dl}{2g} F_1 F_2 \sin\alpha \qquad \text{...(4.40)}$$

Equation (4.40) is the torque per pole pair. If the total number of pole pairs is $P/2$, the torque is

$$T = -\left(\frac{P}{2}\right)\left(\frac{\mu_0 \pi dl}{2g}\right) F_1 F_2 \sin\alpha \qquad \text{...(4.41)}$$

Equation (4.41) indicates that torque is proportional to the peak values of stator and rotor mmfs and sine of electrical angle between them. the negative sign in Eq. (4.40) means that torque is in a direction, so as to decrease the angle $\alpha$. Equal and opposite torques act on the stator and rotor. The stator torque is transmitted to the foundations through the frame of the machine.

From Fig. 4.12(b), it is seen that $F_1 \sin\alpha = F_R \sin\delta$ and $F_2 \sin\alpha = F_R \sin\gamma$. Therefore, Eq. (4.41) can also be written as

$$T = -\left(\frac{P}{2}\right)\left(\frac{\mu_0 \pi dl}{2g}\right) F_2 F_R \sin\delta \qquad \text{...(4.42)}$$

and

$$T = -\left(\frac{P}{2}\right)\left(\frac{\mu_0 \pi dl}{2g}\right) F_1 F_R \sin\gamma \qquad \text{...(4.43)}$$

The above equations can also be expressed in terms of resultant flux per pole (i.e. $\phi_R$)

$$\phi_R = \text{(Average flux density) (area under the pole)}$$

$$= \left(\frac{2}{\pi} B_R\right)\left(\frac{\pi dl}{P}\right) = \frac{2B_R dl}{P} \qquad \text{...(4.44)}$$

$$B_R = \mu_0 F_R / g \qquad \text{...(4.45)}$$

Combining Eqs (4.42), (4.44) and (4.45)

$$T = -\left(\frac{\pi}{2}\right)\left(\frac{P}{2}\right)^2 F_2 \phi_R \sin\delta \qquad \text{...(4.46)}$$

## 4.8 LOSSES

A study of losses in machines is important for the following three reasons: (1) Losses decrease efficiency of machine. (2) The heat energy due to losses has to be dissipated to the surroundings, without exceeding the temperature limits of different parts of the machine. (3) The voltage drop or current components representing the losses has to be taken into account in the equivalent circuit of the machine.

The losses can be classified as (a) copper-losses, (b) core-losses, (c) mechanical losses,(d) stray load losses.

The copper-losses are variable, because they depend on the current. All other losses are more or less constant and are termed as constant losses (or $P_c$).

### 4.8.1 Copper-losses

The copper-losses are the $I^2R$ losses in stator and rotor windings. These losses are calculated with the dc resistance of the winding at 75°C.

### 4.8.2 Core-losses

These losses consist of hysteresis and eddy current losses. The hysteresis loss in an iron core is the energy required for reorientation of domains in each cycle of ac current supplied to the core. It can be calculated as

$$\text{Hysteresis loss} = k_h f B_m^{1.6} \qquad ...(4.47)$$

where $f$ is frequency in Hz, $B_m$ is maximum flux density and $k_h$ is a constant.

The eddy current loss is due to the flow of eddy currents in the core. It can be calculated as

$$\text{Eddy current loss} = k_e B_m^2 f^2 t^2 \qquad ...(4.48)$$

where $t$ is the thickness of laminations and $k_e$ is a constant.

### 4.8.3 Mechanical Losses

These losses are due to brush and bearing friction and windage (or air friction).

When the machine is running at no-load, the input power is used to overcome the mechanical and no-load core losses. Therefore these two losses, lumped together, are known as no-load rotational losses.

### 4.8.4 Stray Load Losses

When a machine is supplying load, additional copper-losses occur due to non-uniform current distribution in the windings. Moreover additional core-losses occur due to distortion of magnetic flux by the load current. It is very difficult to determine these losses. For ac machines they are taken as 0.5% of the rating.

The efficiency of a motor is written as

$$\text{Motor efficiency} = \frac{\text{Input} - \text{Losses}}{\text{Input}} \qquad ...(4.49)$$

For an ac motor, the input can be written as $KVI$, where the constant $K$ equals $\sqrt{3} \times$ power factor for a 3-phase machine. The losses are equal to sum of constant losses $P_c$ and copper-losses $3I^2R$.

Therefore

$$\eta = \text{Motor efficiency} = \frac{KVI - P_c - 3I^2R}{KVI} \quad \text{...(4.50)}$$

For maximum efficiency $d\eta/dI$ should be zero.

$$\frac{d\eta}{dI} = \frac{(KV - 6IR)(KVI) - \left(KVI - P_c - 3I^2R\right)(KV)}{(KVI)^2} = 0$$

or

$$P_c = 3I^2R \quad \text{...(4.51)}$$

$$\eta = \text{Generator efficiency} = \frac{\text{output}}{\text{output + losses}}$$

or

$$\eta = \frac{KVI}{KVI + P_c + 3I^2R} \quad \text{...(4.52)}$$

For maximum efficiency

$$\frac{d\eta}{dI} = \frac{KV\left(KVI + P_c + 3I^2R\right) - KVI(KV + 6RI)}{\left(KVI + P_c + 3I^2R\right)^2} = 0$$

or

$$P_c = 3I^2R \quad \text{...(4.53)}$$

In both cases, maximum efficiency is achieved when constant losses are equal to copper losses (This is true for all ac machines, dc machines and transformers).

## 4.9 SLOT OR TOOTH HARMONICS

The use of distributed windings causes an additional problem in ac machines. The reluctance of the flux path through teeth is lesser as compared to that through slots. This causes tufting of flux at tooth tips as shown in Fig. 4.13 and consequent ripples in the flux wave. The effect is the same as though there

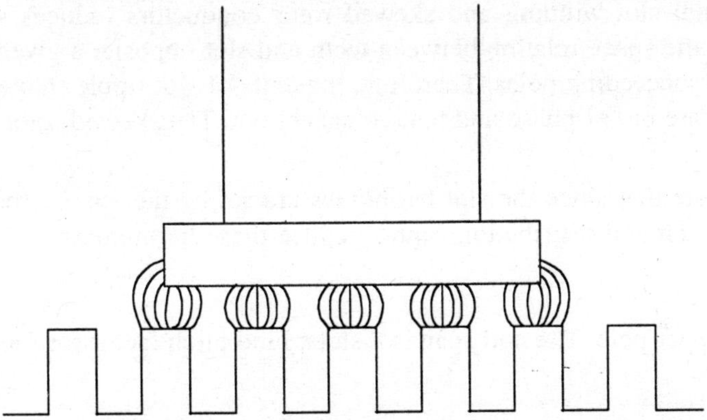

**Fig. 4.13.** Tufting of flux at tooth tips

were superposed, on the main flux, an additional flux $\phi_s$ which pulsates at the frequency with which the teeth pass the pole tips. A complete cycle of this tooth frequency takes place in the time required for the pole to move through an angle equivalent to one tooth pith (or slot pitch). If the machine has $q$ slots per pole per phase and 3-phases, there will be $2 \times 3 \times q$ cycles of tooth frequency for each cycle of fundamental frequency $f$. Thus

$$f_s = f \text{ (Number of slots per pole pair)} = f(2S/P) \qquad \text{...(4.54)}$$

where $f_s$ = tooth (or slot) frequency

$S$ = Total number of stator slots

If the field flux per pole is $\phi$ and an additional flux $\phi_s$ (due to slot harmonics) exists, then

$$\phi_{\text{total}} = (\phi + \phi_s \sin 2\pi f_s t)\cos 2\pi ft \qquad \text{...(4.55)}$$

The induced emf is

$$e = -N\frac{d\phi_{\text{total}}}{dt} = 2\pi fN\phi \sin 2\pi ft$$

$$-\frac{\phi_s N}{2}\left[2\pi(f_s - f)\cos 2\pi(f_s - f)t + 2\pi(f_s + f)\cos 2\pi(f_s + f)t\right] \qquad \text{...(4.56)}$$

The emf contains a component corresponding to fundamental frequency and two other components due to tooth ripples. The frequencies of these components are ($f_s - f$) and ($f_s + f$) or $f(2S/P \pm 1)$. For a machine having 2 slots per pole per phase, the order of these harmonics will be 11th and 13th, corresponding to 550 and 650 Hz.

Slot harmonics may cause many problems, as follows:

(1) If the generated voltage of synchronous machines contains these harmonics, it will cause interference in telecommunication lines.

(2) They introduce vibration and noise in the machine.

(3) The interaction of stator and rotor slot harmonics produces parasitic torques in induction motor. These torques may have a serious effect on the torque-speed curve of the machine.

(4) An increase in core losses by introduction of high frequency components of voltages and currents into the teeth of the stator.

The use of fractional slot winding and skewed rotor conductors reduces slot harmonics. In a fractional slot winding, the space relation between teeth and slot opposite a given pole face is not the same as in the next and succeeding poles. Therefore, the emfs of slot ripple frequencies, (i.e. $f_s \pm f$) in different armature coils are out of phase and tend to cancel out. The skewed rotor conductors are used in induction motors.

It is important to note that since the slot harmonics are set by the spacing between adjacent coil slots, variation in coil pitch and distribution cannot reduce these harmonics.

*Example 4.7*

An alternator has 9 slots per pole. The coil span is 8 slots. Find pitch factor for fundamental frequency.

*Solution*

$$\text{Slot angle} = \frac{180}{9} = 20°$$

Since the coil span is 8 slots, the winding is short pitched by 1 slot

$$\sigma = 1 \times 20° = 20°$$

Pitch factor for fundamental = $\cos(\sigma/2)$

$$= \cos 10° = 0.985.$$

*Example 4.8*

A 3-phase, 8-pole, 750 rpm synchronous alternator has 72 slots. Each slot has 12 conductors and winding is short pitched by 2 slots. Find pitch factor and breadth factor. If flux per pole is 0.06 Wb, find induced emf per phase.

*Solution*

$$\text{Number of slots per pole} = \frac{72}{8} = 9$$

$$\text{Slot angle} = \frac{180}{9} = 20°$$

$$q = \text{Number of slots per pole per phase} = \frac{9}{3} = 3$$

$$\text{Breadth factor } k_b = \frac{\sin(q\alpha/2)}{q\sin(\alpha/2)} = \frac{\sin(3 \times 20/2)}{3\sin(20/2)}$$

$$= 0.96$$

Angle by which coils are short pitched = $2 \times 20 = 40°$

Pitch factor $k_p = \cos(40/2) = 0.94$

Total number of conductors = $72 \times 12 = 864$

$$N = \text{Number of coils per phase} = \frac{864}{3 \times 2} = 144$$

$$E = 4.44 k_b k_p f \phi N$$

$$= 4.44 \times 0.96 \times 0.94 \times 50 \times .06 \times 144$$

$$= 1730.8 \text{ volts per phase.}$$

*Example 4.9*

A 16-pole 3-phase star connected alternator has 144 slots. The coils are short pitched by one slot. The flux per pole is

$$\phi = 100 \sin\theta + 30 \sin 3\theta + 20 \sin 5\theta$$

Find harmonics as percentage of phase voltage and line voltage. [*K.U. 2002*]

*Solution*

$$\text{Number of slots per pole} = \frac{144}{16} = 9$$

$$\text{slot angle } \alpha = \frac{180}{9} = 20°$$

$$q = 3 \text{ and } \sigma = 20°$$

$$k_{b1} = \frac{\sin(q\alpha/2)}{q\sin(\alpha/2)} = \frac{\sin(3\times 20/2)}{3\sin(20/2)} = 0.96$$

$$k_{p1} = \cos\left(\frac{20}{2}\right) = 0.985$$

$$k_{b3} = \frac{\sin(3\times 3\times 20/2)}{3\sin(3\times 20/2)} = \frac{\sin 90°}{3\sin 30°} = 0.667$$

$$k_{p3} = \cos\left(3\times\frac{20}{2}\right) = 0.866$$

$$k_{b5} = \frac{\sin(5\times 3\times 20/2)}{3\sin(5\times 20/2)} = \frac{\sin 150°}{3\sin 50°} = 0.218$$

$$k_{p5} = \cos\left(5\times\frac{20}{2}\right) = 0.643$$

Using Eqs. (4.6) and (4.7)

$$\frac{E_3}{E_1} = \frac{3\phi_3 k_{\omega 3}}{\phi_1 k_{\omega 1}}$$

$$= \frac{3\times 30\times 0.667\times 0.866}{100\times 0.96\times 0.985} = 0.55$$

$$\frac{E_5}{E_1} = \frac{5\phi_3 k_{\omega 5}}{\phi_1 k_{\omega 1}} = \frac{5\times 20\times 0.218\times 0.643}{100\times 0.96\times 0.985} = 0.15$$

Phase emf $=[1^2 + 0.55^2 + 0.15^2]^{0.5} = 1.15$

Third harmonic as percentage of phase voltage $= \frac{0.55}{1.15}\times 100 = 47.83\%$

Fifth harmonic as percentage of phase voltage $= \frac{0.15}{1.15}\times 100 = 13.04\%$

Third harmonic will not appear in line voltage

$$\text{Line voltage} = 1.732\left(E_1^2 + E_5^2\right)^{0.5}$$

$$= 1.732(1^2 + 0.15^2)^{0.5} = 1.751$$

Third harmonic as percentage of line voltage = 0

$$\text{Fifth harmonic as percentage of line voltage} = \frac{\sqrt{3}\times 0.15}{1.751}\times 100 = 14.84\%.$$

*Example 4.10*

The flux′distribution curve of a smooth core 50 Hz generator is

$$B = \sin\theta + 0.2 \sin 3\theta + 0.2 \sin 5\theta + 0.2 \sin 7\theta \text{ Wb/m}^2$$

where θ is the angle measured from neutral axis. The pole pitch is 35 cm, the core length is 32 cm and stator coil span is four-fifth of pole pitch. Find equation for emf induced in one turn and its rms value.

*Solution*

$$k_{p1} = \cos\left[\frac{180(1/5)}{2}\right] = 0.951$$

$$k_{p3} = \cos\left(\frac{3\times 180}{10}\right) = 0.588$$

$$k_{p5} = \cos\left(\frac{5\times 180}{10}\right) = 0$$

$$k_{p7} = \cos\left(\frac{7\times 180}{10}\right) = -0.588$$

Since the coil has only one turn, $k_b = 1$

Radius $r$ at air gap is given by

$$\frac{2\pi r}{P} = 0.35 \text{ or } r = \frac{0.35P}{2\pi}$$

Using Eq. (4.3)

$$\phi_1 = \frac{4}{P}B_1 lr = \frac{4}{P}\times 1\times 0.32\times\frac{0.35P}{2\pi}$$

$$= 0.0713 \text{ Wb}$$

$$E_1 = 4.44\, fN\phi_1\, k_{b1}\, k_{p1}$$

$$= 4.44 \times 50 \times 1 \times 0.0713 \times 1 \times 0.951$$

$$= 15.053 \text{ V}$$

Using Eq. (4.9)

$$E_3 = E_1\left[\frac{k_{\omega 3}B_3}{k_{\omega 1}B_1}\right] = 15.053\left[\frac{1\times 0.588\times 0.2}{1\times 0.951\times 1}\right]$$

$$= 1.86 \text{ V}$$

$$E_5 = 0 \text{ because } k_{p5} = 0$$

$$E_7 = E_1\left[\frac{k_{\omega 7}B_7}{k_{\omega 1}B_1}\right] = 15.053\left[\frac{1\times(-0.588)\times 0.2}{1\times 0.951\times 1}\right]$$

$$= -1.86 \text{ V}$$

$$e = \sqrt{2}\left[E_1 \sin\theta + E_3 \sin 3\theta + E_5 \sin 5\theta + E_7 \sin 7\theta\right]$$

$$= 21.3 \sin\theta + 2.63 \sin 3\theta - 2.63 \sin 7\theta.$$

*Example 4.11*

A 3-phase star connected alternator has 81 slots, 6 poles and a double layer, narrow spread winding having a coil span of 13 slot pitches. The flux density distribution in the air gap is

$$B(\theta) = \sin\theta + 0.4 \sin 3\theta + 0.25 \sin 5\theta$$

Find (a) rms values of third and fifth harmonic phase voltages in terms of fundamental frequency phase voltage, (b) ratio of line voltage to phase voltage. [*P.U. 2002*]

*Solution*

(a)

$$q = \frac{81}{3\times 6} = \frac{9}{2}$$

$$\alpha = \text{slot angle} = \frac{60}{(9/2)} = \frac{40°}{3}$$

$\sigma$ = Angle by which coils are short pitched

$$= 180 - 13\left(\frac{40}{3}\right) = \frac{20°}{3}$$

$$k_{b1} = \frac{\sin(q\alpha/2)}{q\sin(\alpha/2)} = \frac{\sin(30°)}{\frac{9}{2}\sin\left(\frac{20}{3}\right)} = 0.957$$

$$k_{p1} = \cos\left(\frac{\sigma}{2}\right) = \cos\left(\frac{10}{3}\right) = 0.998$$

$$k_{\omega 1} = k_{b1}k_{p1} = 0.957 \times 0.998 = 0.955$$

$$k_{b3} = \frac{\sin(3\times 30°)}{\frac{9}{2}\sin\left(\frac{3\times 20}{3}\right)} = 0.65$$

$$k_{p3} = \cos\left(3\times\frac{10}{3}\right) = 0.985$$

$$k_{\omega 3} = k_{b3}k_{p3} = 0.65 \times 0.985 = 0.64$$

$$k_{b5} = \frac{\sin(5 \times 30°)}{\frac{9}{2}\sin\left(\frac{5 \times 20}{3}\right)} = 0.202$$

$$k_{p5} = \cos\left(5 \times \frac{10}{3}\right) = 0.958$$

$$k_{\omega 5} = k_{b5} \times k_{p5} = 0.202 \times 0.958 = 0.194$$

Using Eq. (4.9)

$$\frac{E_3}{E_1} = \frac{k_{\omega 3}B_3}{k_{\omega 1}B_1} = \frac{0.64 \times 0.4}{0.955} = 0.268$$

$$\frac{E_5}{E_1} = \frac{k_{\omega 5}B_5}{k_{\omega 1}B_1} = \frac{0.194 \times 0.25}{0.955} = 0.051$$

(b) Resultant phase voltage $= \left[E_1^2 + E_3^2 + E_5^2\right]^{0.5}$

$$= E_1\left[1 + 0.268^2 + 0.051^2\right]^{0.5} = 1.0365E_1$$

The third harmonic component will not be present in the line voltage.

Resultant line voltage $= \sqrt{3}\left[E_1^2 + E_5^2\right]^{0.5}$

$$= \sqrt{3}\,E_1\left[1 + 0.051^2\right]^{0.5} = 1.734E_1$$

Ratio of line voltage to phase voltage

$$= \frac{1.734E_1}{1.0365E_1} = 1.673.$$

### *Example 4.12*

The flux density distribution in the air gap of a synchronous alternator is as under:

$$B = B_1 \sin\theta + B_3 \sin 3\theta + B_5 \sin 5\theta$$

where $B_3 = 0.3\ B_1$ and $B_5 = 0.2\ B_1$.

The total flux per pole is 0.08 Wb. The coil span is 0.8 of pole pitch. Find rms emf induced in one turn.

### *Solution*

Since the emf induced in only turn is to be calculated, $k_{b1} = k_{b3} = k_{b5} = 1$

$$\sigma = 180(1 - 0.8) = 36°$$

$$k_{p1} = \cos\left(\frac{\sigma}{2}\right) = \cos 18° = 0.951$$

$$k_{p3} = \cos\left(\frac{3\sigma}{2}\right) = \cos 54° = 0.588$$

$$k_{p5} = \cos\left(\frac{5\sigma}{2}\right) = \cos 90° = 0$$

Using Eq. (4.8)

$$\phi_3 = \frac{1}{3} \times 0.3\phi_1 = 0.1\phi_1$$

$$\phi_5 = \frac{1}{5} \times 0.2\phi_1 = 0.04\phi_1$$

$$\text{Total flux per pole} = \phi = \phi_1 + \phi_3 + \phi_5$$
$$= \phi_1(1 + 0.1 + 0.04) = 1.14\ \phi_1$$

Since $\phi = 0.08$ Wb

$$\phi_1 = \frac{0.08}{1.14} = 0.07\,\text{Wb}$$

$$E_1 = 4.44\ fN\phi_1 k_{p1} k_{b1}$$
$$= 4.44 \times 50 \times 1 \times 0.07 \times 0.915 = 14.78\ \text{V}$$

Using Eq. (4.9)

$$E_3 = \frac{k_{p3}B_3}{k_{p1}B_1} E_1 = \frac{0.588 \times 0.3 \times 14.78}{0.951}$$
$$= 2.74\ \text{V}$$

$$E_5 = 0$$

$$E = \left(E_1^2 + E_3^2\right)^{0.5} = \left(14.78^2 + 2.74^2\right)^{0.5} = 15.03\,\text{V}.$$

### *Example 4.13*

A star connected 3-phase alternator has an induced emf of 400 V between the lines. Due to the presence of third harmonic component, the phase voltage is 244 V. (a) Find the value of third harmonic voltage in the machine. (b) A 3-phase 10 ohm resistance connected in star are connected across the lines with neutrals tied together. Find line current. (c) If in part (b) the neutrals are not connected, find the line current.

### *Solution*

(a) Phase voltage $E = 244$ V

$E_1$ = Fundamental component of phase voltage $= 400/\sqrt{3} = 230.95\,\text{V}$

If $E_3$ is the third harmonic voltage, then

$$E = \left(E_1^2 + E_3^2\right)^{0.5}$$

or
$$244 = \left(230.95^2 + E_3^2\right)^{0.5}$$

or
$$E_3 = 78.73 \text{ V}$$

(b) When neutrals are connected together, the voltage across each 10 ohm resistance will be the phase voltage $E$.

$$\text{Line current} = \frac{244}{10} = 24.4 \text{ A}$$

(c) When neutrals are not connected, the voltage across each 10 ohm resistance will be $400/\sqrt{3}$

$$\text{Line current} = \frac{400\sqrt{3}}{10} = 23.095 \text{ A.}$$

## SUMMARY

1. AC machines are mostly 3 phase machines except small motors used in domestic appliances.
2. AC winding are open windings.
3. The windings of the three phases are spaced 120 electrical degrees apart.
4. AC windings are generally fractional slot windings. In these windings the number of slots per pole per phase is a fraction.
5. AC windings are generally double layer. The coils are short pitched.
6. Induced emf in ac machine = $4.44\, fN\phi k_b k_p$
7. The emfs in adjacent slots are out of phase. Breadth factor is the ratio of phasor sum of component emfs to arithmetic sum of component emfs of a phase.
8. In a short pitch coil the emfs of the two coil sides are not in phase. Pitch factor is the ratio of phasor sum of coil side emfs to the arithmetic sum of coil side emfs.
9. The resultant mmf wave of a distributed winding can be found by phasor addition of slot mmfs.
10. When 3 phase ac is fed to 3 phase distributed winding a rotating field is produced. The speed of this field is called synchronous speed and is equal to 120 $f$/p.
11. Torque is due to the fact that stator and rotor fields tend to align themselves.
12. Losses in ac machines are copper losses, iron losses, mechanical losses and stray load losses.

## TEST POINT QUESTIONS

### A. State whether the following statements are true (T) or false (F).

4.1 AC windings are generally single layer.

4.2 AC windings are generally fractional slot.

4.3 Phase spread of ac windings is generally 60°.

4.4 $k_b = \dfrac{\sin(q\alpha/2)}{q\sin(\alpha/2)}$ where $q$ is the number of slots per pole per phase and $\alpha$ electrical degrees is the slot angle.

4.5 Pitch factor of ac winding is about 0.5

4.6 The mmf of a slot is displaced from that of adjacent slot by slot angle α.

4.7 The mmf of 3 phase winding has a constant amplitude and speed.

4.8 If any two leads of a 3 phase ac machine are reversed, the direction of rotation of ac field reverses.

4.9 Stray load losses do not occur in ac machines.

4.10 Slot harmonics cause vibration and noise in the machine.

## B. Fill in the blanks with suitable words

4.11 AC windings are generally ...............

4.12 In fractional slot windings the ............... is a fraction.

4.13 The windings of 3 phases are ...............

4.14 Short pitch winding results in saving in ...............

4.15 In fractional slot winding the higher order harmonics in emf and mmf are ...............

4.16 Phase spread of ac windings is generally ...............

4.17 Breadth factor is about ...............

4.18 Pitch factor, for a short pitch winding is always ...............

4.19 Synchronous speed = ...............

4.20 Slot harmonics cause ............... in core losses.

## ANSWERS

4.1 F 4.2 T 4.3 T 4.4 T 4.5 F 4.6 T 4.7 T
4.8 T 4.9 F 4.10 T 4.11 Short pitched
4.12 number of slots per pole per phase 4.13 similar 4.14 copper 4.15 reduced 4.16 60°
4.17 0.95 4.18 less than 1 4.19 $\dfrac{120f}{P}$ 4.20 increase

## JOB INTERVIEW QUESTIONS

4.1 What is a fractional slot winding?

4.2 What is the difference between single layer and double layer winding?

4.3 Why are ac windings generally short pitched?

4.4 What is meant by breadth factor?

4.5 What is meant by synchronous speed?

4.6 Why is 3 phase field rotating?

4.7 How can direction of rotating field be reversed.

4.8 What is slot harmonic?

4.9 Do slot harmonics affect the torque speed curve of a 3 phase induction motor? How?

4.10 Write torque equation of an ac machine.

## REVIEW QUESTIONS

4.1 Discuss why

(a) The phase spread of stator windings is 60° and not 120°.

(b) The windings are generally short pitched.

(c) The emf induced in a distributed ac winding is lower than that in a concentrated ac winding.

(d) Fractional slot windings are very commonly used.

4.2 Explain, using suitable diagrams, the production of rotating field in 3-phase machine. Show that an interchange of two connections causes a reversal in direction of rotation.

4.3 Explain the difference between (a) integral slot and fractional slot windings, (b) single and double layer windings, (c) full pitch and short pitch coils.

4.4 Derive the emf equation for an ac machine. What are breadth and pitch factors? Derive expressions for these.

4.5 What are slot harmonics? How can they be reduced?

4.6 Derive an equation for torque in cylindrical rotor ac machines. What assumptions are made in deriving this equation?

4.7 Prepare a table showing the various slot positions and phases which will occupy these slots for a 108-slot 10-pole winding. Choose a suitable coil span.

## PROBLEMS

4.1 A 6-pole 50-Hz 3-phase star connected alternator has 972 conductors distributed in 54 slots. The coils are short pitched by one slot. the flux per pole is 0.01 Wb. Calculate breadth factor, pitch factor and line voltage at no-load. [0.959, 0.985, 588 V]

4.2 A 4-pole 3-phase 50-Hz star connected alternator has a single layer winding in 36 slots with 30 conductors per slot. The flux per pole is 0.05 Wb and winding is full pitched. Find synchronous speed and line voltage on no-load. [1500 rpm, 3320 V]

4.3 A 3-phase 10-pole 50-Hz star connected alternator has 120 stator slots with 8 conductors per slot. All the conductors of a phase are in series. Flux per pole is 56 mWb. Find phase and line emfs.[1900 V, 3290 V]

4.4 An 8-pole alternator has 72 slots and is driven at 750 rpm. Two 100-turn coils *A* and *B* in the stator are as under:

Coil *A*: coil sides lie in slots 1 and 11

Coil *B*: coil sides lie in slots 2 and 10

Find the resultant emfs of the two coils, when they are connected in (a) series aiding, (b) series opposing. $\phi = 0.01$ Wb. [(a) 437.25 V (b) 0]

4.5 The flux density distribution in a 3750 kVA, 3-phase 50Hz 10-pole alternator is as under:

$B = 100 \sin \theta - 11.13 \sin 5\theta + 2.78 \sin 7\theta$

The alternator has 144 slots with $2 \times 5$ conductors per slot. The coil span is 12 slots and flux per pole is 0.116 Wb/m$^2$. Find fundamental, third and fifth harmonic components of phase voltage. [5750, 34, 6 V]

4.6 An ac machine has 6 poles and 96 slots. The coils are wound with 13/16 fractional pitch. Find pitch factor for fundamental. [0.9569]

4.7 A 3-phase induction motor has 96 stator slots with 4 conductors per slot and 120 rotor slots with 2 conductors per slot. The stator is fed from 400 V, 3-phase supply. Both rotor and stator are star connected. Find (a) number of turns per phase on stator and rotor, (b) voltage across sliprings when rotor is open circuited and is at standstill. [(a) 64, 60(b) 250 V]

4.8 The stator of a 3-phase ac machine has four poles, 48 slots and a double layer winding. Find breadth factor. [0.958]

4.9 A 3-phase 4-pole synchronous machine has 84 slots on stator. Find breadth factor. [0.955]

4.10 A 3-phase 16-pole synchronous machine has a star connected stator winding accommodated in 144 slots and 10 conductors per slot. The flux per pole is 0.03 Wb. The speed is 375 rpm and flux distribution is sinusoidal. Find frequency, phase and line values of emf. [50 Hz, 1530 V, 2650 V]

4.11 A 6-pole synchronous machine has 72 slots, a double layer winding with 20 turns per coil. The coil span is 5/6 of the pole pitch. The speed is 1000 rpm and flux per pole is $4.8 \times 10^{-2}$ Wb. Find (a) frequency, (b) number of turns per phase, (c) emf per phase. [(a) 50 (b) 480 (c) 4733.45 V]

4.12 Find the rms values of different harmonic components and total emf per phase for an alternator having following data: 50 Hz, 3-phase, 10 poles, 2 slots per pole per phase, double layer winding, 4 conductors per slot, coil span 150°, flux per pole (fundamental) 0.08, third harmonic component 15%. All coils of a phase are in series. [662.91 V, 53.265 V, 665.05 V]

Chapter 5

# Synchronous Generators

## 5.1 INTRODUCTION

A synchronous machine has a rotor and a stator. The rotor of this machine always moves at synchronous speed. It may be a synchronous generator or synchronous motor. A synchronous generator receives mechanical energy, from a prime mover to which it is mechanically coupled, and converts it into electrical energy. It is also known as alternator, since it delivers ac output. A synchronous motor converts ac electrical energy into mechanical energy. Since ac systems are always 3-phase systems, synchronous machines are 3-phase machines. The frequency of output ac voltage of a synchronous generator is directly proportional to the rotor speed. To maintain the frequency constant; the rotor must always move at synchronous speed.

Synchronous generators are fabricated in a number of ratings from a few kVA to very large sizes. The biggest size used in India has a rating of 500 MVA used in super thermal stations.

A dc machine has a stationary field and revolving armature. This construction is suitable for only very small size synchronous machines (having rating upto about 10 kW). Majority of synchronous machines have very high ratings and armature winding is designed for 6.6 kV or 11 kV or 33 kV. Synchronous machines have a rotating field-stationary armature construction. This construction offers the following advantages. (a) Betters insulation, (b) Ease of current collection, (c) Increased armature tooth strength, (d) More rigid construction, (e) Reduced armature leakage reactance, (f) Lesser number of sliprings, (g) Lesser rotor weight and inertia, (h) Improved ventilation and heat dissipation.

(a) Better insulation. Modern synchronous machines have armature windings designed for 6.6 kV or 11 kV or 33 kV. It is easier to insulate such a winding if it is on stator. Moreover the shaft of every machine is electrically earthed, through metallic bearings, to the stationary frame of the machine. It is easier to insulate a low voltage dc field winding on rotor than a rotating high voltage ac winding.

(b) Ease of current collection. Modern synchronous machines are high current machines. The armature current of a 210 MW, 11 kV, 3-phase, alternator is about 13 kA. It is easier to collect such large currents if armature is stationary. The collection of such large currents through slip rings is very difficult.

(c) Increased armature tooth strength. High armature currents require more copper and hence deep slots to accommodate the winding. When the slots of a stationary armature are made deeper, the teeth become wider and stronger. When the slots of a rotating armature are made deeper, the teeth become narrower and weaker. Strong teeth of the stationary armature construction are likely to be less damaged during fabrication and use.

(d) More rigid construction. The high voltage, high current armature windings of modern 3-phase synchronous machines are rather complex and have many end connections and interconnections. Such windings are more rigid and secure when housed on stator.

(e) Reduced armature leakage reactance. The stationary armature construction leads to lesser leakage flux and hence reduced armature leakage reactance. This leads to improved voltage regulation.

(f) Lesser number of sliprings. If rotating armature construction is used, the machine would require 3 sliprings for the three phases. These sliprings would have to be properly insulated from shaft and also spaced suitably to avoid flashover. Rotating field structure requires only 2 sliprings insulated for low voltage dc (about 250 V or so).

(g) Lesser rotor weight and inertia. Because of low voltage field winding on rotor, the amount of copper and insulation is pretty less. Therefore rotor has lesser weight and inertia. Lesser weight of rotor leads to efficient high speed operation and cheaper bearings. Due to lesser inertia the machine can be brought up to full speed, from rest, in a very small time.

(h) Improved ventilation and heat dissipation. Most of the heat is generated in armature. When armature is on stator, the ventilation ducts and holes for cooling can be wider leading to better ventilation and heat dissipation.

In view of above advantages nearly all modern synchronous machines have rotating field stationary armature construction.

dc current has to be supplied to the filed current on rotor, which is rotating. Two common approaches for this purpose are:

(1) An external dc source supplying the dc power to the rotor winding through sliprings and brushes.

(2) A special dc source mounted directly on the shaft of machine.

Since the field current and voltage are low, the use of sliprings and brushes does not pose any problem. However an external dc source is used only for small synchronous generators. Large size machines used in power plants have brushless exciters mounted on the shaft of the machine. Such an exciter is a small ac generator with its field on stator and armature winding on rotor. The 3-phase output of this exciter is rectified by a solid state rectifier circuit (also mounted on the shaft of the generator) and then fed to the main field winding of the machine.

Two types of constructions are used for rotor and this feature classifies synchronous generators into two categories, i.e. salient pole machine and cylindrical rotor machine. The word salient means protruding or sticking out. Thus in a salient pole machine, the poles stick out from the surface of the rotor and the field winding is concentrated. This construction is suitable for low speed (less than 1000 rpm) machines with water turbines or reciprocating engines as the prime mover. The non-salient or cylindrical rotor machine has distributed field winding in rotor slots. These machines are used in thermal and nuclear power plants and are high speed machines (3000 rpm or 1500 rpm).

## 5.2 FREQUENCY

The emf induced in a conductor passes through one complete cycle when the conductor travels a distance occupied by a pole pair. Therefore, in a two-pole machine one cycle of frequency is generated in one revolution of the rotor. In a machine having $P$ poles, the number of cycles in one revolution is $P/2$. If the rotor is moving at a speed $n_s$ revolutions per second, the frequency is $(P/2)n_s$ Hz. Thus

$$f = (P/2)n_s \text{ Hz} \qquad ...(5.1)$$

The electricity regulations require the frequency of supply to remain constant. Therefore a synchronous generator must always rotate at the synchronous speed $n_s$.

The actual space occupied by a pole, around the periphery of the machine, depends on the number of poles of the machine. However, one complete cycle of frequency is generated when the relative motion between conductor and field is by an angle equal to double pole pitch. In view of this the electrical angle spanned by a double pole pitch is $2\pi$ electrical radians. In a $P$ pole machine, the total number of electrical radians, corresponding to one revolution or $2\pi$ mechanical radians, is $2\pi\ (P/2)$. Thus each degree of actual angular space displacement (or each mechanical degree) is equivalent to $P/2$ electrical degrees.

The standard frequency in India and many other countries is 50 Hz. Therefore synchronous speeds $N_s$ (i.e. revolutions per minute) for different pole numbers are

| $P$ | 2 | 4 | 6 | 8 | 10 | 12 | 14 | 16 | 18 | 20 |
|---|---|---|---|---|---|---|---|---|---|---|
| $N_s$ | 3000 | 1500 | 1000 | 750 | 600 | 500 | 428.57 | 275 | 333.33 | 300 |

## 5.3 CONSTRUCTION

The two main parts of synchronous generators are the stator and rotor. The different types of rotor (i.e. salient pole and cylindrical) result into two types of machines, whose construction and outward appearance are very different from each other.

### 5.3.1 Cylindrical Rotor Machines

They are also known as turbo alternators and are usually designed for 2-poles (i.e. 3000 rpm) or 4 poles (i.e. 1500 rpm). They are suitable for being coupled to steam turbines, because the steam turbines are high speed machines. The centrifugal stress in the rotor of these high speed machines is very large and in order to obtain greater strength and at the same time to reduce the windage loss and to improve flux distribution, cylindrical rotor construction is adopted. The main constructional features of these machines are as under:

(1) The machines have horizontal configuration.

(2) They are wound for 2 or 4 poles. The outside diameter is limited by maximum peripheral speed which should not exceed 175 m/sec. Thus the rotor of a turbo alternator is of small diameter and is very long. The length may be about 5 times the diameter. the maximum diameter for 2-pole machines is around 1 m. The length of rotor between the bearing is limited by the considerations of critical speed. Modern turbo alternators are designed such that the first critical speed lies between 1250 and 1400 rpm and second critical speed is not less than 3700 rpm.

(3) The machines are built in a variety of sizes from 10 MW to over 1500 MW. The maximum size installed in India is 500 MW.

(4) The rotor is made from solid steel forgings of high grade nickel-chrome-molybdenum steel. The outer periphery of rotor has slots in which field winding is placed. About two-third of rotor pole pitch is slotted, leaving the one-third unslotted for the pole centre. Heavy wedges of non-magnetic material are forced into the groves in the teeth outside the field coils, to keep the field coils in position. It is necessary to provide passages in the rotor body for flow of cooling medium. These passage are provided either in the teeth or below the rotor slots.

(5) The stator is built from high permeability sheet steel laminations. One side of each lamination

is treated with insulating material (varnish or enamel or paper) to prevent the flow of eddy currents. The laminations have slots on their inner periphery to house the armature conductors, which are properly insulated from the slots as well as from each other. To reduce skin effect and to obtain a uniform distribution of current in the conductors, the stator windings are divided into a number of thin conductors, each wrapped with insulating material. Double layer fractional slot windings are used. The number of slots per pole per phase, coil pitch and disposition of conductors are so selected as to make the terminal voltage as nearly sinusoidal as possible. The stator windings are connected in star. Figure 5.1 shows a cylindrical rotor machine.

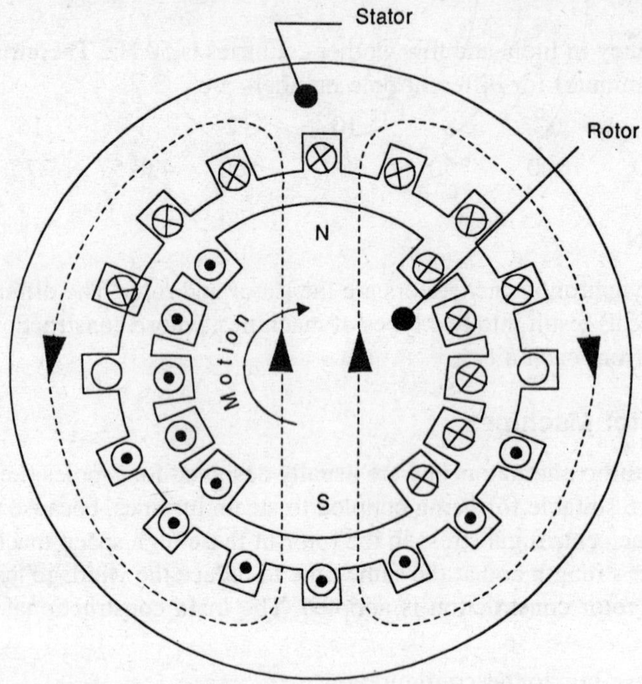


**Fig. 5.1.** Cylindrical rotor synchronous generator

### 5.3.2 Salient Pole Machines

These machines are used in hydro-electric plants and have the following main features:

(1) Speed is low but the inherent danger of run away speeds has to be guarded against. For vertical machines the speeds range from around 50 rpm to 500 rpm. For horizontal machines the speed range is 100-1000 rpm.

(2) The machines may have horizontal or vertical configuration. The alternators used in conjunction with impulse turbines have usually horizontal configuration, while those used with Francis and Kaplan turbines have vertical configuration.

(3) They have large diameters (to accommodate the large number of poles) and small lengths. The number of poles varies from 6 to 150.

(4) They are built in a variety of sizes from 0.5 MW onwards. The largest size used in India is 250 MW. It has been reported that very large size water wheel generators (more than 1000 MW) have been installed in Russia.

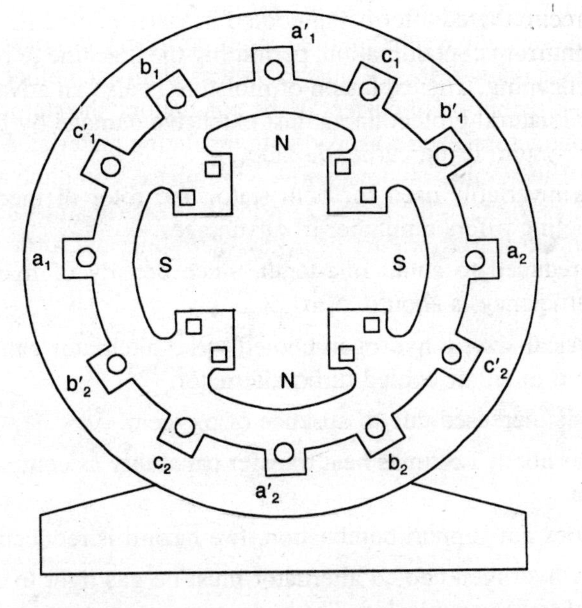


**Fig. 5.2.** Salient pole synchronous generator

(5) The rotor consists of laminated poles projecting radially from a steel core. The core consists of mild steel stampings about 2 mm thick. The poles may be integral with rotor core or bolted or keyed to the core. The stampings are clamped axially under heavy pressure between thick steel end plates by rivets or bolts. The field winding is usually of copper strip, the turns lying edge-wise relative to the pole body. The turns are insulated with asbestos paper inpregnated with varnish. To damp out oscillations, damper windings are housed in pole shoes. This winding consists of low resistance copper bars with their projecting ends connected to short circuiting end rings.

(6) the stator is built from low carbon, high resistivity, silicon steel stampings around 0.5 mm thick, insulated with varnish. The stator core has slots on the inner periphery, to house the armature conductor. Double layer fractional slot winding is used. Each coil may consist of single or more turns wound continuously in series according to the voltage and current rating of the machine. Generators of large ratings have two or more circuits in parallel. The number of slots per pole per phase, coil pitch and winding disposition are so selected as to make the terminal voltage as nearly sinusoidal as possible. The stator windings are connected in star. Figure 5.2 shows a salient pole generator.

## 5.4 COOLING

Salient pole machines used in hydro plants are generally air cooled, either through open ventilation system or closed air circuit. In open ventilation system fresh air from atmosphere enters from one or both ends, flows through ducts between windings and between core sections and discharges at the periphery. It is necessary to clean the incoming air. Passages are provided so that air can be taken from out of doors and discharged out of doors. The air intake is located preferably near the tail race, but sufficiently above the water level in it. The air requirement is about 3 cubic metres per minute per kW loss and air velocity is around 400 metres per minute. Large machines mostly use closed air circuit. The same air is continuously recirculated after being cooled in water cooled heat exchanger. An advantage of closed circuit is freedom from contamination, permitting the machine to run for much longer periods without the necessity of cleaning. The exclusion of moisture is also an advantage. The limited amount of available air provides a natural protection against extensive damage by fire. The ventilation system also include fire fighting system using carbondioxide.

Hydrogen cooling is invariably used for both stator and rotor of medium and large size turbo-alternators. Hydrogen cooling offers a number of advantages.

(1) Windage loss is reduced to about one-tenth, since density of hydrogen is 1/14 of air. The improvement of efficiency is about 0.5 to 1%.

(2) For the same physical size, a hydrogen cooled turbo-alternator can develop about 25% more power, as compared to an air cooled turbo-alternator.

(3) Life of insulation is increased due to absence of oxygen.

(4) Since hydrogen has about 1.5 times heat transfer capability as compared to air, cooling is more efficient and faster.

(5) Since hydrogen does not support combustion, fire hazard is reduced.

The stator frame of a hydrogen cooled alternator must be gas tight to avoid the possibility of air mixing with hydrogen and causing explosion. The hydrogen pressure is kept higher than outside atmospheric pressure to eliminate the leakage of air into the machine. The stator is made sufficiently strong, not only to withstand the gas pressure, but also to withstand the pressure due to explosion.

## 5.5 EXCITATION SYSTEM

A dc current has to be supplied to the field winding of a synchronous machine. Since the field winding is on rotor, a special arrangement is necessary to connect the dc source to the field. In small size synchronous machines, the field winding is excited from a separate dc source through sliprings and brushes. Sliprings are metal rings completely encircling the shaft of the machine, but insulated from it. A brush rides on each slipring. The positive end of a dc voltage source is connected to one brush and negative end is connected to another brush.

### 5.5.1 Brushless Excitation System

Sliprings and brushes create the problems of maintenance and brush voltage drop. The field current of a large size modern turbo-alternator may be about 5 kA or so and invariably a brushless excitation system is used. Figure 5.3 shows a simplified diagram of a brushless excitation system. The portion enclosed by dashed lines is the rotating portion of the system.

The excitation system consists of an alternator rectifier main exciter and a permanent magnet generator (PMG) pilot exciter. Both the main exciter and pilot exciter are driven directly from the main shaft. The main exciter has a stationary field and a rotating armature, which is directly connected, through silicon rectifiers, to the main alternator field. Thus the sliprings and brushes are eliminated. The main exciter's field is fed from a shaft driven PMG, having rotating permanent magnets attached to the shaft and a stationary 3-phase armature. The ac output of PMG is rectified by 3-phase full-wave phase controlled thyristor bridges. The thyristor assembly is usually housed in removable drawers, which can be taken out easily for repair. Redundancy is provided to increase reliability. The thyristors are force cooled through main and reserve fans. The thyristor bridges are controlled by a set of dual firing circuits operating in parallel.

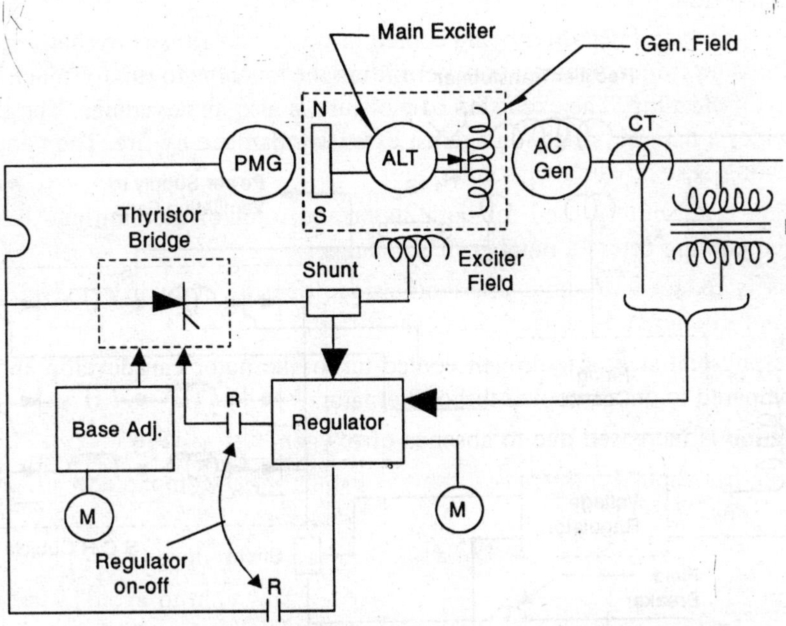


**Fig. 5.3.** Brushless excitation system for synchronous generator

The base excitation is controlled by an input setting to the thyristor gating circuits. This control signal is derived from the PMG via a regulated dc supply, which also serves the regulator logic circuitry. The regulator controls excitation by supplying a buck-boost control signal, which adds algebraically to the base setting. The regulator elements also comprise of solid state circuits.

This excitation system has a short time constant and a response time of less than 0.1 second.

### 5.5.2 Static Excitation System

A static excitation system draws excitation power from the main alternator terminals, through a 3-winding step-down transformer and a rectifier system using thyristors as shown in Fig. 5.4

The rectifier transformer is connected to main alternator bus directly (without a breaker or fuse in between). This transformer is 3-phase indoor type immersed in oil, with forced air cooling. The star connected primary is connected to alternator bus. The delta connected secondary feeds the excitation system. The delta connected tertiary supplies power to grid control circuits and other auxiliary equipment. The rectifiers are connected in parallel to provide sufficient current carrying capacity. Each leg of rectifier is provided with series fuses, surge protection and fault indicating light. The rectifiers are forced air cooled.

The start up is accomplished using the station battery to provide the initial field current. The crowbar circuit is to protect the rectifiers and alternator against the effect of an excessive negative voltage, which can occur if the alternator loses synchronism. Protections are also provided to shut down the alternator unit in the event of a short circuit in rectifier circuit, fault in step-down transformer, excessive current in crowbar circuit and excessive ac current being drawn by excitation circuit.

Thyristors are ideally suited for a static excitation system, because they have high speed of response and high power gain.

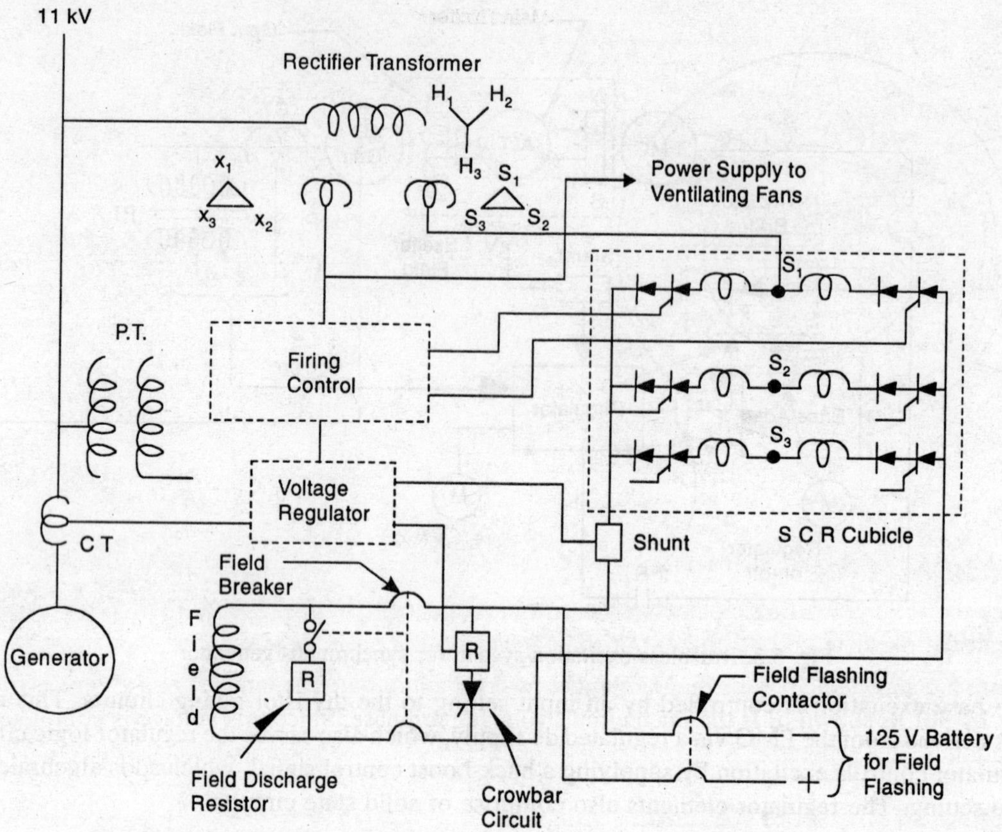


**Fig. 5.4.** Static excitation system for synchronous generator

The advantages of static excitation are as follows:

(1) A very small response time, about 20 m sec.

(2) Elimination of exciter windage loss and commutator, bearing and winding maintenance.

(3) Since the excitation energy is being taken from alternator terminal, the excitation voltage is directly proportional to speed (In shaft driven exciters the excitation voltage is proportional to square of speed). This fact affords a major advantage in load rejection.

## 5.6 PHASOR DIAGRAM OF CYLINDRICAL ROTOR ALTERNATOR

The phasor diagram of an alternator can be developed by first considering the operation at no-load. In the second step, the effect of armature reaction is included in the phasor diagram. Finally the effect of armature resistance and armature leakage reactance are considered.

### 5.6.1 No-Load Condition

When the alternator is operating at no-load (i.e. no current flowing through the armature winding), the field current sets up a magnetic field having an mmf $F$ and flux $\phi$ per pole. Figure 5.5 shows a 2-pole machine with 2 stator coils per pole per phase. The magnetic flux due to field current is an shown.

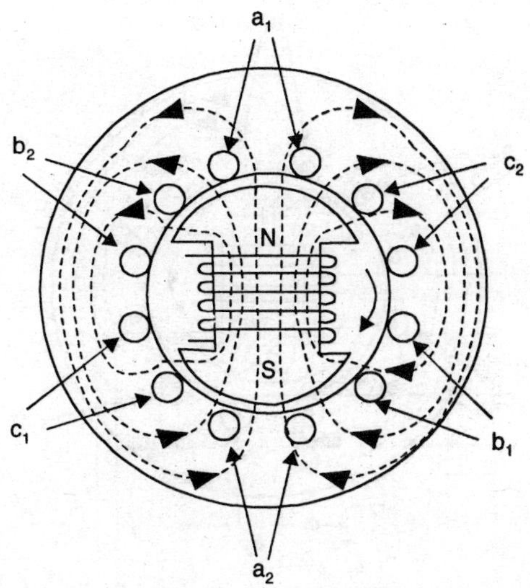


Fig. 5.5. Magnetic field in an alternator on no-load

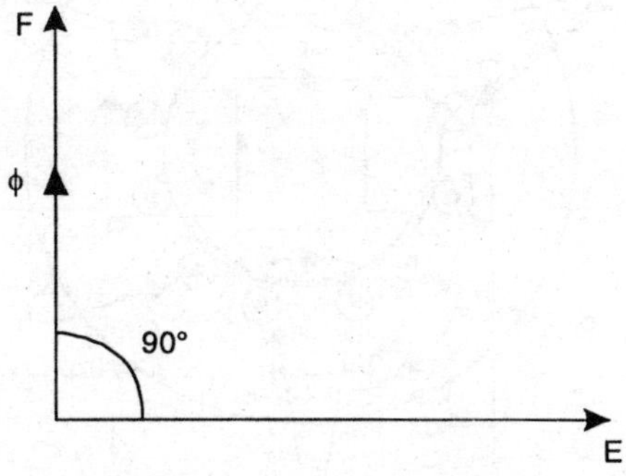


Fig. 5.6. Phasor diagram of alternator on no-load

Since the rotor is rotating clockwise, the emf induced in phase $a$ is maximum at the instant shown and is into the paper in coil $a_1$ and out of the paper in coil $a_2$.

Figure 5.6 shows the phasor diagram under no-load condition. The induced emf $E$ lags is the mmf and flux by 90° as shown. $E$ is the open circuit voltage per phase. Since this voltage is completely due to excitation of the generator, $E$ is known as excitation voltage or generated voltage.

### 5.6.2 Effect of Armature Reaction

When a balanced 3-phase load is connected at the alternator terminals, an rms current $I$ flows through each phase of armature winding. The three phase currents flowing in the 3-phase winding produce a rotating magnetic field. The armature mmf rotates at synchronous speed with respect to stator winding. Since the rotor is also moving at synchronous speed with respect to stator, the rotor and stator mmfs are stationary with respect to each other. Both these mmfs act upon magnetic circuits which have equal reluctance in all axes and may be combined as phasors.

Figure 5.7 shows the distribution of flux due to armature currents at the instant the current in phase $a$ is at maximum value and is at unity power factor. By right hand rule, the armature flux is clockwise due to currents in top conductors and anticlockwise due to currents in the bottom conductors.

The resultant flux due to super imposition of rotor and stator fluxed is shown in Fig. 5.8. A comparison of Figs. 5.5 and 5.7 shows that the two fluxes are in opposition over the leading half of each pole face and are in the same direction over the trailing half of each pole face. Therefore, the effect is to distort the magnetic fluxes as shown in Fig. 5.8. It is seen that the direction of most of the lines of flux in the air gap has been skewed and lengthened. This causes a backward pull on the rotor. To overcome the tangential component of this pull, the prime mover driving the generator has to exert a larger torque than that required at no-load. As seen in Fig. 5.8, the armature mmf has a cross-magnetising effect on field poles.

Figure 5.9 shows the phasor diagram showing the effect of armature reaction at unity power factor. The armature current $I$ creates an armature mmf $F_A$ in phase with $I$. The phasor sum of $F$ and $F_A$ gives the resultant mmf $F_R$. This resultant mmf is responsible for the resultant flux, which induces an emf $E'$

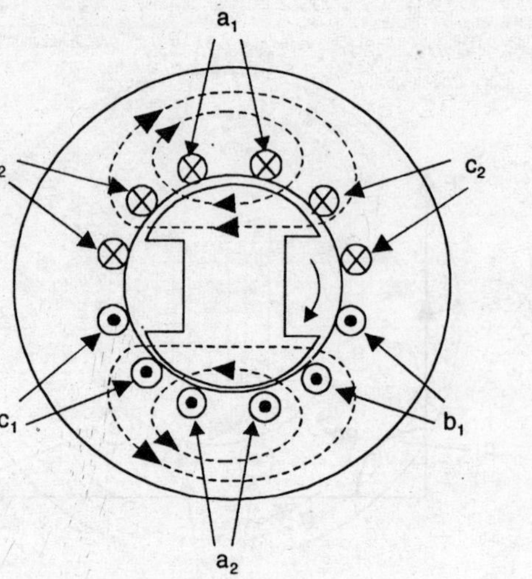


**Fig. 5.7.** Field due to stator current in an alternator

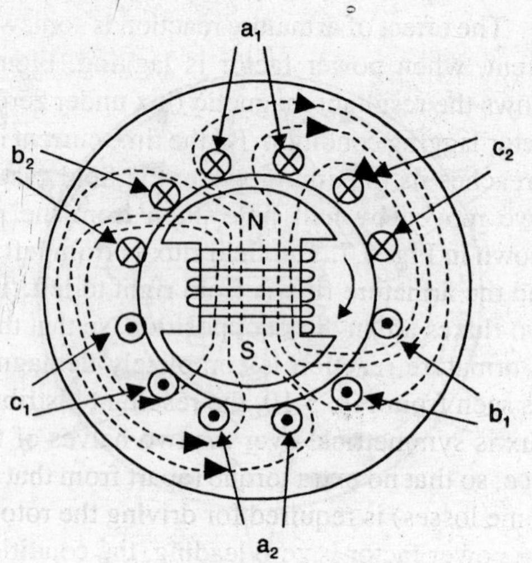


**Fig. 5.8.** Superposition of stator and rotor fields in an alternator

lagging $F_R$ by 90°. In analysing the effect of $F_A$ we can resort to superposition and consider, that $F_A$ sets up a flux $\phi_R$ which is synchronously rotating and cuts the armature winding, there by inducing an armature reaction emf $E_{ar}$ per phase. This emf lags behind mmf $F_A$ and flux $\phi_R$, by 90°. The resultant sum of $E$ and $E_{ar}$ gives emf $E'$. The armature reaction voltage can be considered as a voltage drop across a fictitious inductive reactance $x_a$, such that

$$E_{ar} = -jIx_a \qquad ...(5.2)$$

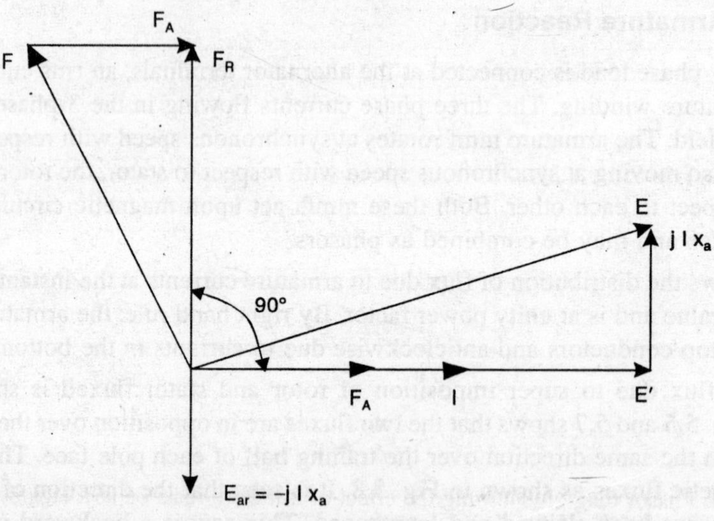


**Fig. 5.9.** Phasor diagram showing effect of armature reaction at unity power factor

The minus sign in Eq. (5.2) has been used for defining $E_{ar}$, so that an equal and opposite quantity can be handled with a plus sign (i.e. plus sign is used for a voltage drop and minus sign for a voltage rise). It should always be remembered that $x_a$ is a fictitious reactance equivalent in its effect to the armature reaction. Evidently

$$E = E' + jIx_a \qquad ...(5.3)$$

The effect of armature reaction is somewhat different, when power factor is lagging. Figure 5.10 shows the resultant magnetic flux under zero power factor lagging condition. By the time current in phase *a* reaches its maximum value, the field system will have moved by half pole pitch from the position shown in Fig. 5.7. The field flux is from left to right and the armature flux is from right to left. Thus the two fluxes are in direct opposition, so that the effect of armature reaction is completely demagnetising. As seen from Fig. 5.10, the resultant distribution of flux is symmetrical over the two halves of the pole face, so that no extra torque (apart from that to overcome losses) is required for driving the rotor. When the power factor is zero leading, the conditions will be similar to that in Fig. 5.10, but the directions of the two fluxes will be the same, so that the effect of armature reaction will be magnetising resulting in an increase in flux.

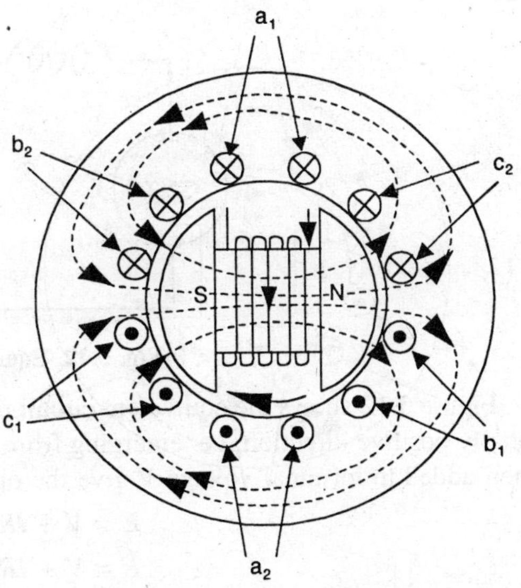


**Fig. 5.10.** Field due to stator current at zero power factor lagging

When power factor has a value between zero lagging and unity, the armature reaction will be partly demagnetising and partly cross-magnetising. Figure 5.11 shows the phasor diagram. The armature mmf $F_A$ is in phase with $I$ and the phasor sum of field mmf $F$ and armature mmf $F_A$ gives the resultant mmf $F_R$. The effect of $F_A$ can be replaced by a fictitious reactance $x_a$. The phasor sum of $E'$ and $jIx_a$ gives the excitations emf $E$.

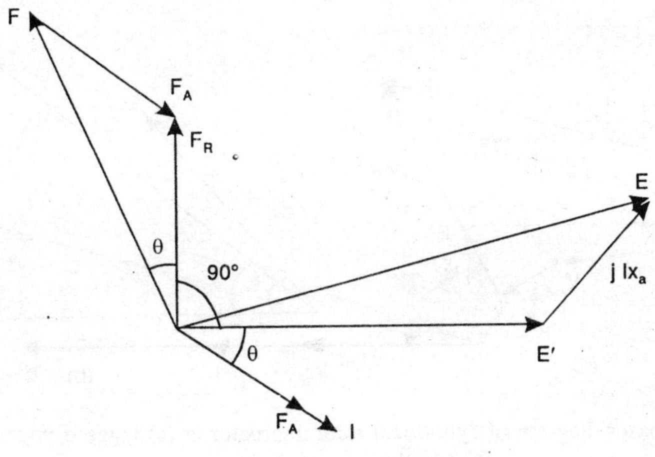


**Fig. 5.11.** Phasor diagram showing the effect of armature reaction at a lagging power factor

### 5.6.3 Effect of Armature Resistance and Leakage Reactance

When the alternator is loaded, a voltage drop takes place in the stator winding, due to its resistance $R_a$ and leakage reactance $x_l$. When these voltage drops are subtracted from $E'$, the result is $V$, i.e. the terminal voltage per phase. In other words, the voltage drops $IR_a$ and $jIx_l$, when added to $V$ give the emf $E'$.

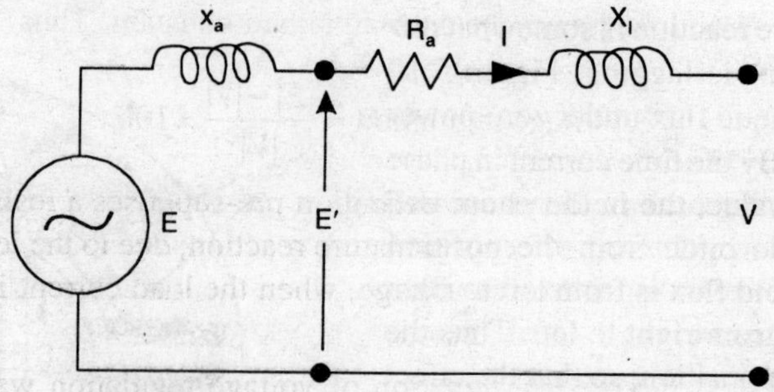


**Fig. 5.12.** Equivalent circuit of alternation

Figure 5.12 shows the equivalent circuit of the alternator. The current flow is shown in the conventionally positive direction, i.e. emerging from the positive terminal. The voltage drops $IR_a$, $jIx_l$ and $jIx_a$ when added to terminal voltage $V$ give the open circuit emf $E$. Thus

$$E = V + IR_a + jIx_l + jIx_a \quad ...(5.4)$$

or

$$E = V + IR_a + jI(x_l + x_a)$$

or

$$E = V + IR_a + jIx_s \quad ...(5.5)$$

or

$$E = V + IZ_s$$

The quantity $x_s$ is the sum of leakage reactance $x_l$ and fictitious armature reactance $x_a$ and is known as synchronous reactance. The quantity $Z_s$ is known as synchronous impedance.

Figures 5.13 (a and b) show the phasor diagram under lagging and leading power factor conditions, respectively. The phasor sum of $V$, $I\,R_a$, $j\,I\,x_l$ and $j\,I\,x_a$ gives the emf $E$.

$E$ and $V$ have a phase difference of $\delta$, which is known as power angle of the machine.

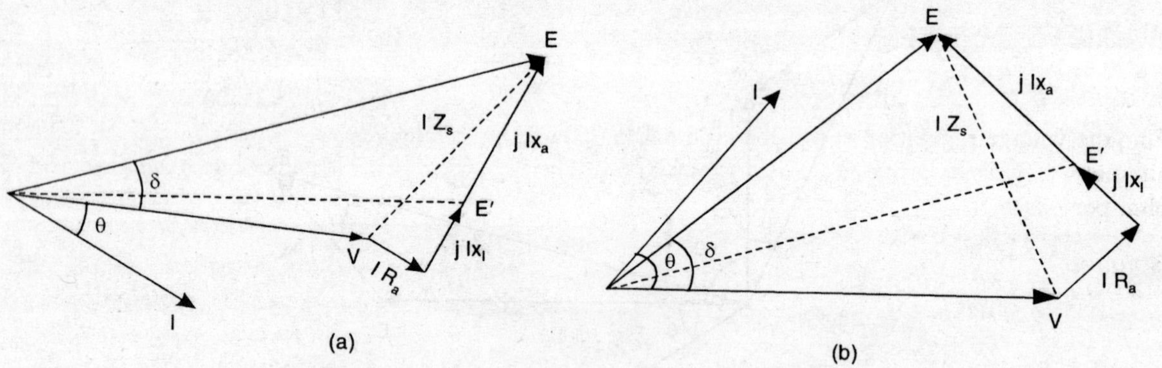


**Fig. 5.13.** Phasor diagram of cylindrical rotor alternator at (a) lagging power factor, (b) leading power factor

## 5.7 VOLTAGE REGULATION

The terminal voltage of an alternator undc. load conditions is different from the open circuit voltage due to the effects of armature resistance, leakage reactance and armature reaction. Voltage regulation is defined as the rise in voltage, expressed as per cent of rated voltage, when the load current is reduced to zero, the field excitation and frequency being maintained constant. Thus

$$\text{Voltage regulation} = \frac{|E| - |V|}{|V|} \times 100 \qquad \text{...(5.6)}$$

The term rise in voltage used in the above definition pre-supposes a resistive or inductive load. If the load is capacitive, the magnetising effect of armature reaction, due to the leading current, may cause $V$ to be higher than $E$, thus causing a drop in voltage, when the load current is reduced to zero. In that case the regulation is negative.

Till about 20 years back, the predetermination of voltage regulation was of great importance to designers and operating engineers. Many methods were devised to calculate the voltage regulation. The advent of automatic voltage regulators transferred attention from voltage regulation to the closely related topic of excitation control to maintain constant terminal voltage. The importance of the subject is further emphasised by the fact that voltage regulation and excitation control are closely related to the current in the alternator under short circuit condition.

## 5.8 SYNCHRONOUS IMPEDANCE METHOD

The terminal voltage under, load conditions is different from that under no-load conditions due to following three factors:

(1) Effect of armature resistance

(2) Effect of armature leakage reactance

(3) Effect of armature reaction.

In the synchronous impedance method, the effect of armature reaction is replaced by a fictitious reactance $x_a$, so that the total voltage drop can be represented as an impedance drop $IZ_s$ where $Z_s$ is the synchronous impedance. As seen from phasor diagram of Fig. 5.13, the phasor sum of $V$, $IR_a$, and $jIx_s$ gives the open circuit emf $E$. From the values of $E$ and $V$, voltage regulation can be calculated.

### *Example 5.1*

Find the voltage regulation at full-load 0.8 p.f. lagging for a 3-phase 1500 kVA, 6600V, star connected alternator having an armature resistance of 0,093 ohm per phase and a synchronous reactance of 8.5 ohm per phase. [*B.U. 2001*]

### *Solution*

$$\text{Phase voltage } V = 6600/\sqrt{3} = 3810.5 \text{ V}$$

$$\text{Full-load current} = \frac{1500 \times 1000}{\sqrt{3} \times 6600} = 131.22\,A$$

The phasor diagram is shown in Fig. 5.13 (a). Let $V$ be the reference phasor

$$V = 3810.5 \angle oV$$

$$I = 131.22(0.8 - j0.6)A$$

$$E = V + I(R_a + jx_s)$$

$$= 3810.5 + 131.22\ (0.8 - j0.6)\ (0.093 + j8.5)$$

$$= 4489.5 + j884.97$$

$$|E| = \left[4489.5^2 + 884.97^2\right]^{0.5} = 4575.9\text{ V}$$

$$\text{Regulation} = \frac{4575.9 - 3810.5}{3810.5} \times 100 = 20.09\%.$$

*Example 5.2*

**A 3-phase star connected alternator has an open circuit line voltage of 6599V. The armature resistance and synchronous reactance are 0.6 ohm and 6 ohm per phase, respectively. Find terminal voltage and voltage regulation and δ if load current is 180 A at power factor of (a) 0.9 lagging, (b) 0.8 leading.**

*Solution*

(a) Figure 5.13(a) shows the phasor diagram. Let $V$ be the reference phasor. $E$ leads $V$ by $\angle\delta$. Thus

$$E = |E|\angle\delta = \frac{6599}{\sqrt{3}}\angle\delta = 3810\angle\delta V$$

$$= 3810 \cos \delta + j3810 \sin \delta$$

$$I = 180(0.9 - j0.436) = 162 - j78.48 \text{ A}$$

$$E = V + I(R_a + jx_s)$$

$$3810 \cos \delta + j3810 \sin \delta = V + (162 - j78.48)(0.6 + j6)$$

$$= V + 568.08 + j924.9$$

Therefore $3810 \sin \delta = 924.9$

or $\delta = 14.05°$

$$V = 3810 \cos \delta - 568.08 = 3127.94 \text{ V}$$

$$\text{Voltage regulation} = \frac{3810 - 3127.94}{3127.94} \times 100$$

$$= 21.8\%$$

(b) Figure 5.13(b) shows the phasor diagram

$$I = 180(0.8 + j0.6) = 144 + j108 \text{ A}$$

$$3810 \cos \delta + j3810 \sin \delta = V + (144 + j108)(0.6 + j6)$$

or $3810 \cos \delta + j3810 \sin \delta = V + j928.8 - 561.6$

or $3810 \sin \delta = 928.8$

or $\delta = 14.1°$

$$V = 3810 \cos \delta + 561.6 = 4256.65 \text{ V}$$

$$\text{Voltage regulation} = -10.49\%.$$

## 5.9 EXPERIMENTAL DETERMINATION OF SYNCHRONOUS IMPEDANCE

**The synchronous impedance can be found from the results of open circuit and short circuit tests on an alternator.**

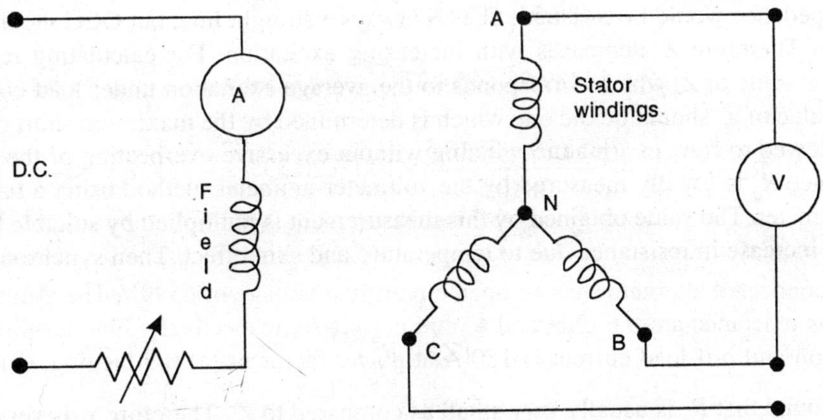


**Fig. 5.14.** Connections for open circuit test on alternator

Figure 5.14 shows the circuit for the open circuit test (or no-load test). The machine is driven at its rated synchronous speed. By varying the resistance in the field circuit, the field current is varied from the minimum to the maximum value. The values of field current and the corresponding values of terminal voltage are noted. The open circuit characteristics shows the relation between open circuit voltage per phase and the field current. For low values of field current, this characteristics is a straight line. However, for higher values of filed current, the characteristics departs from the straight line due to the saturation of magnetic circuit (Fig. 5.15).

The circuit for determination of short circuit characteristics is almost similar to Fig. 5.14, except that an ammeter is connected between terminals *R* and *Y* and the terminals *Y* and *B* are short circuited. The field current is varied from zero to a value sufficient for armature current to be about 1.2 times full-load current. The short circuit characteristics shows the relation between short circuit armature current and field current. This characteristics is a straight line, because the air gap flux, under short circuit conditions, is only about 15% of that under rated voltage conditions. This is due to two reasons. Firstly the field current is very small. Secondly, the power factor of the machine is about zero lagging, so that armature reaction is demagnetising in nature. Therefore there is no saturation of the magnetic circuit and the characteristics is a straight line. Figure 5.15 shows the open circuit and short circuit characteristics.

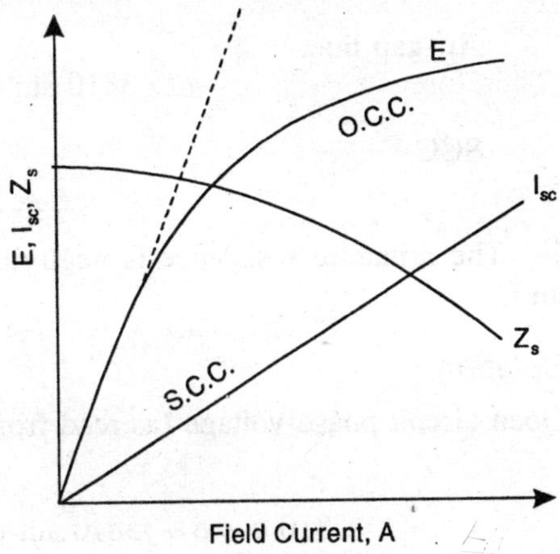


**Fig. 5.15.** Open circuit and short circuit characteristics of alternator

It is evident that when the machine is short circuited the emf generated is consumed as a voltage drop in the synchronous impedance (because the terminal voltage is zero).

$$Z_s = \frac{\text{Open circuit phase voltage for a certain excitation}}{\text{Short circuit current for the same excitation}} \qquad ...(5.7)$$

Using Eq. (5.7) we can find synchronous impedance for different values of excitation. The variation of $Z_s$ with field current is also shown in Fig. 5.15. If OCC and SCC were both straight lines synchronous impedance would be constant. SCC is always a straight line, but OCC includes a partially saturated region. Therefore $Z_s$ decreases with increasing excitation. For calculating regulation, it is logical to select a value of $Z_s$ which corresponds to the average excitation under load conditions. This means that the value of $Z_s$ should be the one which is determined by the maximum short circuit current, which can be allowed to flow in armature winding without excessive overheating of the machine. The armature resistance $R_a$ is usually measured by the voltmeter-ammeter method using a battery, dc voltmeter and dc ammeter. The value obtained by this measurement is multiplied by suitable factors to take into account the increase in resistance due to temperature and skin effect. Then synchronous reactance can be found as

$$x_s = \left(Z_s^2 - R_a^2\right)^{0.5} \quad ...(5.8)$$

It has been found that $R_a$ is usually very small as compared to $Z_s$. Therefore $x_s$ is very nearly equal to $Z_s$.

$Z_s$ is the ratio of open circuit voltage to short circuit current. If the value of open circuit voltage is read from the air gap line (i.e. the extended portion of the initial straight line portion of OCC), the resulting value of $Z_s$ is known as unsaturated synchronous impedance.

*Example 5.3*

A 100 kVA, 400 V 3-phase star connected alternator gave the following test results:

| | | |
|---|---|---|
| OCC: | Line-to-line voltage | 400 V |
| | Field current | 3 A |
| Air-gap line: | Line-to-line voltage | 300 V |
| | Field current | 2 A |
| SCC | Field current | 2 A |
| | Armature current | 144.34 A |

The armature resistance is negligible. Find unsaturated synchronous reactance in ohms and per unit.

*Solution*

Open circuit phase voltage (as read from air gapline for excitation of 2 A)

$$= \frac{300}{\sqrt{3}} = 173.21\text{ V}$$

Short circuit current for excitation of 2 A = 144.34 A

$$\text{Unsaturated synchronous reactance} = \frac{173.21}{144.34} = 1.2\text{ ohm}$$

$$\text{Rated phase voltage} = \frac{400}{\sqrt{3}} = 230.95\text{ V}$$

$$\text{Rated phase current} = \frac{100 \times 1000}{\sqrt{3} \times 400} = 144.34\text{ A}$$

$$\text{Base impedance} = \frac{230.95}{144.34} = 1.6\,\text{ohm}$$

$$\text{Unsaturated } x_s = \frac{1.2}{1.6} = 0.75\text{ pu}.$$

## 5.10 MMF METHOD

Figure 5.16 shows the phasor diagram of a synchronous generator, supplying load at a lagging power factor, based on mmf method. In this method the effect of armature leakage reactance $x_l$ is replaced by an equivalent mmf $F_x$ which is combined with $F_A$ to compute the resultant mmf.

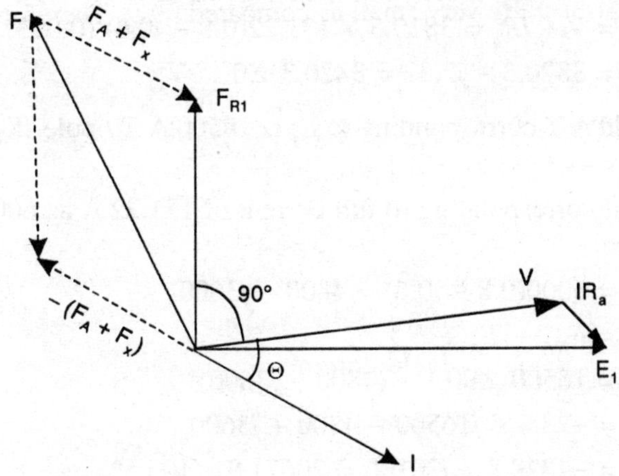


**Fig. 5.16.** Phasor diagram of an alternator as per mmf method

*F* is the field mmf. The phasor sum of *F* and $(F_A + F_x)$ gives the resultant mmf $F_{R1}$, which is responsible for the emf $E_l$. The phasor $IR_a$ is armature resistance drop, which is added to *V* to give $E_l$. When the alternator terminals are short circuited, *V* is zero. Since $IR_a$ very small, $E_1$ is also nearly equal to zero and hence $F_{R1}$ is zero. The mmf $(F_{A.} + F_x)$ is, therefore, the excitation required to circulate full-load current when armature terminals are short circuited.

The phasor sum of *V* and $IR_a$ gives $E_l$ . $F_{Rl}$, excitation corresponding to $E_1$ can be found from OCC. Then

$$F = F_{Rl} - (F_A + F_x) \qquad \text{...(5.9)}$$

*E* is the emf as read from OCC corresponding to excitation *F*.

It is seen that the synchronous impedance and mmf method are complementary to each other. The synchronous impedance method replaces the armature reaction mmf by a fictitious reactance $x_a$, whereas the mmf method replaces the leakage reactance by an mmf $F_x$. However the results of voltage regulation as found by using the two methods are very different from each other. The voltage regulation, as calculated by synchronous impedance method is higher than the actual. Thus this method gives pessimistic results. On the other hand, the mmf method gives results which are less than actual and this method is known as optimistic method. This difference is due to the effect of saturation on OCC. If OCC were a straight line, both the methods would give the same result.

*Example 5.4*

The OCC and SCC of the alternator of example 5.1 are as under:

| Field A T/pole | 0 | 3000 | 6000 | 16500 | 20672 |
|---|---|---|---|---|---|
| O C Voltage per phase | 0 | 900 | 1780 | 3820 | 4080 |
| S C Current | 0 | 65.6 | 131.22 | – | – |

Find voltage regulation at full-load 0.8 p.f. lagging using mmf method.

*Solution*

$$V = 3810.5 \angle 0 \text{ V}$$

$$I = 131.22(0.8 - j0.6) \text{ A}$$

$$E_1 = V + IR_a = 3810.5 + 131.22(0.8 - j0.6)(0.093)$$

$$= 3820.3 - j7.32 = 3820.3 \angle 0.1^\circ \text{ V}$$

From O C C, $F_{R1}$ the field A T corresponding to $E_1$ is 16500 A T / pole. $F_{R1}$ leads $E_1$ by 90°. Thus $F_{R1} = 16500 \angle 90.1°$.

The S C C gives the mmf corresponding to full current of 131.22 A as 6000 A T / pole. This is in phase with current. Therefore

$$F_A + F_x = 6000(0.8 - j0.6) = 4800 - j3600$$

$$F = F_{R1} - (F_A + F_x)$$

$$= 16500 \angle 90.1 - (4800 - j3600)$$

$$= -28.8 + j16500 - 4800 + j3600$$

$$= -4828.8 + j20100 = 20671.9 \angle 103.5°$$

From O C C the emf $E$, corresponding to an mmf of 20671.9 A T / pole is 4080 V

$$\text{Regulation} = \frac{4080 - 3810.5}{3810.5} \times 100 = 7.07\%.$$

## 5.11 SATURATED SYNCHRONOUS IMPEDANCE METHOD

The use of unsaturated synchronous reactance in the calculation of regulation gives very unrealistic results. In the saturated synchronous impedance method, the value of synchronous reactance is calculated in a way which is consistent with the degree of saturation present in the machine. The steps in the calculation of saturated synchronous reactance ($x_{ss}$) are as under:

(1) Find $E_r$ as the phasor sum of $V$ and $IR_a$ and $jIx_l$.

Thus $$E_r = V + I(R_a + jx_l) \quad \text{...(5.10)}$$

(2) From OCC find the mmf (or field current) corresponding to emf $E_c$. In Fig. 5.17 this is shown as $OD$. If there were no saturation, the voltage $E_r$ would correspond to excitation $OB$ as read from air gapline. The saturation factor $k$ is defined

as $$k = \frac{OD}{OB} \quad \text{...(5.11)}$$

The geometry of similar triangles in Fig. 5.17 indicates that $k$ is also given by the ratio $E_{r(ag)}/E_r$.

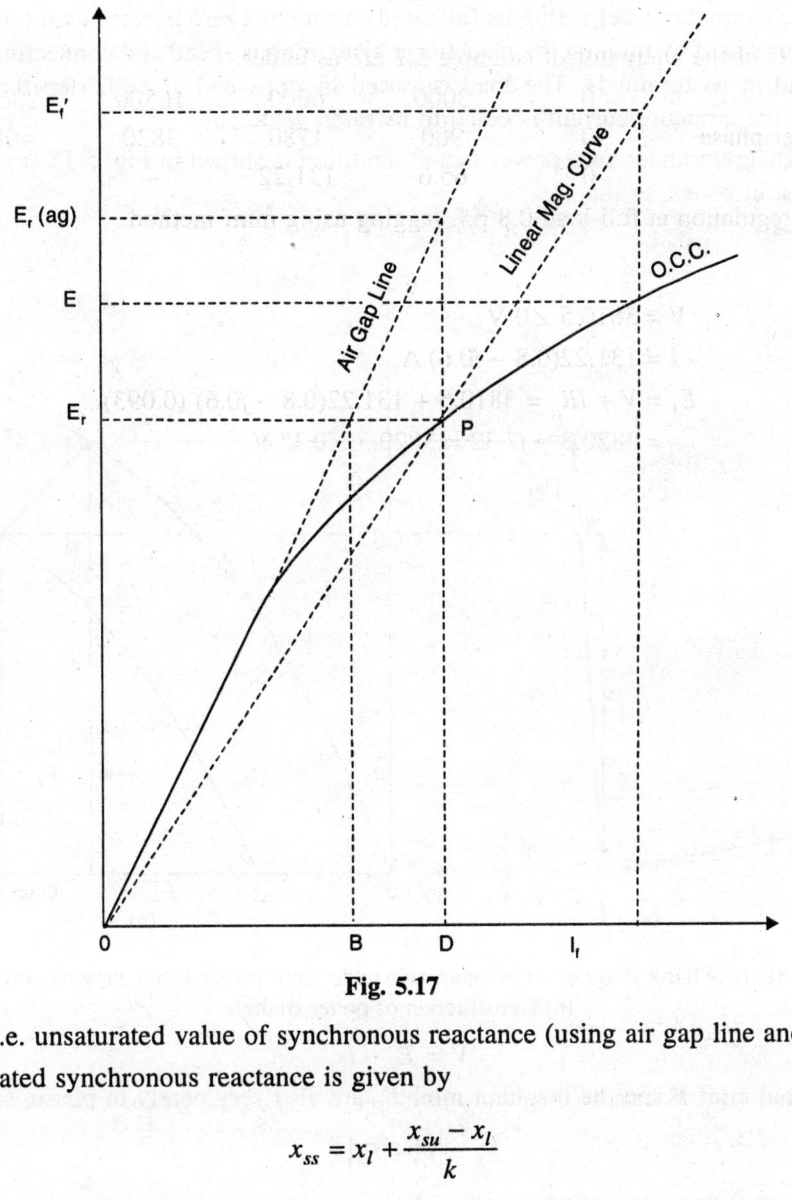


**Fig. 5.17**

(3) Find $x_{su}$ i.e. unsaturated value of synchronous reactance (using air gap line and S.C.C)

(4) The saturated synchronous reactance is given by

$$x_{ss} = x_l + \frac{x_{su} - x_l}{k} \qquad ...(5.12)$$

(5) $$E_f' = V + I(R_a + jx_{ss})$$

Find $E$ from OCC corresponding to above value of $E_f'$. Then find regulation. It is seen from Fig. 5.17 that this method is based on the assumption that degree of saturation remains constant corresponding to emf $E_r$ and the operating point moves, along the dotted line $OP$, which has been designated as linear magnetisation curve.

## 5.12 ZERO POWER FACTOR (POTIER REACTANCE) METHOD

Zero power factor characteristics of an alternator gives the variation of terminal voltage with field current, when the alternator is delivering its full rated current to a zero power factor (lagging) load. This characteristic is obtained by running the machine at synchronous speed and connecting a purely inductive 3-phase load to its terminals. The load is varied in steps and at each step the field current is adjusted, so that the armature current is equal to its rated value.

The phasor diagram under zero power factor condition is shown in Fig. 5.18 (a). It is seen that $V$ and $E_r$ are almost in phase, so that

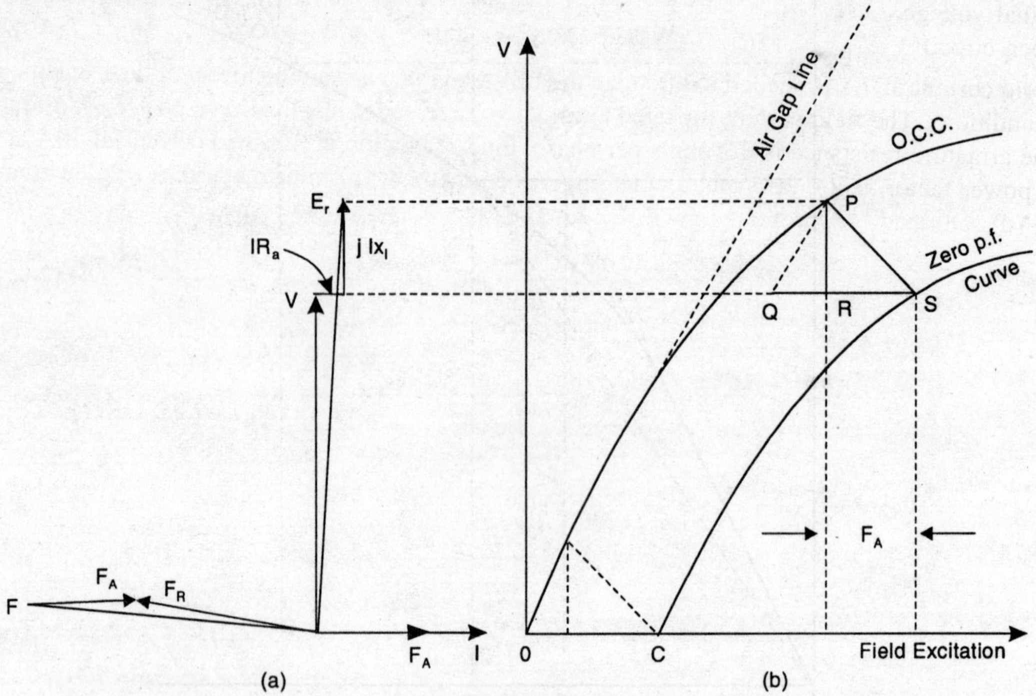


**Fig. 5.18.** (a) Phasor diagram of an alternator under zero power factor lagging condition, (b) Construction of potier triangle

$$V = E_r - Ix_l \qquad ...(5.13)$$

Moreover, the field mmf $F$ and the resultant mmf $F_R$ are also very nearly in phase, so that

$$F_R = |F| - |F_A| \qquad ...(5.14)$$

The zero power factor curve can be used to determine leakage reactance $x_l$ and armature mmf $F_A$. It is not necessary to plot full curve. Only two points $S$ and $C$ (Fig. 5.18 b) are sufficient. The point $C$ corresponds to short circuit condition and full rated armature current flowing. This point can be located from the short circuit characteristic. The point $S$ corresponds to rated terminal voltage condition and can be obtained by loading the alternator by an under excited synchronous motor (An under excited synchronous motor can be made to run at a power factor which is almost zero lagging). The procedure to find $x_l$ and $F_A$ is as under:

(a) Draw horizontal line $SQ$ equal to $C\,O$.

(b) From point $Q$ draw a line $QP$ parallel to air gap line. This line meets OCC at point $P$.

(c) From point $P$ draw $PR$ perpendicular to line $QS$ and meeting it at point $R$.

(d) The triangle *PRS* is known as Potier triangle. The intercept *PR* is equal to $Ix_l$ and intercept *RS* is equal to armature mmf $F_A$. As seen from Fig.5.18(b), Potier triangle can also be located at the base line.

### Example 5.5

The open circuit characteristic of a 6-pole, 440V, 50 Hz, 3-phase, star connected alternator is as under:

| | | | | | | | | | |
|---|---|---|---|---|---|---|---|---|---|
| Field current | A | 2 | 4 | 6 | 7 | 8 | 10 | 12 | 14 |
| Terminal voltage on open circuit | V | 156 | 288 | 396 | 440 | 474 | 530 | 568 | 592 |

A field current of 7A is needed to circulate the full-load rated armature current of 40A under short circuit conditions. The field current for rated terminal voltage under full-load zero power conditions is 15A. The armature resistance is 0.2 ohms per phase. Find regulation at full-load current of 40A at 0.8 lagging power factor, using (a) synchronous impedance method, (b) mmf method, (c) Potier triangle method, (d) saturated synchronous impedance method. Comment on the results.

### Solution

*(a) Synchronous Impedance method*

When the field current is 7A, the O C voltage is $440\sqrt{3} = 254.04$ V per phase and the short circuit current is 40 A.

$$Z_s = \frac{254.04}{40} = 6.35\Omega$$

$$x_s = \left(Z_s^2 - R_a^2\right)^{0.5} = \left(6.35^2 - 0.2^2\right)^{0.5} \cong 6.35\Omega$$

$$V = 254.04\angle 0$$

$$I = 40(0.8 - j0.6)A$$

$$E = V + I(R_a + jx_s)$$

$$= 254.04\angle 0 + 40(0.8 - j0.6)(0.2 + j6.35)$$

$$= 254.04 + 158.8 + j198.4$$

$$= 412.84 + j198.4 = 458.04\angle 25.67°$$

$$\text{Voltage Regulation} = \frac{458.04 - 254.04}{254.04} \times 100 = 80.3\%$$

*(b) MMF method*

The O C C is drawn in Fig. 5.19. Let *I* be the reference phasor

$$I = 40\angle 0\text{A}$$

$$V = 254.04(0.8 + j0.6) = 203.23 + j152.42 \text{ V}$$

The phasor diagram is shown in Fig. 5.16.

$$E_1 = V + IR_a = 203.23 + j152.42 + 40 \times 0.2$$

$$= 211.23 + j152.42 = 260.48\angle 35.81° \ V$$

From OCC, the field current corresponding to 260.48 V per phase is 7.3 A, leading $E_1$ by 90°.

$$\dot{F}_{R1} = 7.3\angle(35.81° + 90°) = 7.3\angle 125.81°$$
$$= -4.27 + j5.92 \text{ A}$$

Under short circuit conditions, the field current required to circulate an armature current of 40A is 7A. This is in phase with current

$$F_A + F_x = 7\angle 0$$
$$F = F_{R1} - (F_A + F_x) = -4.27 + j5.92 - 7$$
$$= -11.27 + j5.92 = 12.73 \angle 152.3A$$

From OCC the emf $E$ corresponding to field current of 12.73 A is 332 V.

$$\text{Voltage Regulation} = \frac{332 - 254.04}{254.04} \times 100$$
$$= 30.69\%$$

*(c) Potier Triangle Method*

The potier triangle is shown in Fig. 5.19. The point $C$ is the point corresponding to short circuit condition. Point $S$ corresponds to rated voltage under full-load zero p.f. condition. PRS is the potier triangle. From this triangle

$$PR = Ix_l = 22 \text{ V}$$

or $$x_l = 22/40 = 0.55\Omega$$

and $$F_A = 6.2A$$

Taking $I$ as reference

$$E_r = V + I(R_a + jx_l)$$
$$= 254.04(0.8 + j0.6) + 40(0.2 + j0.55)$$
$$= 211.23 + j152.42 + j22 = 211.23 + j\,174.42$$
$$= 273.93\angle 39.55°$$

From OCC, the field current ($F_R$) corresponding to 273.93 V is 8 A and $F_A$ = 6.2 A

$F_A$ is in phase with current and $F_R$ leads $E_r$ by 90°

$$F_R = 8\angle(39.55 + 90°) = 8\angle 129.55°$$
$$F = F_R - F_A = 8\angle 129.55 - 6.2$$
$$= -11.29 + j6.17 = 12.86\angle 151.34°\text{A}$$

From OCC, corresponding to a field current of 12.86 A, $E$ = 336 V

$$\text{Voltage regulation} = \frac{336 - 254.04}{254.04} \times 100$$
$$= 32.26\%$$

*(d)* From Fig. 5.19, for $E_r$ = 273.93 V,

$$E_{r(ag)} = 363 \text{ V}$$
$$k = E_{r(ag)}/E_r = 363/273.93 = 1.325$$

As determined in part (c), $x_l = 0.55\ \Omega$

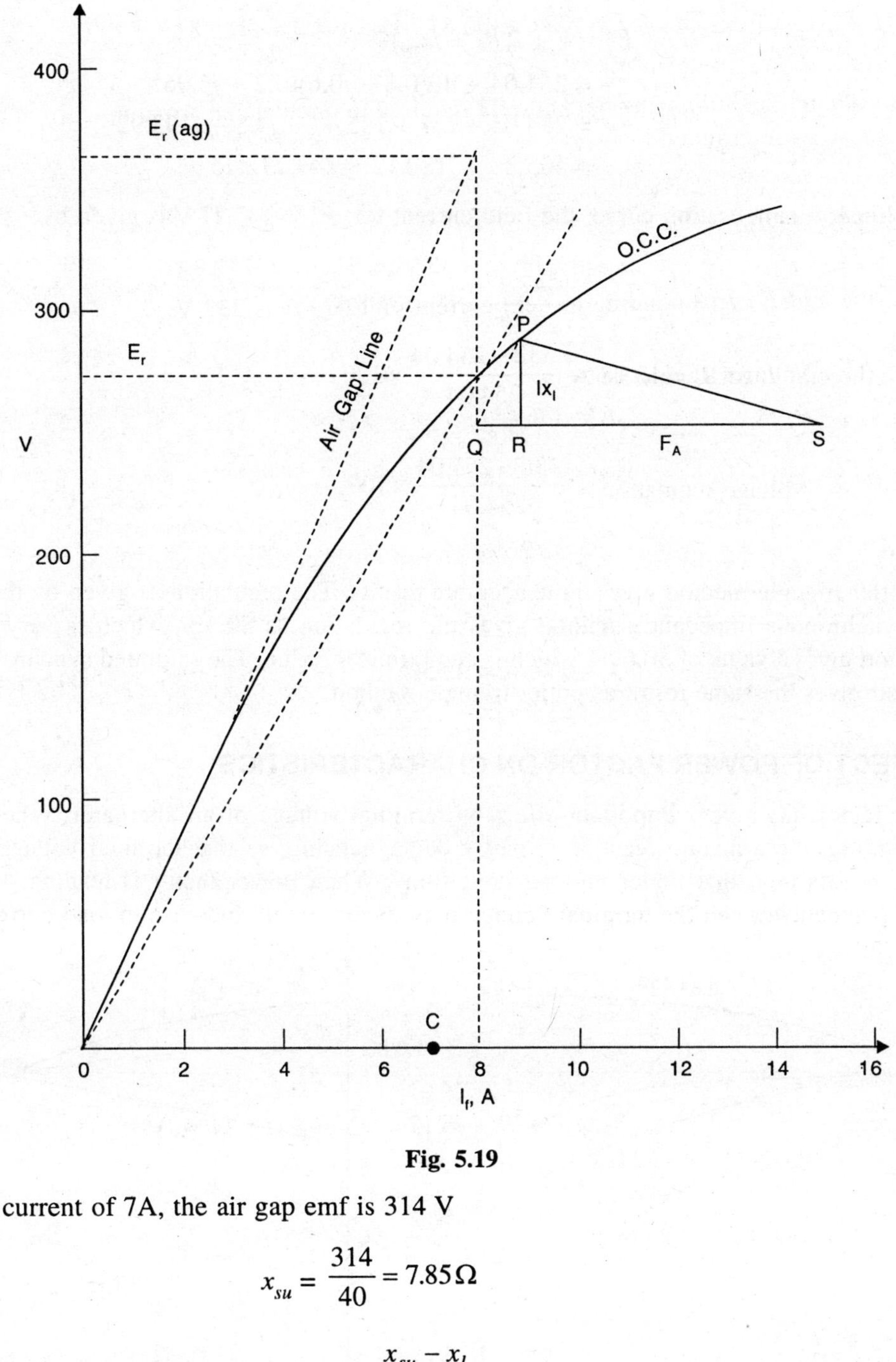


**Fig. 5.19**

For a field current of 7A, the air gap emf is 314 V

$$x_{su} = \frac{314}{40} = 7.85\,\Omega$$

saturated synchronous reactance $x_{ss} = x_l + \dfrac{x_{su} - x_l}{k}$

$$= 0.55 + \frac{7.85 - 0.55}{1.325} = 6.06\,\Omega$$

$$E'_f = V + I(R_a + jx_{ss})$$
$$= 254.04 + 40(0.8 - j0.6)(0.2 + j6.06)$$
$$= 254.04 + 6.4 + 145.44 - j4.8 + j193.92$$
$$= 405.88 + j189.12 = 447.77\angle 25°V$$

Using linear magnetisation curve, the field current for $E'_f = 447.77$ V is given by

$$I_f = \frac{E'_f}{E_r} \times 8 = \frac{447.77}{273.93} \times 8 = 13.08A$$

From OCC, the emf $E$ for $I_f = 13.08$ A is

$$E = 336V$$

$$\text{Voltage regulation} = \frac{336 - 254.04}{254.04} \times 100$$
$$= 32.26\%.$$

The potier triangle method gives quite accurate results. The regulation as given by this method is 32.26%. Synchronous impedance method gives the regulation as 80.3%, which is very pessimistic. MMF method gives a value of 30.69% which is an optimistic value. The saturated synchronous impedance method gives the same result as potier triangle method.

## 5.13 EFFECT OF POWER FACTOR ON CHARACTERISTICS

The power factor has a very important effect on terminal voltage of an alternator. When the power factor is lagging, the armature reaction is partly demagnetising, so that terminal voltage under load conditions, is less than that under no-load conditions. When power factor is leading, the armature reaction is magnetising and the terminal voltage may increase with increase in load current.

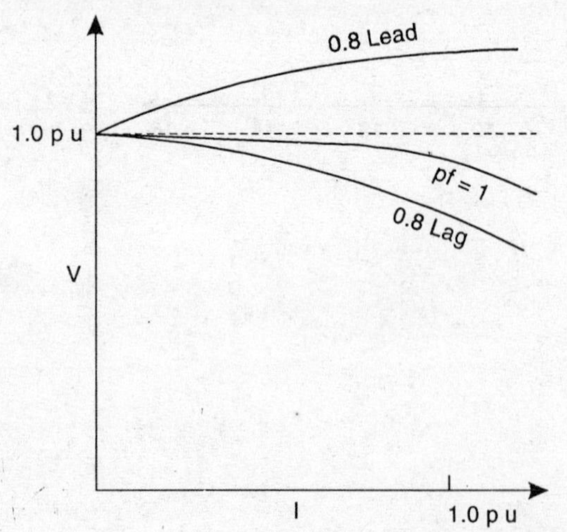


**Fig. 5.20.** Variation of terminal voltage with armature current of an alternator

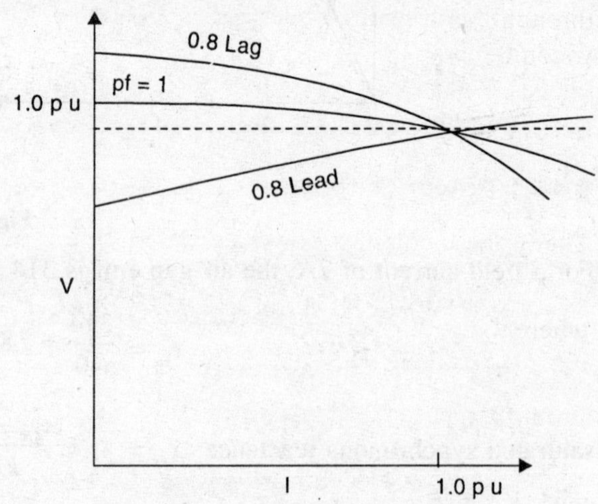


**Fig. 5.21.** Change in terminal voltage when load current is decreased in an alternator

Figure 5.20 shows the variation of terminal voltage with load current for three values of power factor.

When an alternator is delivering rated load current $I$, at rated terminal voltage $V$, a decrease in load current causes a change in terminal voltage. If the power factor is lagging, the terminal voltage will rise (because the demagnetising effect of armature reaction is reduced). If the power factor is leading, the terminal voltage would decrease. This is shown in Fig. 5.21.

The alternator compounding curves indicate the change in field current required to maintain constant terminal voltage as the load current increases. Evidently the alternator will need higher field current, if power factor is lagging. Figure 5.22 shows this variation.

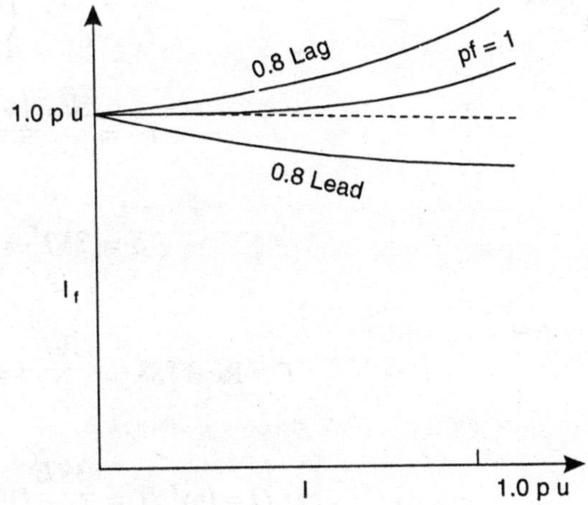


**Fig. 5.22.** Compounding curves for alternator

## 5.14 SHORT CIRCUIT RATIO (SCR)

Short circuit ratio of an alternator is defined as

$$\text{SCR} = \frac{\text{Field current required to produce rated voltage on open circuit}}{\text{Field current required to produce rated current on 3-phase short circuit}} \quad ...(5.15)$$

If there were no saturation, the short circuit ratio would be the reciprocal of the pu value of synchronous reactance. It can be shown that SCR is the reciprocal of the pu value of saturated synchronous reactance.

The short circuit ratio has an important effect on the performance of the machine and the cost. A lower value of SCR means a greater change in field current to maintain constant terminal voltage and a lower value of steady state stability limit. Lesser the SCR, lesser is the size, weight and cost of the machine. Evidently, the short circuit ratio has an important role to play in determining the current through the armature under fault conditions. Modern alternators are built with short circuit ratio between 0.5 and 1.5.

## 5.15 POWER ANGLE CHARACTERISTICS OF CYLINDRICAL ROTOR ALTERNATOR

### 5.15.1 Power Output

The 3-phase complex power delivered by a cylindrical rotor alternator is

$$S = P + jQ = 3VI^* \quad ...(5.16)$$

where $V$ = terminal voltage per phase

$I$ = armature current per phase

From the phasor diagram of Fig. 5.13

$$I = \frac{E\angle\delta - V}{Z_s\angle\phi_s} = \frac{E\angle(\delta - \phi_s) - V\angle - \phi_s}{Z_s}$$

where $$\phi_s = \tan^{-1}\left(\frac{x_s}{R_a}\right)$$

$$I^* = \frac{E\angle(\phi_s - \delta) - V\angle\phi_s}{Z_s} \quad ...(5.17)$$

$$S = 3VI^* = \frac{3V}{Z_s}\left[E\angle(\phi_s - \delta) - V\angle\phi_s\right] \quad ...(5.18)$$

$$P = \text{Real}\,[S] = \frac{3VE}{Z_s}\cos(\phi_s - \delta) - \frac{3V^2}{Z_s}\cos\phi_s \quad ...(5.19a)$$

$$Q = \text{Im}[S] = \frac{3VE}{Z_s}\sin(\phi_s - \delta) - \frac{3V^2}{Z_s}\sin\phi_s \quad ...(5.19b)$$

In large generators the resistance is small as compared to synchronous reactance. If resistance is neglected,

$$Z_s = x_s \text{ and } \phi_s = 90°$$

Then Eq. (5.19) reduces to

$$P = \frac{3VE}{x_s}\sin\delta \quad ...(5.20a)$$

$$Q = \frac{3VE}{x_s}\cos\delta - \frac{3V^2}{x_s} \quad ...(5.20b)$$

The graph of $P$ versus $\delta$ is known as power angle characteristics of the synchronous machine.

The reactance $x_s$ is constant. If $V$ and $E$ are also constant, the real power output varies as sin $\delta$. Therefore angle $\delta$ is known as power angle or load angle. The power angle characteristics is shown in Fig. 5.23. The characteristics for negative values of $\delta$ refers to the operation of machine as motor. $P$ also gives torque in synchronous watts and $P/\omega_s$ gives torque in N-m.

The maximum power output can be found by differentiating Eq. (5.19a) and equating $dP/d\delta = 0$. Thus

$$\frac{dP}{d\delta} = \frac{3VE}{Z_s}\sin(\phi_s - \delta) = 0$$

or $$\phi_s = \delta$$

Thus for maximum power output, the power angle $\delta$ is equal to impedance angle $\phi_s$. The maximum power output is found by putting $\delta = \phi_s$ in Eq. (5.19a)

$$P_{max} = \frac{3VE}{Z_s} - \frac{3V^2}{Z_s}\cos\phi_s$$

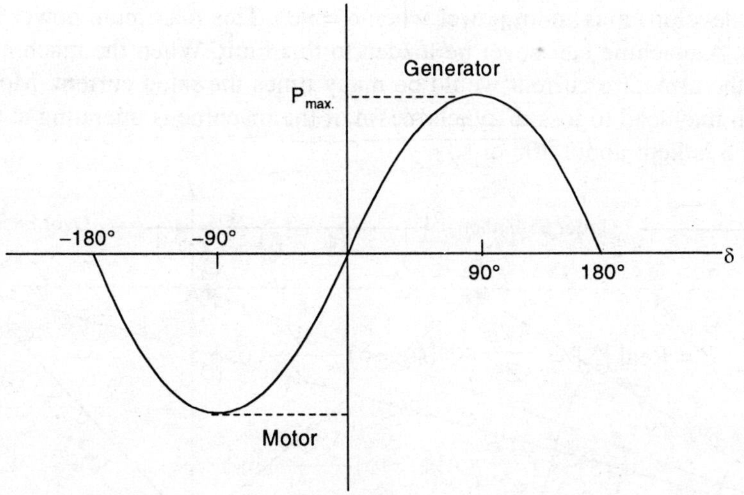


**Fig. 5.23.** Power angle characteristics of cylindrical rotor machine

$$= \frac{3VE}{Z_s} - \frac{3V^2}{Z_s^2} R_a$$

When $P$ is maximum, the reactive power output from Eq. (5.19b) is

$$Q = -\frac{3V^2}{Z_s} \sin \phi_s = -\frac{3V^2}{Z_s^2} x_s$$

The maximum value of reactive power is found by differentiating Eq. (5.19b) and equating $dQ/d\delta$ to zero. Thus

$$\frac{dQ}{d\delta} = \frac{3VE}{Z_s} \cos(\phi_s - \delta) = 0$$

or

$$\phi_s - \delta = \pm 90°$$

or

$$\delta = \phi_s \mp 90°$$

Substituting this value of δ in Eq. (5.19b)

$$Q_{max} = -\frac{3VE}{Z_s} - \frac{3V^2}{Z_s} \sin \phi_s$$

$$= -\frac{3VE}{Z_s} - \frac{3V^2}{Z_s^2} x_s$$

If armature resistance is neglected, $\phi_s = 90°$ and maximum real power output occurs when $\delta = 90°$. This fact is also evident from Eq. (5.20 a).

Equation (5.20) pertains to the power output of the alternator at its terminals. In case the alternator is feeding an infinite bus of voltage $V$ through a transmission line of reactance $x$, the denominator in these equations will be $(x_s + x)$. This total reactance is known as transfer reactance $X$.

The generator develops maximum power when $\delta = 90°$. This maximum power is known as steady state stability limit. A machine can never be loaded to this limit. When the machine is delivering this maximum power, the armature current would be many times the rated current. Moreover, any disturbance in the system may lead to loss of synchronism, if the machine is operating at this limit. In actual practice, the angle $\delta$ is kept about 40° or so.

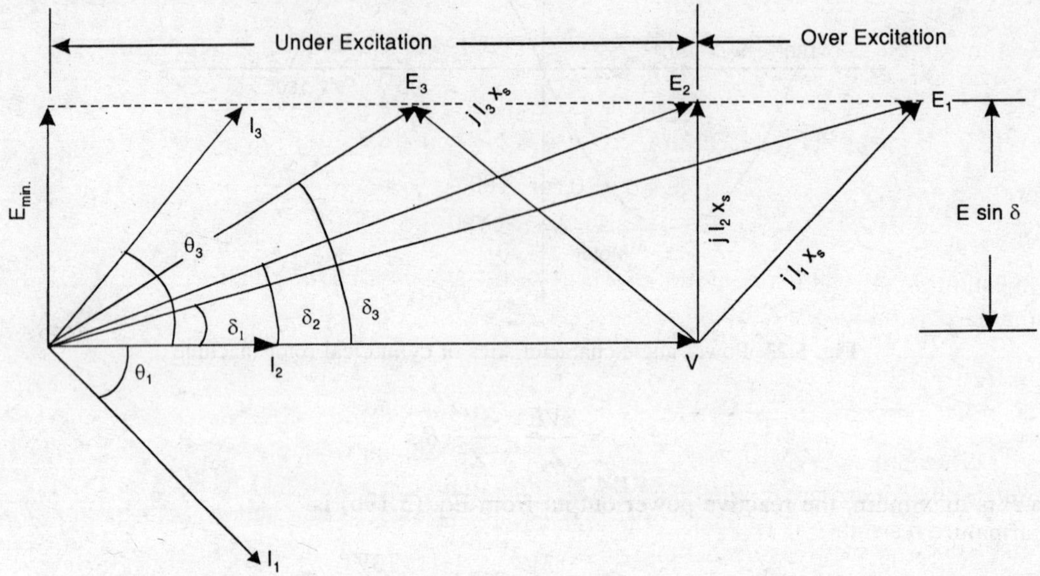


**Fig. 5.24.** Effect of excitation on the power factor of an alternator supplying a constant load

Figure 5.24 shows the operation of the machine with constant load, but varying excitation. Since $V$ and $x_s$ are constant, operation at constant load means that $E \sin \delta$ is constant. The operation is shown for three values of current, viz. $I_1$ at a lagging power factor $\cos\theta_1$, $I_2$ at unity power factor and $I_3$ at a leading power factor $\cos\theta_3$. In every case, the phasor sum of $V$ and $jIx_s$ gives the emf $E$. The operation at a lagging power factor leads to higher values of $E$ (i.e. over excitation) and operation at a leading power factor leads to lower values of field current (and therefore $E$) i.e. under excitation. The maximum value of field current (and therefore $E$) is limited by rotor heating. Therefore an alternator cannot be operated at very low lagging power factors. The minimum value of excitation (shown as $E_{min}$) occurs when $\delta = 90°$. This is the stability limit. If the alternator is operating at or near $\delta = 90°$, a small disturbance is enough to cause instability and the alternator will lose synchronism.

### 5.15.2 Power Input

The power input per phase is the product of $E\angle\delta$ and $I^*$. As found above

$$I^* = \frac{E\angle(\phi_s - \delta) - V\angle\phi_s}{Z_s}$$

So that the three-phase complex power input is

$$S_{\text{input}} = 3E\angle\delta \left[\frac{E\angle(\phi_s - \delta) - V\angle\phi_s}{Z_s}\right]$$

$$= \frac{3E^2}{Z_s}\angle\phi_s - \frac{3EV}{Z_s}\angle(\phi_s + \delta)$$

$$P_{\text{input}} = \frac{3E^2}{Z_s}\cos\phi_s - \frac{3VE}{Z_s}\cos(\delta + \phi_s) \qquad \text{...(5.21)}$$

The maximum power input is given by the condition $dP_{\text{input}}/d\delta = 0$

$$\frac{dP_{\text{input}}}{d\delta} = \frac{3VE}{Z_s}\sin(\delta + \phi_s) = 0$$

or $\delta + \phi_s = 0$ or $180°$

or $\delta = -\phi_s$ or $180 - \phi_s$

Substituting this value of δ in Eq. (5.21) gives the maximum power input. Evidently, $\delta = 180 - \phi_s$ gives the maximum power input condition

$$\text{Max. } P_{\text{input}} = \frac{3E^2}{Z_s}\cos\phi_s + \frac{3VE}{Z_s} \qquad \text{...(5.22)}$$

When armature resistance is neglected, $\phi_s = 90°$ and maximum power input $= 3VE/x_s$. It is seen that under this condition the maximum power input and output are equal. This has to be so, because when armature resistance is neglected, the armature copper-loss is zero and input and output become equal.

*Example 5.6*

A 480V, 50Hz, 3-phase, 4-pole, delta connected cylindrical rotor generator has $R_a = 0.015$ ohm and $x_s = 0.1$ ohm. The rated current and power factor are 1200 A and 0.8 lagging, respectively. At full-load the mechanical losses are 40 kW and core-losses are 30 kW. The resistance of field winding is 500 ohm. The OCC of generator is as under:

| $I_f$ | 1.0 | 2.0 | 3.0 | 4.0 | 5.0 | 6.0 | 7.0 A |
|---|---|---|---|---|---|---|---|
| OC Voltage | 120 | 250 | 360 | 451 | 505 | 532 | 550 V |

(a) What is the speed of this machine ?

(b) What is the field current at no-load, if terminal voltage is 480 V ?

(c) What is the field current if the generator is delivering rated current at rated terminal voltage and power factor ? What is the power output, power angle and efficiency ?

(d) When the generator is supplying full load, the load is suddenly disconnected. What will happen to terminal voltage.

(e) Find the field current for a load current of 1200 A, at 480 V and 0.8 power factor leading.

*Solution*

(a) $n = n_s = 2f/p = 2 \times 50/4 = 25$ rps or 1500 RPM.

(b) Since $I = 0$ ; $E = V = 480$ V

Since the machine is delta connected, $E$ per phase is also 480 V. From OCC, the value of field current for this value of $E$ is 4.5 A.

(c) Since the generator is delta connected, armature current per phase = $I = 1200/\sqrt{3} = 692.8$ A

$$E = V + I(R_a + jx_s)$$

Let $V = 480\angle 0$

$$I = 692.8(0.8 - j0.6)$$

$$E = 480\angle 0 + 692.8(0.8 - j0.6)\ (0.015 + j0.1)$$

$$= 529.9 + j49.2 = 532.0\angle 5.3°$$

From OCC, the field current corresponding to $E = 532.0$ V is 6A

er angle = 5.3°

$$\text{Power output} = 3VI \cos\theta \times 10^{-3} \text{ kW}$$

$$= 3(480)\ (692.8)\ (0.8) \times 10^{-3}$$

$$= 798.1 \text{ kW}$$

$$\text{Armature copper losses} = 3I^2R_a \times 10^{-3} \text{ kW}$$

$$= 3(692.8)^2\ (0.015) \times 10^{-3} = 21.6 \text{ kW}$$

$$\text{Field copper-losses} = I_f^2 R_f \times 10^{-3} \text{ kW}$$

$$= (6)^2(500) \times 10^{-3} = 18 \text{ kW}$$

$$\text{Total losses} = 21.6 + 18 + 30 + 40 = 109.6 \text{ kW}$$

$$\text{Efficiency} = \frac{\text{Output}}{\text{Output} + \text{Losses}} \times 100$$

$$= \frac{798.1 \times 100}{798.1 + 109.6} = 87.93\%$$

(d) Since the load current has been reduced to zero, $V = E$. Because the field current has not been changed, the terminal voltage will rise to 532.0 V.

(e)
$$I = 692.8(0.8 + j0.6)$$

$$E = V + I(R_a + jx_s)$$

$$= 480\angle 0 + 692.8(0.8 + j0.6)(0.015 + j0.1)$$

$$= 446.7 + j61.7 = 451\angle 7.9° V$$

From OCC, the field current for $E = 451$ V is 4 A.

### *Example 5.7*

A 480 V, 6-pole, star connected 50 Hz cylindrical rotor synchronous generator has negligible armature resistance and 1.0 ohm per phase synchronous reactance. The full-load armature current is 60 A. Total losses at full-load are 4.2 kW. The field current has been so adjusted that terminal voltage at no-load is 480 V. Find (a) speed, (b) terminal voltage and regulation at full-load p.f. 0.8 lagging, (c) terminal voltage and regulation at full-load unity p.f., (d) terminal voltage and regulation at full-load 0.8 p.f. leading, (e) efficiency at full-load 0.8 power factor lagging, (f) shaft torque applied by prime mover at full-load, 0.8 p.f. lagging, (g) torque developed by the machine, at full-load 0.8 p.f. lagging.

*Solution*

(a) $$n = n_s = 2f/P = 2 \times 50/6$$
$$= 16.666 \text{ rps or } 1000 \text{ rpm}$$

(b) $$E = 480/\sqrt{3} = 277.14\,\text{V}$$

The phasor diagram for 0.8 lagging p.f. is shown in Fig. 5.25(a)

$$V + jx_s I = E$$

or $$V\angle 0 + j(1)[60(0.8 - j0.6)] = (277.14)^2$$
$$(V + 36)^2 + (48)^2 = (277.14)^2 \text{ or } V = 236.95 \text{ V}$$

$$\text{Line voltage at terminals} = \sqrt{3} \times 236.95 = 410.4\,\text{V}$$

$$\text{Voltage regulation} = \frac{277.14 - 236.95}{236.95} \times 100 = 16.96\%$$

(c) The phasor diagram for unity p.f. is shown in Fig. 5.25 (b)

$$E = 277.14 \text{ V}$$
$$V^2 + (x_s I)^2 = E^2$$

or $$V^2 + 60^2 = 277.14^2 \text{ or } V = 270.57 \text{ V}$$

$$\text{Line voltage at terminals} = \sqrt{3} \times 270.57 = 468.6\,\text{V}$$

$$\text{Voltage regulation} = \frac{277.14 - 270.57}{270.57} \times 100 = 2.4\%$$

(d) The phasor diagram for 0.8 leading power factor is shown in Fig. 5.25 (c).

$$E = 277.14 \text{ V}$$
$$V + jx_s I = E$$

or $$V\angle 0 + j(1)[60(0.8 + j0.6)] = 277.14$$

or $$(V - 36)^2 + 48^2 = 277.14 \text{ V}$$

or $$V = 308.95 \text{ V}$$

$$\text{Line voltage at terminals} = \sqrt{3}(308.95) = 535.1\,\text{V}$$

$$\text{Voltage regulation} = \frac{277.14 - 308.95}{308.95} \times 100 = -10.3\%$$

As expected, the voltage regulation is positive at lagging and unity power factor and negative at leading power factor.

(e) Power output when I = 60 A and p.f. is 0.8 lagging is

$$P = 3VI\cos\theta = 3(236.95)(60)(0.8)$$
$$= 34120 \text{ W or } 34.12 \text{ kW}$$

$$\text{Losses} = 4.2 \text{ kW}$$

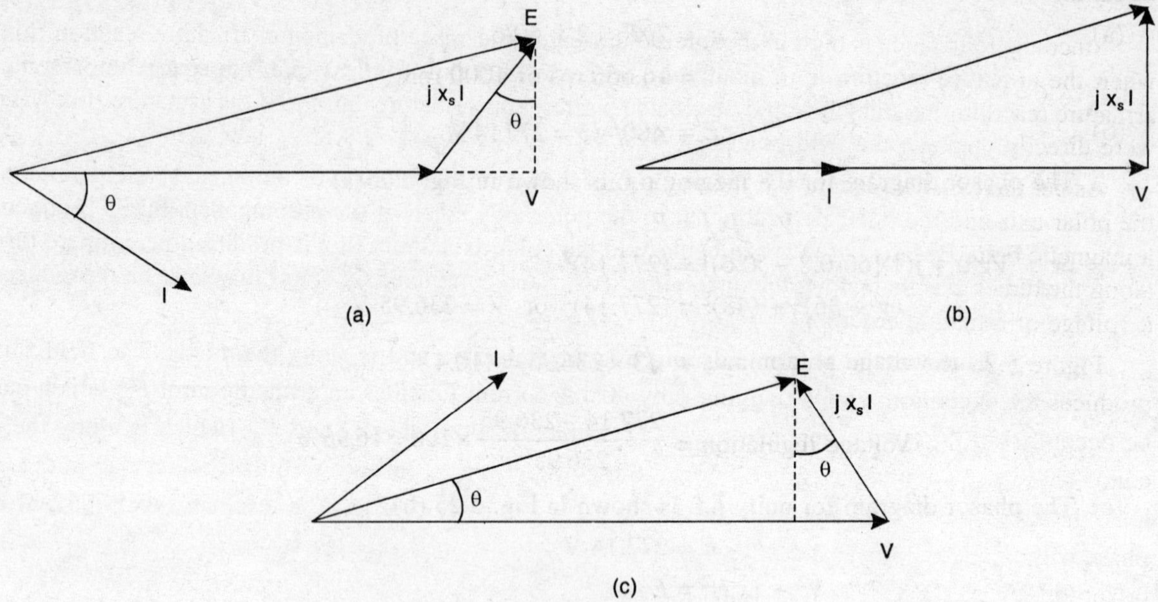


**Fig. 5.25**

$$\text{Efficiency} = \frac{34.12}{34.12+4.2} \times 100 = 89.04\%$$

(f) $$(\text{shaft torque})\ (\omega_m) = \text{Power Input}$$

$$\omega_m = \frac{1000}{60} \times 2\pi = 104.72 \text{ rad/sec.}$$

$$\text{Shaft torque} = \frac{(34.12+4.2)1000}{104.72}$$

$$= 365.93 \text{ N-m}$$

(g) $$(\text{torque developed})(\omega_m) = \text{Power Output}$$

or $$\text{Torque developed} = \frac{34.12 \times 1000}{104.72}$$

$$= 325.82 \text{ N-m.}$$

## 5.16 SALIENT POLE MACHINES: TWO-REACTION THEORY

The equivalent circuit, phasor diagram and the equations developed in the preceding sections are applicable only for cylindrical rotor machines. In these machines, the reluctance of the magnetic path is the same in all directions, so that the field mmf and armature mmf can be combined directly. However, in a salient pole machine, the reluctance of the magnetic circuit along the polar (or direct) axis is much less than the reluctance along the interpolar (or quadrature) axis. The analysis of salient pole machines was proposed by Andre Blondel and is known as two-reaction theory.

Since the reluctance is high in the interpolar region, the space fundamental armature reaction flux, when the armature reaction is in quadrature with the field poles is less than the space fundamental armature reaction flux, which would be created by the same armature current, if the armature flux wave were directly opposite the field poles.

As per the two-reaction theory, the stator mmf is decomposed into two components, one parallel to the polar axis and the other perpendicular to the polar axis. Each of these component mmfs produces a magnetic field. Because of lower reluctance in the polar axis, more flux is produced per ampere turn along the direct axis than along the quadrature axis. Each component of stator magnetic field produces a voltage of armature reaction.

Figure 5.26 shows the phasor diagram. $\phi$ is the field flux and is along the *d*-axis. The field flux produces the excitation, emf *E*, lagging $\phi$ by 90°. A current *I* creates an armature mmf $F_a$, which can be decomposed into two components, viz. $F_{ad}$ (which is along *d*-axis) and $F_{aq}$ (which is along the *q* axis). These two component mmfs produce fluxes $\phi_{ad}$ and $\phi_{aq}$. Because of higher reluctance in the *q*-axis, the ratio $\phi_{ad}/F_{aq}$ is lower than the ratio $\phi_{ad}/F_{ad}$. As a result of this the resultant flux $\phi_{ar}$ is not in phase with armature mmf $F_a$ but is lagging behind $F_a$. In a cylindrical rotor machine, the flux $\phi_{ar}$ would be in phase with $F_a$. The flux $\phi_{ad}$ is opposing the field flux $\phi$ in this case, because the power factor is lagging. For a leading power factor, $\phi_{ad}$ would be in the same direction as field flux $\phi$. The *q*-axis flux $\phi_{aq}$ is cross-magnetising flux.

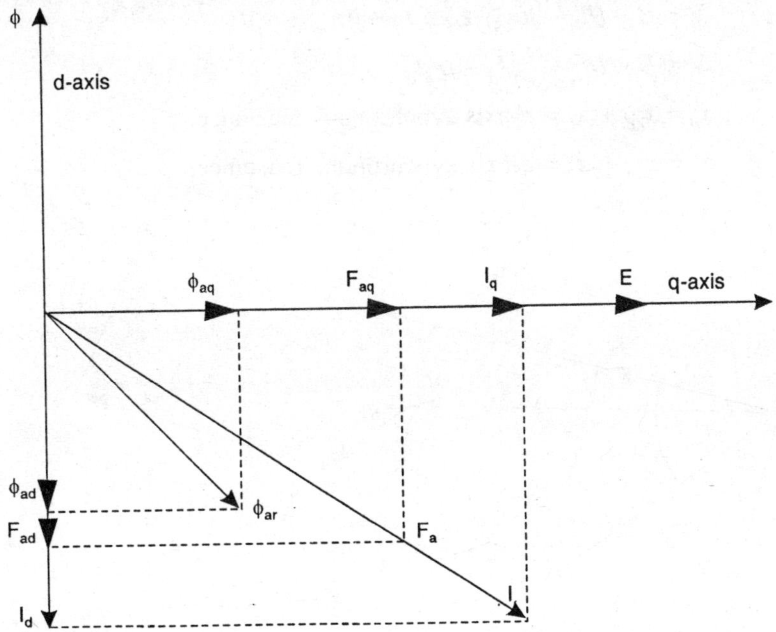


**Fig. 5.26.** Decomposition of stator mmf and flux into direct and quadrature axis components

Since the armature mmf $F_a$ is proportional to armature current *I*, decomposing this mmf into direct and quadrature components in effect means the decomposition of armature current into the direct axis component $I_d$ and quadrature axis component $I_q$.

The component fluxes $\phi_{ad}$ and $\phi_{aq}$ are rotating at synchronous speed and induce armature reaction emfs in each phase of stator winding. These emfs are $E_d$ and $E_q$, respectively. The total voltage $E'$ in each phase of stator winding is then

$$E' = E + E_d + E_q \quad ...(5.23)$$

where $E$ is the emf due to field flux, $E_d$ is the emf due to direct axis component of armature reaction and $E_q$ is the emf due to the quadrature axis component to armature reaction.

As in the case of cylindrical rotor machine, the armature reaction emf can be considered as a voltage drop across a fictitious reactance, so that $E_d$ and $E_q$ can be written as

$$E_d = -jx_{ad} I_d \quad ...(5.24a)$$

$$E_q = -jx_{aq} I_q \quad ...(5.24b)$$

Therefore the total stator $E'$ is

$$E' = E - jx_{ad} I_d - jx_{aq} I_q \quad ...(5.25)$$

In addition to above, voltage drops take place in armature resistance $R_a$ and leakage reactance $x_l$, so that the terminals voltage $V$ is

$$V = E' - IR_a - jIx_l \quad ...(5.26)$$

Equations (5.25 and 5.26) can now be combined. Since the leakage reactance is independent of rotor position, the voltage drop $jIx_l$ can be decomposed into $jI_d x_l$ and $jI_q x_l$ to give the final voltage equation of a salient pole machine, i.e.

$$V = E - IR_a - j(x_{ad} + x_l) I_d - j(x_{aq} + x_l)I_q$$

or

$$V = E - IR_a - jx_d I_d - jx_q I_q \quad ...(5.27)$$

where $x_d = x_{ad} + x_l$ = $d$-axis synchronous reactance.

$x_q = x_{aq} + x_l$ = $q$-axis synchronous reactance.

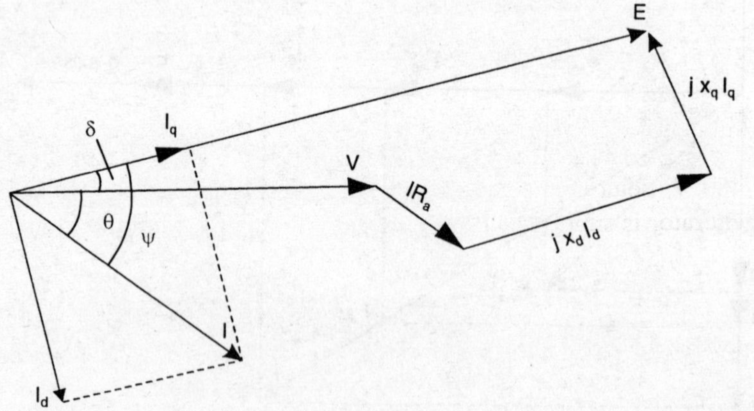


**Fig. 5.27.** Phasor diagram of salient pole alternator as per two-reaction theory

Figure 5.27 shows the phasor diagram corresponding to Eq. (5.27). $\theta$ is the power factor angle. The armature current is resolved into two components, $I_d$ which is perpendicular to $E$ and $I_q$ in phase with $E$. The phasor sum of $V$, $IR_a$, $j I_d x_d$ and $jI_q x_q$ gives the emf $E$.

In using the phasor diagram of Fig. 5.27, the armature current $I$ has to be resolved into $I_d$ and $I_q$. This is possible only if ($\angle\theta + \delta$) is known. Initially, the $\angle\theta$ is known, but the power angle $\delta$ is not known. ($\angle\theta + \delta$) can be found by redrawing the phasor diagram, as in Fig. 5.28. It is seen that triangles *ode* and *abc* are similar.

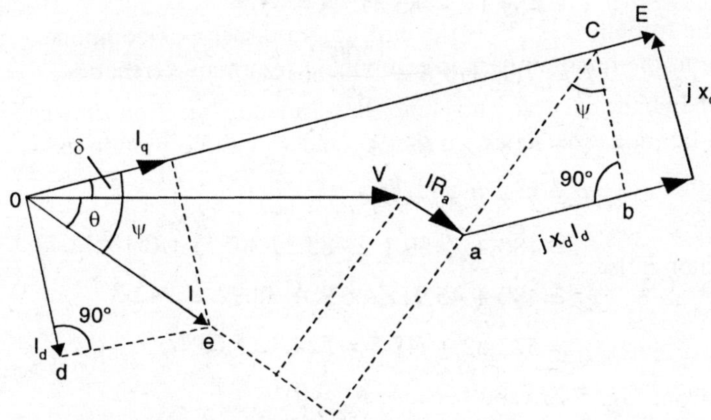


**Fig. 5.28.** Modified phasor diagram of salient pole alternator

$$\frac{ac}{oe} = \frac{bc}{de}$$

or
$$ac = \frac{(bc)(oe)}{de} = \frac{(jx_q J_q)(I)}{I_q} = jx_q I$$

Therefore phasor $oc = V + IR_a + j\,x_q\,I$

Thus point $c$ is located and $\angle\psi$ can be found, as under

$$\tan\psi = \frac{V\sin\theta + Ix_q}{V\cos\theta + IR_a}. \qquad ...(5.28)$$

### *Example 5.8*

A 50 Hz, 3-phase, 480 V, delta connected salient pole synchronous generator has $x_d = 0.1\ \Omega$ and $x_q = 0.075\ \Omega$. The generator is supplying 1200 A at 0.8 p.f. lagging. Find the excitation emf. Neglect armature resistance.

### *Solution*

$$V = 480 \text{ V}$$

$$I = 1200/\sqrt{3} = 692.8 \text{ A}$$

Using Eq. (5.28)

$$\tan\psi = \frac{480 \times 0.6 + (692.8)(0.075)}{480 \times 0.8} = 0.885$$

$$\psi = 41.5°, \; \delta = 41.5 - 36.87° = 4.63°$$

$$I_d = I \sin \psi \angle -(90 - \delta) = \frac{1200}{\sqrt{3}} \sin(41.5°) \angle -85.37°$$

$$= 459.1 \angle -85.37° \text{ A}$$

$$I_q = I \cos \psi \angle \delta = \frac{1200}{\sqrt{3}} \cos(41.5°) \angle 4.63°$$

$$= 518.9 \angle 4.63° \text{ A}$$

$$E = V + jI_d x_d + jI_q x_q$$

$$= 480 + j(459.1 \angle -85.37°)(0.1) + j(518.9 \angle 4.63°(0.075)$$

$$= 480 + 45.91 \angle 4.63° + 38.92 \angle 94.63°$$

$$= 522.62 + j42.5 = 524.3 \angle 4.63° \text{V}$$

Alternatively, as seen from Fig. 5.27 (and because $R_a = 0$)

$$E = V \cos \delta + I_d x_d$$

$$= 480(\cos 4.63°) + 459.1(0.1) = 524.3 \text{ V.}$$

## 5.17 POWER ANGLE CHARACTERISTICS OF SALIENT POLE GENERATOR

Figure 5.29 shows a salient pole synchronous generator feeding an infinite bus. V is the bus voltage, $E$ is the excitation voltage of generator, $V_t$ is the terminal voltage of generator, and $X$ is the reactance of the line connecting generator to the bus. The resistance of generator and line are negligible.

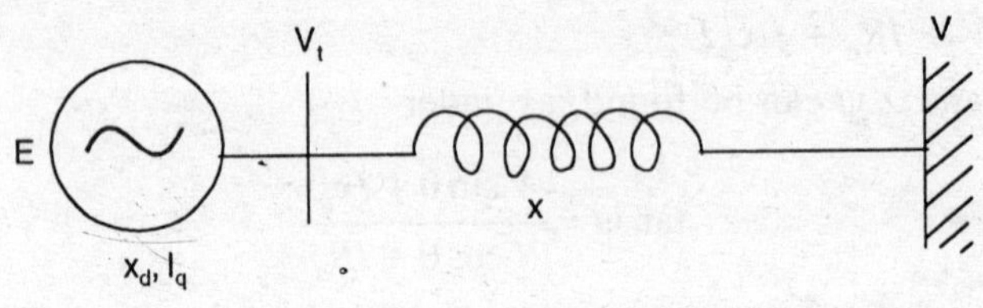


**Fig. 5.29.** Salient pole alternator feeding an infinite bus

The current $I$ can be resolved into $I_d$ and $I_q$. The phasor sum of $V$, $j I_d X$ and $j I_q X$ gives the generator terminal voltage $V_t$. The phasors $j I_d x_d$, and $j I_q x_q$ when added to $V_t$ give the excitation emf $E$ as shown in Fig. 5.30.

The voltage drops $j i_d X$ and $j I_d x_d$ can be combined as $j I_d X_d$ and the voltage drops $j I_q X$ and $j I_q x_q$ can be combined as $j I_q X_q$. Thus the total $d$- and $q$-axis reactances are

$$X_d = X + x_d$$

$$X_q = X + x_q$$

The phasor diagram after combining these voltage drops is shown in Fig. 5.31. The total power delivered by the generator can be found by resolving the bus voltage $V$ into $V \sin \delta$ (which is in-phase with $I_d$) and $V \cos \delta$ (which is in-phase with $I_q$). thus

$$P = 3I_d V \sin\delta + 3I_q V \cos\delta \qquad \text{...(5.29)}$$

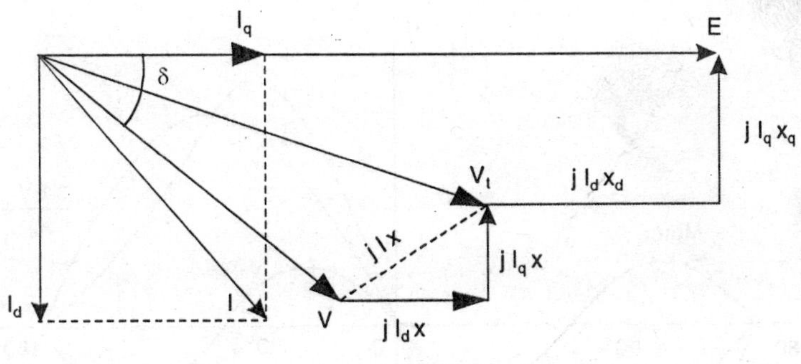


**Fig. 5.30.** Phasor diagram of salient pole alternator feeding an infinite bus

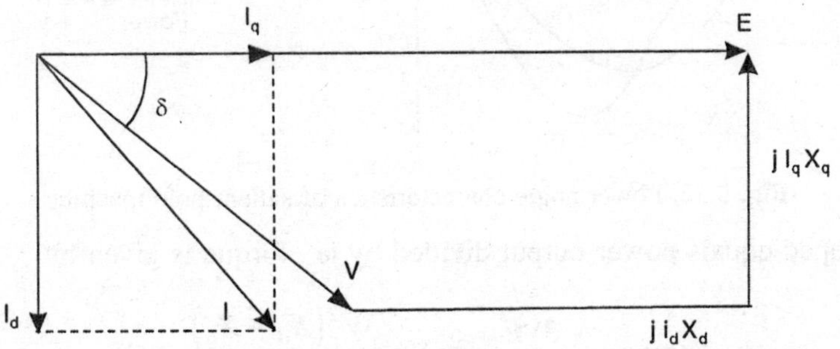


**Fig. 5.31**

From Fig. 5.31, the currents $I_d$ and $I_q$ are given by

$$I_d = \frac{E - V\cos\delta}{X_d} \qquad ...(5.30)$$

and

$$I_q = \frac{V\sin\delta}{X_q} \qquad ...(5.31)$$

Substituting Eqs. (5.30) and (5.31) into Eq. (5.29)

$$P = 3(V)\left[\frac{E - V\cos\delta}{X_d}\right]\sin\delta + 3(V)\left[\frac{V\sin\delta}{X_q}\right]\cos\delta$$

or

$$P = \frac{3VE}{X_d}\sin\delta + 3V^2\left(\frac{X_d - X_q}{2X_dX_q}\right)\sin 2\delta \qquad ...(5.32)$$

The first term in Eq. (5.32) is the same as that in a cylindrical rotor generator. The second term is the additional power due to saliency and is known as reluctance power (or torque). This second term does not depend on field excitation and is present even when field current is zero. Its maximum value occurs at $\delta = 45°$. The power angle curve is shown in Fig. 5.32. The maximum power occurs at an $\angle\delta$ which is less than 90°.

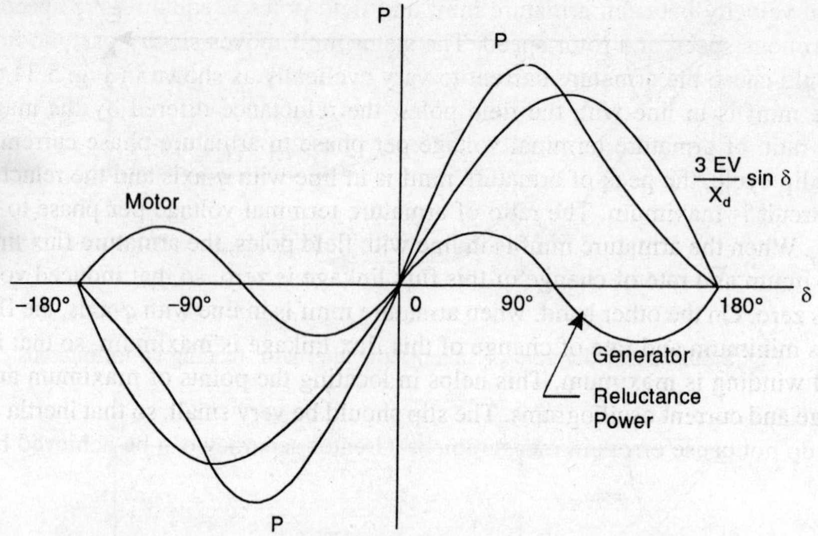


**Fig. 5.32.** Power angle characteristics of salient pole machine

Since torque developed equals power output divided by $\omega_s$, torque is given by

$$\text{Torque} = \frac{3VE}{\omega_s (X_d)} \sin\delta + \frac{3V^2 (X_d - X_q)}{2\omega_s X_d X_q} \sin 2\delta \qquad ...(5.33)$$

## 5.18 MEASUREMENT OF $x_d$ AND $x_q$: SLIP TEST

The direct and quadrature axis reactances can be measured by slip test. The machine is driven by an auxiliary motor (preferably a dc motor) at a speed slightly less or slightly more than synchronous speed. The field winding is kept open circuited and a low voltage 3-phase supply (about 20 % of the rated voltage) is applied to the armature terminals (Fig. 5.33). The direction of rotation should be the same as the direction of rotating field. If this condition is fulfilled, a small ac voltage would be indicated by the voltmeter across the field winding.

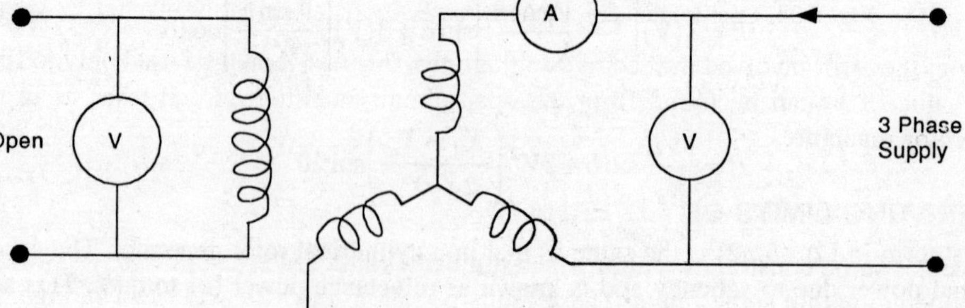


**Fig. 5.33.** Connections for slip test

The relative velocity between armature mmf and field poles is equal to slip speed, i.e. difference between synchronous speed and rotor speed. The stator mmf moves slowly past the field poles at slip speed. This would cause the armature current to vary cyclically as shown in Fig. 5.34. When the peak of the armature mmf is in line with the field poles, the reluctance offered by the magnetic circuit is minimum. The ratio of armature terminal voltage per phase to armature phase current gives $x_d$. After one quarter of slip cycle, the peak of armature mmf is in line with $q$-axis and the reluctance offered by the magnetic circuit is maximum. The ratio of armature terminal voltage per phase to armature phase current gives $x_q$. When the armature mmf is in line with field poles, the armature flux linkage with field winding is maximum and rate of change of this flux linkage is zero, so that induced voltage across the field winding is zero. On the other hand, when armature mmf is in line with $q$-axis, the flux linkage with field winding is minimum and rate of change of this flux linkage is maximum, so that induced voltage across the field winding is maximum. This helps in locating the points of maximum and minimum on armature voltage and current oscillograms. The slip should be very small, so that inertia of moving parts of instruments do not cause errors in measurement. Greater accuracy can be achieved by using recording oscillogram.

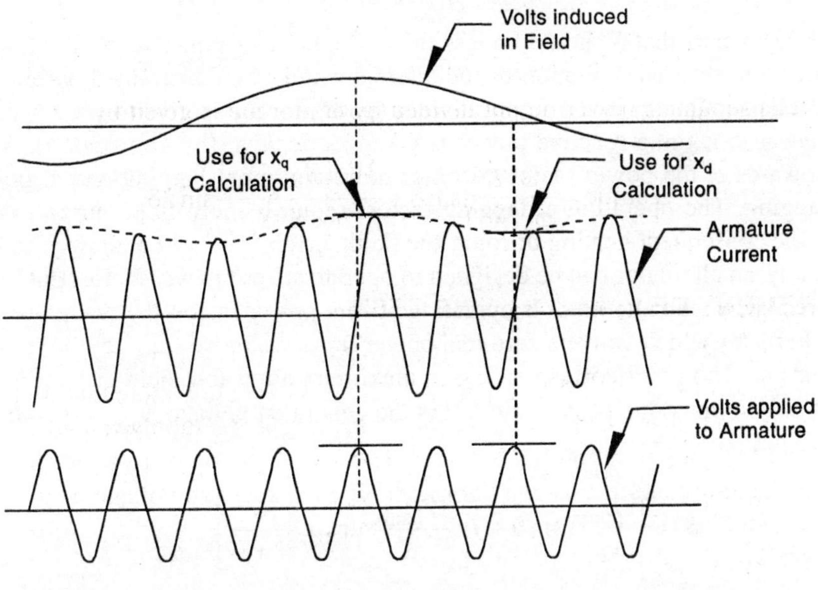


**Fig. 5.34.** Oscillograms of armature voltage and current during slip test

In view of the error involved in reading oscillograms, this test should be used only to find the ratio $x_d/x_q$. The value of $x_d$ can be found from the open circuit and short circuit tests, as in the case of cylindrical rotor machine.

## 5.19 OPERATING LIMITS ON ALTERNATORS

Every device has some constraints within which it must work. An alternator delivering power to a grid has also to operate within some limits. These limits pertain to the MW and MVA rating of alternator, the maximum field current and the stability limit.

The power systems are operated as constant voltage systems, i.e. the voltage at any bus in the system is held within ±10 per cent of nominal value. At constant voltage, power is proportional to current and this brings in the question of current carrying capacity. The stator winding has resistance $R_a$. When a current $I$ flows in the winding, copper-loss equal to 3 $I^2R_a$ takes place. This loss is converted into heat and this heat must be removed, or it will raise the temperature of the winding and the immediate surroundings of the winding. No matter how efficient the alternator cooling system is, there is a current $I$ which if exceeded by a large margin, will cause excessive heating and may cause damage to the machine. This current $I$ is referred to as rated current, i.e. $I_{rated}$. The extent to which the actual current may be allowed to be more than the rated current depends on the time for which the excess current is likely to exist, (the shorter the time, the greater is the permissible excess current). Associated with $V_{rated}$ and $I_{rated}$ is the 3-phase volt-ampere rating of the alternator given by

$$S_{rated} = \sqrt{3}(V_{rated})(I_{rated}) \qquad ...(5.34)$$

$V_{rated}$ and $I_{rated}$ in Eq. (5.24) refer to their line values (The 3-phase rating is usually specified in MVA). The actual 3-phase volt-amperes $S$ delivered by the alternator should always be less than or equal to $S_{rated}$, i.e.

$$|S| \leq S_{rated} \qquad ...(5.35)$$

Equation (5.25) means that $S$ should be less than or equal to 1 p.u. This limit when applied to $PQ$ diagram (Fig. 5.35) means that operation of the alternator must be constrained within the circle $a$ $b$ $c$ $d$ $e$. The part of this semicircle in the first quadrant (portion $c$ $d$ $e$) refers to the operation at lagging power factors, because lagging reactive power is taken as positive (by convention). As the operation moves from $c$ towards $e$, the power factor becomes more and more lagging and at point $e$ the power factor is zero lagging. The operation at lagging factors requires more field current. Because of field copper-loss and the consequent heating of rotor, the field current can not be allowed to increase indefinitely. Theoretically, an alternator can be designed to operate at zero power factor. But in actual practice this is not required (when the alternator is operating at zero power factor, the real power output is zero. Who needs an alternator which delivers zero real power?). As we move for $c$ to $e$ we require increasing value of field current. The point corresponding to maximum allowable field current is (say) $d$. Corresponding to this maximum value of field current is the generated voltage $E_{max}$. Substituting $E_{max}$ for $E$ in Eq. (5.18), we get

$$S = \frac{3VE_{max}}{x_s}\sin\delta + j\left[\frac{3VE_{max}}{x_s}\cos\delta - \frac{3V^2}{x_s}\right] \qquad ...(5.36)$$

This the equation of a circle in $PQ$ plane with centre at $(O, -V^2/x_s)$ and radius $VE_{max}/x_s$. a segment of this circle $f$ $g$ $d$ $h$ is shown in Fig. 5.35. The operation of the alternator is constrained to be within this circle, because of rotor field winding heating.

Now consider the portion of the semicircle in the 4th quadrant (i.e. portion $abc$). In this region $Q$ is negative, which means that the alternator is operating at a leading power factor, which requires less excitation. As we move from point $c$ to $a$, the excitation requirement decreases. For a particular value of leading power factor (usually around 0.5), the power angle becomes 90°. To keep the operation well within stability limits, the maximum value of δ should not be more than 60°. Corresponding to $\delta_{max}$, the real and reactive power (Eq. 5.20) are

$$P = \frac{3VE}{x_s}\sin\delta_{max} \qquad ...(5.37a)$$

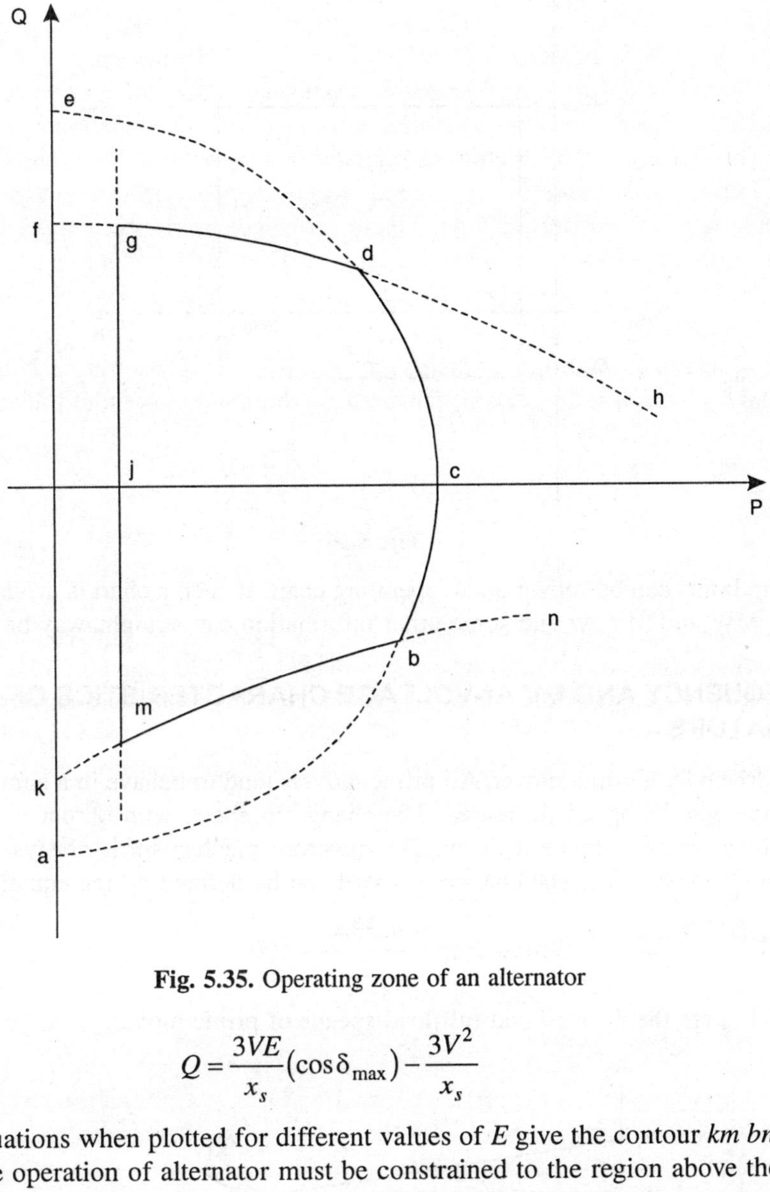


**Fig. 5.35.** Operating zone of an alternator

$$Q = \frac{3VE}{x_s}(\cos \delta_{max}) - \frac{3V^2}{x_s} \quad ...(5.37b)$$

These two equations when plotted for different values of $E$ give the contour *km bn*. From stability considerations, the operation of alternator must be constrained to the region above the curve *km bn.*

For some alternators it is not desirable to allow the real power output to become zero, unless it is necessary to take the unit off the bus bars, e.g. in a steam station a constant boiler temperature is required to prevent slagging problems. Moreover, if a unit is shut down, it would take a long time to restart it and bring it to bus bars. Thus the real power output should not be allowed to become less than a value $P_{min}$. Therefore, the operation of the alternator should be constrained within the region *m b c d g j m*. This constraint can be shown in a simplified was by fitting the largest possible rectangle inside the region *m b c d g j m*. This is shown in Fig. 5.36. This simplified characteristics is particularly useful for computer control, because only four values $P_{max}$, $P_{min}$, $Q_{max}$ and $Q_{min}$ completely define the operating area.

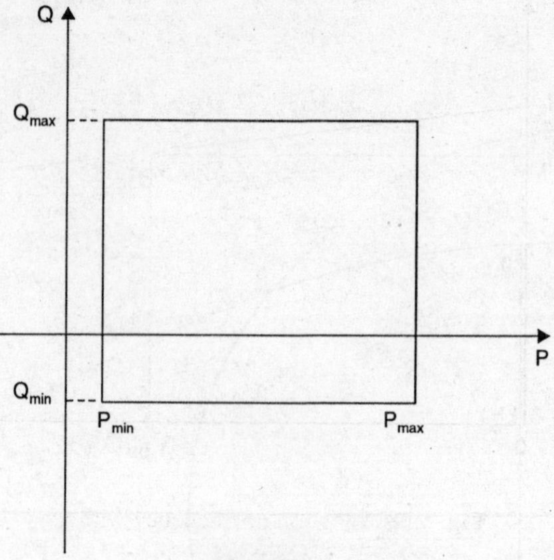


**Fig. 5.36**

The operating limits can be shown on an operating chart. If such a chart is available, the excitation for any value of MW and MV Ar and some other information can straightaway be obtained.

## 5.20 MW-FREQUENCY AND MV Ar-VOLTAGE CHARACTERISTICS OF ALTERNATORS

An alternator is driven by a prime mover. All prime movers tend to behave in a similar fashion. As the power output increases the speed decreases. The change in speed with output is nonlinear, but the governor mechanism tends to make it linear. The governor mechanism is always adjusted to give a slightly drooping characteristics. The change in speed can be defined by the equation

$$\text{Speed drop} = \frac{n_o - n_f}{n_f} \times 100 \qquad ...(5.38)$$

Where $n_o$ and $n_f$ are the no-load and full-load speeds of prime mover.

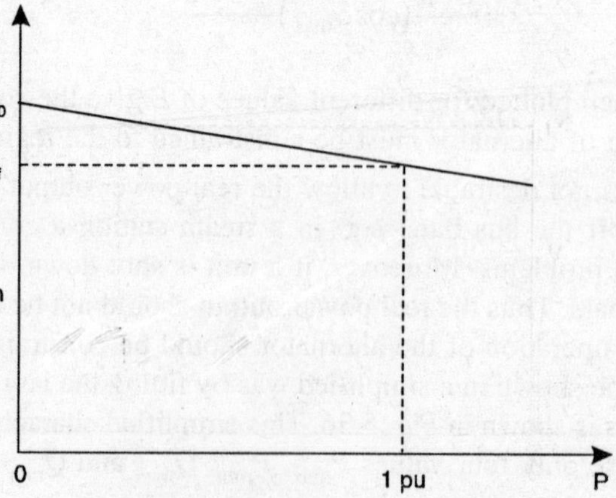


**Fig. 5.37.** Variation of speed with load

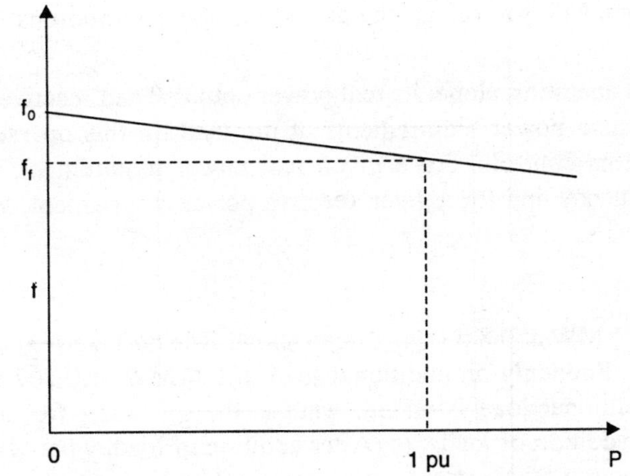


**Fig. 5.38.** Variation of frequency with load

Figure 5.37 shows the speed-power curve. The speed drop from no-load to full-load is around 2-4%. All governors have some type of set point adjustment to allow the no-load speed adjustment. Since frequency is directly proportional to speed, the real power is related to frequency as shown in Fig. 5.38. The change in power output $\Delta P$ is related to change in frequency $\Delta f$ by the equation

$$\Delta P = -\frac{1}{R}\Delta f \qquad ...(5.39)$$

The constant $R$ is known as speed regulation of governor and has the units of Hz per MW or Hz/pu MW. Evidently 1/R is the slope of load-frequency curve.

The terminal voltage varies with reactive power. A lagging reactive power leads to a drop in terminal voltage and a leading reactive power leads to a rise in terminal voltage. This is shown in Fig. 5.39. Generally this curve is nonlinear, but voltage regulators include a feature to make it linear. This characteristic curve can be moved up and down by changing the no-load terminal voltage set point on the voltage regulator.

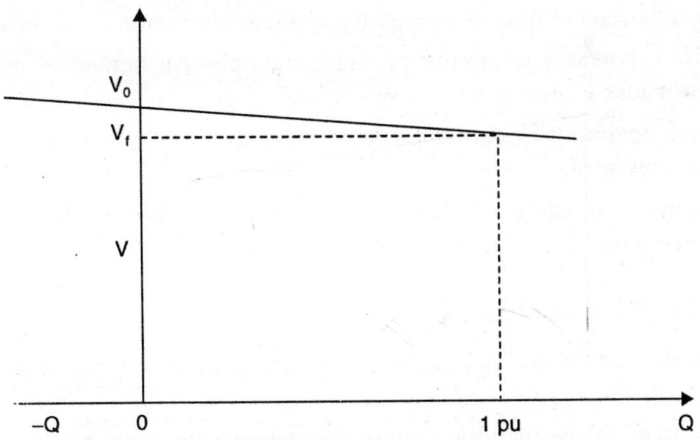


**Fig. 5.39.** Variation of terminal voltage with reactive power

The MW-frequency and MV Ar-voltage characteristics play an important role in parallel operation of alternators.

When an alternator is operating alone, its real power output $P$ and reactive power output $Q$ will be equal to the real and reactive power requirements of the load. In this operation $P$ and $Q$ cannot be controlled by the alternators controls. For a given real power requirement, the governor set points control the operating frequency and for a given reactive power requirement, the field current controls the terminal voltage.

*Example 5.9*

An alternator is supplying 1 MW at 0.8 p.f. lagging to a load. The no-load frequency is 51 Hz and speed regulation $R$ is 1 Hz/MW. Suddenly an additional load of 0.8 MW at 0.707 lagging power factor is added. (a) Before the additional load is added, what is the operating frequency ? (b) What is the operating frequency after addition of load ? (c) After addition of load, what action should the operator take to restore the frequency to 50 Hz ?

*Solution*

(a) From Eq. (5.39)

$\Delta f = (-R)(\Delta P) = (-1)(1) = -1$

Operating frequency = 51 − 1 = 50 Hz

(b) $\Delta f = (-R)(\Delta P) = (-1)(0.8) = -0.8$

Operating frequency = 50 − 0.8 = 49.2 Hz

(c) The total drop is frequency from no-load to 1.8 MW load is 1.8 Hz. To restore the operating frequency to 50 Hz, the operator should adjust the no-load set points of governor to 51.8 Hz.

## 5.21 PARALLEL OPERATION OF ALTERNATORS

### 5.21.1 Need for Parallel Operation

In modern power systems alternators are operated in parallel to supply the total load. The need for parallel operation arises due to the following reasons:

(i) The total load requirement cannot be met by a single alternator.

(ii) Parallel operation increases reliability of electric supply. An outage of one alternator will not cause total power loss to the load.

(iii) If alternators are operating in parallel, one or more of them can be shut down for preventive maintenance turn by turn.

(iv) Parallel operation of alternators leads to economy in operating costs. The less efficient machines can be shut down when the load requirement is less.

### 5.21.2 Conditions for Parallel Operation

The following conditions must be fulfilled before an incoming alternator can be put in parallel with the bus bars:

(i) The terminal voltage of the incoming alternator must be the same as that of bus bars.

(ii) The frequency of incoming alternator must be the same as that of bus bars.

(iii) The phase of the voltage of incoming alternator must be identical with the phase of the bus bar voltage, with respect to the external circuit, i.e. opposite in phase with respect to the local circuit through the armature and the bus bars.

(iv) The phase sequence of the voltage of incoming alternator must be the same as that of bus bar voltage.

The condition of same phase sequence is checked by the phasing out test during the commissioning of the alternator. The remaining three conditions have to be checked each time the alternator is to be put in parallel with bus bars.

### 5.21.3 Synchronising

To avoid severe shock to the incoming and existing alternators, a definite procedure known as synchronising procedure has to be followed. Figure 5.40 shows an alternator $G_1$ to be put in parallel with bus bars.

The condition of correct phase sequence has already been checked at the time of initial commissioning of $G_1$. The remaining three conditions (i.e. equal voltage, equal frequency and same phase) have to be checked before switch $S_1$ can be closed.

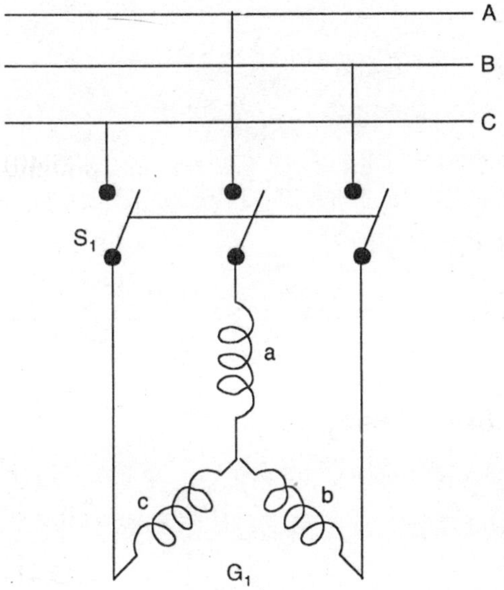


**Fig. 5.40.** Alternator in parallel with bus bars

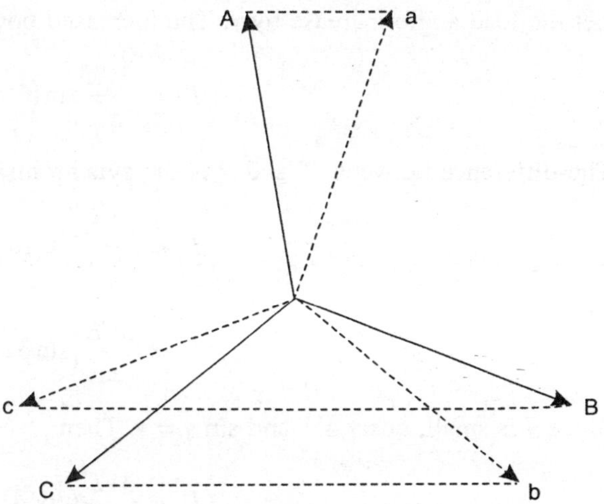


**Fig. 5.41.** Voltages of bus bar and incoming alternator

First of all the field current of $G_1$ is adjusted, so that the terminal voltage of $G_1$ is equal to bus bars voltage. Then the speed of $G_1$ is adjusted, so that its frequency is slightly higher than the bus bar frequency [All alternators have a reverse power relay connected to them. If the frequency of $G_1$ is less than system frequency, it will start drawing power from the system as soon as it is put in parallel. This will cause the reverse power relay to trip].

Figure 5.41 shows the line-to-neutral voltages *a b c* of the incoming alternator $G_1$ and the line-to-neutral voltage *A, B, C* of the bus bar. As the two frequencies are not exactly equal, these two sets of phasors change phase with respect to each other at $2\pi(\Delta f)$ rad/sec, where $\Delta f$ is the difference is frequencies. The exact instant of switching is when the two systems are in phase. This is done by using a synchronoscope. A synchronoscope is an instrument which measures the phase difference between voltage of one of the phases (say *a*) of incoming alternator $G_1$ and the corresponding phase (i.e. *A*) of bus bar. The dial of synchronoscope shows 0° (i.e. in phase) at the top and 180° at the bottom. The switch $S_1$ is closed when synchronoscope indicates zero phase difference.

Before the use synchronoscope became common, the instant of switching was determined by two bright and one dark lamp methods. For this purpose one lamp is connected across *a A*, one across *b C* and one across *B c*. It is seen from Fig. 5.41 that when the voltages are in phase, lamp across *a A* is dark, while the other two lamps are equally bright. Since the voltage across each lamp can be twice the system voltage the lamps should have a voltage rating of twice the system voltage. For modern generators operating at 11 kV, the synchronoscope (or the lamps) have to be connected through potential transformers. For large generators used in power systems the process of synchronising is automated using a computer.

## 5.22 SYNCHRONIZING POWER, TORQUE

The power output of a cylindrical rotor alternator is

$$P = \frac{3VE}{x_s}\sin\delta$$

Let the load angle increase by $\gamma$. The increased power $P'$ is

$$P' = \frac{3VE}{x_s}\sin(\delta+\gamma) \quad \text{...(5.40)}$$

The difference between $P'$ and $P$ is the synchronising power $P_s$

$$P_s = \frac{3VE}{x_s}[\sin(\delta+\gamma)-\sin\delta]$$

$$= \frac{3VE}{x_s}[\sin\delta\cos\gamma + \cos\delta\sin\gamma - \sin\delta]$$

Since $\gamma$ is small, $\cos\gamma \cong 1$ and $\sin\gamma = \gamma$. Then

$$P_s = \frac{3VE}{x_s}(\cos\delta)(\gamma) \quad \text{...(5.41)}$$

Synchronising power per electrical radian

$$= \frac{3VE}{x_s}\cos\delta \quad \text{...(5.42)}$$

Alternatively, for a particular value of excitation, the relationship between a small change in load angle $\Delta\delta$ and the corresponding change in additional power $\Delta P$ is given by

$$\frac{\Delta P}{\Delta\delta} \cong \frac{dP}{d\delta} \cong \frac{3VE}{x_s}\cos\delta \quad \text{...(5.43)}$$

The slope $\frac{dP}{d\delta}$ is the synchronising power per electrical radian. It is also known as stiffness ratio of the machine. $P_s$ is maximum at no-load and zero at $\delta = 90°$. For $\delta > 90°$, $P_s$ is negative, indicating unstable operating region. To ensure that the generator has reasonable synchronising power, δ is kept about 40° or so.

For a salient pole machine having $x_d$ and $x_q$ as direct and quadrature axis reactances, the power output is

$$P = \frac{3VE}{x_d}\sin\delta + \frac{3V^2(x_d - x_q)}{2x_d x_q}\sin 2\delta$$

$$\frac{dP}{d\delta} = \frac{3VE}{x_d}\cos\delta + \frac{3V^2(x_d - x_q)}{x_d x_q}\cos 2\delta \qquad ...(5.44)$$

Synchronising torque = $P_s$ synchronous watts

$$= \frac{P_s}{2\pi n_s}\text{ N-m} \qquad ...(5.45)$$

where $n_s$ is synchronous speed in revolutions per second.

Equations (5.42) and (5.44) indicate that synchronising power varies directly as excitation voltage $E$. Therefore an overexcited machine has higher synchronising power than an underexcited one. Moreover a machine with large air gap has lesser $x_s$ (or $x_d$). Thus a machine with larger air gap is stiffer than a machine with smaller air gap.

### Example 5.10

A 5 MVA, 10000 V, 1500 rpm, 3-phase, 50 HZ alternator is operating on infinite bus bar. Find synchronising power per mechanical degree of angular displacement at (a) no-load, (b) full-load at rated voltage and 0.8 power factor lagging. Also find synchronising torque for a 0.5° mechanical displacement in each case. $x_s$ is 20%.

### Solution

(a) No-load, $\delta = 0$

$V = E = 1$ p.u $\quad x_s = 0.2$ pu

$$P_s = \frac{1 \times 1}{0.2} = 5 \text{ pu} = 5 \times 5000 \text{ kW/elect. radian}$$

Since it is a 4-pole machine

$$1 \text{ elect. rad.} = \frac{180}{\pi \times 2} = \frac{90}{\pi} \text{ mech. degree}$$

$$\text{Synchronising power} = \frac{5 \times 5000}{90/\pi} = 873 \text{ kW/mech. deg}$$

Synchronising torque for 0.5° displacement

$$= \frac{873 \times 1000 \times 0.5}{2\pi n_s} = \frac{873000 \times 0.5}{2\pi(25)}$$

$$= 2777 \text{ N-m}$$

(b) Full-load

$$E = V + jIx_s = 1 + j(0.8 - j0.6)(0.2)$$

$$= 1.1313 \angle 8.13° \text{ pu}$$

$$P_s = \frac{1 \times 1.1313}{0.2} \cos(8.13°) = 5.6 \text{ pu}$$

Synchronising power per mech. degree

$$= \frac{5.6 \times 5000}{90/\pi} = 977.38 \text{ kW per mech. degree}$$

Synchronising torque of 0.5° displacement

$$= \frac{977.38 \times 1000 \times 0.5}{2\pi \times 25}$$

$$= 3111 \text{ N-m.}$$

*Example 5.11*

A 2 MVA, 3-phase, star connected, 8-pole, 750 rpm alternator is operating on 6000 V bus bars. $x_s$ is 6 Ω per phase. Find synchronising power and torque per mechanical degree of displacement for full-load, 0.8 power factor lagging.

*Solution*

$$I = \frac{2000 \times 1000}{\sqrt{3} \times 6000} = 192.45 \text{ A}$$

$$E = \frac{6000}{\sqrt{3}} + 192.45\,(0.8 - j0.6)(j6)$$

$$= 4258.32 \angle 12.53° \text{ V}$$

$$P_s = \frac{3VE}{x_s}\cos\delta = \frac{3\left(6000/\sqrt{3}\right)(4258.32)}{6}\cos(12.53°) \text{ watt/elect. radian}$$

$$= \frac{3\left(6000/\sqrt{3}\right)(4258.32)}{1000 \times 6} \times \frac{4\pi}{180}\cos(12.53°) \text{ kW per mech. degree}$$

$$= 502.8 \text{ kW per mech. degree}$$

$$\text{Synchronising torque} = \frac{502.8 \times 1000}{2\pi(750/60)}$$

$$= 6400 \text{ N-m per mech. degree.}$$

## 5.23 TWO ALTERNATORS IN PARALLEL

### 5.23.1 Effect of change in Excitation

We discuss the effect of change of excitation by considering two identical alternators sharing the load equally. In this system, the sum of real power outputs of the two must be equal to real power demand of load and the sum of reactive power outputs of the two must be equal to reactive power demand of the load.

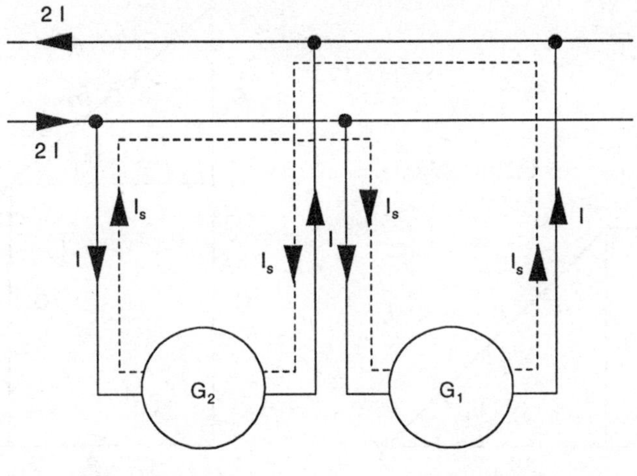


**Fig. 5.42.** Circulating current of two alternators in parallel

Figure 5.42 shows the two alternators in parallel. Initially each machine is operating at the same power factor angle θ. Let excitation of machine 1 be increased. The induced emf $E_1$ of machine 1 becomes more than the induced emf $E_2$ of machine 2 and a circulating current $I_s$ is set up. The magnitude of this current is

$$I_s = \frac{E_1 - E_2}{2Z_s} \quad \text{...(5.46)}$$

$I_s$ will lag behind $E_1$ by an angle $\phi_s$ (= $\tan^{-1} x_s/R_a$). This current will be superposed on the original current distribution as shown in Fig. 5.43(b).

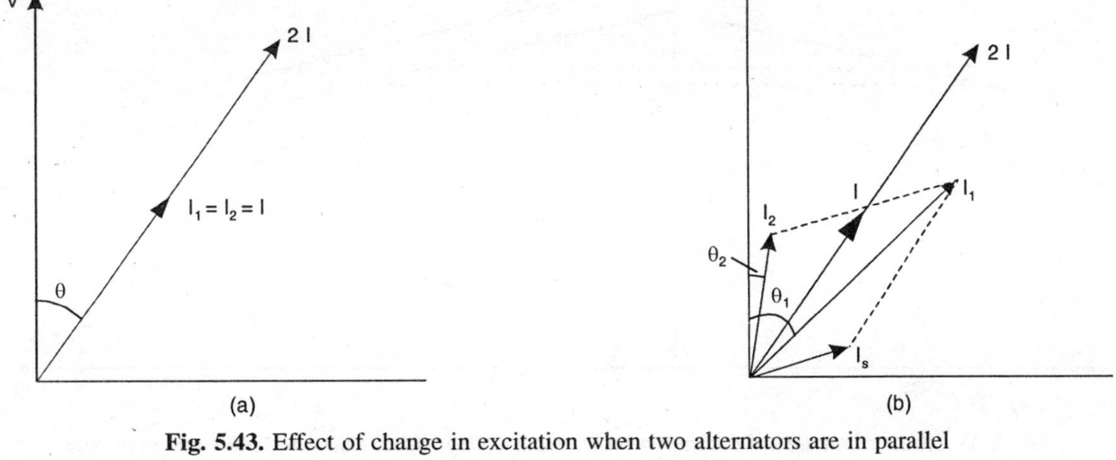


**Fig. 5.43.** Effect of change in excitation when two alternators are in parallel

$I_s$ is added to load current of machine 1 and subtracted from the load current of machine 2. The new currents $I_l$ and $I_2$ are as shown. Since the angle $\phi_s$ is almost 90°, there is almost no change in active power division between the two machines. The power factor of machine 1 is decreased and that of machine 2 is increased. The change in power factors evidently leads to a change in reactive power division. Figure 5.44(a) shows initial load sharing and Fig. 5.44 (b) shows the load sharing after increase in excitation of machine 1.

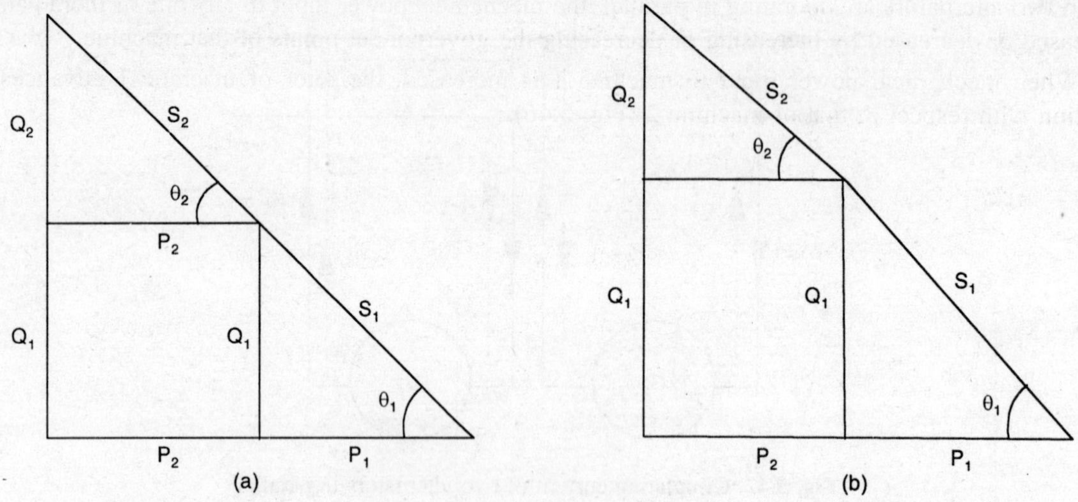


**Fig. 5.44.** Sharing of active and reactive power by two alternators in parallel

The reactive power supplied by machine 1 is increased while that supplied by machine 2 is decreased.

The effect of change in excitation on terminal voltage is shown in Fig. 5.45. $G_1$ and $G_2$ are the initial characteristics. $Q_1$ and $Q_2$ are initial reactive powers. An increase in field current of machine 1

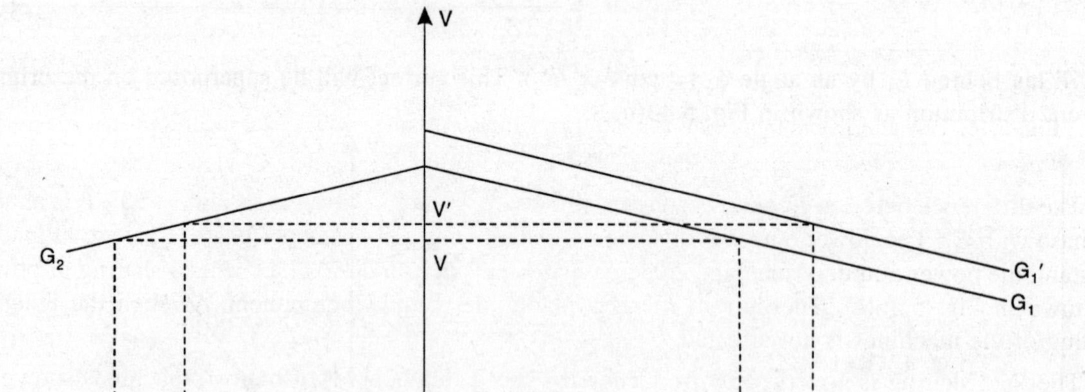


**Fig. 5.45.** Effect of change of excitation on terminal voltage when two alternators are in parallel

raises its characteristics to $G'_1$. The new reactive powers are $Q'_1$ and $Q'_2$. The terminal voltage has also increased from initial value $V$ to $V'$. If it is desired to keep the terminal voltage constant, it would be necessary to decrease the field current of machine 2. To change the terminal voltage without changing the reactive power sharing, it is necessary to increase or decrease both the field currents simultaneously.

### 5.23.2 Effect of Change in Mechanical Power Input (Steam Supply)

When two alternators are operating in parallel, the mechanical power input to any one of them can be increased or decreased by increasing or decreasing the governor set points of that machine.

When mechanical power input to machine 1 is increased, the rotor of machine 1 advances in position with respect to that of machine 2 (Fig. 5.46).

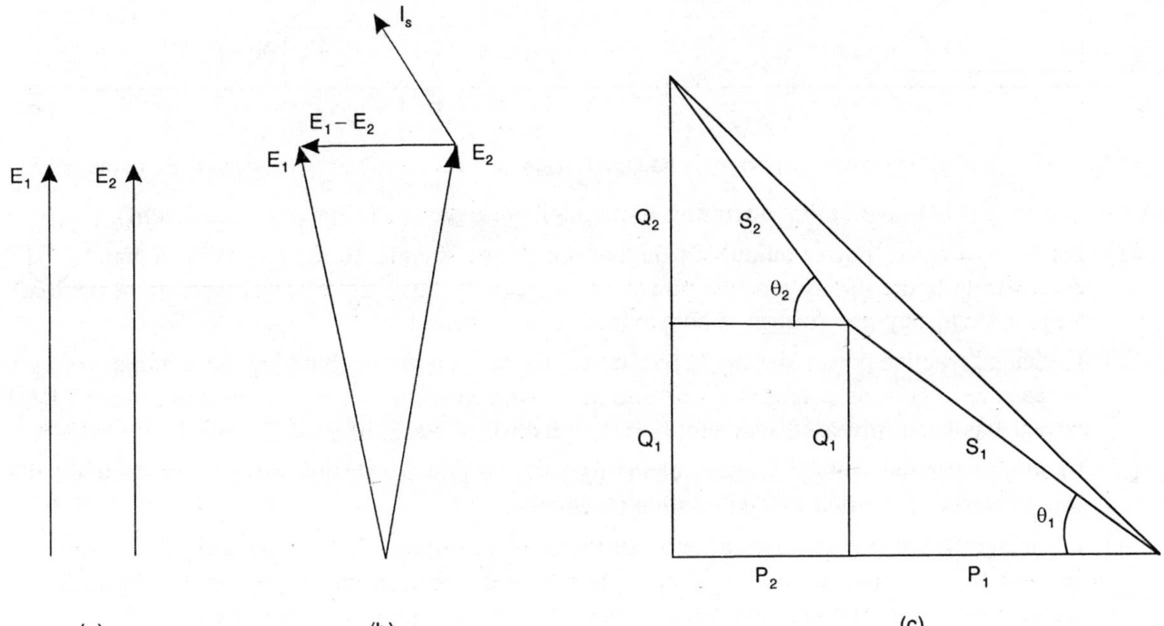


**Fig. 5.46.** Two alternators in parallel (a) initial conditions, (b) after mechanical input to machine 1 is increased, (c) load sharing

The difference between $E_1$ and $E_2$ causes a circulating current $I_s$. Since $\phi_s$ is almost 90°, $I_s$ is almost in phase with $E_1$. The power output of machine 1 increases by $E_1 I_s$ watts per phase. Since total load is constant, the power output of machine 2 decreases by the same amount. The changed sharing of power is shown in Fig. 5.46(c). The current $I_s$ has a negligible reactive component. As such the reactive loading of the machines is not affected.

The load sharing is governed by frequency (or speed) load curves of the two machines shown in Fig. 5.47. $G_1$ and $G_2$ are initial characteristics. $f$ is the operating frequency and $P_1$, $P_2$ are the real power outputs of the two machines. Increasing the governor set points of machine 1 raises its characteristics to $G'_1$. The new real power outputs are $P'_1$ and $P'_2$. The operating frequency has also increased to $f'$. If it is desired to keep the frequency constant, it would be necessary to decrease the governor set points of machine 2. To adjust the frequency without changing real power sharing, it is necessary to simultaneously increase or decrease the set points of both governors.

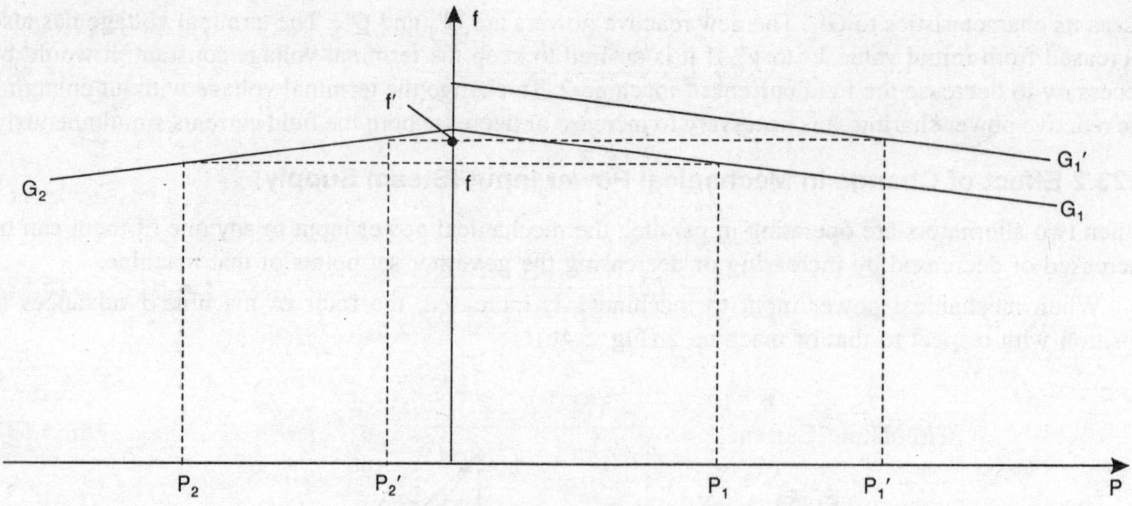


**Fig. 5.47.** Effect of increasing governor set points of machine 1 on load sharing between two alternators

To summarise, for two alternators operating in parallel, we have the following conclusions:

(1) The sum of active power outputs of the two machines is equal to active power demand of the load. Similarly the sum of reactive power outputs equals the reactive power demand of the load. Neither frequency nor voltage is constrained to be constant.

(2) To change reactive power sharing between two machines without changing the terminal voltage, increase field current of one machine and decrease that of the other. The machine whose field current has been increased will supply more reactive power.

(3) To adjust terminal voltage without changing reactive power sharing, increase or decrease the field currents of the two machines simultaneously.

(4) To change active power sharing between the two machines without changing the frequency, increase the governor set points of one machine and decrease that of the other. The machine whose governor set points have been increased, will supply more active power.

(5) To adjust frequency without changing the real power sharing, increase or decrease the governor set points of both the machines simultaneously.

### 5.23.3 Sharing of Load Currents

Let $E_1$, $E_2$ be the induced emfs per phase and $Z_{s1}$, $Z_{s2}$ be the synchronous impedances per phase of the two machines. $Z$ is the load impedance, $V$ the voltage across load and $I_1$, $I_2$ are the currents supplied by the two machines. Then

$$E_1 = I_1 Z_{s1} + (I_1 + I_2)\,Z \qquad ...(5.47)$$

$$E_2 = I_2 Z_{s2} + (I_1 + I_2)\,Z \qquad ...(5.48)$$

Solving Eqs. (5.47) and (5.48)

$$I_1 = \frac{E_1 Z_{s2} + (E_1 - E_2)\,Z}{Z_{s1} Z_{s2} + (Z_{s1} + Z_{s2})\,Z} \qquad ...(5.49)$$

$$I_2 = \frac{E_2 Z_{s1} + (E_2 - E_1)Z}{Z_{s1}Z_{s2} + (Z_{s1} + Z_{s2})Z} \quad ...(5.50)$$

$$I = \frac{E_1 Z_{s2} + E_2 Z_{s1}}{Z_{s1}Z_{s2} + (Z_{s1} + Z_{s2})Z} \quad ...(5.51)$$

$$V = IZ = \frac{E_1 Z_{s2} + E_2 Z_{s1}}{\dfrac{Z_{s1}Z_{s2}}{Z} + Z_{s1} + Z_{s2}} \quad ...(5.52)$$

$$\text{Circulating Current } I_s = \frac{(E_1 - E_2)Z}{Z_{s1}Z_{s2} + (Z_{s1} + Z_{s2})Z} \quad ...(5.53)$$

The above equations are similar to those for transformers [Eqs. (2.93) to (2.99)]. A solution for more than two alternators can be obtained by using parallel generator theorem [Eq. (2.100)], which in this case can be written as

$$V\left[\frac{1}{Z} + \frac{1}{Z_{s1}} + \frac{1}{Z_{s2}} + ...\right] = \frac{E_1}{Z_{s1}} + \frac{E_2}{Z_{s2}} + ... \quad ...(5.54)$$

### *Example 5.12*

Two similar 4000 kVA alternators operate in parallel. The governor of the first machine is such that frequency drops from 50 Hz at no-load to 47.5 Hz on full-load. The corresponding drop for the second machine is 50 Hz to 48 Hz. (a) How will they share a load of 6000 kW ? (b) How much maximum unity power factor load can they carry without any one of them getting overloaded?

### *Solution*

(a) Let the load supplied by the first machine be $L_1$ kW. The load supplied by second machine is $(6000 - L_1)$ kW

$$50 - \frac{(50 - 47.5)}{4000} L_1 = 50 - \frac{(50 - 48)}{4000}(6000 - L_1)$$

Whence $L_1 = 2666.66$ kW, $L_2 = 3333.33$ kW

(b) Machine 2 will be able to supply its rated load of 4000 kW. The frequency drop will be from 50 Hz to 48 Hz.

$$\text{Load Supplied by machine } 2 = \left[\frac{50 - 48}{50 - 47.5}\right] 4000 = 3200\,\text{kW}$$

Total load = 4000 + 3200 = 7200 kW.

### *Example 5.13*

Two 3-phase star connected alternators supply a load of 3000 kW at 0.8 p.f. lagging and share the load equally. The excitation of second machine is so adjusted that it supplies 150 A at a lagging p.f. The synchronous impedance are $0.4 + j\,12$ and $0.5 + j\,10$ ohm. Find current, power factor, induced emf and load angle of each machine. The terminal voltage is 6.6 kV.

*Solution*

$$P_1 = P_2 = 1500 \text{ kW}$$

For second machine

$$\sqrt{3} \times 6.6 \times 150 \cos\theta_2 = 1500$$

or $\cos\theta_2 = 0.875$, $\sin\theta_2 = 0.485$, $\theta_2 = 29°$

$$I_2 = 150(0.875 - j0.485) = 131.2 - j\,72.75 \text{ A}$$

$$\text{Total Current} = \frac{3000}{\sqrt{3} \times 6.6 \times 0.8} = 328 \text{ A}$$

$$I_1 + I_2 = 328(0.8 - j0.6) = 262.4 - j\,196.8 \text{ A}$$

$$I_1 = 262.4 - j196.8 - I_2 = 131.2 - j124.05 \text{ A}$$

$$= 180.6\angle{-43.4°}$$

$$\cos\theta_1 = \cos(-43.4°) = 0.726 \text{ lagging}$$

$$E_1 = \frac{6.6}{\sqrt{3}} + \frac{(131.2 - j124.05)(0.4 + j12)}{1000}$$

$$= 5.35 + j1.52 \text{ kV per phase} = 5.56 \text{ kV per phase}$$

$$\text{Load angle } \delta_1 = \tan^{-1}\frac{1.52}{5.35} = 15.9°$$

$$E_2 = \frac{6.6}{\sqrt{3}} + \frac{(131.2 - j72.75)(0.5 + j10)}{1000}$$

$$= 4.6 + j1.27 \text{ kV per phase}$$

$$= 4.77\angle 15.4° \text{ kV per phase}$$

$$\text{Load angle } \delta_2 = \tan^{-1}\frac{1.27}{4.6} = 15.4°.$$

*Example 5.14*

The emfs of two alternators are 3000 ∠20° and 2900 ∠0 V. Their synchronous impedances are 2 + *j* 20 and 2.5 + *j* 30 ohm. The load impedance is 10 + *j* 4 Ω. Find circulating current.

*Solution*

$$I_s = \frac{(E_1 - E_2)Z}{Z_{s1}Z_{s2} + (Z_{s1} + Z_{s2})Z}$$

$$= \frac{(3000\angle 20 - 2900)(10 + j4)}{(2 + j20)(2.5 + j30) + (2 + j20 + 2.5 + j30)(10 + j4)}$$

$$= 10.37 - j4.56 = 11.33 \text{ A.}$$

### *Example 5.15*

Two alternators are sharing a load of 2.5 MW at 0.8 p.f. lagging. Machine 1 has a no-load frequency of 51.5 Hz and a frequency drop of 1 Hz/MW. The second machine has no-load frequency of 51 Hz and a frequency drop of 1 Hz/MW. (a) At what frequency is the system operating and how much load is supplied by each machine? (b) Load is increased by 1 MW. Find the new frequency and load sharing. (c) The governor set points of machine 2 are increased by 0.5 Hz. Find the new frequency and load sharing for the total load of 3.5 MW.

### *Solution*

(a) Let load on machine 1 = $P_1$ MW

$$\text{Load on machine 2} = (2.5 - P_1) \text{ MW}$$

$$\text{Operating frequency of machine 1} = 51.5 - 1 \times P_1$$

$$\text{Operating frequency of machine 2} = 51 - 1\,(2.5 - P_1)$$

Since operating frequency must be the same, we have

$$51.5 - P_1 = 51 - (2.5 - P_1) \text{ or } P_1 = 1.5 \text{ MW}$$

$$P_2 = 2.5 - P_1 = 1 \text{ MW}$$

$$\text{Frequency} = 51.5 - 1 \times 1.5 = 50 \text{ Hz}$$

(b)

$$\text{Total load} = 3.5 \text{ MW}$$

$$\text{Let load on machine 1} = P_1'$$

$$\text{Load on machine 2} = P_2' = 3.5 - P_2'$$

$$51.5 - 1 \times P_1' = 51 - 1(3.5 - P_1')$$

or

$$P_1' = 2 \text{ MW}, P_2' = 1.5 \text{ MW}$$

$$\text{Frequency} = 51.5 - 1 \times 2 = 49.5 \text{ Hz}$$

(c)

$$\text{Total load} = 3.5 \text{ MW}$$

$$\text{Let load on machine 1} = P_1''$$

$$\text{Load on machine 2} = 3.5 - P_1'' = P_2''$$

$$\text{No-load frequency of machine 2} = 51.5 \text{ Hz}$$

$$51.5 - 1 \times P_1'' = 51.5 - 1(3.5 - P_1'')$$

or

$$P_1'' = 1.75 \text{ MW}, P_2'' = 1.75 \text{ MW}$$

$$\text{Frequency} = 51.5 - 1 \times 1.75 = 49.75 \text{ Hz.}$$

## 5.24 OPERATION OF ALTERNATOR ON INFINITE BUS BARS

Most of the alternators feed into power systems which are composed of a large number of interconnected alternators. The size of the system is so large that the addition or removal of one alternator does not affect the system voltage and frequency. Such a large system, operating at constant voltage and frequency, is known as an infinite bus bar system.

The operation of an alternator on infinite bus bar is shown in Fig. 5.48. $E$ is the induced emf of alternator and $V$ is the bus bar voltage. When an alternator has just been put in parallel with the bus, $V$

and $E$ would be equal, alternator current would be zero and alternator power output would be zero (Fig. 5.48a). The power input to the machine would be just sufficient to supply the losses of the alternator. The machine would be electrically floating, neither generating nor motoring. Let the governor set points be raised to increase the power input. The increase in power input would tend to accelerate the rotor and $E$ would move ahead of $V$. However, the speed of rotor is fixed by the system frequency. Therefore the speed change is only temporary. Once angle $\delta$ assumes such a value that current $I$ has a power component corresponding to input, steady state condition would be restored, (Fig. 5.48 b). The alternator is now operating at a leading power factor. The magnitude of $E$ depends on field current. By adjusting the field current, power factor can be adjusted to any value (Fig. 5.48 c). When excitation is changed, the active power output remains unchanged. Thus excitation controls the reactive power output, while the governor speed setting controls the active power output.

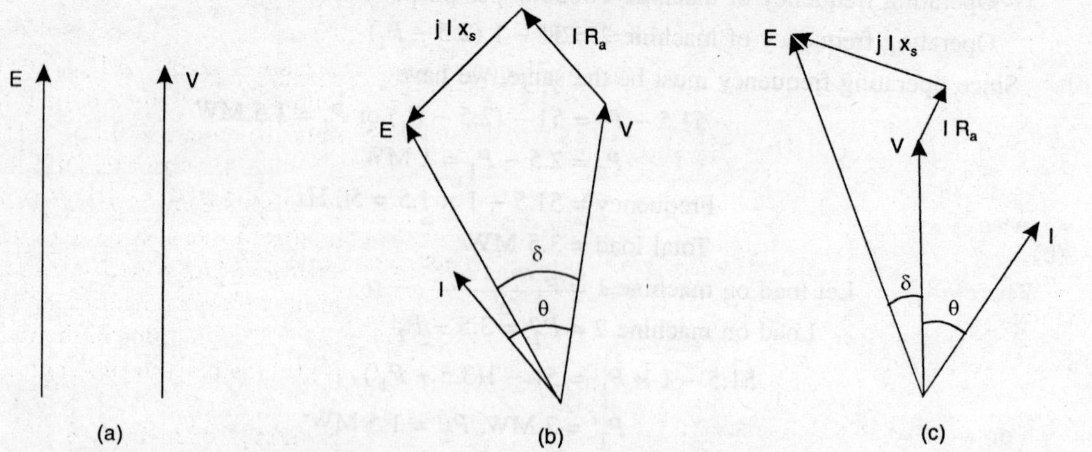


**Fig. 5.48.** Alternator in parallel with infinite bus (a) floating, (b) after increasing input to alternator, (c) after increasing excitation of alternator

An alternator working on infinite bus bar will become a motor if its excitation is maintained, but the prime mover is replaced by a mechanical load.

The power angle $\delta$ plays an important role in the operation of the machine. When $\delta$ is made to increase excessively, the machine becomes unstable. When operation alters from generator to motor, $\delta$ changes from positive to negative.

After synchronising, an alternator is loaded by adjusting the governor setting to admit more steam (in case of turbo alternators) or water (in case of water wheel generators). This tends to advance the rotor and load current flows from the machine to the bus bars maintaining an exact balance between the mechanical input and electrical output. The governor setting can be altered by remote control from the switchboard. In modern power systems having on-line control, the governor settings are adjusted by computer. Excitation is adjusted to provide necessary reactive power output.

To summarise, when an alternator is operating in parallel with an infinite bus system, the following conditions prevail:

(1) The terminal voltage and frequency remain constant at the system voltage and frequency.

(2) The governor set points of the alternator control the real power output of the alternator.

(3) The field current of the alternator controls its reactive power output and power factor.

*Example 5.16*

An alternator delivers 100 A at 0.8 lagging p.f. to a 3-phase 11 kV infinite bus bar system. Synchronous reactance of alternator is 4 ohm per phase. (a) Find open circuit emf and load angle, (b) The steam supply is increased and δ increases by 10°. Find the new load current and power factor, (c) The excitation is now changed till power factor becomes 0.8 lagging. Find new value of load current.

*Solution*

(a)
$$E = \frac{11000}{\sqrt{3}} + 100\,(0.8 - j0.6)(j4)$$
$$= 6600\angle 2.78°\text{V}$$

Hence
$$E = 6600 \text{ V per phase}$$
$$\delta = 2.78°$$

(b)
$$\delta' = 10 + 2.78 = 12.78°$$

$$P = \frac{3VE}{x_s}\sin\delta = \frac{3V \times 6600 \sin 12.78°}{4}$$

Also
$$P = 3\,VI\cos\theta$$

Therefore
$$I\cos\theta = \frac{6600 \sin 12.78°}{4} = 365$$

$$\frac{11000}{\sqrt{3}} + I(\cos\theta - j\sin\theta)(j4) = 6600$$

or
$$\frac{11000}{\sqrt{3}} + j4I\cos\theta + 4I\sin\theta = 6600$$

or
$$\left(\frac{11000}{\sqrt{3}} + 4I\sin\theta\right)^2 + (4 \times 365)^2 = 6600^2$$

Whence
$$I\sin\theta = 21.37$$

$$\tan\theta = \frac{21.37}{365}$$

$$\cos\theta = 0.998$$

$$I = \frac{365}{0.998} = 365.7 \text{ A}$$

(c) Since steam supply is kept unchanged, the active component of current remains the same.

$$I'\cos\theta' = 365.7 \times 0.998$$
$$I' \times 0.8 = 365$$
$$I' = \frac{365}{0.8} = 456.25 \text{ A.}$$

## 5.25 SYMMETRICAL SHORT CIRCUIT TRANSIENT

A sudden short circuit of all the three terminals (known as symmetrical short circuit) causes very severe transient currents.

Consider a 3-phase alternator running at synchronous speed and with the field excited by a dc source. The terminals are open circuited (i.e. alternator is not supplying any load). When a 3-phase short circuit is applied to the terminals, the resulting currents in the three phases are shown in Fig. 5.49.

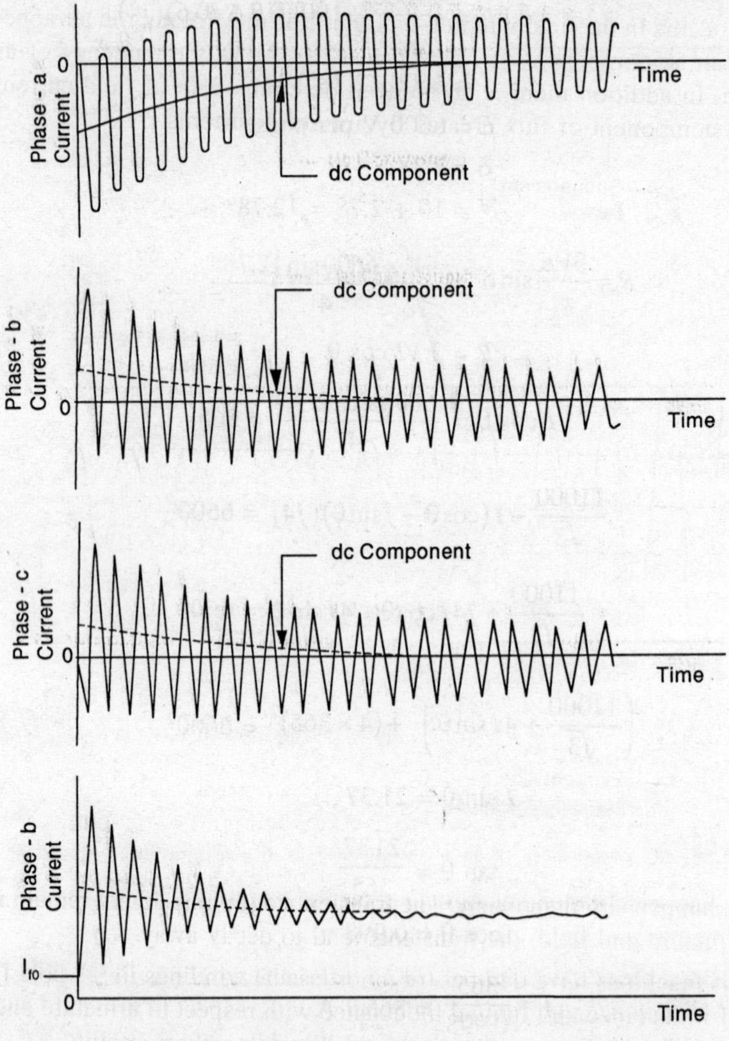


**Fig. 5.49.** Armature and field currents in an alternator after a sudden 3-phase short circuit

The current in each phase consists of an ac component and a dc component. The ac component is shown in Fig. 5.50. The transient field current is also shown in Fig. 5.49.

The transients are due to two fundamental facts: (i) the current in an inductive circuit cannot change instantaneously, (ii) the theorem of constant flux linkage which states that flux linkages with a closed circuit having zero resistance and no voltage source cannot change.

Before the application of short circuit, the only current is the field current. Each armature phase has a time varying flux linkage (due to field flux) as the rotor rotates.

When short circuit is applied, currents flow in armature winding to maintain their flux linkages constant at the values which existed at the instant of short current. Each armature current has an ac component and a dc component. The ac component corresponds to the armature current required to oppose the time varying flux produced by the field winding as it rotates. The dc component corresponds to the initial flux linkage which existed at the instant of short circuit. The net results of these currents is an armature flux linkage. Each armature phase keeps its initial flux linkage constant.

A similar situation occurs in the field winding on rotor. It is now rotating in a trapped armature flux wave and the result is an ac component of field current to oppose the tendency of armature flux to change its own linkage. In addition there is an induced dc component of field current to oppose the synchronously rotating component of flux created by armature currents.

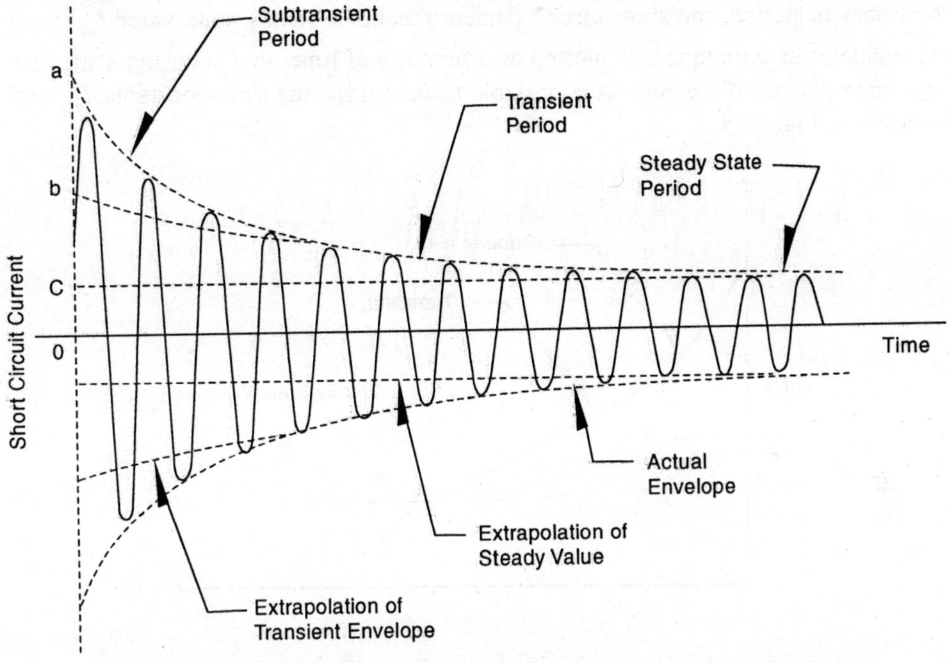


**Fig. 5.50.** AC component of short circuit current in a synchronous machine

The above actions happen simultaneously. The transient conditions do not remain indefinitely. Due to the resistance of armature and field, the transients tend to decay away.

Many synchronous machines have damper (or amortisseur) windings in the pole faces. This winding has resistance, self-inductance and mutual inductance with respect to armature and field windings. This winding also affects the currents in the machine following a short circuit.

Since the voltage of the three phases are out of phase by 120°, the short circuit occurs at different point on the voltage wave of each phase. Therefore the ac components of current, at the instant of short circuit, are different in each phase. As a result of this the dc component of armature current is also different in each phase. These dc components decay fairly quickly, but may be about 50% - 60% of ac component at the instant of short circuit. The total initial current is, therefore, about 1.5 times ac component of current.

The ac component of current (Fig. 5.50) can be divided into three distinct periods. During the first one cycle or so, the ac current is very large and falls very rapidly. This period of time is known as subtransient period and the ac rms current flowing during this period is called sub-transient current (denoted by $I''$). This current is due to the effect of both the damper winding and the field winding. It decays with a time constant $T_d''$. Since the resistance of damper winding is high, the decay of sub-transient current is very fast.

After the sub-transient period is over, the current continues to fall at a slower rate till it reaches a steady value. This period of time when the current decays at a slower rate is known as transient period and the ac rms current during this period is known as transient current (denoted by $I'$). It is caused by the dc component of current induced in the field winding at the instant of short circuit. Since the time constant of field circuit is much longer than the time constant of damper circuit, the transient period lasts much longer than the sub-transient period. The time constant during transient period is denoted by $T_d'$. The rms current during transient period may be about 5 times the steady state short circuit current.

After the transient period, the short circuit current reaches a steady state value $I_{ss}$.

If the rms value of ac component is plotted as a function of time on a semi-log scale, it is possible to observe the three periods of current. It is possible to determine the time constants $T_d''$ and $T_d'$ from such a plot shown in Fig. 5.51.

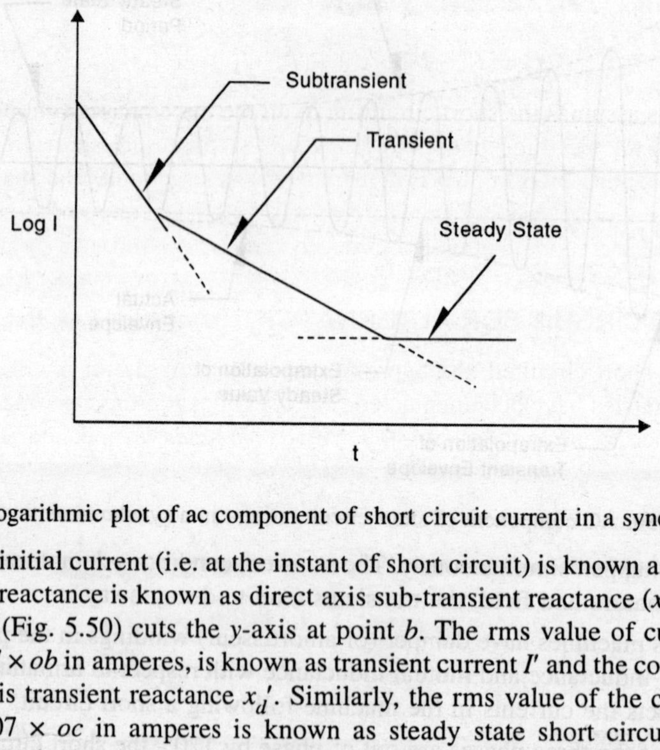


**Fig. 5.51.** Semi-logarithmic plot of ac component of short circuit current in a synchronous machine

The rms value of initial current (i.e. at the instant of short circuit) is known as sub-transient current $I''$ and corresponding reactance is known as direct axis sub-transient reactance ($x_d''$). The extrapolation of transient envelope (Fig. 5.50) cuts the *y*-axis at point *b*. The rms value of current, represented by intercept *ob*, i.e. 0.707 × *ob* in amperes, is known as transient current $I'$ and the corresponding reactance is known as direct axis transient reactance $x_d'$. Similarly, the rms value of the current represented by intercept *oc*, i.e. 0.707 × *oc* in amperes is known as steady state short circuit current $I_{ss}$ and the corresponding reactance is the direct axis reactance $x_d$. Thus

$$|I_{ss}| = \frac{oc}{\sqrt{2}} = \frac{|E|}{x_d} \quad ...(5.55)$$

$$|I'| = \frac{ob}{\sqrt{2}} = \frac{|E|}{x'_d} \quad \text{...(5.56)}$$

$$|I''| = \frac{oa}{\sqrt{2}} = \frac{|E|}{x''_d} \quad \text{...(5.57)}$$

where $|E|$ = rms value of open circuit voltage per phase

$|I_{ss}|$ = rms value of steady state short circuit current per phase

$|I'|$ = rms value of transient current per phase

$|I''|$ = rms value of sub-transient current per phase

$x_d$ = direct axis synchronous reactance per phase

$x_d'$ = direct axis transient reactance per phase

$x_d''$ = direct axis sub-transient reactance per phase

The rms current $I(t)$ at any time $t$ after short circuit occurs can be written as

$$I(t) = (I'' - I')e^{-t/T_d''} + (I' - I_{ss})e^{-t/T_d'} + I_{ss} \quad \text{...(5.58)}$$

The above analysis assumes the short circuiting of all the three phases simultaneously. Such a short circuit which involves all the three phases is known as symmetrical short circuit. If a short circuit involves only one or two phase, it is known as unsymmetrical short circuit. The analysis of unsymmetrical short circuits is more complex and can be carried out using the method of symmetrical components. This analysis is beyond the scope of this book and the reader should refer to [15] in the references.

## 5.26 EQUIVALENT CIRCUIT FOR ALTERNATOR TRANSIENT CONDITIONS

When an alternator is short circuited, the approximate equivalent circuit for first few cycles is shown in Fig. 5.52. The resistances have been neglected. The conditions in the machine are similar to those in a 3-winding transformer, the armature being the primary and the two secondaries being (i) field winding having reactance $x_f$, (ii) damper winding having reactance $x_D$. As discussed in Sec 5.16, $x_l$ is

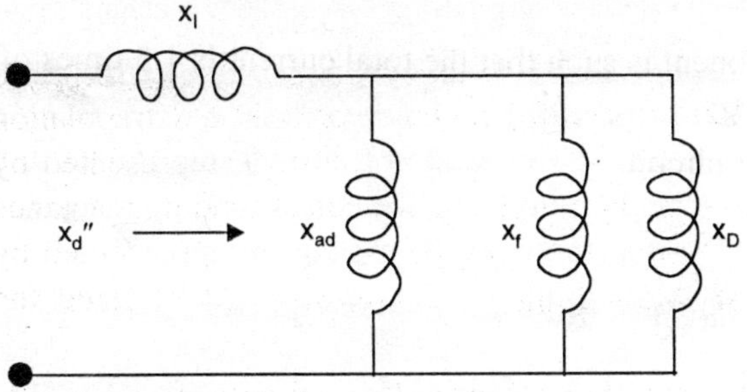


**Fig. 5.52.** Approximate equivalent circuit of alternator at the instant of short circuit (sub-transient conditions)

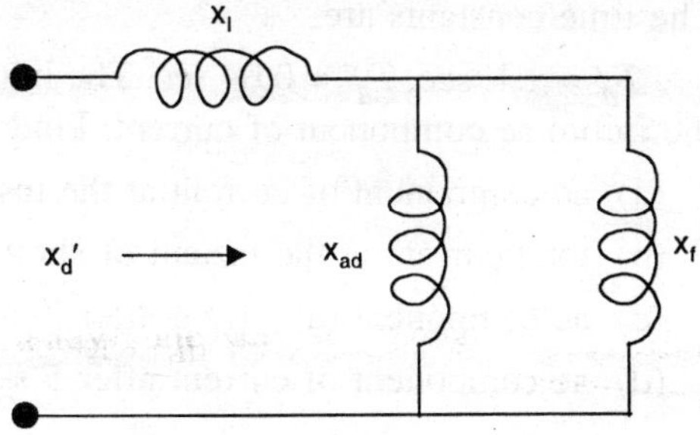


**Fig. 5.53.** Approximate equivalent circuit of alternator during transient conditions

leakage reactance and $x_{ad}$ is reactance which represents demagnetising effect of $I_d$. The equivalent reactance of this circuit is the direct axis sub-transient reactance $x_d''$. Evidently

$$x_d'' = x_l + \frac{1}{(1/x_{ad}) + (1/x_f) + (1/x_D)} \quad ...(5.59)$$

The effect of damper winding disappears after the first few cycles, because its resistance is much larger than resistance of field winding. Therefore the equivalent circuit during transient conditions takes the form shown in Fig. 5.53. The equivalent reactance of this circuit is the direct axis transient reactance $x_d'$. Evidently

$$x_d' = x_l + \left(\frac{1}{1/x_{ad} + 1/x_f}\right) \quad ...(5.60)$$

A comparison of Eqs. (5.59) and (5.60) indicates that $x_d'' < x_d'$.

After the transient period is over, the transformer action between armature and field winding disappears and steady state short circuit conditions start. The equivalent circuit then takes the form shown in Fig. 5.54. The effect is the same as though the secondary of transformer (represented by field winding in this case) has been opened.

The direct axis reactance $x_d$, is the sum of $x_l$ and $x_{ad}$, i.e.

$$x_d = x_l + x_{ad} \quad ...(5.61)$$

A comparison of Eqs. (5.60) and (5.61) shows that $x_d' < x_d$.

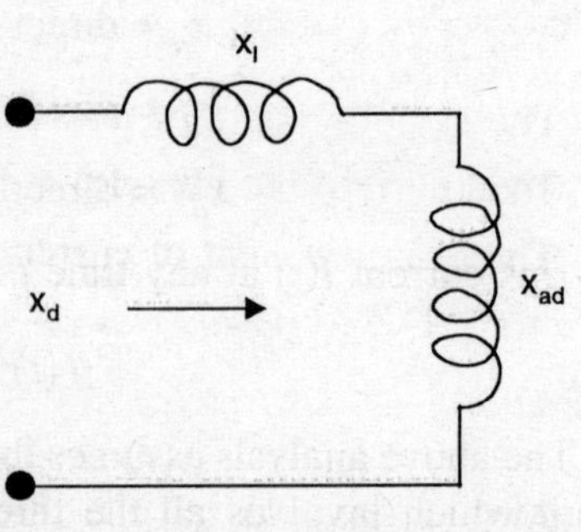


**Fig. 5.54.** Approximate equivalent circuit of alternator during steady state short short circuit conditions

*Example 5.17*

A 100 MVA, 11 kV, 3-phase star connected alternator is running at synchronous speed with rated voltage. Its terminals are open circuited. A sudden 3-phase fault develops at its terminals. The per unit reactances are

$$x_d = 1.0, x_d' = 0.25, x_d'' = 0.12$$

The time constants are

$T_d' = 1.1$. sec, $T_d'' = 0.04$ sec. The initial dc component is such that the total current is 1.5 times of the initial ac component of current. Find the following:

(a) ac component of current at the instant of short circuit.

(b) total current at the instant of short circuit.

(c) ac component of current after 2 cycles.

(d) ac component of current after 5 sec.

*Solution*

$$\text{Base current} = \frac{100 \times 10^6}{\sqrt{3} \times 11 \times 10^3} = 5248.8 \text{ A}$$

$$I'' = \text{Sub-transient current} = \frac{E}{x''_d} = \frac{1.0}{0.12} = 8.333\,\text{pu}$$

$$= 8.333 \times 5248.8 = 43738.3 \text{ A}$$

$$I' = \text{Transient current} = \frac{E}{x'_d} = \frac{1.0}{0.25} = 4.0\,\text{pu}$$

$$= 4 \times 5248.8 = 20995.2 \text{ A}$$

$$\text{Steady state short circuit current} = I_{ss} = \frac{E}{x_d}$$

$$= \frac{1.0}{1.0} = 1 \text{ pu} = 5248.8\,\text{A}$$

(a) The ac component of initial current = $I''$ = 43738.3 A

(b) Total current (ac plus dc) at the instant of fault = 1.5 $I''$ = 65607.5 A

(c) The ac component of current is given by Eq. (5.48)

$$I(t) = (I'' - I')e^{-t/T''_d}(I' - I_{ss})e^{-t/T'_d} + I_{ss}$$

$$= 22743.1e^{-t/0.04} + 15746.4e^{-t/1.1} + 5248.8 \text{ A}$$

At 2 cycles, $t$ = 2/50 = 1/25 sec

$$I(t) = 22743.1e^{-1/25\times 0.04} + 15746.4e^{-1/25\times 1.1} + 5248.8$$

$$= 8366.72 + 15184.1 + 5248.8$$

$$= 28799.62 \text{ A}$$

(d) At $t$ = 5 sec

$$I(t) = 22743.1\,e^{-5/0.04} + 15746.7\,e^{-5/1.1} + 5248.8$$

$$= 0 + 167.15 + 5248.8 = 5415.95 \text{ A.}$$

## 5.27 SYNCHRONOUS MACHINE REACTANCES

### 5.27.1 Direct Axis Synchronous Reactance $x_d$

The flow of 3-phase armature current sets up a rotating field. This field is stationary with respect to field structure. When the direct axis (polar axis) is in line with crest of armature mmf wave, a path of least reluctance exists and the fundamental flux wave has greatest possible magnitude for a given armature current. The flux linkage of an armature phase per ampere of armature current under there conditions is the direct axis synchronous inductance $L_d$. The direct axis synchronous reactance $x_d = \omega L_d$.

### 5.27.2 Direct Axis Transient Reactance $x_d'$

Sudden short circuit causes abrupt application of armature currents and the consequent armature mmf. This mmf tends to establish flux through pole core. By the theorem of constant flux linkage, at the instant immediately after short circuit occurs, the field flux linkage is still zero. Therefore the only flux which can be established immediately is that which does not link the field winding, but rather passes through high reluctance leakage paths largely in air. Under this condition, the flux per ampere is small and is defined as direct axis a transient inductance $L_d'$. The direct axis transient reactance $x_d' = \omega L_d'$. It is always less than direct axis synchronous reactance, i.e. $x_d' < x_d$.

### 5.27.3 Direct Axis Sub-transient Reactance $x_d''$

In defining $x_d'$ we have considered only one rotor circuit, i.e. field winding. Many salient pole machines have damper windings. In round rotor machines, the solid steel core furnishes significant paths for eddy currents.

When sudden short circuit is applied, transient currents are induced in the field winding as well as in additional direct axis rotor circuits. These transient currents oppose armature mmf to keep the flux linkage of every rotor circuit constant. The additional rotor circuit is near the air gap and, therefore, the flux set up by armature current is initially forced into leakage paths of smaller cross-sectional area and higher reluctance. Under these circumstances, the armature flux linkage per ampere of armature current is direct axis sub-transient inductance $L_d''$ and corresponding reactance ($= \omega L_d''$) is the direct axis sub-transient reactance $x_d''$. It is always less than direct axis transient reactance, i.e. $x_d'' < x_d'$.

### 5.27.4 Potier Reactance

In Sec 5.12 we discussed the construction of Potier triangle. The armature leakage reactance determined by this method is known as potier reactance.

### 5.27.5 Quadrature Axis Synchronous Reactance

We discussed the two-reaction theory for salient pole machines in Sec 5.16. When the quadrature axis is in line with crest of armature mmf wave, the reluctance of magnetic circuit is more. The flux linkage of an armature phase per ampere of armature current under this condition is the quadrature axis synchronous inductance $L_q$ and the corresponding reactance ($= \omega L_q$) is the quadrature axis synchronous reactance $x_q$. For salient pole machines $x_q$ is about 60% of $x_d$.

A small salient pole effect is present in turbo alternators also, even though they are cylindrical rotor machines. This is due to the effect of rotor slots on the quadrature axis reluctance. The reluctance in the quadrature axis is slightly more than that in direct axis. Therefore $x_d$ is slightly more than $x_q$ in turbo alternators.

### 5.27.6 Quadrature Axis Transient Reactance $x_q'$ and Sub-transient Reactance $x_q''$

These reactances are defined in the same way as $x_d'$ and $x_d''$ respectively, except that the suddenly applied armature current is in such time phase that the crest of space fundamental mmf wave is in line with quadrature axis of the rotor instead of direct axis.

In salient pole machines $x_q'$ is equal to $x_q$ but $x_q''$ is much less than $x_q'$. In turbo generators, the value of $x_q'$ lies between the values of $x_d'$ and $x_q$ and $x_q''$ is less than $x_q'$.

### 5.27.7 Negative Sequence Reactance $x_2$

The analysis for unsymmetrical faults is carried out by the method of symmetrical component. As per this method, the 3-phase unsymmetrical currents can be transformed into three balanced sets (1) positive sequence set having the same phase sequence as the original set of unbalanced currents, (2) negative sequence set having phase sequence opposite to that of original set, and (3) a zero sequence set, having three currents equal in magnitude and in phase with each other.

When negative sequence currents flow in stator winding, they produce a rotating field having a direction of rotation opposite to that of rotor. Thus this field rotates backward at twice synchronous speed with respect to rotor. Currents of twice the rated frequency are induced in all rotor circuits, keeping the flux linkages of these circuits constant at zero value. The flux due to armature current is forced into path of high reluctance. The armature flux linkage per ampere of armature current is the negative sequence inductance $L_2$ and corresponding reactance ($= \omega L_2$) is the negative sequence reactance $x_2$.

### 5.27.8 Zero Sequence Reactance $x_0$

When zero sequence currents flow in armature, there is no rotating field. However a pulsating mmf is produced. This mmf is opposed by currents induced in the closed rotor circuits and air gap flux is very small. Therefore, the flux per ampere of armature current is small. The corresponding zero sequence reactance $x_0$ is also very small and is the lowest of all reactances.

**Table 5.1:** Typical values of synchronous machine reactances (pu) and time constants (sec)

| | *Turbo generator* | *Salient pole generator* | *Synchronous condenser* | *Synchronous motor* |
|---|---|---|---|---|
| $x_d$ | 1.9 | 1.7 | 2.2 | 1.45 |
| $x_d'$ | 0.25 | 0.3 | 0.5 | 0.40 |
| $x_d''$ | 0.2 | 0.18 | 0.3 | 0.20 |
| $x_q$ | 1.85 | 1.0 | 1.3 | 1.0 |
| $x_q'$ | 0.50 | 1.0 | 1.3 | 1.0 |
| $x_q''$ | 0.20 | 0.25 | 0.35 | 0.25 |
| $x_2$ | 0.2 | 0.22 | 0.32 | 0.22 |
| $x_0$ | 0.08 | 0.10 | 0.10 | 0.08 |
| $x_p$ | 0.15 | 0.3 | 0.3 | 0.3 |
| $T_{do}$ | 4.0 | 3.0 | 6.5 | 1.5 |
| $T_d'$ | 0.55 | 0.50 | 1.5 | 0.4 |
| $T_{do}''$ | 0.1 | 0.08 | 0.13 | 0.04 |
| $T_d''$ | 0.02 | 0.02 | 0.03 | 0.01 |
| $T_a$ | 0.16 | 0.05 | 0.25 | 0.4 |

Zero sequence reactance can be measured by connecting the three armature windings in series and passing a single phase current through them. The ratio of voltage to current is $3\,x_0$.

Table 5.1 gives some typical values of reactances and time constants.

## 5.28 SYNCHRONOUS MACHINE TIME CONSTANTS

During short circuit, a number of currents and voltages exist. They decay with different time constants. Some of these time constants are as under:

### 5.28.1 Direct Axis Transient Open Circuit Time Constant $T'_{do}$

If the armature circuit is open and there is no damper winding, the change in field current is governed only by the inductance to resistance ratio of field circuit. This ratio is known as direct axis transient open circuit time constant $T'_{do}$.

$$T'_{do} = L_f/R_f \qquad \text{...(5.62)}$$

Where $L_f$ is the inductance of field winding and $R_f$ is its resistance.

### 5.28.2 Direct Axis Transient Short Circuit Time Constant $T_d'$

When both field and armature circuits are closed, the current in either circuit induces a current in the other. Because of rotation of one circuit with respect to the other, the two currents are not of the same frequency. Direct current in the field winding is associated with alternating current in the armature winding and direct current in armature winding is associated with alternating current in the field. The direct axis transient short circuit time constant $T_d'$ controls the rate of change of direct current in the field and of alternating current in the armature. It is equal to the ratio of apparent inductance of field (when coupled to closed armature circuit) to the apparent resistance of field under the same conditions.

The apparent inductance of field (with armature short circuited) is much less than the field inductance with armature open circuited. However, the resistance of field circuit is the same under the two conditions. Therefore, $T_d'$ is much less than $T'_{do}$.

$$T_d' = \frac{x'_d}{x_d} T'_{do} \qquad \text{...(5.63)}$$

### 5.28.3 Armature Short Circuit Time Constant $T_a$

This time constant applies to direct current in the armature winding and to alternating currents in the field and damper windings, both of which are closed. It is equal to the ratio of armature inductance to armature resistance under above conditions. It can be shown that this armature inductance is equal to negative sequence inductance $L_2$. Therefore

$$T_a = L_2/R_a \qquad \text{...(5.64)}$$

### 5.28.4 Direct Axis Sub-transient Time Constants $T''_{do}$ and $T''_d$

In addition to field winding, damper winding is also present in the direct axis. This winding rotates with respect to armature winding.

When armature winding is open, the ratio of damper winding inductance to its resistance is $T''_{do}$.

When armature winding is short circuited, a direct current in the damper winding is associated with an alternating current in the armature. The time constant $T_d''$ governors the rate of change of this current.

## 5.29 ELECTRO-MECHANICAL TRANSIENTS, HUNTING

A generator receives mechanical torque input $T_i$ via shaft from the prime mover and develops electro-magnetic torque $T_e$, which oppose the mechanical torque $T_i$. The algebraic difference between $T_i$ and $T_e$ is the net torque $T_a$, which causes acceleration of rotor.

$$T_a = T_i - T_e \quad \text{...(5.65)}$$

When the machine is operating in steady state, $T_a$ is zero. Whenever the balance between $T_i$ and $T_e$ is upset, there is acceleration (if $T_a$ is positive) or retardation (if $T_a$ is negative)

From laws of mechanics

$$I\alpha = T_a$$

or

$$I\frac{d^2\theta}{dt^2} = T_a \quad \text{...(5.66)}$$

where $I$ is moment of inertia in kg–m$^2$ and $\alpha$ or $\frac{d^2\theta}{dt^2}$ is angular acceleration is electrical degree/sec$^2$.

The angular position $\theta$ of the rotor continuously varies with time. It is convenient to measure the angular position with respect to a synchronously rotating reference axis. Let

$$\delta = \theta - \omega_s t \quad \text{...(5.67)}$$

where $\omega_s$ = synchronous speed in elect.degree/sec

$\delta$ = angular displacement of rotor from synchronously rotating reference axis in electric degrees.

Differentiating Eq. (5.67) twice with respect $t$

$$\frac{d^2\delta}{dt^2} = \frac{d^2\theta}{dt^2} \quad \text{...(5.68)}$$

Combining Eqs. (5.65), (5.66) and (5.68)

$$I\frac{d^2\delta}{dt^2} = T_a = T_i - T_e \quad \text{...(5.69)}$$

Multiplying Eq. (5.69) by rotor speed $\omega$

$$M\frac{d^2\delta}{dt^2} = P_a = P_i - P_e \quad \text{...(5.70)}$$

where $M = I\omega$ = angular momentum

$P_i = T_i\omega$ = mechanical power input corrected for windage and friction losses.

$P_e = T_e\omega$ is the electrical power output, corrected for electrical losses.

$P_a = P_i - P_e$ is the accelerating power.

Eq. (5.70) is known as swing equation. The electrical power output $P_e$ is given by Eq. (5.19). In this equation $V$ and $E$ are line-to-neutral voltages. If $V$ and $E$ represent line-to-line voltages, the factor 3 is automatically taken care of and $P_e$ can be written as

$$P_e = \frac{VE}{X}\sin\delta \qquad ...(5.71)$$

$$\frac{\Delta P_e}{\Delta\delta} \cong \frac{dP_e}{d\delta} \cong \frac{VE}{X}\cos\delta = P_r \qquad ...(5.72)$$

where $P_r$ is the synchronising power per electrical radian or synchronising coefficient.

Assume that the generator is operating in steady state at load angle $\delta_0$. Let the load angle change by $\Delta\delta$. Then

$$\frac{dP_e}{d\delta}(\Delta\delta) = P_r(\Delta\delta) \qquad ...(5.73)$$

where $P_r$ is the synchronising coefficient at angle $\delta_0$ and equals $\frac{VE}{X}\cos\delta_0$ and is a constant corresponding to its value at $\delta = \delta_0$. The equation of motion is

$$M\frac{d^2\delta}{dt^2} + P_r(\Delta\delta) = 0 \qquad ...(5.74)$$

Substituting $p^2$ for $d^2\delta/dt^2$, the characteristic equation is

$$Mp^2 + P_r = 0 \qquad ...(5.75)$$

or

$$p = \pm[-P_r/M]^{0.5} = \pm j\omega_d \qquad ...(5.76)$$

where $\omega_d = [P_r/M]^{0.5}$ is the undamped frequency of oscillations in rad/sec. The losses in the system damp the oscillations and the system returns to the original state. The system is stable so long as $P_r > 0$, because under this condition, both roots of Eq. (5.75) are imaginary. When $P_r < 0$, the roots of Eq. (5.75) are real and the system is unstable. The condition $P_r > 0$ means that $0 < \delta < 90°$.

The above oscillations of the synchronous machine are known as hunting. During this phenomenon $\delta$ is changing and the position of phasor $E$ is changing with respect to $V$. Therefore, this phenomenon is also known as phase swinging. It produces mechanical stresses and fatigue in the shaft.

## 5.30 DAMPER WINDING

The electro-mechanical transient should be reduced to a tolerable level. This can be done by introducing a damping term (proportional to $d\delta/dt$) in the swing equation (Eq. 5.70). A small amount of damping is provided by system losses. Damper windings are used in salient pole machine to provide additional damping. The damper winding consists of low resistance copper or aluminium bars embedded in notches carved in rotor pole shoes and short circuited at both ends by rings as in the case of squirrel cage induction motor. This winding is also known as amortisseur winding.

When the rotor is turning at synchronous speed, there is no induced emf in these bars. However, whenever the speed of rotor is different from synchronous speed, there is relative motion between stator magnetic field (rotating at synchronous speed) and rotor and a voltage is induced in the bars. Since the bars are short circuited at the ends, a current flows and this current produces a magnetic field. The interaction of the two fields produces a torque, which tends to make the rotor speed equal to synchronous speed. Thus the addition of damper winding helps in damping out the rotor oscillations.

It is not necessary to use damper winding in turbo generators. The solid steel rotor cores of these machines provide path for eddy currents, which provide the damping torque.

*Example 5.18*

For a cylindrical rotor alternator, answer the following: (a) When is reactive power maximum and what is its value ? (b) When is reactive power positive ? When is it negative ? (c) What is the reactive power when active power is maximum ?

*Solution*

As derived is Section 5.15

$$Q = 3\frac{EV}{x_s}\cos\delta - \frac{3V^2}{x_s} \quad \text{...}(i)$$

(a) $Q$ is maximum when $\frac{dQ}{d\delta}$ is zero. This occurs when $\delta = 0$. Then

$$Q_{\max} = \frac{3EV}{x_s} - \frac{3V^2}{x_s}$$

(b)
$$Q = \frac{3V}{x_s}(E\cos\delta - V)$$

$Q$ is positive when $E \cos \delta$ is greater than $V$. Under this condition the generator is over excited and reactive power output is positive (i.e. generator is delivering inducture reactive power to the load).

$Q$ is negative when $E \cos \delta$ is less than $V$. Under this condition, the generator is under excited and generator is absorbing reactive power (i.e. delivering capacitive reactive power to the load).

When $E \cos \delta = V$, the reactive power is zero. The generator is working at unity power factor and is not delivering or absorbing any reactive power.

(c) $P$ is maximum when $\delta = 90°$. From Eq. (i) the reactive power under this condition is

$$Q = -\frac{3V^2}{x_s}.$$

*Example 5.19*

Find an expression for reactive power divered by a salient pole generator to an infinite bus. The bus voltage is $V$ and excitation voltage of generator is $E$. The direct and quadrature axis reactances are $X_d$ and $X_q$, respectively. Hence find $Q_{\max}$.

*Solution*

The reactive power is given by the product of voltage and component of current lagging the voltage by 90°.

From the phasor diagram of Fig. 5.30 it is seen that $I_d$ lags $V \cos \delta$ by 90° and $I_q$ leads $V \sin \delta$ by 90°. Therefore

$$Q = 3I_dV\cos\delta - 3I_qV\sin\delta \quad \text{...}(i)$$

From Fig. 5.30, the currents $I_d$ and $I_q$ are given by

$$I_d = \frac{E - V\cos\delta}{X_d}$$

and
$$I_q = \frac{V\sin\delta}{X_q}$$

Substituting the value of $I_d$ and $I_q$ in Eq. (i), we get

$$Q = 3V\left[\frac{E - V\cos\delta}{X_d}\right]\cos\delta - 3V\left[\frac{V\sin\delta}{X_q}\right]\sin\delta$$

$$= \frac{3EV}{X_d}\cos\delta - \frac{3V^2}{X_d}\cos^2\delta - \frac{3V^2}{X_q}\sin^2\delta$$

$$= \frac{3EV}{X_d}\cos\delta - \frac{3V^2}{X_d}\left(1 - \sin^2\delta\right) - \frac{3V^2}{X_q}\sin^2\delta$$

$$= \frac{3EV}{X_d}\cos\delta - \frac{3V^2}{X_d} - 3V^2\left(\frac{I}{X_q} - \frac{I}{X_d}\right)\sin^2\delta \qquad \text{...(ii)}$$

$Q$ is maximum when $\frac{dQ}{d\delta} = 0$. Thus

$$\frac{dQ}{d\delta} = -\frac{3EV}{X_d}\sin\delta - 3V^2\left(\frac{1}{X_q} - \frac{1}{X_d}\right)2\sin\delta\cos\delta = 0$$

or
$$\cos\delta = \frac{-EX_q}{2V\left(X_d - X_q\right)} \qquad \text{...(iii)}$$

Substituting the value of cos δ form Eq. (iii) into Eq. (ii), we get

$$Q = \frac{-3E^2X_q}{2X_d\left(X_d - X_q\right)} - \frac{3V^2}{X_d} - 3V^2\left(\frac{X_d - X_q}{X_dX_q}\right)\left[1 - \frac{E^2X_q^2}{4V^2\left(X_d - X_q\right)^2}\right]$$

or
$$Q = \frac{-3V^2}{X_q} - \frac{3E^2X_q}{4X_d\left(X_d - X_q\right)}.$$

### *Example 5.20*

Find the synchronous impedance and reactance of a single-phase alternator, in which a given field current produces an armature current of 250 A on short circuit and a generated impedance of 1500 V on open circuit. The armature resistance is 2Ω. Hence calculate the terminal voltage when a load of 250 A at 6.6 kV at 0.8 p.f. lagging is switched off.

*Solution*

$$Z_s = \frac{\text{OC voltage for a certain field current}}{\text{SC current for the same field current}}$$

$$= \frac{1500}{250} = 6 \text{ ohms}$$

$$x_s = \sqrt{Z_s^2 - R_a^2} = \sqrt{6^2 - 2^2} = 5.66\,\Omega$$

$$E = V + I(R_a + jx_s)$$

Let

$$V = 6600 \angle 0 \text{ V}$$

$$I = 250(0.8 - j0.6)$$

$$E = 6600\angle 0 + 250(0.8 - j0.6)(2 + j5.66)$$

$$= 7849 + j832 = 7893\angle 6.05°$$

When load is switched off, the terminal voltage rises to 7893 V or 7.893 kV.

*Example 5.21*

A 3-phase alternator having a reactance of 10 ohm has an armature current of 220 A at unity p.f., when running on 11 kV constant frequency bus bars. If the steam admission is unchanged and the emf is raised by 25 %, find graphically or otherwise the new value of machine current and power factor.

*Solution*

Assume that alternator is star connected.

The initial value of $E$ is

$$E = V + I(jx_s) = \frac{11000}{\sqrt{3}} + (220\angle 0)\,(j10)$$

$$= 6721.3\angle 19.1°\text{V}$$

New value of emf = $E'$ = 6721.3 × 1.25 = 8401.6

Let the new current be $I$ and new power factor be cos θ. Since the steam supply is not changed, the active component of current, i.e. $I$ cos θ will remain unchanged. Then

$$8401.6 = 6351 + (I \cos\theta - jI \sin\theta)(j10)$$

or

$$8401.6 = 6351 + (220 - jI \sin\theta)\,(j10)$$

or

$$8401.6^2 = (6351 + 10I \sin\theta)^2 + 2200^2$$

or

$$I \sin\theta = 175.74$$

$$\tan\theta = \frac{I\sin\theta}{I\cos\theta} = \frac{75.74}{220} = 0.7988$$

or

$$\theta = 38.62°$$

New power factor = cos(38.62°) = 0.78 lagging

$$I = \frac{220}{0.78} = 282.05\,\text{A}.$$

*Example 5.22*

Two generators rated 400 MW and 600 MW are operating in parallel. The drop characteristics of their governors are 3 % and 4 %, respectively, from no-load to full-load. If they are operating at 50 Hz at no-load, how will they share 1000 MW load and what will be the system frequency ?

*Solution*

Let $L_1$ be load on 400 MW generator. Load on 600 MW generator = $1000 - L_1$

Let $f$ be the frequency at load

$$f = 50 - 50 \times \frac{3}{100} \times \frac{L_1}{400} = 50 - \frac{4}{100} \times 50 \times \frac{(1000 - L_1)}{600}$$

Whence $$L_1 = 470.6 \text{ MW}$$

$$\text{Load on 600 MW generator} = 10000 - L_1 = 529.4 \text{ MW}$$

$$f = 50 - 50 \times \frac{3}{100} \times \frac{470.6}{400} = 48.235\,\text{Hz}$$

*Example 5.23*

A number of alternators are working in parallel with terminal voltage equal to rated value. One of the machines has a synchronous reactance of 50 % and resistance of 1 % and delivers power output in kW equal to 70% of its rated kVA. If the emf of this machine is 1.2 times terminal voltage, find the power factor at which this machine is working.

*Solution*

$$V = 1 \text{ pu}, \quad E = 1.2 \text{ pu}$$

$$x_s = 0.5 \text{ pu}, \; R_a = 0.01 \text{ pu}$$

$$P = 0.7 \text{ pu}$$

Therefore $$I \cos\theta = 0.7 \text{ pu}$$

$$E = V + (I\cos\theta - jI\sin\theta)(0.01 + j0.5)$$

or $$1.2 = 1 + (0.7 - jI\sin\theta)(0.01 + j0.5)$$

or $$1.2^2 = (1 + 0.007 + 0.5I\sin\theta)^2 + (0.35 - 0.01I\sin\theta)^2$$

Neglecting the term $0.01\, I \sin\theta$, we get $I \sin\theta = 0.28$

$$\tan\theta = \frac{I\sin\theta}{I\cos\theta} = \frac{0.28}{0.7} = 0.4$$

$$\theta = 21.8°$$

$$\text{Power factor} = \cos\theta = 0.93.$$

*Example 5.24*

A 3 MVA, 50 Hz, 11 kV, 3-phase star connected alternator supplies 100A at zero power factor leading.

The line voltage is 12370 V. When load is removed, the line voltage is 11000 V. Find regulation at full-load 0.8 power factor lagging. $R_a = 0.4\ \Omega$/phase.

*Solution*

At zero power factor leading

$$I = j100\,\text{A},\ V = \frac{12370}{\sqrt{3}} = 7142\,\text{V}$$

$$E = \frac{11000}{\sqrt{3}} = 6351\,\text{V},\ R_a = 0.4\,\Omega,\ x_s = ?$$

$$E = V + I(R_a + jx_s)$$

or
$$6351 = 7142 + (j100)(0.4 + jx_s)$$

or
$$6351^2 = (7142 - 100\,x_s)^2 + 40^2$$

or
$$x_s = 7.91\ \text{ohm}$$

At full-load, 0.8 p.f. lagging

$$I = \frac{3 \times 10^6}{\sqrt{3} \times 11 \times 10^3} = 157.46\ \text{A}$$

$$6351 = V + 157.46(0.8 - j0.6)\,(0.4 + j7.91)$$

or
$$6351 = (V + 797.67) + j958.6$$

or
$$6351^2 = (V + 797.67)^2 + 958.6^2$$

or
$$V = 5480.6\ \text{V}$$

$$\text{Regulation} = \frac{E - V}{V} \times 100 = \frac{6351 - 5480.6}{5480.6} \times 100$$

$$= 15.88\%.$$

*Example 5.25*

A 25 MVA, 6.6 kV, 50 Hz, 4-pole alternator has pu armature resistance of 0.004 and synchronous reactance 0.67 p.u. When the machine is supplying rated current at rated voltage, the induced emf is 1.5 pu. Find torque angle and power factor.

*Solution*

$$I = 1\ \text{pu}, \qquad V = 1\ \text{pu}$$

$$R_a = 0.004\ \text{pu} \qquad x_s = 0.67\ \text{pu}$$

From the phasor diagram of Fig. 5.13 (a)

$$E^2 = (V\cos\theta + IR_a)^2 + (V\sin\theta + Ix_s)^2$$

or
$$1.5^2 = (\cos\theta + 0.004)^2 + (\sin\theta + 0.67)^2$$

or
$$2.25 = \cos^2\theta + 0.004^2 + 0.008\cos\theta + \sin^2\theta + 0.67^2 + 2(0.67)\sin\theta$$

or $$0.801 = 1.34 \sin \theta + 0.008 \cos \theta$$

This equation can be solved either by hit and trial or by using an iterative method.

The solution is power factor = $\cos \theta = 0.805$ lagging

From the phasor diagram of Fig. 5.13(a)

$$E \cos \delta = V + IR_a \cos \theta + Ix_s \sin \theta$$

or $$1.5 \cos \delta = 1 + 0.004 \times 0.805 + 0.67 \times 0.593$$

or $$\cos \delta = 0.9337$$

or $$\delta = 21°.$$

*Example 5.26*

A 4 kVA, 3-phase, 110 V, 50 Hz, star connected alternator has $x_d = 3$ ohm and $x_q = 2$ ohm. The machine is delivering full-load current at 0.8 p.f. lagging at rated voltage. Find the induced emf, load angle and maximum power output of the alternator.

*Solution*

$$x_d = 3 \text{ ohm}, \quad x_q = 2 \text{ ohm}, \quad R_a = 0$$

$$V = \frac{110}{\sqrt{3}} = 63.51 \text{V}$$

$$\text{I} = \frac{4 \times 1000}{\sqrt{3} \times 110} = 21 \text{A}$$

Using Eq. (5.28)

$$\tan \psi = \frac{V \sin \theta + Ix_q}{V \cos \theta + IR_a}$$

$$= \frac{63.51 \times 0.6 + 21 \times 2}{63.51 \times 0.8} = 1.577$$

$$\psi = 57.62°$$

$$\delta = \psi - \theta = 57.62 - \cos^{-1}(0.8) = 20.75°$$

From Fig. 5.28.

$$E = V \cos \delta + I_d x_d$$

$$I_d = I \sin \psi = 21 \sin(57.62°)$$

$$= 27 \times 0.845 = 17.73 \text{ A}$$

$$E = 63.51 \cos(20.75°) + 17.73 \times 3$$

or $$E = 112.58 \text{ V per phase}$$

The power output of a salient pole machine is given by Eq. (5.32)

$$P = 3\frac{VE}{x_d} \sin \delta + \frac{3V^2(x_d - x_q)}{2x_d x_q} \sin 2\delta$$

or
$$P = \frac{3(63.51)(112.58)}{3}\sin\delta + \frac{3(63.51)^2(1)}{2(3)(2)}\sin 2\delta$$

or
$$P = 7150 \sin\delta + 1008.38 \sin 2\delta$$

For maximum power, $\frac{dP}{d\delta} = 0$

$$\frac{dP}{d\delta} = 7150\cos\delta + 2\,(1008.38)\cos 2\delta = 0$$

or
$$7150 \cos\delta + 2016.76 \cos 2\delta = 0$$

or
$$7150 \cos\delta + 2016.76(2\cos^2\delta - 1) = 0$$

or
$$4033.52 \cos^2\delta + 7150 \cos\delta - 2016.76 = 0$$

or
$$\cos^2\delta + 1.773 \cos\delta - 0.5 = 0$$

The solution of this equation is

$$\cos\delta = 0.247 \text{ or } \delta = 75.7°$$

$$P_{max} = 7150 \sin(75.7°) + 1008.38 \sin(2 \times 75.7°)$$
$$= 6928.46 + 482.7 = 7411.16 \text{ W.}$$

### *Example 5.27*

A salient pole alternator is operated at 0.8 p.f. lagging. It is delivering rated current at rated voltage. $R_a = 0.02$ pu, $x_d = 1.2$ pu and $x_q = 0.8$ pu. Find direct and quadrature axis currents and load angle.

### *Solution*

$$V = 1 \text{ pu}, \quad I = 1 \text{ pu}$$

Using Eq. (5.28)

$$\tan\psi = \frac{V\sin\theta + Ix_q}{V\cos\theta + IR_a} = \frac{1\times 0.6 + 1\times 0.8}{1\times 0.8 + 1\times 0.02}$$

$$\frac{1.4}{0.82} = 1.707$$

$$\psi = 59.64°, \ \delta = 59.64 - 36.87 = 22.77°$$

$$I_d = I \sin\psi = 0.863 \text{ pu}$$

$$I_q = I \cos\psi = 0.505 \text{ pu.}$$

### *Example 5.28*

For the salient pole generator of Example 5.27, find no-load emf. Neglect $R_a$.

### *Solution*

From Fig. 5.28

$$E = V\cos\delta + I_d x_d$$

$$= 1 \cos(59.64 - 36.87) + (0.863)(1.2)$$

$$= 0.922 + 1.036 = 1.958 \text{ pu.}$$

### *Example 5.29*

Two identical round rotor alternators supply inductive load of 1.414 pu at 0.707 p.f. lagging and at rated voltage. Pu synchronous reactance of each is 0.6. Find emf and power angle. State assumption made.

### *Solution*

Since the alternators are identical, it is assumed that they share the load equally. Moreover, it is assumed that each of them operates at the same p.f., i.e. 0.707 lagging.

$$V = 1 \text{ pu}, \; I = 0.5 \times 1.414 = 0.707 \text{ pu}$$

$$\cos\theta = 0.707 \text{ lagging}, \quad \sin\theta = 0.707$$

$$x_s = 0.6 \text{ pu}, \quad R_a = 0$$

Let $V = 1\angle 0$

$$I = 0.707(0.707 - j0.707)$$

$$= 0.5 - j0.5$$

$$E = V + IR_a + jIx_s$$

$$= 1\angle 0 + (0.5 - j0.5)(j0.6)$$

$$= 1 + j0.3 + 0.3 = 1.3 + j0.3$$

or $E = 1.334\angle 13° \text{ pu}$

Hence $E = 1.334$ pu and $\delta = 13°$.

### *Example 5.30*

A 3-phase star connected cylindrical rotor alternator has synchronous impedance of $1 + j\,10$ ohm per phase. It is operating at a constant voltage of 6.6 kV and the field current is adjusted to give an excitation voltage of 6.4 kV. Find power output, armature current and power factor under the condition of maximum power input.

### *Solution*

As derived in Section 5.15, maximum power input occurs when

$$\delta = 180 - \phi_s$$

$$Z_s = 1 + j10 = 10.05\angle 84.3° \text{ ohm}$$

$$\phi_s = 84.3°$$

$$E = \frac{6400}{\sqrt{3}} = 3695.15 \text{ V}$$

$$V = \frac{6600}{\sqrt{3}} = 3810.62 \text{ V}$$

Under this condition, the real power output is given by Eq. (5.19 a).

$$\delta = 180 - \phi_s = 180 - 84.3 = 95.7°$$

$$P = \frac{3VE}{Z_s}\cos(\phi_s - \delta) - \frac{3V^2}{Z_s}\cos\phi_s$$

$$= \frac{3(3810.62)(3695.15)}{10.05}\cos(84.3 - 95.7) - \frac{3(3810.62)^2}{10.05}\cos(84.3°)$$

$$= 3689794 \text{ W or } 3689.794 \text{ kW}$$

$$P = 3VI\cos\theta = 3689794$$

or $$I\cos\theta = \frac{3689794}{3(3810.62)} = 322.76 \text{ A}$$

Since $E$ is less than $V$, power factor must be leading.

$$V = 3810.62\angle 0 \text{V}$$

$$E = 3695.15\angle\delta = 3695.15\angle 95.7°$$

$$E = V + IZ_s$$

or $$3695.15\angle 95.7 = 3810.62 + (I\cos\theta + jI\sin\theta)(1 + j10)$$

or $$-367 + j3676.88 = 3810.62 + (322.76 + jI\sin\theta)(1 + j1)$$

Equating the real parts

$$-367 = 3810.62 + 322.76 - 10\,I\sin\theta$$

or $$I\sin\theta = 450.04 \text{ A}$$

$$\tan\theta = \frac{I\sin\theta}{J\cos\theta} = \frac{450.04}{322.76} = 1.394$$

$$\text{Power factor} = \cos\theta = 0.583 \text{ leading}$$

$$\text{Armature current } I = \frac{322.76}{0.583} = 553.62 \text{ A}.$$

## SUMMARY

1. A synchronous generator (alternator) converts mechanical energy into ac electrical energy. To keep the output frequency constant it must run at synchronous speed.
2. Synchronous generator has armature winding on stator and field winding on rotor. The field winding is excited by dc.
3. It may be a cylindrical rotor machine or salient pole machine. Cylindrical rotor machines have horizontal configuration, have 2 or 4 poles, (i.e. 3000 or 1500 rpm) and are used in thermal plants. Salient pole machines have large number of poles, run at slow speed and are used in hydro electric plants.
4. The excitation system may be a brushless excitation system or static excitation system. Thyristors are used in static excitation system.

5. The effect of armature reaction can be represented by an inductive reactance $x_a$. The sum of $x_a$ and leakage reactance $x_l$ is called synchronous reactance $x_s$. Synchronous impedance $Z_s = R_a + jx_s$.
6. Voltage regulation is the change in terminal voltage when load current is reduced to zero. It is expressed as percentage of full load terminal voltage.
7. Voltage regulation can be estimated by synchronous impedance method, MMF method, saturated synchronous method and zero power factor method.
8. Synchronous impedance can be determined from open circuit and short circuit tests.
9. Power output $P = \dfrac{3VE}{x_s} \sin\delta$
10. The analysis of salient pole machine can be done by two reaction theory. As per this theory the stator mmf (armature reaction mmf) can be decomposed into two components, one parallel to polar axis and the other perpendicular to polar axis.
11. For salient pole machine, power output $P$ is

$$P = \frac{3VE}{X_d}\sin\delta + 3V^2\left[\frac{X_d - X_q}{2X_dX_q}\right]\sin 2\delta$$

12. $X_d$ and $X_q$ can be measured by slip test.
13. Every alternator has limits of MW, MVA, maximum field current and stability.
14. Alternators have to be operated in parallel to meet the total load requirements. The conditions for parallel operation are: same voltage, same frequency, same phase sequence and same phase.
15. An incoming alternator has to be synchronised with the bus bars to ensure that conditions for parallel operation are satisfied.
16. When two alternators are operating in parallel and supplying a certain load the frequency and terminal voltage are not constrained to be constant. The increase in field current of one machine increases its reactive power output.
17. When an alternator is feeding a bus bar, the terminal voltage and frequency remain constant at system voltage and frequency. The governor set points control real power output and field current controls the reactive power output and power factor of alternator.
18. Short circuit transient consists of sub-transient period, transient period and steady state period.
19. A number of machine reactances and time constants are defined to analyse the machine behaviour.
20. Damper winding is used in salient pole machines to damp out the electromechanical transients. It consists of low resistance copper or aluminium bars embedded in rotor pole shoes and short circuited at both ends by rings.

## TEST POINT QUESTIONS

### A. State whether the following statements are true (T) or false (F)

5.1 The field winding of alternator is on rotor.

5.2 Alternators used in hydro electric plants have low speed.

5.3 Alternators used in thermal plants have large diameters.

5.4 The generation voltage is generally 132 kV.

5.5 $f = (P/2)(n_s)$

5.6 The maximum peripheral speed of cylindrical rotor machines should not exceed 175 m/s.

5.7 Hydrogen does not support combustion.

5.8 Use of hydrogen cooling lowers efficiency of alternator.

5.9 The maximum size of cylindrical rotor alternator, used in India, is lesser than the maximum size of salient pole alternator.

5.10 Static excitation system has vary small response time.

5.11 $x_a$ is a fictitious reactance.

5.12 Armature mmf distorts the magnetic flux.

5.13 At leading power factor armature reaction has a magnetising effect in an alternator.

5.14 Voltage regulation of an alternator at leading power factor may be negative.

5.15 Synchronous impedance method of finding voltage regulation gives pessimistic results.

## B. Fill in the blanks with suitable words.

5.16 MMF method of calculating voltage regulation of an alternator gives .............. results.

5.17 Zero power factor method of calculating voltage regulation of an alternator is also called .............. method.

5.18 For a cylindrical rotor alternator P = ..............

5.19 If the field current of an alternator is increased, its reactive power ..............

5.20 Salient pole machine can be analysed using ..............

5.21 For a salient pole alternator, $x_d$ and $x_q$ can be found by ..............

5.22 As the power output of a prime mover increases, the speed ..............

5.23 The method of putting an alternator in parallel with bus bars is called ..............

5.24 When an alternator is feeding an infinite bus the .............. and .............. remain constant.

5.25 The initial period following a sudden short circuit of an alternator is called ..............

5.26 Out of $x_d$, $x_d'$ and $x_d''$, the reactance .............. has the lowest value.

5.27 When zero sequence currents flow in armature there is no ..............

5.28 Out of all the reactances of a synchronous machine the .............. has the minimum value.

5.29 Damper winding is located in ..............

5.30 Damper winding is .............. at both ends.

## ANSWERS

| | | | | |
|---|---|---|---|---|
| 5.1 T | 5.2 T | 5.3 F | 5.4 F | 5.5 T |
| 5.6 T | 5.7 T | 5.8 F | 5.9 F | 5.10 T |
| 5.11 T | 5.12 T | 5.13 T | 5.14 T | 5.15 T |

5.16 Optimistic    5.17 Potier reactance

5.18 $\frac{3VE}{x_s}$ 5.19 increases 5.20 two reaction theory
5.21 Slip test 5.22 decreases 5.23 synchronising
5.24 Voltage, frequency 5.25 sub-transient period 5.26 $x_d''$
5.27 Rotating field 5.28 Zero sequence reactance
5.29 rotor pole shoes 5.30 short circuited

## JOB INTERVIEW QUESTIONS

5.1 What is a synchronous generator? Why is it so called?
5.2 What is synchronous speed?
5.3 What is the difference between salient pole machine and cylindrical rotor machine?
5.4 What is meant by voltage regulation?
5.5 What are the advantages of hydrogen cooling?
5.6 What is synchronous impedance?
5.7 How is synchronous impedance found?
5.8 What is zero power factor characteristics?
5.9 What is synchronising?
5.10 What is an infinite bus bar?
5.11 What is a damper winding? Why is it used?
5.12 What are the operating limits on an alternator?
5.13 What is sub-transient period?
5.14 Name the conditions for parallel operation?
5.15 Why is parallel operation of alternators needed?
5.16 Write the equation for synchronising power.
5.17 What is parallel generator theorem?
5.18 What is the difference between $x_d$ and $x_q$?
5.19 Why is hydrogen cooling used?
5.20 what is an exciter?

## REVIEW QUESTIONS

5.1 Give reasons:
(a) The speed of a synchronous generator should be kept constant at synchronous speed.
(b) Cylindrical rotor alternators have small diameters and large lengths of core.
(c) Generally cylindrical rotor construction is used for high speed machines and salient pole construction for low speed machines.
(d) Open circuit characteristics of a synchronous machine has curved shape.
(e) The short circuit characteristics of a synchronous generator is a straight line.

(f) The voltage regulation for unity and lagging power factors is positive but may be negative for leading power factors.

5.2 Discuss the main constructional features of cylindrical rotor and salient pole alternators.

5.3 What is breadth factor? What is its effect. Derive the equation for breadth factor of a winding having $q$ slots per pole per phase and slot angle $\alpha$?

5.4 What is pitch factor? Is it always less than or more than 1? Derive an expression for the pitch factor of a winding using coils short pitched by angle $\sigma$. What is the effect of fractional pitch on the magnitude and wave shape of voltage induced in an alternator?

5.5 Derive the phasor diagram of a cylindrical rotor alternator. What is the effect of armature reaction and how is it included in phasor diagram? Draw phasor diagrams for lagging, unity and leading power factors?

5.6 Discuss the synchronous impedance and mmf methods of calculating regulation of an alternator. Why do these methods give different results?

5.7 Discuss the Potier triangle method of finding regulation of an alternator.

5.8 What is synchronizing power? Derive equations for synchronising power of cylindrical rotor and salient pole alternators. Discuss the role of this quantity in successful operation of an alternator. When is it maximum? When is it negative?

5.9 Bring out the characteristics of two alternators working in parallel. What is the effect of change in excitation and change in mechanical power input on load sharing?

5.10 In what respect is the operation of an alternator on infinite bus bar different from parallel operation of two alternators? What is the effect of change is excitation and change in mechanical power input on the operation of an alternator on infinite bus?

5.11 Discuss the phenomenon of sudden 3-phase short circuit of an alternator. Draw a typical wave shape of current and mark the different regions. Write an expression for the current.

5.12 Define the various reactances and time constants of a synchronous machine.

5.13 Explain the two-reaction theory of salient pole machines. Draw the phasor diagram as per this theory. How can the regulation of an alternator be calculated using this theory?

5.14 Describe the procedure of slip test. How can $x_d$ and $x_q$ found from this test?

## PROBLEMS

5.1 A 3-phase star connected alternator is rated at 1600 kVA 13.5 kV. The effective armature resistance and synchronous reactance are 1.5 ohms and 30 ohm per phase, respectively. Find regulation for a load of 1280 kW, rated voltage and power factor of (a) 0.78 leading, (b) unity, (c) 0.8 lagging.

[(a) –11.8% (b) 3.2% (c) 18.6%]

5.2 A 3-phase 11 kV star connected alternator has $R_a$ = 1 ohm and $x_s$ = 20 ohm per phase. Find regulation for a load of 1500 kW at power factors of (a) 0.8 lagging, (b) unity, (c) 0.8 leading.

[(a) 22% (b) 4.25% (c) –13.4%]

5.3 The effective resistance of a 2200 V, 50 Hz, 440 kVA, single-phase alternator is 0.5 ohm. On short circuit a field current of 4 A gives the full-load current. The emf on open circuit for the same field current is 1160 V. Find $Z_s$ and $x_s$. [5.80 Ω, 5.78 Ω]

5.4 A 5000 kVA, 2-pole, 50 Hz alternator has a rated line voltage of 4160 V. The open circuit characteristics is

| Field current | 20, | 40, | 60, | 80, | 100, | 120, | 140, | 160, | 180, | 200 A. |
|---|---|---|---|---|---|---|---|---|---|---|
| Line Voltage | 1250, | 2500, | 3650, | 4450, | 4950, | 5150, | 5300, | 5440, | 5530, | 5600 V. |

When the alternator terminals are short circuited, a field current of 84A is required to circulate full-load current. Use mmf method to find regulation at full-load, rated voltage and power factors of (a) 1, (b) 0.8 lagging. The alternator is star connected. Neglect armature reaction. [(a) 21.4% (b) 34%]

5.5 Two identical 3000 kVA alternators are running in parallel. The frequency drop from no-load to full-load for the two alternators are 50 Hz to 47 Hz and 50 Hz to 48 Hz, respectively.

(a) How will they share a load of 4000 kW?

(b) What is the maximum unity power factor load, which they can jointly supply without any one of them getting over loaded? [(a) 1600 kW, 2400 kW (b) 5000 kW]

5.6 Two similar star connected 3-phase alternators share a load of 7500 kW equally at 6000 V and 0.8 p.f. lagging. The synchronous impedance of each is $2.5 + j50$ ohm. The excitation of second machine is changed, so that it delivers 40 A at a lagging p.f. Find (a) armature current of first machine, (b) emf of each machine, (c) power factor of each machine. [(a) 51.4 A (b) 9780 V, 8270 V (c) 0.7 lagging, 0.9 lagging]

5.7 Two single-phase alternators together supply a load of $7.5 + j10$ ohm. The synchronous impedance of each is $0.5 + j\,2.5$ ohm. Emf of first machine is $230 \angle 0$ and that of second machine is $250 \angle 0$. Find (a) terminal voltage, (b) power output of each machine. [(a) 218 V (b) 1.32 kW, 0.91 kW]

5.8 Two three-phase star connected identical alternators share a load of 25 MW at0.9 p.f. lagging. The terminal voltage is 11 kV and synchronous reactance of each machine is 4.8 ohm, the armature resistance being negligible. Alternators 2 delivers 10 MW and its induced emf is 15 kV. Find (a) load angle of each machine, (b) current supplied by each machine, (c) power factor of each machine, (d) kVA of each machine, (e) induced emf of first machine.

[(a) 27.6°, 16.9°, (b) 820 A, 660 A (c) 0.96 lagging, 0.8 lagging (d) 15.6 MVA, 12.6 MVA, (e) 14.6 kV]

5.9 Two 3-phase alternators running in parallel share a load of 6 MW at 0.8 p.f. lagging. The frequency of machine *A* falls from 51 Hz to 49.5 Hz, when loaded to 2 MW. The frequency drop of machine *B* is from 51 Hz to 49.75 Hz at 10 MW load. Find (a) power supplied by each machine, (b) power factor of second machine if that of first machine is 0.71 lagging.

[$P_A$ = 857 kW, $P_B$ = 5143 kW p.f. of machine *B* = 0.815 lagging]

5.10 Two alternators rated 100 MW and 200 MW are operating in parallel. Their governor characteristics are such that their speed drops form no-load to full-load are 4% and 5%, respectively. The machines are operating at 50 Hz and sharing a load of 300 MW in the ratio of their ratings. (a) If the load reduces to 200 MW, how will the load be shared and what will be the system frequency. Assume free governor operation. (b) The speed change of the governors are reset such that they share the 200 MW load in the ratio of their rating and the system frequency at 200 MW is 50 Hz. What are the no-load frequencies of the two machines? [(a) 66.66, 133.33 MW, 50.7 Hz (b) 51.37 Hz, 51.72 Hz]

5.11 A 10 MVA 3-phase alternator has a short circuit reactance of 20%. Calculate the synchronizing power per mechanical degree of displacement when the machine is running at 1500 rpm on 10 kV infinite bus bar. [1744 kW]

5.12 A 2000 kVA, 3-phase, 8-pole, star connected machine is running on 6000 V, 50 Hz, infinite bus bar. Find the synchronizing power and torque per mechanical degree of displacement at (a) no-load with excitation adjusted, so as to give 6000 V on open circuit, (b) full-load, 0.8 p.f. lagging. (c) What are the values of open circuit emf and load angle for full-load 0.8 p.f. lagging. $Z_s = (1 + j20)\%$.

[(a) 698 kW, 8887 N-m (b) 790 kW, 10058 N-m (c) 1.139 p u, 7.8°]

5.13 The speed of a 500 MW generator drops by 5% from no-load to full-load. (a) Find speed regulation *R*. Express the results in Hz per unit MW and Hz per MW. (b) If frequency drops by 0.1 Hz find the change in power output. [(a) 2.5 Hz/pu MW, $0.5 \times 10^{-2}$ Hz/MW (b) 20 MW]

5.14 A salient pole synchronous generator is rated at 200 MVA and 11 kV. Its armature winding is star connected, having negligible resistance. $x_d = 0.8$ pu, $x_q = 0.5$ pu. The generator is supplying rated MVA at rated voltage at 0.8 p.f. lagging. Find load angle and open circuit emf in p.u. [17.1°. 1.603 pu]

5.15 A 3 phase salient pole synchronous generator is rated at 3.5 MVA 6.6 kV. Its direct and quadrature axis reactances are 9.6 ohm and 6 ohm per phase. The winding are star connected and resistance is negligible. If the generator is supplying 2.5 MW at rated voltage and at 0.8 p.f. lagging to an infinite bus, find regulation. [50.9%]

5.16 A 200 MVA, 11 kV, 3-phase, star connected alternator is operated at rated speed and voltage and at no-load. A sudden 3-phase short circuit occurs at its terminals. The p.u. reactances are $x_d = 1.3$ pu, $x_d' = 0.25$ p.u, $x_d'' = 0.15$ pu. The initial dc component is 60% of the initial ac component of current. Find (a) ac component of current at the instant of short circuit, (b) total current at the instant of short current, (c) ac component of current after 1 cycle, (d) ac component of current after 2 sec. The time constants are $T_d' = 1$ sec, $T_d'' = 0.05$ sec. [(a) 69984 A (b) 111974.4 A (c) 60083.5 A (d) 12665 A]

5.17 A 1 MVA, 3-phase, 50 Hz, star connected 4600 V alternator has a stator impedance of 2 + *j* 20 ohm per phase. Find full load generated emf at (a) unity p.f. (b) 0.75 lagging p.f.

[(a) 3636 V/phase (b) 4814 V/phase]

5.18 A 100 kVA, 50 Hz, 3-phase star connected, 2400 V alternator has an armature reactance of 2.5 Ω/phase and synchronous reactance of 30 Ω/phase. Fine (a) voltage regulation at full load unity p.f., (b) actual generated line voltage at full load unity p.f. if leakage reactance per phase is 10 ohms. [(a) 16.7% (b) 2539 V]

5.19 Find the synchronising torque per mechanical degree of displacement for a 6000 kVA, 50 Hz, 1500 rpm alternator having a generated emf of 10 kV and synchronous reactance of 0.25 pu. Assume E = V.

[5333.3 N-m]

5.20 For the alternator of Problem 5.17, find generated voltage at full load (a) 0.75 p.f. leading, (b) 0.4 p.f. leading. [(a) 2364 V/phase (b) 1315 V/phase]

5.21 A 750 kVA, 50 Hz, 3000 rpm, 1380 V 3-phase alternator gave the following test results:
O.C. test: field current 17.5 A, line voltage 1020 V.
S.C. test: field current 17.5 A, line current rated value.
D.C. resistance test: Applied voltage 15 V, current 100 A.

The alternator is delta connected. Assume that effective ac resistance is 40% higher than dc resistance. Find $x_s$ [5.62 ohm]

5.22 Two alternators are operating in parallel. For the first alternator the induced emf is (2000 + *j* 0) V per phase and for the second machine it is (2200 + *j* 100) V per phase. The synchronous impedance of each is (0.2 + *j* 2) ohm per phase. The load impedance is (3 + *j* 4) ohm per phase. Find kW output and p.f. of each machine.

[Alternator 1: 399.2 kW, 0.606 lagging
Alternator 2: 746.9 kW, 0.596 lagging]

5.23 A 3-phase, star connected, 50 Hz, 100 kVA, 1100 V synchronous generator gave an O.C. voltage (line value) of 420 V for a field current of 12.5 A. The short circuit armature current for the same value of field current was equal to rated value. A 6 V battery connected across two terminals gave a current of 10 A. Find (a) armature resistance, synchronous impedance and synchronous reactance per phase, (b) voltage regulation at 0.8 p.f leading. Assume that ac resistance is 1.5 times dc resistance.

[(a) 0.45 Ω, 4.62 Ω, 4.6 Ω (b) – 13.4%]

5.24 Two identical alternators are being driven by a common prime mover. Their induced emf are in phase and synchronous impedance of each is (0.3 + *j* 4) ohm per phase. Their emfs are 250 V and 350 V per phase respectively. Find (a) circulating current, (b) terminal voltage, (c) electrical output of generating machine, (d) power asupplied by prime mover if mechanical and core losses total 2 kW. Both the machines are delta connected.

[(a) 12.4 A (b) 300 V (c) .84 kW (d) 2.28 kW]

5.25 Two similar 3-phase, 50 Hz, 6.6 kV synchronous generators share a total load of 10 MW at 0.8 p.f lagging equally. The excitations of the two machines are equal and each has a synchronous impedance of

(0.05 + *j* 0.5) ohm per phase. The excitation of one machine is changed so that its armature current is 438 A. The steam input to the turbine is not changed. Find armature current, emf and p.f. of second machine. The machines are star connected. [788.8 A, 7200 V, 0.556 lagging]

5.26 Two 3-phase, 50 Hz alternators A and B are to be put in parallel to a common bus. Each has synchronous impedance of (0.1 + *j* 1) ohm per phase. Their line voltages are adjusted to 2500 V and 2300 V and are in phase with respect to bus bar. Both the machines are star connected. Find (a) resultant voltage between the machines, (b) synchronising current per phase, (c) phase angle between resultant voltage and synchronising current, (d) power developed by alternator A per phase and mode of operation, (e) power developed per phase by alternator B and mode of operation, (f) power loss per phase, (g) bus voltage per phase.

[(a) 115.5 V per phase (b) 57.5 A (c) 84.3° (d) 8.243 kW, generating (e) 7.583 kW, motoring (f) 660 W (g) 1386 V]

5.27 A 3-phase star connected alternator is operating on 10 kV infinite bus. It's synchronous impedance is (0.4 + *j* 6) ohm per phase and it is delivering 300 A at 0.8 p.f. lagging. If steam input to prime mover is kept constant, find the percentage change in emf to increase p.f. to unity. [15%]

Chapter 6

# Synchronous Motors

## 6.1 PRINCIPLE OF OPERATION

A synchronous motor is a synchronous machine which converts 3-phase electrical power to mechanical power. Its rotor winding is fed with a dc current which sets up a steady magnetic field. Since the rotor is moving at synchronous speed, the rotor field also moves at synchronous speed with respect to inertial space. The stator is fed with 3-phase ac current. As discussed in Chapter 4, a field rotating at synchronous speed is set up by the stator currents. Thus two magnetic fields are present in the machine, both revolving at synchronous speed and therefore stationary with respect to each other. The rotor field tends to line up with the stator field. The larger the angle between the two fields, greater is the amount of torque produced. The basic principle of this motor is that rotor chases the rotating field of stator around the circle, but is never able to catch up with it.

## 6.2 STEADY STATE CONSTANT SPEED OPERATION

A synchronous motor is a constant speed motor. It develops torque only when running at synchronous speed. At all other speeds its average torque is zero. Moreover, it has no starting torque.

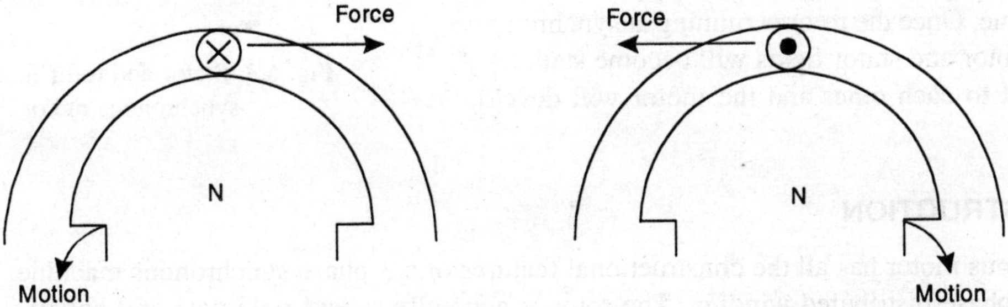


**Fig. 6.1.** Force on a conductor in positive and negative half cycles of current

Figure 6.1 (a) shows a conductor carrying current into the page. The conductor is under the influence of a north pole. As per Fleming's left hand rule, the force on the conductor will be towards right. Since the conductor is on stator and the pole is mechanically coupled to the shaft, the shaft moves in the counter clockwise direction. After one-half cycle, the current in the conductor will reverse in direction (i.e. coming out of page) as depicted in Fig. 6.1 (b). Now the force on the conductor will be towards left and, therefore, the shaft will tend to move in clockwise motion. Thus the direction of torque on the rotor reverses every half-cycle and the average torque is zero.

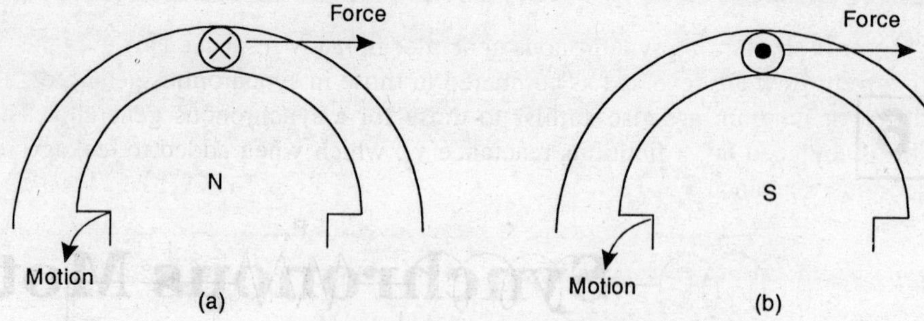


**Fig. 6.2.** Force on a conductor in positive and negative half cycles of current if rotor is moving at synchronous speed

However the direction of torque is the same if the rotor is already moving at synchronous speed. This is depicted in Fig. 6.2. In Fig. 6.2 (a) the conductor is under the influence of a north pole and the direction of torque on the shaft will be counter clockwise. After one-half cycle the direction of current in the conductor will reverse. If the rotor is moving at synchronous speed, the conductor will now be under the influence of a south pole (Fig. 6.2(b)). The direction of torque will again be counter clockwise. Thus the motor will develop unidirectional torque, if it is already running at synchronous speed.

The action of motor is illustrated in Fig. 6.3. The stator and rotor fields are stationary with respect to each other. The stator field attracts a rotor field of opposite polarity and continues to pull it around at synchronous speed. Thus the motor develops positive torque. However, if it is at standstill or running at a speed other than synchronous speed, its average torque is zero. Therefore it is essential to bring the rotor to the synchronous speed by an auxiliary starting method before it can develop a torque. Once the rotor is running at synchronous speed, the rotor and stator fields will become stationary with respect to each other and the motor will develop torque.

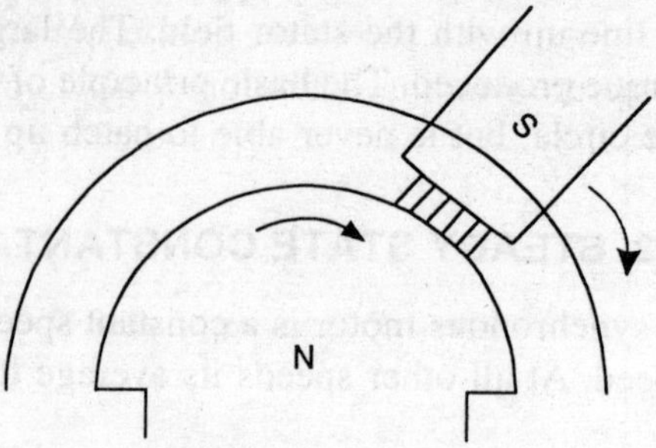


**Fig. 6.3.** Stator and rotor fields in a synchronous motor

## 6.3 CONSTRUCTION

A synchronous motor has all the constructional features of a 3-phase synchronous machine. The stator carries a 3-phase distributed winding. The rotor is generally salient pole type and has two windings. The dc winding takes the form of field coils around the pole cores and takes in the dc current from an external source for setting up the magnetic field. The second rotor winding is the amortisseur winding, which consists of copper (or aluminium or bronze) bars embedded in slots in the pole faces. These bars are short circuited at the two ends by conducting metal rings. The function of this winding is to develop starting torque by induction motor action. There is no current in these bars when the motor is running at synchronous speed. However, whenever the speed tends to be slightly different from the synchronous speed, due to a change in mechanical load on the shaft, currents are induced in these bars and oscillations of the speed are damped out.

## 6.4 EQUIVALENT CIRCUIT AND PHASOR DIAGRAM: CYLINDRICAL ROTOR THEORY

A synchronous motor is similar to a synchronous generator in many respects. However the direction of power flow and current flow are reversed as compared to those in synchronous generator. The equivalent circuit and phasor diagram are also similar to those for a synchronous generator. The effect of armature reaction is replaced lay a fictitious reactance $x_a$, which when added to leakage reactance $x_l$ gives the synchronous reactance $x_s$.

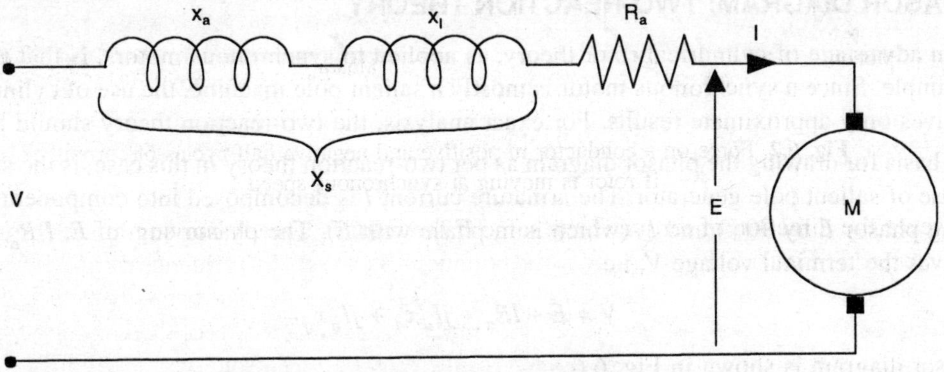


**Fig. 6.4.** Equivalent circuit of synchronous motor

Figure 6.4 shows the equivalent circuit. The terminal voltage *V* has to send a current through the armature winding. Therefore *V* has to overcome the voltage drop across the synchronous impedance $Z_s$ and the counter emf *E*. Thus the voltage equation for a synchronous motor takes the form.

$$V = E + IR_a + jIx_s \qquad ...(6.1)$$

All quantities in Eq. (6.1) are in per phase values.

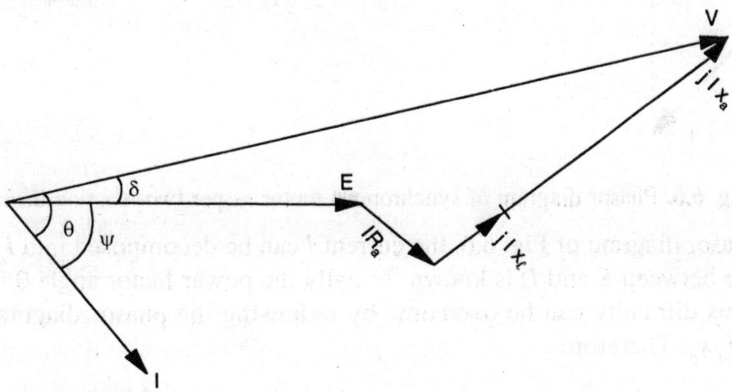


**Fig. 6.5.** Phasor diagram of cylindrical rotor synchronous motor

Figure 6.5 shows the phasor diagram corresponding to Eq. (6.1). It is seen that phasor *V* is ahead of phasor *E* (In an alternator, phasor *E* is ahead of phasor *V*). This is again due to the fact that, in a synchronous motor, the direction of power flow is into the motor. In an alternator the driving action of the prime mover causes the centreline of rotor poles to be ahead of the centreline of stator poles by $\angle\delta$. In a synchronous motor, the rotor falls behind the stator due to the mechanical load on the shaft. In this case the centreline of stator poles is ahead of the centreline of rotor poles by $\angle\delta$. This phase difference

$\delta$ between the stator and rotor magnetic axis governs the flow of power into the motor. Greater this $\angle\delta$, greater is the mechanical power developed by the motor.

As seen in the phasor diagram of Fig. 6.5, the phasor sum of $E$, $I\,R_a$ and $jIx_s$ gives the terminal voltage $V$. The phase angle between $V$ and $I$ is the power factor angle $\theta$ and the motor is operating at power factor cos $\theta$.

## 6.5 PHASOR DIAGRAM: TWO-REACTION THEORY

The main advantage of cylindrical rotor theory, as applied to synchronous motors, is that the analysis is very simple. Since a synchronous motor is mostly a salient pole machine, the use of cylindrical rotor theory gives only approximate results. For exact analysis, the two-reaction theory should be used.

The basis for drawing the phasor diagram as per two-reaction theory in this case, is the same as that in the case of salient pole generator. The armature current $I$ is decomposed into components $I_d$ (which is lagging phasor $E$ by 90 °) and $I_q$ (which is in phase with $E$). The phasor sum of $E$, $I\,R_a$, $j\,I_d\,x_d$ and $j\,I_q\,x_q$ gives the terminal voltage $V$, i.e.

$$V = E + IR_a + jI_d x_d + jI_q x_q \qquad ...(6.2)$$

The phasor diagram is shown in Fig. 6.6.

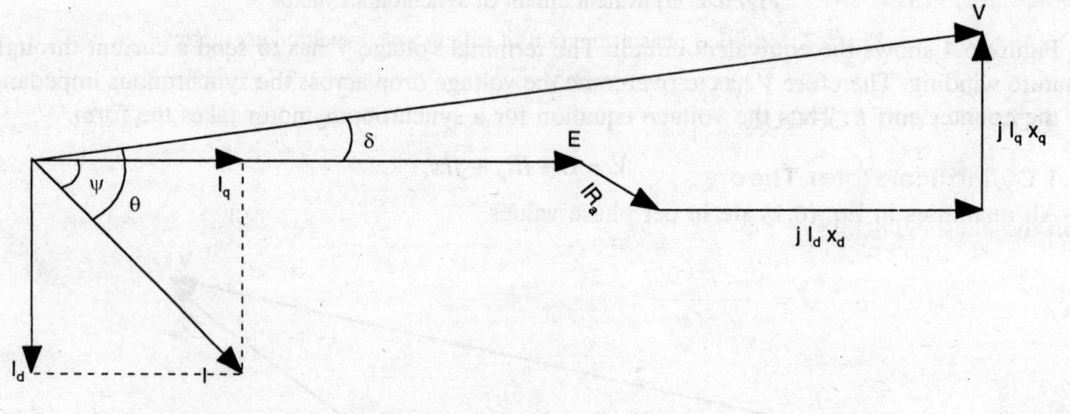


**Fig. 6.6.** Phasor diagram of synchronous motor as per two-reaction theory

In using the phasor diagram of Fig. 6.6, the current $I$ can be decomposed into $I_d$ and $I_q$, only if $\angle\psi$ (i.e. the phase angle between $E$ and $I$) is known. Initially the power factor angle $\theta$ is known, but angle $\psi$ is not known. This difficulty can be overcome by redrawing the phasor diagram as in Fig. 6.7. In triangle $abc$, $bc = I_q\,x_q$. Therefore

$$ac = \frac{bc}{\cos\psi} = \frac{I_q x_q}{\cos\psi} = Ix_q$$

Then,

$$\tan\psi = \frac{V\sin\theta - Ix_q}{V\cos\theta - IR_a} \qquad ...(6.3)$$

Angle $\psi$ can be found from Eq. (6.3). Then $I_d = I \sin\psi$ and $I_q = I\cos\psi$.

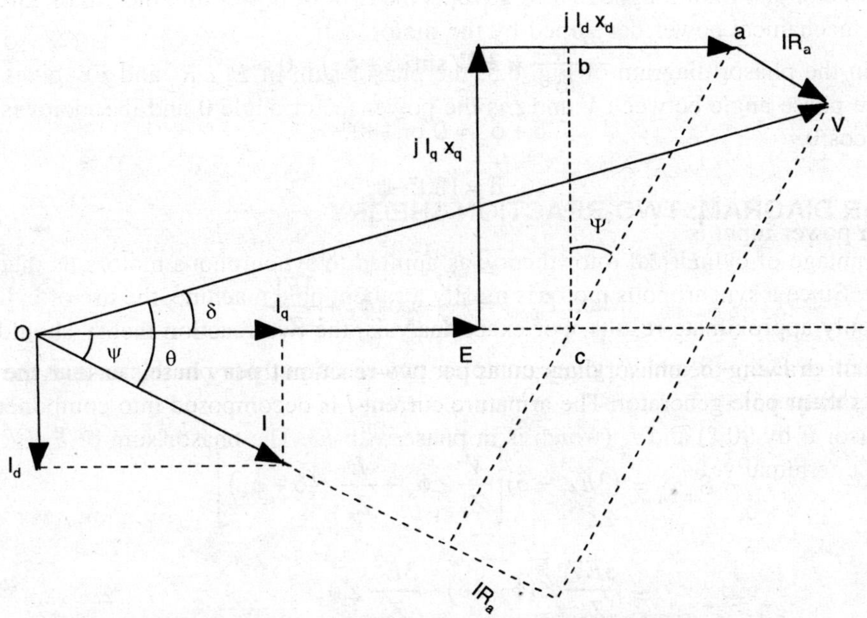


**Fig. 6.7.** Modified phasor diagram of salient pole synchronous motor

## 6.6 POWER AND TORQUE

### 6.6.1 Cylindrical Rotor Theory

From the phasor diagram of Fig. 6.5

$$I = \frac{V - E\angle -\delta}{Z_s \angle \phi_s} = \frac{V}{Z_s}\angle -\phi_s - \frac{E}{Z_s}\angle -(\delta + \phi_s)$$

$$I^* = \frac{V}{Z_s}\angle \phi_s - \frac{E}{Z_s}\angle (\delta + \phi_s) \qquad \text{...(6.4)}$$

The product of $V$ and $I^*$ gives complex power input per phase, so that the 3-phase complex power input is

$$S_{\text{input}} = 3VI^* = \frac{3V^2}{Z_s}\angle \phi_s - \frac{3EV}{Z_s}\angle (\delta + \phi_s)$$

The real part of $S_{\text{input}}$ gives $P_{\text{input}}$

$$P_{\text{input}} = \frac{3V^2}{Z_s}\cos\phi_s - \frac{3EV}{Z_s}\cos(\delta + \phi_s)$$

Maximum power input is given by the condition

$$dP_{\text{input}} / d\delta = 0$$

$$\frac{dP_{\text{input}}}{d\delta} = 3EV\sin(\delta + \phi_s) = 0$$

or
$$\delta + \phi_s = 0 \text{ or } 180°$$

or
$$\delta = 180° - \phi_s$$

and maximum power input is

$$\text{Max. } P_{\text{input}} = \frac{3V^2}{Z_s}\cos\phi_s + \frac{3EV}{Z_s}$$

The product of $E\angle - \delta$ and $I^*$ gives complex power output per phase, so that the three-phase complex power output is

$$S_{\text{input}} = (3E\angle - \delta)\left[\frac{V}{Z_s}\angle\phi_s - \frac{E}{Z_s}\angle(\delta + \phi_s)\right]$$

$$= \frac{3EV}{Z_s}\angle(\phi_s - \delta) - \frac{3E^2}{Z_s}\angle\phi_s \quad \text{...(6.5)}$$

$$P_{\text{output}} = \text{Re}\left[S_{\text{output}}\right] = \frac{3EV}{Z_s}\cos(\phi_s - \delta) - \frac{3E^2}{Z_s}\cos\phi$$

The condition of maximum power output is obtained by putting $dP_{\text{output}}/d\delta = 0$

$$\frac{dP_{\text{ouput}}}{d\delta} = \frac{3EV}{Z_s}\sin(\phi_s - \delta) = 0$$

or
$$\delta = \phi_s$$

The maximum power output is

$$\text{Max.} P_{\text{output}} = \frac{3EV}{Z_s} - \frac{3E^2}{Z_s}\cos\phi_s$$

Power output equals the torque is synchronous watts

$$T = P_{\text{output}} \text{ synchronous watts}$$

$$= \frac{P_{\text{output}}}{2\pi n_s}, \text{N-m} \quad \text{...(6.6)}$$

The power output less the core loss and mechanical losses gives the power available at the shaft.

If armature resistance is neglected, $Z_s = x_s$ and $\phi_s = 90°$. Then power output and power input become equal and the expression reduces to familiar power angle characteristics. In this case the simplified expression for $I$ is

$$I = \frac{V - E\angle - \delta}{jx_s} = \frac{V\angle - 90° - E\angle(-\delta - 90°)}{x_s}$$

$$I^* = \frac{V\angle 90° - E\angle(\delta + 90°)}{x_s}$$

$$S = 3VI^* = 3V\left[\frac{V\angle 90° - E\angle(\delta + 90°)}{x_s}\right] \qquad ...(6.7)$$

The real part of $S$ in Eq. (6.7) gives the real power $P$

$$P = \frac{3VE}{x_s}\sin\delta \qquad ...(6.8)$$

Equation (6.8) is the power angle curve (Fig. 5.23). Power is maximum when $\delta = 90°$.

Synchronising power is $dP/d\delta$.

The imaginary part of $S$ in Eq. (6.7) gives the reactive power $Q$

$$Q = \frac{3V^2}{x_s} - \frac{3VE}{x_s}\cos\delta \qquad ...(6.9)$$

Equation (6.9) can be rewritten as

$$Q = \frac{3V}{x_s}(V - E\cos\delta) \qquad ...(6.10)$$

Equation (6.10) depicts a very important property of synchronous motor. When $E \cos \delta < V$, reactive power is positive, i.e. the machine is operating at a lagging power factor. When $E \cos \delta > V$, reactive power is negative, i.e. the machine is working at a leading power factor. Thus an underexcited synchronous motor operates at a lagging power factor, whereas an overexcited one operates at a leading power factor. This characteristic is used to correct the power factor of installations having lagging power factor loads.

### 6.6.2 Two-Reaction Theory

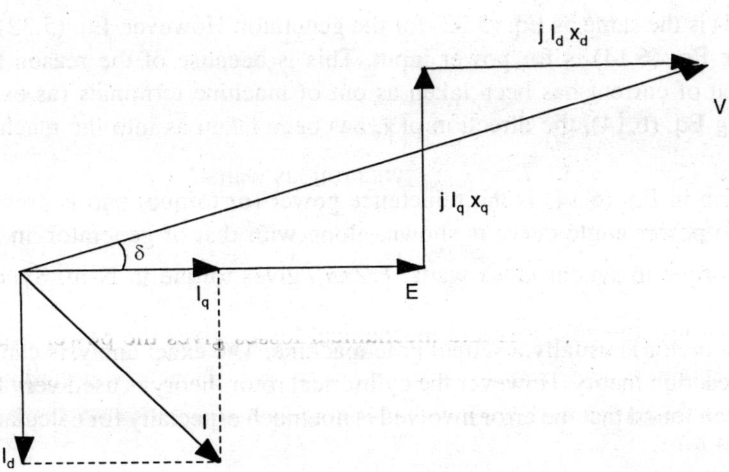


**Fig. 6.8.** Phasor diagram of salient pole synchronous motor fed from infinite bus

Figure 6.8 shows the phasor diagram of a salient pole synchronous motor being fed by an infinite bus of voltage $V$. The excitation voltage of motor is $E$. The reactances $X_d$ and $X_q$ are

$$X_d = x + x_d$$

$$X_q = x + x_q$$

where $X_d$ and $X_q$ are total direct and quadrature axis reactances, $x_d$ and $x_q$ are direct and quadrature axis reactances of motor and $x$ is the reactance of the line connecting infinite bus to motor. The armature resistance has been neglected. As in Sec. 5.17, the total power can be found by resolving the bus voltage $V$ into $V \sin \delta$ (which is in phase opposition with $I_d$) and $V \cos \delta$ (which is in phase with $I_q$). Thus

$$P = -3I_d V \sin\delta + 3I_q V \cos\delta \qquad ...(6.11)$$

In Eq. (6.11) minus sign has been used with the first term, because $V \sin \delta$ and $I_d$ are in phase opposition.

From Fig. 6.8

$$I_d = \frac{V\cos\delta - E}{X_d} \qquad ...(6.12)$$

and

$$I_q = \frac{V\sin\delta}{X_q} \qquad ...(6.13)$$

Substituting Eqs. (6.12) and (6.13) into Eq. (6.11)

$$P = -3V\sin\delta \frac{(V\cos\delta - E)}{X_d} + 3V\cos\delta\left(\frac{V\sin\delta}{X_q}\right)$$

or

$$P = \frac{3VE}{X_d}\sin\delta + \frac{3V^2(X_d - X_q)}{2X_d X_q}\sin 2\delta \qquad ...(6.14)$$

Equation (6.14) is the same as Eq. (5.32) for the generator. However, Eq. (5.32) is for power output of generator, while Eq. (6.14) is for power input. This is because of the reason that in deriving Eq. (5.32), the direction of current has been taken as out of machine terminals (as exists in a generator), whereas in deriving Eq. (6.14), the direction of $I$, has been taken as into the machine (as exists in the motor).

The second term in Eq. (6.14) is the reluctance power (or torque) and is present even when field current is zero. The power angle curve is shown, along with that of generator, in Fig. 5.32.

$P$ is equal to torque in synchronous watts. $P/2\pi n_s$, gives torque in N-m. Synchronising power is $dP/d\delta$:

A synchronous motor is usually a salient pole machine. The exact analysis can be carried out only by the use of two-reaction theory. However the cylindrical rotor theory is used very frequently due to its simplicity. It has been found that the error involved is not much especially for calculations in steady state.

## 6.7 CHANGE IN LOAD

When a synchronous motor is operating at synchronous speed with a certain mechanical load on the shaft, the rotor and stator pole axes are out of phase by angle δ. If the load on the shaft is increased, the first reaction is to slow down the rotor. This causes an increase in angle δ. Thus the difference between $V$ and $E$ increases. Since this difference equals $x_s I$ (neglecting $IR_a$ drop) and $x_s$ is constant, the armature current $I$ increases. Thus the power input to the motor increases till the power output matches the power input. The motor continues to run at synchronous speed with a new increased angle δ. A reverse action takes place if the load is decreased. The change in power angle δ can be measured by a stroboscope. The stroboscope furnishes a convincing demonstration of the physical character of power angle.

## 6.8 SPEED-TORQUE CHARACTERISTICS

A synchronous motor is constant speed motor. The speed of rotation depends on the number of poles and supply frequency. It is usually fed from an infinite bus system which operate at constant voltage and frequency. Figure 6.9 shows the torque-speed curve. Using cylindrical rotor theory, the torque is given by

$$\text{Torque} = \frac{3VE\sin\delta}{(x_s)2\pi n_s} \quad \text{...(6.15)}$$

The speed remains constant from no-load up to the maximum torque (known as pull-out torque), which occurs at δ = 90°. The full-load torque is generally about 40% of the pull-out torque.

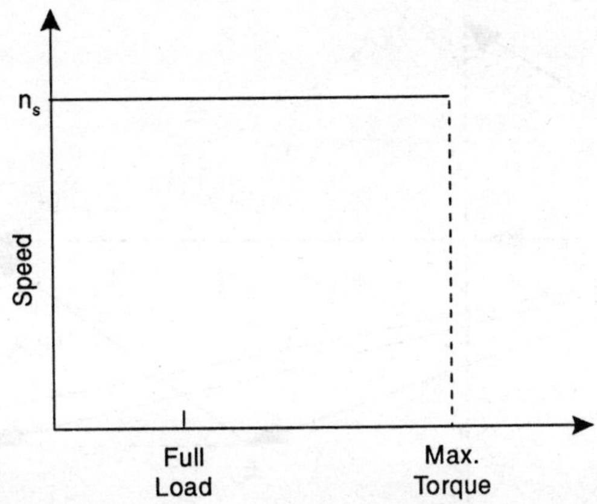


**Fig. 6.9.** Speed torque characteristics of synchronous motor

When the load on the shaft is more than the pull-out torque, the rotor cannot remain locked to the stator and starts slipping behind it. As the rotor slows down, the stator magnetic field laps it repeatedly and the direction of the induced torque reverses with each pass. Thus the torque acts in one direction and then in the opposite direction and this is repeated with each pass. This results in severe vibrations and consequent loss of synchronism. Since torque is proportional to excitation emf $E$, there is an advantage, from stability point of view, in operating the motor with higher excitation.

## 6.9 POWER FACTOR CONTROL: EFFECT OF CHANGE OF EXCITATION

The terminal voltage $V$ of a synchronous motor is constant. If the mechanical load is constant, a change in excitation (i.e. field current) causes a change in power factor, because the phasor relation between

$E$ and $V$ (Fig. 6.5) must be satisfied at all values of $E$. This characteristics of a synchronous motor makes it as a very suitable device for improvement of power factor is power systems.

A constant applied voltage means a constant resultant flux. This resultant flux is due to the action of both dc source and ac source. If excitation is large, the field produced by rotor is too high. Consequently the armature mmf becomes demagnetising, so as to keep the resultant flux constant. If excitation is small, the field produced by rotor is too small. Consequently the armature mmf becomes magnetising, so as to keep the resultant flux constant.

Figure 6.10 shows the phasor diagram of a synchronous motor supplying a constant mechanical load at three different values of excitation $E_1$, $E_2$ and $E_3$. The power developed is $(3\ VE/x_s)\sin\delta$. As $E$ changes, $\delta$ also changes, so as to keep the power developed at a constant value. The locus of $E$ is the dotted line. When emf is $E_1$, the field flux is too high. The current $I_1$ is now at a leading power factor $\cos\theta_1$. The leading reactive current exerts a demagnetising effect to restore the resultant flux to the constant value. The phasor sum of $E_1$ and $jI_1 x_s$ gives the terminal voltage $V$ (the armature resistance $R_a$ has been neglected). When the excitation emf is $E_2$, there is no excess flux produced by the rotor winding. Therefore the input current $I_2$ has no reactive component and the power factor is 1. When the excitation voltage is $E_3$, the motor is underexcited. The armature current now assumes the position $I_3$ and the power factor is lagging. The lagging current has a magnetising effect, which helps to establish the air gap flux demanded by the terminal voltage.

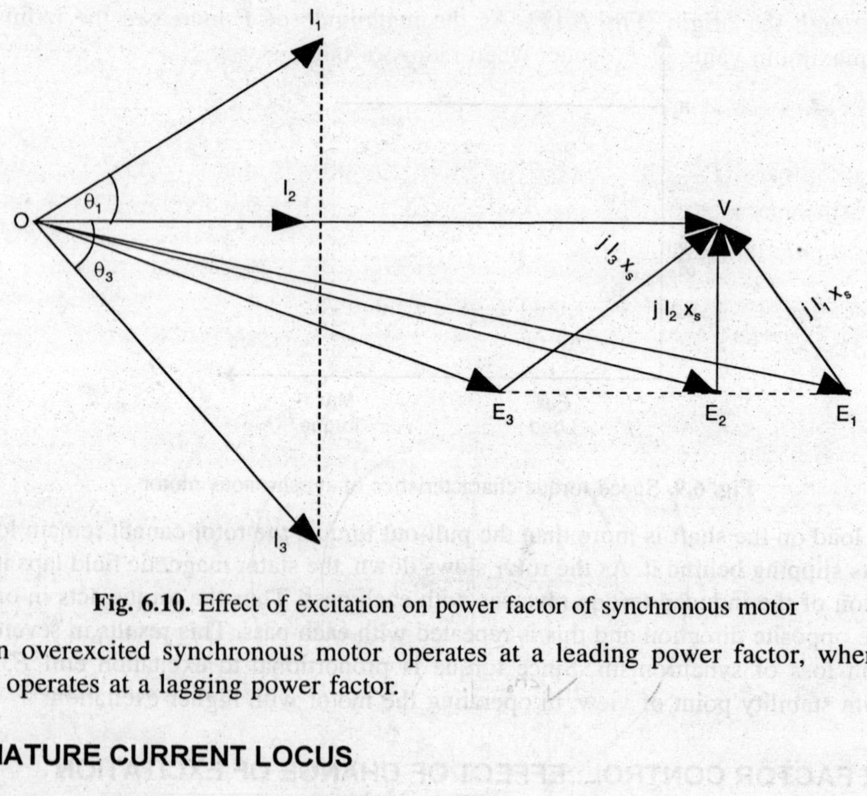


**Fig. 6.10.** Effect of excitation on power factor of synchronous motor

Thus an overexcited synchronous motor operates at a leading power factor, whereas an under excited one operates at a lagging power factor.

## 6.10 ARMATURE CURRENT LOCUS

### 6.10.1 Constant Power

The power input is $VI\cos\theta$ per phase. $I^2R_a$ is the stator copper-loss per phase. Therefore mechanical developed is

$$P = VI\cos\theta - I^2 R_a \quad ...(6.16)$$

Equation (6.16) can be written as

$$I^2 - \frac{V}{R_a} I\cos\theta = -\frac{P}{R_a} \quad ...(6.17)$$

Adding $(V/2R_a)^2$ to both sides of Eq. (6.17)

$$I^2 + \left(\frac{V}{2R_a}\right)^2 - 2\left(\frac{V}{2R_a}\right) I\cos\theta = \left(\frac{V}{2R_a}\right)^2 - \frac{P}{R_a} \quad ...(6.18)$$

The quantity $V/2R_a$ is a current in phase with $V$. Equation (6.18) depicts that if $I$ and $V/2R_a$ are two sides of a triangle with angle $\theta$ between them, the third side is $\left[(V/2R_a)^2 - P/R_a\right]^{0.5}$. If $P$ and $V$ are constant, two sides of the triangle are constant and as $\angle\theta$ changes, the locus of $I$ must be a circle with centre at (0, $V/2R_a$) and $\left[(V/2R_a)^2 - P/R_a\right]^{0.5}$ as radius. When $P = 0$ (i.e. at no-load) the radius is $V/2R_a$ and the circle passes through the origin (Fig. 6.11). As the magnitude of $P$ increases the radius of circle decreases. The maximum value of $P$ occurs when radius of circle is zero, i.e.

$$P_{max} = V^2/4R_a \quad ...(6.19)$$

Under the condition of $P_{max}$, the current $I$ is $V/2R_a$ and power input is $V^2/2R_a$, i.e. power input is twice $P_{max}$ and efficiency is 50%. This maximum power is much above the rating of machine, which is limited by maximum temperature rise.

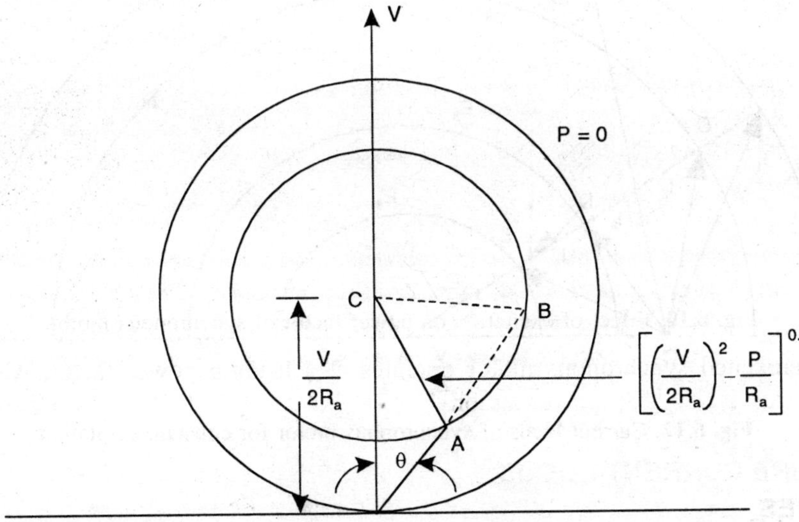


**Fig. 6.11.** Current locus of synchronous motor for constant power

For a given value of *P*, the current *I* is given by Eq. (6.18). For one value of *P* and cos θ, this equation gives two values of *I* as shown in Fig. 6.11 by phasors *OA* and *OB*.

### 6.10.2 Constant Excitation

$$I = \frac{V - E}{Z_s} = \frac{V\angle 0 - E\angle -\delta}{Z_s\angle\phi_s}$$

or

$$I = \frac{V\angle -\phi_s}{Z_s} - \frac{E}{Z_s}\angle -(\delta + \phi_s)$$

or

$$I = \frac{V}{Z_s}\cos\phi_s - \frac{E}{Z_s}\cos(\delta + \phi_s) + j\left[-\frac{V}{Z_s}\sin\phi_s + \frac{E}{Z_s}\sin(\delta + \phi_s)\right]$$

or

$$I^2 = \left(\frac{V}{Z_s}\right)^2 + \left(\frac{E}{Z_s}\right)^2 - 2\frac{V}{Z_s}\times\frac{E}{Z_s}\cos\delta \qquad ...(6.20)$$

Equation (6.20) indicates that if quantities $V/Z_s$ and $E/Z_s$ with $\angle\delta$ between them are two sides of a triangle, the third side is current *I*. $V/Z_s$ is constant. If *E* is also constant, the extremities of $E/Z_s$, and *I* must lie on a circle of radius $E/Z_s$ as shown in Fig. 6.12. As *E* takes all possible values from zero to the maximum possible (permissible by rotor heating) the locus of *I* is a family of concentric circles as shown is Fig. 6.12.

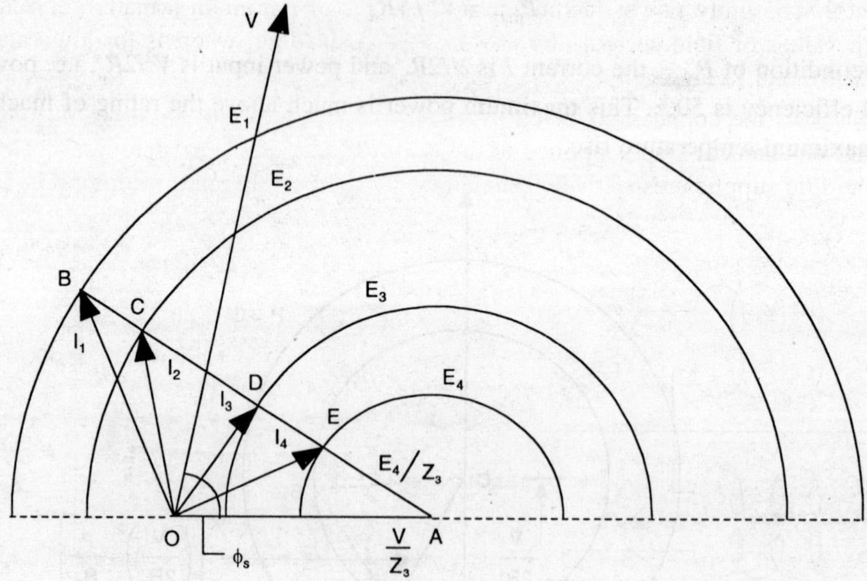


**Fig. 6.12.** Current locus of synchronous motor for constant excitation

## 6.11 V CURVES

A change in field current of a synchronous motor, having a constant mechanical load, affects the power factor and magnitude of current. As seen in Fig. 6.10 a change in excitation voltage from $E_1$ to $E_2$ causes the armature current to decrease from $I_1$ to $I_2$. For an excitation voltage $E_3$, the current is $I_3$. Thus the current is large for under excitation and over excitation and is minimum at some intermediate point. A plot of armature current $I$ for different values of field current $I_f$ resembles a $V$ shape as shown in Fig. 6.13.

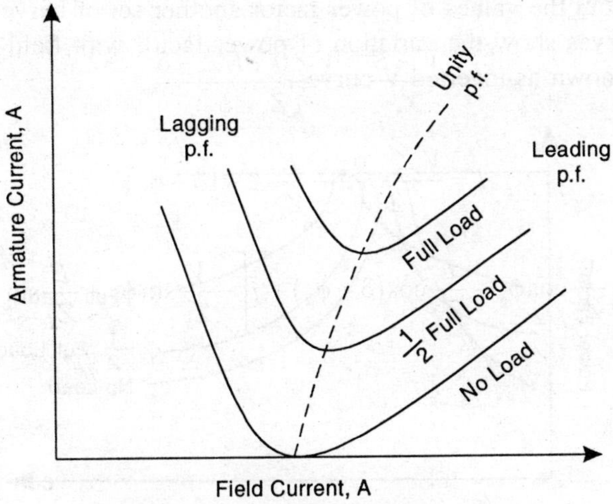


**Fig. 6.13.** V curves of synchronous motor

Since a higher mechanical load calls for a higher armature current, the family of curves for different loads appears as shown in Fig. 6.13. For each value of load, the minimum armature current occurs at unity power factor. The unity power factor line is the locus of minimum armature current points as shown. For high values of field current, the power factor is leading, whereas for low values of field current, the power factor is lagging.

Figure 6.14 shows the connections of an experimental set up to plot V curves of a synchronous motor. A constant mechanical load is applied to the shaft of the motor. The stator and rotor are fed by the 3-phase ac and dc supplies respectively. The ammeters in the stator and rotor circuits indicate the

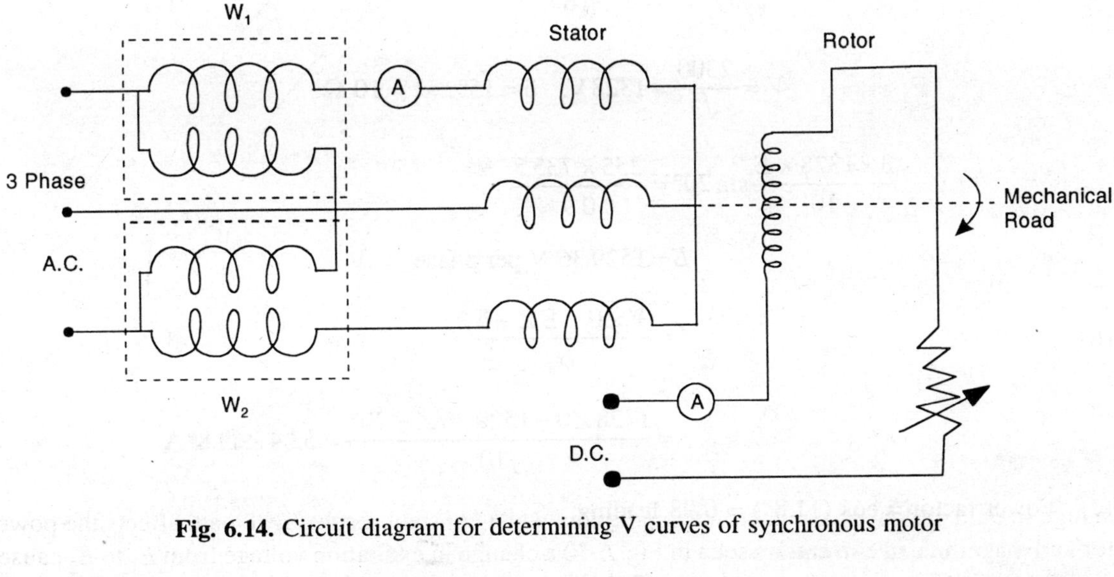


**Fig. 6.14.** Circuit diagram for determining V curves of synchronous motor

armature current and the field current respectively. The additions of two wattmeters in the stator supply circuit enables the calculation of power factor. By applying a constant mechanical load to the shaft and varying the dc field current from under excitation to over excitation, the V curves be plotted. The experiment can be repeated for a number of loads. The power factor angle is given by the equation (B–41) of appendix B. From the values of power factor another set of curves shown in Fig. 6.15 can also be plotted. These curves show the variation of power factor with field current for constant mechanical load. They are known as inverted V curves.

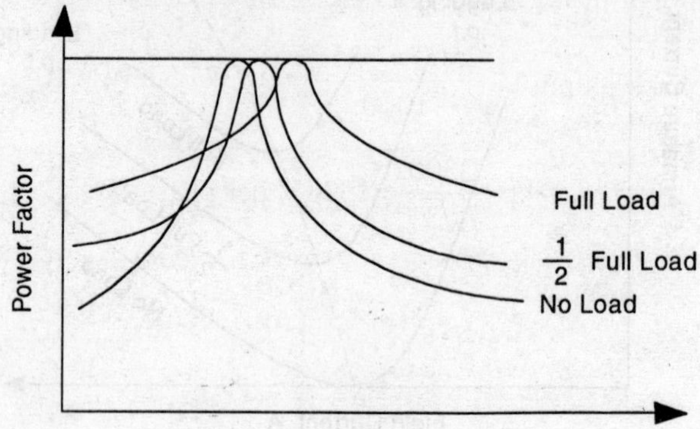


**Fig. 6.15.** Inverted V Curves

### *Example 6.1*

A 2300 V, 3-phase star connected synchronous motor has a synchronous reactance of 10 ohm per phase. When the motor delivers 255 hp (metric) the efficiency is 90% (exclusive of field loss). The power angle is 20°. Calculate (a) *E* per phase, (b) *I*, (c) power factor. Neglect resistance.

### *Solution*

(a)
$$\frac{3VE}{x_s}\sin\delta = \frac{225 \times 735.5}{0.9}$$

$$V = \frac{2300}{\sqrt{3}} = 1328\,\text{V}, \quad \delta = 15°, \; x_s = 10\,\Omega$$

$$\frac{3 \times 1328 \times E}{10}\sin 20° = \frac{255 \times 735.5}{0.9}$$

or
$$E = 1529.36 \text{ V per phase}$$

(b)
$$I = \frac{V\angle 0 - E\angle -\delta}{jx_s}$$

$$= \frac{1328\angle 0 - 1529.36\angle -20°}{j10} = 53.4\angle 11.8°\,\text{A}$$

(c) Power factor = cos (11.8°) = 0.98 leading.

*Example 6.2*

A 2000 hp (metric), 2300 V, 20-pole, 50 Hz synchronous motor has a reactance of 1.85 ohm per phase. The motor is supplied from a constant voltage system at 2300 V. Its field excitation is so adjusted that the power factor is unity at rated load. Calculate the maximum torque which the motor can develop. Neglect losses.

*Solution*

The phasor diagram is shown in Fig. 6.16.

$$\text{Three-phase output} = \frac{2500 \times 735.5}{1000} = 1838.75\,\text{KW}$$

$$V = \frac{2300}{\sqrt{3}} = 1328\,\text{V per phase}$$

$$3VI\cos\theta = 1838.75 \times 10^3$$

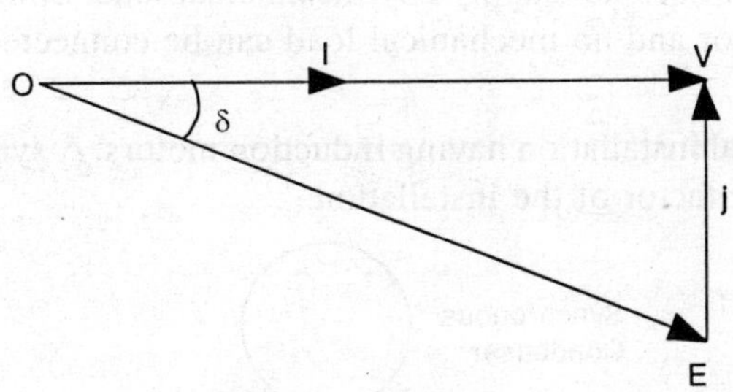


**Fig. 6.16.**

or
$$3(1328)(I)(1) = 1838.75 \times 10^3$$

or
$$I = 442.67\,\text{A}$$

$$Ix_s = 442.67 \times 1.85 = 818.9\,\text{V}$$

As seen from Fig. 6.16, the phasors $V$ and $Ix_s$ are at right angles. Therefore

$$E = \left[V^2 + (Ix_s)^2\right]^{0.5} = \left[1328^2 + 818.9^2\right]^{0.5} = 1560.2\,\text{V}$$

$$\text{Maximum power output} = \frac{3VE}{x_s} = \frac{3 \times 1328 \times 1560.2}{1.85} = 3360 \times 10^3\,\text{W}$$

$$\text{Maximum torque} = 3360 \times 10^3 \text{ synchronous watts}$$

$$n_s = 2f/P = 2 \times 50/20 = 5$$

$$\text{Maximum torque} = \frac{3360}{2\pi(5)} = 106.95 \times 10^3\ N\text{-}m.$$

## 6.12 SYNCHRONOUS CONDENSER

Industrial installations usually operate at lagging power factors. This is mainly due to wide use of induction motors in such installations. The full-load power factor of an induction motor is about 0.8. This is because of a large magnetising current drawn by an induction motor. At part loads, the power factor is still poorer.

Miscellaneous equipments like arc lamps, welding equipment, etc. also contribute to low power factor. Transformers also draw magnetising current and cause the power factor to be low.

The power utilities penalise the industrial installations for low power factor. Therefore, these installations have to install power factor improvement devices at their premises. One such device is a static capacitor. A synchronous motor is another such device.

The power factor of a synchronous motor can be controlled very easily by controlling its field current. When the field current is low, the power factor is lagging. When the field current is high, the power factor is leading.

An over excited synchronous motor, used specifically for power factor correction, is known as synchronous condenser. Usually it does not supply any mechanical load. Sometimes its shaft does not come out of the frame of the motor and no mechanical load can be connected, even if one wanted to do so.

Figure 6.17 shows an industrial installation having induction motors. A synchronous condenser has been added to improve the power factor of the installation.

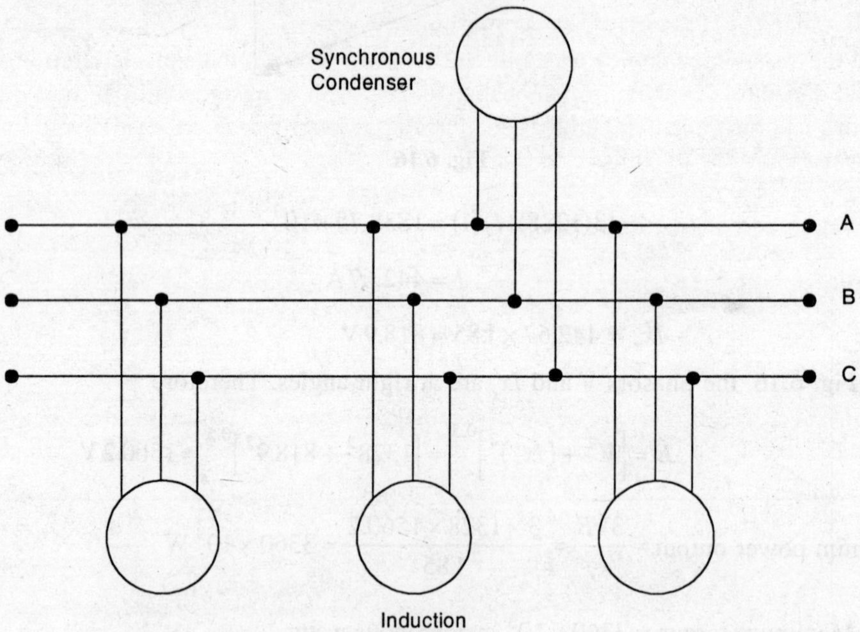


**Fig. 6.17.** Synchronous condenser to improve the power factor of an installation having induction motors

### 6.12.1 Capacity of Synchronous Condenser Running at No-load

Such a synchronous condenser is installed only for the purpose of power factor correction. Since it is not carrying any load, its capacity is expressed in kVAr. The kVAr capacity of such a synchronous condenser is equal to the total lagging kVAr of the installation.

### 6.12.2 Capacity of Synchronous Condenser Supplying Mechanical Load

Sometimes a synchronous motor is installed in some installations to drive constant speed loads. In addition it is operated as over excited motor to improve the power factor of the installation.

Figure 6.18 shows the phasor diagram in this case. The subscripts *I*, *s* and *T* refer to induction motor load, synchronous motor load and total load. Before the installation of synchronous motor, the kVA, active and reactive power are $kVA_I$, $kW_I$ and $kVAr_1$. Addition of synchronous motor with $kVA_s$, $kW_s$ and $kVAr_s$ as the kVA, active and reactive power improves the overall power factor from $\cos \theta_I$ to $\cos \theta_T$. The overall kVA, active and reactive power are $kVA_T$, $kW_T$ and $kVAr_T$. $\cos \theta_s$ is the power factor of synchronous motor.

## 6.13 SYNCHRONOUS REACTOR

The load on a transmission line varies between wide limits. At times of peak load, the line is loaded to nearly its full capacity. During off-peak conditions the load at the receiving end is nearly zero. The transmission line has to transfer power with a voltage drop of less than 10%. The major cause of voltage drop, in the line, is the flow of reactive power. Moreover reactive current causes $I^2R$ losses in the system but produce no revenue.

Most loads absorb positive vars. During peak load conditions the lagging vars demanded by loads greatly exceed the vars which can be transmitted over the line from the consideration of voltage drop. Therefore static capacitors or synchronous capacitors have to be installed at the receiving end of the line. They absorb leading vars and thereby generate lagging vars which are used by the loads operating at lagging power factor.

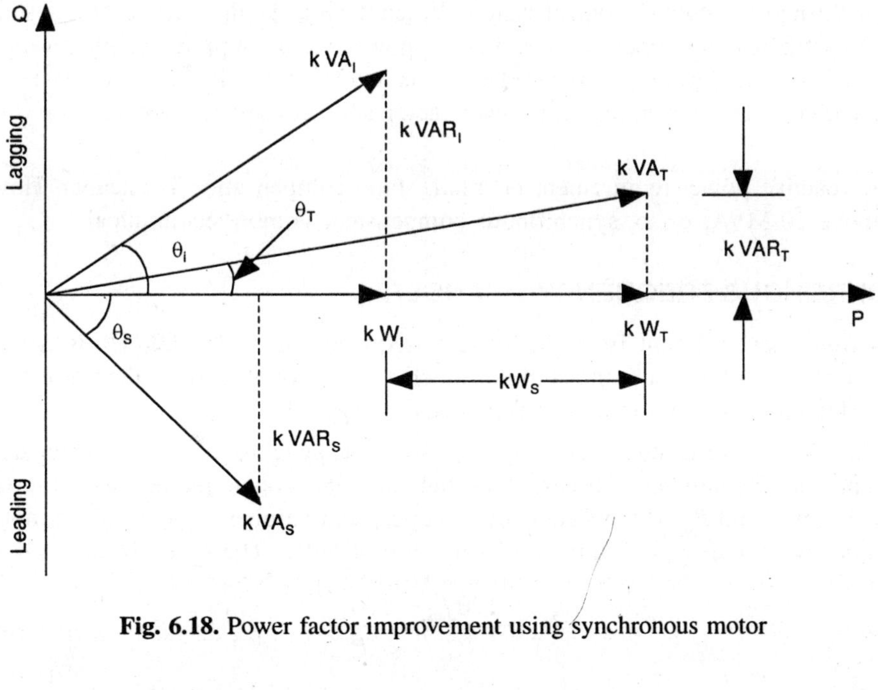


**Fig. 6.18.** Power factor improvement using synchronous motor

The shunt capacitance of the line absorb leading vars (i.e. generate lagging vars). During off-peak conditions the lagging vars produced by the lines are much larger than that required by loads. This causes the voltage at the receiving end to become higher than the sending end voltage (This phenomenon is known as Ferranti effect). It is necessary to connect shunt inductive reactance at the receiving end, during off-peak conditions to prevent the voltage rise due to Ferranti effect. Thus a transmission line needs reactive power compensation to generate or absorb vars. Application of compensation equipment in a system has the following effects:

1. Reduction in reactive component of line current.
2. Maintenance of voltage profile within limit.
3. Reduction of $I^2R$ losses in the line and other equipment due to reduction in current.
4. Reduction in investment in the system per kW of load supplied.
5. Decrease in kVA loading of generators, transformers and other equipments. This decrease in kVA loading may relieve an over load condition or release capacity for load growth.
6. Reduction in kVA demand charges for large consumers.
7. Improvement in power factor of generators.

Reactive compensation equipment consists of shunt capacitors (for peak load conditions) and shunt reactors (for off-peak conditions). Alternatively synchronous compensators can also be used.

A synchronous compensator is essentially a synchronous motor with no mechanical output. The excitation of the machine is varied to provide the necessary amount of compensation. It delivers lagging vars (or absorbs leading vars) when its excitation is large and delivers leading vars (or absorbs lagging vars) when its excitation is small. Thus it is a very flexible source of reactive power because its reactive power can be varied continuously in magnitude and direction (from lagging to leading) by varying the excitation. Thus it serves the purpose of a capacitor as well as a reactor.

When used as a synchronous compensator the dc excitation of the machine is controlled by the system voltage through a feedback control system. When the load is high, the control system adjusts the excitation so that the machine operates at a leading power factor, supplies lagging vars to loads and prevents the voltage from decreasing. When the load is low the control system decreases the excitation so that the machine operates at a lagging power factor and prevents the voltage rise due to Ferranti effect.

When the reactive power requirement is small, static compensation is cheaper. However for requirements above 10 MVAr or so, synchronous compensator is more economical.

## 6.14 SYNCHRONOUS FREQUENCY CHANGER

Sometimes a frequency different from the system frequency of 50 (or 60) Hz is needed. A motor generator set consisting of a synchronous motor mechanically coupled to an alternator can be used for this purpose. This motor generator set is a frequency changer.

The synchronous motor is fed by the ac system. It runs at synchronous speed which depends on the system frequency and its number of poles. Since the alternator is coupled to the motor, it also runs at the same speed. Let $f_m$ and $P_m$ be the frequency of supply and number of poles of the motor and $f_a$ and $P_a$ be the frequency of alternator output and its number of poles. The speed $N_s$ is

$$N_s = \frac{120 f_m}{P_m} = \frac{120 f_a}{P_a}$$

or
$$f_a = f_m\left(\frac{P_a}{P_m}\right) \qquad ...(6.21)$$

We can select the number of poles $P_a$ and $P_m$ so that we get the desired frequency $f_a$.

*Example 6.3*

It is desired to change 50 Hz frequency to 400 Hz. Select the suitable number of poles for the motor and generator of a synchronous frequency changer.

*Solution*

Using Eq. (6.21)

$$400 = 50\left(\frac{P_a}{P_m}\right)$$

or
$$P_a = 8\,P_m$$

Let the motor have 4 poles. The alternator should have 32 poles. Other possible combination is $P_m$ = 2 and $P_a$ = 16 or $P_m$ = 6 and $P_a$ = 48.

*Example 6.4*

An industrial installation having induction motors takes in 2000 kW from a 3-phase 3300 V system at 0.75 p.f. lagging. It is desired to improve the power factor to 0.95 lagging. (a) Find the rating of synchronous condenser. Assume that it is not to supply any active load. (b) If the armature of synchronous condenser is star connected, find the armature current.

*Solution*

(a) kW = 2000

Initial p.f. = 0.75 lagging = $\cos\theta_1$

$$\text{Initial kVA} = \frac{2000}{0.75} = 2666.67 \text{ kVA}$$

Initial kVAr = 2666.67 sin θ = 1763.8 kVAr

Final p.f. = 0.95 lagging = $\cos\theta_2$

$$\text{Final kVA} = \frac{2000}{0.95} = 2105.26 \text{ kVA}$$

Final kVAr = 2105.26 $\sin\theta_2$ = 657.37 kVAr

Rating of synchronous motor = 1763.8 − 657.37 = 1106.43 kVAr

(b) $\sqrt{3}\,VI = 1106.43 \times 10^3$

or
$$I = \frac{1106.43\times 10^3}{\sqrt{3}\times 3300} = 193.58 \text{ A.}$$

*Example 6.5*

A 400 V, 3-phase system feeds a 300 kVA load at 0.7 p.f. lagging. A synchronous motor supplying a 80 b.h.p. mechanical load is connected to improve the power factor. If the synchronous motor is operating at a 0.8 p.f. leading, find total kVA, total active power and overall power factor. Efficiency of synchronous motor is 0.8.

*Solution*

Referring to symbols used in Fig. 6.18

$$\cos\theta_l = 0.7 \text{ lag}, \sin\theta_l = 0.714$$

$$\cos\theta_s = 0.8 \text{ lead}, \sin\theta_s = -0.6$$

$$kW_l = 300 \times 0.7 = 214 \text{ kW}$$

$$kVAr_l = 300 \times 0.714 = 214.2 \text{ kVAr}$$

$$kW_s = \frac{80 \times 0.746}{0.8} = 74.6 \text{ kW}$$

$$kVA_s = \frac{74.6}{0.8} = 93.25 \text{ kVA}$$

$$kVAr_s = (93.25)(-0.6) = -55.95 \text{ k VAr}$$

(As per convention leading kVAr is taken as negative and lagging kVAr is taken as positive)

$$\text{Total kW} = 210 + 74.6 = 284.6 \text{ kW}$$

$$\text{Total kVAr} = 214.2 - 55.95 = 158.25 \text{ k VAr lagging}$$

$$\text{Total kVA} = \left(284.6^2 + 158.25^2\right)^{0.5} = 325.64 \text{ kVA}$$

$$\text{Overall p.f.} = \frac{284.6}{325.64} = 0.874 \text{ lagging}$$

## 6.15 STARTING OF SYNCHRONOUS MOTOR

Since a synchronous motor does not have starting torque, a special method is needed to start it and bring it to the synchronous speed. Once it is brought to the synchronous speed, it will continue running. Two methods used for starting are:

(a) Auxiliary motor starting

(b) Induction motor starting.

### 6.15.1 Auxiliary Motor Starting

The function of the auxiliary motor is to bring the synchronous motor to synchronous speed or to a speed near this speed. The auxiliary motor has to be mechanically coupled to the synchronous motor. No load is put on the synchronous motor during starting. Therefore the auxiliary motor has to overcome only the inertia of the synchronous motor and may have a rating much smaller than the rating of synchronous motor.

The auxiliary motor may be a dc motor or an induction motor. Many large synchronous motors have brushless excitation systems mounted on their shafts and it is often possible to use these exciters as starting motors.

The induction motor used for starting may have the same number of poles as synchronous motor or two poles less than the synchronous motor.

If the induction motor has the same number of poles, it brings the synchronous motor to a speed close to the synchronous speed. When this is achieved, the 3-phase supply is switched on to the armature and dc supply to the field circuit of synchronous motor. The synchronous motor pulls into step, its speed rises to synchronous speed and it continues to run at this speed.

If the induction motor has two poles less than synchronous motor, the induction motor speed is higher than the synchronous speed of the synchronous motor. Once this higher speed is attained, the induction motor is switched off and 3-phase supply is switched on to armature of synchronous motor. When the speed of synchronous motor becomes almost equal to synchronous speed, its field winding is energized. The synchronous motor pulls into step and starts running at synchronous speed.

### 6.15.2 Induction Motor Starting

Synchronous motor are generally provided with an amortisseur winding consisting of copper or aluminium or bronze bars embedded in slots in pole faces and short circuited at both ends by conducting rings. When 3-phase supply is switched on to stator, the rotating field induces an emf in these bars. This emf causes a current to flow in the bars. The reaction of this current on the stator field produces a torque which accelerates the rotor in the direction of rotating field.

The amortisseur winding cannot accelerate the motor to synchronous speed because as synchronism is approached, the torque (due to induction motor action) decreases. During the period the motor is accelerating, the field poles (not yet excited) are slipping backward, more and more, through the synchronously rotating armature field. They tend to get magnetised first and one polarity, then with the other, so that if the field circuit is energized as synchronism is approached, the stator and rotor will lock together, provided the impressed polarity of field is of proper sign at the instant the field switch is closed. In case the impressed polarity happens to lie opposite of what it should be, the machine suddenly experiences generator action and its field system slips back. When the field system has slipped back by one pole pitch, the armature and field are in locking position and the machine pulls into step. It is necessary that the field system is energized only when the speed is near synchronous speed.

During the starting period the field winding is also cut by the revolving stator field and a dangerously high voltage may be induced in the field winding. Some of the methods to limit this high voltage include: (1) sectionalising the field winding into sections, so that the voltages induced are not cumulative, (2) short circuiting the field winding through an external resistor (about 6 times the resistance of field winding) during the starting period and reducing this resistance in steps as motor speeds up, and (3) starting at reduced ac voltage.

## 6.16 HUNTING

The problem of hunting in a synchronous motor is similar to that in a synchronous generator. A change of load calls for a change in power angle. The power angle adjusts to the new value corresponding to the new load. However the inertia of the rotor and its connected load prevents an instantaneous adjustment of the power angle. When the load is increased, the rotor slows down and power angle increases. The inertia causes the rotor to fall back to a position where it develops more torque than that demanded by the load. Because of this the rotor speeds up attempting to attain the proper power angle. It may however move past this angle because of large torque impulse. A similar situation may happen when the load is reduced. These oscillations of the rotor, about mean position, are known as hunting and may cause the motor to lose synchronism and come to a stop.

The equations for electro-mechanical oscillations are the same as that for generator (Sec. 5.29) except that direction of power flow is reversed (i.e. input is electrical power and output is mechanical power). In Eq. (5.65) if both $T_e$ and $T_i$ are taken as negative, this equation is valid for synchronous motor also.

The amortisseur winding inserted in the pole faces for the purposes of starting has the added advantage of suppressing hunting. During the period of hunting a torque due to induction motor action is developed due to this winding. This torque opposes the synchronous motor torque and damps out the oscillations.

## 6.17 APPLICATIONS

The main advantage of a synchronous motor is that its power factor can be controlled by the control of field current. In certain application the fact that its speed is independent of load and applied voltage, may be important. Its efficiency is slightly higher than that of induction motor. For low speeds and high rating (above 500 kW) a synchronous speed is cheaper than induction motor.

The main disadvantages are: the need of dc source for excitation (usually an exciter is mounted on the shaft to provide the excitation), starting and synchronising, instability and hunting and expensive control devices.

If view of the above advantages and disadvantages, synchronous motors are used only for low speed, high hp requirements. The applications include large low-head pumps, rubber mills and mixers, crushers, paper mill drives, compressers, rolling mills, ball mills, etc. Its use as a synchronous phase modifier for voltage regulation of transmission lines is another important application.

## 6.18 SYNCHRONOUS INDUCTION MOTOR

A synchronous induction motor has 3-phase ac winding on stator and a wound rotor (similar to that of a wound rotor 3-phase induction motor). The rotor windings are connected to slip rings (Fig. 6.19). During starting external resistances are connected across the slip rings to give high starting torque. As the rotor speeds up, the external resistance is reduced to zero. When the speed is near synchronous speed, the connections of the sliprings are switched over, so as to connect dc supply to the rotor winding. This dc supply is obtained from an exciter mounted on the motor shaft (Fig. 6.19). During running, one phase winding of rotor carries full excitation current and other two windings carry half the excitation current each, because they are connected in parallel.

Figure 6.20 shows another type of synchronous induction motor. It has two separate windings on rotor. One is the dc field winding and the other is 3-phase damper winding for starting. The 3-phase damper rotor winding is housed in the pole faces. Both the rotor winding are brought out to sliprings as shown. Thus the motor has five sliprings. The sliprings connected to dc field winding are fed by dc supply for field current. The sliprings connected to damper winding are connected to variable resistances.

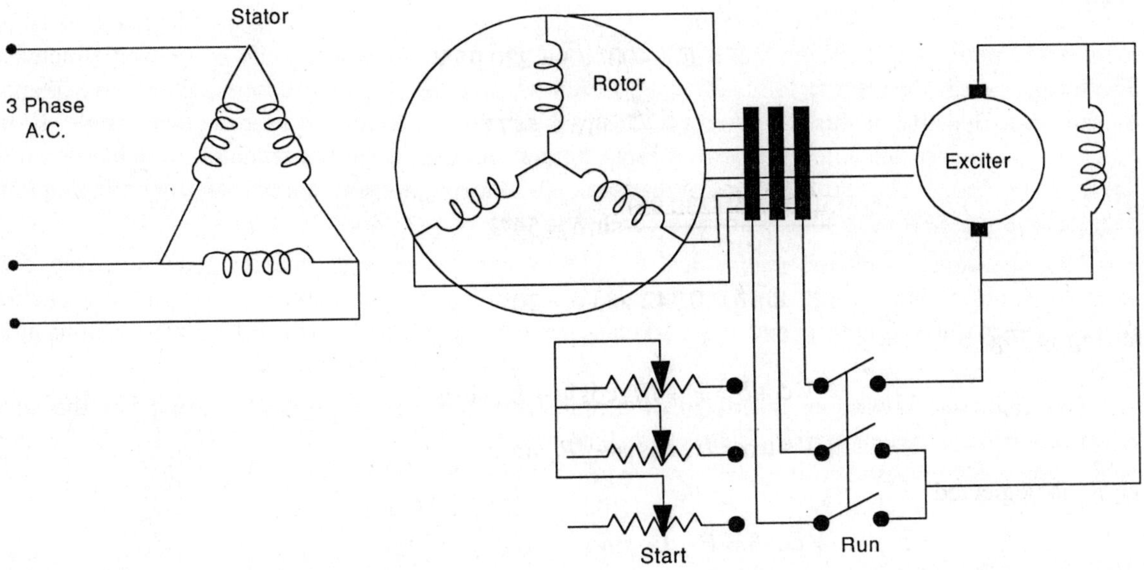


**Fig. 6.19.** Synchronous induction motor

At the time of starting these resistances are adjusted to full value to provide high starting torque to motor. As the motor speeds up these resistances are cut out. When the motor attains a speed nearly equal to synchronous speed, the dc field supply is switched on and the motor pulls into synchronism. When the motor is running at synchronous speed, the emf induced and hence current in damper windings is zero.

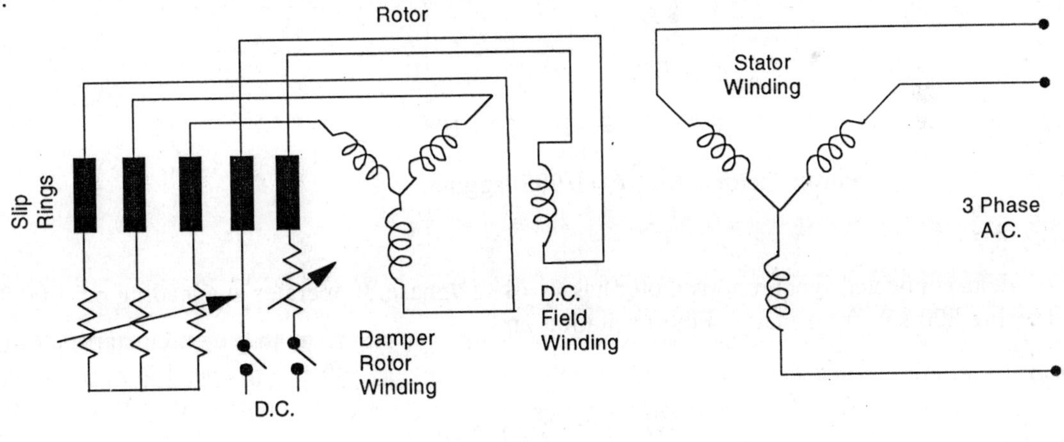


**Fig. 6.20.** Synchronous induction motor

*Example 6.6*

A 3-phase star connected induction motor has a power input of 5472 W at rated voltage, $x_s$ = 10 ohm. If excitation voltage is adjusted equal to the rated voltage of 400 V, find power angle, armature current and power factor. Neglect resistance.

*Solution*

$$V = E = 400/\sqrt{3} = 230.95\,\text{V}$$

$$\frac{3VE}{x_s}\sin\delta = 5472$$

or
$$\frac{3(230.95)(230.95)}{10}\sin\delta = 5472$$

or
$$\sin\delta = 0.342 \text{ and } \delta = 20°$$

Referring to Fig. 6.5

$$V\cos\delta = E + IR_a\cos\psi + Ix_s\sin\psi$$

$$V\sin\delta = Ix_s\cos\psi - IR_a\sin\psi$$

Since $R_a$ is neglected

$$V\cos\delta = E + Ix_s\sin\psi$$

$$V\sin\delta = Ix_s\cos\psi$$

Substituting the values

$$230.95\cos 20° = 230.95 + 10\,I\sin\psi$$
$$230.95\sin 20° = 10\,I\cos\psi$$

From above equations

$$I = 8\,\text{A}$$

$$\psi = -10°$$

$$\theta = \psi + \delta = -10 + 20 = 10°$$

$$\text{Power factor} = \cos 10° = 0.985 \text{ lagging.}$$

*Example 6.7*

A 3300 V delta connected synchronous motor has $x_s$ = 18 Ω/phase. It operates at a leading p.f. of 0.707 when drawing 800 kW from mains. Find excitation emf.

*Solution*

$$3VI\cos\theta = \text{Total input}$$

or
$$3 \times 3300 \times I \times 0.707 = 800 \times 10^3$$

or
$$I = 114.3\,\text{A}$$

Since p.f. is 0.707 leading

$$I = 114.3(0.707 + j0.707)$$

$$V = E + I(jx_s)$$

$$3300\angle 0 = E\angle -\delta + 114.3(0.707 + j0.707)(j18)$$

or $$3300\angle 0 = E\angle(-\delta) - 1454.5 + j1454.5$$

or $$3300\angle 0 = (E\cos\delta - 1454.5) + j(1454.5 - E\sin\delta)$$

or $$E\cos\delta = 4754.5$$

$$E\sin\delta = 1454.5$$

whence $$\delta = 17° \text{ and } E = 4972 \text{ V/phase}.$$

### *Example 6.8*

A 2000 hp (metric), unity power factor, 3-phase star connected, 2300 V, 50 Hz synchronous motor has $x_{sm}$ = 1.95 Ω/phase. It is supplied from a 2300 V, 1750 kVA, star connected generator with $x_{sg}$ = 2.6 ohm/phase. For rated load on motor, find (a) back emf of motor, (b) induced emf of generator, (c) maximum power which can flow from generator to motor. The power factor is 1.

### *Solution*

The equivalent circuit is shown in Fig. 6.21 (a) and phasor diagram is as shown in Fig. 6.21 (b).

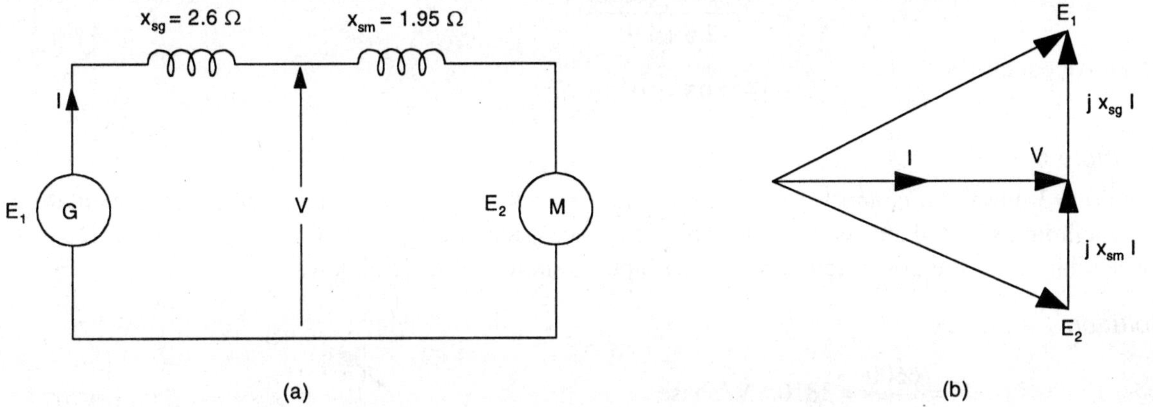


**Fig. 6.21.**

$$\text{Motor Input} = 2000 \times 735.5 = 1471 \times 10^3 \text{ W}$$

$$3VI\cos\theta = \text{Motor Input}$$

or $$3 \times \frac{2300}{\sqrt{3}} \times I \times 1 = 1471 \times 10^3$$

$$I = 369.2 \text{ A}$$

Let $E_1$ and $E_2$ be the induced emf of generator and back emf of motor. Synchronous reactance of motor = $x_{sm}$ = 1.95 ohm.

Synchronous reactance of generator = $x_{sg}$ = 2.6 Ω

(a) As seen from the phasor diagram of Fig. 6.21(b)

$$E_2^2 = V^2 + (x_{sm} I)^2$$

$$= \left(\frac{2300}{\sqrt{3}}\right)^2 + (1.95 \times 369.2)^2$$

or Back emf of motor = $E_2$ = 1510.6 volts/phase

(b) As seen from the phasor diagram (Fig. 6.21b)

$$E_1^2 = V^2 + (x_{sg} I)^2$$

$$= \left(\frac{2300}{\sqrt{3}}\right)^2 + (2.6 \times 369.2)^2$$

or Induced emf of generator = $E_1$ = 1638.5 V/phase

(c) Maximum power transfer = $\dfrac{3E_1E_2}{x_{sg} + x_{sm}}$

$$= \frac{3(1638.5)(1510.6)}{2.6 + 1.95}$$

$$= 1631.95 \times 10^3 \text{ W.}$$

*Example 6.9*

A 6600 V, 3-phase star connected synchronous motor works at constant voltage and constant excitation. Its synchronous impedance is 2 + *j* 20 ohm per phase. When the input is 1000 kW, the power factor is 0.8 leading. Find the power factor when the input is increased to 1500 kW.

*Solution*

$$V = \frac{6600}{\sqrt{3}} = 3810.6 \text{ V/phase}$$

$$3VI_1 \cos\theta = \text{Input}$$

or $3(3810.6)(I_1)(0.8) = 1000 \times 10^3$

or $I_1 = 109.34$ A

Phasor diagram is shown in Fig. 6.22.

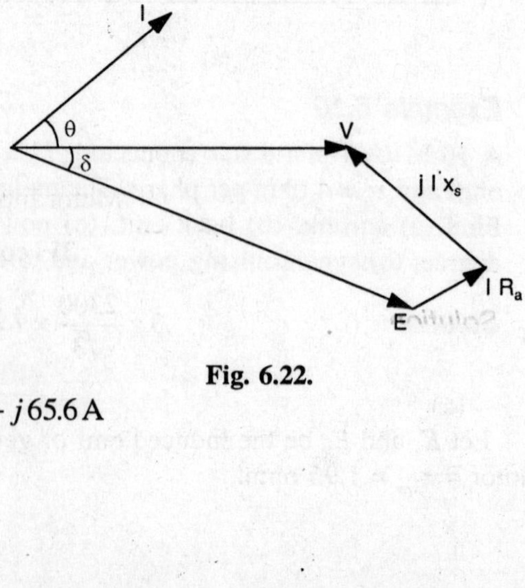


Fig. 6.22.

Since p.f. is 0.8 leading

$$I_1 = 109.34(0.8 + j0.6) = 87.47 + j65.6 \text{ A}$$

$$V\angle 0 = E\angle -\delta_1 + I_1(R_a + jx_s)$$

or $$3810.6 = E\cos\delta_1 - jE\sin\delta_1 + (87.47 + j65.6)\,(2 + j20)$$

$$= (E\cos\delta_1 - 1137.06) - j(E\sin\delta_1 - 1880.6)$$

or $$E\sin\delta_1 = 1880.6$$

and $$E\cos\delta_1 = 4947.66$$

$$\delta_1 = 20.81° \text{ and } E = 5293.4\text{ V}$$

When input is 1500, let power angle be $\delta_2$, current $I_2$ and power factor cos $\theta_2$.

$$Z_s = 2 + j20 = 20.1\angle 84.29°$$

$$I_2 = \frac{V - E\angle - \delta_2}{Z_s} = \frac{3810.6}{20.1\angle 84.29°} - \frac{5293.4\angle - \delta_2}{20.1\angle 84.29°}$$

$$= 189.58\angle - 84.29° - 263.35\angle(-\delta_2 - 84.29°)$$

$$I_2\cos\theta_2 = 189.59\cos(-84.29°) - 263.35\cos(\delta_2 + 84.29°) \quad \text{...(i)}$$

Also $$3VI_2\cos\theta_2 = 1500 \times 10^3$$

or $$I_2\cos\theta_2 = \frac{1500 \times 10^3}{3 \times 3810.6} = 131.2 \quad \text{...(ii)}$$

From Eqs. (i) and (ii), $\delta_2 = 30.96\ °$

$$V\angle 0 = E\angle - \delta_2 + (I_2\cos\theta_2 + jI_2\sin\theta_2)(2 + j20)$$

or $$3810.6 = 5293.4\angle - 30.96° + (131.2 + jI_2\sin\theta_2)(2 + j20) \quad \text{...(iii)}$$

From Eq. (iii)

$$5293.4\sin(30.96°) = 2624 + 2I_2\sin\theta_2$$

or $$I_2\sin\theta_2 = 49.565$$

$$\tan\theta_2 = \frac{I_2\sin\theta_2}{I_2\cos\theta_2} = \frac{49.565}{131.2} = 0.378$$

$$\text{Power factor} = \cos\theta_2 = 0.935 \text{ leading.}$$

*Example 6.10*

A 10 MVA 3-phase star connected, 11 kV, 16-pole, 50 hz, salient pole synchronous motor has $x_d = 6$ ohm and $x_q = 4$ ohm per phase. The motor is working at full-load unity power factor and rated voltage. Find (a) current, (b) back emf, (c) power angle, (d) synchronising power and torque per electrical degree, (e) synchronising power and torque per mechanical degree.

*Solution*

(a) $$V = \frac{11000}{\sqrt{3}} = 6351.04\text{ V/phase}$$

$$3VI = 10 \times 10^6$$

or $$I = \frac{10 \times 10^6}{3 \times 6351.04} = 524.85\text{ A}$$

Since p.f. is unity, $\cos\theta = 1, \sin\theta = 0.$

$$n_s = \frac{2 \times 50}{16} = \frac{25}{4}\text{ r.p.s}$$

(b) The phasor diagram is shown in Fig. 6.23. From the phasor diagram

$$\tan\delta = \frac{I_q x_q}{V\cos\delta} = \frac{I\cos\delta x_q}{V\cos\delta} = \frac{Ix_q}{V}$$

$$= \frac{524.85 \times 4}{6351.04} = 0.33$$

$$\delta = 18.26° = \psi$$

$$I_d = I\sin\psi = 524.85\sin 18.26°$$

or $$I_d = 164.45\ A$$

$$|E| - |I_d x_d| = |V|$$

or $$|E| = 6351.04 + 164.45 \times 6$$

$$= 7337.74 \text{ volts per phase}$$

(c) $$\delta = 18.26°$$

(d) $$P = \frac{3VE}{x_d}\sin\delta + \frac{3V^2(x_d - x_q)}{2x_d x_q}\sin 2\delta$$

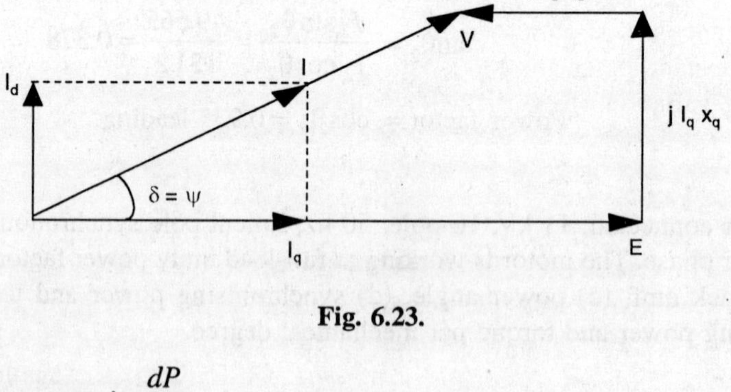


**Fig. 6.23.**

Synchronizing power is $\frac{dP}{d\delta}$

$$\frac{dP}{d\delta} = \frac{3VE}{x_d}\cos\delta + \frac{3V^2(x_d - x_q)}{x_d x_q}\cos 2\delta$$

$$= \frac{3(6351.04)(7337.74)}{6}\cos(18.26°) + \frac{3(6351.04)^2(6-4)}{6\times 4}\cos(2\times 18.26°)$$

$$= (22127.8 + 8103.9)10^3 \text{ W} = 30231.7 \text{ kW per radian}$$

Synchronizing power per electrical degree

$$= \frac{30231.7}{180/\pi} = 527.6 \text{ kW per elect. degree}$$

$$\text{Synchronizing torque} = \frac{\text{Synchronizing power}}{2\pi n_s}$$

$$= \frac{527.6\times 10^3}{2\pi(25/4)} = 13435.2 \text{ N-m/elect. degree}$$

(e) Since it is a 16-pole machine, 1 mech. degree $= \frac{16}{2} = 8$ elect. degree

Synchronizing power per mech. degree = 526.6 × 8

= 4212.8 kW per mech. degree

Synchronizing torque per mech. degree = $13435.2 \times 8$

= 107481.6 N-m/per mech. degree.

*Example 6.11*

A 2 MW, 6.6 kV, 3-phase, 50 Hz, 16-pole star connected cylindrical rotor synchronous motor has the following OC characteristics:

| $I_f$ | 10 | 15 | 20 | 25 | 30 | 40 |
|---|---|---|---|---|---|---|
| OC voltage per phase | 1300 | 2500 | 3750 | 4590 | 5280 | 5850 |

The armature resistance is one ohm. A current of 1500 A is produced on short circuit with a field current of 20 A. The constant losses are 200 kW. It is operating on rated load, voltage and frequency. Plot its *V* curve and the variation of power factor with field current.

*Solution*

When

$$I_f = 20 \text{ A}, E = 3750 \text{ V/phase}$$

$$Z_s = \frac{3750}{1500} = 2.5 \text{ ohm}$$

$$\phi_s = \cos^{-1} 1/2.5 = 66.4°$$

$$\text{Terminal voltage } V \text{ per phase} = \frac{6600}{\sqrt{3}} = 3810.6 \text{ V}$$

$$\frac{V}{2R_a} = \frac{3810.6}{2} = 1905.3 \text{ A}$$

$$\frac{V}{Z_s} = \frac{3810.6}{2.5} = 1524.2 \text{ A}$$

$$P = \frac{2000 + 200}{3} = 733.33 \text{ kW/phase}$$

$$\left[\left(\frac{V}{2R_a}\right)^2 - \frac{P}{R_a}\right]^{0.5} = \left[(1905.3)^2 - \frac{733.33 \times 10^3}{1}\right]^{0.5}$$

$$= 1702 \text{ A}$$

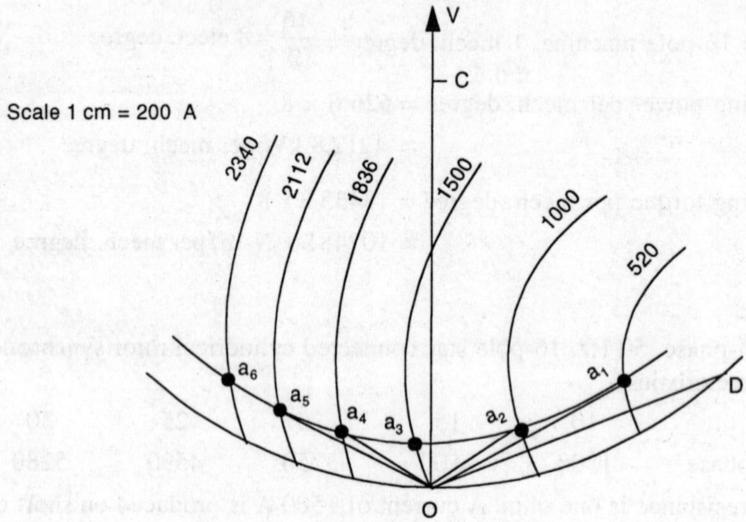


**Fig. 6.24.**

With $C$ as centre and $CO$ = 1905.3/200 as radius draw circle. $OC$ is the vertical axis. With $C$ as centre and 1702/200 cm as radius draw another circle. This is the circle for the rated power output. $OD$ = 1524.2/200 cm and is inclined at 66.4° to OC (Fig. 6.24).

| $I_f$ | 10 | 15 | 20 | 25 | 30 | 40 | A |
|---|---|---|---|---|---|---|---|
| $E$ | 1300 | 2500 | 3750 | 4590 | 5280 | 5850 | V |
| $E/Z_s$ | 520 | 1000 | 1500 | 1836 | 2112 | 2340 | A |

With $D$ as centre draw arcs for above values of $E/Z_s$ (to the scale of 1 cm = 200 A). The points of intersection of these arcs and the power circle give the armature current for different values of $E$ (and therefore $I_f$). $Oa_1$ is the armature current for a field current of 10 A. The power factor angle is $\angle COa_1$. Similarly, measure other values of armature current and power factor angle.

Figure 6.25 shows the *V* curve and variation of power factor with field current. The power factor is lagging for low values of excitation (i.e. under excitation) and leading for high values of excitation (i.e. over excitation).

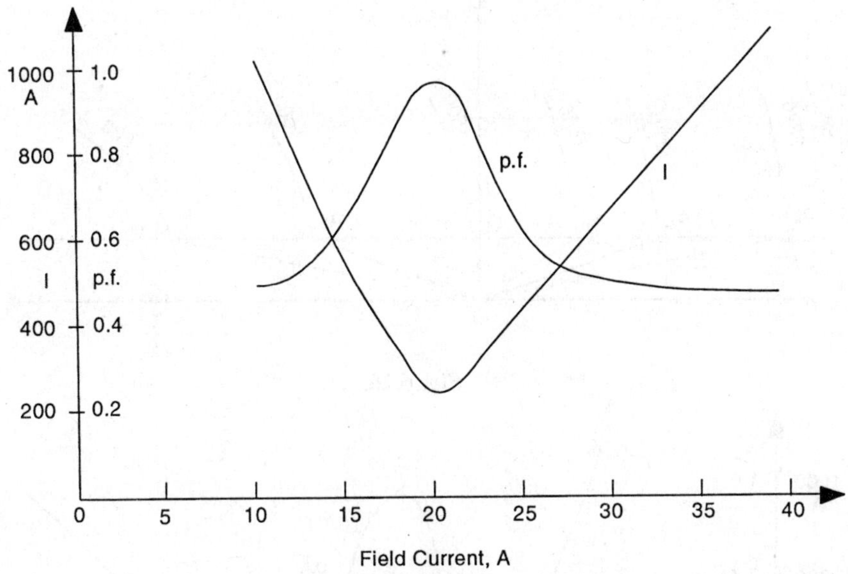


**Fig. 6.25**

*Example 6.12*

Repeat example 6.11, if the armature resistance is negligible.

*Solution*

Since $R_a$ is neglected

$$x_s = Z_s = 2.5\,\text{ohm}$$

Since $\dfrac{V}{2R_a}$ is infinite, the power circles become straight lines.

$$P = 733.33 \text{ kW/phase}$$

$$P = \text{VI} \cos \theta$$

or $$733.33 \times 10^3 = 3810.6 \times I \cos \theta$$

or $$I \cos \theta = 192.4 \text{ A}$$

Scale 1 cm = 200 A

As shown in Fig. 6.26, *OV* is the voltage axis. Take *OG* = 192.4 A and draw a straight line parallel to *x*-axis and passing through *G*. This line represents the full-load output. Take $OD = V/x_s = 1524.24$ A along *x*-axis (i.e. at 90 ° to voltage axis). With *D* as center draw arcs corresponding to different values of $E/x_s$. The points of intersections of these arcs with the straight line through *G* are the various operating points $a_1$, $a_2$, etc., ....$Oa_1$, $Oa_2$, etc. are the armature currents for different values of $E/x_s$ (and hence $I_f$). The angles which the lines $Oa_1$, $Oa_2$, ....., make with voltage axis are the power factor angles. The $V$ curve and variation of power factor are shown in Fig. 6.27.

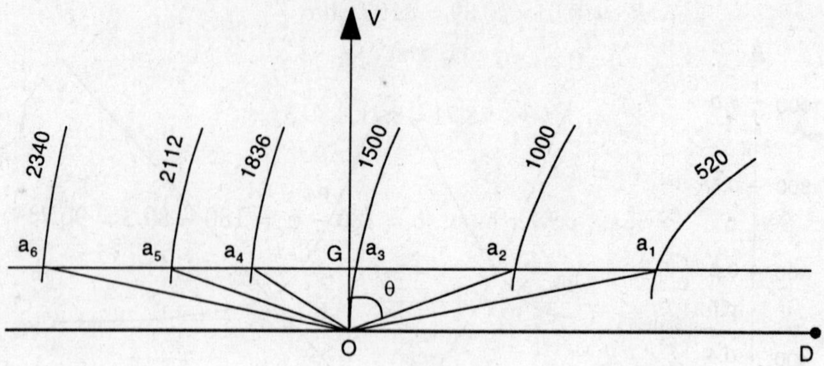


Fig. 6.26.

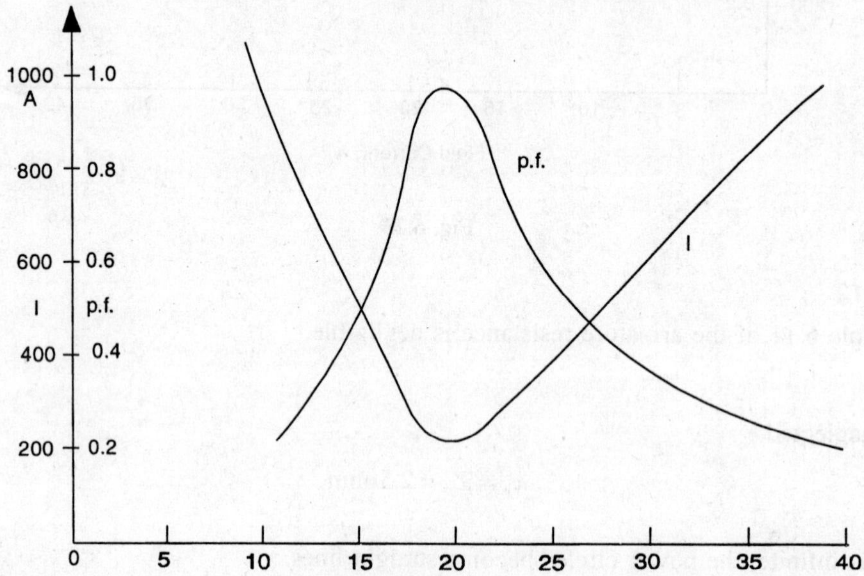


Fig. 6.27.

*Example 6.13*

A 1 MVA, 3.3 kV, 50 Hz, 6-pole 3-phase round rotor synchronous motor has a p.u. resistance of 0.01 and pu synchronous reactance of 0.8. If $E$ = 1.4 pu, find the value of maximum power input and corresponding armature current and power factor.

*Solution*

$$\text{Base current} = \frac{1 \times 10^6}{\sqrt{3} \times 3300} = 174.96\ \text{A}$$

$$\text{Base impedance} = \frac{3300}{\sqrt{3} \times 174.96} = 10.89 \text{ ohm}$$

$$R_a = 0.01 \times 10.89 = 0.109 \text{ ohm}$$

$$x_s = 0.8 \times 10.89 = 8.71 \text{ ohm}$$

$$Z_s = 0.109 + j8.71 = 8.71\angle 89.3°$$

$$\phi_s = 89.3°$$

As derived in Sec. 6.6, for max. power input, $\delta = 180 - \phi_s = 180 - 89.3 = 90.7°$

$$\text{Max. } P_{input} = \frac{3V^2}{Z_s}\cos\phi_s + \frac{3EV}{Z_s}$$

$$V = \frac{3300}{\sqrt{3}} = 1905.3 \text{ V per phase}$$

$$E = \frac{1.4 \times 3300}{\sqrt{3}} = 2667.4 \text{ V per phase}$$

$$\text{Max. } P_{input} = \frac{3 \times 1905.3^2}{8.71}\cos 89.3° + \frac{3 \times 1905.3 \times 2667.4}{8.71} \text{ W}$$

$$= 1.765 \text{ MW}$$

$$3VI\cos\theta = 1.765 \times 10^6$$

or

$$I\cos\theta = \frac{1.765 \times 10^6}{3 \times 1905.3} = 308.84 \text{ A}$$

$$V = E\angle -\delta + (I\cos\theta + jI\sin\theta)(R_a + jx_s)$$

or

$$1905.3 = 2667.4\angle -90.7° + (308.84 + jI\sin\theta)(0.109 + j8.71)$$

Which gives

$$I\sin\theta = -215.3$$

$$\text{Therefore armature current} = 308.84 - j215.3 = 376.48\angle -34.88° \text{ A}$$

$$\text{Power factor} = 0.82 \text{ lagging.}$$

*Example 6.14*

The motor of example 6.13 is drawing rated current at rated voltage. $E = 1.4$ pu. Find the gross torque developed by the motor.

*Solution*

$$I = 174.96 \text{ A}$$

$$IZ_s = 174.96 \times 8.71 = 1523.9 \text{ V}$$

$$V = 1905.3 \text{ V per phase}$$

$$E = 2667.4 \text{ V per phase}$$

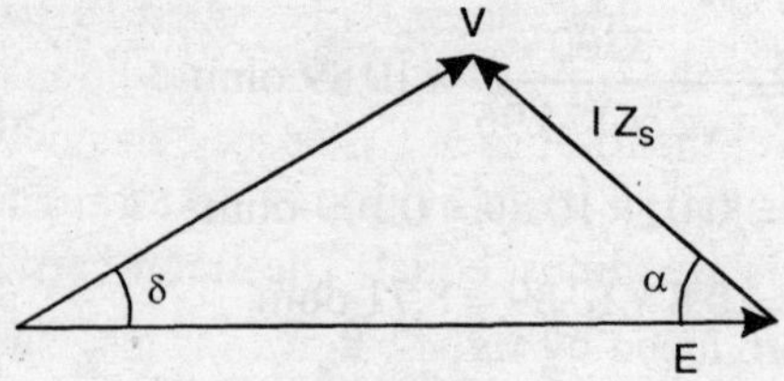


**Fig. 6.28.**

The phasor diagram is shown in Fig. 6.28. Let the phase angle between phasors $E$ and $I\,Z_s$ be $\alpha$. From the phasor diagram.

$$V^2 = E^2 + (IZ_s)^2 - 2EIZ_s\cos\alpha$$

or
$$(1905.3)^2 = (2667.4)^2 + (1523.9)^2 - 2(2667.4)(1523.9)\cos\alpha$$

which gives
$$\cos\alpha = 0.714 \text{ and } \alpha = 44.4°$$

Moreover
$$V\sin\delta = I\,Z_s\sin\alpha$$

or
$$\sin\delta = \frac{1523.9\sin 44.4°}{1905.3} = 0.56$$

and
$$\delta = 34.02°$$

The power output $P$ is

$$P = \frac{3EV}{Z_s}\cos(\phi_s - \delta) - \frac{3E^2}{Z_s}\cos\phi_s$$

$$= \frac{3(2667.4)(1905.3)}{8.71}\cos(89.3 - 34.02) - \frac{3(2667.4)^2}{8.71}\cos(89.3°)$$

$$= 967 \times 10^3 \text{ W}$$

$$\text{Gross torque} = 967 \times 10^3 \text{ synch. watts}$$

$$n_s = \text{synchronous speed} = \frac{2 \times 50}{6} - \frac{50}{3}\text{rps}$$

$$\text{Gross torque} = \frac{967 \times 10^3}{2\pi(50/3)} = 9234 \text{ N-m.}$$

## 6.19 PERMANENT MAGNET SYNCHRONOUS MOTOR

Synchronous motor has very high efficiency and runs at absolutely constant speed. If the ac input to the stator is through a solid state rectifier-inverter combination, the input frequency can be changed and thus the speed of the motor can be adjusted at a desired value.

The main disadvantage of a synchronous motor is that it requires dc supply for exciting the field winding. During the past two decades samarium-cobalt and neodymium-iron-cobalt magnetic materials have been developed. These materials have very high retentivity and are, thus, used for making permanent magnets. The use of these materials for rotor poles gives a permanent magnet synchronous motor.

Fig. 6.29 shows the cross-section of such a motor. It is called surface magnet synchronous motor because the rotor magnets are glued to the surface of the rotor. Additional stainless steel (non-magnetic) cylindrical shell may be used to cover the rotor so as to keep the magnets in position against centrifugal force in high speed motors. The air gap in these motors in uniform and large. Consequently the synchronous inductance in *d* and *q* axis are almost equal. The stator carries 3 phase distributed winding..

The air gap flux distribution produced by rotor magnets is also sinusoidal due to special shaping of rotor magnets. Because of large air gap the inductance is low and armature reaction is also low. Since the excitation is fixed (due to permanent rotor magnets) there is no scope for control of power factor.

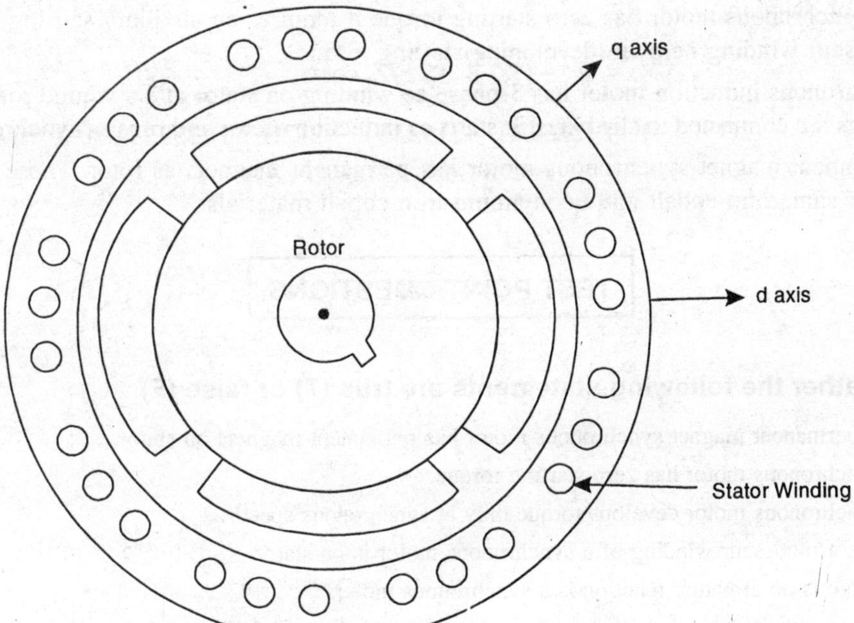


**Fig. 6.29.** Cross-section of surface magnet synchronous motor

## SUMMARY

1. A synchronous motor converts 3-phase ac energy into mechanical energy. It develops torque only at synchronous speed.
2. The stator of a synchronous motor has 3-phase distributed winding. The rotor is, generally, salient pole and has field winding and amortisseur winding. The amortisseur winding helps in developing a starting torque.
3. The exact phasor diagram and analysis requires the use of two reaction theory. However the analysis assuming the motor to have cylindrical rotor is simple and generally fairly accurate.
4. Torque $= \dfrac{P_{\text{output}}}{2\pi n_s}$ Newton metres
5. An overexcited synchronous motor operates at a leading power factor whereas an underexcited

one operates at a lagging power factor. Thus a synchronous motor can be used for power factor control.

6. V curves of synchronous motor shows the variation of armature current in the field current for different load conditions.
7. A synchronous motor used for power factor correction is called synchronous condenser.
8. A motor generator set consisting of synchronous motor and alternator coupled mechanically can be used as frequency changer. The output frequency depends on the input frequency and the number of poles of synchronous motor and alternator.
9. Since synchronous motor has zero starting torque it requires an auxiliary starting method. The amortisseur winding helps in developing starting torque.
10. A synchronous induction motor has 3-phase ac winding on stator and a wound rotor. The rotor windings are connected to slip rings. It starts as induction motor and runs as synchronous motor.
11. A permanent magnet synchronous motor has permanent magnets as rotor. These magnets are made of samarium-cobalt and neodymium-iron-cobalt materials.

## TEST POINT QUESTIONS

### A. State whether the following statements are true (T) or false (F)

6.1 A permanent magnet synchronous motor has permanent magnets on stator.

6.2 Synchronous motor has zero starting torque.

6.3 Synchronous motor develops torque only at synchronous speed.

6.4 The armotisseur winding of a synchronous motor is on stator.

6.5 There is no armature reaction is a synchronous motor.

6.6 An overexcited synchronous motor operates at a leading power factor.

6.7 In *V* curves the region nearer the origin corresponds to leading power factor.

6.8 Synchronous condenser can also be used as synchronous reactor.

6.9 Amortisseur winding helps in suppressing hunting.

6.10 A synchronous induction motor has a 3 phase wound rotor.

### B. Fill in the blanks with suitable words

6.11 For a constant load current, as the field current of a synchronous motor is increased, its power factor .........

6.12 A synchronous induction motor has ......... winding on rotor.

6.13 A synchronous motor is a ......... motor.

6.14 The rotor of a synchronous motor is generally .........

6.15 V curves of synchronous motor depict the relation between ......... and .........

6.16 For a synchronous motor the graph between power factor and field current for a constant load is called .........

6.17 Improvement of power factor ......... line losses.

6.18 To provide starting torque an additional winding called ......... is provided in a synchronous motor.

6.19 When field current is zero the torque developed by synchronous motor is called .........

6.20 For a synchronous motor the reactance ......... is always more than the reactance .........

## ANSWERS

6.1 F 6.2 T 6.3 T 6.4 F 6.5 F 6.6 T 6.7 F

6.8 T 6.9 T 6.10 T 6.11 improves 6.12 3 phase 6.13 constant speed

6.14 salient pole 6.15 armature current, field current

6.16 inverted V curve 6.17 reduces 6.18 amortisseur winding

6.19 reluctance torque 6.20 $x_d$, $x_q$

## JOB INTERVIEW QUESTIONS

6.1 Why does synchronous motor develop torque only at synchronous speed?

6.2 Name the main application of synchronous motor.

6.3 What is amortisseur winding?

6.4 What is a permanent magnet synchronous motor?

6.6 What is a doubly excited machine?

6.7 What is a phase advancer?

6.8 What is the difference between $x_d$ and $x_q$.

6.9 How is a synchronous motor started?

6.10 What is V curve?

## REVIEW QUESTIONS

6.1 Is it possible to vary the speed of synchronous motor by varying the field excitation or by any other method? Explain what happens when the field current is (a) increased, (b) decreased.

6.2 Why is it so that a synchronous motor develops torque only at synchronous speed, whereas an induction motor develops torque at all speeds, except synchronous speed? Discuss.

6.3 In what respects is the phasor diagram of a synchronous motor similar to that of alternator? In what respects is it different?

6.4 Draw and explain the phasor diagram of a synchronous motor operating at (a) lagging power factor, (b) leading power factor.

6.5 Why does the power factor of industrial installation tend to be low? How can it be improved?

6.6 Explain the sequence of events which take place when the load on a synchronous motor is changed? What is hunting? How can it be avoided?

6.7 Briefly discuss (a) construction, (b) *V* curves, (c) starting, and (d) applications of synchronous motors.

6.8 Derive the power angle characteristic of a synchronous motor having a cylindrical rotor.

6.9 Draw the phasor diagram of a salient pole synchronous motor. Derive the power angle characteristics.

6.10 What is a synchronous condenser? What is its application?

6.11 Explain the operation of a synchronous induction motor.

## PROBLEMS

6.1 What frequency must be applied to the stator of a 20-pole synchronous motor, so that it may have a speed of 300 rpm? (50 Hz)

6.2 The input to an 11 kV 3-phase star connected synchronous motor is 60 A. The effective resistance and reactance are 1 ohm and 30 ohm per phase, respectively. Find (a) power input and induced emf for a power factor of 0.8 leading, (b) induced emf for a power factor of 0.8 lagging.

[(a) 914.5 kW, 13 kV, (b) 9.36 kV]

6.3 A 3-phase 8-pole, 50 Hz, 6.6 kV star connected synchronous motor has a synchronous impedance of 0.66 + $j$ 6.6 ohm per phase. When excited to give an emf of 4500 V per phase, it takes an input of 2500 KW. Find (a) torque developed, (b) input current, (c) power factor, (d) power angle.

[(a) 30435 N-m (b) 235.2 A (c) 0.93 leading (d) 19.5°]

6.4 A 2000 V, 3-phase, star connected synchronous motor has effective armature resistance and synchronous reactance of 0.2 ohm and 2.2 ohm, respectively. The input is 800 kW at rated voltage. The induced emf is 1440 V per phase. Find line current and power factor. [254 A, 0.91 leading]

6.5 A 3-phase, 220 V, 1500 rpm, delta connected synchronous motor has synchronous reactance of 4 ohm per phase. Its input current is 30 A and power factor 0.8 leading. Find the induced emf and load angle expressed in mechanical degrees. If the mechanical load is thrown off without change of excitation, find the line current under new conditions. [267.4 V, 6°, 20.6 A]

6.6 A 2000 V 3-phase 4-pole star connected synchronous motor has a resistance of 0.2 ohm per phase and a synchronous reactance of 1.9 ohm per phase. Find the emf and rotor displacement when the machine has as input of 800 kW at power factors of (a) 0.8 lagging, (b) 0.8 leading.

[(a) 878 V/phase, 13.7° mech. (b) 1514 V/phase, 9° mech.]

6.7 A synchronous motor is over excited to take 600 kVA. It develops 360 hp at 0.9 efficiency. It runs in parallel with a load of 1 MVA at 0.8 p.f. lagging. Find overall power factor. [0.997 lagging]

6.8 An alternator supplying 800 kW at 0.8 p.f. lagging has its power factor raised to unity by an over excited synchronous motor. With the armature current unchanged, how much power can alternator supply to synchronous motor? How much hp can synchronous motor develop when working at 90% efficiency. What is the power factor of synchronous motor? [200 kW, 244.73 hp (metric), 0.316 leading]

6.9 A 1200 kW load has a power factor 0.6 lagging. Find the rating of a synchronous condenser to raise the power factor to 0.9 lagging. Also find the total kVA supplied at the new power factor.

[1020 kVAr, 1333 kVA]

6.10 A synchronous frequency changer is required to change frequency from 50 to 300 Hz. Suggest suitable number of poles on the motor and alternator. [2 and 12 or 4 and 24 or 6 and 36]

6.11 A 20-pole, 40 hp, 660 V, 50 Hz, 3 phase, star connected synchronous motor is operating at no-load with $V = E$. At no-load the rotor is retarded by 0.5° mechanical from its synchronous position. The armature impedance per phase is (1 + $j$ 10) ohm. Find (a) rotor shift in electrical degrees (b) resultant emf across armature per phase, (c) armature current, (d) power drawn from supply.

[(a) 5 °, (b) 33.2 $\angle 87.5°$ V (c) 3.31 $\angle 3.2°$ A (d) 3777 W]

6.12 A 50 Hz, 3-phase, 100 hp, 440 V, 1000 rpm synchronous motor has a star connected stator At full load the line current is 106 A at unity p.f. The armature impedance is (0.09 + $j$ 2.25) ohms per phase. Find (a) generated emf per phase, (b) torque angle, (c) h.p (metric) developed.

[(a) 342 V (b) 44.2° (c) 105.71 hp]

6.13 Repeat Problem 6.12 for 0.8 lagging p.f. [(a) 212 V (b) 60.8° (c) 83.71 hp]

6.14 Repeat Problem 6.12 for 0.8 leading p.f. [(a) 436 V (b) 26.7 ° (c) 83.71 hp]

6.15 A 3-phase, 50 hp, 440 V, 6-pole, 50 Hz star connected synchronous motor has an armature resistance of 0.1 ohm and synchronous reactance of 2.4 ohm per phase. The motor is operating under excited with a power angle of 20°. The generated emf is 240 V per phase. Find armature current, p.f. and hp (metric) developed.

[36.17 A, 0.9576 lagging, 35.35 hp (metric)]

6.16 Repeat Problem 6.15 if generated voltage is 265 V per phase. [37.79 A, unity, 38.57 hp (metric)]

6.17 A 550 V, 3-phase system supplies 500 kVA load at 0.5 lagging p.f. A synchronous motor operating at 0.8 leading p.f. and supplying 74.6 kW load at 0.9 efficiency is added to the system. Find overall p.f., total kW and total kVA. [0.667 lagging, 333 kW, 499 kVA]

6.18 An industrial installation takes in 2000 kW at 0.6 p.f. lagging from a 6000 V, 3-phase ac system. A new additional dc motor generator set is to be added to the system. The motor may be a 746 kW, 6000 V synchronous motor operating at 0.8 leading p.f. and 0.92 efficiency or a 746 kW, 6000 V, 3 phase induction motor operating at 0.8 lagging p.f. and 0.92 efficiency. Find (a) total load current and p.f. if synchronous motor is used, (b) total load current and p.f. if induction motor is used.

[(a) 338.9 A, 0.8 lagging (b) 415.3 A, 0.65 lagging]

6.19 A 1000 metric hp, 2.2 kV, 3-phase induction motor load is connected to a 3-phase ac system. The power factor is 0.8 lagging and efficiency of induction motor is 0.9. (a) Find kW input and line current. (b) If 400 metric hp load is converted from induction motor to synchronous motor, find the new line current, kVA, and overall power factor. The synchronous motor is operating at a power factor of 0.85 leading and has an efficiency of 0.9. Assume that efficiency and power factor of induction motor remain unchanged.

[(a) 817.22 kW, 268.1 A; (b) 218 A, 833.74 kVA, 0.98 lagging]

Chapter 7

# Polyphase Induction Motors

## 7.1 CONSTRUCTION

An induction motor is a singly excited machine. The power is supplied only to the stator. It is known as induction motor because voltage (and current) are induced in the rotor. The rotor is not connected to any external circuit. As such it does not require any dc field current.

### 7.1.1 Stator

The stator of a 3-phase induction motor is very similar to that of a synchronous machine. The stator core is laminated and is slotted on its inner cylindrical surface. These slots house the 3-phase stator winding. The large size motors use open slots, so that form-wound and properly insulated coils can be placed in them. Small size motors use semi-enclosed slots, so that the effective length of air gap (between stator and rotor) may be small.

The air gap between the stator and rotor should be as small as is mechanically possible, so that the primary and secondary leakage fluxes are low. In very small size motors, the air gap is about 0.5 mm and in large ones it is about 1 mm.

### 7.1.2 Rotor

The rotor core is also laminated. The rotor winding may be in the form of bars or a complete 3-phase winding. These two types are known as squirrel cage rotor and wound rotor.

A squirrel cage rotor has a number of conducting bars (made of copper or aluminum) laid in the slots of the rotor core. These bars are short circuited at both ends by conducting end rings. If these end rings are not provided, current would not be able to flow in the rotor bars and the motor would not develop torque. It is known as squirrel cage rotor, because it looks like one of the exercise wheels of a squirrel. The cage winding is adaptable to any number of poles.

The cage rotor motor is cheap and robust. However its starting torque is low.

A wound rotor has a laminated core with slots on its outer surface. These slots carry 3-phase rotor winding, which is similar to the stator winding. Both the stator and rotor windings are designed for the same number of poles. The 3-phase rotor winding is usually star connected. The ends of three phases are tied to sliprings mounted on the motor shart. The rotor windings are shorted through brushes, which ride on the sliprings. Thus the rotor currents are accessible at these brushes. Extra resistance can be connected to the sliprings. This extra resistance is usually necessary to give a high starting torque.

### 7.1.3 Frame

The frame and the end shields serve to carry the bearings and supports (and protects) the core. The choice of a frame depends on the operating conditions and environment. Some commonly used frames are as under:

*(1) Open Type Frame*

This frame presents minimum obstruction to the flow of ventilating air consistent with the requirement of mechanical strength. These motors should be operated in clean environment and should not be accessible to unskilled labour.

*(2) Totally Enclosed Frame*

This frame is not necessarily air tight. However no exchange of air with ambient can take place.

*(3) Totally Enclosed Fan Cooled (TEFC)*

In these motors the frame is totally enclosed. One or more fans are used for cooling. The fans are an integral part of the motor. Most of the motors, up to a few kW rating, have this frame.

*(4) Enclosed Self-Ventilated*

These frames have openings for admission and discharge of air. The air is circulated through ducts. Small motors have radial ducts. Motors of moderate output and high speed have axial ducts. Both radial and axial ventilation is used for large motors. Circulation of air is effected by fans integral with the rotor.

*(5) Drip Proof Frame*

It is also known as splash proof frame. This frame has openings in the end shield, so designed as to exclude drops of liquid or solid particles approaching at an angle not greater than 100° from vertical, or which strike and run along the surface.

## 7.2 PRINCIPLE OF OPERATION

The stator winding is connected to 3-phase ac supply. The flow of currents in the stator winding produces a rotating field as discussed in Chapter 4. This field rotates at synchronous speed. The relative motion between this field and rotor conductors induces an emf in the rotor conductors. As the rotor is a closed one, currents flow in the rotor winding. The rotor currents produce a rotor mmf. The interaction of stator and rotor fields produces a torque and the rotor starts moving in the same direction as the stator rotating field. When the rotor is standstill, the frequency of rotor emf (proportional to relative speed between stator rotating field and rotor speed) is equal to the supply frequency. As the rotor starts moving, the frequency of rotor emf and the magnitude of rotor emf are reduced (because relative speed between stator rotating field and rotor is lower). However, the rotor speed cannot become equal to synchronous speed, because if it does so the emf induced in rotor conductors would become zero and torque would become zero. Therefore, the speed will always remain a bit lower than the synchronous speed. At no-load, the difference between rotor speed and synchronous speed is only about 1%. When the rotor shaft is loaded, the rotor slows down. The emf induced in the rotor conductors and rotor current increase, there by increasing the torque. Under steady state conditions, the electro-magnetic torque is equal to the load torque (if losses are neglected). Under full-load conditions, the difference between synchronous speed and rotor speed is about 3-5%.

In an induction motor, the rotor speed cannot be equal to synchronous speed. Therefore this machine is also known as asynchronous machine.

## 7.3 SLIP

The difference between the synchronous speed and rotor speed is known as slip. This term is descriptive of the manner is which the rotor slips back from the synchronous speed and is given by

$$s = \frac{n_s - n}{n_s}$$

or

$$n = n_s (1 - s) \qquad ...(7.1)$$

Where $n_s$ = synchronous speed, rps

$s$ = slip

$n$ = rotor speed, rps

When the rotor is at standstill, s = 1. At no-load slip is about 1% and at full-load it is about 3-5%. Slip cannot be zero, because when $s = 0$, $n = n_s$ and the electro-magnetic torque is zero. If the rotor is driven by external prime mover such that $n$ is greater than $n_s$, slip is negative and the machine acts as induction generator.

The variation in slip, and therefore speed, from no-load to full-load is very small. Thus a 3-phase induction motor is known as a constant speed motor.

## 7.4 PRODUCTION OF TORQUE

Figure 7.1(a) shows a cage motor operating at no-load. $B_1$ and $B_2$ are fields of stator and rotor. $B_R$ is the net field produced by the magnetising current. The magnitude of magnetising current and $B_R$ is directly proportional to stator emf $E_1$. If $E_1$ is constant, the magnetising current and $B_R$ are constant. At no-load the rotor slip is very small. Therefore the rotor frequency (which is equal to slip × stator frequency) is small. The rotor emf $E_2$ and rotor current $I_2$ are also small. Since the rotor frequency is small, the reactance of rotor is negligible and the rotor current $I_2$ is nearly in phase with the rotor emf $E_2$. The rotor current produces a rotor field $B_2$ at an angle which is slightly more than 90° behind the net field $B_R$. As per Eq. (4.46) the torque is

$$T = K B_2 B_R \sin \delta \qquad ...(7.2)$$

where $\delta$ is the angle between $B_2$ and $B_R$.

When the machine is loaded, the speed decreases. The higher relative motion between the stator magnetic field and rotor causes an increase in rotor emf $E_2$ and rotor current $I_2$. The rotor field $B_2$ also increases. Since slip is more, the rotor reactance is also more and the phase angle between $E_2$ and $I_2$ is also more. Figure 7.1 (b) shows this condition. Angle $\delta$ also increases. The motor develops more torque to supply the load.

It is seen from Fig. 7.1(b) that when the motor is supplying load, the angle $\delta$ is more than 90°. As the load is increased from no-load to full-load, the rotor current (and hence rotor field $B_2$) increases and this causes an increase in torque. The increase in angle $\delta$ tends to decrease the torque (since torque $\propto$ $\sin \delta$ and $\delta$ is more than 90°). The first effect is larger than the second one and the torque increases. As the motor is loaded still further (beyond full load), a condition is reached when $\sin \delta$ term decreases more than the increase in $B_2$. At this point a further increase in load decreases the torque produced and the motor stops. The torque produced at this condition, is known as pull-out torque.

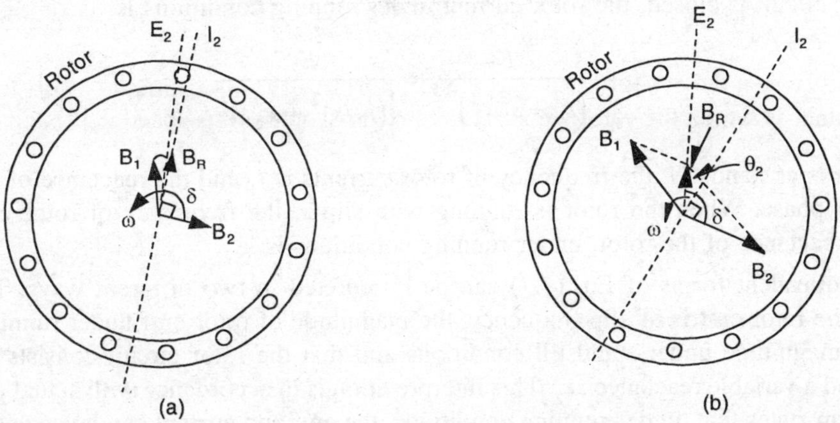


**Fig. 7.1.** Production of torque in cage motor (a) at no load, (b) under load conditions

## 7.5 EMF AND CURRENT

The analysis of induction motor is usually carried out under the assumptions that applied voltage $V_1$ is constant, purely sinusoidal and alternating at a frequency $f_1$, the flux per pole $\phi$ is sinusoidally distributed in space around the air gap and is rotating at synchronous speed and the rotor winding has the same number of poles and phases as the stator.

When the motor is at standstill, the emf induced in each phase of stator and rotor is

$$E_1 = 4.44\ K_{b1}\ K_{p1}\ f_1 N_1 \phi \qquad ...(7.3)$$

$$E_2 = 4.44\ K_{b2}\ K_{p2}\ f_1 N_2 \phi \qquad ...(7.4)$$

where $E_1$ = emf induced per phase in stator, volts
$E_2$ = emf induced per phase in rotor, volts
$f_1$ = supply frequency, Hz
$\phi$ = flux per pole, Wb
$N_1$ = number of turns in series per phase of stator
$N_2$ = number of turns in series per phase of rotor.
$K_{b1}$, $K_{b2}$ = breadth factors for the stator and rotor windings, respectively
$K_{b1}$, $K_{b2}$ = pitch factors for the stator and rotor windings, respectively

When the rotor is stationary, the frequency of the emf induced in rotor is the same as the supply frequency.

When the rotor speed is $n$ revolutions per second, the relative speed between the rotor and stator rotating field is $(n_s - n)$ revolutions per second. Then the frequency of emf induced in the rotor conductors is

$$f_2 = \frac{1}{2} P(n_s - n) = \frac{1}{2} P n_s \frac{n_s - n}{n_s} = s f_1 \qquad ...(7.5)$$

The emf induced in the rotor winding under running conditions is $sE_2$ and equals

$$sE_2 = 4.44\ K_{b2}K_{p2}sf_1N_2\phi \qquad \text{...(7.6)}$$

Since the rotor circuit is closed, the rotor current under running conditions is

$$I_2 = \frac{sE_2}{\left(r_2^2 + s^2x_2^2\right)^{0.5}} = \frac{E_2}{\left[\left(r_2/s\right)^2 + x_2^2\right]^{0.5}} \qquad \text{...(7.7)}$$

When the rotor is at standstill, the frequency of rotor currents is $f_1$ and the reactance of rotor winding is $x_2$ ohms per phase. When the rotor is running with slip $s$, the frequency of rotor currents is $sf_1$. Therefore the reactance of the rotor, under running conditions is $sx_2$.

The two equivalent forms of Eq. (7.7) can be interpreted in two different ways. The first form indicates that the rotor emf is of slip frequency, the magnitude of rotor emf under running conditions is $s$ times the magnitude under standstill conditions and that the rotor circuit consists of a constant resistance $r_2$ and a variable reactance $sx_2$. This interpretation is in accordance with actual physical facts. The second form states that, under running conditions, the emf and current can be considered to be of line frequency, the magnitude of rotor emf is the same as that at standstill, the rotor reactance is constant (equal to $x_2$) but the rotor resistance is variable and is $r_2/s$. This interpretation is useful in drawing the equivalent circuit and analysing the behaviour of the machine under running condition.

The rotor currents produce a rotor mmf. The frequency of rotor currents is $sf_1$. Therefore the speed of rotating mmf of rotor is $sn_s$ with respect to the rotor winding. The rotor winding is rotating at speed $n$ with respect to inertial space. Therefore speed of rotor mmf in space.

= speed of rotor mmf with respect to rotor winding + speed of rotor relative to space or speed of rotor mmf in space

$= sn_s + n = sn_s + n_s(1 - s)$

$= n_s$ rev/sec. ...(7.8)

It is seen that both stator and rotor mmfs rotate at synchronous speed and are stationary with respect to each other. Therefore we can combine the stator mmf and rotor mmf (as in the case of transformer) to analyse the behaviour of the motor.

### *Example 7.1*

A 400 V, 4-pole, 50Hz, 3-phase, 10 hp (metric), star connected induction motor has a no-load slip of 1% and full-load slip of 4%. Find (a) synchronous speed, (b) speed at no-load, (c) speed at full-load, (d) frequency of rotor currents at full-load, (e) full-load torque.

### *Solution*

(a) $$N_s = \frac{120f}{P} = \frac{120 \times 50}{4} = 1500 \text{ rpm}$$

(b) $$N = N_s(1 - s) = 1500(1 - 0.01) = 1485 \text{ rpm}$$

(c) $$N = N_s(1 - s) = 1500(1 - 0.04) = 1440 \text{ rpm}$$

(d) $$f_2 = sf_1 = 0.04 \times 50 = 2 \text{ Hz}$$

(e) $$T = \frac{\text{Power output}}{\omega_m}$$

$$= \frac{10 \times 735.5}{1440 \times \frac{1}{60} \times 2\pi} = 48.77 \text{ N-m.}$$

*Example 7.2*

A 12-pole 3-phase synchronous generator is driven by an engine running at 500 rpm. The generator feeds a 3-phase induction motor having a full-load speed of 1425 rpm. Find the slip and number of poles of motor.

*Solution*

$$\frac{PN_s}{120} = \frac{12 \times 500}{120} = 50\,\text{Hz}$$

The stator of induction motor is fed by 50 Hz supply. The synchronous speed of induction motor can be 3000 rpm, 1500 rpm, 1000 rpm, etc. Since full-load slip is about 5%, the synchronous speed of induction motor is 1500 rpm.

$$s = \frac{1500 - 1425}{1500} = 0.05$$

$$P = \frac{120 \times 50}{1500} = 4 \text{ poles.}$$

*Example 7.3*

A 3-phasor, 50 Hz, 4-pole, 400 V induction motor has a delta connected stator winding and star connected rotor winding. The stator winding has 240 conductors per phase and the rotor winding has 48 conductors per phase. The breadth factor is 0.95 and pitch factor is 1 for each winding. The rotor resistance and standstill reactance are 0.02 ohm/phase and 0.08 ohm/phase. Neglect stator impedance. Find (a) flux per pole, (b) rotor emf per phase at standstill, (c) rotor current and power factor at standstill with sliprings short circuited, (d) rotor emf per phase at 4% slip, (e) rotor current and power factor at 4% slip.

*Solution*

(a) $$400 = 4.44\, k_{b1}\, k_{p1}\, f_1 N_1 \phi$$

or $$400 = 4.44 \times 0.95 \times 1 \times 50 \times \frac{240}{2} \phi$$

or $$\phi = 0.0158 \text{ Wb}$$

(b) $$E_2 = 4.44\, k_{b2}\, k_{p2}\, f_1 N_2 \phi$$

$$= 4.44 \times 0.95 \times 1 \times 50 \times \frac{48}{2} \times 0.0158$$

$$= 80 \text{ volts/phase}$$

(c) $$I_2 = \frac{E_2}{\left(r_2^2 + x_2^2\right)^{0.5}} = \frac{80}{\left(0.02^2 + 0.08^2\right)^{0.5}}$$

$$= 970.14 \text{ A}$$

$$\text{Rotor p.f.} = \frac{r_2}{\left(r_2^2 + x_2^2\right)^{0.5}} = \frac{0.02}{\left(0.02^2 + 0.08^2\right)^{0.5}}$$

$$= 0.24 \text{ lagging}$$

It is seen that rotor current at standstill (i.e. at starting) is very large. Therefore stator current will also be very large and must be limited by external means.

(d) Rotor emf at 4% slip

$$sE_2 = 0.04 \times 80 = 3.2 \text{ V/phase}$$

(e)

$$I_2 = \frac{sE_2}{\left[r_2^2 + (sx_2)^2\right]^{0.5}}$$

$$= \frac{0.04 \times 80}{\left[0.02^2 + (0.04 \times 0.08)^2\right]^{0.5}} = 158 \text{ A}$$

$$\text{Rotor p.f.} = \frac{0.02}{\left[0.02^2 + (0.04 \times 0.08)^2\right]^{0.5}}$$

$$= 0.987 \text{ lagging.}$$

## 7.6 INDUCTION MOTOR AS TRANSFORMER

It is possible to compare the operation of induction motor to that of a transformer. The induction motor (like a transformer) is singly energized device. The stator and rotor of an induction motor can be compared to primary and secondary of transformer.

When a 3-phase supply is fed to the stator, a rotating field is set up and emfs are induced in stator and rotor [Eqs (7.3) and (7.4)]. These equations are similar to Eqs. (2.3b, c) for transformer. The slight difference in these equations is due to the fact that in induction motor the windings are distributed in slots, so that breadth and pitch factors have to be included in the equations (in contrast the windings in transformer consist of concentrated coils). An induction motor at standstill exhibits the characteristics of a transformer on no-load. In induction motor also the leakage fluxes exist as in a transformer.

The behaviour of both the devices is similar under load conditions also. The flow of rotor current causes rotor mmf. The rotor and stator mmfs can be combined because both of them are rotating at synchronous speed (and are therefore stationary with respect to each other). This is similar to what happens in a transformer, where primary and secondary mmfs can be combined. When the load on secondary is increased, the primary current increases in a transformer. In an induction motor also, an increase in mechanical load causes an increased rotor and stator currents. A transformer under short circuit is similar to an induction motor with blocked rotor. Moreover, the stator and rotor have winding resistance and leakage reactances just as transformer windings have resistances and leakage reactances. In view of these similarities, the phasor diagram and equivalent circuit of an induction motor are also similar to those in a transformer.

However these are some very important differences also between the two devices. These are:

(1) A transformer is a static device whereas and induction motor is a rotating device. Therefore an air gap exists in an induction motor. Due to the presence of air gap, the magnetising current is pretty high in an induction motor.

(2) Because of the presence of air gap, the leakage reactances in an induction motor are higher than in a transformer.

(3) The losses are higher and the efficiency is lower in an induction motor than in transformer.

(4) A transformer winding consists of concentrated coils. In induction motor the windings are distributed in slots.

## 7.7 PHASOR DIAGRAM

The behaviour of an induction motor is similar to that of a transformer. Therefore the phasor diagrams of the two devices are also similar. Figure 7.2 shows the phasor diagram of an induction motor.

The mutual flux $\phi_m$ induces emfs in stator and rotor. These emfs, under running conditions are $E_1$ and $sE_2$ respectively. Treating these emfs as voltage drops (as in the case of transformer), these emfs lead $\phi_m$ by 90°. $I_0$ is the no-load current. The rotor current $I_2$ is reflected as $I_2'$ in the stator. The phasor sum of $I_2'$ and $I_0$ gives stator current $I_1$. the phasor sum of $E_1$, $I_1r_1$ and $I_1x_1$ gives the applied voltage $V_1$, cos $\theta_1$ is the power factor at stator terminals. In the phasor diagram of Fig. 7.2 all quantities are expressed in per phase values.

$$V_1 = E_1 + I_1 (r_1 + jx_1) \quad \text{...(7.9)}$$

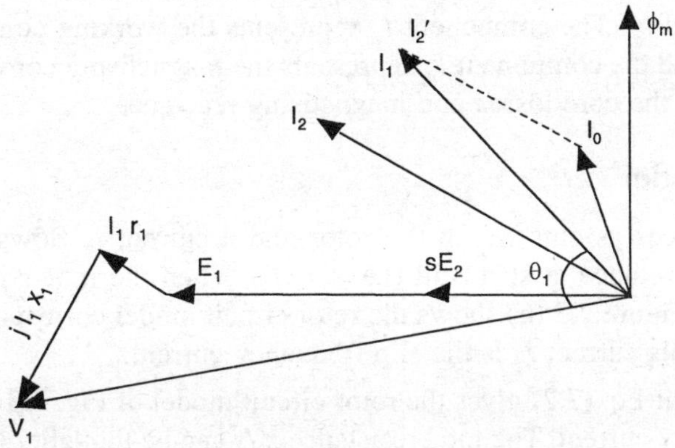


**Fig. 7.2.** Phasor diagram of induction motor

## 7.8 EQUIVALENT CIRCUIT

In developing the equivalent circuit of an induction motor, we have to develop circuit models for stator and rotor and then combine them, so as to give the equivalent circuit.

### 7.8.1 Stator Circuit Model (Fig. 7.3 a)

When a voltage $V_1$ is applied to stator terminals, a flux $\phi_m$ is set up. This flux induces emf $E_1$ in the stator winding. The flow of current $I_1$ causes voltage drops $I_1r_1$ and $I_1x_1$ in the stator winding $E_1$, $V_1$, $I_1r_1$ and $I_1x_1$ are related through Eq. (7.9).

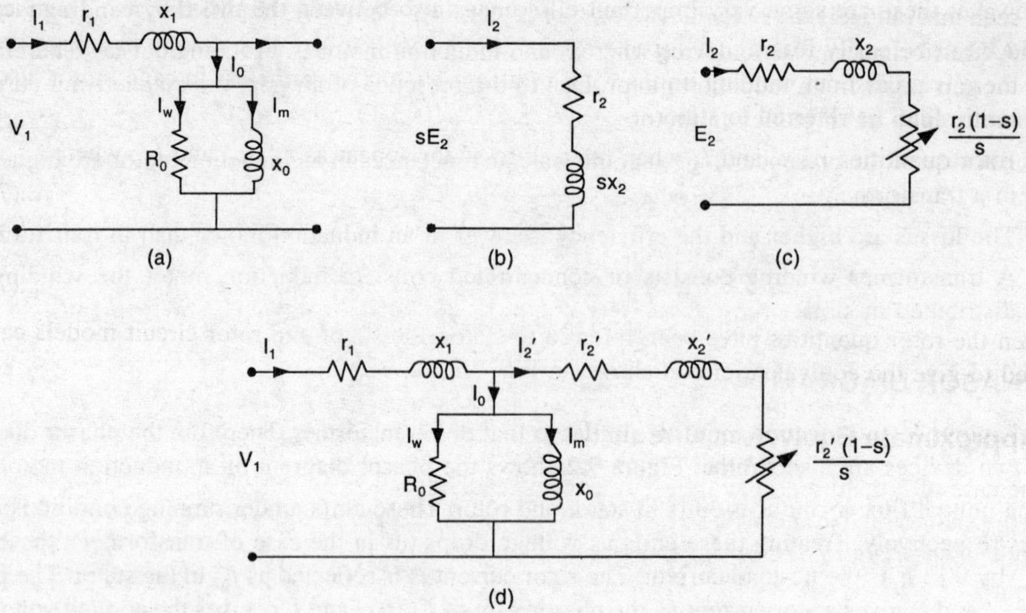


**Fig. 7.3.** Development of equivalent circuit of 3-phase induction motor (a) stator circuit model, (b and c) rotor circuit models, (d) complete equivalent circuit

Even when the motor is at no-load, a current $I_o$ flows in the stator winding. This current consists of two components $I_w$ and $I_m$. The component $I_w$ represents the working component of $I_o$ to take into account the core losses and the component $I_m$ represents the magnetising component (for setting up the flux). $R_o$ and $X_o$ simulate the core losses and magnetising reactance.

### 7.8.2 Rotor Circuit Model

The mutual flux $\phi_m$ induces as emf $sE_2$ in the rotor and a current $I_2$ flows. $r_2$ and $x_2$ are the rotor resistance and leakage reactance at standstill (i.e. $s = 1$). When the motor is operating at slip $s$, the leakage reactance is $sx_2$. Figure 7.3 (b) shows the rotor circuit model corresponding to above facts and first form of Eq. (7.7). This current $I_2$ is the slip frequency current.

The alternative form of Eq. (7.7) gives the rotor circuit model of Fig. 7.3 (c). The current $I_2$ in this model is the line frequency current. The total resistance $r_2/s$ can be thought of as two components, viz. $r_2$ and $r_2(1 - s)/s$. The component $r_2$ is the actual rotor winding resistance, while the component $r_2$ $(1 - s)/s$ represents the mechanical load. The current $I_2$ in Fig. 7.3(c) is line frequency current.

### 7.8.3 Exact Equivalent Circuit

It is necessary to combine the circuit models of Fig. 7.3 (a and c) to obtain the final equivalent circuit. In order to do so it is necessary to transfer rotor quantities to stator side. This can be done through the effective transformation ratio obtained from Eqs. (7.3) and (7.4), i.e.

$$k = \frac{E_1}{E_2} = \frac{k_{b1} k_{p1} N_1}{k_{b2} k_{p2} N_2} \qquad ...(7.10)$$

It is seen that ratio $k$ is the ratio of stator turns to rotor turns per phase, modified by winding factors. This ratio $k$ can be easily found in a slipring motor. It is difficult to define $k$ exactly in a cage motor, because there is no distinct winding on the rotor of a cage motor. Nevertheless in cage motor also the rotor quantities can be referred to stator.

The rotor quantities $r_2$, $x_2$ and $I_2$ when referred to stator appear as $r_2'$, $x_2'$ and $I_2'$, where

$$r_2' = k^2 r_2 \quad ...(7.11)$$

$$x_2' = k^2 x_2 \quad ...(7.12)$$

$$I_2' = I_2/k \quad ...(7.13)$$

When the rotor quantities have been referred to stator, the stator and rotor circuit models can be combined to give the equivalent circuit shown in Fig. 7.3(d).

### 7.8.4 Approximate Equivalent Circuit

As in the case of transformer, the approximate equivalent circuit of an induction motor is obtained by shifting the shunt branch (consisting of $R_0$ and $X_0$) to the input terminals. Figure 7.4 shows the approximate equivalent circuit. However, it is important to remember that in a transformer the no-load current $I_0$ is about 2–5% of the full-load primary current. However, in an induction motor the no-load current $I_0$ is about 25%–40% of rated full-load stator current. Therefore the computations using approximate equivalent circuit of an induction motor involve considerable error.

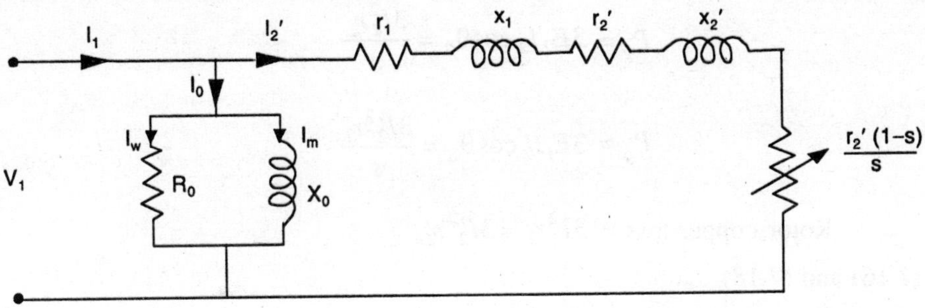


**Fig. 7.4.** Approximate equivalent circuit of 3-phase induction motor

## 7.9 LOSSES, POWER OUTPUT

The 3-phase power input $P_i$ to an induction motor is

$$P_i = 3V_1 I_1 \cos \theta_1 \quad ...(7.14)$$

where $V_1$, $I_1$ and $\cos \theta_1$ are the per phase values of stator voltage and current and stator power factor, respectively.

A part of this power is consumed in stator core-losses and stator copper-losses. The remainder is $P_g$, the air gap power.

$$\text{Stator copper-losses} = 3I_1^2 r_1$$

$$P_g = P_i - 3I_1^2 r_1 - \text{stator core-losses} \quad ...(7.15)$$

$P_g$ is the power transferred, across the air gap, to the rotor. After rotor copper-losses are subtracted from $P_g$, we get the mechanical power developed $P_m$. A part of $P_m$ is consumed as friction and windage and stray losses and the remainder is the power output $P_0$. Figure 7.5 shows the power flow diagram.

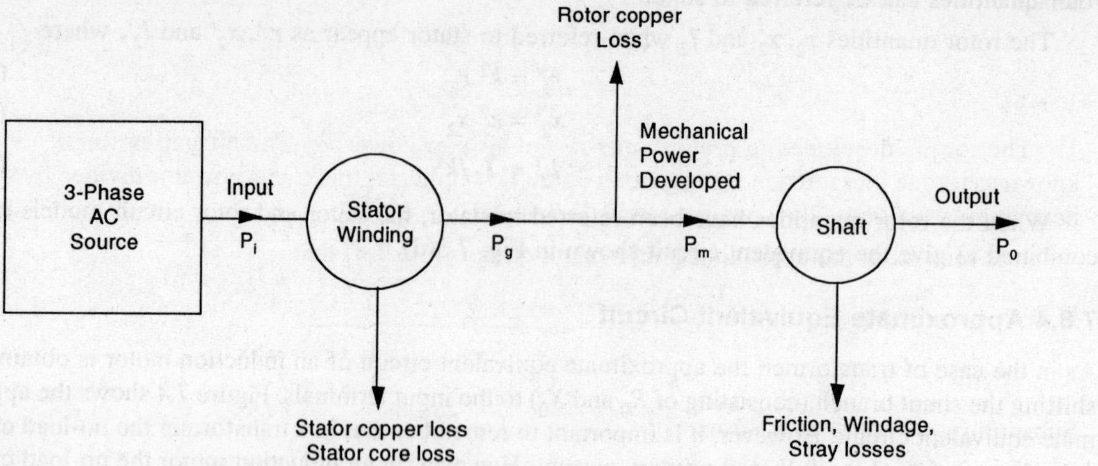


**Fig. 7.5.** Power flow diagram of 3-phase induction motor

Now $$P_g = 3E_2 I_2 \cos\theta_2 = \frac{3I_2^2 r_2}{s} \qquad ...(7.16)$$

Also $$P_g = 3E_1 I_2' \cos\theta_2 = \frac{3I_2'^2 r_2'}{s} \qquad ...(7.17)$$

$$\text{Rotor copper-loss} = 3I_2^2 r_2 = 3I_2'^2 r_2' \qquad ...(7.18)$$

From Eqs. (7.16) and (7.18)

$$\text{Rotor copper-loss} = sP_g \qquad ...(7.19)$$

$$P_m = P_g - \text{Rotor copper-loss} = P_g(1 - s) \qquad ...(7.20)$$

$$P_0 = P_m - \text{friction, windage and stray losses} \qquad ...(7.21)$$

The frequency of rotor currents is very low, Therefore rotor core-loss is very small and hence negligible.

## 7.10 TORQUE

The torque developed $T$ is defined as the torque generated due to mechanical power developed by electro-mechanical power conversion process in the motor. The actual load torque available at the shaft is the torque developed less the windage and friction and stray losses.

Thus $$T = \frac{P_m}{\omega_m} = \frac{P_m}{2\pi n} \qquad ...(7.22)$$

or $$T = \frac{3I_2^2 r_2}{2\pi n}\left[\frac{1-s}{s}\right] \qquad ...(7.23)$$

or
$$T = \frac{3I_2'^2 r_2'}{2\pi n}\left[\frac{1-s}{s}\right] \qquad ...(7.24)$$

Since
$$n = n_s\,(1 - s)$$

$$T = \frac{3I_2'^2 r_2'}{(2\pi n_s)s} = \frac{P_g}{2\pi n_s} \qquad ...(7.25)$$

The torque developed is proportional to the air gap power $P_g$. The air gap power $P_g$ is usually known as torque measured in synchronous watts. This indicates that if this power is divided by synchronous angular velocity $2\pi n_s$, we get torque developed.

It is seen from the approximate equivalent circuit (Fig. 7.4) that

$$I_2' = \frac{V_1}{\left[(r_1 + r_2'/s)^2 + (x_1 + x_2')^2\right]^{0.5}} \qquad ...(7.26)$$

Substituting the value of $I_2'$ from Eq. (7.26) into Eq. (7.25)

$$T = \left[\frac{3}{2\pi n_s}\right]\left[\frac{r_2'}{s}\right]\left[\frac{V_1^2}{(r_1 + r_2'/s)^2 + (x_1 + x_2')^2}\right] \qquad ...(7.27)$$

*Example 7.4*

The power input to the rotor of a 3-phase, 50 Hz, 6-pole induction motor is 80 kW. The rotor emf makes 100 complete alternations per minute. Find (a) slip, (b) motor speed, (c) mechanical power developed, (d) rotor copper-loss per phase, (e) rotor resistance per phase if rotor current is 65 A, (f) torque developed.

*Solution*

$$n_s = \frac{2f_1}{P} = \frac{2 \times 50}{6} = \frac{50}{3}\text{ rps}$$

(a)
$$sf_1 = \frac{100}{60}$$

or
$$s = \frac{100}{60 \times 50} = \frac{1}{30}$$

(b)
$$n = n_s\,(1 - s) = \frac{50}{3}\left(1 - \frac{1}{30}\right)$$

$$= 16.111 \text{ rev/sec.}$$

$$= 966.66 \text{ rev/min.}$$

(c)
$$P_g = 80 \text{ kW}$$

$$P_m = P_g(1 - s) = 80\left(1 - \frac{1}{30}\right)$$

$$= 77.333 \text{ kW}$$

(d) $$\text{Rotor copper-loss} = sP_g = \frac{1}{30} \times 80 = \frac{8}{3} \text{ kW}$$

$$\text{Rotor copper-loss per phase} = \frac{1}{3} \times \frac{8}{3} \text{ kW} = 888.9 \text{ W}$$

(e) $$\text{Rotor copper-loss per phase} = I_2^2 r_2$$

or $$I_2^2 r_2 = 888.9$$

$$r_2 = \frac{888.9}{I_2^2} = \frac{888.9}{65^2} = 0.21 \text{ ohm/phase}$$

(f) $$T = \frac{P_g}{2\pi n_s} = \frac{80 \times 1000}{2\pi(50/3)} = 763.94 \text{ N-m.}$$

*Example 7.5*

A 400 V, 25 hp, 4-pole, 50 Hz 3-phase, star connected induction motor has the following impedances per phase as referred to stator:

$$r_1 = 0.6 \text{ ohm}, \qquad x_1 = 1.1 \text{ ohm}$$

$$r_2' = 0.3 \text{ ohm}, \qquad x_2' = 0.5 \text{ ohm}$$

$$X_0 = 25 \text{ ohm}$$

The mechanical losses are 1000 W and stator core-losses are 500 W. The slip is 3%. Find (a) speed, (b) stator current, (c) power factor, (d) mechanical power developed, air gap power and power output, (e) torque developed and load torque, (f) efficiency, (g) rotor copper-loss per phase. Neglect $R_0$.

*Solution*

(a) $$n_s = \frac{2f_1}{P} = \frac{2 \times 50}{4} = 25 \text{ rev/sec.}$$

$$n = n_s(1 - s) = 25\,(1 - 0.03)$$

$$= 24.25 \text{ rev/sec.}$$

$$= 1455 \text{ rev/min.}$$

(b) $$V_1 = \frac{400}{\sqrt{3}} = 230.95 \text{ V/phase}$$

From Fig. 7.4

$$I_2' = \frac{V_1}{\left(r_1 + \frac{r_2'}{s}\right) + j(x_1 + x_2')}$$

$$= \frac{230.95}{\left(0.6 + \dfrac{0.3}{0.03}\right) + j(1.1 + 0.5)}$$

$$= \frac{230.95}{10.6 + j1.6} = 21.54\angle -8.6^\circ \text{ A}$$

$$I_0 = \frac{230.95}{j25} = 9.24\angle -90^\circ \text{ A}$$

Stator current $I_1 = I_2' + I_0$

$$= 21.54\angle -8.6^\circ + 9.24\angle -90^\circ$$

$$= 21.16 - j12.46 \text{ A}$$

$$= 24.56\angle -30.5^\circ \text{ A}$$

(c) Power factor $= \cos(-30.5^\circ)$

$$= 0.86 \text{ lagging}$$

(d) Stator copper-losses $= 3I_1^2 r_1 = 3(24.56)^2(0.6)$

$$= 1085.75 \text{ W}$$

Power input $P_i = 3V_1 I_1 \cos\theta_1$

$$= 3(230.95)(24.56)(0.86)$$

$$= 14634.1 \text{ W}$$

Air gap power $P_g = P_i - 1085.75 - 500$

$$= 14634.1 - 1085.75 - 500$$

$$= 13048.35 \text{ W}$$

Mechanical power developed $P_m = P_g(1 - s)$

or $P_m = 13048.35(1 - 0.03) = 12656.9$ W

$$P_0 = P_m - \text{mechanical losses}$$

$$= 12656.9 - 1000 = 11656.9 \text{ W}$$

(e) Torque developed $= \dfrac{P_s}{2\pi n_s} = \dfrac{13048.35}{2\pi \times 25} = 83.07$ N-m

Load torque $= \dfrac{P_0}{2\pi n} - \dfrac{11656.9}{2\pi \times 24.25} = 76.51$ N-m

(f) Effciency $= \dfrac{P_0}{P_i} = \dfrac{11656.9}{14634.1} = 0.797$ or 79.7%

(g) Rotor copper-losses $= sP_g = 0.03 \times 13048.35 = 391.45$ W

$$\text{Rotor copper-loss per phase} = \frac{391.45}{3} = 130.48\,\text{W}.$$

*Example 7.6*

The motor of Example 7.5 is fed at 300 V. Find (a) stator current, (b) torque developed, (c) power factor.

*Solution*

(a) It is seen from the equivalent circuit (Fig. 7.4) that $I_1 \propto V_1$

$$\text{Stator current} = 24.56 \times \frac{300}{400} = 18.42\ \text{A}$$

(b)
$$\text{Torque} \propto V_1^2$$

$$\text{Torque developed} = 83.07\left(\frac{300}{400}\right)^2 = 46.73\ \text{N-m}$$

(c) Since the parameters of the equivalent circuit are the same, the power factor remains unaffected by change of voltage.

Power factor = 0.86 lagging.

## 7.11 TORQUE-SLIP CURVE

The torque-slip curve can be derived from Eq. (7.2) which is

$$T = KB_2 B_R \sin\delta$$

Each term can be studied independently to find the variation of torque with slip.

(1) $B_2$. The rotor field $B_2$ is directly proportional to the rotor current (assuming the magnetic circuit to be unsaturated). The rotor current increases with increase is slip [Eq. (7.7)] when slip is low. However for high values of slip, $I_2$ becomes almost constant. This variation is shown in Fig. 7.6(a).

(2) $B_R$. The resultant field $B_R$ is proportional to $E_1$, which decreases slightly with increase in current. However, this decrease is very small and $E_1$ can be assumed to remain constant. Thus $B_R$ remains constant (Fig. 7.6 b).

(3) sin δ. It is seen from Fig. 7.1 (b) that angle δ is equal to the power factor angle of rotor (i.e. $\theta_2$) plus 90°. Therefore

$$\sin\delta = \sin(\theta_2 + 90°) = \cos\theta_2 \qquad ...(7.28)$$

Thus sin δ is equal to rotor power factor.

$$\theta_2 = \tan^{-1}\left(\frac{sx_2}{r_2}\right) \qquad ...(7.29)$$

and
$$\cos\theta_2 = \cos\left[\tan^{-1}(sx_2/r_2)\right] \qquad ...(7.30)$$

The variation of $\cos \theta_2$ with slip is shown in Fig. 7.6 (c).

Since torque is proportional to product of the above three terms, the torque-speed (or torque-slip) curve can be constructed by a point-by-point multiplication of the three curves (Fig. 7.6 a, b, c) and is shown in Fig. 7.6(d). This curve can be divided into three regions. In the low slip region, the slip increases approximately linearly with increase in load (i.e. the speed decreases approximately linearly with load). In this region, the rotor reactance is negligible and rotor power factor is almost 1. The normal steady state operation of induction motor lies in this region.

In the region with moderate slip values, the rotor reactance ($sx_2$) is of the same order as rotor resistance. The rotor current does not increase as rapidly as in the low slip region and rotor power factor begins to decrease. The maximum torque (or pull-out torque) occurs when increase in rotor current is exactly balanced by decrease in rotor power factor.

In the region with high values of slip, the torque decreases with increase in slip, because the increase in rotor current is over-shadowed by the decrease in rotor power factor.

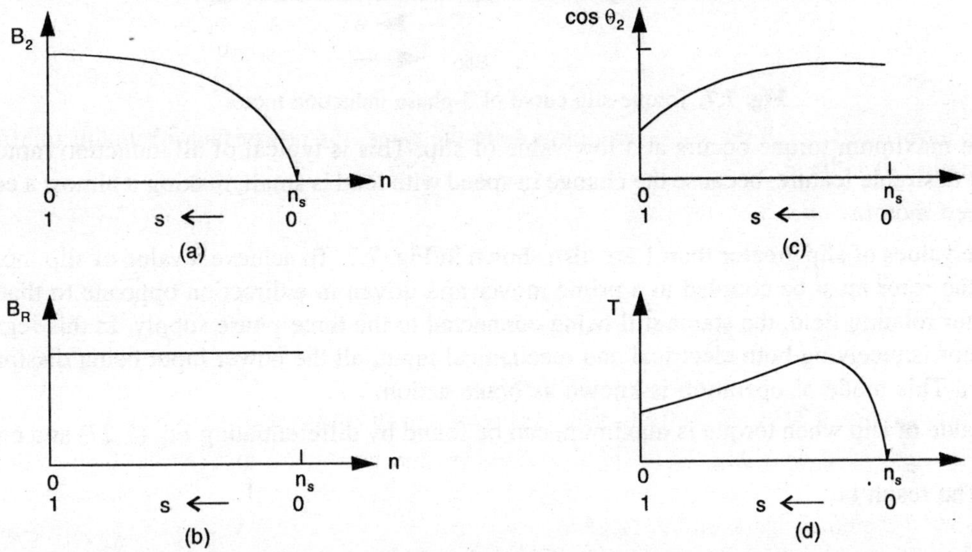


**Fig. 7.6.** Development of torque slip curve of a 3-phase induction motor (a) Variation of rotor field and rotor current with slip. (b) Variation of resultant field with slip, (c) Variation of rotor power with slip, (d) torque slip curve

The torque-slip curve is again drawn in Fig. 7.7. This curve has the following characteristic features:

(1) When $n = n_s$ and $s = 0$, torque is zero. This is because at synchronous speed, there is no relative motion between rotor conductors and stator rotating field, so that rotor emf and current are zero.

(2) the hatched area shows the normal operating range for motor action. In this region torque increases linearly with slip.

(3) *CD* represents the torque at $s = 1$, i.e. at standstill conditions. If load torque is less than *CD*, the motor will accelerate till the torque developed by motor is equal to load torque. This happens at a speed which is close to but less than synchronous speed.

(4) *AB* is the maximum or pull-out torque. If the load torque is more than this torque, the motor will come to standstill. The region *BD* is the unstable region.

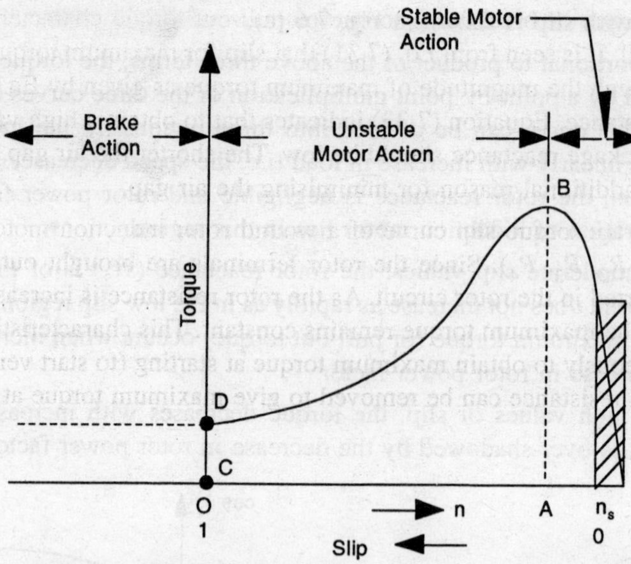


**Fig. 7.7.** Torque-slip curve of 3-phase induction motor

(5) The maximum torque occurs at a low value of slip. This is typical of all induction motors and is a desirable feature, because the change in speed with load is small, making it almost a constant speed motor.

(6) The values of slip greator than 1 are also shown in Fig. 7.7. To achieve a value of slip more than 1, the rotor must be coupled to a prime mover and driven in a direction opposite to that of the stator rotating field, the stator still being connected to the three-phase supply. In this region the motor is receiving both electrical and mechanical input, all the power input being dissipated as loss. This mode of operation is known as brake action.

The value of slip when torque is maximum can be found by differentiating Eq. (7.27) and equating $\frac{dT}{dx} = 0$. The result is

$$s = \pm \frac{r_2'}{\left[r_1^2 + (x_1 + x_2')^2\right]^{0.5}} \quad ...(7.31)$$

In Eq. (7.31),positive sign corresponds to motor action and negative sign to the generator action.

The value of slip for maximum torque can also be obtained by examining Fig. 7.4. Torque will be maximum when there is maximum power transfer into the load $r_2'/s$. The source impedance is $r_1 + j(x_1 + x_2')$. Applying maximum power transfer principle, the condition for maximum power transfer is

$$\frac{r_2'}{s} = \sqrt{r_1^2 + (x_1 + x_2')^2} \quad ...(7.32)$$

Eqs. (7.31) and (7.32) are identical. Substituting this value of $s$ is Eq. (7.27), the maximum torque is

$$T_{max} = \frac{3V_1^2}{2\pi n_s (2)} \left[\frac{1}{r_1 + \left[r_1^2 + (x_1 + x_2')^2\right]^{0.5}}\right] \quad ...(7.33)$$

The maximum torque or breakdown torque or pull-out torque characterises the load which will cause the motor to stall. It is seen from Eq. (7.31) that slip for maximum torque is directly proportional to rotor resistance though the magnitude of maximum torque as given by Eq. (7.33) is independent of the value of rotor resistance. Equation (7.33) indicates that to obtain a high value of maximum torque, the stator and rotor leakage reactance should be low. The shorter the air gap, the lesser is the leakage reactance. This is an additional reason for minimising the air gap.

Figure 7.8 shows the torque-slip curve of a wound rotor induction motor for different values of rotor resistances ($R_1$, $R_2$, $R_3$, $R_4$). Since the rotor terminals are brought out to slip rings, additional resistance can be inserted in the rotor circuit. As the rotor resistance is increased, the slip increases (or speed decreases) but the maximum torque remains constant. This characteristics of wound rotor motor can be used advantageously to obtain maximum torque at starting (to start very heavy loads). After the motor starts, the extra resistance can be removed to give maximum torque at a low value of slip. This is shown in Fig. 7.9.

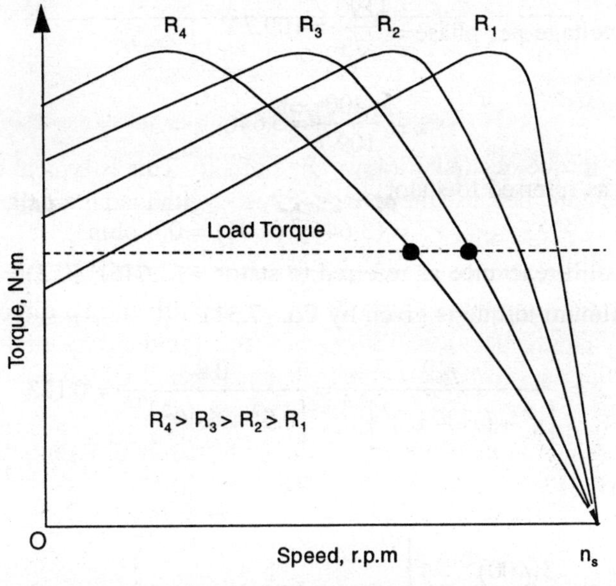


Fig. 7.8. Torque-slip curves of 3-phase induction motor for different values of rotor resistance

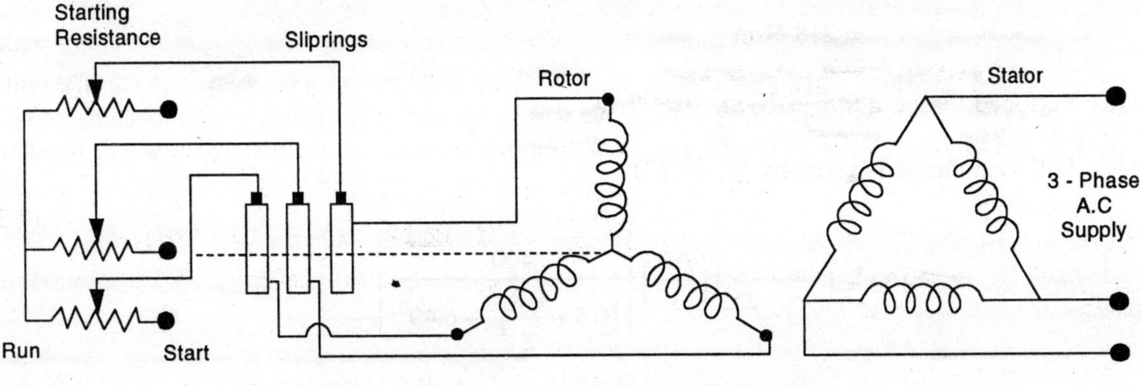


**Fig. 7.9.** Starting of wound rotor induction motor

### *Example 7.7*

A 400 V 3-phase 4-pole 50 Hz induction motor has a delta connected stator winding and a star connected rotor winding. At standstill the voltage between sliprings is 190 V. The stator impedance is 0.5 + *j*2.5 ohm. The rotor resistance and standstill rotor reactance are 0.06 ohm and 0.3 ohm, respectively. Determine (a) $n_s$, (b) slip at which maximum torque occurs, (c) maximum torque, (d) torque, slip and power output at full-load if maximum torque is twice full load torque, (e) frequency of rotor currents at full-load.

### *Solution*

(a) $$n_s = \frac{2 \times 50}{4} = 25 \text{ rev/sec.}$$

(b) Stator voltage per phase = 400 V

$$\text{Rotor voltage per phase} = \frac{190}{\sqrt{3}} = 109.7 \text{ V}$$

$$k = \frac{400}{109.7} = 3.646$$

Rotor resistance as referred to stator

$$r_2' = (3.646)^2\,(0.06) = 0.8 \text{ ohm}$$

$x_2'$ = Rotor standstill reactance as referred to stator = $(3.646)^2$ (0.3) = 3.99 ohm

The slip for maximum torque is given by Eq. (7.31).

Therefore $$s = \frac{r_2'}{\left[r_1^2 + (x_1 + x_2')^2\right]^{0.5}} = \frac{0.8}{\left[0.5^2 + 6.49^2\right]^{0.5}} = 0.123$$

(c) Using Eq. (7.33)

$$T_{\max} = \frac{3(400)^2}{2\pi \times 25} \times \frac{1}{2}\left[\frac{1}{0.5 + \left(0.5^2 + 6.49^2\right)^{0.5}}\right]$$

$$= 218 \text{ N-m}$$

(d) Full-load torque $$= \frac{218}{2} \text{ N-m} = 109 \text{ N-m}$$

Full-load torque is given by Eq. (7.27)

or $$109 = \left(\frac{3}{2\pi \times 25}\right)\left(\frac{0.8}{s}\right)\left[\frac{400^2}{\left(0.5 + \frac{0.8}{s}\right)^2 + 6.49^2}\right]$$

$$109 = \frac{2444.6}{s\left[\left(0.5 + \frac{0.8}{s}\right)^2 + 6.49^2\right]}$$

or $$42.37s^2 - 21.628s + 0.64 = 0$$

or $$s = 0.0315 \text{ or } 0.479$$

Since slip for maximum torque is 0.123, the value of $s = 0.479$ is in the unstable region. Hence $s = 0.0315$.

$$\text{Full-load speed } n = 25\,(1 - 0.0315) = 24.2 \text{ rev/sec.}$$

$$\text{Full-load power output } P_0 = 2\pi nT$$

or $$P_0 = 2\pi(24.2)(109) = 16573.8 \text{ W}$$

(e) frequency of rotor currents at full-load

$$sf_1 = 0.0315 \times 50 = 1.755 \text{ Hz.}$$

*Example 7.8*

A slipring induction motor has a full-load speed of 288 rpm. If the supply frequency is 50 Hz, find (a) number of poles, (b) slip, (c) slip for full-load if rotor resistance is doubled, (d) new value of rotor copper-loss if initial value of copper-loss is 275 W.

*Solution*

(a) $$N = 288$$

Slip for full-load condition is from 0.03 to 0.05. Therefore

$$N_s = 300 \text{ rpm}$$

$$P = \frac{120f}{N_s} = \frac{120 \times 50}{300} = 20$$

(b) $$s = \frac{N_s - N}{N_s} = \frac{300 - 288}{300} = 0.04$$

(c) From Eq. (7.27) it is seen that if torque is constant, slip $\propto r_2' \propto r_2$. Since rotor resistance has been doubled, slip also becomes twice.

Hence $s = 0.08$.

(d) Since torque is constant, rotor current is also constant. The rotor resistance has been doubled. Therefore rotor copper-loss = 2 × 275 = 550 W.

## 7.12 STARTING OF INDUCTION MOTOR

From the torque-slip curve of Fig. 7.7, it is seen that a 3-phase induction motor has a positive finite starting torque at $s = 1$. Therefore it starts by itself when connected to a 3-phase supply.

However an induction motor, when at standstill, is like a transformer with short circuited secondary. It takes a high current (about 5 to 8 times full-load current) if started at full voltage. This large line current (at low power factor), drawn by motors at starting is objectionable because of the possible drop in the voltage of the supply circuit and the consequent dimming of lamps as well as undesirable effects on other devices connected to the same supply circuit. Special methods are, therefore needed to reduce the starting current. In addition short circuit, overload and under voltage protections are also provided.

### 7.12.1 Wound Rotor Induction Motor

A wound rotor induction motor is started by introducing external resistance across the sliprings (Fig. 7.9). This possibility makes the wound rotor induction motors especially suitable for starting loads which have large initial friction or unusually large inertia, without drawing excessive starting current. This method has the added advantage that a high starting torque can be obtained. The external resistance is kept at such a value as to give maximum or near maximum torque at starting. From Eq. (7.32) it is seen that to have maximum torque at $s = 1$,

$$r_2' = \left[r_1^2 + (x_1 + x_2')^2\right]^{0.5} \qquad \text{...(7.34)}$$

As the motor accelerates, the external resistance is gradually reduced to zero. This method not only limits the starting current but also raises the power factor at starting, thereby reducing the line disturbance.

### 7.12.2 Cage Rotor Induction Motor

The squirrel cage motors are usually started by applying a reduced voltage to the motor initially and then increasing the voltage to the ful-line value, when the motor has picked up speed. Three methods used are: direct starting, star-delta starting and auto-transformer starting.

*(a) Direct Starting*

This method means switching the motor directly on to the supply without any device for reducing the starting current. This method is restricted to small motors upto 2 hp. For these small motors. the starting torque is about twice the full-load torque and the starting period lasts only a few seconds.

Figure 7.10 shows a starter for direct starting with in-built short circuit, over load and under voltage protection. When the motor is to be started, the main switch is put on and start button is pressed. This energizes the relay coil $M$ causing the normally open contacts $M_1$, $M_2$, $M_3$ to close. Power is supplied to the motor and it starts. The contact $M_4$ also shuts, thus shorting out the starting switch allowing the operator to release it without removing power from the $M$ relay. When the stop button is pressed, the $M$ relay is deenergized and the $M$ contacts open, thus stopping the motor. The protective devices work as under: Short circuit protection is provided by fuses $F_1$, $F_2$ and $F_3$. If a short circuit occurs within the motor, and causes a current many times larger than the rated current, these fuses blow and disconnect the motor from supply. However, these fuses are so selected that they can withstand the starting current. Overload protection is provided by the device $OL$ and overload heaters. Under normal conditions the overload contacts are shut. When a high overload occurs and persists, the temperature of the overload heating elements rises far enough, the contact $OL$ opens thus deenergizing the $M$ relay, which opens the $M$ contacts, thus disconnecting the motor from the mains.

The starter also provides under-voltage protection. It is seen that control power for the $M$ relay comes from the mains supply. If the voltage is too low, the $M$ relay is deenergized, the $M$ contacts open and the motor is disconnected from the mains.

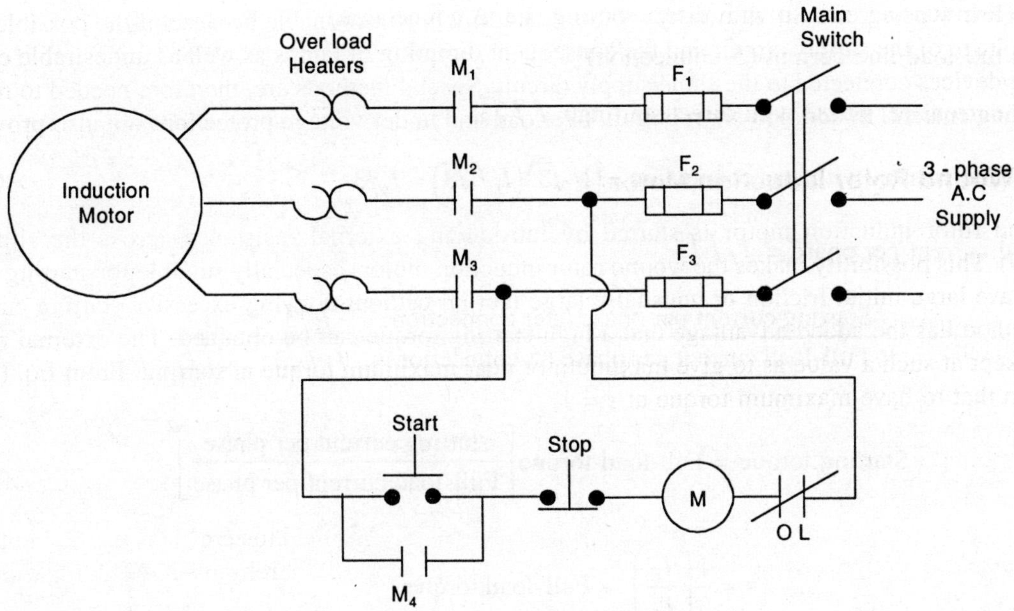


**Fig. 7.10.** Starter for 3-phase induction motor

### *(b) Star-Delta Starting*

This method requires the motor to be delta connected for normal operation. All the six terminals (of the three phases) are brought out. For starting the phases are connected in star, thereby reducing the voltage of each phase to $1/\sqrt{3}$ of its normal value.

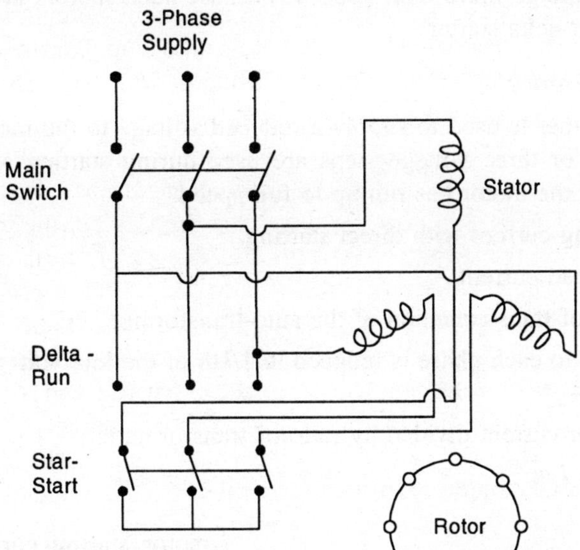


**Fig. 7.11.** Star-delta starter

Let $I_s$ = line starting current with direct starting (i.e. Δ connection)

$I_f$ = full-load line current (Δ connection)

Starting current per phase with direct starting $= I_s/\sqrt{3}$

Starting line current with star connection $= (1/\sqrt{3})(I_s/\sqrt{3}) = I_s/3$ ...(7.35)

Full-load current per phase $= I_f/\sqrt{3}$

$$\frac{\text{Starting current per phase (star connection)}}{\text{Full-load current per phase } (\Delta \text{ connection})} = \frac{I_s/3}{I_f/\sqrt{3}} = \frac{I_s}{\sqrt{3}\,I_f}$$

$$\text{Starting torque} = \text{Full-load torque}\left[\frac{\text{starting current per phase}}{\text{Full-load current per phase}}\right]^2$$

$$= \frac{1}{3}\left(\frac{I_s}{I_f}\right)^2 \times \text{Full-load torque}$$

$$= \frac{1}{3} \times \text{starting torque with direct starting} \qquad \text{...(7.36)}$$

Thus this method reduces the starting line current to one-third, but the starting torque is also reduced by the same amount. This method is cheap but limited to applications where high starting torque is not necessary, e.g. machine tools, pumps, motor generator sets. This method is not possible for motors with rated voltage more than 3000 V, because such motors are generally star connected. Figure 7.11 shows a star-delta starter.

### *(c) Auto-transformer Starter*

A 3-phase auto-transformer is used to supply a reduced voltage to the motor at starting. Figure 7.12 shows the circuit. Two or three voltage steps are used during starting and the auto-transformer is completely cut out after the motor has run up to full speed.

Let $I_s$ = starting current with direct starting,

$I_f$ = full-load current,

$k$ = ratio of transformation of the auto-transformer.

The voltage applied to each phase is reduced to 1/*k*th of the line voltage. Current is each phase = $I_s/k$.

Line current is the motor current divided by ratio of transformation.

Therefore line current $= I_s/k^2$ ...(7.37)

$$\text{Starting torque} = \text{Full-load torque}\left(\frac{\text{motor starting current}}{\text{full-load current}}\right)^2$$

$$= \text{Full-load torque}\left(\frac{I_s/k}{I_f}\right)^2$$

$$= \frac{1}{k^2}[\text{Starting torque with direct starting}]. \qquad ...(7.38)$$

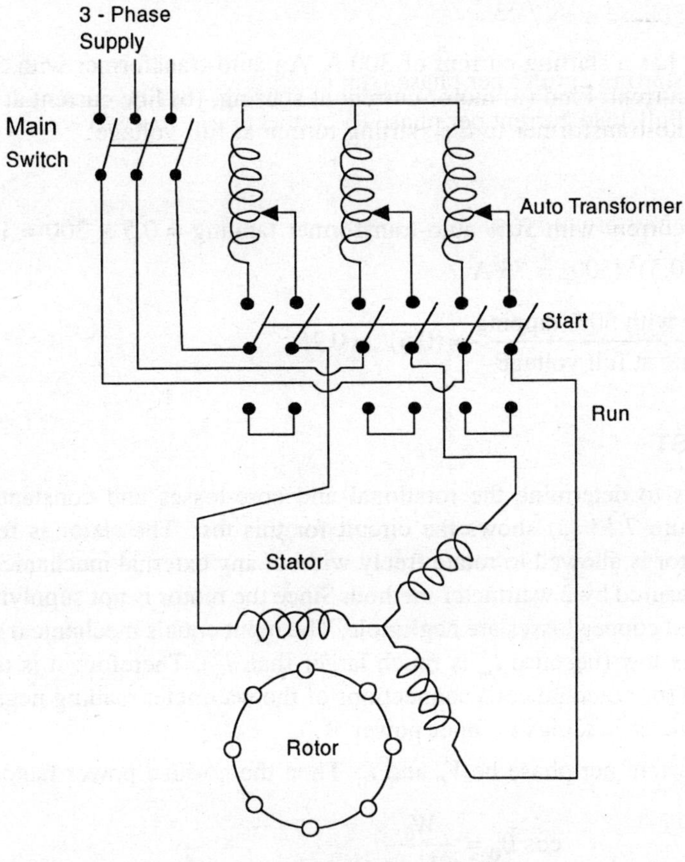


**Fig. 7.12.** Auto-transformer starter

*Example 7.9*

A 3-phase induction motor takes 4 times full-load current on direct starting and develops 1.5 times full-load torque. What must be the applied voltage to obtain full-load torque at starting and what will be the starting current?

*Solution*

$$\frac{\text{Full-load torque}}{\text{Starting torque with direct starting}} = \frac{1}{1.5} = \frac{1}{k^2}$$

or $\qquad k = \sqrt{1.5} = 1.2247$

$$\text{Applied voltage} = \frac{1}{1.2247} \text{ or } 0.816 \times \text{line voltage}$$

$$\text{Line current at starting} = \frac{1}{k^2} \times 4 \times \text{ full-load current}$$

$$= 2.67 \times \text{full-load current.}$$

*Example 7.10*

A squirrel cage motor has a starting current of 300 A. An auto-transformer with 50% tapping is used to reduce the starting current. Find (a) motor current at starting, (b) line current at starting, (c) ratio of starting torque with auto-transformer to the starting torque at full voltage.

*Solution*

(a) Motor starting current with 50% auto-transformer tapping = 0.5 × 300 = 150 A

(b) line current = $(0.5)^2$ (300) = 75 A

(c) $$\frac{\text{Starting torque with 50\% tapping}}{\text{Starting torque at full voltage}} = (0.5)^2 = 0.25.$$

## 7.13 NO-LOAD TEST

The aim of this test is to determine the rotational and core-losses and constants $R_0$ and $X_0$ of the equivalent circuit. Figure 7.13 (a) shows the circuit for this test. The stator is fed by 3-phase rated voltage supply. The rotor is allowed to rotate freely without any external mechanical load on the shaft. The power input is measured by 2 wattmeter method. Since the motor is not supplying any load, the no-load current is small and copper-losses are negligible. The input equals mechanical and core losses. The no-load power factor is low (because $I_m$ is much larger than $I_w$). Therefore it is usually necessary to reverse the current coil (or potential coil) connections of the wattmeter reading negative (in such a case the difference of wattmeter readings is input power $W_0$).

Let voltage and current per phase be $V_0$ and $I_0$. Then the no-load power factor cos $\theta_0$ is

$$\cos\theta_0 = \frac{W_0}{3V_0I_0} \quad \text{...(7.39)}$$

$$R_0 = \frac{V_0}{I_w} = \frac{V_0}{I_0\cos\theta_0} \quad \text{...(7.40)}$$

$$X_0 = \frac{V_0}{I_m} = \frac{V_0}{I_0\sin\theta_0} \quad \text{...(7.41)}$$

It is evident that this test is similar to the no-load test on a transformer.

## 7.14 BLOCKED ROTOR TEST

The aim of this test is to determine the winding resistance, i.e. $(r_1 + r_2')$ and leakage reactances $(x_1 + x_2')$. It is similar to the short circuit test on a transformer. Figure 7.13(b) shows the connection

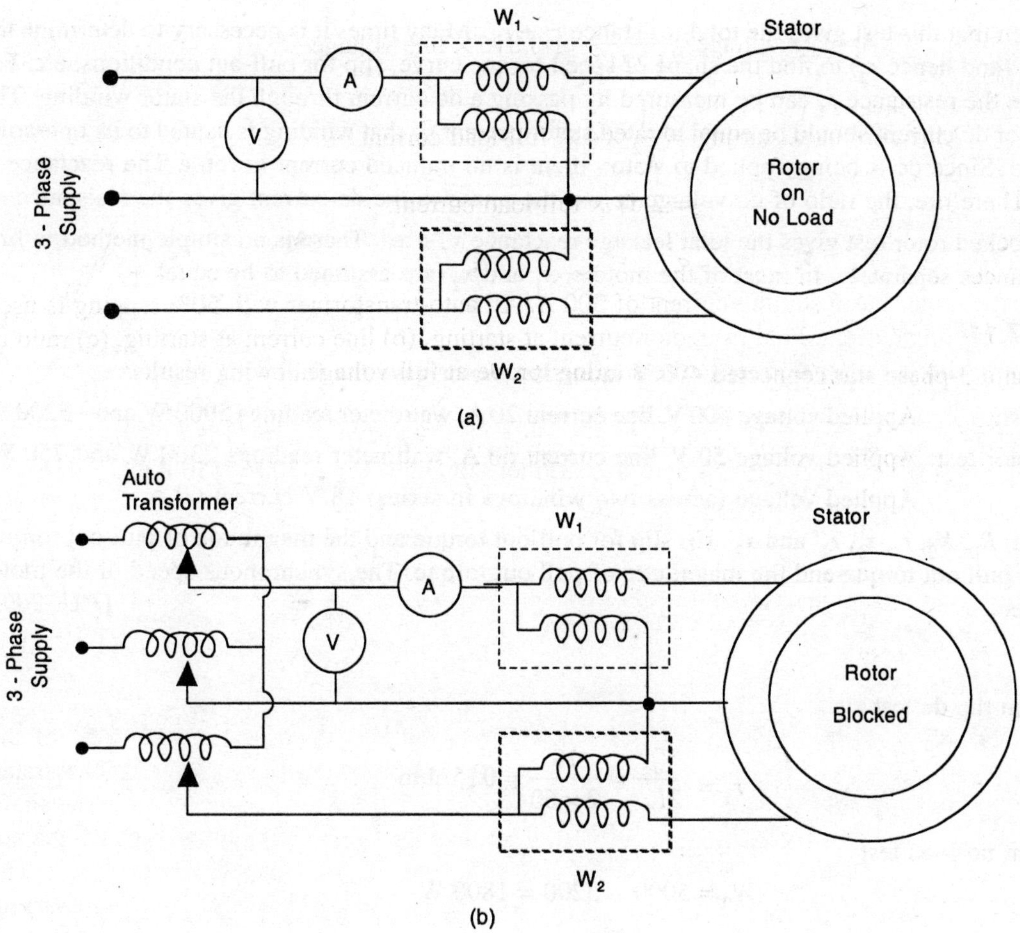


**Fig. 7.13.** Connections for (a) no-load test, (b) blocked rotor test on 3 phase induction motor

diagram. The motion of the rotor is blocked by a brake. A low voltage 3-phase supply (obtained from a 3-phase auto-transformer) is applied to the stator. The magnitude of applied voltage is adjusted, so that the ammeter reads rated line current of stator. The power input is measured by the two wattmeters (In this case also, the power factor is usually less than 0.5 and it may be necessary to reverse either the current coil or potential coil connections of the watt-meter reading negative). Let $W_{sc}$ be the total power input and $V_{sc}$ and $I_{sc}$ be the per phase value of applied voltage and current.

Then
$$\cos\theta_{sc} = \frac{W_{sc}}{3V_{sc}I_{sc}} \quad ...(7.42)$$

$$r_1 + r_2' = \frac{V_{sc}}{I_{sc}}\cos\theta_{sc} \quad ...(7.43)$$

$$x_1 + x_2' = \frac{V_{sc}}{I_{sc}}\sin\theta_{sc} \quad ...(7.44)$$

It is seen that this test gives the total resistance $r_1 + r_2'$. Many times it is necessary to determine the value of $r_2'$ (and hence $r_2$) to find the shape of speed-torque curve, slip for pull-out conditions, etc. For this purpose the resistance $r_1$ can be measured by passing a dc current through the stator winding. The magnitude of dc current should be equal to rated stator current so that winding is heated to its operating temperature. Since dc is being applied to stator, there is no induced current in rotor. The reactance is also zero. Therefore, the ratio of dc voltage across the phase to the dc current gives the resistance $r_1$.

The blocked rotor test gives the total leakage reactance $x_1 + x_2'$. There is no simple method to find these reactances separately. In most of the motors, $x_1$ and $x_2'$ are assumed to be equal.

### Example 7.11

The tests on a 3-phase star connected 400 V induction motor gave the following results:

No-load test: Applied voltage 400 V, line current 20 A, wattmeter readings 5000 W and –3200 W.

Blocked rotor test: Applied voltage 50 V, line current 60 A, wattmeter readings 2300 W and 750 W.

dc test: Applied voltage (across two windings in series) 18 V current 60 A.

Find (a) $R_0$, $X_0$, $r_1$, $x_1$, $r_2'$ and $x_2'$. (b) slip for pull-out torque and the magnitude of pull-out torque. (b) slip for pull out torque and the magnitude of pull out torque. The synchronous speed of the motor is 25 rev/sec. *[P.U. 2003]*

### Solution

(a) From the dc test

$$r_1 = \frac{V_{dc}}{2I_{dc}} = \frac{18}{2 \times 60} = 0.15\,\text{ohm}$$

From no-load test

$$W_0 = 5000 - 3200 = 1800\text{ W}$$

$$V_0 = 400\sqrt{3} = 230.95\text{ V}$$

$$I_0 = 20\text{ A}$$

$$\cos\theta_0 = \frac{1800}{3 \times 230.95 \times 20} = 0.13$$

$$R_0 = \frac{230.95}{20 \times 0.13} = 88.83\,\text{ohm}$$

$$X_0 = \frac{230.95}{20\sin\theta_0} = 11.65\,\text{ohm}$$

From blocked rotor test

$$W_{sc} = 2300 + 750 = 3050\text{ W}$$

$$V_{sc} = 50/\sqrt{3} = 28.87\text{ V}$$

$$I_{sc} = 60\text{ A}$$

$$\cos\theta_{sc} = \frac{3050}{3\times 28.87\times 60} = 0.587$$

$$r_1 + r_2' = \frac{28.87}{60}\times 0.587 = 0.282\,\text{ohm}$$

$$r_2' = 0.282 - 0.15 = 0.132\text{ ohm}$$

$$x_1 + x_2' = \frac{28.87}{60}\sin\theta_{sc} = 0.39\,\text{ohm}$$

Assume $x_1 = x_2'$

Therefore $x_1 = x_2' = \dfrac{0.39}{2} = 0.195\,\text{ohm}$

(b) Using Eq. (7.31), the slip for maximum (or pull-out) torque is

$$s = \frac{r_2'}{\left[r_1^2 + (x_1 + x_2')^2\right]^{0.5}}$$

$$= \frac{0.132}{\left(0.15^2 + 0.39^2\right)^{0.5}} = 0.316$$

Using Eq. (7.33)

$$T_{\max} = \frac{3V_1^2}{2\pi n_s}\cdot\frac{1}{2}\left[\frac{1}{r_1 + \left[r_1^2 + (x_1 + x_2')^2\right]^{0.5}}\right]$$

$$= \frac{3(230.95)^2}{4\pi\times 25}\left[\frac{1}{0.15 + \left(0.15^2 + 0.39^2\right)^{0.5}}\right]$$

$$= 896.96\text{ N-m.}$$

## 7.15 BRAKE TEST

This is a very convenient method to determine the operating characteristics directly. However, it can be used only for small motors. The method is essentially similar, to the one used for dc shut motor (Sec. 3.36). The desired load is put by tightening the belt or rope. The input to the induction motor is measured by two wattmeter method.

Let $W_1$ and $W_2$ be the tensions is kilograms on the tight and slack side of the brake, $r$ be the radius of pulley in metres and $n$ be the speed in rev/sec.

$$\text{Torque } T = (W_1 - W_2)\, r \times 9.81\text{ N-m}$$

$$\text{Output} = 2\pi n T\text{ watts.}$$

The input and power factor can be calculated from the wattmeter readings. The graphs of power factor, efficiency, torque and speed versus output can be drawn.

## 7.16 COMPUTATION OF PERFORMANCE BY CIRCLE DIAGRAM

The performance of an induction motor can be computed from the equivalent circuit. A graphical solution is also possible by drawing the circle diagram. The results obtained by graphical method are not very accurate. However, this method has the advantage of providing a complete panoramic view of the effect of changes in machine parameters, voltage, frequency, etc. on the operating characteristics.

### 7.16.1 Theory

This method aims at plotting the locus of input stator current $I_1$ for various values of slip. $I_1$ is the phasor sum of $I_2'$ (i.e. rotor current as referred to stator) and $I_0$.

$$I_2 = \frac{E_2}{\frac{r_2}{s} + jx_2} \qquad \text{...(7.45)}$$

or

$$-j\frac{E_2}{x_2} = I_2 - j\frac{r_2}{sx_2}I_2 \qquad \text{...(7.46)}$$

The phasor $E_2/x_2$ denotes a constant current. As seen from Eq. (7.46), this current lagging $E_2$ by 90°, is composed of two components, viz. $I_2$ and $-j\,r_2I_2/sx_2$. These two components are at right angles to each other. Therefore the extremely of $I_2$ lies on the circumference of a circle. Since $I_2$ and $I_2'$ are related by a constant, the locus of $I_2'$ is also a circle. Since $I_1$ is the phasor sum of $I_2'$ and $I_0$, we can displace this circle of $I_2'$ from origin by the magnitude of $I_0$. Then the extremity of $I_1$ will lie on the circumference of the circle.

### 7.16.2 Construction

The data available from the no-load and blocked rotor tests is used for drawing the circle diagram. All quantities are per phase value.

Figure 7.14 shows the construction. Select a suitable scale for current. The vertical line represents voltage axis. Draw *OA*, lagging *V* by angle $\theta_0$, equal to no-load current $I_0$ to the selected scale. At a voltage $V_{sc}$, the short circuit (i.e. blocked rotor) current is $I_{sc}$. Therefore this current at rated voltage is $(V_{rated}/V_{sc})\,I_{sc}$. Draw *OB* lagging *V* by angle $\theta_{sc}$ and equal to $(V_{rated}/V_{sc})\,I_{sc}$ to the selected scale. Join *AB*. Draw right bisector of *AB*. This right bisector intersects the horizontal line through *A* at point *P*, which is the centre of the circle. With *P* as centre and *PA* as radius, draw the semicircle *A G S T B F*. Draw vertical *BD* and divide it at point *C*, so that $\frac{BC}{CD} = \frac{r_2'}{r_1}$. Join *AC*.

### 7.16.3 Computation of Performance

Let the stator input current, for a certain load, be *OG*. The power input per phase is proportional to the component of *OG* in phase with voltage. This is equal to *Ge*. Therefore *Ge* is power input per phase to a certain scale.

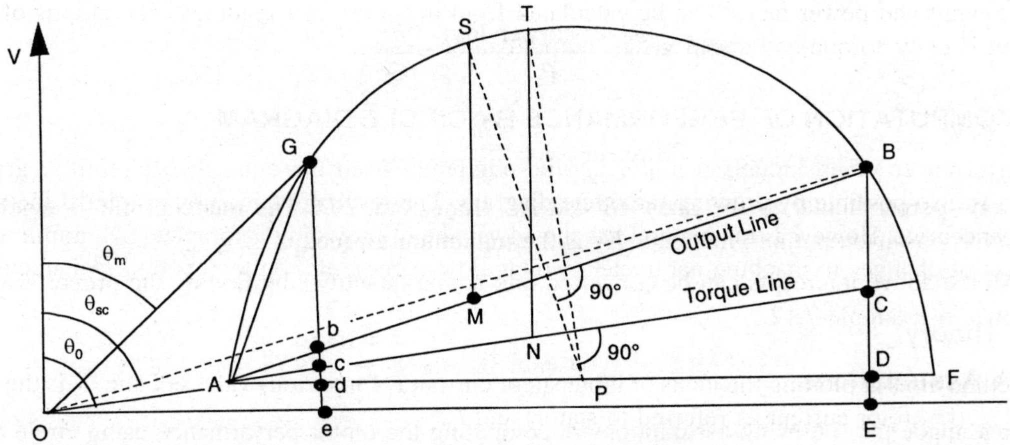


**Fig. 7.14.** Circle diagram of 3-phase induction motor

The component of $I_0$ (i.e. *OA*) in phase with voltage is *de*. This is proportional to no-load, losses, i.e. core-loss, windage and friction loss. The remaining component *Gd* represents power output plus copper-losses in stator and rotor.

Under blocked rotor conditions at rated voltage, the power input per phase is *BE*. This represents core-loss and copper-losses (because when the rotor is blocked, friction and windage losses are absent). The intercept *DE* (= *de*) is approximately equal to core-loss, because though the friction and windage losses are absent, the core-loss in rotor is slightly increased under blocked rotor conditions. Hence the intercept *BD* represents the sum of stator and rotor copper-losses.

The triangles *ABD* and *Abd* are similar. Therefore

$$\frac{bd}{BD} = \frac{Ad}{AD} = \frac{AG\cos\angle GAF}{AB\cos\angle BAF} = \frac{AG(AG/AF)}{AB(AB/AF)} = \frac{AG^2}{AB^2}$$

$$= \frac{AG^2(r_1 + r_2')}{AB^2(r_1 + r_2')} = \frac{\text{copper-losses at full-load}}{\text{copper-losses under blocked rotor condition}}$$

Therefore *bd* is copper-loss when the stator current is *OG*. Evidently the remaining intercept *Gb* is the mechanical power output per phase. The line *AB* is the locus of all such points as *b* and is known as output line. A vertical intercept from any point on the circle to this line *AB* is the output per phase for the corresponding value of stator current.

Torque is given by $P_g/2\pi n_s$ and is thus proportional to power input to rotor. The rotor copper-loss at standstill is *BC*. Therefore rotor copper-loss, corresponding to stator current *OG*, is *bc*. Hence *Gc* is the power output plus rotor copper-loss, i.e. the power input to rotor. The locus of all points similar to *c* is the torque line *AC*.

$$\text{Starting torque} = BC$$

$$\text{Efficiency} = \frac{Gb}{Ge}$$

$$\text{slip} = \frac{\text{Rotor copper-loss}}{P_g} = \frac{bc}{Gc}$$

$$\text{Power factor} = \frac{Ge}{OG}$$

Draw perpendicular to the lines *AB* and *AC* from point *PD*. The intercept *SM* is evidently the maximum power output and intercept *TN* is the maximum torque.

All the above intercepts can be converted into actual quantities by finding the proper scales. This is shown in Example 7.12.

### 7.16.4 Assumptions

We have made the following assumptions in computing the motor performance using circle diagram:

(1) Approximate equivalent circuit has been used.

(2) The leakage reactance have been assumed to be constant.

(3) Saturation of the magnetic circuit has been neglected. This assumption affects the maximum output and maximum torque especially.

*Example 7.12*

The results of no-load and blocked rotor tests on a 400 V, 50 hp, 50 Hz, 3-phase, 4-pole induction motor are as under.

No-load test : $V_0 = 400$ V, $I_0 = 30$ A, $W_0 = 1800$ W.

Blocked rotor test : $V_{sc} = 110$ V, $I_{sc} = 80$ A, $W_{sc} = 4000$ W

The motor has a star connected stator having a resistance of 0.1 ohm per phase. Draw circle diagram and find out (a) full-load current, power factor, slip, torque and efficiency, (b) starting torque, maximum torque and maximum output.

*Solution*

Now $$\cos\theta_0 = \frac{1800}{\sqrt{3} \times 400 \times 30} = 0.0866$$

$$\theta_0 = 85°$$

$$\cos\theta_{sc} = \frac{4000}{\sqrt{3} \times 110 \times 80} = 0.262$$

$$\theta_{sc} = 74.8°$$

$$n_s = \frac{2 \times 50}{4} = 25\,\text{rev/sec.}$$

Blocked rotor current at rated voltage $= 80 \times \dfrac{400}{110} = 290.9\,\text{A}$

Power input at rated voltage under blocked rotor condition $= 4000\left(\dfrac{400}{110^2}\right)$

$$= 52892.6\ W$$

Stator copper-loss at rated voltage under blocked rotor condition $= 3(0.1)(290.9)^2$

$= 25386.8$ W

Rotor copper-loss at rated voltage under blocked rotor condition

$$= 52892.6 - 25386.8 = 27505.8 \text{ W}$$

The circle diagram is drawn to scale in Fig. 7.15. *OA* is 20 *A* and *OB* is 290.9 A. Divide *BD* at point *C* so that $\frac{BC}{CD} = \frac{27505.8}{225386.8}$. Draw output line *AB* and torque line *AC* as shown.

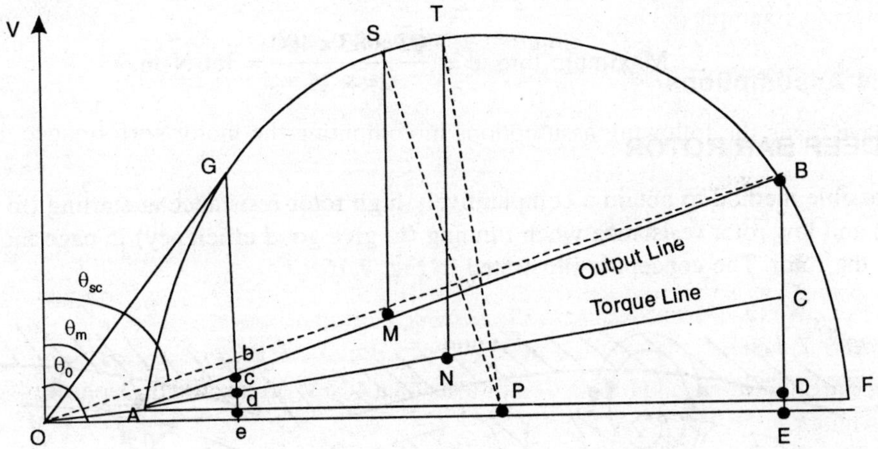


**Fig. 7.15**

(a) The full-load point *G* is located such that the vertical intercept *Gb* is equal to full-load output.

$$Gb = \frac{50 \times 735.5}{\sqrt{3} \times 400} = 53.08 \text{ A to scale}$$

Full-load stator current = *OG* = 87 A

Full-load power factor = $\cos \theta_m = \cos 42° = 0.74$

$$\text{Full-load slip} = \frac{bc}{Gc} = 0.05$$

$$\text{Full-load efficiency} = \frac{Gb}{Ge} = 0.82$$

Full-load torque = *Gc* = 58 A

$$= \frac{\sqrt{3} \times 400 \times 58}{2\pi \times 25} = 255.8 \text{ N-m}$$

(b) Starting torque = *BC* = 28 A

$$= \frac{\sqrt{3} \times 400 \times 28}{2\pi \times 25} = 123.5 \text{ N-m}$$

Draw perpendicular to output line passing through point $P$. This perpendicular meets the circle at point $S$. The maximum output is $SM$.

$$SM = 70 \text{ A}$$

$$\text{Maximum output} = \frac{\sqrt{3} \times 400 \times 70}{1000} = 48.5 \text{ kW}$$

Draw perpendicular to torque line passing through point $P$. This perpendicular meets the circle at point $T$. The maximum torque is $TN$.

$$TN = 83 \text{ A}$$

$$\text{Maximum torque} = \frac{\sqrt{3} \times 83 \times 400}{2\pi \times 25} = 366 \text{ N-m.}$$

## 7.17 DEEP BAR ROTOR

One possible method to obtain a comparatively high rotor resistance at starting (to give good starting torque) and low rotor resistance when running (to give good efficiency) in cage motors is to use deep bars in the rotor. The concept is illustrated in Fig. 7.16.

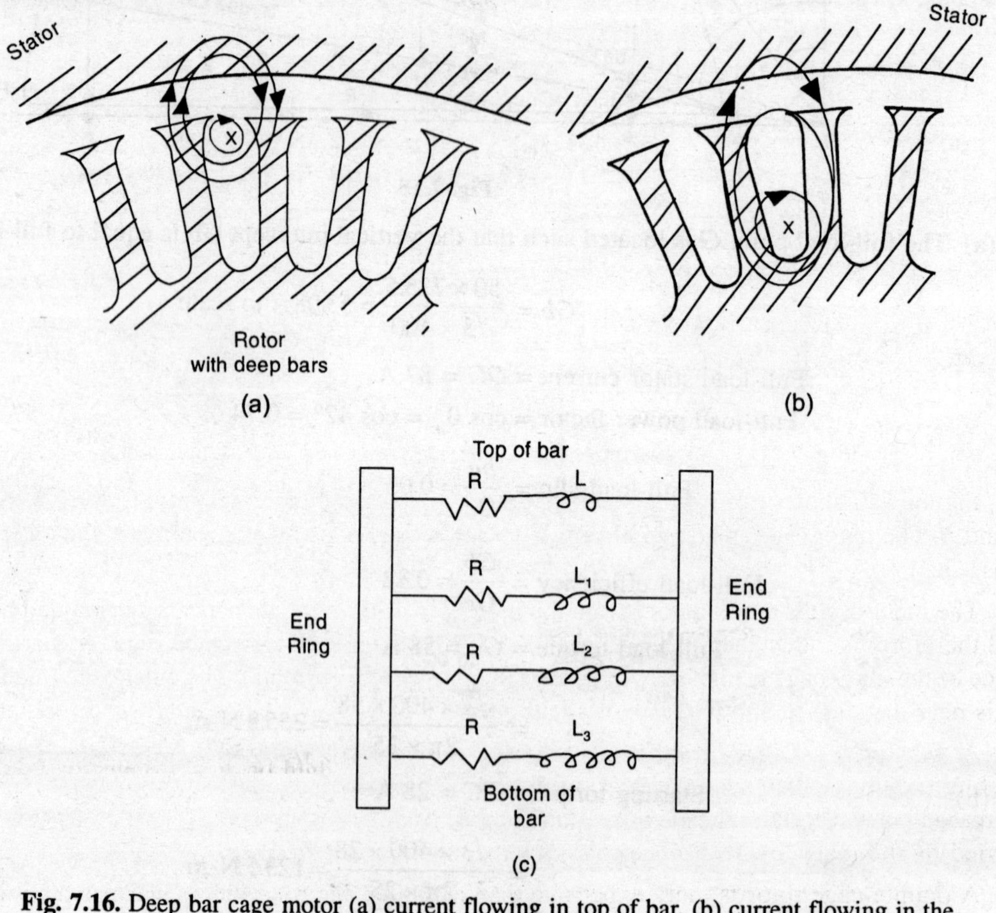


**Fig. 7.16.** Deep bar cage motor (a) current flowing in top of bar, (b) current flowing in the bottom of bar, (c) equivalent circuit of the bar as a function of depth of the bar

Figure 7.16(a) shows current in the upper part of the bar. The current in this part of the bar is tightly coupled to the stator. Therefore the leakage inductance is small for this part of the bar. Figure 7.16 (b) shows current in the lower part of the bar. The leakage inductance of this part of the bar is higher. All parts of the bar are in parallel and the equivalent circuit consists of a number of parallel branches, the upper ones having a lower inductance as compared to the lower branches, (Fig. 7.16 c).

When slip is small, (i.e. during running conditions) the rotor frequency is small (about 2 Hz or so) and the reactance of all the parallel branches are small as compared to resistance. The impedances of all branches are nearly equal and the current in different parts of the bars are equal. The resulting large cross-sectional area makes the rotor resistance quite small. Therefore efficiency is high.

When slip is high (i.e. at the time of starting) the rotor frequency is high and the reactances are large as compared to resistance of the different parallel branches. Therefore, almost the whole of rotor current is forced to flow in the low reactance part of bar (i.e. top part of the bar). This results in lower effective cross-section and higher rotor resistance. The high rotor resistance gives good starting torque.

## 7.18 DOUBLE CAGE ROTOR

Another economical method of obtaining a comparatively high rotor resistance at starting and low rotor resistance at running is to use a double-cage rotor.

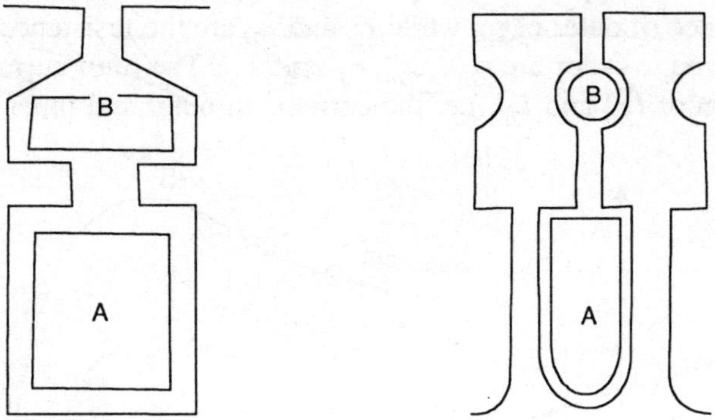


**Fig. 7.17.** Rotor punchings for double cage rotor of a 3-phase induction motor

Figure 7.17 shows two rotor punchings for double cage motors. Each rotor is divided into two parts *A* and *B*. The inner cage (made up of bars *A*) is deeply embedded in the rotor iron and has bars of large cross-section.

The inner cage is much farther from the periphery of the core, its bars are surrounded by more iron and therefore its inductance is higher than that of the outer high resistance cage. At starting the reactance of the inner cage is high (due to line frequency of rotor currents). The outer cage (made up of bars *B*) is near the surface and has bars of small cross-section. Its reactance is low and resistance is high.

At starting most of the rotor current is carried by the high resistance outer cage (bar *B*) and high starting torque is obtained. As motor speeds up the rotor frequency decreases, permitting the flow of increased current in the inner low resistance cage. At full running speed, most of the rotor current is carried by the inner low resistance cage, thus giving a high efficiency.

A double-cage rotor is more expensive than a single-cage rotor, but is cheaper than wound rotor. However, it permits the combination of high starting torque and good efficiency (during running conditions) at a lower cost (as compared to wound rotor) and without the need of maintaining sliprings and brushes.

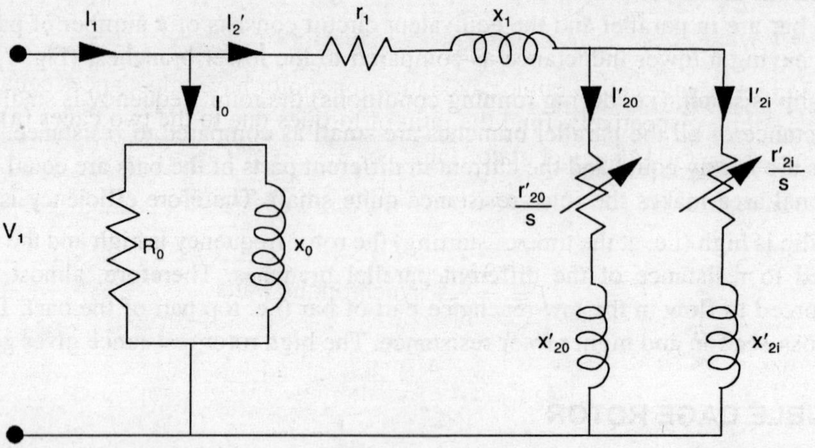


**Fig. 7.18.** Approximate equivalent circuit of a double cage 3-phase induction motor

Figure 7.18 shows the approximate equivalent circuit of a double cage motor. $r_{2o}$ and $x_{2o}$ are the resistance and reactance of outer cage, while $r_{2i}$ and $x_{2i}$ are the resistance and reactance of inner cage. These values as referred to stator are $r_{2o}'$, $x_{2o}'$, $r_{2i}'$ and $x_{2i}'$. The rotor current (as referred to stator) is $I_2'$ and is the phasor sum of $I_{2o}'$ and $I_{2i}'$, i.e. the currents in outer and inner cage, respectively.

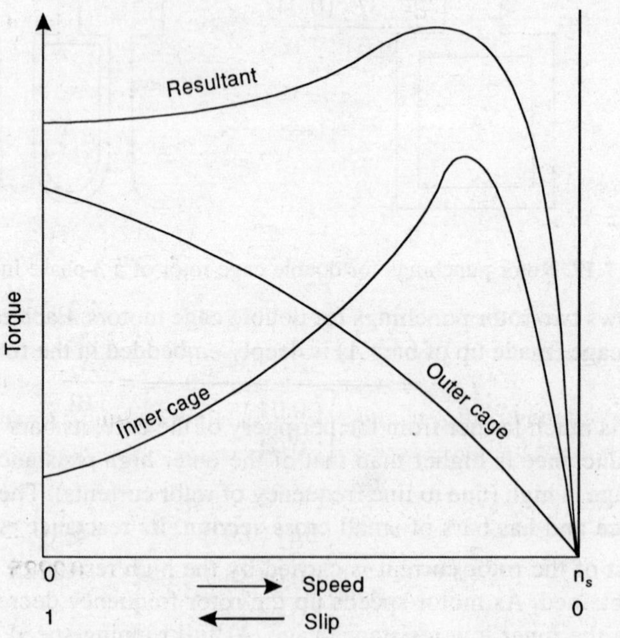


**Fig. 7.19.** Torque slip curve for a double cage 3-phase induction motor

Figure 7.19 shows the torque-slip curve of a double cage motor. When slip is 1 (i.e. at starting) a greater portion of torque is contributed by outer cage. During running, the outer cage contributes only a small portion of the total torque.

*Example 7.13*

The outer and inner cages of a double cage induction motor have impedances equal to $(0.04 + j\,0.1)$ and $(0.01 + j\,0.4)$ ohm, respectively. Find the ratio of torques due to the two cages (a) at starting, (b) when running with slip = 0.04. Neglect stator impedance.

*Solution*

Now
$$T = \frac{3I_2^2 r_2}{s} \text{ synchronous watts}$$

$$I_2 = \frac{sE_2}{\left(r_2^2 + s^2 x_2^2\right)^{0.5}}$$

or
$$T = \frac{3s^2 E_2^2 r_2}{s\left(r_2^2 + s^2 x_2^2\right)} = \frac{3E_2^2 r_2}{s\left(\dfrac{r_2^2}{s^2} + x_2^2\right)}$$

(a) At starting $s = 1$

$$T_{\text{outer}} = \frac{3E_2^2(0.04)}{0.04^2 + 0.1^2}$$

$$T_{\text{inner}} = \frac{3E_2^2(0.01)}{0.01^2 + 0.4^2}$$

$$\frac{T_{\text{outer}}}{T_{\text{inner}}} = \frac{0.04}{0.04^2 \times 0.1^2} \times \frac{0.01^2 + 0.4^2}{0.01} = 55.2$$

(b) At $s = 0.04$

$$T_{\text{outer}} = \frac{3E_2^2(0.04)}{0.04\left[\left(\dfrac{0.04}{0.04}\right)^2 + 0.1^2\right]} = \frac{3E_2^2}{1.01}$$

$$T_{\text{inner}} = \frac{3E_2^2(0.01)}{0.04\left[\left(\dfrac{0.01}{0.04}\right)^2 + 0.4^2\right]} = \frac{3E_2^2(0.25)}{0.2225}$$

$$\frac{T_{\text{outer}}}{T_{\text{inner}}} = \left(\frac{3E_2^2}{1.01}\right)\left[\frac{0.2225}{3E_2^2(0.25)}\right] = 0.88.$$

## 7.19 COGGING AND CRAWLING

A cage motor may show peculiar behaviour at starting, if the motor has a certain relationship between the number of poles and the stator and rotor slots. When the number of stator and rotor slots are equal, the machine may not start, a phenomenon known as cogging. For some other ratio of rotor-to-stator slots, the machine may run stably at low speed, e.g. one-seventh of normal, a phenomenon known as crawling. These phenomena are almost absent in slipring machines having high starting torque.

In addition to the magnetic field corresponding to fundamental frequency, a cage motor possesses harmonic fields. These parasitic magnetic fields are due to (a) winding arrangement, (b) slotting, (c) saturation, and (d) gap length irregularity. The $n$th order harmonic has an amplitude $1/n^2$ of the fundamental. The movement of the fields due to these harmonics is in accordance with the expression $n = 6m \pm 1$, where $m$ is any integer. If $n = 6m + 1$, the motion is in the same direction as fundamental, whereas if $n = 6m - 1$, the motion is in reverse direction. The fifth harmonic has an amplitude of about 4% of fundamental and rotates in a direction opposite to that of fundamental. The seventh harmonic has an amplitude equal to 2% of fundamental and rotates in the same direction as fundamental. Certain combinations of stator and rotor poles augment the effect of these space harmonics.

The harmonic fields also produce rotor emfs, rotor currents and torques in the same way as fundamental. These torques have the same general shape as that due to fundamental field, but with synchronous speed one-fifth (in reverse direction) and one-seventh (in forward direction) for the fifth and seventh harmonics, respectively. The resultant torque-slip curve is a combination of fundamental, fifth and seventh harmonic torques and is shown in Fig. 7.20.

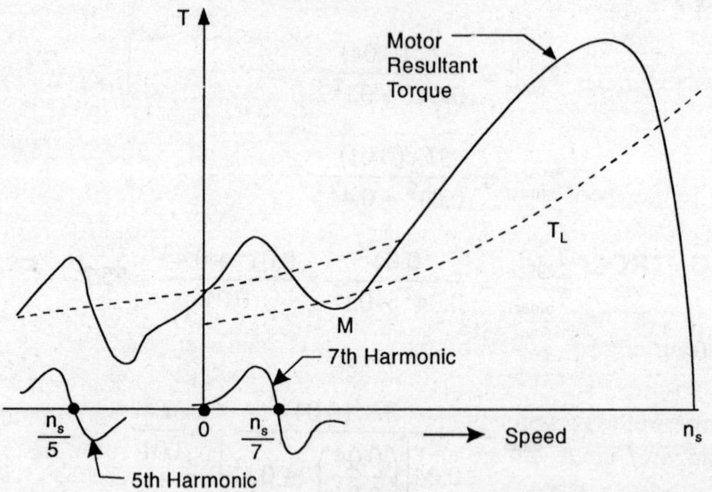


**Fig. 7.20.** Torque slip curve of a 3-phase induction motor as affected by harmonics

It is seen that this curve has a prominent saddle effect. A stable operating region exists around one-seventh of normal speed. The load-torque curve ($T_L$) intersects the motor torque-slip curve at point $M$ and the motor may continue to crawl at one-seventh of the normal speed.

Certain stator and rotor slot ratios may produce a synchronising torque (similar to that in synchronous motor). For a 4-pole machine having 24 stator slots and 28 rotor slots, the stator develops a reversed eleventh and forward thirteenth harmonic and the rotor develops a reversed thirteenth and forward fifteenth harmonic. The thirteenth harmonics of stator and rotor have opposite directions of rotation. Let $n_s$ be the synchronous speed of fundamental and $n$ the rotor speed. Then speed of thirteenth stator harmonic is $n_s/13$ with respect to stator and that of thirteenth rotor harmonic is $-(n_s - n)/13$ with respect to rotor. The stator and rotor thirteenth harmonics fall into step and produce a synchronising torque, when

$$\frac{n_s}{13} = n - \frac{n_s - n}{13}$$

or

$$n = n_s/7 \qquad \text{...(7.47)}$$

Therefore the thirteenth harmonic produces a discontinuity in torque-slip characteristics at one-seventh the synchronous speed. This is shown in Fig. 7.21. The motor may continue to operate at point $M$ corresponding to one-seventh synchronous speed.

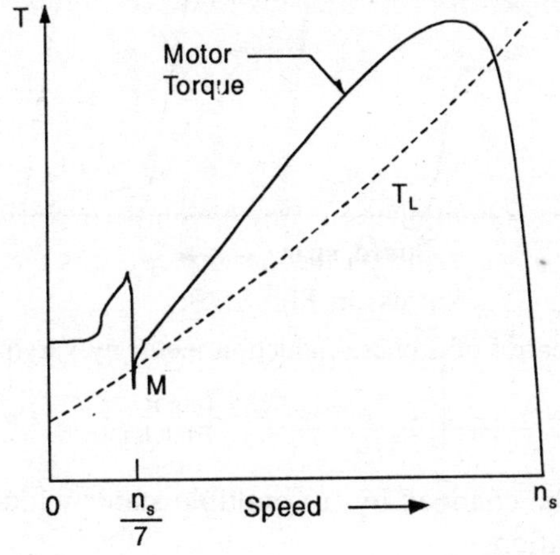


**Fig. 7.21.** Torque speed curve of 3-phase induction motor showing sub synchronous crawling

## 7.20 SPEED CONTROL

The slip of an induction motor is very small (less than 5%). As such this motor operates at almost constant speed. Speed control is possible only by elaborate methods.

Since the synchronous speed depends on line frequency and number of poles, the two basic methods of speed control are: (1) frequency control, (2) pole changing. In addition to this, limited variation is possible by change of input voltage and addition of resistance in rotor circuit.

### 7.20.1 Frequency Control

A change of input frequency changes the synchronous speed and provides a direct method of speed control. To keep the magnetisation current within limits, the applied voltage must be reduced in direct proportation to the frequency. Otherwise the magnetic circuit will become saturated resulting in excessive magnetisation current. If voltage is changed in direct proportion to frequency, the resulting torque-speed curves will be as shown in Fig. 7.22.

The starting torque at reduced frequency is not reduced in the same proportion, because rotor power factor improves with reduction in frequency. The torque that can be produced by maximum permissible rotor current is equal to that at rated conditions. Since power is product of torque and speed, operation at reduced speed results in lesser permissible output.

In earlier days, the variable frequency was obtained from a motor-generator set or mercury arc inverter. The present day inverters use SCRs.

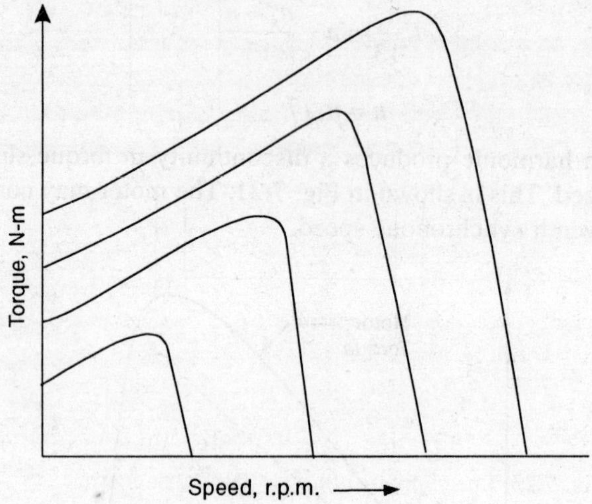


**Fig. 7.22.** Speed control of 3-phase induction motor by varying line frequency

### 7.20.2 Pole Changing

The number of stator poles can be changed by (a) multiple stator winding, (b) method of consequent poles, (c) pole amplitude modulation.

#### *(a) Multiple Stator Winding*

The stator has two different windings with different number of poles and only one winding is energized at a time. A motor may have two windings (say) for 6 and 4 poles and the synchronous speeds will be 1000 and 1500 rpm, respectively. If the full-load slip is 5% in each case the operating speeds will be 950 rpm and 1425 rpm. Such a machine tends to be more costly and less efficient. Therefore this method is used only when absolutely necessary.

#### *(b) Method of Consequent Poles*

In this method the number of poles can be changed in the ratio of 2:1 by changing the connection of coils.

Figure 7.23 shows four coils of one phase of a stator winding with alternate coils in series and terminals $x_1$, $y_1$, $x_2$, $y_2$ brought out as shown. If these terminal are connected in the order $x_1\ y_1\ y_2\ x_2$, there will be a total of 4 poles giving a synchronous speed of 1500 rpm. If these terminals are connected in the order $x_1\ y_1\ x_2\ y_2$, all the four coils will magnetise in the same direction and form *N* poles. Consequent poles of opposite polarity (i.e. *S*) will be formed in between the *N* poles (as shown) giving a total of 8 poles and a synchronous speed of 750 rpm. It is also possible to connect these coils in parallel for the low and high speeds. The series and parallel connections are illustrated in Fig. 7.24 (a and b).

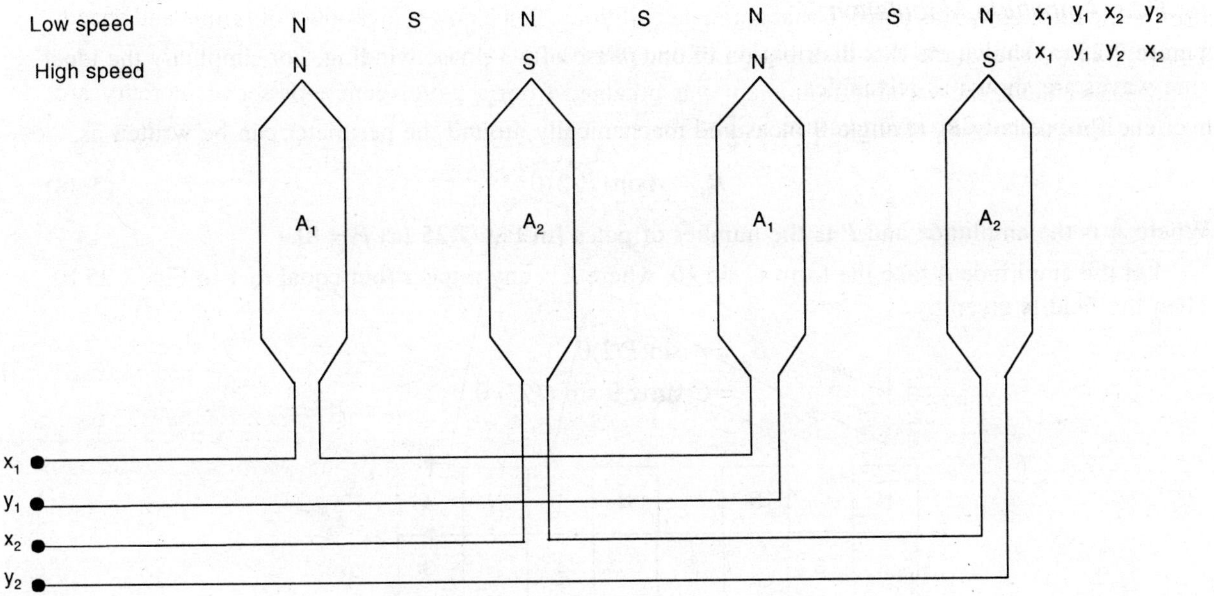


**Fig. 7.23.** Speed control through pole changing

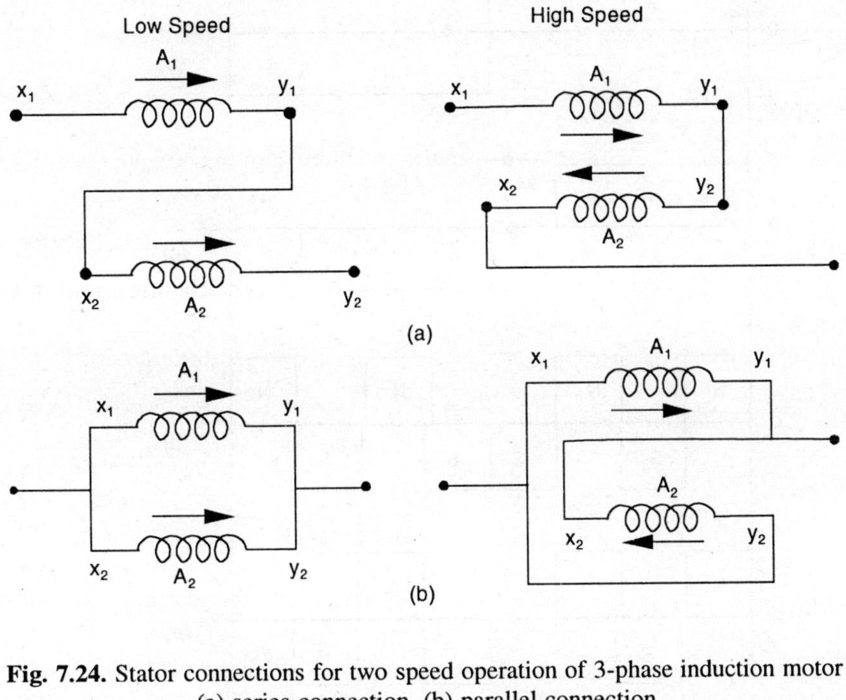


**Fig. 7.24.** Stator connections for two speed operation of 3-phase induction motor (a) series connection, (b) parallel connection

It is important to note that slot angle (i.e. electrical degrees), phase spread, breadth factor and pitch factors will be different for the low and high speed connections. The three phases can be connected in star or delta, thus giving a number of connections.

### (c) Pole Amplitude Modulation

Figure 7.25 (a) shown the flux distribution of one phase of a 3-phase winding. For simplicity the ideal sine waves are shown as rectangles.

The flux density $B_\theta$ at angle θ measured mechanically around the perimeter can be written as

$$B_\theta = A \sin(P/2)\theta \qquad ...(7.48)$$

Where $A$ is the amplitude and $P$ is the number of poles [in Fig. 7.25 (a) $P = 8$].

Let the amplitude $A$ take the form $C \sin k\theta$, where $k$ is any integer (but equal to 1 in Fig. 7.25 b). Then the field is given by

$$B_\theta = A \sin(P/2)\theta$$
$$= C \sin k\,\theta \sin(P/2)\,\theta$$

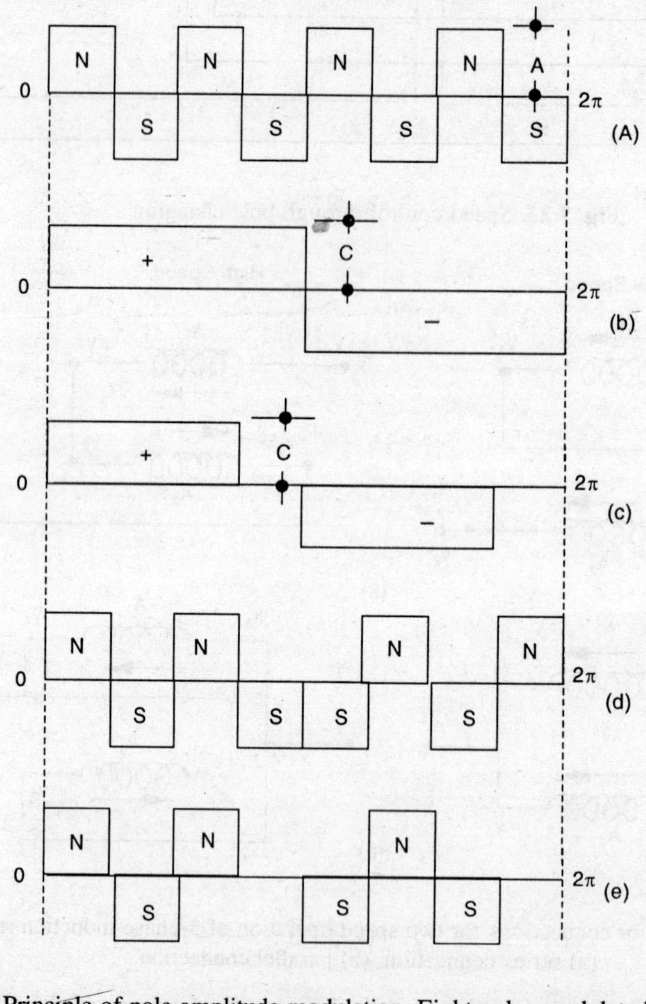


**Fig. 7.25.** Principle of pole amplitude modulation. Eight poles modulated to 6 poles (a) original 8-pole stator magnetic field, (b and c) space modulating waves, (d and e) final 6 pole stator magnetic field

$$= \frac{C}{2} \cos\left(\frac{P}{2} - k\right)\theta - \cos\left(\frac{P}{2} + k\right)\theta \qquad ...(7.49)$$

Equation (7.49) indicates a double field of pole numbers ($P - 2k$) and ($P + 2k$), respectively super imposed on each other. In the case shown in Fig. 7.25, the resultant field $B_\theta$ will be a mixture of two fields, i.e. a 6-pole field and 8-pole field.

It is necessary to remove one of the above two pole numbers and retain the other. It has been found that if the windings of the three phases, before modulation, are spaced electrically by ($2\pi r$)/3, where $r$ is given by any of the following equations, one of the pole numbers is eliminated.

Thus

$$\frac{m}{r} = \frac{1}{3}\left(1 + \frac{2k}{P}\right) \qquad ...(7.50a)$$

or

$$\frac{m}{r} = \frac{1}{3}\left(1 - \frac{2k}{P}\right) \qquad ...((7.50b)$$

where $m$ is any integer.

If $P = 8$ and $k = 1$, the number of poles after modulation is 6 and 10. If $m = 1$ and $r = 4$, Eq. (7.50 b) is satisfied and the lower pole number 6 vanishes. Thus an 8-pole field is modulated to 10-pole field.

There are two methods of connections by which the desired modulation can be achieved in practice. In the first method, the second half of each phase winding is reversed with respect of the first half [Fig. 7.25 (b)]. In the second method a section is first omitted from each half and half of the remaining portion is then reversed with respect to the first half (in Fig. 7.25 (c), 4th and 8th coils are omitted and 5th, 6th and 7th coils are then reversed with respect to the first three). These two changes are equivalent to multiplication of the original ware (Fig. 7.25 a) by one or the other rectangular waves shown in Fig. 7.25 (b and c). The resultant waves obtained are shown in Fig. 7.25 (d and e).

The methods of speed control by pole changing (multiple stator winding, method of consequent poles and pole amplitude modulation) are suitable for squirrel cage motors only because, a cage rotor has as many poles induced in it as there are in the stator and can thus adopt when the number of stator poles changes.

### 7.20.3 Line Voltage Control

The torque developed by an induction motor is proportional to square of voltage. As shown in Fig. 7.26 speed control, over a limited range, can be achieved by change of line voltage. The three torque-speed curves correspond to input-voltages 1 pu, 0.75 pu and 0.5 pu.

The operating speeds in the three cases are also shown. This method is sometimes used on small motors driving fans, whose torque requirement is proportional to square of speed.

### 7.20.4 Rotor Resistance Control

The shape of torque-speed curve of a wound rotor induction motor can be adjusted by addition of resistance in the rotor circuit. As seen in Fig. 7.27 speed control over a limited range is possible by this method. However addition of resistance in the rotor circuit reduces the efficiency. Therefore this method should be used only when speed control is required for a short period.

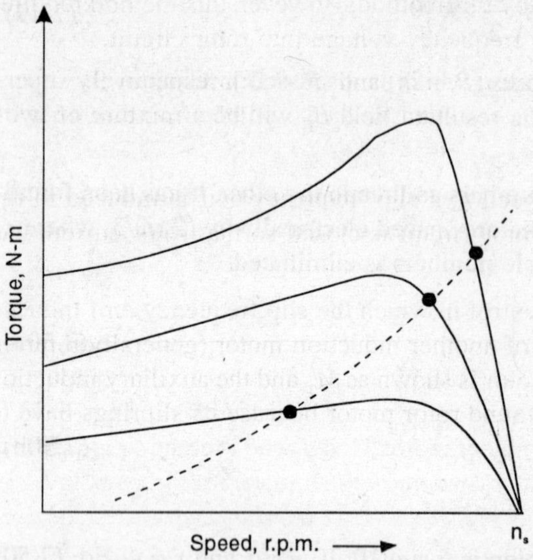


**Fig. 7.26.** Speed control of 3-phase induction motor by line voltage control

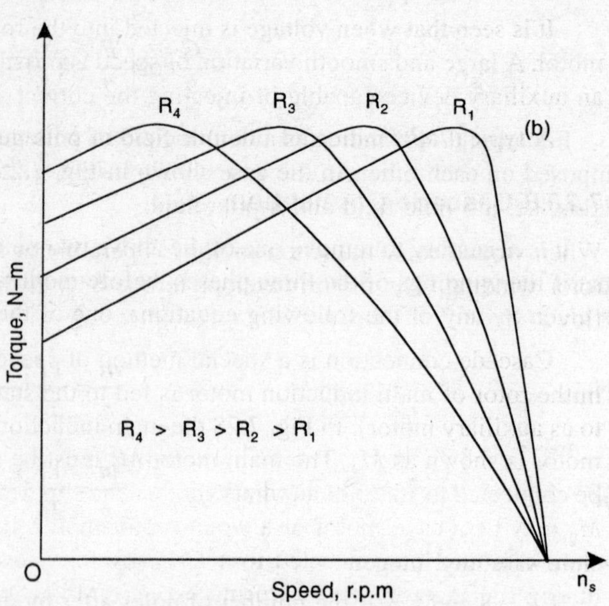


**Fig. 7.27.** Speed control of 3-phase induction motor by changing rotor resistance

### 7.20.5 Voltage Injection in Rotor Circuit

The speed control of a wound rotor induction motor can also be achieved by injection of a slip frequency voltage into the rotor winding (i.e. across the slip rings).

Assume that rated torque is to be delivered to a constant torque load at different speeds. At any slip $s$ the rotor reactance $sx_2$ is small as compared to rotor resistance and is therefore negligible. The rotor power factor is, therefore, almost unity.

The torque developed by induction motor is given by Eq. (7.2), which is

$$T = KB_2B_R \sin \delta \qquad ...(7.51)$$

The resultant field $B_R$ is proportional to the applied voltage and is almost constant. As derived in Eq. (7.28), $\sin \delta = \cos \theta_2$, i.e. rotor power factor and is almost unity.

The rotor field $B_2$ is directly proportional to rotor current $I_2$. To maintain constant torque, $I_2$ should be constant. Under normal operation with slip $s$ and negligible rotor reactance, the rotor current and voltage are related by

$$I_2 = sE_2/r_2 \qquad ...(7.52)$$

If a 3-phase voltage $V_i$ per phase is injected into the rotor circuit, the slip changes to a new value $s'$, such that

$$I_2r_2 = sE_2 = s'E_2 + V_i \qquad ...(7.53)$$

It is evident that if $V_i$ is phase opposition to $sE_2$, the new slip $s'$ is higher than $s$, i.e. speed of the machine has decreased. If $V_i$ is in phase with $sE_2$, the new slip $s'$ is lower than $s$, so that the speed is higher. If $V_i$ is in phase with the equal to $sE_2$, then $s'$ is zero and the machine runs at synchronous speed. If the value of $V_i$ is increased still further, slip becomes negative and the machine operates at speed above synchronous speed.

It is seen that when voltage is injected into the rotor circuit, the machine becomes a doubly excited motor. A large and smooth variation of speed is possible by this method. However, this method requires an auxiliary device capable of injecting the correct slip frequency voltage into rotor circuit.

A typical application of this principle is schrage motor, which is discussed in Chapter 9.

### 7.20.6 Cascade Connection

When wound rotor induction motor is operating at a speed $n$, a slip frequency emf is induced in the rotor winding. The motor develops torque only if the rotor circuit is closed so that rotor current can flow.

Cascade connection is a special method of speed control in which the slip frequency emf induced in the rotor of main induction motor is fed to the stator of another induction motor (generally referred to as auxiliary motor). In Fig. 7.28 the main induction motor is shown as $M_1$ and the auxiliary induction motor is shown as $M_2$. The main motor $M_1$ must be a wound rotor motor because its sliprings have to be connected to stator of auxiliary motor. The slip frequency of motor $M_1$ is $f_2$ and is equal to $s_1 f_1$. Motor $M_2$ may be a cage motor or a wound rotor motor. If $M_2$ is a wound rotor motor, its sliprings must be short circuited or connected to a resistance as shown in Fig. 7.28. This resistance is generally used during starting conditions. During operation the slip rings of $M_2$ are short circuited.

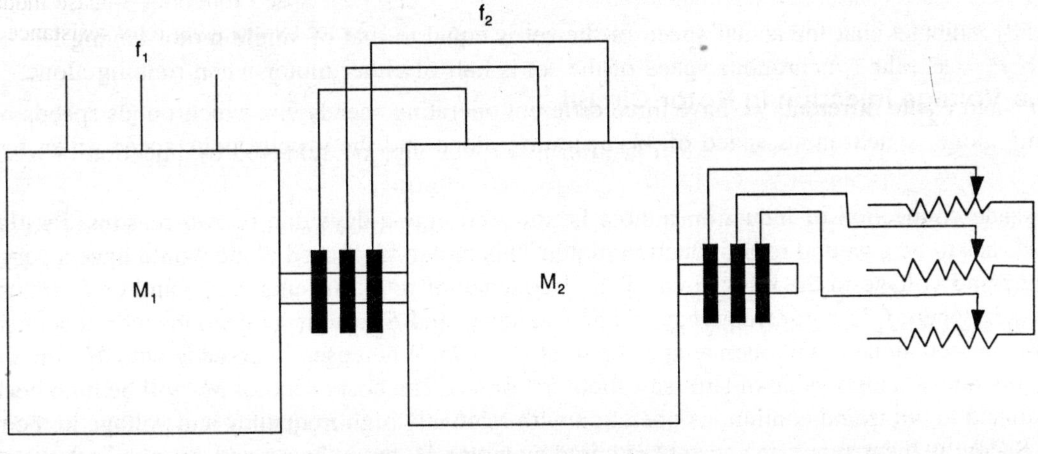


**Fig. 7.28.** Cascade connection

Let $P_1$ and $P_2$ be the number of poles of motors $M_1$ and $M_2$ and $f_1$ and $f_2$ be the frequencies of supply to stators of motors $M_1$ and $M_2$ respectively.

Then,

$$N_1 = \text{Synchronous speed of } M_1 = \frac{120 f_1}{P_1} \qquad ...(7.54)$$

$$N_2 = \text{Synchronous speed of } M_2 = \frac{120 f_2}{P_2} \qquad ...(7.55)$$

Let $s_1$ and $s_2$ be the slips of motors $M_1$ and $M_2$.

$$\text{Actual speed of } M_1 = N_1(1-s_1) = \frac{120 f_1}{P_1}(1-s_1) \qquad \text{...(7.56)}$$

$$\text{Actual speed of } M_2 = N_2(1-s_2) = \frac{120 f_2}{P_2}(1-s_2) \qquad \text{...(7.57)}$$

Since the two motors are mechanically coupled, their actual speeds are equal. Moreover $f_2 = s_1 f_1$. Therefore,

$$\frac{120 f_1}{P_1}(1-s_1) = \frac{120 s_1 f_1}{P_2}(1-s_2)$$

The sliprings of $M_2$ are shortcircuited during normal operation. Therefore $s_2 \cong 0$. Therefore,

$$s_1 = \frac{P_2}{P_1 + P_2}$$

and
$$N = \frac{120 f_1}{P_1}\left(1 - \frac{P_2}{P_1 + P_2}\right) = \frac{120 f_1}{P_1 + P_2} \qquad \text{...(7.58)}$$

Eq. (7.58) indicates that the actual speed of the set is equal to that of single motor having $(P_1 + P_2)$ poles. If $P_1 = P_2$, the synchronous speed of the set is half of either motor when running alone.

If $P_1$ and $P_2$ are different, we have three different operating speeds viz. synchronous speeds of $M_1$ operating alone, synchronous speed of $M_2$ operating alone and the synchronous speed given by Eq. (7.58).

Cascade connection of induction motors is not used now-a-days due to two reasons. Firstly the motor $M_1$ has to be a wound rotor induction motor. This motor $M_1$ if used alone would have a very low frequency and voltage in the rotor circuit [The frequency of rotor currents is $s_1 f_1$ and emf induced in rotor is $s_1 E_2$ where $f_1$ is stator frequency or line frequency and $E_2$ is emf induced in rotor at standstill]. When $M_1$ is used alone it will operate at a slip of about 0.05. When used in cascade with $M_2$, the motor $M_1$ will operate at a high value of slip (say about 0.4 or so). The cost of motor $M_1$ will be high because it is insulated to withstand continuous operation with relatively high frequency and voltage in the rotor circuit. Secondly the magnetising current required by motor $M_2$ has to be carried through both the stator and rotor of $M_1$. Therefore the cascade set operates at a lower power factor as compared to the power factor of a single motor with $(P_1 + P_2)$ poles.

*Example 7.14*

A 4-pole, 400 V, 3-phase, 15 hp induction motor has rotor resistance equal to 0.16 ohm per phase and standstill reactance equal to 0.27 ohm per phase. The voltage across sliprings at standstill is 141 V per phase. the motor develops a torque of 62.4 N-m at $s = 0.035$ and at a rotor current of 31.1 A. Find the magnitude and phase of emf to be injected into rotor circuit, so that motor may operate at a speed 40% above synchronous speed and develop the same torque.

*Solution*

Let $I_2$ and $I_2'$ be the initial and new value of rotor current and $\theta_2$ and $\theta_2'$ be the initial and new value of rotor power factor angle. Since torque is constant

$$I_2 \cos \theta_2 = I_2' \cos \theta_2'$$

$$\theta_2 = \tan^{-1}(sx_2/r_2) = \tan^{-1}(0.035 \times 0.27/0.16) = 3.4°$$

Since the new speed is 40% above synchronous speed

$$s' = -0.4$$

$$\theta_2' = \tan^{-1}(0.4 \times 0.27/0.16) = 34°$$

$$I_2' = 31.1 \cos(3.4°)/\cos(34°) = 37.45 \text{ A}$$

$$I_2'(r_2 + js' x_2) = s'E_2 + V_i$$

$$(37.45)[0.16 + j(0.4)(0.27)] = (-0.4 \times 141) + V_i$$

or
$$7.23 = -56.4 + V_i$$

or
$$V_i = 63.63 \text{ V.}$$

## 7.21 OPERATION WITH UNBALANCED SUPPLY VOLTAGE

In the above analysis we have assumed that the three phase windings of the induction motor are identical and the supply voltages are balanced. The similarity in the three windings is a necessary consequence of proper design and manufacture. As such the three windings are always similar. However the supply voltages may be unbalanced due to disturbances in the power supply network feeding the motor. The analysis with unbalanced supply voltages can be carried out using symmetrical components.

### 7.21.1 Symmetrical Components in Three Phase Systems

Any set of unbalanced three phase voltages (or currents) can be transformed into three balanced sets. These are:

(1) A positive sequence set of three symmetrical voltages (i.e. all numerically equal and displaced from each other by 120°) having the same phase sequence abc as the original set and denoted by $V_{a1}$, $V_{b1}$ and $Vc_1$ as shown in Fig. 7.29(a).

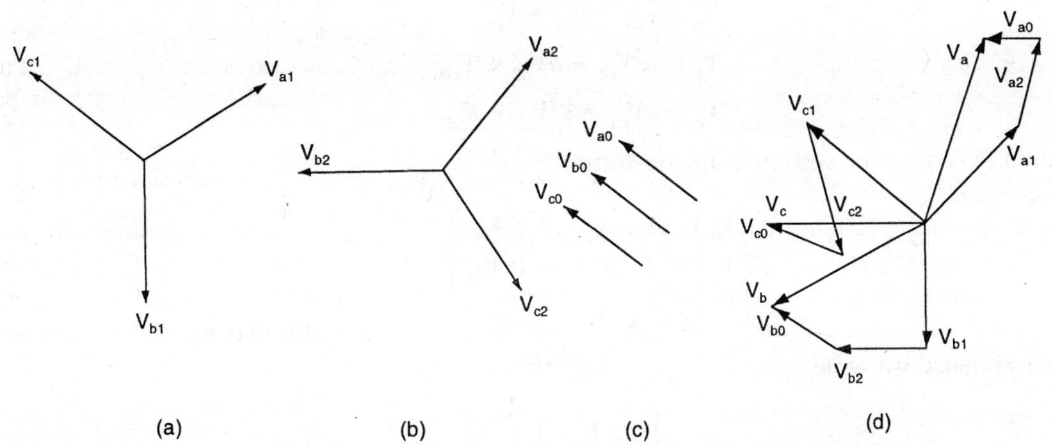


**Fig. 7.29.** Symmetrical components (a) Positive sequence set, (b) Negative sequence set, (c) Zero sequence set, (d) Summation of positive, negative and zero sequence voltages

(2) A negative sequence set of three symmetrical voltages having a phase sequence opposite to that of the original set and denoted by $V_{a2}$, $V_{b2}$ and $V_{c2}$ as shown in Fig. 7.29(b).

(3) A zero sequence set of three voltages, all equal in magnitude and in phase with each other and denoted by $V_{a0}$, $V_{b0}$ and $V_{c0}$ (Fig. 7.29c).

The positive, negative and zero sequence components are known as symmetrical components. Figure 7.29(d) shows the addition of symmetrical components to give the three phase voltages $V_a$, $V_b$ and $V_c$. Thus

$$\begin{aligned} V_a &= V_{a1} + V_{a2} + V_{a0} \\ V_b &= V_{b1} + V_{b2} + V_{b0} \\ V_c &= V_{c1} + V_{c2} + V_{c0} \end{aligned} \qquad ...(7.59)$$

Equation (7.59) can be put in a better form by using an operator '$a$' as under

$$a = 1 \angle 120° = 1e^{2\pi/3} = -0.5 + j\,0.866 \qquad ...(7.60)$$

Any phasor when multiplied by '$a$' gets rotated by 120° in positive (anticlock wise direction) without any change in magnitude. If the phasor is multiplied by '$a$' twice it gets rotated in positive direction by 240°. Thus

$$\begin{aligned} a^2 &= 1\angle 240° = -0.5 - j\,0.866 \\ a^3 &= 1\angle 360° = 1 \\ a^4 &= 1\angle 480° = 1\angle 120° = a \end{aligned} \qquad ...(7.61)$$

It is seen from Fig. 7.29 that components of voltages $V_b$ and $V_c$ can be written in terms of components of voltage $V_a$ by using operator '$a$'. Thus

$$\begin{aligned} V_{b1} &= a^2 V_{a1} & V_{c1} &= aV_{a1} \\ V_{b2} &= aV_{a2} & V_{c2} &= a^2 V_{a1} \\ V_{b0} &= V_{a0} & V_{c0} &= V_{a0} \end{aligned} \qquad ...(7.62)$$

Substituting Eq. (7.62) into Eq. (7.59) we get

$$\begin{aligned} V_a &= V_{a1} + V_{a2} + V_{a0} \\ V_b &= a^2 V_{a1} + aV_{a2} + V_{a0} \\ V_c &= aV_{a1} + a^2 V_{a2} + V_{a0} \end{aligned} \qquad ...(7.63)$$

Equation (7.63) can be written in matrix form

$$\begin{bmatrix} V_a \\ V_b \\ V_c \end{bmatrix} = \begin{bmatrix} 1 & 1 & 1 \\ 1 & a^2 & a \\ 1 & a & a^2 \end{bmatrix} \begin{bmatrix} V_{a0} \\ V_{a1} \\ V_{a2} \end{bmatrix} \qquad ...(7.64)$$

For convenience we write

$$[A] = \begin{bmatrix} 1 & 1 & 1 \\ 1 & a^2 & a \\ 1 & a & a^2 \end{bmatrix} \qquad ...(7.65)$$

It can be easily verified that

$$[A]^{-1} = \frac{1}{3}\begin{bmatrix} 1 & 1 & 1 \\ 1 & a & a^2 \\ 1 & a^2 & a \end{bmatrix} \quad \text{...(7.66)}$$

Multiplying both sides of Eq. (7.64) by $A^{-1}$ we get

$$\begin{bmatrix} V_{a0} \\ V_{a1} \\ V_{a2} \end{bmatrix} = \frac{1}{3}\begin{bmatrix} 1 & 1 & 1 \\ 1 & a & a^2 \\ 1 & a^2 & a \end{bmatrix}\begin{bmatrix} V_{a0} \\ V_b \\ V_c \end{bmatrix} \quad \text{...(7.67)}$$

Eqs. (7.64) and (7.67) are valid for currents also. Thus using these equations we can transform phase voltages and currents into symmetrical components and vice versa.

The advantage of using symmetrical components is that each sequence voltage causes a current of the same sequence. Thus each sequence current can be assumed to flow in an independent circuit composed of impedances to the currents of that sequence only. If a positive sequence set of voltages is applied to the terminals of induction motor, a magnetic field rotating in a certain direction is set up. This field interacts with the rotor winding to produce a torque $T_1$. The negative sequence set of voltages sets up a magnetic field rotating in an opposite direction to that of positive sequence. The interaction of this field with the rotor winding produces a torques $T_2$. The resultant torque is $T_1 - T_2$.

The stator windings of the induction motor may be connected in star or delta. Even when connected in star, the neutral is seldom brought out and the motor is fed by a 3-phase 3-wire system. Therefore the zero sequence impedance is infinite. However zero sequence voltages may exist at the terminals and at the neutral point of star connected windings.

### 7.21.2 Positive Sequence Equivalent Circuit

The positive sequence equivalent circuit is similar to that shown in Fig. 7.3(d) or 7.4. The only difference is that the voltage applied is positive sequence voltage $V_{a1}$. Accordingly the stator current is $I_{11}$ (where first subscript 1 refers to stator and the second subscript 1 refers to positive sequence) and rotor current is $I_{21}$. The rotor current as referred to stator is $I_{21}'$.

$$\text{Power input} = 3\,V_{a1}\,I_{11}\cos\theta_1 \quad \text{...(7.68)}$$

where $\theta_1$ is the phase angle between positive sequence voltage $V_{a1}$ and positive sequence stator current $I_{11}$.

Positive sequence air gap power $P_{g1}$

= Positive sequence power input–stator copper-losses–stator core-losses ...(7.69)

Also
$$P_{g1} = \frac{3I_{21}^2 r_2}{s} = \frac{3I_{21}'^2 r_2'}{s} \quad \text{...(7.70)}$$

where $I_{21}$ is rotor current, $I_{21}'$ is rotor current as referred to stator, $r_2$ is rotor resistance, $r_2'$ is rotor resistance as referred stator and $s$ is the slip.

$$\text{Rotor copper-losses} = 3I_{21}^2 r_2 = 3I_{21}'^2 r_2' \quad \text{...(7.71)}$$

$$T_1 = \frac{P_{g1}}{2\pi n_s} \qquad \text{...(7.72)}$$

Eqs. (7.68 to 7.72) are similar to the corresponding equations in Section 7.9.

### 7.21.3 Negative Sequence Equivalent Circuit

The slip $s$ is defined in terms of the speed of the rotor in the positive direction with respect to the rotating field set up by positive sequence voltages. The rotating field due to negative sequence voltages has a direction opposite to that of positive sequence field. Therefore the slip $s_2$ for negative sequence equivalent circuit can be found as under

$$s = \frac{n_s - n}{n_s} \qquad \text{...(7.73)}$$

where $n$ and $n_s$ are the actual rotor speed and synchronous speed in revolutions per second.

$$s_2 = \frac{n_s - (-n)}{n_s} = 1 + \frac{n}{n_s} \qquad \text{...(7.74)}$$

Combining Eqs. (7.73) and (7.74) we get

$$s_2 = 2 - s \qquad \text{...(7.75)}$$

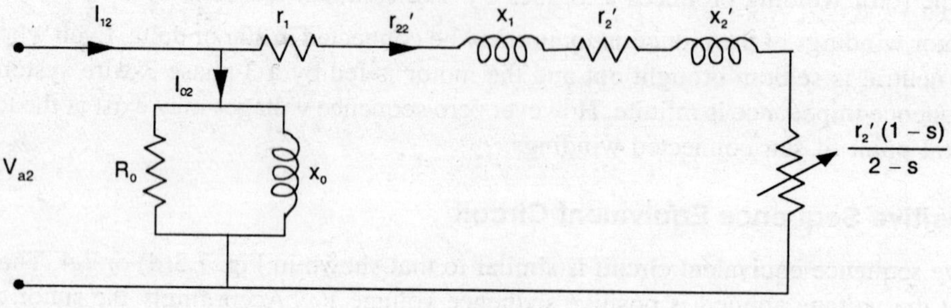


**Fig. 7.30**

The approximate equivalent circuit for negative sequence is shown in Fig. 7.30. The applied voltage is $V_{a2}$. The stator and rotor currents are $I_{12}$ and $I_{22}$. The rotor current as referred to stator is $I_{22}'$. The resistance which replaces the mechanical load is $r_2(1 - s)/(2 - s)$ and this resistance when referred to stator becomes $r_2'(1 - s)/(2 - s)$. $I_{02}$ denotes no-load current for negative sequence

$$\text{Negative Sequence Power input} = 3V_{a2}I_{12}\cos\theta_2 \qquad \text{...(7.76)}$$

where $\theta_2$ is the phase angle between $V_{a2}$ and $I_{12}$

$$\text{Negative sequence air gap power } P_{g2} = \frac{3I_{22}^2 r_2}{2-s} = \frac{3I_{22}'^2 r_2'}{2-s} \qquad \text{...(7.77)}$$

$$\text{Negative sequence rotor copper-losses} = 3I_{22}^2 r_2 = 3I_{22}'^2 r_2' \qquad \text{...(7.78)}$$

$$T_2 = \frac{P_{g2}}{2\pi n_s}$$

$$\text{Net torque} = T_1 - T_2 = \frac{P_{g1} - P_{g2}}{2\pi n_s} \qquad ...(7.79)$$

If Eq. (7.79) is multiplied by $(1 - s)$ we get the shaft output torque.

The zero sequence voltages, even if present, do not produce any resultant field and do not affect the torque or mechanical power output.

### 7.21.4 Single Phasing

An extreme case of unbalance supply voltage can occur if one phase of supply gets disconnected when the motor is running. It could happen due to blowing of fuse in one phase. If the motor is operating on full load and single phasing occurs, the motor may even stall. If the load is less the motor may continue operating but with poor performance. Both stator and rotor will get over heated. The temperature rise in stator will be very prominent in two phases if it is star connected and in one phase if it is delta connected.

The performance of the motor under the condition of single phasing can also be evaluated using symmetrical components. Figure 7.31 shows some characteristics under this condition. For comparison the characteristics for balanced 3-phase operation are also shown.

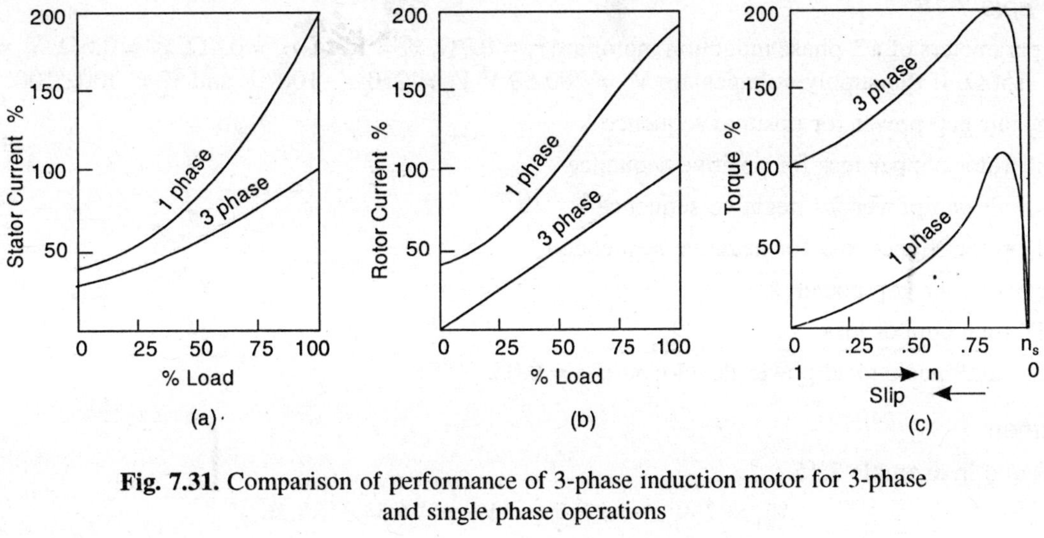


**Fig. 7.31.** Comparison of performance of 3-phase induction motor for 3-phase and single phase operations

*Example 7.15*

In an unbalanced 3-phase system the phase voltages are $V_a = 200\angle 0$ *V*, $V_b = 250\angle{-100°}$ *V*, $V_c = 300\angle 100°$ *V*. Find the symmetrical components $V_{a1}$, $V_{a2}$ *and* $V_{a0}$.

*Solution*

Using Eq. (7.67)

$$V_a = 200 + j0,\ V_b = -43.4 - j246.2 \text{ and } V_c = -52.1 + j295.4$$

$$V_{a0} = \frac{1}{3}(V_a + V_b + V_c) = \frac{1}{3}(+104.5 + j49.2)$$

$$= 34.83 + j16.4V = 38.5\angle 25.2°V$$

$$V_{a1} = \frac{1}{3}\left(V_a + aV_b + a^2 V_c\right)$$

$$V_{a1} = \frac{1}{3}[200\angle 0 + (1\angle 120°)(250\angle -100°) + (1\angle 240°)(300\angle 100°)]$$

$$= \frac{1}{3}[200\angle 0 + 250\angle 20° + 300\angle 340°]$$

$$= 238.94 - j5.7 = 239\angle -1.37°V$$

$$V_{a2} = \frac{1}{3}\left(V_a + a^2 V_b + aV_c\right)$$

$$= \frac{1}{3}[200\angle 0 + 250\angle 140° + 300\angle 220°]$$

$$= -73.77 - j10.71 = 74.54\angle 188.26°V.$$

*Example 7.16*

The parameters of a 3-phase induction motor are $r_1 = 0.7\Omega$, $x_1 = 1.3\Omega$, $r_2' = 0.4\Omega$, $x_2' = 0.6\Omega$, $X_0 = 35\Omega$, $R_0 = 150\Omega$. If the supply voltages are $V_a = 200\angle 0\ V$, $V_b = 250\ \angle -100°\ V$ and $V_c = 300\angle 100° V$ find

(a) air gap power for positive sequence
(b) rotor copper loss for positive sequence
(c) air gap power for negative sequence
(d) rotor copper loss for negative sequence
(e) total air gap power
(f) total copper loss
(g) total mechanical power developed if $s = 0.05$.

*Solution*

As found in example 7.15

$$V_{a1} = 239\ \angle -1.37°V, \qquad V_{a2} = 74.54\angle 188.26°V$$

(a) Positive sequence rotor current (as referred to stator)

$$I'_{21} = \frac{V_{a1}}{\left(r_1 + \frac{r'_2}{s}\right) + j\left(x_1 + x'_2\right)}$$

$$= \frac{239\angle -1.37°}{\left(.7 + \frac{.4}{.05}\right) + j(1.3 + .6)} = \frac{239\angle -1.37°}{8.74\angle 13.36°}$$

or $$I'_{21} = 26.73\angle -14.73°\ A$$

$$P_{g1} = 3I_{21}'^2 \frac{r_2'}{s} = 3(26.73)^2\left(\frac{.4}{.05}\right) = 17147.83 \text{ W}$$

(b) Positive sequence rotor copper losses $= 3I_{21}'^2 r_2'$

$$= 3(26.73)^2(.4) = 857.39\ W$$

(c)

$$I_{22}' = \frac{V_{a2}}{\left(r_1 + \frac{r_2'}{2-s}\right) + j(x_1 + x_2')}$$

$$= \frac{74.54\angle 188.26°}{0.7 + \frac{0.4}{2-0.5} + j(1.3 + .6)} = \frac{74.54\angle 188.26°}{.905 + j1.9}$$

$$= \frac{74.54\angle 188.26°}{2.1\angle 64.53°} = 35.49\angle 123.73° \text{ A}$$

$$P_{g2} = 3I_{22}'^2 \frac{r_2'}{2-s} = 3(35.49)^2\left(\frac{.4}{1.95}\right) = 775.1 \text{ W}$$

(d) Negative sequence rotor copper-losses

$$= 3I_{22}'^2 r_2' = 3(35.49)^2 (0.4)$$

$$= 1511.45 \text{ W}$$

(e) Total air gap power $= 17147.83 + 775.1$

$= 17922.93$ W

(f) Total copper-losses $= 857.39 + 1511.45$

$= 2368.84$ W

(g) Total mechanical power developed

$= 17922.93 - 2368.84 = 15554.09$ W.

## 7.22 INDUCTION GENERATOR

Figure 7.32 shows the complete torque-speed curve of an induction machine for all ranges of speed. If the machine is driven at a speed greater than synchronous speed by a prime mover, the direction of induced torque reverses and it acts as an induction generator. As seen by the torque-speed curve, there is a maximum torque, which can be induced when it is acting as a generator (just as in motoring mode the maximum torque or pull out torque exists). If the prime mover applies a torque greater than this maximum value, the machine will overspeed.

The rotating magnetic field in an induction motor is set up by the magnetising current drawn from the mains. Even when the speed of the machine is above synchronous speed and it is acting as a generator, this magnetising current must be available. Thus an induction generator is not self-exciting, but must be operated in parallel with another generator which can supply its magnetising current.

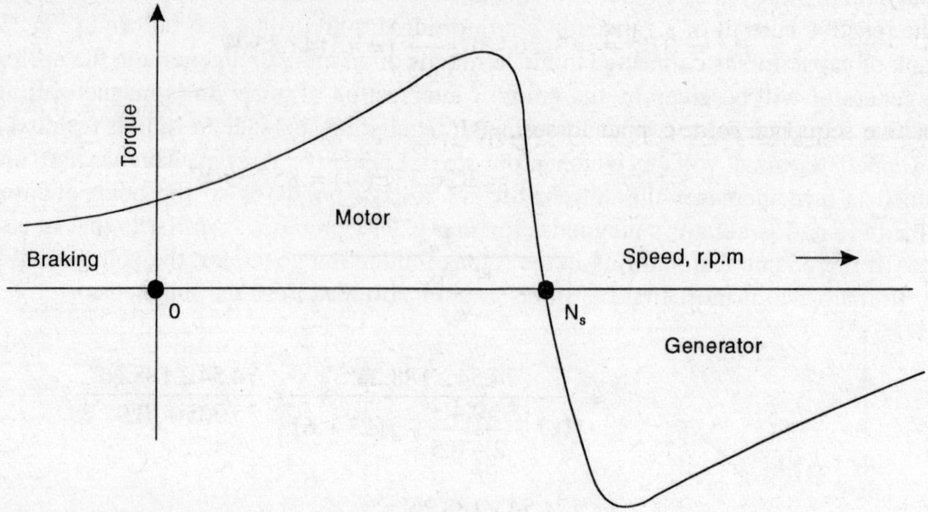


**Fig. 7.32.** Speed torque curve of induction machine showing braking, motoring and generating regions

An induction generator has many limitations. Since it does not have a field circuit, is cannot produce reactive power. Rather it consumes reactive power because the magnetising current must be supplied to it. It cannot control its output voltage. Normally the generator's output voltage is maintained by the external power system connected to it.

## 7.22.1 Induction Generator in Parallel with Capacitor Bank

It is possible to operate an isolated induction generator, if a capacitor bank is connected across its terminals as shown in Fig. 7.33. The capacitor bank supplies the reactive power of the generator and the other loads.

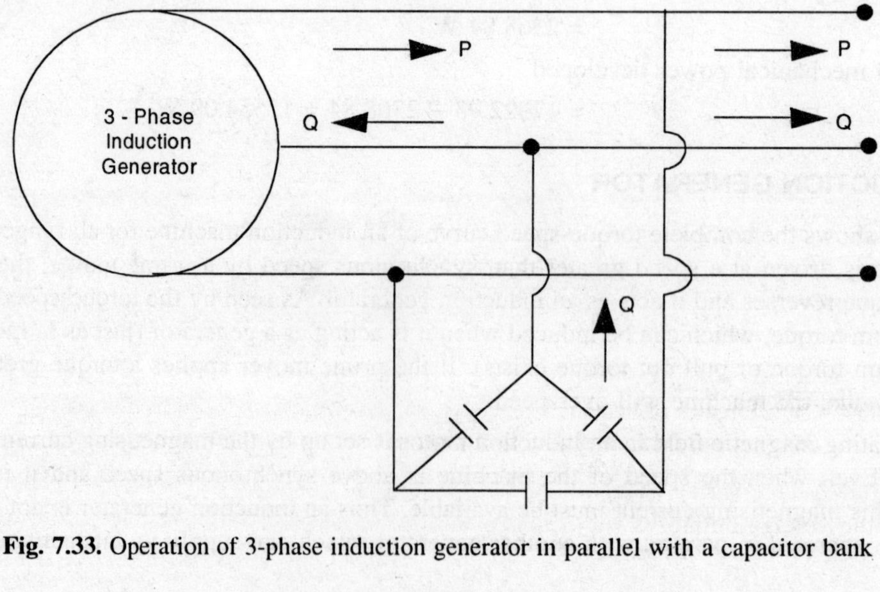


**Fig. 7.33.** Operation of 3-phase induction generator in parallel with a capacitor bank

The magnetizing current of an induction machine is a function of terminal voltage as shown in Fig. 7.34(a). The reactive current of a capacitor is proportional to the voltage as shown in Fig. 7.34(b). If a 3-phase bank of capacitors is connected to the terminals of an induction generator, the no-load terminal voltage of generator will be given by the point of intersection of generator's magnetization curve and capacitor's load line (Fig. 7.34c). When the induction generator is made to run, its residual magnetism produces a small terminal voltage which produces a capacitive current. This current increases the voltage which is turn increases the current till the voltage builds up to the point of intersection of capacitor loadline and generator's magnetisation curve. The process is similar to that in a self-excited dc generator. If there is no residual flux in the rotor of induction generator, the voltage build-up cannot take place. In that case, it must first be run as a motor to create residual magnetism.

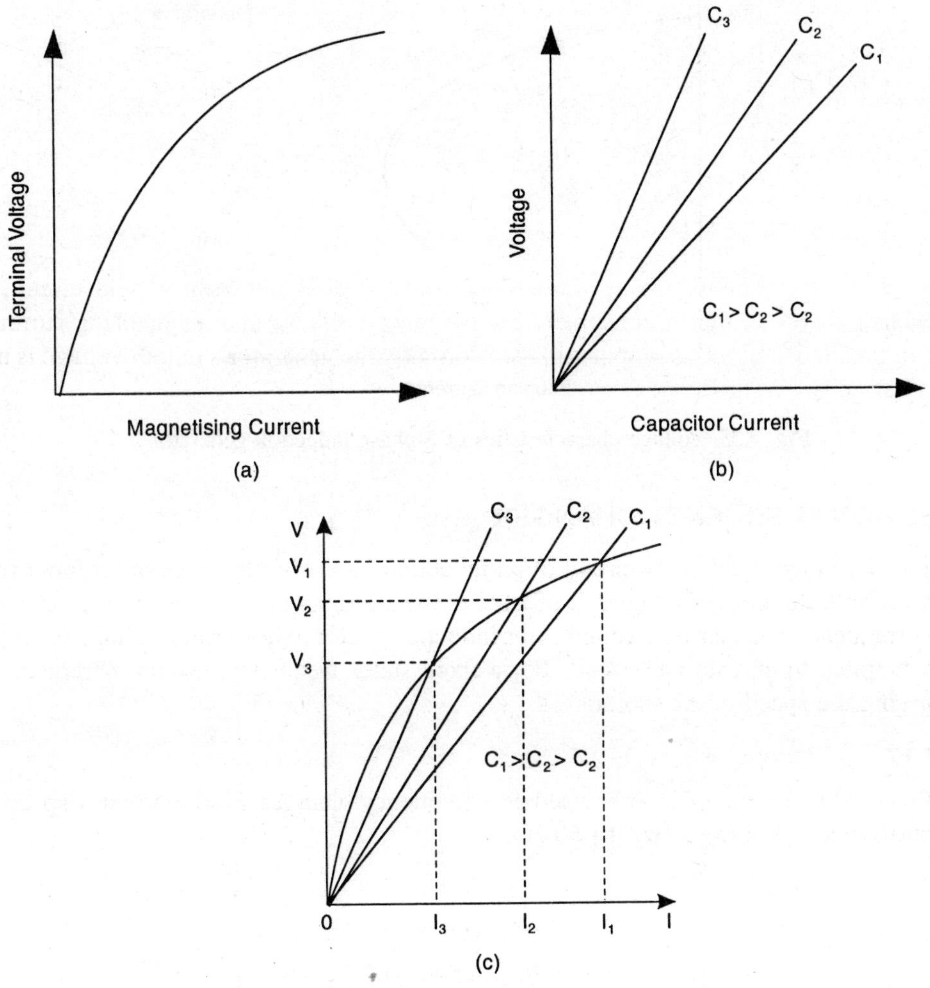


**Fig. 7.34.** Characteristics of induction generator and capacitor bank (a) induction generator alone, (b) capacitor bank alone, (c) induction generator in parallel with capacitors

The output voltage varies widely with changes in load, as shown in Fig. 7.35. For a lagging power factor load, the voltage collapses very rapidly and this is a serious disadvantage of an induction generator. Because of its torque-speed characteristics, its speed and therefore, the output frequency varies by about 5% from no-load to full-load.

Induction generators were used till early twentieth century. With increasing fuel prices, it may stage a come-back, because energy recovery has become an important aspect of industrial processes. Induction generator is ideal for such applications, because its control and maintenance are very simple. Because of its simplicity and small size per kW of power output, an induction generator is very suited for windmills. Many commercial windmills are operated in parallel with large power systems. In such an operation, the power system provides the voltage and frequency control and static capacitors can be used for power factor correction.

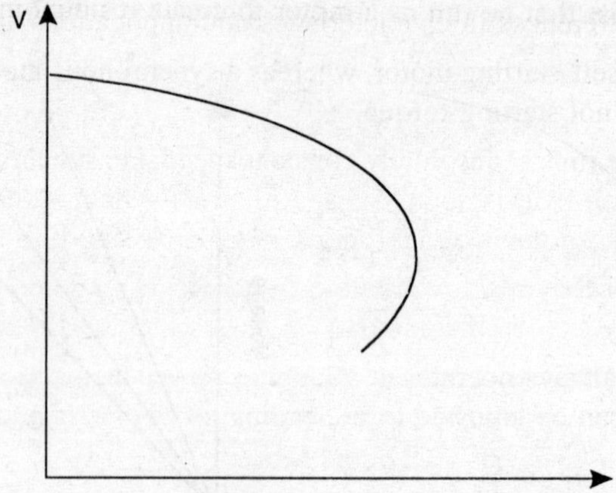


**Fig. 7.35.** Voltage characteristics of 3-phase induction generator

## 7.23 INDUCTION FREQUENCY CHANGER

The frequency of voltage in the rotor circuit of an induction motor is slip times stator input frequency. Therefore, a variable frequency voltage is available at the sliprings of a wound rotor induction motor. An induction frequency changer is a wound rotor induction motor, whose rotor voltage is tapped off at brushes and supplied to an external circuit. For a fixed stator frequency, the rotor frequency can be varied by varying the speed of the motor.

*Example 7.17*

A 4-pole, 50 Hz induction motor is to be used as a frequency changer. Find the motor speed, so as to give an output frequency of (a) 5 Hz, (b) 80 Hz.

*Solution*

(a)
$$f_2 = sf_1$$

or
$$s = f_2/f_1 = 5/50 = 0.1$$

$$N_s = \frac{120 \times 50}{4} = 1500 \text{ rpm}$$

$$s = \frac{N_s - N}{N_s} \text{ or } 0.1 = \frac{1500 - N}{1500} \text{ or } N = 1350 \text{ rpm}$$

(b)

$$80 = s\ (50) \text{ or } s = 1.6$$

$$1.6 = \frac{1500 - N}{1500} \text{ or } N = -900 \text{ rpm}$$

The speed –900 means that the direction of rotation of the shaft is opposite to the direction of rotation of magnetic field.

## 7.24 COMPARISON OF INDUCTION MOTOR AND SYNCHRONOUS MOTOR

A three-phase induction motor and synchronous motor have different characteristics and their own field of application. The main differences in characteristics and applications are as under:

(a) Induction motor is a self-starting motor, whereas a synchronous motor requires a special starting method, since it has not starting torque.

(b) A synchronous motor runs at absolutely constant speed, i.e. synchronous speed. The speed of an induction motor at full-load is about 5% less than its synchronous speed. An induction motor does not develop any torque at synchronous speed and cannot run at this speed.

(c) A synchronous motor requires dc excitation (it is a doubly excited machine), whereas an induction motor is a singly excited machine and does not require any dc excitation.

(d) An induction motor always operates at a lagging power factor. However, the power factor of a synchronous motor can be adjusted to be leading as well as lagging, by variation of excitation current.

(e) An induction motor is used only for supplying mechanical loads, whereas a synchronous motor is used for supplying mechanical loads as well as for power factor control in power systems.

## 7.25 APPLICATIONS OF INDUCTION MOTORS

The 3-phase induction motor is the most widely used motor in industry and work-shops. It is manufactured in a variety of ratings from a few kW to 5000 kW or so. Its simple and rugged construction, almost maintenance-free operation and the fact that it does not require any dc excitation current make it an ideal choice for many applications. The squirrel cage motor is generally built in small sizes (upto a few kW). For bigger sizes wound-rotor motors are preferred.

Typical applications of 3-phase induction motors include fans, blowers, centrifugal pumps, machine tools wood working machines, conveyors, compressors, crushers, agitators, reciprocating pumps, shears, punch presses, bull dozers, boilers, hoists, cranes, etc.

### *Example 7.18*

The standstill impedance of the outer cage of a double-cage induction motor is $0.3 + j\,0.4$ ohm and that of inner cage is $0.1 + j\,1.5$ ohm. Compare the relative currents and torques of the two cages: (a) at standstill, (b) at 5% slip. Neglect stator impedance. [*P.U. 2002*]

### *Solution*

(a) At standstill $s = 1$

$$I_2 = \frac{sE_2}{\left(r_2^2 + s^2 x_2^2\right)^{0.5}}$$

$$I_{2,\ \text{outer}} = \frac{(1)(E_2)}{(0.3^2 + 0.4^2)^{0.5}} = \frac{E_2}{0.5}$$

$$I_{2,\ \text{inner}} = \frac{(1)(E_2)}{(0.1^2 + 1.5^2)^{0.5}} = \frac{E_2}{1.5033}$$

$$\frac{I_{2,\text{outer}}}{I_{2,\text{inner}}} = \frac{(E_2)}{0.5}\left(\frac{1.5033}{E_2}\right) = 3$$

$$T = \frac{3I_2^2 r_2}{s} = \frac{3s^2 E_2^2 r_2}{s(r_2^2 + s^2 x_2^2)} = \frac{3E_2^2 r_2}{s\left(\frac{r_2^2}{s^2} + x_2^2\right)}$$

$$T_{\text{outer}} = \frac{3E_2^2(0.3)}{0.3^2 + 0.4^2} = 3.6\,E_2^2$$

$$T_{\text{inner}} = \frac{3E_2^2(0.1)}{0.1^2 + 1.5^2} = 0.133\,E_2^2$$

$$\frac{T_{\text{outer}}}{T_{\text{inner}}} = \frac{3.6E_2^2}{0.133\,E_2^2} = 27.07$$

(b) At $s = 0.05$

$$I_{2,\ \text{outer}} = \frac{0.05\,E_2}{\left[0.3^2 + (0.05 \times 0.4)^2\right]^{0.5}} = 0.166\,E_2$$

$$I_{2,\ \text{inner}} = \frac{0.05\,E_2}{\left[0.1^2 + (0.05 \times 1.5)^2\right]^{0.5}} = 0.4\,E_2$$

$$\frac{I_{2,\text{outer}}}{I_{2,\text{inner}}} = \frac{0.166E_2}{0.4E_2} = 0.415$$

$$T_{\text{outer}} = \frac{3\,E_2^2(0.3)}{0.05\left[\left(\frac{0.1}{0.05}\right)^2 + 0.4^2\right]} = 0.498\,E_2^2$$

$$T_{\text{inner}} = \frac{3\,E_2^2(0.1)}{0.05\left[\left(\frac{0.1}{0.05}\right)^2 + 1.5^2\right]} = 0.96\,E_2^2$$

$$\frac{T_{\text{outer}}}{T_{\text{inner}}} = \frac{0.498E_2^2}{0.96E_2^2} = 0.519.$$

*Example 7.19*

A 400 V, 3-phase, 50 Hz, 6-pole star connected induction motor when running light at normal voltage and frequency takes an input of 1.2 kW and line current is 8.25 A. With rotor blocked and 180 V applied to stator, the input power is 5.25 kW and line current is 45 A. The stator resistance measured between two line terminals when hot is 0.9 ohm. Find parameters of approximate equivalents circuit of the motor and find the net mechanical power output and torque at a slip of 4%.

*Solution*

No-load test

$$V_0 = 400/\sqrt{3} = 230.95 \text{ V/phase}$$

$$I_0 = 8.25 \text{ A per phase}$$

$$\cos\theta_0 = \frac{W_0}{3V_0I_0} = \frac{1.2 \times 1000}{3 \times 230.95 \times 8.25} = 0.21$$

$$R_0 = \frac{V_0}{I_0 \cos\theta_0} = \frac{230.95}{8.25 \times 0.21} = 133.3 \text{ ohm}$$

$$X_0 = \frac{V_0}{I_0 \sin\theta_0} = \frac{230.95}{8.25 \times 0.977} = 28.65 \text{ ohm}$$

Blocked rotor test

$$V_{sc} = 180/\sqrt{3} = 103.93 \text{ V per phase}$$

$$I_{sc} = 45 \text{ A per phase}$$

$$\cos\theta_{sc} = \frac{5.25 \times 1000}{3 \times (103.93)(45)} = 0.375$$

$$r_1 + r_2' = \frac{V_{sc}}{I_{sc}} \cos\theta_{sc} = \frac{103.93}{45} \times 0.375$$

$$= 0.866 \text{ ohm per phase}$$

$$x_1 + x_2' = \frac{V_{sc}}{I_{sc}} \sin\theta_{sc} = \frac{103.93}{45} \times 0.927$$

$$= 2.14 \text{ ohm per phase}$$

$$r_1 = 0.9/2 = 0.45 \text{ ohm}$$

$$r_2' = 0.866 - 0.45 = 0.416 \text{ ohm}$$

When
$$s = 0.05$$

$$I_2' = \frac{230.95}{r_1 + r_2' + r_2'\left(\frac{1-s}{s}\right) + j(x_1 + x_2')}$$

$$= \frac{230.95}{0.866 + 0.416\left(\frac{1-0.05}{0.05}\right) + j2.14}$$

$$= \frac{230.95}{8.77 + j2.14} = 25.58 \text{ A}$$

$$\text{Mech. power output} = 3(I_2')^2\, r_2'\left(\frac{1-s}{s}\right)$$

$$= 3(25.58)^2(0.416)\left(\frac{0.95}{0.05}\right) \text{ W}$$

$$= 15.52 \text{ kW}$$

$$N_s = 120 \times 50/6 = 1000 \text{ rpm}$$

$$N = N_s(1-s) = 1000(1-0.05) = 950 \text{ rpm}$$

$$\frac{2\pi NT}{60} = 15.52 \times 1000$$

$$T = \frac{15.52 \times 1000 \times 60}{2\pi \times 950} = 156 \text{ N-m.}$$

### *Example 7.20*

A 100 hp induction motor is coupled to a drive whose torque varies as square of speed. Rotor emf at standstill is 100 V per phase. Full-load slip is 0.05%. Find the resistance to be added in the rotor circuit to reduce the speed by 20%. Neglect mechanical losses and rotor reactance.

#### *Solution*

At full-load,
$$s = 0.05$$

$$\text{Output} = 100 \text{ hp (metric)}$$

$$= 100 \times 735.5 = 73550 \text{ W}$$

If rotor resistance is $r_2$

$$\text{Rotor current } I_2 = \frac{sE_2}{r_2} = \frac{0.05 \times 100}{r_2} = \frac{5}{r_2} \text{ A}$$

$$\text{Power Output} = \frac{3I_2^2 r_2(1-s)}{s} = 3\left[\frac{5}{r_2}\right]^2 r_2\left(\frac{0.95}{0.05}\right)$$

$$= \frac{3 \times 25 \times 0.95}{0.05\, r_2} = 73550\,\text{W}$$

or

$$r_2 = 0.0194 \text{ ohm}$$

$$N = N_s(1-s) = 0.95\, N_s$$

Let external resistance = $r$ ohm/phase

$$\text{New value of speed} = 0.8 \times 0.95\, N_s = 0.76\, N_s$$

$$\text{Torque} \propto N^2$$

$$\text{Since power output} \propto T \times N \propto N^3$$

$$\text{New power output} = 73550\left(\frac{0.76 N_s}{0.95 N_s}\right)^3 = 37657.6\,\text{W}$$

$$\text{New value of slip} = 0.24$$

$$\text{New value of rotor current} = \frac{0.24 \times 100}{r + r_2} = \frac{24}{r + r_2}$$

$$3\left(\frac{24}{r + r_2}\right)^2 (r + r_2)\left(\frac{1 - 0.24}{0.24}\right) = 37657.6$$

or

$$\frac{3 \times 24^2 \times 0.76}{(r + r_2)(0.24)} = 37657.6$$

or

$$r + r_2 = 0.145\ \Omega$$

$$r = 0.145 - 0.0194 = 0.1256 \text{ ohm/phase.}$$

*Example 7.21*

A 3-phase 6-pole, 50 Hz, 400 V induction motor develops maximum torque at a slip of 10%. In a particular application it runs at 5% slip at rated voltage driving a load whose torque demand is proportional to the square speed. What is the minimum value to which the supply voltage can drop, if the speed of the motor is not to decrease below 900 rpm. Neglect stator impedance drop.

*Solution*

Neglecting $r_1$ and $x_1$ and using Eq. (7.32)

$$r_2' = 0.1\, x_2'$$

$$N_s = 120 \times 50/6 = 1000\,\text{rpm}$$

Let $T_1$, $N_1$, $s_1$, $V_1$ denote the initial values of torque, speed, slip and voltage and $T_2$, $N_2$, $s_2$, $V_2$ denote the new values of these quantities.

Assuming delta connected motor $V_1 = 400$ V, $s_1 = 0.05$

$$N_1 = 1000(1-0.05) = 950 \text{ rpm}$$

$$N_2 = 900 \text{ rpm}$$

$$s_2 = (1000-900)/1000 = 0.1$$

Since load torque ∝ square of speed

$$\frac{T_2}{T_1} = \left(\frac{900}{950}\right)^2$$

Using Eq. (7.27)

$$T_1 = \frac{3}{2\pi n_s}\left[\frac{400^2}{\left(\frac{r_2'}{0.05}\right)^2 + x_2'}\right]\frac{r_2'}{0.05}$$

$$T_2 = \frac{3}{2\pi n_s}\left[\frac{V_2^2}{\left(\frac{r_2'}{0.1}\right)^2 + x_2'}\right]\frac{r_2'}{0.1}$$

Therefore
$$\frac{T_2}{T_1} = \left(\frac{900}{950}\right)^2 = \left[\left(\frac{V_2^2}{400^2}\right)\left(\frac{400r_2'^2 + x_2'^2}{100r_2'^2 + x_2'^2}\right)\right]\frac{0.05}{0.1}$$

or
$$\left(\frac{900}{950}\right)^2 = \frac{V_2^2}{400^2}\left[\frac{5x_2'^2}{2x_2'^2}\right]\frac{0.05}{0.1}$$

or
$$\left(\frac{900}{950}\right)^2 = \frac{V_2^2}{400^2}(1.25)$$

$$V_2 = 338.95 \text{ V.}$$

*Example 7.22*

A 4-pole, 400 V, 3-phase induction motor has a standstill rotor emf of 100 V per phase. The rotor has resistance of 50 milli ohm per phase and standstill reactance of 0.5 ohm per phase. Calculate the maximum torque and the slip at which it occurs. Neglect stator impedance. [*K.U. 2001*]

*Solution*

Since stator impedance is neglected, the slip for maximum torque is

$$s = r_2'/x_2' = r_2/x_2 = \frac{50 \times 10^{-3}}{0.5} = 0.1$$

$$n_s = 2 \times 50/4 = 25 \text{ rps}$$

$$\text{Rotor current} = \frac{sE_2}{\left(r_2^2 + s^2 x_2^2\right)^{0.5}}$$

$$= \frac{0.1(100)}{\left[\left(50 \times 10^{-3}\right)^2 + (0.1 \times 0.5)^2\right]^{0.5}}$$

$$= 141.42 \text{ A per phase}$$

$$\text{Torque} = \frac{3I_2^2 r_2}{s(2\pi n_s)} = \frac{3(141.42)^2\left(50 \times 10^{-3}\right)}{2\pi(25)(0.1)}$$

$$= 190.98 \text{ N-m.}$$

*Example 7.23*

Show that in a 3-phase induction motor, torque $T$ at any slip $s$ is

$$T = \frac{2T_{max}}{s/s_m + s_m/s}$$

where $T_{max}$ = maximum torque

$s_{max}$ = slip for maximum torque

Make simplifying assumptions. [*P.U. 2002*]

*Solution*

$$T = 3I_2^2 \frac{r_2}{s}$$

Neglecting stator impedance

$$I_2 = \frac{sE_2}{\left(r_2^2 + s^2 x_2^2\right)^{0.5}}$$

or
$$T = \frac{3s^2 E_2^2}{r_2^2 + s^2 x_2^2}\left(\frac{r_2}{s}\right) = \frac{3sE_2^2 r_2}{r_2^2 + s^2 x_2^2}$$

or
$$T_{max} = \frac{3s_m E_2^2 r_2}{r_2^2 + s_m^2 x_2^2}$$

or
$$\frac{T}{T_{\max}} = \frac{s\left(r_2^2 + s_m^2 x_2^2\right)}{s_m\left(r_2^2 + s^2 x_2^2\right)}$$

But
$$s_m = r_2/x_2$$

or
$$\frac{T}{T_{\max}} = \frac{s\left(r_2^2 + r_2^2\right)}{s_m\left(r_2^2 + \frac{s^2 r_2^2}{s_m^2}\right)} = \frac{2r_2^2 s s_m}{r_2^2\left(s^2 + s_m^2\right)}$$

$$= \frac{2 s s_m}{s^2 + s_m^2} = \frac{2}{\frac{s}{s_m} + \frac{s_m}{s}}$$

or
$$T = \frac{2T_{\max}}{s/s_m + s_m/s}.$$

*Example 7.24*

A 3-phase squirrel cage motor has maximum torque equal to twice full-load torque. $r_2 = 0.2\Omega$ and $x_2 = 2\Omega$. Determine the ratio of starting torque to full-load torque, if it is started by (a) direct on-line stator, (b) star-delta starter. (c) auto-transformer starter with 7% tapping.

*Solution*

As derived in example 7.23 torque $T$ at any slip $s$ is

$$\frac{T}{T_{\max}} = \frac{2}{s/s_m + s_m/s}$$

$$s_m = \text{slip at maximum torque} = \frac{0.2}{2} = 0.1$$

(a) Direct on-line starting

At starting $s = 1$

$$\frac{T_s}{T_{\max}} = \frac{2}{1/0.1 + 0.1/1} = \frac{2}{10.1}$$

$$T_s = 0.198\ T_{\max}$$

Since
$$T_{\max} = 2T_{fL}$$

$$T_s = (0.198 \times 2) \text{ full-load torque}$$

$$= 0.396 \times \text{full-load torque}$$

(b) When star-delta starter is used, the starting torque is 1/3 of the starting torque, with direct on-line starting.

$$T_s = \frac{1}{3}(0.396) \times \text{full-load torque}$$

$$= 0.132 \times \text{full-load torque}$$

(c) When auto-transformer with 70% is used

$$T_s = (0.7)^2\ (0.396)\ (\text{full-load torque})$$

$$= 0.194 \times \text{full-load torque.}$$

*Example 7.25*

An 11 kW, 220 V, 3-phase induction motor has a full-load efficiency of 0.87 and a power factor of 0.85. At standstill the motor draws 6 times full-load current and develops 1.5 times full-load torque. An auto-transformer is used to reduce the starting current and to give full-load torque at starting. Neglecting no-load current, determine (a) voltage applied at starting, (b) full-load line current, (c) current drawn by motor at starting, (d) line current in the primary of auto-transformer at starting.

*Solution*

(a) Assume motor to be delta connected. Let $V_1$ be the voltage applied at starting.

$$T \propto (\text{Voltage})^2$$

$$\left(\frac{V_1}{220}\right)^2 = \frac{T_{\text{full-load}}}{(1.5)\,T_{\text{full-load}}} = \frac{1}{1.5}$$

or

$$V_1 = \frac{220}{\sqrt{1.5}} = 179.63\text{ V}$$

(b)

$$\text{Full-load line current} = \frac{11 \times 1000}{\sqrt{3} \times 220 \times 0.87 \times 0.85}$$

$$= 39.04\text{ A}$$

(c) When applied voltage is 220 V, the current is 6 times full-load current. When applied voltage is 170.63 V, motor starting current

$$= \frac{6 \times 39.04 \times 179.63}{220}$$

$$= 191.26\text{ A}$$

(d) Current on primary side of auto-transformer $= \dfrac{191.26 \times 179.63}{220} = 156.16\text{ A}.$

*Example 7.26*

A 3-phase star connected 400 V, 50 Hz, 4-pole induction motor has the following parameters as referred to stator $r_1 = 0.15$ ohm, $x_1 = 0.45$ ohm, $r_2' = 0.12$ ohm, $x_2' = 0.45$ ohm, $X_0 = 28.5$ ohm.

Neglect $R_0$. Find stator current, p.f. at rated voltage and at $s = 0.04$.

*Solution*

$$V_1 = 400/\sqrt{3} = 230.95\text{ V/phase}$$

Referring to approximate equivalent circuit of Fig. 7.4

$$I_2' = \frac{V_1}{r_1 + \frac{r_2'}{s} + j(x_1 + x_2')}$$

$$= \frac{230.95}{0.15 + \frac{0.12}{0.04} + j(0.45 + 0.45)}$$

or $$I_2' = \frac{230.95}{3.15 + j0.9} = 70.5\angle -15.95^\circ$$

$$I_0 = I_m = \frac{230.95}{j28.5} = -j8.1\text{ A}$$

Stator current $$I_1 = I_2' + I_0 = 70.5\angle -15.95^\circ - j8.1$$

$$= 67.79 - j27.47 = 73.15\angle -22.06^\circ\text{ A}$$

Power factor $$= \cos(-22.06)^\circ = 0.93 \text{ lagging.}$$

*Example 7.27*

An 8-pole, 50 Hz, 3-phase induction motor develops a maximum torque of 150 N-m at 650 rpm. The rotor resistance is 0.5 ohm per phase. Find torque at 4% slip. Neglect stator impedance.

*Solution*

$$N_s = 120 \times 50/8 = 750 \text{ rpm}$$

For maximum torque

$$s = r_2/x_2$$

$$s = \frac{750 - 650}{750} = \frac{1}{7.5}$$

$$x_2 = r_2/x_2 = 0.5 \times 7.5 = 3.75\,\Omega$$

Torque at any slip $s$ is

$$T = \frac{3I_2^2 r_2}{2\pi n_s(s)} = \frac{3E_2^2}{\left(\frac{r_2}{s}\right)^2 + x_2^2}\left[\frac{r_2}{2\pi n_s(s)}\right]$$

$$T_{\max} = 150 = \frac{3E_2^2}{(3.75^2 + 3.75^2)}\left[\frac{0.5 \times 7.5}{2\pi \times 15}\right]$$

or $$E_2 = 188 \text{ V}$$

At $s = 0.04$

$$T = \frac{3(188)^2}{\left(\frac{0.5}{0.04}\right)^2 + (3.75)^2}\left[\frac{0.5}{2\pi \times 15 \times 0.04}\right] = 82.57 \text{ N-m}.$$

*Example 7.28*

A 3-phase cage rotor induction motor has a full-load slip of 4%. The standstill current at rated voltage is 6.5 times full-load current. Find auto-transformer tapping to give full-load torque at starting and also find line current at starting. [*P.U. 2001*]

*Solution*

If $T_s$ is starting torque, $I_s$ standstill current, $T_{fL}$ full-load torque, $I_{fL}$ full-load current and $s_{fL}$ full-load slip, then

$$\frac{T_s}{T_{fL}} = \left(\frac{I_s}{I_{fL}}\right)^2 s_{fL} = 6.5^2(0.04) = 1.69$$

$$\text{Torque} \propto (\text{Voltage})^2$$

If $$T_s = T_{fL}$$

$$\text{Auto-transformer tapping} = (1/1.69)^{0.5} .= 0.77$$

$$\text{Line current with 0.77 tapping} = (0.77)^2\ (6.5)(\text{full-load})(\text{current})$$
$$= 3.85 \times \text{full-load current.}$$

*Example 7.29*

A 6-pole, 50 Hz, 3-phase induction motor has rotor resistance = 0.2 ohm/per phase. Maximum torque is 200 N-m at 850 rpm. Find (a) torque at 4% slip, (b) additional rotor resistance to get two-third of maximum torque at starting. Neglect stator impedance.

*Solution*

$$N_s = 120 \times 50/6 = 1000 \text{ rpm}$$

$$n_s = \frac{1000}{60} = \frac{50}{3} \text{ rps}$$

$$\text{Slip for maximum torque} = \frac{r_2}{x_2}$$

$$\frac{1000 - 850}{1000} = 0.75 = \frac{0.2}{x_2}$$

or $$x_2 = \frac{0.2}{0.15} = 1.333\,\text{ohm}$$

Torque at any slip $s$ is

$$T = \frac{3I_2^2 r_2}{2\pi n_s (s)} = \frac{3E_2^2}{\left(\frac{r_2}{s}\right)^2 + x_2^2}\left[\frac{r_2}{2\pi n_s (s)}\right]$$

$$T_{\max} = 200 = \frac{3E_2^2}{\left(\frac{0.2}{0.15}\right)^2 + (1.333)^2}\left[\frac{0.2}{2\pi \frac{50}{3}(0.15)}\right]$$

or $$200 = \frac{3E_2^2}{(1.333)^2 + (1.333)^2}\left(\frac{0.6}{15\pi}\right)$$

or $$E_2 = 136.41$$

At $s = 0.04$

$$T = \frac{3(136.41)^2}{\left(\frac{0.2}{0.04}\right)^2 + 1.333^2}\left[\frac{0.2}{2\pi\left(\frac{50}{3}\right)(0.04)}\right] = 99.54\,\text{N-m}$$

Let $r$ be the extra rotor resistance to get two-third maximum torque at starting (i.e. $s = 1$)

$$T_s = \frac{2}{3}(200) = 133.3\,\text{N-m}$$

$$133.33 = \frac{3(136.41)^2}{(0.2 + r)^2 + 1.333^2}\left[\frac{(0.2 + r)}{2\pi\left(\frac{50}{3}\right)}\right]$$

or $$r = 0.31\text{ ohm.}$$

## SUMMARY

1. Three phase induction motor is a singly excited motor. Its stator is fed with 3 phase ac but the rotor is not given any supply.
2. The rotor may be a wound rotor or squirrel cage rotor.
3. The wound rotor has 3-phase winding and has the same number of poles as stator. The terminals of rotor winding are brought out to sliprings.

4. A squirrel cage rotor has a number of conducting bars (of copper or aluminium) laid in slots on rotor core. The ends of these bars are short circuited by conducting rings on both sides.
5. slip $s = \dfrac{n_s - n}{n_s}$
6. Three phase induction motor develops torque at all speeds except synchronous speed.
7. The speed of a 3-phase induction motor is always less than synchronous speed by about 3-5%.
8. Both stator and rotor mmfs rotate at synchronous speed.
9. In some respects the operation of a 3-phase induction motor is similar to that of transformer.
10. The equivalent circuit helps in analysing the performance.
11. Losses in a 3-phase induction motor include stator copper losses, stator core losses, rotor copper losses, friction and windage losses and stray losses. Rotor core losses are negligible because frequency of rotor currents is very low.
12. Rotor copper loss = $s\ P_g$
13. Three phase induction motor has positive starting torque at zero speed. The maximum torque occurs at about 3–5% slip.
14. A wound rotor motor is started by adding additional resistances across slip rings. These resistances are reduced to zero as the speed increases to full load speed.
15. A cage induction motor is started by star-delta starter. At the time of starting the stator windings are connected in star. When the motor picks up speed the connections are changed so as to reconnect the stator windings in delta.
16. The parameters of equivalent circuit can be determined by no load test and blocked rotor test.
17. Circle diagram is convenient method to analyse the performance of a 3-phase induction motor. Parameters like starting torque, efficiency, slip, power factor, maximum torque, etc. can be determined from the circle diagram.
18. A double cage rotor has two sets of rotor bars. It has high starting torque.
19. Methods for speed control of 3-phase induction motor include frequency control (using rectifier — inverter), pole changing, line voltage control, rotor resistance control and cascade connection.
20. When the supply voltage is unbalanced, the performance of motor can be found by the use of symmetrical components.
21. If a 3-phase induction motor is run at a speed more than synchronous speed (by a prime mover) it becomes an induction generator. However an induction generator has many limitations and is not suitable for large scale energy conversion.
22. These phase induction motor is the most widely used motor in many industries. Typical applications include fans, blowers, machine tools, centrifugal pumps, shears, punch presses, bulldozers, compressors etc.

## TEST POINT QUESTIONS

### A. State whether the following statements are true (T) or false (F)

7.1 A 3-phase induction motor has zero torque at synchronous speed.

7.2 A 3-phase induction motor has no starting torque.

7.3 Induction motor operates at a lagging power factor.

7.4 The rotor core losses of a 3-phase induction motor are negligible.

7.5 The frequency of rotor currents in a 3-phase induction motor is the same as line frequency.

7.6 A 3-phase induction motor rotates at a speed slightly less than synchronous speed.

7.7 Stator and rotor mmfs, in a 3-phase induction motor, rotate at different speeds.

7.8 The magnetising current of a 3-phase induction motor is more than that of the same kW rating transformer.

7.9 Speed control of a 3-phase induction motor by stator voltage control is possible if load torque increases with speed.

7.10 In a double cage motor both the sets of rotor bars are similar.

7.11 If external resistance is added in rotor circuit of a 3-phase induction motor the starting torque increases.

7.12 The airgap in a 3-phase induction motor of medium size is about 1 cm.

7.13 In a squirrel cage motor the rotor bars are made of iron.

7.14 Most of 3-phase induction motors have a totally enclosed fan cooled frame.

7.15 Full load efficiency of a 3-phase induction motor of medium size is about 0.5.

### B. Fill in the blanks with suitable words

7.16 Rotor copper loss = ( ) ($P_g$)

7.17 In a 3-phase induction motor the maximum torque occurs at a small value of .........

7.18 In a star-delta starter the stator windings are connected in ......... at the instant of starting.

7.19 A 3-phase induction motor has a starting current of 300 A. If an auto transformer with 50% tapping is used the line current at starting will be .........

7.20 Blocked rotor test of 3-phase induction motor is similar to ......... on a transformer.

7.21 Full load power factor of a medium size 3 phase induction motor is about .........

7.22 In a 3-phase induction motor the power factor at the time of starting is generally about .........

7.23 If $n$ is the speed in rev/s and $T$ is the torque in N-m, outputs in watts = .........

7.24 If a 3-phase induction motor runs at about one-seventh of rated speed, the phenomenon is called .........

7.25 The best method to control speed of 3 phase induction motor is .........

7.26 For a 3-phase induction motor speed control by stator input voltage control is possible only if load torque is .........

7.27 Squirrel cage motor has the advantage of ......... and .........

7.28 When the 3-phase induction motor is operating at rated speed the rotor reactance is .........

7.29 The no load current of a 3-phase induction motor is about .........of full load current.

7.30 The resistance which is equivalent to mechanical load in a 3-phase induction motor = .........

## ANSWERS

| | | | | |
|---|---|---|---|---|
| 7.1 T | 7.2 F | 7.3 T | 7.4 T | 7.5 F |
| 7.6 T | 7.7 F | 7.8 T | 7.9 T | 7.10 F |
| 7.11 T | 7.12 F | 7.13 F | 7.14 T | 7.15 F |
| 7.16 $s$ | 7.17 slip | 7.18 star | 7.19 75 A | 7.20 short circuit test |
| 7.21 0.8 lagging | 7.22 0.5 lagging | 7.23 $2\pi n$ T | 7.24 crawling | |
| 7.25 line frequency control | | 7.26 variable | 7.27 low cost and rugged construction | |
| 7.28 very small | 7.29 20% | 7.30 $\frac{r_2(1-s)}{s}$ | | |

## JOB INTERVIEW QUESTIONS

7.1 Why is a 3-phase induction motor called singly excited motor?
7.2 What is meant by slip?
7.3 Why is squirrel cage motor so popular?
7.4 What is a rotating field?
7.5 What is blocked rotor test.
7.6 Why are rotor core losses negligible in 3-phase induction motor.
7.7 Draw equivalent circuit of 3-phase induction motor.
7.8 What is meant by crawling?
7.9 Why is the power factor of a 3-phase induction motor lagging?
7.10 Why is the rotor power factor of a 3-phase induction motor nearly unity?
7.11 Draw torque slip curve of 3-phase induction motor.
7.12 What is a double cage rotor?
7.13 What is the difference between cage rotor and wound rotor.
7.14 What is a star-delta stator.
7.15 How can speed of 3-phase induction motor be controlled?

## REVIEW QUESTION

7.1 Give reasons:

(a) The speed of an induction motor cannot be equal to synchronous speed.
(b) The reactance of rotor varies widely between starting and running conditions.
(c) The induction motor is the most commonly used motor.
(d) Induction motor can be called a generalised transformer.
(e) In a 3-phase induction motor, the rotor core-loss is negligible.

7.2 When the mechanical load on a 3-phase induction motor is increased, how does the input current increase to meet with the new conditions?

7.3 Explain why?

(a) The no-load test on a 3-phase induction motor is conducted at rated voltage.
(b) Only a small voltage is applied to the stator during blocked rotor test.
(c) The core-loss, windage and friction-loss can neglected for the blocked rotor condition.
(d) The locus of the rotor current for different operating condition of a 3-phase induction motor is a circle.

7.4 The rotating fields of the stator and rotor are stationary with respect to each other. Justify the statement.

7.5 Discuss (a) construction, (b) principle of operation of 3-phase induction motor.

7.6 Justify the statement: Induction motor can be regarded as a generator feeding a fictitious load resistance.

7.7 What is meant by slip of an induction motor? Explain the importance of slip in a 3-phase induction motor.

7.8 Deduce the equation for rotor current of a 3-phase induction motor. What determines the value of current in the rotor conductors at standstill and under running condition?

7.9 In what respects is the operation of an induction motor similar to that of a transformer? In what respects are they different?

7.10 Starting from first principles, develop the equivalent circuit of a 3-phase induction motor.

7.11 Draw and explain the phasor diagram of a 3-phase induction motor.

7.12 Draw and explain the power flow diagram of a 3-phase induction motor. Show that rotor copper-loss is slip times the power input to rotor.

7.13 Derive the equation for torque developed by an induction motor. Draw a typical torque-slip curve and deduce the condition for maximum torque.

7.14 What is the effect of introducing resistance in the rotor circuit of an induction motor? In which type of motor is it possible? Under what conditions is it done and why?

7.15 Discuss the methods of starting (a) wound rotor, (b) cage rotor induction motor.

7.16 Discuss the procedure of no-load and blocked rotor test on a 3-phase induction motor. How are the parameters of equivalent circuit determined from test results?

7.17 Explain the procedure of drawing the circle diagram of an induction motor. What information can be obtained from the circle diagram and how?

7.18 What is induction generator? Discuss its principle of operation. What are its limitations?

7.19 Discuss briefly the various methods of speed control of 3-phase cage motor and wound rotor induction motor.

7.20 What is pole amplitude modulation? Discuss the application of this principle for speed control of 3-phase induction motor.

7.21 Discuss the working of (a) deep rotor bar, (b) double cage induction motor.

7.22 Explain the phenomenon of cogging and crawling in a 3-phase induction motor.

7.23 What is the effect of injecting a voltage in the rotor circuit of a 3-phase induction motor? What practical use can be derived from this method?

## PROBLEM

7.1 A 12-pole 3-phase alternator is coupled to an engine running at 500 rpm. It supplies a 3-phase induction motor having a full-load speed of 1440 rpm. Find the percentage slip and number of poles of motor. [4%, 4 poles]

7.2 A 4-pole induction motor is run at 1450 rpm from a 50 Hz supply. Find the percentage slip and the frequency of rotor current. [3.33%, 1.67 Hz]

7.3 If the emf in the stator of a 4-pole, 3-phase induction motor has a frequency of 50 Hz and that in the rotor is 1.5 Hz, at what speed is the motor running and what is the slip? [1455 rpm, 0.03]

7.4 The rotor of a 3-phase, 440 V, 50 Hz, 4-pole induction motor has a resistance of 0.3 ohm per phase and an inductance of 0.008 H per phase. The ratio of stator-to-rotor turns is 2:1. The stator is delta connected and rotor is start connected. Determine the voltage between the slip-rings when the motor is at standstill. Calculate the standstill rotor current and rotor current when the motor is operating at 3% slip.

[381 V, 87 A, 22 A]

7.5 An induction motor has a efficiency of 0.9 when the output is 50 hp. At this load the stator and rotor copper-loss each equals the iron-loss. The windage and friction losses are one-third of no-load losses. Calculate the slip. [3.12%]

7.6 A 500 V, 3-phase, 6-pole, 50 Hz induction motor develops 20 hp inclusive of windage and friction losses, when running at 975 rpm. The power factor is 0.87. Find (a) slip, (b) rotor copper-loss, (c) total input if stator losses are 1500 W, (d) line current, (e) frequency of rotor current.

[(a) 2.5% (b) 382.6 W (c) 16.8 kW (d) 22.3 A (e) 1.25 Hz]

7.7 The power input to the rotor of a 3-phase, 440 V, 50 Hz, 6-pole induction motor is 80 kW. The rotor emf is observed to make 100 complete alternations per minute. Calculate (a) slip, (b) motor speed, (c) mechanical power developed, (d) rotor copper-loss per phase, (e) rotor resistance per phase if rotor current is 65 A.

[(a) 0.033 (b) 967 rpm (c) 104.98 hp (d) 880 W (e) 0.208 ohm]

7.8 A 6-pole 3-phase induction motor develops 30 hp including 2 hp mechanical losses at a speed of 950 rpm when connected to 550 V 3-phase mains. The power factor is 0.88. Find (a) slip, (b) rotor-copper-loss, (c) total input if the stator losses are 2 kW, (d) efficiency, (e) line current.

[(a) 0.05 (b) 1175 W (c) 25.6 kW (d) 0.82 (e) 30.4 A]

7.9 A 3-phase induction motor has a 4-pole star connected stator and runs on a 220 V, 50 Hz supply. The rotor resistance and standstill reactance are 0.1 ohm and 0.9 ohm, respectively. The ratio of stator-to-rotor turns is 1.75 and full-load slip is 5%. Find full-load torque, hp output, maximum torque and speed at maximum torque. Neglect stator impedance. [42 N-m, 8.52 hp (metric), 50 N-m, 1330 rpm]

7.10 A 440 V, 3-phase, 50 Hz, 8-pole, star connected induction motor has $r_1 = r_2' = 0.1$ ohm, $x_1 = x_2' = 0.7$ ohm, $R_0 =$ 100 ohm, $X_0 = 25$ ohm. Calculate the rotor current as referred to stator, stator current, input power factor, torque and efficiency at 4% slip and starting torque. Mechanical losses are 1 kW. Use approximate equivalent circuit. [86 A, 93.3 A, 0.838, 706 N-m, 87.5%, 123.25 N-m]

7.11 A 3-phase induction motor has a synchronous speed of 250 rpm and a full load slip of 0.04. The resistance and stand still reactance of rotor are 0.02 ohms and 0.15 ohms per phase. Find (a) ratio of maximum torque to full load torque, (b) speed at which maximum torque occurs. Neglect stator impedance. The frequency is 50 Hz. [(a) 1.82 (b) 217 rpm]

7.12 A 3-phase, 50 Hz, 4-pole, 50 hp, 480 V star connected induction motor has the following parameters: $r_1 =$ 0.1 Ω, $x_1 = 0.35$ Ω, $r_2' =$ 0.12 Ω, $x_2' = 0.4$ Ω. At no-load the motor draws a current of 18 A at 0.089 lagging power factor. The stator core losses are 1200 W and rotational losses are 950 W. If slip is 2.5%, find (a) input line current, (b) power factor, (c) developed torque, (d) efficiency.

[(a) 63.6A (b) 0.895 lagging (c) 287.18 N-m (d) 0.905]

7.13 A 3-phase, 60 Hz, 4-pole induction motor has a full load speed of 1730 rpm with slip rings short circuited. The rotor resistance and stand still reactance per phase are 0.1 Ω and 0.4 Ω respectively. The motor develops 1.5 times full load torque when started with sliprings short circuited. If 0.25 Ω per phase resistance is added to rotor circuit find (a) new speed, (b) new starting torque.

[(a) 1555 rpm (b) 3.19 × full load torque]

7.14 A 3-phase, 400 V induction motor has 6-poles. Find the rotor speed if

(a) frequency is 60 Hz and slip is 0.05

(b) frequency is 50 Hz and slip is 0.07 [(a) 1140 rpm (b) 930 rpm]

7.15 A 3-phase, 6-pole, 400 V, 50 Hz induction motor develop maximum torque at a speed of 792 rpm. The rotor resistance is 0.2 Ω per phase. Find rotor reactance per phase at stand still. [0.9615 Ω]

7.16 For the 3-phase induction motor of problem 7.12, find (a) slip at which maximum torque occurs, (b) input current and p.f. when maximum torque occurs. [(a) 0.159 (b) 258.33 A, 0.72 lagging]

7.17 A 3-phase, 50 Hz, 4-pole induction motor has a full load slip of 0.06. Find frequency of rotor current at the instant of starting and at full load. [50 Hz, 3 Hz]

7.18 A 3-phase, 50 Hz, star connected, 208 V induction motor has an effective stator to rotor turn ratio of 3:1. It requires 52 V to produce rated stator line current at stand still. The rotor resistance and stand still reactance per phase are 0.1 Ω and 0.5 Ω. Find rotor current with rotor blocked (and 52 V input) and when operating at 5% slip. [19.62 A, 19.43 A]

7.19 A 3-phase, 50 Hz, 6-pole induction motor has a star connected rotor. When rated voltage is applies to stator, the voltage across sliprings is 55 V with sliprings open circulated. The rotor impedance is $0.7 + j5$ ohm per phase. Find rotor current (a) when motor is running at 0.05 slip, (b) at starting with slip rings short circulated, (c) at starting with an impedance of $4 + j3$ ohm per phase inserted in rotor circuit. [(a) 2.14 A (b) 6.29 A (c) 3.42 A]

7.20 A 3-phase, 6-pole, 60 Hz induction motor has a no-load speed of 1196 rpm and full load speed of 1150 rpm. At stand still the rotor induced emf per phase is 55 V. Find (a) slip at no-load and full-load (b) rotor induced emf at no-load and full-load. [(a) 0.3333% 4.167% (b) .1833 V, 2.29 V]

7.21 A 3-phase, 4-pole, 50 Hz, 208 V induction motor has a starting line current of 700 A and a starting torque of 225 N-m. If a reduced voltage of 120 V is applied to stator at the time of starting, find starting torque and starting line current. [75 N-m, 403 A]

7.22 A 3-phase, 60 Hz, 10-pole induction motor has a full-load speed of 690 rpm. The rotor resistance and standstill reactance per phase are 0.3 Ω and 1.1 Ω respectively. Find (a) speed at pull out torque, (b) additional rotor resistance per phase to obtain maximum torque at starting, (c) full load speed with added rotor resistance. [(a) 523.64 rpm (b) 0.8 Ω (c) 610 rpm]

7.23 A 200 hp, 3-phase, 25 Hz, 5000 V induction motor has star connected stator and rotor windings with 576 and 56 turns per phase respectively. The stator impedance is $3.6 + j16$ ohms per phase. The rotor impedance at stand still is $0.03 + j0.07$ ohms per phase. No load current is 6A at 0.07 lagging power factor. Draw circle diagram and find (a) start torque on full voltage and rotor short circuited, (b) value of rotor resistance per phase to be added to give twice full load torque at starting, (c) starting current and p.f. on full voltage with rotor short circuited, (d) starting current and p.f. with added rotor resistance. [(a) 135000 sync. watts (b) 0.05 Ω (c) 118 A, 0.28 p.f. (d) 112 A, .45 p.f.]

7.24 A 3-phase 4-pole, 500 V, 25 hp induction motor has a rotor impedance of $0.2 + j0.35$ ohm at standstill. The voltage across sliprings at standstill is 161 V per phase. The motor develops a torque of 73 N-m at 4% slip with rotor short circuited. The rotor current under this condition is 51 A. Find the emf which should be injected in each rotor phase so that the motor may develop the same torque at speed 30% above synchronous speed. [61.28 V]

7.25 A 3-phase cage motor has maximum torque equal to 2.5 times full load torque. Its rotor impedance at standstill is $0.3 + j3$ ohm per phase. Find the ratio of starting torque to full load torque if started by (a) direct on line starter, (b) star delta starter, (c) auto-transformer with 60% tapping. [(a) 0.495 (b) 0.165 (c) 0.1782]

7.26 In an unbalanced 3-phase system the phase voltages are $V_a = 150 \angle 0°$, $V_b = 250 \angle -90°$ and $V_c = 350 \angle 110°$V. Find sequence voltages $V_{a1}$, $V_{a2}$ and $V_{a0}$. [238 ∠5.16 V, 108.25 ∠ 206.15V, 28.17 ∠ 69°V]

7.27 The parameter of a 3-phase induction motor are $r_1 = 0.8$ Ω, $x_1 = 1.4$ Ω, $r_2' = 0.4$ Ω and $x_2' = 0.7$ Ω, $X_0 = 40$ Ω, $R_0 = 160$ Ω. The supply voltages are $V_a = 150 \angle 0°$V, $V_b = 250 \angle -90°$V and $V_c = 350 \angle 110°$V. If slip is 0.04 find (a) airgap power for positive and negative sequence, (b) rotor copper-loss for positive and negative sequence, and (c) total mechanical power developed. [(a) 14043.5 W, 1324.12 W (b) 561.74 W, 2595. 25 W (c) 12210.63 W]

Chapter 8

# Single-Phase Induction Motor

## 8.1 INTRODUCTION

Single-phase induction motors are used very widely in many equipments in houses, offices and factories. They are used only in small sizes, less than 2 hp and mostly in the fractional horse power range. They are relatively simple in construction but considerably more difficult to analyse than 3-phase induction motors. Single-phase induction motors in the 1/8 to 3/4 hp range are very widely used for fans, washing machines, refrigerators, blowers, centrifugal pumps, etc. They are used in very small sizes (1/300 to 1/20 hp) in toys, hair dryers, vending machines, etc. Single-phase induction motors operate at low power factors and are less efficient than 3-phase induction motors.

## 8.2 CONSTRUCTION

The construction of a single-phase induction motor is very similar to that of a 3-phase induction motor. The rotor is always squirrel cage, while the stator carries a single phase winding. The stator winding is disposed in slots around the inner periphery of a laminated ring as shown in Fig. 8.1. In addition the stator also carries an auxiliary winding for providing the starting torque.

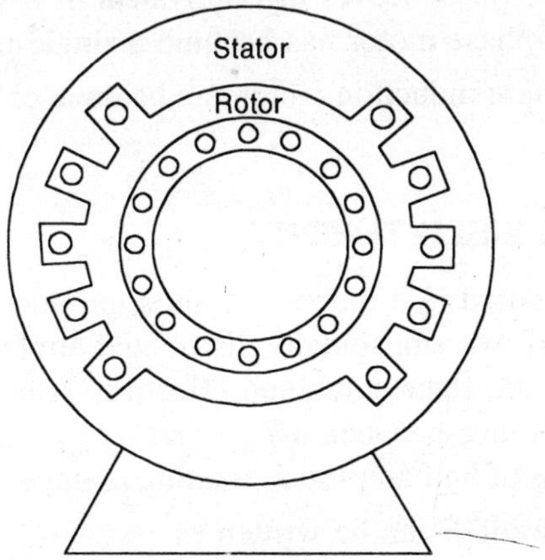


**Fig. 8.1.** Construction of single-phase induction motor

## 8.3 CONDITIONS AT STARTING

A single-phase induction motor has a special feature. It does not develop any torque when at standstill but does develop a torque when running.

Figure 8.2 shows the conditions at starting. The stator winding is actually distributed in slots but is shown as a concentrated field acting on horizontal axis. As the applied voltage varies sinusoidally, the flux also varies sinusoidally but always acts along $x$-axis. As flux is increasing from left to right, voltages are induced in rotor bars due to transformer action. The resulting rotor currents produce a field which opposes the stator field. The torque is produced due to reaction of rotor currents with stator field. As seen in Fig. 8.2, the forces on the rotor bars in top half are in downward direction, whereas the forces on the rotor bars in bottom half are in upward direction. The net torque is zero when the motor is at standstill.

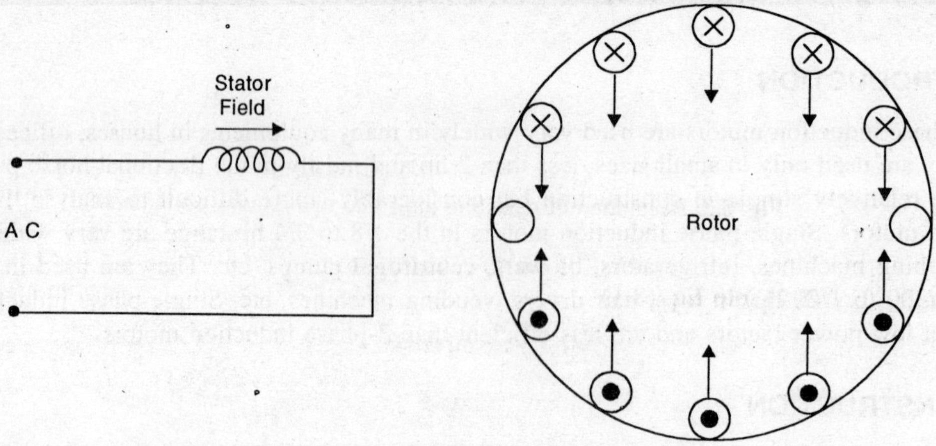


**Fig. 8.2.** Conditions at starting of single-phase induction motor

However, the motor develops a torque if it is in motion. It is known that if one phase of a 3-phase induction motor is opened when it is supplying a moderate or light load, it continues to run at a slightly reduced speed. Opening of one phase leaves the equivalent of a single-phase connected to the stator winding. In other words the 3-phase motor has become a single-phase motor.

The analysis of a single-phase induction motor can be done either by double revolving field theory or by cross-field theory.

## 8.4 DOUBLE REVOLVING FIELD THEORY

A single-phase winding, distributed in a number of slots, produces a sinusoidal space distribution of mmf centred on the coil axis. If two sinusoidally distributed mmfs each of constant magnitude $F_{max}/2$ are rotating in opposite direction, their combined effect, as seen from Fig. 8.3 is equivalent to one pulsating field $F_{max} \cos \omega t$ varying between $+F_{max}$ and $-F_{max}$. Conversely a pulsating field can be resolved into two rotating fields of half amplitude rotating in opposite directions at synchronous speed.

The space wave of stator mmf $F$ can be written as

$$F = F_i \cos \theta \qquad \text{...(8.1)}$$

where θ is the electrical space angle measured from the stator coil axis and $F_i$ is the instantaneous value of mmf at the coil axis and is proportional to instantaneous stator current. Thus

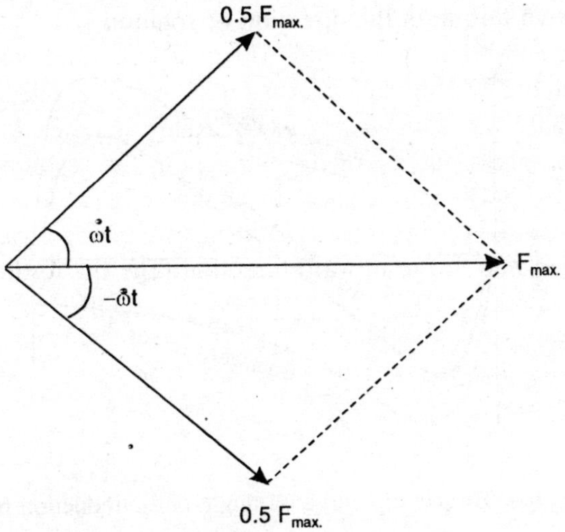


**Fig. 8.3.** Resolution of pulsating mmf into two rotating components

$$F_i = F_{max} \cos \omega t \qquad \text{...(8.2)}$$

Substituting Eq. (8.2) into Eq. (8.1)

$$F = F_{\max} \cos \omega t \cos \theta$$

or

$$F = 0.5 F_{\max} \cos(\theta - \omega t) + 0.5 F_{\max} \cos(\theta + \omega t) \qquad \text{...(8.3)}$$

Each of the cosine terms in Eq. (8.3) describes a sinusoidal function of space angle θ. Each has a peak value of half the amplitude of the pulsating wave and a space phase angle $\omega t$. The angle $\omega t$ provides rotation of each wave around the air-gap at constant angular velocity ω radians per second. The two waves travel in opposite directions. The first wave whose argument is $(\theta - \omega t)$ travels in the forward direction of θ. The second wave whose argument is $(\theta + \omega t)$ travels in the backward direction of θ.

Each of the two rotating mmf waves produces torque due to induction motor action, the two torques acting in the opposite directions. If the rotor is stationary it is subjected to the inductive effects of fields due to these equal and opposite rotating mmfs and naturally the two sets of induced rotor currents will develop equal and opposite torques, so that the net torque is zero. However, if the rotor is given an initial rotation in either direction, the torque due to the rotating field acting in the direction of initial rotation will be more than that due to the other rotating field and the motor will develop a net positive torque in the same direction as the initial rotation. Thus the motor will keep running in the direction of initial rotation. It is convenient to consider the actual motor as a composite of two structures, mechanically coupled together, one of which has a rotor subjected to the influence of a stator carrying balanced 3-phase currents having positive sequence, while the other rotor is in a field produced by another set of balanced 3-phase currents having negative sequence.

The torque-slip curve of the actual motor can be obtained by applying the principle of the superposition to the hypothetical constituent motors. The difference between these two motors is that, when one of them is operating at a slip $s$, the other is operating at a slip $(2 - s)$. Their individual torque-slip curves will have the form shown in Fig. 8.4 and the algebraic sum of their ordinates will give the resultant torque. It is seen that the resultant torque is zero at standstill. At any other value of slip, the motor develops a net positive torque in the direction of rotation.

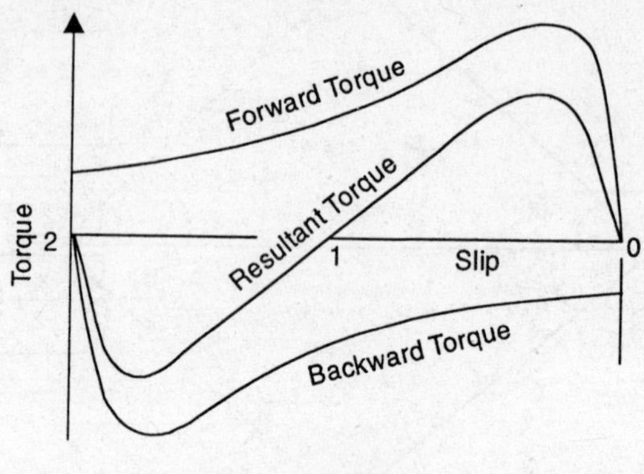


**Fig. 8.4.** Torque-slip curve of single-phase induction motor

In addition to the average net torque shown in Fig. 8.4, there are torque pulsations at twice the stator frequency. These torque pulsations are caused when the forward and backward magnetic fields cross each other twice each cycle. Although these torque pulsations produce no average torque, they do increase vibrations which make a single-phase induction motor more noisy than a three-phase induction motor of the same size.

## 8.5 EQUIVALENT CIRCUIT (CIRCUIT MODEL)

When the motor is at standstill, the equivalent circuit is exactly similar to that of a transformer on short circuit as shown in Fig. 8.5 (a). $r_1$ and $x_1$ are the resistance and leakage reactance of the stator. The currents in $R_0$ and $X_0$ represent the iron-loss and magnetising component of no-load current. $r_2'$ and $x_2'$ represent the rotor resistance and leakage reactance as referred to stator.

At standstill the air gap flux can be resolved into two equal and opposite magnetic fields within the motor. Since these fields are of equal size, each one contributes an equal share to the resistive and reactive voltage drops in the rotor circuit. Therefore $r_2$ and $x_2$ can be split into two parts, each one corresponding to the effects of one of the magnetic fields. This is shown in Fig. 8.5(b). The voltages $E_f$ and $E_b$ are equal at standstill.

When the motor is operating at slip $s$, the resistive term in the forward field group changes to $r_2'/2s$ as in a 3-phase induction motor. The slip for the backward field is $(2 - s)$, so that the resistance in the backward field group changes to $(r_2'/2)/(2 - s)$. The complete equivalent circuit under running conditions is shown in Fig. 8.5 (c). Since the value of $s$ is generally small (around 0.05 or so), $r_2'/2s$ is considerably higher than $r_2'/2\ (s - 2)$. Therefore $E_f$ is greater than $E_b$ and forward flux is greater than the backward flux.

## 8.6 ANALYSIS

The equivalent circuit of a single-phase induction motor is similar to that of 3-phase one. The analysis is also similar to that of 3-phase motor. If the parameters of equivalent circuit (Fig. 8.5c) are known, then $I_{2f}'$ and $I_{2b}'$ can be calculated. Then

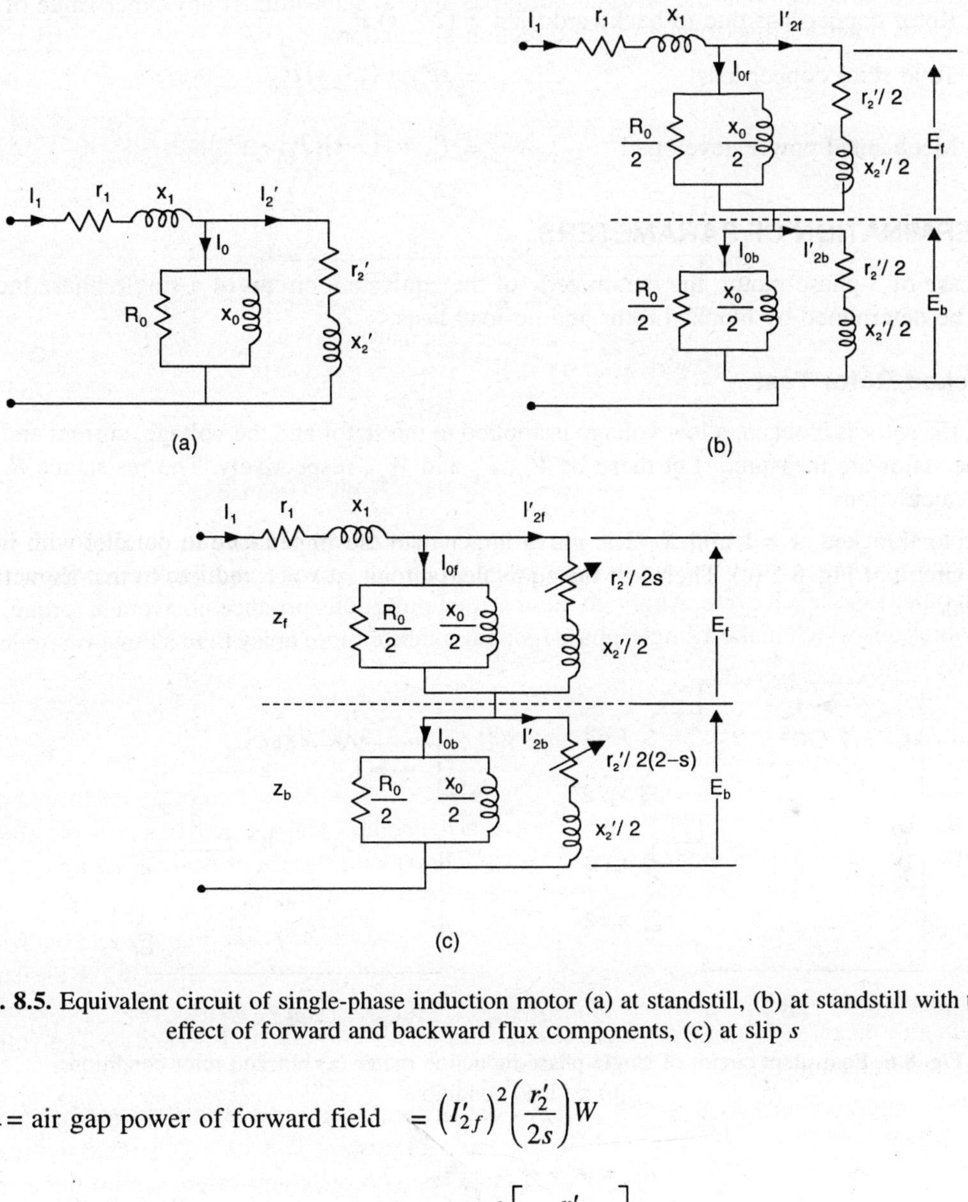


**Fig. 8.5.** Equivalent circuit of single-phase induction motor (a) at standstill, (b) at standstill with the effect of forward and backward flux components, (c) at slip *s*

$$P_{gf} = \text{air gap power of forward field} = \left(I_{2f}'\right)^2\left(\frac{r_2'}{2s}\right)W \quad ...(8.4)$$

$$P_{gb} = \text{air gap power of backward field} = \left(I_{2b}'\right)^2\left[\frac{r_2'}{2(2-s)}\right]W \quad ...(8.5)$$

$$T_f = \text{torque due to forward field} = \frac{P_{gf}}{2\pi n_s}\text{N-m} \quad ...(8.6)$$

$$T_b = \text{torque due to backward field} = \frac{P_{gb}}{2\pi n_s}\text{N-m} \quad ...(8.7)$$

The net torque $T = T_f - T_b$ ...(8.8)

Rotor copper-loss due to forward field $= sP_{gf}$ ...(8.9)

Rotor copper-loss due to backward field $= (2 - s)\,P_{gb}$ ...(8.10)

Total rotor copper-loss $= sP_{gf} + (2 - s)\,P_{gb}$ ...(8.11)

Mechanical power developed $= P_m = (1 - s)\left(P_{gf} - P_{gb}\right).$ ...(8.12)

## 8.7 DETERMINATION OF PARAMETERS

As in the case of 3-phase motor, the parameters of the equivalent circuit of a single-phase induction motor can be determined by blocked rotor and no-load tests.

### 8.7.1 Blocked Rotor Test

In this test, the rotor is blocked, a low voltage is applied to the stator and the voltage, current and power input to the stator are measured. Let these be $V_{sc}$, $I_{sc}$ and $W_{sc}$, respectively. The resistance $R_0$ is neglected in calculations.

With rotor blocked, $s = 1$ and $X_0/2$ is much larger than the impedance in parallel with it in the equivalent circuit of Fig. 8.5 (c). Therefore the equivalent circuit, at $s = 1$, reduces to that shown in Fig. 8.6 (a). Thus

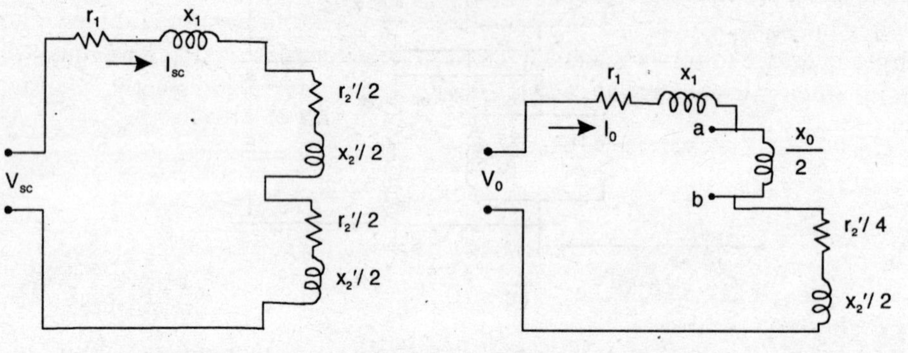


**Fig. 8.6.** Equivalent circuit of single-phase induction motor (a) blocked rotor conditions, (b) no-load conditions.

$$R_{sc} = r_1 + r_2' = \frac{W_{sc}}{I_{sc}^2} \quad ...(8.13)$$

$$Z_{sc} = \frac{V_{sc}}{I_{sc}} \quad ...(8.14)$$

$$x_1 + x_2' = \left(Z_{sc}^2 - R_{sc}^2\right)^{0.5} \quad ...(8.15)$$

In all well designed machines, $r_1 = r_2'$. Moreover, it is assumed that $x_1 = x_2'$. Thus $r_1$, $r_2'$, $x_1$, $x_2'$ can be determined.

### 8.7.2 No-load Test

In this test, the motor is run on no-load and voltage $V_0$, current $I_0$ and power input $W_0$ to the stator are measured.

At no-load $s$ is very small (close to zero). Therefore $r_2'/2s$ is very large as compared to $X_0/2$. Moreover $r_2'/2(2-s) \cong r_2'/4$ and $(r_2'/4 + j\,x_2'/2)$ is very small as compared to $X_0/2$. The resistance $R_0$ is neglected. Therefore, under no-load conditions, the equivalent circuit of Fig. 8.5 (c) simplifies to that shown in Fig. 8.6 (b).

$$\text{No-load power factor, } \cos\theta_0 = \frac{W_0}{V_0 I_0} \qquad ...(8.16)$$

$$V_{ab} = V_0 - [I_0 \angle -\theta_0]\left[\left(r_1 + \frac{r_2'}{4}\right) + j\left(x_1 + \frac{x_2'}{2}\right)\right] \qquad ...(8.17)$$

Since $r_1$, $r_2'$, $x_1$, $x_2'$ are known from the blocked rotor test, $V_{ab}$ can be calculated. Then

$$\frac{X_0}{2} = \frac{V_{ab}}{I_0}. \qquad ...(8.18)$$

*Example 8.1*

A 220 V, 50 Hz, 6-pole, 1/6 hp single-phase induction motor has $r_1$ = 11.4 ohm, $r_2'$ = 13.8 ohm, $x_1 = x_2'$ = 14.3 ohm and $X_0$ = 275 ohm. Using double revolving field theory, find (a) input current and power factor, (b) torques due to forward and backward field components and gross torque, (c) output, (d) input power, (e) efficiency when the motor is operating at a slip of 0.06. Neglect $R_0$. Constant losses are 30.2 W.

*Soiution*

Referring to the equivalent circuit in Fig. 8.5 (c)

(a)
$$\frac{r_2'}{2s} + j\frac{x_2'}{2} = \frac{13.8}{2 \times 0.06} + j\frac{14.3}{2} = 115 + j7.15\,\text{ohm}$$

$$= 115 \angle 3.6° \text{ ohm}$$

$$j\frac{X_0}{2} = 137.5 \angle 90° \,\text{ohm}$$

$$Z_f = \frac{(115\angle 3.6°)(137.5\angle 90°)}{115\angle 3.6° + 137.5\angle 90°} = 63.5 + j\,56.5 \text{ ohm}$$

$$= 85.5 \angle 41.8°$$

$$(2 - s) = 1.94$$

$$\frac{r_2'}{2(2-s)} + j\frac{x_2'}{2} = \frac{13.8}{2(1.94)} + j\frac{14.3}{2} = 3.56 + j7.15 \text{ ohm}$$

$$= 8\angle 63.5° \text{ ohm}$$

$$Z_b = \frac{(8\angle 63.5°)(137.5\angle 90°)}{8\angle 63.5° + 137.5\angle 90°} = 7.6\angle 64.9°$$

$$= 3.22 + j6.9 \text{ ohm}$$

$$\text{Total impedance} = r_1 + jx_1 + z_f + z_b$$

$$= 11.4 + j14.3 + 63.5 + j56.5 + 3.22 + j6.9$$

$$= 78.1 + j77.7$$

$$= 110.2\angle 44.85° \text{ ohm}$$

$$\text{Input current} = \frac{220}{110.2} = 2\text{A}$$

$$\text{Input power factor} = \cos(44.85°) = 0.71 \text{ lagging}$$

(b)

$$E_f = 2\,(85.5) = 171 \text{ V}$$

$$E_b = 2\,(7.6) = 15.2 \text{ V}$$

$$I'_{2f} = \frac{E_f}{\frac{r'_2}{2s} + j\frac{x'_2}{2}} = \frac{171}{115 + j7.15} = 1.49\,\text{A}$$

$$I'_{2b} = \frac{E_b}{\frac{r'_2}{2(2-s)} + j\frac{x'_2}{2}} = \frac{15.2}{8\angle 63.5°} = 1.9\,\text{A}$$

$$n_s = \frac{2 \times 50}{6} = \frac{50}{3} \text{ rps}$$

$$\text{Torque due to forward field} = \frac{(I'_{2f})^2\left(\frac{r'_2}{2s}\right)}{2\pi n_s}$$

$$= \frac{(1.49)^2(115)}{2\pi(50/3)} = 2.438 \text{ N-m}$$

$$\text{Torque due to backward field} = \frac{(I'_{2b})^2\left[\frac{r'_2}{2(s-2)}\right]}{2\pi n_s}$$

$$= \frac{(1.9)^2(3.56)}{2\pi(50/3)} = 0.123 \text{ N-m}$$

$$\text{Gross torque} = 2.438 - 0.123 = 2.315 \text{ N-m}$$

(c) $$\text{Gross output } (2.315)(2\pi n) = 2.315(2\pi n_s)(1-s)$$

$$= 2.315\left(\frac{2\pi 50}{3}\right)(1-0.06) = 227.9 \text{ W}$$

$$\text{Net output} = 227.9 - 30.2 = 197.7 \text{ W}$$

(d) $$\text{Input power} = 220 \times 2 \times 0.71 = 312.4 \text{ W}$$

(e) $$\text{Efficiency} = \frac{\text{output}}{\text{Input}} = \left(\frac{197.7}{312.4}\right)100$$

$$= 63.3\%.$$

*Example 8.2*

A 220 V single-phase induction motor gave the following test results.

| | | | |
|---|---|---|---|
| Blocked rotor test: | 110 V, | 10 A, | 400 W |
| No-load test: | 220V, | 4 A, | 100 W. |

(a) Find the parameters of equivalent circuit. Neglect $R_0$. (b) Find iron-friction and windage losses.

*Solution*

(a) Using the results of Sec. 8.7

$$r_1 + r_2' = \frac{400}{10^2} = 4 \text{ ohm}$$

$$r_1 = r_2' = 2 \text{ ohm}$$

$$Z_{sc} = \frac{110}{10} = 11 \text{ ohm}$$

$$x_1 + x_2' = \left(11^2 - 4^2\right)^{0.5} = 10.25 \text{ ohm}$$

$$x_1 = x_2' = 5.125 \text{ ohm}$$

$$\cos\theta_0 = \text{No-load power factor} = \frac{100}{220 \times 4} = 0.114$$

$$\theta_0 = 83.5°$$

$$V_{ab} = 220 - [4\angle -83.5°][(2.5 + j7.6875)]$$

$$= 220 - (4\angle -83.5°)(8.084\angle 71.99°) = 188.42 \text{ V}$$

$$\frac{X_0}{2} = \frac{V_{ab}}{I_0} = \frac{188.42}{4} = 47.1 \text{ ohm}$$

(b) $\quad$ Iron, friction and windage loss $= W_0 - I_0^2\left(r_1 + \frac{r_2'}{4}\right)$

$$= 100 - 4^2\left(2 + \frac{2}{4}\right) = 60\,\text{W}.$$

## 8.8 SYMMETRICAL COMPONENTS

The method of symmetrical components allows us to represent any unbalanced polyphase system by balanced sets. A set of unbalanced 3-phase voltage (or currents) can be resolved into a positive sequence set, a negative sequence set and a zero sequence set as discussed in Section 7.21.1. Any unbalanced two-phase system can be resolved into a two-phase balanced positive sequence set and a two-phase balanced negative sequence set.

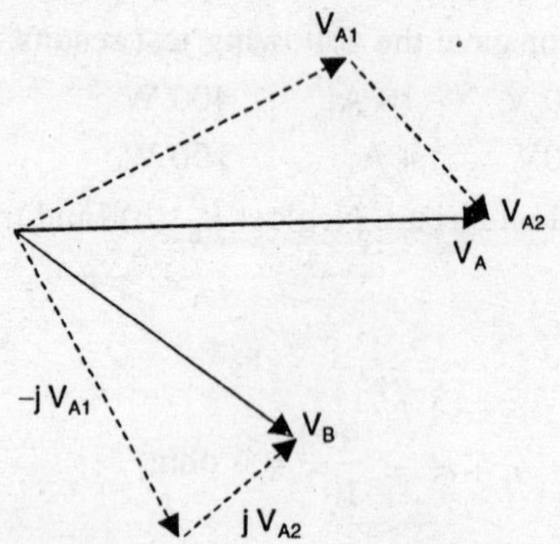


**Fig. 8.7.** Symmetrical components of two-phase system

Fig. 8.7 shows an unbalanced set of two-phase voltages $V_A$ and $V_B$. Each of these can be resolved into two components as shown. Thus we can write

$$V_A = V_{A1} + V_{A2} \qquad \text{...(8.19a)}$$

$$V_B = -jV_{A1} + jV_{A2} \qquad \text{...(8.19b)}$$

or in matrix form

$$\begin{bmatrix} V_A \\ V_B \end{bmatrix} = \begin{bmatrix} 1 & 1 \\ -j & j \end{bmatrix} \begin{bmatrix} V_{A1} \\ V_{A2} \end{bmatrix} \qquad \text{...(8.20)}$$

From Eqs. (8.19) we can write

$$V_{A1} = 0.5(V_A + jV_B) \qquad \text{...(8.21a)}$$

$$V_{A2} = 0.5(V_A - jV_B) \qquad \text{...(8.21b)}$$

or in matrix form

$$\begin{bmatrix} V_{A1} \\ V_{A2} \end{bmatrix} = 0.5\begin{bmatrix} 1 & j \\ 1 & -j \end{bmatrix} \begin{bmatrix} V_A \\ V_B \end{bmatrix} \qquad \text{...(8.22)}$$

If a system has two phase unbalanced currents $I_A$ and $I_B$, they can be decomposed into positive and negative sequence currents $I_{A1}$ and $I_{A1}$ using Eq. (8.22). Similarly Eq. (8.20) can be used to find $I_A$ and $I_B$ if $I_{A1}$ and $I_{A2}$ are known.

## 8.9 ANALYSIS OF SINGLE PHASE INDUCTION MOTOR USING SYMMETRICAL COMPONENTS

A single phase induction motor has only one stator winding and one stator current $I_A$. The current $I_B$ is zero. Using Eq. (8.22) we get

$$I_{A1} = 0.5I_A \quad \text{...(8.23a)}$$

$$I_{A2} = 0.5I_A \quad \text{...(8.23b)}$$

Thus the positive and negative sequence currents, in a single-phase induction motor are the same. This has to be so since the motor has only one stator winding.

The equivalent circuit of a single-phase induction motor consists of two parts one representing the response to the positive sequence voltage set and the other representing the response to the negative sequence voltage set.

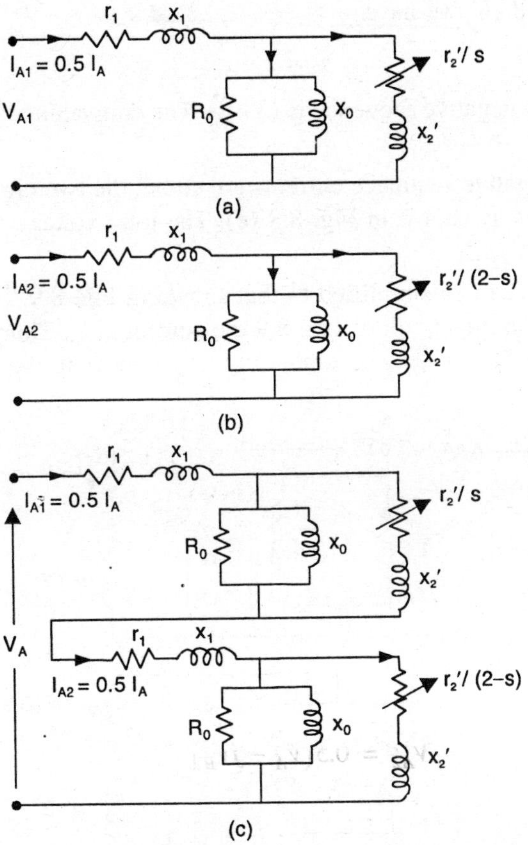


**Fig. 8.8.** Equivalent circuit of single phase induction motor (a) positive sequence (b) negative sequence (c) series connection of positive and negative sequence circuits.

The equivalent circuit corresponding to the positive sequence voltage set is shown in Fig. 8.8(a). $V_{A1}$ and $I_{A1}$ represent the positive sequence voltage and current. This equivalent circuit is similar to that of 3-phase induction motor. $r_1$ and $x_1$ represent the stator winding resistance and leakage reactance. $R_0$ and $X_0$ represent the core-loss and magnetising component of no-load current. As discussed in Chapter 7, the rotor circuit is represented by a resistance $r_2'/s$ and reactance $x_2'$.

The equivalent circuit corresponding to negative sequence set will have an applied voltage $V_{A2}$ and a current $I_{A2}$. However, as seen from Eq. (8.23) the positive and negative sequence currents are equal. The slip of the rotor with respect to positive sequence set is $s$ i.e.

$$s = \frac{n_s - n}{n_s} \quad \text{...(8.24)}$$

where $n_s$ and $n$ are the synchronous and actual rotor speeds in revolutions per second. The negative sequence voltage sets up a field which rotates in a direction opposite to that of the field due to positive sequence voltage. Therefore $s_2$, i.e. slip with respect to negative sequence field is

$$s_2 = \frac{n_s - (-n)}{n_s} = 1 + \frac{n}{n_s} \quad \text{...(8.25)}$$

Combining Eqs. (8.24) and (8.25) we have

$$s_2 = 2 - s \quad \text{...(8.26)}$$

Thus the rotor slip for the negative sequence is $(2 - s)$. The equivalent circuit for negative sequence voltages is shown in Fig. 8.8 (b).

Since the positive and negative sequence currents are equal, the two circuits of Fig. 8.8 (a) and 8.8 (b) can be connected in series as shown in Fig. 8.8 (c). The total voltage applied to the circuit is $V_A$ which is the sum of $V_{A1}$ and $V_{A2}$.

The circuit of Fig. 8.8 (c) can be simplified to that shown in Fig. 8.9. In the circuit of Fig. 8.8 (c) the current is 0.5 $I_A$ where as in the circuit of Fig. 8.9 the current is $I_A$. Therefore it is only natural that all impedances in Fig. 8.9 are half of the corresponding impedances in the circuit of Fig. 8.8 (c). The

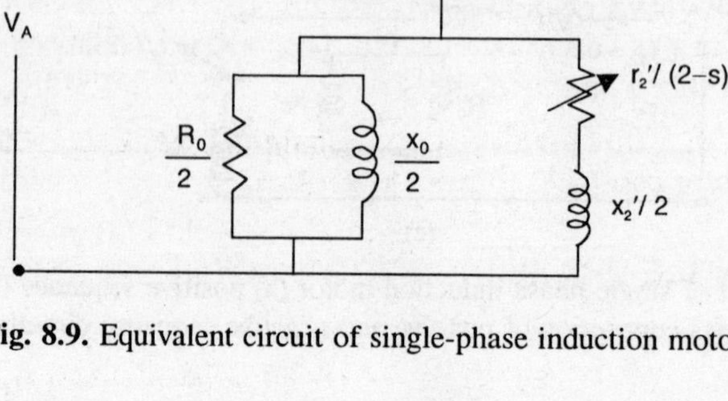


**Fig. 8.9.** Equivalent circuit of single-phase induction motor

total stator impedance in Fig. 8.8 (c) is 2 $(r_1 + jx_1)$. This changes to $(r_1 + jx_1)$ since the current has been increased from $0.5I_A$ to $I_A$. The same is true about $R_o$, $X_o$ as also for the rotor circuit impedance.

The circuit of Fig. 8.9 is the same as that shown in Fig. 8.5 (c). The stator current has been written as $I_1$ in Fig. 8.5 (c). This is the same as $I_A$. In fact this is the current in the main winding of the stator under steady state operating conditions. This current has been denoted by $I_m$ in Section 8.11 to distinguish it from the auxiliary winding current $I_a$.

## 8.10 CROSS-FIELD THEORY

In this theory the flux is resolved into two components. One component acts along the stator winding axis and the other at right angles to it as shown in Fig. 8.10. One group of short circuited rotor conductors links with one component flux and the second group links with the other component flux. In Fig. 8.10, *ab* and *cd* represent these two groups of conductors.

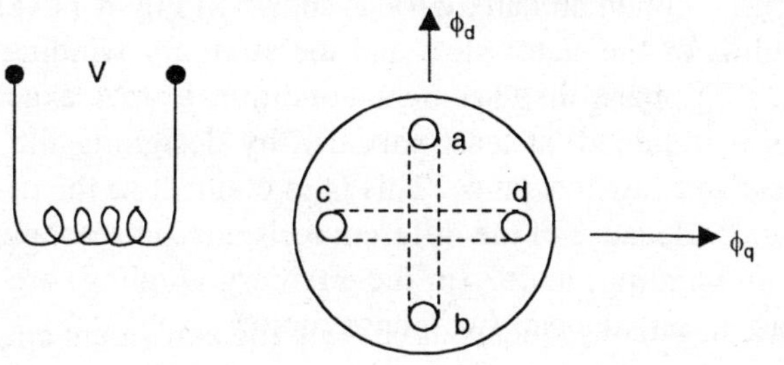


**Fig. 8.10.** Cross-field theory of single-phase induction motor

### 8.10.1 Motor at Standstill

When the motor is at standstill, and voltage $V$ is applied to stator, an mmf is set up in $q$-axis and this mmf causes flux $\phi_q$ in this axis. This flux links with turn *ab* and induces a transformer emf $E_{tq}$ in this turn, which causes a current $I_q$. This current sets up an opposing mmf along $q$-axis. The flux $\phi_q$ is actually set up by the resultant of these two mmfs. As both these mmfs are along the same axis, no torque is produced. The coil *cd* has its axis at 90° to $q$-axis. Therefore no emf is induced in this coil and hence no current. Thus there is no torque on either of the coils.

### 8.10.2 Motor Running

When the motor is running, the stator current will cause voltage drops in stator resistance and leakage reactance. If these voltage drops are neglected, the back emf induced in stator winding is equal to the voltage applied to stator. The flux $\phi_q$ will also be constant. The coil *cd* is moving in the flux $\phi_q$ and a motional emf $E_{rq}$ is set up in this coil. Since the coil is short circuited, a current $I_d$ flows in it. This current sets up an mmf and flux $\phi_d$ in the $d$-axis.

The interaction of flux $\phi_d$ with current $I_q$ (in coil *ab*) and of flux $\phi_q$ with current $I_d$ (in coil *cd*) produces torque and the motor continues to run.

## 8.11 STARTING OF SINGLE-PHASE INDUCTION MOTORS

Since a single-phase induction motor does not have a starting torque, it needs special methods of starting. These motors are classified according to the method of starting and are usually referred to by names descriptive of these methods. All these methods modify the motor, so that it becomes a two-phase motor at the time of starting. The stator is provided with two windings, called main and auxiliary windings, whose axes are space displaced by 90 electrical degrees. The auxiliary windings is excited by a current which is out-of phase with the current in the main winding, both currents derived from the same supply mains. If the phase difference between the two currents is 90° and the mmfs created by them are equal, maximum starting torque is produced. If the phase difference is not 90° and if mmfs are not equal, starting torque will be small, but in many applications it is still sufficient to start the motor. The auxiliary winding is disconnected by a centrifugal switch after the motor has achieved about 75% speed. The various starting methods differ in the arrangement to create a phase difference between the currents in the main and auxiliary winding.

### 8.11.1 Split Phase Motor

The stator has two windings — main and auxiliary as shown in Fig. 8.11 (a). Usually the main winding is made to occupy two-thirds of the stator slots and the auxiliary winding is placed in the remaining one-third slots. Thus the 90° phase displacement condition is met exactly. The time displacement between the two currents is obtained, at least partially, by designing the auxiliary winding for high resistance (using thin wire) and low reactance. This is in contrast to the main winding, which has low resistance and high reactance. Because of the different resistance to reactance ratios in the two winding, the currents $I_m$ (in the main winding) and $I_a$ (in the auxiliary winding) are out-of-phase by about 45°. The motor is equivalent to an unbalanced two-phase motor.

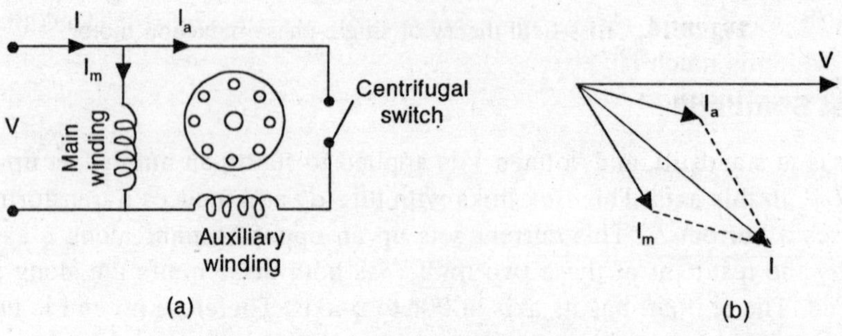


**Fig. 8.11.** Split phase motor (a) circuit, (b) phasor diagram at starting

A revolving field is obtained by this two-phase arrangement and the motor develops a starting torque. Because of the high resistance character of auxiliary winding, this motor is also known as resistance start motor. The auxiliary winding is not designed for continuous operation and is disconnected by the centrifugal switch, when the motor has achieved about 75% speed.

In a split-phase motor the current $I_a$ in the auxiliary winding reaches its peak value earlier than the current $I_m$ in the main winding. Therefore the magnetic field due to $I_a$ also reaches its peak value earlier than that due to $I_m$. The direction of rotation depends on whether space angle of magnetic field from auxiliary winding is 90° ahead or 90° behind the angle of the main winding. This angle can be changed from 90° ahead to 90° behind, by reversing the connections of auxiliary winding. Thus the direction of rotation can be reversed by reversing the connections of auxiliary winding leaving the connections of main winding unchanged.

Figure 8.12 shows the torque speed characteristic of this motor. Split phase motors have moderate starting torque and low starting current. Typical applications include fans, blowers, centrifugal pumps, etc. Typical ratings are 1/20 to 0.5 hp and they are the cheapest motors in this range.

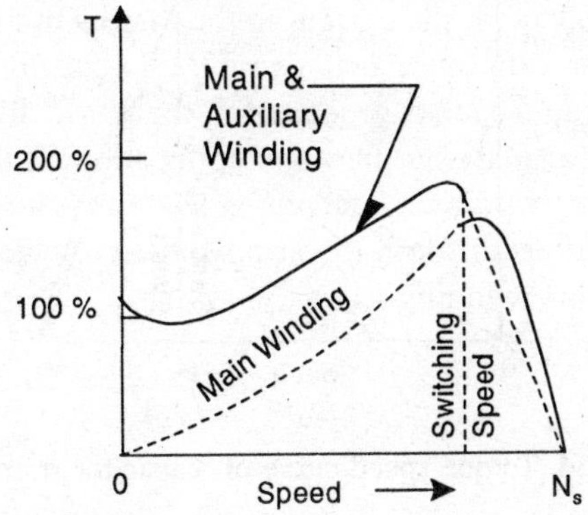


**Fig. 8.12.** Torque-speed curve of split phase motor

### 8.11.2 Capacitor Start Motor

In this motor also, the main and auxiliary windings are space displaced by 90°. The time displacement between the currents in the main and auxiliary winding is achieved by connecting a capacitor in series with auxiliary winding. By using a capacitor of proper value, the current $I_a$ in the auxiliary winding can be made to lead the current $I_m$ in the main winding by 90° at standstill. Thus the motor develops a starting torque which is much larger than that in a split phase motor. Both the capacitor and auxiliary winding are designed for short time duty and are disconnected by centrifugal switch when the motor has reached 75% speed. Figure 8.13 shows the connections and phasor diagram at starting. Figure 8.14 shows the torque-speed characteristic. A high starting torque is an outstanding feature of this motor.

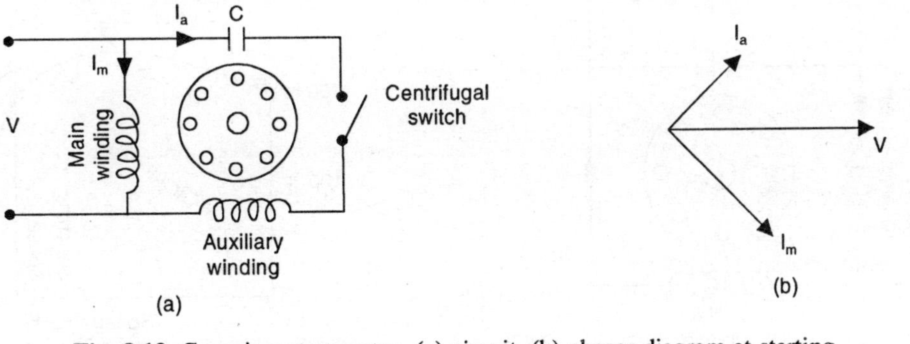


**Fig. 8.13.** Capacitor start motor (a) circuit, (b) phasor diagram at starting

Capacitor start motors are manufactured in a number of sizes from about 1/8 hp to 3/4 hp and are used for many general purpose applications like fans, blowers, compressors, pumps, etc. The size of the capacitor depends on motor rating and has a value of around 50 μF to a few hundred μF.

### 8.11.3 Capacitor Start Capacitor Run Motor

This motor has two capacitors in the auxiliary winding. Both the capacitors $C_1$ and $C_2$ are in the circuit

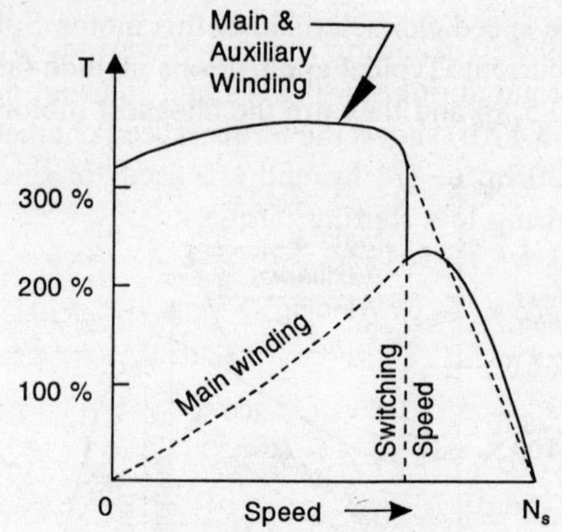


**Fig. 8.14.** Torque-speed curve of capacitor start motor

during starting. After the motor has picked up speed, the centrifugal switch opens and disconnects the capacitor $C_2$. The auxiliary winding and capacitor $C_1$ remain in the circuit during running condition too. Thus the motor is, actually, a two-phase motor. The use of capacitor $C_1$ during running condition improves the power factor. The two-phase operation results in better efficiency. The auxiliary winding and capacitor $C_1$ have to be designed for continuous duty rating and this feature makes this motor more expensive than a capacitor start motor. Two capacitors $C_1$ and $C_2$ are needed, because the size of capacitor needed for good starting torque is very much different from that required for good operating conditions. Capacitor $C_1$ has a value of about a few μF, which $C_2$ is much bigger (about 50 to a few hundred μF). Figure 8.15 shows the connections of this motor and Fig. 8.16 shows the torque-speed characteristic. These motors are manufactured in a number of sizes from 1/8 to 3/4 hp and are used in compressors, conveyors, pumps and other high torque loads.

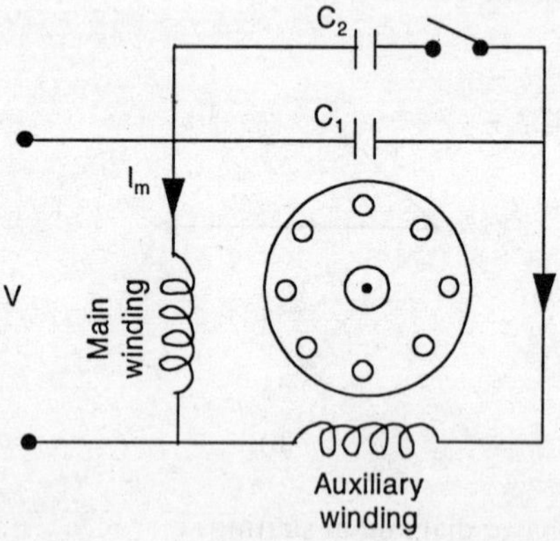


**Fig. 8.15.** Capacitor start capacitor run motor

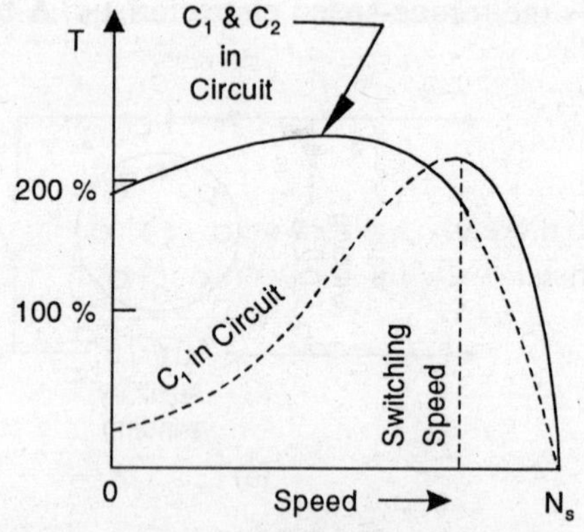


**Fig. 8.16.** Torque-speed curve of capacitor start capacitor run motor

### 8.11.4 Permanent Split Capacitor Motor

In applications where the motor starts under practically zero load, it is possible to avoid the expense of centrifugal switch and the starting capacitor. Thus this motor has a circuit similar to that of Fig. 8.15, except that centrifugal switch and capacitor $C_2$ are not there. Both the windings as well as capacitor $C_1$ remain in circuit during starting as well as during running conditions. Since capacitor $C_1$ is small, the current in the auxiliary winding circuit during starting is small and starting torque is low. Figure 8.17(a) shows the connections, while Fig. 8.17(b) shows the torque-speed characteristic of this motor. They are manufactured in the range of 1/20 hp to 3/4 hp and are used for direct connected fans, blowers, centrifugal pumps and loads requiring low starting torque.

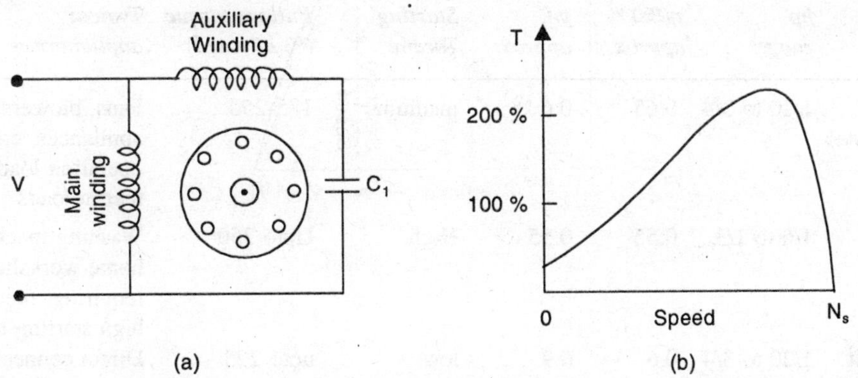


**Fig. 8.17.** Permanent split capacitor motor (a) circuit, (b) torque-speed curve

### 8.11.5 Shaded Pole Motor

Small motors requiring only small starting torque may be made self-starting by using shading bands. A shaded pole motor has a salient pole construction as shown in Fig. 8.18. Low resistance copper bands called shading bands *C* and *D* are placed, so as to surround approximately half of each pole face.

The emf induced in the shading band causes a current to flow in the band. This current opposes the change of flux to which it is due and thus there is a time lag between the fluxes in the shaded and unshaded portion of the pole. The flux in the shaded portion of the pole lags the flux in the other portion. The result is like a rotating field moving in the direction from the unshaded to the shaded portion of the pole and a starting torque is produced. The efficiency and starting torque are poor. Figure 8.19 shows the torque-speed characteristics. Since this motor relies on the shading band for the starting torque, there is no easy way to reverse the direction of rotation of this motor. Perhaps the only method is to install two shading bands on each pole face and to selectively short circuit one or the other.

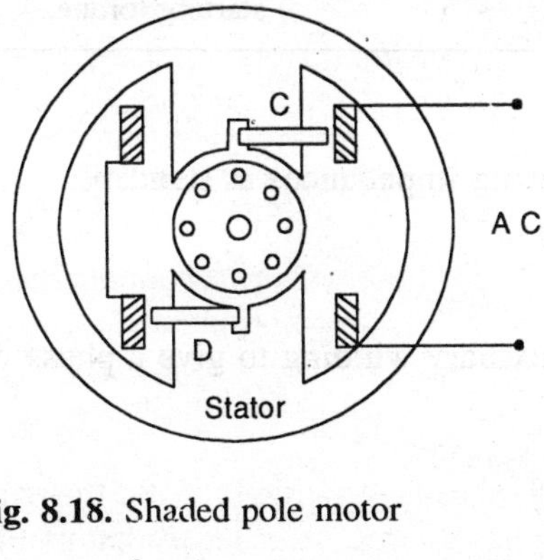


**Fig. 8.18.** Shaded pole motor

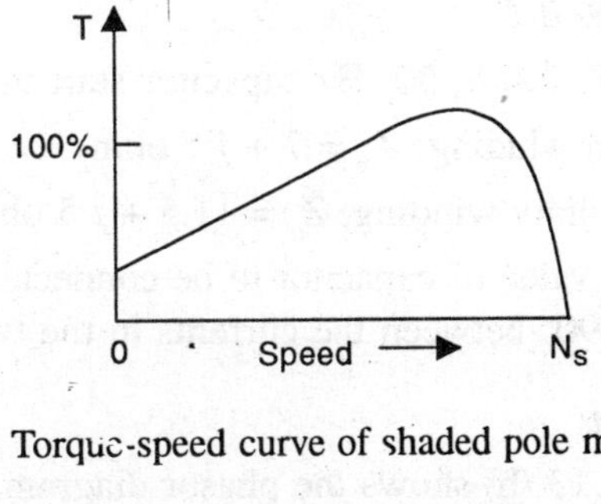


**Fig. 8.19.** Torque-speed curve of shaded pole motor

These motors are manufactured in small sizes from 1/300 and 1/20 hp and are used in toys, hair dryers, desk fans, etc.

Table 8.1 shows the characteristics and applications of the single phase induction motors.

**Table 8.1 Characteristics and Applications of Single Phase Induction Motors**

| *Motor type* | *hp range* | *η(%) approx.* | *p.f. approx.* | *Starting Torque* | *Pullout torque (% of rated)* | *Typical applications* |
|---|---|---|---|---|---|---|
| Split Phase (general purpose) | 1/20 to 3/4 | 0.65 | 0.6 | medium | 185-250 | Fans, blowers, office appliances, continuous operation loads, low inertia loads |
| Split Phase (high torque) | 1/6 to 1/3 | 0.55 | 0.55 | High | Upto 350 | Washing machines, home workshops, loads requiring medium and high starting torque |
| Permanent split capacitor (low torque) | 1/20 to 3/4 | 0.6 | 0.9 | low | upto 225 | Direct connected fans, blowers, loads requiring low starting torque and not driven through belts. |
| Permanent split capacitor (high torque) | 1/6 to 3/4 | 0.6 | 0.9 | high | 260 | Fans and blowers (belt driven or direct driven) centrifugal pumps loads requiring moderate starting torque. |
| Capacitor start | 1/8 to 3/4 | 0.6 | 0.7 | Very high | upto 400 | Conveyors, pumps, compressors. Belt driven loads with high static friction |
| Capacitor start capacitor run | 1/8 to 3/4 | 0.6 | 0.9 | High | upto 260 | Conveyors, pumps, compressors, High torque loads |
| Shaded pole | 1/300 to 1/20 | 0.35 | 0.45 | Very low | 150 | Fans, toys, hair dryers and loads requiring low starting torque. |

### *Example 8.3*

A 250 W, 230 V, 50 Hz capacitor start motor has the following impedances at standstill.

Main winding, $Z_m = 7 + j\,5$ ohm

Auxiliary winding, $Z_a = 11.5 + j\,5$ ohm

Find the value of capacitor to be connected in series with auxiliary winding to give a phase displacement of 90° between the currents in the two windings.

### *Solution*

Figure 8.13 (b) shows the phasor diagram.

$$\text{Phase angle of } I_m = \tan^{-1}\left(\frac{5}{7}\right) = 35.54°$$

After addition of a capacitor $C$ in series, the impedance of auxiliary winding

$$= 11.5 + j\left(5 - \frac{1}{\omega C}\right)$$

$$\text{Phase angle of } I_a = \tan^{-1}\left[\frac{(5 - 1/\omega C)}{11}\right]$$

$$-\tan^{-1}\left[\frac{(5 - 1/\omega C)}{11}\right] + 35.54° = 90°$$

or $$\tan^{-1}\left[\frac{(5 - 1/\omega C)}{11}\right] = -54.46°$$

or $$\frac{(5 - 1/\omega C)}{11} = -1.4$$

or $$\frac{1}{\omega C} = 20.4$$

or $$C = 156\,\mu F.$$

*Example 8.4*

A 0.5 hp, 230 V, single-phase induction motor (split phase) takes a current of 4.2 A lagging the voltage by 10° for the auxiliary winding and a current of 6.2 A lagging the voltage by 40° for its main winding. Find (a) total current and p.f. at the time of starting, (b) total current and p.f. during running, (c) phase angle between the main winding current and auxiliary winding current, (d) power drawn by auxiliary winding, (e) power drawn by main winding, (f) total power drawn during starting, (g) total power drawn during running, (h) motor efficiency if the motor output is 0.5 h.p. (metric) and currents are as above.

*Solution*

(a)
$$I_a = 4.2\angle -10°\,A = 4.136 - j0.73\,A$$

$$I_m = 6.2\angle -40°\,A = 4.75 - j3.986A$$

$$I = I_a + I_m = 8.886 - j4.716$$

$$= 10.06\angle -27.96°\,A$$

Power factor = 0.883 lagging

(b) During running the auxiliary winding is not connected. Hence the current drawn by main winding is the total current

current = 6.2 $A$

$$\text{p.f.} = \cos(-40°) = 0.766 \text{ lagging}$$

(c) $\text{Phase angle} = 40 - 10 = 30°$

(d) $P_a = 230 \times 4.2 \cos(10°) = 951.32 \text{ W}$

(e) $P_m = 230 \times 6.2 \cos(40°) = 1092.38 \text{ W}$

(f) $\text{Total power during starting} = P_a + P_m$

$$= 2043.7 \text{ W}$$

(g) During running only the main winding is connected. Hence power is 1092.38 W

(h) $\text{Output} = 0.5 \times 735.5 = 367.75 \text{ W}$

$$\eta = \frac{367.75}{1092.38} \times 100 = 33.66\%.$$

*Example 8.5*

A capacitor start, single phase, 0.5 hp, induction motor takes the same currents as the split phase motor of example 8.4 but the auxiliary winding current leads the supply voltage by 45°. Find (a) total starting current and its power factor, (b) ratio of starting torque of this motor to the split phase motor of example 8.4.

*Solution*

(a)
$$I = I_a + I_m = 4.2\angle 45° + 6.2\angle -40°$$
$$= 2.97 + j2.97 + 4.75 - j3.986$$
$$= 7.72 - j1.016 = 7.79\angle -7.5° \, A$$

$$\text{Power factor} = 0.99 \text{ lagging}$$

(b) The starting torque $\propto I_a I_m \sin\theta$ where $\theta$ is the phase angle between auxiliary and main winding currents

$$= \frac{\text{Starting torque of capacitor start motor}}{\text{Starting torque of split phase motor}}$$

$$= \frac{4.2 \times 6.2 \sin(45 + 40)°}{4.2 \times 6.2 \sin(30)°} = 1.99.$$

*Example 8.6*

A small motor has an output torque of 0.25 N-m and a speed of 100 rad/s. If the input current is 0.6 A at 230 V and 0.6 lagging p.f. find (a) output power in W, (b) hp (metric) output, (c) efficiency.

*Solution*

(a) $\text{output} = \omega T = 100 \times 0.25 = 25 \text{ W}$

(b) $\text{h.p.} = \dfrac{25}{735.5} = 0.034$

(c) $$\eta = \frac{25}{230 \times 0.6 \times 0.6} = 30.19\%.$$

## 8.12 SPEED CONTROL

The speed control of a single-phase induction motor can be obtained by the methods used for 3-phase induction motors, viz change of frequency, change of number of poles and change of voltage applied to stator. Since these motors are used only in small sizes, the only economical method is to change the voltage applied to stator. This can be achieved in any of the following ways:

(a) Use of auto-transformer. The motor is fed from the variable secondary voltage of auto-transformer. This method is expensive and rarely used.

(b) SCR or Triac circuit may be used to reduce the applied voltage by ac phase control as discussed in Chapter 11.

(c) A resistor in series with motor circuit gives a very cheap method of voltage control. It has the disadvantage of reduced efficiency due to energy loss in the resistor. However it is the most commonly used method.

### *Example 8.7*

A 2-pole, 50 Hz, single-phase induction motor has an effective rotor resistance and leakage reactance of 0.5 ohm and 0.5 ohm, respectively. If it is running at 2600 rpm, find (a) frequencies of rotor current components, (b) relative magnitudes of forward and backward fluxes. Neglect magnetising current and stator impedance.

*Solution*

(a) $$N_s = \frac{120 \times 50}{2} = 3000 \text{ rpm}$$

$$s = \frac{3000 - 2600}{3000} = \frac{2}{15}$$

Frequency of forward component = $sf$

$$= \left(\frac{2}{15}\right)(50) = 6.67 \text{ Hz}$$

Frequency of backward component

$$(2 - s)f = \left(2 - \frac{2}{15}\right)(50) = 93.33 \text{ Hz}$$

(b) From the equivalent circuit of Fig. 8.5 (c) it is seen that if $r_1$, $x_1$, $R_0$, $X_0$ are neglected

$$\frac{E_f}{E_b} = \frac{(r_2'/2s) + j(x_2'/2)}{[r_2'/2(2-s)] + j(x_2'/2)}$$

$$= \frac{(r_2/2s) + j(x_2/2)}{[r_2/2(2-s)] + j(x_2/2)}$$

$$= \frac{\left[(0.5)/\left(2 \times \frac{2}{15}\right)\right] + j0.5/2}{\left[(0.5)/2\left(2 - \frac{2}{15}\right)\right] + j0.5/2} = \frac{1.875 + j0.25}{0.134 + j0.25}$$

$$= 6.65$$

The two fluxes are proportional to $E_f$ and $E_b$. Hence the ratio of forward and backward fluxes is 6.65.

*Example 8.8*

In a 6-pole, 50 Hz , one-phase induction motor the gross power absorbed by the forward and backward fields are 160 W and 20 W, respectively, at a motor speed of 950 rpm. If the no-load frictional losses are 75 W, find the shaft torque.

*Solution*

$$\text{Net Power} = 160 - 20 = 140 \text{ W}$$

$$N_s = \frac{120 \times 50}{6} = 1000 \text{ rpm}$$

$$s = \frac{1000 - 950}{1000} = 0.05$$

$$\text{Power output} = 140(1 - s) - 75$$

$$= 140 \times 0.95 - 75 = 85 \text{ W}$$

$$n = \frac{950}{60} = \frac{95}{6} \text{ rps}$$

$$\text{Shaft torque} = \frac{\text{Power output}}{2\pi n}$$

$$= \frac{58}{2\pi(95/6)} = 0.583 \text{ N-m.}$$

## SUMMARY

1. A single phase induction motor has a stator and rotor. The stator has a single phase ac winding. The rotor is squirrel cage.
2. Single phase induction motors are mostly used in fractional horse power range. They are widely used in domestic fans, room coolers, air conditioners, refrigerators, hair dryers, etc.
3. A single phase induction motor does not have a starting torque. To provide the starting torque, an additional auxiliary winding is provided on stator.
4. A single phase induction motor can be analysed by double revolving field theory. As per this theory the pulsating magnetic field created by single phase stator winding can be decomposed into two rotating fields. These two fields rotate in opposite directions.

5. The analysis is done by calculating the airgap power and torque produced by the forward and backward rotating fields. The equivalent circuit is also drawn using this concept.
6. The parameters of equivalent circuit can be found from the results of no load test and blocked rotor test.
7. The analysis using double revolving field theory is very similar to the analysis using symmetrical components.
8. A single phase induction motor can also be analysed using cross-field theory. As per this theory the flux is resolved into two components, one acting along stator winding axis and the other at right angles to it.
9. Single phase induction motors are classified as per the method used to provide starting torque. Split phase motor, capacitor start motor, capacitor start capacitor run motor, permanent split capacitor motor and shaded pole motor are some commonly used single phase induction motors.

## TEST POINT QUESTIONS

### A. State whether the following statements are true (T) or false (F).

8.1 The auxiliary winding is in the circuit only at the time of starting in all single phase induction motors.

8.2 The two component rotating fields revolve at the same synchronous speed but in opposite directions.

8.3 At starting the torques produced by the two component rotating fields are equal and opposite.

8.4 The efficiency and power factor of a single phase induction motor are poorer than the efficiency and power factor of three phase induction motor.

8.5 Split phase motor requires a capacitor in series with the auxiliary winding.

8.6 A capacitor start motor has higher starting torque than a split phase motor.

8.7 Capacitor start capacitor run motor does not require a centrifugal switch.

8.8 A capacitor start capacitor run motor uses two capacitors.

8.9 A permanent split capacitor motor uses two capacitors.

8.10 Shaded pole motor has high starting torque.

### B. Fill in the blanks with suitable words

8.11 The rotor of a single phase induction motor is .........

8.12 As per double revolving field theory the magnitude of each component field is ......... of that of pulsating field.

8.13 If auxiliary winding is not used, the starting torque of a single phase induction motor is .........

8.14 Under running conditions at any slips, the effective rotor resistance $\frac{r_2'}{2s}$ is ......... than $\frac{r_2'}{2(2-s)}$.

8.15 The most common application of single phase induction motors is in .........

8.16 The axes of main and auxiliary stator windings are space displaced by .........

8.17 A single phase induction motor with main and auxiliary stator windings is equivalent to ......... motor.

8.18 Generally the centrifugal switch opens when the motor speed is ......... of rated speed.

8.19 In a capacitor start capacitor run motor, the auxiliary winding is designed for .........

8.20 The lowest hp range single phase induction motors are .........

## ANSWERS

| | | | | | | |
|---|---|---|---|---|---|---|
| 8.1 F | 8.2 T | 8.3 T | 8.4 T | 8.5 F | 8.6 T | 8.7 F |
| 8.8 T | 8.9 F | 8.10 F | 8.11 squirrel cage | | 8.12 half | 8.13 zero |
| 8.14 higher | 8.15 fans | 8.16 90 electrical degrees | | 8.17 two phase | | 8.18 75% |
| 8.19 continuous duty | | 8.20 shaded pole type | | | | |

## JOB INTERVIEW QUESTIONS

8.1 Name same applications of single phase induction motors.

8.2 Why are efficiency and power factor of single phase induction motor poor.

8.3 How are single phase induction motors classified?

8.4 What is the type of motor used in ceiling fans?

8.5 What is squirrel cage rotor?

8.6 What is auxiliary winding?

8.7 What is double revolving field theory?

8.8 What is a shaded pole motor?

8.9 How can speed of single phase induction motor be controlled?

8.11 What is centrifugal switch?

## REVIEW QUESTIONS

8.1 A 3-phase induction motor develops starting torque, but a one-phase induction motor does not develop a starting torque. Why?

8.2 A single-phase inductions motor has poorer performance as compared to a 3-phase induction motor. Discuss the reasons.

8.3 Briefly discuss the different methods used for starting of one-phase induction motors giving the application of each method.

8.4 Explain the operation of a one-phase induction motor on the basis of (a) double revolving field theory, (b) cross-field theory.

8.5 Draw a typical torque-speed curve of a one-phase induction motor on the basis of double revolving field theory.

8.6 Prepare a table showing the ratings and applications of different types of one-phase induction motors.

8.7 Why is the auxiliary winding in a capacitor-start motor disconnected, after the motor has picked up speed?

8.8 Discuss the differences between capacitor start, capacitor start-capacitor run and permanent split capacitor motors.

8.9 Draw the equivalent circuit of a one-phase induction motor. How can the performance of the motor be analysed?

8.10 Discuss the procedure to determine the parameters of equivalent circuit of one-phase induction motor.

## PROBLEMS

8.1 A 220 V, 50 Hz, 4-pole, one-phase induction motor has rotational losses of 15 W and the following parameters:

$r_1$ = 2.6 ohm, $x_1$ = 5.0 ohm, $X_0$ = 96 ohm

$r_2'$ = 6.4 ohm, $x_2'$ = 4.4 ohm

Find input current, p.f., output and efficiency at a slip of 4%. [4.51 A, 0.53 lagging, 376.1 W, 72%]

8.2 A 2/3 hp, 220 V, 50 Hz, 6-pole induction motor has the following parameters

$r_1$ = 3.04 ohm, $x_1$ = 4.2 ohm, $X_0$ = 105.6 ohm

$r_2'$ = 6.26 ohm, $x_2'$ = 2.12 ohm

The core-losses are 70 W, friction, windage and stray losses are 32 W. The motor is operating at 5% slip. Find (a) motor speed, (b) input current and power factor, (c) air gap power, (d) output power, (e) shaft torque. [(a) 950 rpm (b) 4.66 A, 0.635 lagging (c) 518.8 W (d) 390.9 W (e) 3.93 N-m]

8.3 A 230 V, 4-pole, one-phase, 50 Hz induction motor has the following parameters

$r_1$ = 2.8 ohm, $x_1$ = 3.5 ohm, $X_0$ = 35.5 ohm

$r_2'$ = 4.6 ohm, $x_2'$ = 2.4 ohm

It is operating at 5% slip. Find (a) forward, backward and net torque in synchronous watts, (b) ratio of forward and backward fluxes. [(a) 2100, 700, 1400 synch. watts (b) 2.43]

8.4 For the motor of problem 8.3 assume different values of slip and plot torque-speed curve.

8.5 A 0.25 hp, 110 V, split phase induction motor draws a starting winding current of 4 A lagging the voltage by 15° and a main winding current of 6 A lagging the voltage by 40°. Its output is 0.25 hp (metric). Find (a) starting current and its power factor, (b) current during running and its power factor, (c) phase angle between auxiliary winding and main winding currents, (d) power drawn by auxiliary winding, (e) power drawn by main winding, (f) total power during starting, (g) total power input during running, (h) motor efficiency. [(a) 9.77 A, 0.866 lagging, (b) 6A, 0.766 lagging (c) 25° (d) 425 W (e) 505.6 W (f) 930.6 W (g) 505.6 W (h) 36.37%]

8.6 The motor of Problem 8.5 is converted into a capacitor start motor by connecting a capacitor in series with the auxiliary winding. The magnitude of the two currents remains the same but the auxiliary winding current leads the supply voltage of 42°. Find (a) starting current and its power factor, (b) ratio of starting torque of capacitor start motor to that of split phase motor. [(a) 7.66 A, 0.988 lagging (b) 2.35]

8.7 A small motor has an output torque of 0.35 N-m at 150 rad/s. Its input current is 1.2 A at 230 V and 0.57 lagging power factor. Find (a) output in W, (b) hp (metric) output, (c) efficiency. [(a) 52.5 W (b) 0.0714 hp (c) 33.37%]

8.8 A 60 Hz, 110 V, 4-pole single phase induction motor has the following circuit parameters $r_1 = 1.3\ \Omega$; $x_1 = 2.5\ \Omega$, $r_2' = 3.2\ \Omega$, $x_2' = 2.2\ \Omega$, $x_0 = 48\ \Omega$. Resistance $R_0$ can be neglected. Find input line current, input power, power factor, positive sequence gap power and negative sequence gap power. [4.495 A, 250.2 W, 0.506 lagging, 208.52 W, 15.07 W]

8.9 A 110 V single phase induction motor gave the following test results:

No-load test 110 V, 2.8 A, 60 W

Blocked rotor test 50 V, 6.72 A, 232.6 W

Find (a) parameters of equivalent circuit.

(b) iron, friction and windage loss. Neglect $R_0$.

$$\left[(a)\ r_1 = r_2' = 2.65\ \Omega,\ x_1 = x_2' = 2.68\ \Omega,\ \frac{X_0}{2} = 35.3\ \Omega\ (b)\ 35.2\ \text{W}\right]$$

8.10 In a 230 V, 4-pole single phase induction motor, the gross power of positive sequence and negative sequence fields are 210 W and 30 W respectively at a speed of 1425 rpm. If the no load losses are 90 W find shaft torque. [0.543 N-m]

Chapter 9

# AC Commutator Motors

## 9.1 SCHRAGE MOTOR

### 9.1.1 Construction

The schrage motor is a 3-phase induction motor with an emf injected in the secondary circuit for the purpose of speed control. The injected emf is obtained from a commutator winding mounted on the stator.

Figure 9.1 shows the arrangement of windings in a schrage motor. In contrast with induction motor, the primary winding, in schrage motor, is on rotor and is connected to the mains through sliprings. The secondary winding is on stator. Another winding, known as commutator winding or auxiliary adjusting winding is housed in the tops of rotor slots. This third (or tertiary) winding is tapped to the segments of a commutator. The commutator is provided with two distinct sets of brushes supported on yoke rings mounted in parallel planes. Either set of brushes can be shifted relative to the other in either direction. The brushes *a, c* and *b* have a displacement of 120° with respect to each other. Similarly brushes *a′*, *c′* and *b′* are also displaced by 120° from each other. Each secondary (stator) phase is connected to a pair of movable brushes, which collect, from the commutator, the required emf for injection into the secondary circuit.

### 9.1.2 Operation

Let the brushes of a pair, which are connected to each secondary phase, rest on the same commutator segment. Then the secondary winding is short circuited and the motor behaves as an ordinary 3-phase induction motor. The three-phase currents in the primary (rotor) winding set up a field rotating at synchronous speed, with respect to rotor. Emfs and currents are induced in the secondary winding. The reaction between the induced secondary currents and the rotating magnetic field sets up a torque. The rotor will start running in a direction, so as to tend to reduce the frequency of emf's induced in the secondary winding, i.e. the rotor will rotate in a direction opposite to the direction of rotation of its field. Therefore, as the rotor speeds up in the opposite direction, the speed of magnetic field in space decreases and approaches zero, as the rotor speed approaches synchronous speed. If $n_s$ and $n_r$ are the synchronous and rotor speeds, respectively, the speed of the rotating field relative to fixed secondary winding is $(n_s - n_r)$ i.e. slip speed. Therefore, slip frequency emfs are induced in the secondary winding (as happens in an ordinary 3-phase induction motor). $n_r$ is always less than $n_s$ and slip is positive.

The rotating flux is also cutting the tertiary winding conductors at a speed $n_s$ and induces in them an emf, which is independent of the speed of rotor and is of the same frequency as mains. However, the field is moving at a speed ($n_s - n_r$) relative to the fixed brushes. Therefore the frequency of emf appearing between brushes will be slip frequency. Thus the commutator changes the frequency of tertiary winding conductors from line frequency to slip frequency at the brushes. Thus the emf between the brushes is always of correct frequency for injection in the secondary circuit. This emf can be added to or subtracted from the secondary emf. If the emfs of secondary and tertiary windings are additive, speed is increased, whereas if they are oppositely directed, the speed will be decreased. Let the emfs be additive for connections shown in Fig. 9.1. Interchanging the positions of *a* and *a′* (as also those *bb′* and *cc′*) will bring the two emfs into opposition. The magnitude of emf appearing between brushes (and injected into secondary circuit) is nearly proportional to the chord of the arc subtended by the brush pair.

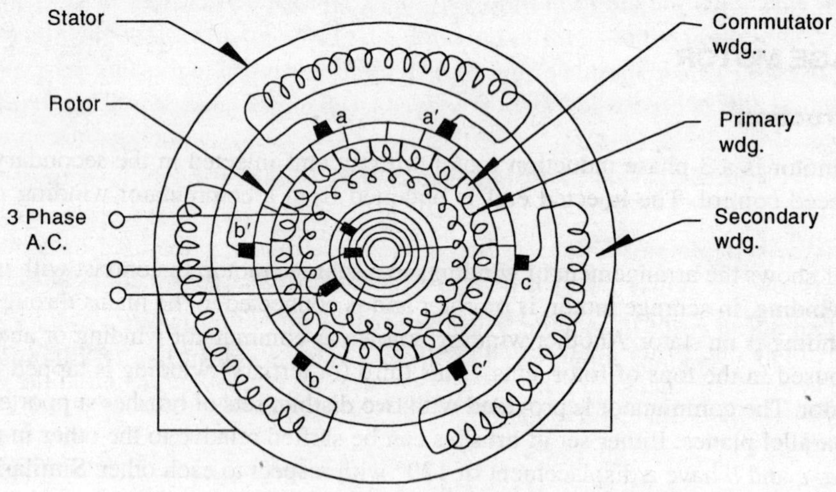


**Fig. 9.1.** Schrage Motor

Figure 9.1 has been drawn, such that the centre line of winding belt *a a′* is in line with the centre of the corresponding phase belt of secondary (and similarly for the other two phases). Under this condition, the emf of auxiliary winding will be either exactly in phase or in phase opposition to secondary emf. However, if the two brushes of a pair are not placed symmetrically with respect to corresponding phase belt of secondary, the secondary and tertiary emfs will have a phase difference. Under this condition, the injection of tertiary emf into the secondary circuit will change the phase of the resultant secondary current. Thus an improvement of power factor can effected.

### 9.1.3 Advantages

A schrage motor has following advantages over an induction motor:

(1) External resistances are not required, so that overall efficiency is improved.

(2) Speed is almost independent of load.

(3) Almost full-load braking torque is available at speeds only slightly above the no-load speed.

(4) Speed can be easily adjusted (increased and decreased).

### 9.1.4 Disadvantages

A schrage motor is more costly as compared to an induction motor. The maintenance cost is also higher. The moment of inertia of rotor is considerably more than that of an induction motor of the same size.

### 9.1.5 Application

A schrage motor is a very suitable variable speed ac motor. In all applications where a variable speed is required, this motor has been used very successfully. The main applications include cranes and hoists, fans and centrifugal pumps (required to be driven at variable speeds), paper making machinery, ring spinning frames, etc.

## 9.2 SINGLE-PHASE SERIES MOTOR

### 9.2.1 Construction

If an ordinary dc series motor is fed from a single-phase ac supply, the flux produced by the field winding and the current in the armature will reverse almost at the same time. Therefore the torque will always be in the same direction, through pulsating in magnitude at twice the line frequency. The average torque will be positive. Thus a series motor has the ability to run on both the direct and alternating current.

However, a dc series motor does not show good performance when used on alternating current. The efficiency is low, the power factor is poor and there is considerable sparking at brushes. The poor efficiency is caused by the increased hysteresis and eddy current losses due to the alternating flux. The poor power factor is due to the high reactance of the field and armature windings. The sparking at brushes is due to poor commutation. The voltage induced by transformer action in the coil undergoing commutation further intensifies commutation difficulties.

To improve the efficiency, both the stator and rotor cores are laminated (the rotor or armature core is laminated in dc machines also). To reduce the reactance of armature winding, a compensating winding $C$ is added in the pole faces and connected in series with armature winding, as shown in Fig. 9.2. The number of turns per pole in the compensating winding is so chosen, that the mmf it produces is equal and opposite to that of the armature, i.e.

$$N_c = \frac{Z}{2PA} \qquad ...(9.1)$$

where $N_c$ = number of turns per pole in compensating winding

$Z$ = number of armature conductors.

$A$ = number of parallel paths in armature

$P$ = number of poles

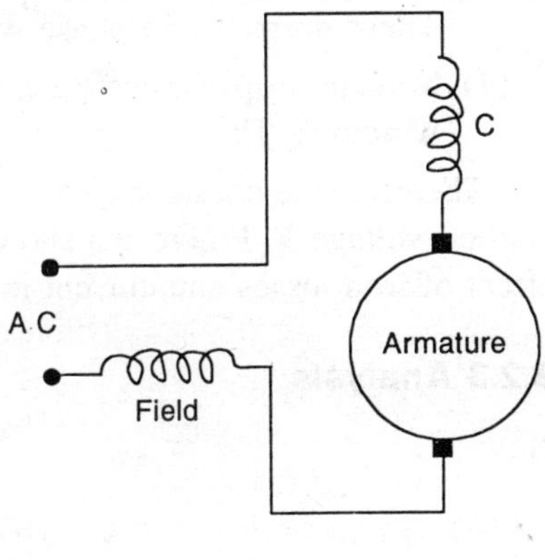


**Fig. 9.2.** AC series motor with compensating winding

The compensating winding cannot neutralise the armature field in the commutation zone, because its space distribution is not identical with that of armature winding. Thus there will be residual armature field owing to the armature conductors lying between the pole tips. To neutralise this flux in the commutating zone, interpoles are necessary.

The flux which is responsible for the reactance of the field winding is the main flux which is essential for the operation of the motor. Evidently this flux should not be neutralised. Nevertheless, the field winding reactance must be kept as low as possible, if the power factor is to be reasonably high. This requirement calls for the minimum possible number of field turns per pole, which means minimum flux per pole. Since the field flux is small, the number of armature conductors has to be high to develop the required torque. Thus the ratio of armature ampere turns per pole to the field ampere turns per pole is high in ac series motors. However, there is a limit to the number of armature ampere turns, which can be accommodated on a given core, this limit being fixed by the consideration of heating. The only way to develop the desired torque in motors of considerable rating is to use numerous poles, thus introducing an additional difference between ac series motor and dc series motor. The use of numerous poles, each with relatively few turns, improves the commutation and reduces the leakage inductance of field winding. The power factor, efficiency and commutation are all improved by lowering the frequency of ac supply.

### 9.2.2 Phasor Diagram

The current in series field windings sets up a flux $\phi_d$ in the direct axis. This current also sets up a quadrature axis flux $\phi_q$, but this flux is neutralised by the compensating winding. The current $I$ and flux $\phi_d$ are in phase. The applied voltage $V$ has to balance the following emfs and voltage drops:

(1) The speed emf $E_r$ produced by the flux $\phi_d$. This emf appears across the brushes which are in neutral or $q$-axis. This emf $E_r$ is in phase with flux $\phi_d$.

(2) The self-induced emf (or transformer emf) $E_f$ is induced in the field winding by the flux $\phi_d$. This emf is induced because flux $\phi_d$ is an alternating flux. If the machine is operated on dc, this emf is absent. This emf $E_f$ is in quadrature with flux $\phi_d$.

(3) Voltage drops across, $r_a$, $r_f$ and $r_c$, i.e. resistance of armature, field and compensating windings. These drops are in phase with current $I$.

(4) Voltage drops across $x_a$, $x_f$ and $x_c$, i.e. leakage reactances of armature, field and compensating windings. These drops are in quadrature with current.

Therefore the phasor sum of $E_r$, $E_f$ and voltage drops in resistances and leakage reactances gives the applied voltage $V$. Figure 9.3 shows the approximate phasor diagram. The power factor is cos θ. The effect of iron losses and current in the coil undergoing commutation have been neglected.

### 9.2.3 Analysis

Now

$$V = I\left[\left(r_a + r_f + r_c\right) + j\left(x_a + x_f + x_c\right)\right] + E_r + E_f \qquad \text{...(9.2)}$$

Let $\quad r_a + r_f + r_c = R'$ and $x_a + x_f + x_c = X'$

The speed voltage $E_r$ is proportional to current $I$ and speed $n$ and can, therefore, be written as $CIn$. The voltage $E_f$ can be written as $jIX$, where $X$ is the constant of proportionality and has the dimensions of reactance. Therefore

$$V = I\,(R' + Cn) + jI(X' + X) \qquad \text{...(9.3)}$$

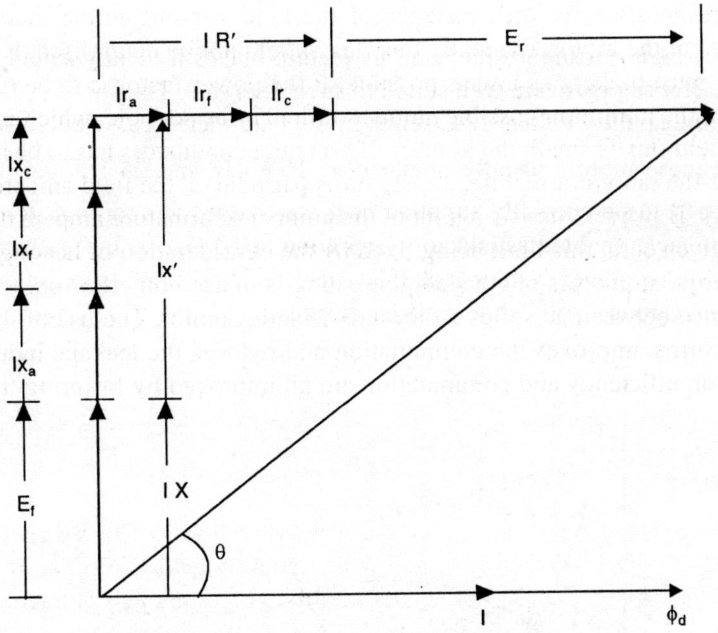


**Fig. 9.3.** Phasor diagram of single-phase series motor

$$\tan\theta = \frac{X + X'}{R' + Cn} \qquad \text{...(9.4)}$$

Power factor angle θ decreases with increase in speed *n*.

From Eq. (9.3)

$$\left(\frac{V}{I}\right)^2 = (R' + Cn)^2 + (X' + X)^2$$

or

$$n = \frac{1}{C}\left[\sqrt{\left(\frac{V}{I}\right)^2 - (X' + X)^2} - R'\right] \qquad \text{...(9.5)}$$

Equation (9.5) gives the typical series characteristic, the speed decreasing with increase in current. As current falls to zero, the speed increases indefinitely. The torque $T$ is proportional to $I^2$ and can be expressed as $kT = I^2$. Hence

$$n \propto \sqrt{\frac{V^2}{kT} - (X' + X)^2} - R' \qquad \text{...(9.6)}$$

When an ac series motor is operated on dc, leakage reactance $X'$ and transformer emf $E_f$ (represented by reactance $X$) is zero. It follows that speed of the motor when operated on dc is higher than that when operated on ac.

Figure 9.4 shows the typical characteristics of a 300 hp ac series motor used for electric traction.

### 9.2.4 Application

The series motor has a high starting torque and this feature makes it ideally suited for use in traction. The speed and torque characteristic are well suited to such a service. The speed is easily and efficiently regulated by varying the applied voltage. AC series motors are therefore, widely used in electric locomotives. In the past, these motors usually operated at $16\frac{2}{3}$ Hz and 25 Hz and the railway electric distribution network had to be specially supplied at these frequencies through frequency converters. Modern developments have enabled satisfactory operation at 50 Hz, so that railway electric network can be fed directly from the power system grid. However, the latest trend is to use dc series motors fed from the power system through solid state rectifier circuits.

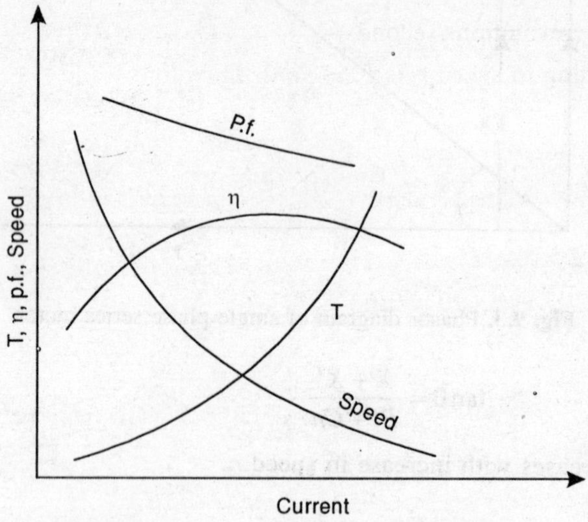


**Fig. 9.4.** Characteristics of ac series motor

## 9.3 UNIVERSAL MOTOR

### 9.3.1 Applications

A series motor designed for operation on dc does not show good performance when operated on ac as discussed in Sec. 9.2. Inspite of this fact, small fractional hp motors known as universal motors are widely used for food mixers, vacuum cleaners, hair driers, electric shavers, portable drills, sewing machines, etc. They are ac series motors, generally without compensating winding and can be used on both ac and dc. They have ratings below 0.75 hp and have operating speed of around 7000 rpm or so. The high speed results in a small weight and size of the motor. The friction and windage losses are sufficient to limit the no-load speed to a safe value. The armature current is small, so that serious commutation problems are eliminated.

### 9.3.2 Phasor Diagram

In a universal motor, the *q*-axis field, created by armature currents is also present in addition to the direct axis field $\phi_d$ created by the series field winding (In an ac series motor with compensating winding the $q$-axis field is neutralised by the effect of compensating winding). Both these fields alternate at supply frequency. Each of these fields sets up rotational (i.e. speed) emf and transformer emf.

The $d$-axis flux $\phi_d$ creates a rotational emf across the brushes (which are in neutral or $q$-axis). In a dc machine this emf $E$ is

$$E = \frac{\phi ZnP}{A} \text{volts} \quad ...(9.7)$$

where $\phi$ = flux per pole

$Z$ = number of armature conductors

$P$ = number of poles

$A$ = number of parallel paths

$n$ = speed, revolutions/second

The frequency corresponding to speed $n$ is $f_r = Pn/2$. Therefore

$$E = 2\phi Z \frac{f_r}{A} \text{ volts} \quad ...(9.8)$$

If flux $\phi$ is alternating, i.e. $\phi = \phi_d \sin \omega t$, the instantaneous value of rotational emf $e_r$ is

$$e_r = \frac{2\phi_d Z f_r}{A} \sin \omega t \quad ...(9.9)$$

and the rms value $E_r$ is

$$E_r = \sqrt{2}\phi_d Z \frac{f_r}{A} \text{ volts} \quad ...(9.10)$$

This emf is in phase with flux $\phi_d$.

Since flux $\phi_d$ is an alternating flux, transformer emf is also induced in armature coils. This emf adds to zero at the brushes.

The $q$-axis flux $\phi_q$ (created by armature coils) also sets up rotational and transformer emfs. The rotational emf due to this flux adds to zero at the brushes. The transformer emf due to this $\phi_q$ [Eq. (2.3)] is

$$E_t = \sqrt{2}\pi f \frac{Z}{2A} \phi_q k_b \quad ...(9.11)$$

where $E_t$ = rms value of transformer emf, volts

$f$ = supply frequency

$k_b$ = breadth factor (since the armature winding is distributed in slots)

$\phi_q$ = $q$-axis flux

If the armature winding is finely distributed, $k_b = 2/\pi$, so that

$$E_t = 2\sqrt{2} f \frac{Z}{2A} \phi_q \quad ...(9.12)$$

This emf is in quadrature with the flux. Since the emf $E_t$ is proportional to and in quadrature with flux, it is proportional to and in quadrature with current $I$.

Since $\phi_d$ is alternating at the supply frequency *f*, it induces a transformer emf in the field winding which is a concentrated winding of $N_f$ turns. This emf $E_f$ is [Eq. (2.3)] given by

$$E_f = \sqrt{2}\pi f N_f \phi_d \text{ volts} \quad ...(9.13)$$

This emf is in quadrature with flux $\phi_d$. Therefore, $E_t$ and $E_f$ are in phase. Both these emfs are proportional to and in quadrature with current *I*. Therefore it is possible to write $E_t + E_f = jXI$, where *X* is a constant of proportionality and has the dimensions of reactance.

Figure 9.5 shows the approximate phasor diagram of a universal motor. The effects of iron-losses and current in the coil undergoing commutation have been neglected. The current *I*, flux $\phi_d$ and flux $\phi_q$ are in phase; $x_a$ and $x_f$ are reactances and $r_a$ and $r_f$ are the resistances of the armature and field windings. The phasor sum $E_f$, $E_t$, $E_r$, $Ix_f$, $Ir_a$, $Ir_f$ gives the applied voltage *V*. The power factor is cos θ.

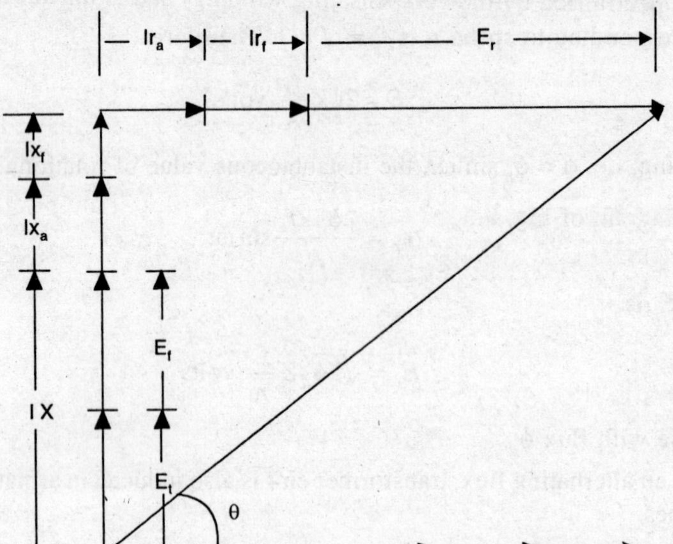


**Fig. 9.5.** Phasor diagram of universal motor

### 9.3.3 Analysis

$$V = I\left(r_a + r_f\right) + jI\left(x_a + x_f\right) + E_t + E_f + E_r \quad ...(9.14)$$

Let $r_a + r_f = R'$ and $x_a + x_f = X'$

The speed emf $E_r$ is proportional to current *I* and speed *n* and can be written as *CIn*. The sum of $E_t$ and $E_f$ can be written as *jXI*. Thus

$$V = I\,(R' + Cn) + jI\,(X' + X) \quad ...(9.15)$$

$$\tan\theta = \frac{X' + X}{R + Cn} \quad ...(9.16)$$

As speed increases, θ decreases, so that power factor improves. From Eq. (9.15)

$$n = \frac{1}{C}\left[\sqrt{\left(\frac{V}{I}\right)^2 - (X' + X)^2} - R'\right] \quad ...(9.17)$$

The torque $T$ is proportional to $I^2$ so that

$$n \propto \left[\sqrt{\frac{V^2}{kT} - (X' + X)^2} - R'\right] \quad ...(9.18)$$

Equation (9.15) to (9.18) are exactly similar to Eqs. (9.3) to (9.6) for ac series motor. However, the definitions of $R'$, $X'$ and $X$ are different in the two cases, because in ac series motor, the emf $E_t$ is absent (because flux $\phi_q$ is neutralised by the compensating winding) and is universal motor the terms $r_c$ and $x_c$ do not appear (because compensating winding is absent).

When a universal motor is operated on dc supply, the reactance $X'$ and emfs $E_f$ and $E_t$ (represented by reactance $X$) are zero. If $V_{dc}$, $I_{dc}$ and $n_{dc}$ denote the dc applied voltage, current and speed, then

$$V_{dc} = I_{dc}(R' + Cn_{dc}) \quad ...(9.19)$$

From the phasor diagram of Fig. 9.5

$$V\cos\theta = I(r_a + r_f) + E_r$$

or

$$V\cos\theta = I(R' + Cn_{ac}) \quad ...(9.20)$$

From Eqs. (9.18) and (9.19)

$$CI_{dc}n_{dc} = V_{dc} - I_{dc}R' \quad ...(9.21)$$

$$CIn_{ac} = V\cos\theta - IR' \quad ...(9.22)$$

Therefore,

$$\frac{n_{dc}}{n_{ac}} = \left(\frac{V_{dc}}{I_{dc}} - R'\right)\Big/\left(\frac{V\cos\theta}{I} - R'\right) \quad ...(9.23)$$

If

$$V_{dc} = V \text{ and } I_{dc} = I, \text{ then}$$

$$\frac{n_{dc}}{n_{ac}} = \frac{1 - IR'/V}{\cos\theta - IR'/V} \quad ...(9.24)$$

where $n_{dc}$ is speed with dc input and $n_{ac}$ is speed, with ac input.

Figure 9.6 shows the typical characteristics of a 1/5 hp universal motor.

*Example 9.1*

A universal motor has an inductance of 0.5 H and a resistance of 30 ohm. When loaded and connected to 250 V dc supply, it takes 0.8 A and runs at 2000 rpm. Find the speed and power factor when it is operating on 250 V, 50 Hz, ac supply and is drawing the same current.

*Solution*

When operating on dc

$$250 = E_{r1} + 0.8(30)$$

or

$$E_{r1} = 250 - 24 = 226 \text{ V}$$

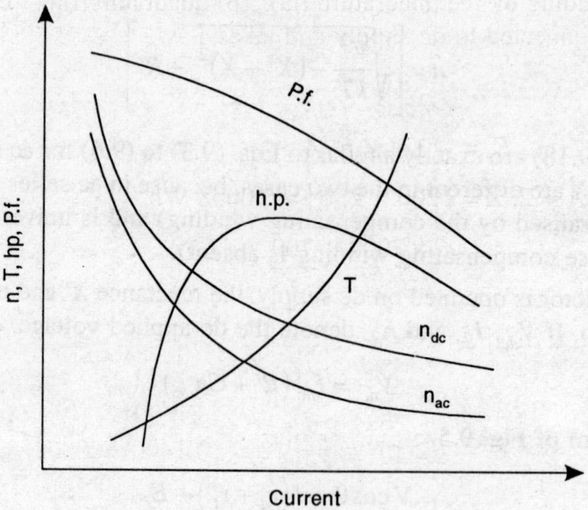


**Fig. 9.6.** Characteristics of universal motor

When operating on ac

$$250 = E_{r2} + 0.8(30 + j2\pi + 50 \times 0.5)$$

or

$$250 = (E_{r2} + 24) + j125.66$$

or

$$(E_{r2} + 24)^2 + (125.66)^2 = 250^2$$

or

$$E_{r2} = 192.12 \text{ V}$$

$$\frac{E_{r1}}{E_{r2}} = \frac{2000}{N_{ac}}$$

or

$$N_{ac} = \frac{192.12 \times 2000}{226} = 1700.2 \text{ rpm}$$

Using Eq. (9.15)

$$\tan\theta = \frac{I(X' + X)}{IR' + CIn_{ac}} = \frac{125.66}{24 + 192.12}$$

$$= 0.58$$

Power factor = cos θ = 0.865 lagging.

*Example 9.2*

A 2-pole, 50 Hz, 230 V universal motor has 1200 armature turns and 400 series field turns. When operated on ac, the full-load current is 1 A, field flux is 0.0008 Wb and speed is 7000 rpm. The total resistance and leakage reactance are 20 ohm and 30 ohm, respectively. Find (a) transformer emf induced in the armature winding by the quadrature flux, (b) quadrature flux, (c) hp output, (d) speed of operation if machine is connected to dc supply and takes 1 A.

*Solution*

(a)

$$E_r = \sqrt{2}\,\phi_d Z f_r / A$$

$$A = 2,\ Z = 1200 \times 2,\ \phi_d = 0.0008 \text{ Wb}$$

$$f_r = \frac{2 \times 7000}{60 \times 2} = \frac{700}{6} \text{ Hz}$$

$$E_r = \sqrt{2}\,(0.0008)(2 \times 12000)\left(\frac{700}{6}\right)\left(\frac{1}{2}\right) = 158.37\text{ V}$$

$$E_f = \sqrt{2}\,\pi f N_f \phi_d = \sqrt{2}\,\pi(50)(400)(0.0008) = 71 \text{ V}$$

From Fig. 9.5

$$V^2 = \left(E_f + E_t + Ix_a + Ix_f\right)^2 + \left(Ir_a + Ir_f + E_r\right)^2$$

or

$$230^2 = (71 + E_t + 30)^2 + (20 + 158.37)^2$$

or

$$E_t = 44.2 \text{ V}$$

(b) Using Eq. (9.12)

$$44.2 = 2\sqrt{2} f \frac{Z}{2A} \phi_q$$

or

$$44.2 = 2\sqrt{2}\,(50)\left(\frac{2400}{2 \times 2}\right)(\phi_q)$$

or

$$\phi_q = 0.00052 \text{ Wb}$$

(c) From Fig. 9.5

$$\tan\theta = \frac{E_f + E_t + Ix_a + Ix_f}{Ir_a + Ir_f + E_r}$$

$$= \frac{71 + 44.2 + 30}{20 + 158.37} = 0.814$$

$$\cos\theta = 0.776 \text{ lagging}$$

$$\text{Power input} = VI\cos\theta = 230 \times 1 \times 0.776$$

$$= 178.48 \text{ W}$$

$$\text{copper-losses} = I^2(r_a + r_f) = 20 \text{ W}$$

Neglecting other losses

$$\text{Power output} = 178.48 - 20 = 158.48 \text{ W}$$

$$= \frac{158.48}{735.5} = 0.215 \text{ hp (metric)}$$

(d) When connected to dc supply

$$V_{dc} = 230 \text{ V}, I_{dc} = 1 \text{ A}$$

$$E_{rdc} = 230 - I_{dc}(r_a + r_f) = 230 - 20 = 210 \text{ V}$$

$$\frac{N_{dc}}{N} = \frac{210}{158.37} \text{ or } N_{dc} = \frac{210}{158.37} \times 7000$$

$$= 9282 \text{ rpm.}$$

### *Example 9.3*

**The motor of Example 9.2 is provided with a compensating winding having a resistance of 2 ohm and reactance of 5 ohm. The compensating winding neutralizes armature flux completely. Find the speeds of operation on ac and dc if $V$ = 230 V and $I$ = 1 A.**

### *Solution*

Since $\phi_q$ has been neutralized fully

$$E_t = 0, \quad x_c = 5\,\Omega, \quad r_c = 2\,\Omega$$

From Fig. 9.3

$$V^2 = (E_f + Ix_a + Ix_f + Ix_c)^2 + (E_r + Ir_a + Ir_f + Ir_c)^2$$

or

$$230^2 = [71 + 1(30+5)]^2 + [E_r + 1\,(20+2)]^2$$

or

$$E_r = 182.12 \text{ V}$$

If $N_{ac}$ is the new speed for ac operation

$$\frac{N_{ac}}{7000} = \frac{182.12}{158.37} \text{ or } N_{ac} = 8050 \text{ rpm}$$

When connected to dc supply

$$E_{rdc} = V_{dc} - I_{dc}(r_a + r_f + r_c)$$

$$= 230 - 1(20+2) = 208 \text{ V}$$

$$\frac{n_{dc}}{8050} = \frac{208}{182.12} \text{ or } N_{dc} = 9194 \text{ rpm.}$$

## 9.4 REPULSION MOTOR

A repulsion motor has its rotor energized inductively. The rotor has a closed type armature winding, similar to that in a dc machine and is provided with a commutator and brushes. The brushes are short circuited along an axis displaced by an angle α from the axis of the single-phase stator winding. The stator winding is housed in slots around the inner periphery of a laminated ring similar to the stator frame of a single-phase induction motor. The arrangement is shown in Fig. 9.7 (a). This single stator winding is electrically and magnetically equivalent to two stator windings shown in Fig. 9.7 (b), the mmfs being related by the equations.

$$IN_c = IN \cos \alpha \text{ and } IN_f = IN \sin \alpha$$

so that

$$N_c = N \cos \alpha \text{ and } N_f = N \sin \alpha.$$

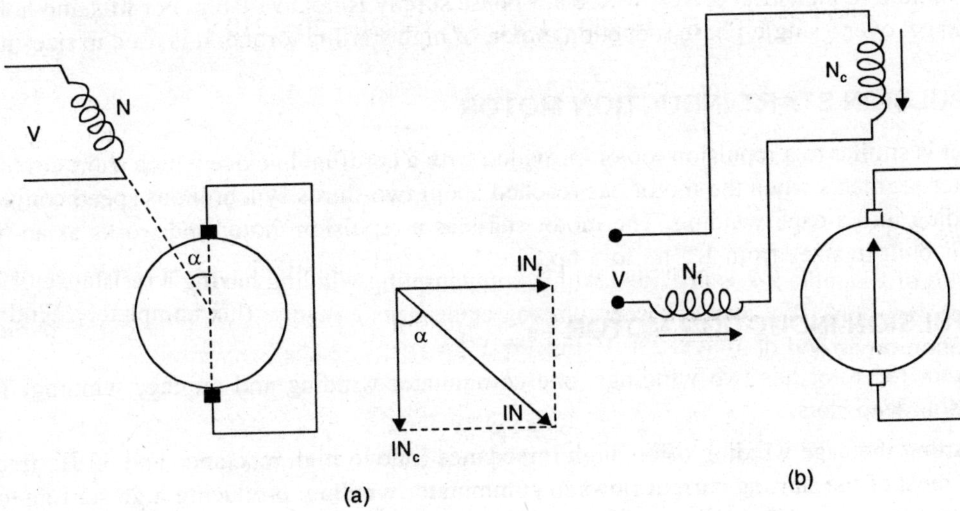


**Fig. 9.7.** Repulsion motor (a) stator winding and resolution of stator mmf into two components, (b) two equivalent stator windings

For production of torque it is necessary that the stator winding is displaced from both the *d*- and *q*- axes. If only winding $N_f$ is present, there will be no mutual induction between stator and rotor, no emf would be induced in the rotor winding and no rotor current would flow. Since there is no rotor current, there will be no torque. The conditions will be similar to that of a transformer on open circuit. If only winding $N_c$ is present, the mutual induction will be maximum and rotor current would exist. However no torque would be produced because the rotor and stator mmfs are coincident. The conditions are similar to that in a transformer on short circuit. For production of torque both the windings $N_f$ and $N_c$ are necessary. The single winding of Fig. 9.7 (a) is equivalent in effect to the two stator windings of Fig. 9.7 (b). For analysis it is convenient to treat the single stator winding of Fig. 9.7 (a) as two separate windings of Fig. 9.7 (b).

The current $I$ in the winding $N_f$ produces a flux $\phi_f$ in phase with itself. The rotation of armature through this flux causes rotational emf $E_r$ in phase with $\phi_f$. The alternating flux created by winding $N_c$ induces a transformer emf $E_t$ in the armature which is in quadrature with the flux. The total armature voltage $E_a$ is phasor sum of $E_r$ and $E_t$. The voltage $E_a$ causes a current $I_a$ to flow in short circuited armature, the current $I_a$ lagging the voltage $E_a$ by an angle corresponding to armature leakage impedance $Z_a$. This current $I_a$ sets up an armature reaction mmf $F_a$ on the brush axis. This mmf $F_a$ combines with mmf of winding $N_c$ to produce $F_0$, the resultant mmf. Torque is produced by the interaction of flux $\phi_f$ and armature current $I_a$.

The current input $I$ from the mains varies with the load and the flux due to winding $N_f$ is proportional to $I$. Thus the motor has variable field flux and displays series motor characteristic. The magnetic field due to armature reaction is largely neutralized by mmf of winding $N_c$, which serves as a compensating winding functioning by induction, instead of by conduction as in ac series motor.

The name repulsion motor is a misnomer and appears to have come from the torque observed on two current carrying coils at an angle to each other.

A repulsion motor is self-starting and has high starting torque (because torque is proportional to $I^2$). Its use is limited to industrial drives, where a 3-phase supply is not available. For lifts and hoists it has the advantage, over a single-phase induction motor, of high starting torque. It is used in sizes upto 5 hp.

## 9.5 REPULSION START-INDUCTION MOTOR

This motor is similar to a repulsion motor, provided with a certifugal device which short circuits all the commutator segments when the motor has reached about two-thirds synchronous speed converting the rotor winding into a cage winding. The motor starts as a repulsion motor and works as an induction motor. It is built in sizes from 1/5 hp to 5 hp.

## 9.6 REPULSION-INDUCTION MOTOR

In this motor the rotor has two windings, one commutator winding and on cage winding. The cage winding is in deep slots.

At starting the cage winding offers high impedance (due to high reactance and 50 Hz frequency). Therefore most of the starting current flows in commutator winding, producing high starting torque by repulsion motor action. The resistance of cage winding is considerably lower than that of commutator winding. Therefore, during running conditions, most of the rotor current flows in cage winding, thus giving induction motor running characteristics.

Nowadays repulsion start induction motor and repulsion induction motors are rarely used. Instead capacitor start and capacitor start-capacitor run induction motors are preferred.

## SUMMARY

1. A Schrage motor is a 3-phase induction motor with an emf injected in the secondary circuit for the purpose of speed control.
2. The primary winding is on rotor and is connected to 3-phase ac supply through sliprings. The secondary winding is an stator.
3. A commutator winding is housed in the tops of rotor slots. This third winding is tapped to the segments of a commutator.
4. The rotating field induces an emf in the third winding. This emf is injected in the secondary winding. If this emf, is added to the secondary emf speed is increased. If the tertiary emf is subtracted from secondary emf, speed is decreased. Thus schrage motor is an adjustable speed motor.

5. Single phase series motor is somewhat similar to dc series motor. However it has a compensating winding (in addition to field winding) connected in series with armature. The purpose of compensating winding is to reduce the reactance of armature winding. This compensating winding is housed in pole faces.
6. Single phase series motor has a high starting torque and it is very suited for electric traction.
7. Universal motor can be used for dc as well as single phase ac. It is a series motor and is used only in fractional horse power range. It does not have a compensating winding.
8. Applications of universal motor include food mixers, vacuum cleaners, hair dryers etc. It has the advantage of small size and weight.
9. A repulsion motor is also a single phase ac motor. The stator winding is housed in slots around the inner periphery of a laminated ring and is similar to stator winding of single phase induction motor. The rotor has a closed armature winding (similar to that on the rotor of dc machine) and is provided with commutator. The rotor is energised inductively. The brushes are short circuited along an axis displaced by an angle $\alpha$. It has high starting torque.

## TEST POINT QUESTIONS

### A. State whether the following statements are true (T) or false (F)

9.1 Schrage motor has three windings.

9.2 Schrage motor is an adjustable speed motor.

9.3 The commutator winding, in a schrage motor, is in the tops of rotor slots.

9.4 The speed of schrage motor is independent of load.

9.5 Single phase ac series motor has a compensating winding housed in pole faces.

9.6 The purpose of compensating winding, a single phase series motor, is to strengthen the armature mmf.

9.7 Single phase ac series motor has speed emf as well as transformer emf.

9.8 Universal motor is a constant speed motor.

9.9 Universal motor is used only in fractional hp range.

9.10 Repulsion motor has an armature winding similar to that in dc machines.

### B. Fill in the blanks with suitable words

9.11 Schrage motor is an ......... speed motor.

9.12 Schrage motor has ......... windings.

9.13 The primary winding of schrage motor is on .........

9.14 Schrage motor is ......... as compared to induction motor.

9.15 If a dc series motor is connected to single phase ac the torque is .........

9.16 Single phase series motor has ......... number of field turns per pole.

9.17 Single phase series motor has ......... number of field poles.

9.18 Universal motor can be used on ......... as well as .........

9.19 As the current increases the speed of universal motor .........

9.20 In repulsion motor, the rotor is energised .........

## ANSWERS

| | | | | | |
|---|---|---|---|---|---|
| 9.1 T | 9.2 T | 9.3 T | 9.4 T | 9.5 T | 9.6 F |
| 9.7 T | 9.8 F | 9.9 T | 9.10 T | 9.11 adjustable | 9.12 three |
| 9.13 rotor | 9.14 more costly | 9.15 positive | 9.16 small | 9.17 large | 9.18 dc, ac |
| 9.19 decreases | 9.20 inductively | | | | |

## JOB INTERVIEW QUESTIONS

9.1 How many windings are there in a schrage motor?

9.2 What is the principle of Schrage motor.

9.3 What are the applications of Schrage motor.

9.4 What is the function of compensating winding in a single phase series motor.

9.5 Why has a single phase series motor a large number of poles.

9.6 What is the difference between speed emf and transformer emf?

9.7 What are the applications of universal motor?

9.8 Why is universal motor used only in fractional hp range?

9.9 What is a repulsion motor?

9.10 How many windings are these in a repulsion motor?

## REVIEW QUESTIONS

9.1 Discuss why
- (a) A dc series motor gives poor performance when operated on ac.
- (b) An ac series motor has high ratio of armature ampere turns per pole to field ampere turns per pole.
- (c) An ac series motor is commonly used for electric traction.
- (d) In a schrage motor the frequency of emf induced in tertiary winding is the mains frequency but the emf across the brushes is of slip frequency.

9.2 Draw a neat diagram of a schrage motor and explain its working. Discuss its applications.

9.3 What modifications are necessary in a dc series motor, so that it may work satisfactorily on ac?

9.4 Draw the phasor diagram of an ac series motor. How can its performance be analysed? Draw its typical characteristics.

9.5 What is a universal motor? Draw its phasor diagram and discuss its operation. Bring out the effects of various emf induced in its armature. Find out an expression for ratio of speeds when operated on dc and ac.

9.6 Explain the principle of operation of a repulsion motor.

## PROBLEMS

9.1 A universal motor has a resistance of 20 ohm and total inductance of 0.5 H. It runs at 2400 rpm, taking 1 A from 230 V dc mains. Find the speed and power factor, if it is fed from 230 V ac mains and takes 1A. [1409.5 rpm, 0.73 lagging]

9.2 A 220 V, 2-pole, 50 Hz universal motor has the following data for ac operation. Input current 1.5 A, input power 228.75 W, total windings resistance 13 ohm, speed 4800 rpm, number of field winding turns 125, number of armature turns, 600, leakage reactance drop is 15% of total reactive drop in motor. Find (a) speed emf, (b) field flux per pole, (c) quadrature axis flux per pole, (d) speed of the motor if connected to 220 V dc supply and taking 1.5 A. [(a) 133 V (b) 0.00196 Wb (c) .0019 Wb (d) 7236 rpm]

9.3 The universal motor of problem 9.2 is fitted with a compensating winding having a resistance of 2 ohm. The addition of this winding neutralizes the quadrature axis flux completely, but increases the leakage reactance drop by 20%. Find speeds on ac and dc operation. In each case $V = 220$ V and $I = 1.5$ A. [6540 rpm, 7128 rpm]

9.4 An ac series motor without compensating winding has a reactance of 40 ohms at 50 Hz and a resistance of 0.5 ohm. At 500 V dc the current is 90 A and speed is 780 rpm. Find the speed and power factor when operating on 500 V ac at (a) 50 Hz, (b) 25 Hz and taking the same current. [(a) 517.7 rpm, 0.694 lagging (b) 722.5 rpm, 0.933 lagging]

9.5 A single-phase series commutator motor has one turn per armature coil. The maximum permissible voltage in the coil short circuited by brushes is 4 V. Find the maximum value of flux per pole if frequency is 50/3 Hz. [0.038 Wb]

9.6 A single-phase commutator motor was subjected to dynamometer brake test. The data obtained was: rated voltage 230 V, input current 1 A, speed 1550 rpm, length of brake arm 0.35m, dyanamometer reading 0.2 kg. Find efficiency, power factor, motor hp (metric) and torque. The power input to motor was 160 W. [0.697, 0.696 lagging, 0.1515 hp, 0.6867 N-m]

Chapter 10

# Special Machines

## 10.1 RELUCTANCE MOTOR

A reluctance torque is the torque produced in a motor, in which the reluctance of the air gap is a function of the angular position of rotor, with respect to stator coils. In a salient pole synchronous motor, the torque has two components, one component being a function of $(x_d - x_q)$, i.e. difference in the motor reactance in the direct and quadrature axis. This component is reluctance torque. A motor which develops torque only due to difference in reluctance in the two axes is known as reluctance motor.

The reluctance torque is proportional to sin 2δ, where δ is the angle between the rotor and stator magnetic field. Therefore reluctance torque is maximum when δ = 45°. The reluctance torque arises due to the tendency of the rotor to align itself in the minimum reluctance position, with respect to the stator field revolving at synchronous speed.

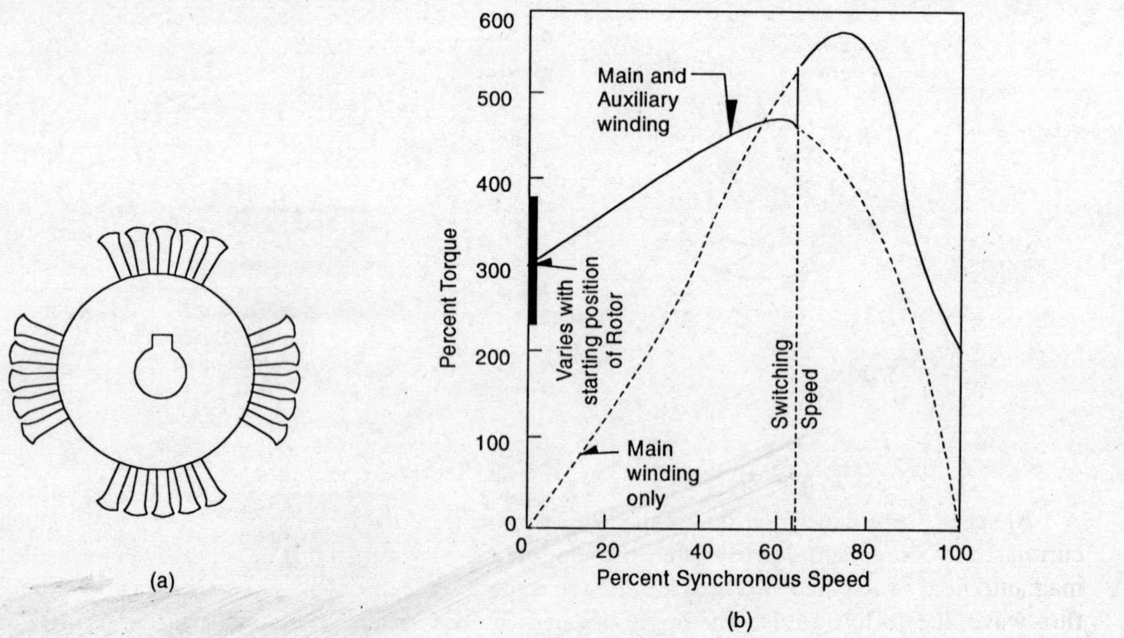


**Fig. 10.1.** Reluctance motor (a) rotor punching, (b) torque-speed curve

Figure 10.1 shows the rotor punching of a self-starting reluctance motor. It is similar to the punching of a cage rotor, except that some of the teeth have been removed, leaving the bars and end rings intact. That stator may be either single-phase (having main and auxiliary windings as in single-phase induction motor) or three-phase. The motor starts due to induction motor action. At speeds less than synchronous speed, the torque alternates in positive and negative half-cycles. If the rotor has small inertia and the motor is lightly loaded, the rotor will be pulled into synchronism during the positive half-cycle and continue to run at synchronous speed.

Figure 10.1 (b) shows the torque-speed curve of a reluctance motor. As mentioned above, it is basically a salient pole synchronous motor, but has a split-phase stator winding for providing the starting torque. The starting torque is a function of the rotor position. This phenomenon is known as cogging. The starting torque is pretty high, because to obtain satisfactory synchronous motor characteristics, it is necessary to use frames bigger than that for an induction motor of the same rating.

It is used in recording instruments, timing devices, control apparatus, phonograph twin tables, etc.

## 10.2 HYSTERESIS MOTOR

In a hysteresis motor, the torque is produced due to the hysteresis phenomenon. The stator has a 3-phase or permanent split capacitor single-phase winding, thus producing rotating field. It is necessary that stator field has sinusoidal space distribution, so that the losses are minimum. The rotor is a smooth cylinder of hard steel and does not have any teeth or winding as shown in Fig. 10.2(a).

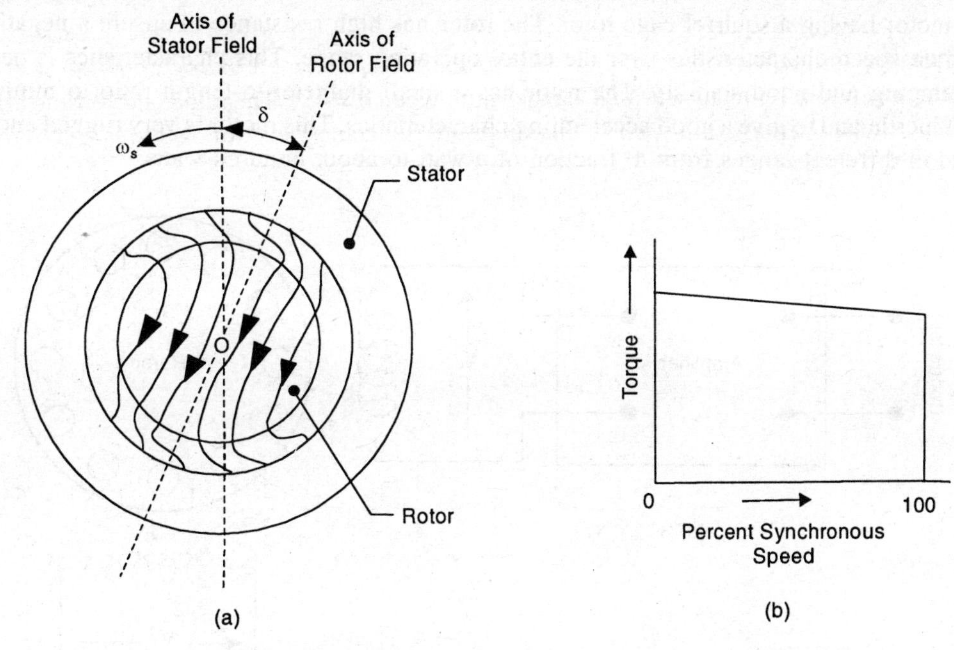


**Fig. 10.2.** Hysteresis motor (a) magnetic fields, (b) torque-speed curve

At speeds less than synchronous speed, the torque is partly due to hysteresis and partly due to eddy currents, the component due to hysteresis being much greater than that due to eddy currents. When the magnetic field of the stator sweeps around the rotor surface, the rotor flux wave lags behind the stator flux wave due to hysteresis. The angle of lag is $\delta$. The greater the intrinsic hysteresis loss of rotor material, the greater the angle by which the rotor magnetic field lags the stator field. The torque is proportional to sin δ. In addition to above, the stator magnetic field induces eddy currents in the rotor. These eddy currents produce a magnetic field of their own and torque is produced. The greater the relative motion between the rotor and stator fields, the greater the eddy currents and eddy current torque.

When motor reaches synchronous speed, the eddy current torque vanishes and only hysteresis torque is present. The hysteresis torque is constant from zero to synchronous speed. Since the total torque at any speed, less than synchronous speed, is slightly more than the torque at synchronous speed, this motor can accelerate any load up to its full-load. After reaching synchronous speed, the motor adjusts its torque angle, so as to develop the torque required by the load and continues to run at synchronous speed. Since the eddy current torque is much less than hysteresis torque, the total torque is nearly constant from starting to synchronous speed (Fig. 10.2b).

A very small hysteresis motor can be built with shaded pole stator construction also. This gives a small, self-starting single-phase synchronous motor. It is used as the driving mechanism in electric clocks.

## 10.3 TWO-PHASE SERVO MOTOR

### 10.3.1 Construction

A two-phase servo motor is commonly used in feed back control systems. It is basically a two-phase induction motor having a squirrel cage rotor. The rotor has high resistance to ensure a negative slope for the torque-speed characteristics over the entire operating range. This characteristics is needed for positive damping and good stability. The rotor has a small diameter-to-length ratio to minimize the moment of inertia and to give a good accelerating characteristics. This motor is very rugged and reliable and is used in different ranges from a fraction of a watt to about hundred watts.

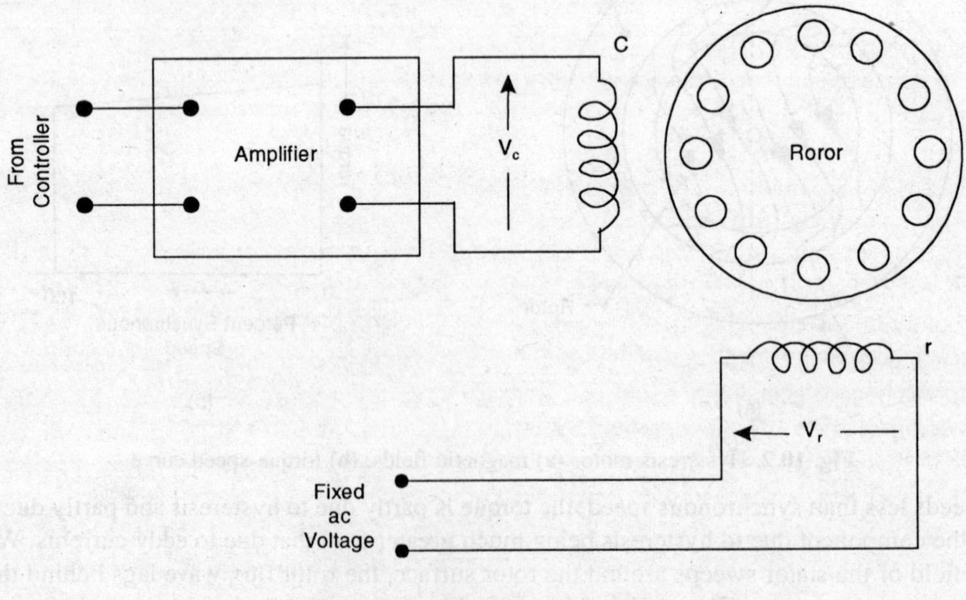


**Fig. 10.3.** Two-phase servo motor

Figure 10.3 shows the connection diagram. One of the stator windings (known as reference winding) is excited by a fixed ac voltage $V_r$, while the second winding $c$ (usually known as control winding) is excited by the control voltage $V_c$. The two voltages must be in synchronium, which is a necessary condition for production of torque. Therefore, the two voltages must be derived from the same source. The control voltage is fed to the motor through an amplifier. Moreover, the two voltages must have a phase difference of 90°. This phase shift is usually obtained by using a phase shifting network in the amplifier circuit. The 90° phase shift can also be obtained by adding a capacitor in series with the reference winding. When $V_c$ leads $V_r$ by 90°, rotation in one direction is obtained and when $V_c$ lags $V_r$ by 90°, rotation in reverse direction is obtained. Since the torque is a function of both $V_r$ and $V_c$, changing the magnitude of $V_c$ changes the torque developed.

### 10.3.2 Analysis Using Symmetrical Components

We have used symmetrical components to develop the equivalent circuit and analyse the performance of single phase induction motor. Symmetrical components can also be used for the analysis of a two-phase servomotor. The voltages $V_r$ and $V_c$ can be resolved into symmetrical components $V_{r1}$ and $V_{r2}$ using Eqs. (8.20) and (8.21).

$$\begin{bmatrix} V_r \\ V_c \end{bmatrix} = \begin{bmatrix} 1 & 1 \\ -j & j \end{bmatrix} \begin{bmatrix} V_{r1} \\ V_{r2} \end{bmatrix} \quad ...(10.1)$$

and

$$\begin{bmatrix} V_{r1} \\ V_{c2} \end{bmatrix} = \frac{1}{2} \begin{bmatrix} 1 & j \\ 1 & -j \end{bmatrix} \begin{bmatrix} V_r \\ V_c \end{bmatrix} \quad ...(10.2)$$

Thus each of the two voltages, $V_r$ and $V_c$ consists of components of a balanced two-phase set, one positive sequence and the other negative sequence. The positive sequence components $V_{r1}$ and $-j\,V_{r1}$ set up a rotating magnetic field which interacts with the rotor winding to produce a torque $T_1$. The negative sequence components $V_{r2}$ and $+j\,V_{r2}$ set up a rotating field in the reversed direction. This rotating field also interacts with rotor winding to produce a torque $T_2$. The resultant torque is $(T_1 - T_2)$. Both the positive and negative sequence sets are balanced. Thus all the techniques of analysis of 3-phase induction motor can be used to analyse the performance of a two phase servomotor.

Fig. 10.4 (a) shows the per phase equivalent circuit of a two-phase servomotor for the positive sequence set. This is similar to the equivalent circuit of a 3-phase induction motor (Fig. 7.3d). The applied voltage is the positive sequence voltage $V_{r1}$. It is assumed that the main winding (or reference winding) is similar to the control winding. Therefore equivalent circuit for both the windings is similar. In case the two windings are not similar, a suitable transformer can be connected to bring about the balanced condition. $r_1$ and $x_1$ denote the resistance and leakage reactance of the stator. $X_0$ is the magnetising reactance. The core-loss resistance $R_0$ has been neglected. $r_2'$ and $x_2'$ denote the equivalent resistance and leakage reactance of rotor as referred to stator. As discussed in Chapter 7 the total resistance of rotor (including the resistance which replaces the mechanical load) as referred to stator is $r_2'/s$.

Fig. 10.4 (b) shows the perphase equivalent circuit for the negative sequence set. The applied voltage is the negative sequence voltage $V_{r2}$. Since the negative sequence rotating field has a direction of rotation opposite to that of positive sequence field, this direction of rotation is also opposite to the rotation of rotor. Therefore the slip, in this case, in $(2 - s)$. The total rotor resistance as referred to stator is $r_2'/(2 - s)$. The performance can be evaluated from the equivalent circuits shown in Fig. 10.4 (a and b).

For the positive sequence circuit (Fig. 10.4 (a) the input impedance $Z_1$ is

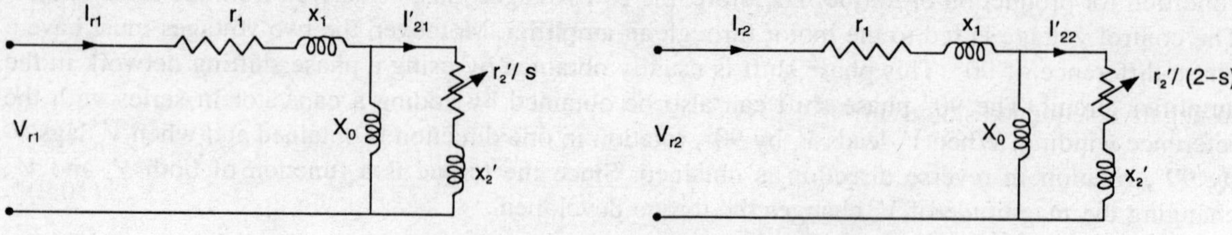


(a) (b)

**Fig. 10.4.** Equivalent circuit of two phase servomotor
(a) Positive sequence, (b) Negative sequence

$$Z_1 = r_1 + jx_1 + \frac{jX_0\left[\dfrac{r_2'}{s} + jx_2'\right]}{\dfrac{r_2'}{s} + j(x_2' + X_0)} \qquad \text{...(10.3)}$$

The positive sequence stator current is

$$I_{r1} = \frac{V_{r1}}{Z_1} \qquad \text{...(10.4)}$$

The positive sequence rotor current (as referred to stator) is

$$I_{21}' = \frac{(I_{r1})(jX_0)}{\dfrac{r_2'}{s} + j(X_0 + x_2')} \qquad \text{...(10.5)}$$

In the notation $I_{21}'$, the first subscript 2 refers to rotor and the second subscript 1 refers to positive sequence. The air gap power $P_{g1}$ for the positive sequence can be found from Eq. (7.17). Thus

$$P_{g1} = 2\left(I_{21}'\right)^2\left(\frac{r_2'}{s}\right) \qquad \text{...(10.6)}$$

In Eq. (10.6), the factor 2 arises because it is a two-phasor motor.

The positive sequence torque $T_1$ is

$$T_1 = \frac{P_{g1}}{2\pi n_s} \qquad \text{...(10.7)}$$

where $n_s$ is the synchronous speed in revolutions per second.

For the negative sequence circuit (Fig. 10.4b) the input impedance $Z_2$ is

$$Z_2 = r_1 + jx_1 + \frac{jX_0\left[\dfrac{r_2'}{2-s} + jx_2'\right]}{\dfrac{r_2'}{2-s} + j(x_2' + X_0)} \quad \text{...(10.8)}$$

The negative sequence stator current $I_{r2}$ is

$$I_{r2} = \frac{V_{r2}}{Z_2} \quad \text{...(10.9)}$$

The negative sequence rotor current (as referred to stator) is

$$I_{22}' = \frac{(Ir_2)(jX_0)}{\dfrac{r_2'}{s} + j(X_0 + x_2')} \quad \text{...(10.10)}$$

The negative sequence air gap power $P_{g2}$ is

$$P_{g2} = 2(I_{22}')^2\left(\frac{r_2'}{2-s}\right) \quad \text{...(10.11)}$$

and negative sequence torque $T_2$ is

$$T_2 = \frac{P_{g2}}{2\pi n_s} \quad \text{...(10.12)}$$

The resultant torque is

$$T = T_1 - T_2 = \frac{P_{g1} - P_{g2}}{2\pi n_s} \quad \text{...(10.13)}$$

From the values of $I_{r1}$ and $I_{r2}$, the currents in the two windings $I_r$ and $I_c$ can be found from Eq. (10.1).

In application of servomotor, efficiency is not of much significance and hence need not be calculated.

An alternative expression of air gap power and torque can be found by considering that the two voltages $V_r$ and $V_c$ are balanced (i.e. $|V_r| = |V_c|$ and their phase difference is 90°). Then the phase current in stator is

$$I_r = \frac{V_r}{Z_1} \quad \text{...(10.14)}$$

where $Z_1$ is given by Eq. (10.3).

The rotor current as referred to stator is

$$I_2' = \frac{(I_r)(jX_0)}{\dfrac{r_2'}{s} + j(x_2' + X_0)} \quad \text{...(10.15)}$$

The air gap power $P_g$ for balanced operation is

$$P_g = 2(I_2')^2\left(\frac{r_2'}{s}\right) \qquad \text{...(10.16)}$$

Comparing Eqs. (10.6) and (10.16) we get

$$\frac{P_{g1}}{P_s} = \left(\frac{I_{21}'}{I_2'}\right)^2 = \left(\frac{V_{r1}}{V_r}\right)^2 \qquad \text{...(10.17)}$$

Similarly for the negative sequence we get

$$\frac{P_{g2}}{P_g'} = \left(\frac{\dot{V}_{r2}}{V_r}\right)^2 \qquad \text{...(10.18)}$$

where $P_g'$ denotes the air gap power for balanced operation at slip $(2 - s)$. The resultant torque is

$$T = T_1 - T_2 = \frac{1}{2\pi n_s}\left[\left(\frac{V_{r1}}{V_r}\right)^2 P_g - \left(\frac{V_{r2}}{V_r}\right)^2 P_g'\right] \qquad \text{...(10.19)}$$

The values of $P_g$ and $P_g'$ are generally supplied by the manufacturer. Therefore the net torque T can be calculated directly from Eq. (10.19) without referring to the equivalent circuit. Hence torque slip or torque-speed characteristic can be plotted.

### 10.3.3 Torque-speed Curve

Figure 10.5 shows the torque-speed characteristics of a two-phase servo motor for different values of control voltage and unity reference voltage. It is necessary that torque should be high at speeds near zero. The slope of torque-speed curve should be negative for positive damping and good stability. The maximum torque is developed at a reverse speed of one-half the synchronous speed. This ensures that normal operation near zero speed is in stable region. The analysis of a two-phase motor can be done by the method of symmetrical components.

### 10.3.4 Transfer function

The torque developed is a function of shaft angular speed ω and control voltage $V_c$.

Thus

$$T = K_c V_c - k_n \omega \qquad \text{...(10.1)}$$

where $K_c$ and $K_n$ are positive constants.

Moreover

$$T = J\dot{\omega} + f\omega \qquad \text{...(10.2)}$$

where $J$ is moment of inertia of motor and load, $f$ is viscous friction coefficient of motor and load and ω is acceleration.

Thus

$$J\dot{\omega} + f\omega = K_c V_c - K_n \omega$$

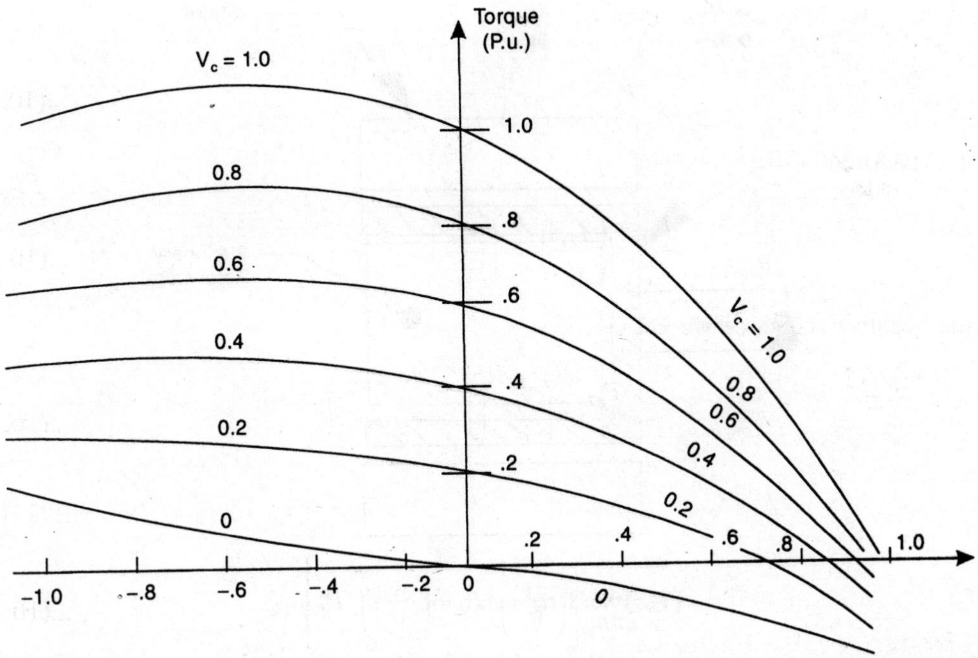


**Fig. 10.5.** Torque-speed curve of two phase servo motor

or
$$J\dot{\omega} + \omega(f + K_n) = K_c V_c \quad \text{...(10.3)}$$

Taking Laplace transform we get

$$\text{Transfer function} = \frac{\omega(s)}{V_c(s)} = \frac{K_c}{Js + (f + K_n)}$$

$$= \frac{K_m}{\tau_m s + 1} \quad \text{...(10.4)}$$

where
$$K_m = \frac{K_c}{f + K_n} = \text{motor gain constant}$$

$$\tau_m = \frac{J}{f + K_n} = \text{motor time constant.}$$

### 10.3.5 Drag Cup Rotor

When the rating of a servo motor is only a few watts or less, inertia can be decreased by using a drag cup rotor. This rotor (Fig. 10.6) is in the form of a thin metallic cup. The iron core is stationary. Since only the metallic cup is the rotating member, the inertia is very small.

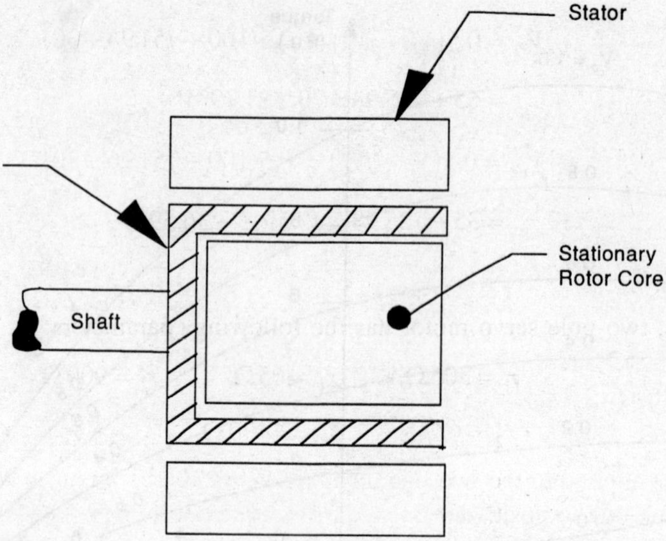


**Fig. 10.6.** Drag cup rotor

### 10.3.6 Special Features

A servomotor has the following special features

1. High torque to inertia ratio
2. High torque to volume ratio
3. Low inductance of windings
4. Low cogging torque at low speed
5. Efficient heat dissipation

High torque to inertia ratio enables fast acceleration and deceleration. This is made possible by low rotor inertia and low winding inductance. If the cogging torque is appreciable the rotor may tend to have preferential positions. This may affect position accuracy. Efficient heat dissipation is necessary because a servomotor is often operated with frequent start and stop duty coupled with fast acceleration and deceleration. Due to this the motor current may be about twice the rated current leading to high $I^2R$ losses. Therefore adequate cooling measures are incorporated in the motor housing for efficient heat dissipation.

*Example 10.1*

The voltage applied to a servomotor are $V_r = 100\angle 0° V$ and $V_c = 60 \angle{-30°} V$. Find the positive and negative sequence voltages.

*Solution*

$$V_r = 100\angle 0° = 100 + j0\,V$$

$$V_c = 60\angle -30° = 51.96 - j30\,V$$

Using Eq. (10.2)

$$V_{r1} = 0.5(V_r + jV_c) = 0.5(100 + j51.96 + 30)$$

$$= 65 + j25.98 = 70\angle 21.79°\, V$$

$$V_{r2} = 0.5(V_r - jV_c) = 0.5(100 - j51.96 - 30)$$

$$= 35 - j25.98 = 43.59\angle -36.59°\, V.$$

*Example 10.2*

A 230 V, 50 Hz, two-pole servo motor has the following parameters:

$$r_1 = 305\,\Omega, \quad x_1 = 65\,\Omega, \quad r_2' = 900\,\Omega,$$

$$x_2' = 65\,\Omega, \quad X_0 = 1100\,\Omega.$$

The voltages applied to the two windings are $V_r = 230\angle 0° V$, and $V_c = 150\angle -30° V$. If the slip is 0.5 find (a) torque due to positive sequence, (b) torque due to negative sequence, (c) resultant torque, (d) stator currents $I_r$ and $I_c$.

*Solution*

Using the notations of section 10.3.2

$$V_r = 230 + j0V,\ V_c = 150\angle -30° = 129.9 - j75V$$

$$V_{r1} = 0.5(V_r + jV_c) = 0.5(230 + j129.9 + 75)$$

$$= 152.5 + j64.95 = 165.76\angle 23.07°\, V$$

$$V_{r2} = 0.5(V_r - jV_c) = 0.5(230 - j129.9 - 75)$$

$$= 77.5 - j64.95 = 101.1\angle -39.97°\, V$$

(a) $$Z_1 = 305 + j65 + \frac{j1100\left[\dfrac{900}{0.5} + j65\right]}{\dfrac{900}{0.5} + j(65 + 1100)} = 305 + j65 + 473.71 + j793.4$$

$$= 778.71 + j858.4 = 1158.98\angle 47.79°\, ohm$$

$$I_{r1} = \frac{V_{r1}}{Z_1} = \frac{165.76\angle 23.07}{1158.98\angle 47.79} = 0.143\angle -24.72°\, A$$

$$I_{21}' = \frac{(0.143\angle -24.72°)(1100\angle 90°)}{1800 + j1165}$$

$$= \frac{(0.143\angle -24.72°)(1100\angle 90°)}{2144.1\angle 32.91°} = 0.07336\angle 32.37°\, A$$

$$P_{g1} = 2(0.07336)^2\left(\frac{900}{0.5}\right) = 19.374\ W$$

$$T_1 = 19.374 \text{ synch. watts.}$$

$$= \frac{19.374}{2\pi\left(\frac{3000}{60}\right)} = \frac{19.374}{2\pi(50)} = 0.06167 \text{ N-m}$$

(b)

$$Z_2 = 305 + j65 + \frac{j1100\left[\frac{900}{1.5} + j65\right]}{\frac{900}{1.5} + j(1100+65)}$$

$$= 305 + j65 + \frac{(j1100)(600+j65)}{600+j1165}$$

$$= 305 + j65 + 422.78 + j279.09$$

$$= 727.78 + j344.09 = 805.02\angle 25.3° \text{ ohm}$$

$$I_{r2} = \frac{V_{r2}}{Z_2} = \frac{101.1\angle -39.97°}{805.02\angle 25.3°} = 0.1256\angle -65.27°\ A$$

$$I'_{22} = \frac{(0.1256\angle -65.27°)(1100\angle 90°)}{600+j1165} = 0.105\angle -38.02°\ A$$

$$P_{g2} = 2(0.105)^2\left(\frac{900}{1.5}\right) = 13.23\ W$$

$$T_2 = 13.23 \text{ synch. watts} = \frac{13.23}{2\pi(50)} = 0.0421\ N\text{-}m$$

(c)

$$T = T_1 - T_2 = 0.06167 - 0.0421$$
$$= 0.01957\ N\text{-}m$$

(d)

$$I_r = I_{r1} + I_{r2} = 0.143\angle -24.72° + 0.1256\angle -65.27°$$
$$= 0.1824 - j0.1739 = 0.252\angle -43.63°\ A$$

$$I_c = -jI_{r1} + jI_{r2} = 0.143\angle -114.72° + 0.1256\angle 24.73°$$
$$= 0.0543 - j0.0774 = 0.0945\angle -54.95°\ A.$$

*Example 10.3*

**In a two phase, 50 Hz, 2 pole servomotor the air gap powers for the balanced positive sequence and**

balanced negative sequence operation for $V_r = V_c = 100$ V and $s = 0.4$ are 7 W and 5 W respectively. Find the resultant torque in synchronous watts and N-m if $V_r = 200\angle 0°$ V and $V_c = 120\ \angle -60°$ for $s = 0.4$

*Solution*

$$V_{r1} = 0.5(V_r + jV_c) = 0.5[200 + j(60 - j103.92)]$$

$$= 151.96 + j30 = 154.89\angle 11.17\, V$$

$$V_{r2} = 0.5(V_r - jV_c) = 0.5[200 - j(60 - j103.92)]$$

$$= 48.04 - j30 = 56.64\angle -31.98°\, V$$

$$n_s = \frac{2 \times 50}{2} = 50\ rps$$

Using Eq. (10.19)

$$T = \left(\frac{154.89}{100}\right)^2 \times 7 - \left(\frac{56.64}{100}\right)^2 \times 5 = 16.79 - 1.6$$

$$= 15.19 \text{ synch. watts}$$

$$= \frac{15.19}{2\pi \times 50} = 0.04835 \text{ N-m.}$$

## 10.4 STEPPER MOTOR

It is a form of synchronous motor and is designed to rotate a specific number of degrees for each electrical pulse applied to its control winding. Typical steps are 2°, 2.5°, 5°, 7.5° and 1.5° for each pulse.

This motor is designed with a multipole multi-phasor stator winding. The number of phases is 3 or 4 and number of poles depends on the required degrees of rotation per pulse. The rotor can be either permanent magnet type or variable reluctance type. The permanent magnetic field provides the motor with constant dc excitation, so that it behaves as a synchronous motor.

Figure 10.7 shows a 4-phase stepper motor, with permanent magnet rotor. A stepper motor always operates with an external drive logic circuit. When a train of pulses is applied to the drive circuit, it delivers appropriate currents to the stator windings of the motor to make the axis of air-gap field step around in accordance with the input pulses. Depending on the pulse rate and load torque, the rotor follows the axis of the air-gap magnetic field due to the torque acting on it. In Fig. 10.7 the rotor assumes the angles 0, 45°, 90°, ... as the windings are excited in the sequence $N_a$, $N_a + N_b$, $N_b$, .... If the windings are excited in the sequence $N_a$, $N_b$, $N_c$, ..., the rotor moves in steps of 90°. Different winding patterns will result in smaller steps of rotation.

The construction of Fig. 10.8 can be used for small steps of rotation. The stator has a 2-phase winding, while the rotor has five projecting poles. When winding *a* is excited, the position of the rotor is as shown. If current is transferred to winding *b*, the rotor will shift by $\theta = 90 - 72 = 18°$, so as to align

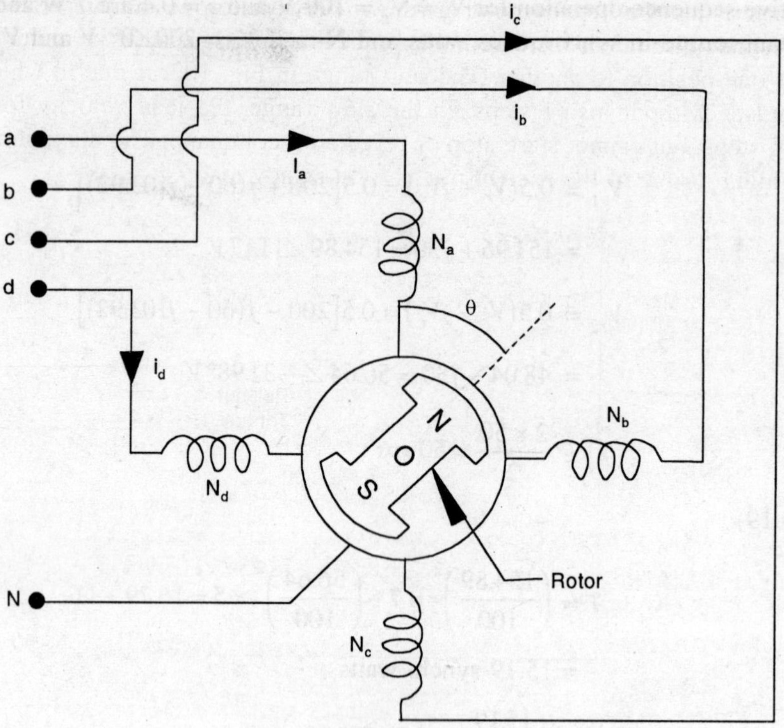


**Fig. 10.7.** 4-Phase stepper motor

pole 2 with $N_b$ axis. This construction is known as differential construction. Both permanent magnet and variable reluctance rotor can be used. However, a permanent magnet rotor provides smooth response.

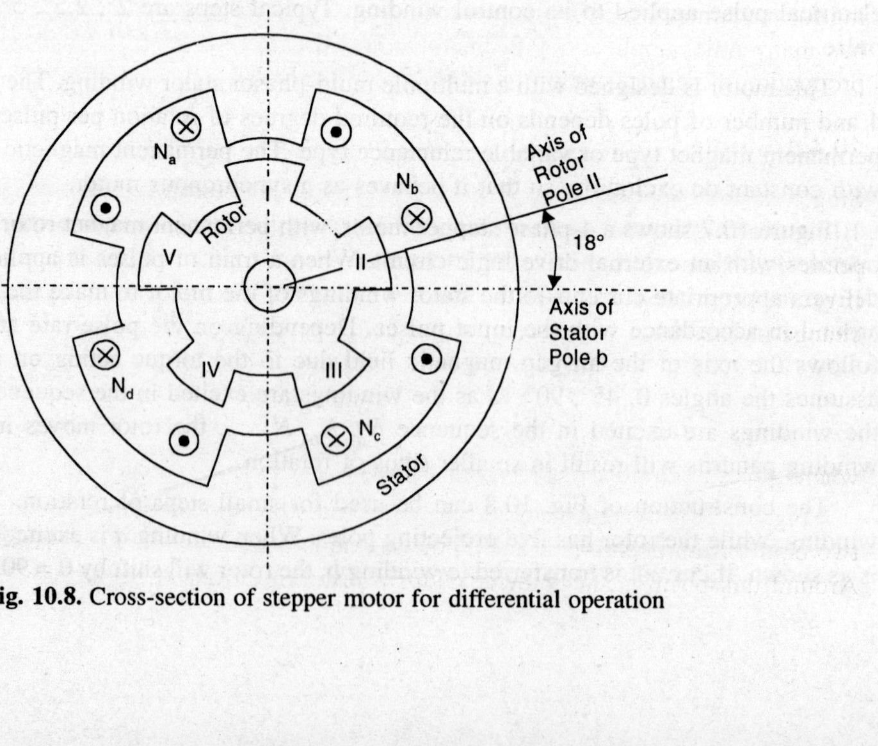


**Fig. 10.8.** Cross-section of stepper motor for differential operation

The characteristics of a stepper motor are drawn between torque and pulse rate (i.e. number of steps per second). As the pulse rate is increased, the torque is reduced because the motor has less time to shift the load from one position to another. The start range in Fig. 10.9 is one in which the load position follows the pulses without losing steps. In the slew range, the load velocity follows the pulse rate without losing steps, but cannot start, stop or reverse on command. The maximum torque point is the maximum holding torque of the excited motor for a steady load.

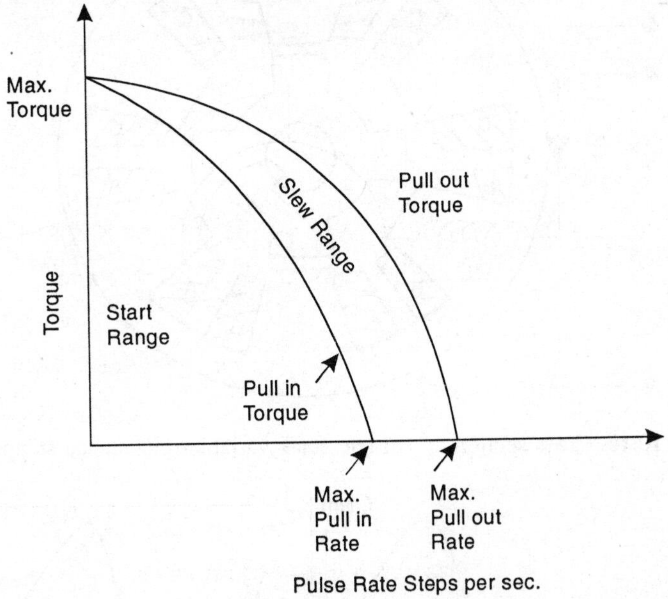


**Fig. 10.9.** Torque versus pulse rate characteristics of stepper motor

A stepper motor has a small size and is a cheap drive unit. Typical applications are: table positioning for machine tools, *X-Y* plotters, tape drivers, line printers and other computer peripheral devices. It has many other applications in manufacture of packages food stuffs, commercial end products and production of science fiction movies.

### 10.4.1 Single Stack Variable Reluctance Stepper Motor

Single stack stepper motor is generally of variable reluctance type and has no excitation on rotor. Fig. 10.10 shows the cross-section of a 3-phase motor having two stator poles per phase and four rotor poles. The steps are clockwise if the phase winding is excited in the sequence ABCA and anti clockwise if excitation sequence is ACBA. The step angle is the angle through which the rotor moves for each change in excitation sequence of motor and is given by

$$\text{Step angle} = \frac{360°}{Np}$$

where $N$ is the number of phases in the stator and $p$ is the number of poles in the stator.

Fig. 10.11 shows the static torque characteristics. It is drawn between torque $T$ and angle $\theta$. In the two equilibrium positions $X$ and $Y$, the torque is zero. The position $X$ is called stable detent position. Around this position the rotor develops a restoring torque if displaced. As the rotor moves from its position $X$, the restoring torque increases and reaches $T_{max}$ on either side. The level of excitation controls the slope of T – θ characteristics and maximum torque. For forward stepping the excitation sequence is AB-BC-CA. For reverse stepping the excitation sequence is AB-CA-BC.

**Fig. 10.10.** Cross section of a single-stack variable-relucatance stepper motor

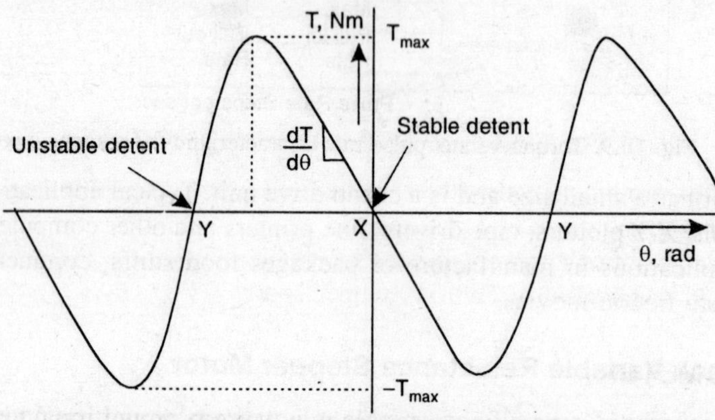


**Fig. 10.11.** Static torque characteristic of a stepper motor

## 10.4.2 Multistack Variable Reluctance Stepper Motor

In this motor the stator windings are stacked along the shaft. Each stack section has the same number of poles in the stator and rotor. Each stator stack is staggered with respect to its neighbour by one Nth of a pole pitch where N is the number of stator/rotor phases or sections. In a 6 phase multistack motor with 32 stator and rotor poles, the step size is 360°/(6 × 32) = 1.875°. The excitation sequence is similar to that in section in 10.4.1 but more excitation sequences are possible. When a stator winding is excited, the rotor poles of that section tend to align with those defined by the stator excitation. In the other sections the stator and rotor teeth are not aligned. The rotor is made to move one step by changing the combination of excited phases to the next in sequence.

### 10.4.3 Hybrid Stepper Motor

This motor has axially oriented permanent magnet sandwiched between two sections of stator and rotor as shown in Fig. 10.12. The magnetic flux distributes radially through the two sections of stator and rotor which are toothed. Moreover the magnetic flux distributes axially through the back iron of the stator and the shaft. The stator has two phase windings each of which creates alternate polarities of magnetic poles in both sections of the stator.

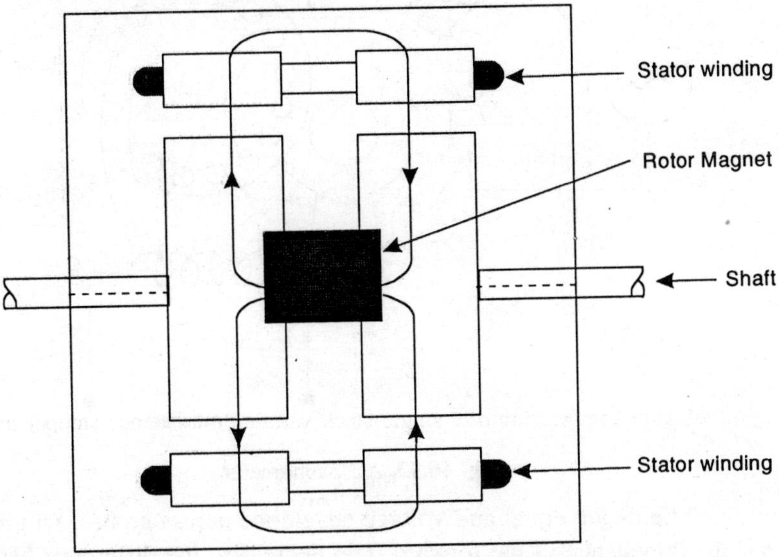


**Fig. 10.12.** Axial Section of Hybrid Stepper Motor

The magnetic flux produced by the stator winding is circumferential in each stator and rotor section and crosses the air gap axially. The rotor magnet causes the stator and rotor teeth to settle at the minimum reluctance position. The stepping angle of a hybrid motor is 90°/p where $p$ is the number of rotor poles.**Fig. 10.9.** Axial view of Hybrid Stepper Motor

## 10.5 DC TACHOMETER

In many control systems, it is necessary to feedback a voltage proportional to speed of the shaft. In a dc servo mechanism, this can be achieved by using a dc tachometer, which is a permanent magnet dc generator. The permanent magnet ensures a constant air-gap flux. Therefore its output voltage is directly proportional to speed of the shaft only.

A dc tachometer can be used in ac servomechanism by converting the dc output voltage to an ac voltage by using an inverter circuit.

## 10.6 AC TACHOMETER

An ac tachometer is used in feedback control systems, to feedback an ac voltage proportional to the speed of the shaft. It is basically a two-phase induction motor as shown in Fig. 10.13. One stator phase winding ($r$) is excited by a suitable ac voltage of constant magnitude and frequency. A voltage of the same frequency is generated across the other winding $c$, known as auxiliary winding or control winding. This output voltage is applied to the high impedance input circuit of an amplifier, so that the control winding can be considered as open circuited. It is necessary that the voltage developed across the control winding is linearly proportional to shaft speed and the phase of this voltage be fixed with respect to voltage applied to reference winding *r*.

The operation of an ac tachometer can be explained by the double revolving theory. As seen from winding *r*, the tachometer is equivalent to a small single-phase induction motor. At standstill the

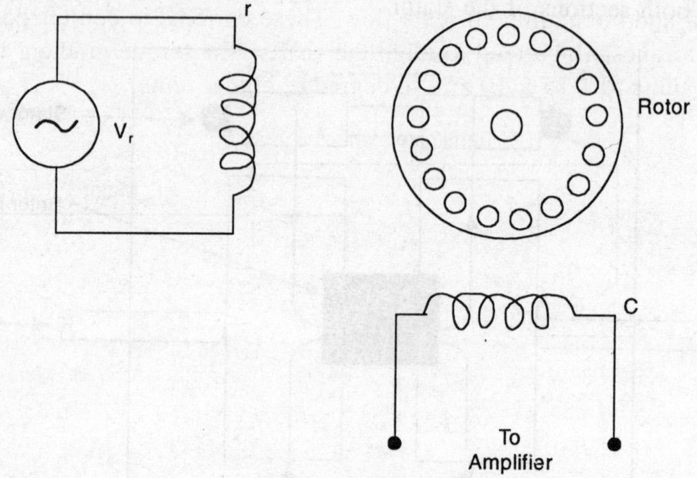


**Fig. 10.13.** AC Tachometer

forward and backward fields are equal and voltage developed across control winding *c* is zero. When rotor is revolving, the impedance of the forward field decreases, the difference between them being a function of speed. Therefore, the voltage developed across control winding is a function of speed. Reversal of direction of rotation reverses the phase of output voltage.

For a constant phase angle of output voltage and linear relationship between output voltage and speed, the ratio $x_2/r_2$ of the rotor (where $x_2$ is rotor reactance and $r_2$ is rotor resistance) should be either small (i.e. less than 0.1) or high (i.e. more than 10). If $x_2/r_2$ is low, the sensitivity (i.e. volts per revolution per minute) is sacrificed but linear speed range is wide. If $x_2/r_2$ is high, the speed range is limited to a fairly small fraction of synchronous speed to meet the condition of linearity of voltage and constancy of phase angle. Generally an intermediate value of $x_2/r_2$ ratio gives satisfactory performance.

An ac tachometer should have low inertia when rapid speed variations are encountered as in automatic control systems. The drag cup construction (Fig. 10.6) gives low inertia and is used many times.

## 10.7 SYNCHROS AND CONTROL TRANSFORMERS

A synchro (also known as selsyn) is a self-synchronising device used widely in servomechanisms as a position indicator. It transmits shaft position information, so that synchronism between two or more shafts be maintained. A control transformer is a particular type of synchro. It provides an error signal proportional to the deviation of its shaft from the synchronous position. A synchro is a miniature ac synchronous machine. It has a distributed 3-phase winding on stator and a 2-pole single-phase winding on rotor.

Figure 10.14 shows a circuit configuration, using two synchros, for maintaining synchronism between two shafts. The constructions of synchro transmitter and synchro motor are similar. The stator has 3-phase star connected distributed winding. The rotor has single-phase winding. Both rotor windings are connected to the same ac source. When rotor windings are excited, emfs are induced by transformer action in the stator windings of transmitter and motor. If the two shafts are in the same position (relative to stator windings), equal emfs are induced in the two stator windings and no circulating current exists and hence no torque. If the two shaft positions do not correspond, the emfs induced in the two stator windings are unequal and circulating currents flow. These currents in conjunction with air-gap magnetic fields produce torques which tend to align the shafts. The torque gradient varies from $5 \times 10^{-4}$ N-m/degree in small units to $1.25 \times 10^{-2}$ N-m/degree in bigger units.

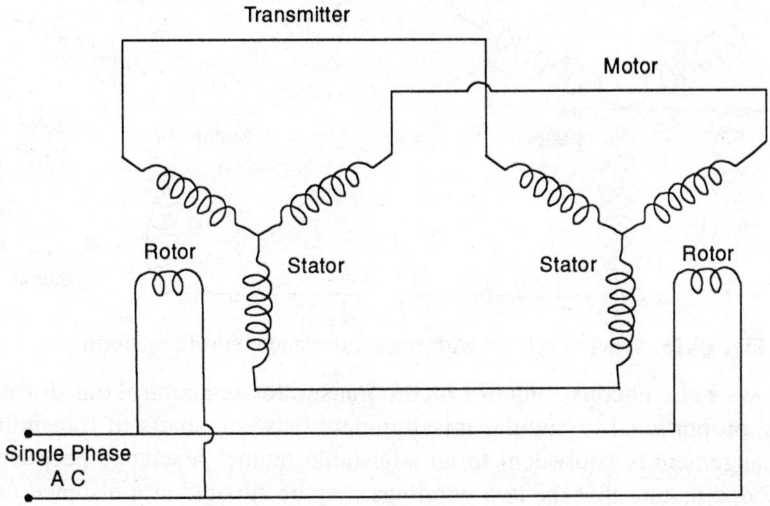


**Fig. 10.14.** Single-phase synchro: transmitter motor system

Figure 10.15 shows a synchro systems with transmitter, differential and motor. The function of this system is to permit the rotation of a shaft to be a function of sum or difference of the rotation of two other shafts. The differential has 3-phase distributed windings on both stator and rotor. The voltages impressed on its stator windings induce corresponding voltages in its rotor windings. It can be shown that

$$\alpha_r = \alpha_t - \alpha_d \tag{10.5}$$

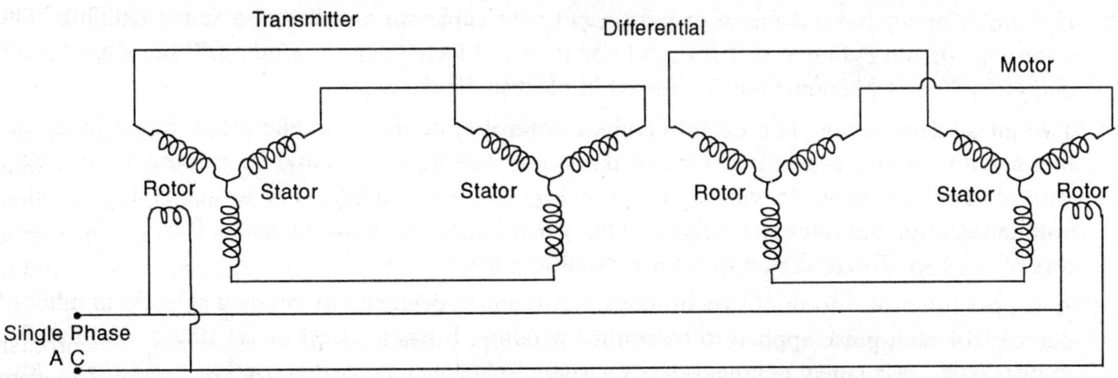


**Fig. 10.15.** Synchro system with transmitter, differential and motor

where $\alpha_r$ is the displacement of receiver shaft, $\alpha_t$ is the displacement transmitter shaft and $\alpha_d$ is displacement of differential shaft. If the phase sequence of stator and rotor windings of differential are reversed, the minus sign in Eq. (10.5) changes to plus sign.

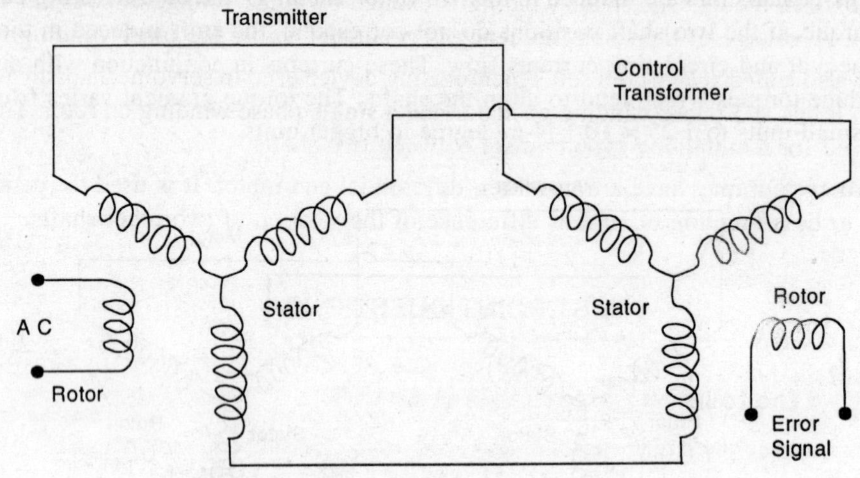


**Fig. 10.16.** Synchro system with transmitter and control transformer

Figure 10.16 shows a circuit consisting of synchro transmitter and control transformer. This system gives an error signal proportional to angular misalignment between shafts of transmitter and control transformer. The arrangement is equivalent to an adjustable mutual reactance between the two rotor windings with the added feature that the two windings may be situated at a distance from each other. When the angle is 90° corresponding to 90° displacement of two shafts, error voltage is zero. When the angle is different from 90° (and 270°), a voltage whose magnitude is proportional to angular displacement between the two shafts, is produced.

## SUMMARY

1. Reluctance motor develops torque due to difference in reluctance in the two axes. Its stator may have a single phase winding or 3-phase winding. It starts due to induction motor action and runs at synchronous speed.
2. Hysteresis motor has a 3-phase or permanent split capacitor single phase stator winding. The rotor is a smooth cylinder of hard steel and does not have teeth or winding. It develops torque due to hysteresis phenomenon. It is used in electric clocks etc.
3. Two phase servomotor is used in feedback control systems. It is like a two phase induction motor with squirrel cage rotor. One of the stator windings is excited by a fixed ac reference voltage. The second stator winding is excited by a control voltage. The torque is a function of both the control and reference voltage. The two voltages are out of phase by 90°. For servomotors of very small size drag cup rotor is used to reduce inertia.
4. A stepper motor is a form of synchronous motor and is designed to rotate a specific number of degrees for each pulse applied to its control winding. It has a 3 or 4 phase stator winding. The rotor is either permanent magnet type or variable reluctance type. It is used in *X-Y* plotters, tape drives, line printers and other computer peripheral devices.

5. A dc tachometer is a permanent magnet dc generator. Its output voltage is directly proportional to speed. It is used to feedback a voltage proportional to speed of shaft.
6. An AC tachometer generates an ac voltage proportional to speed of shaft. It is basically a two phase induction motor. One stator winding is excited by an ac voltage of constant voltage and frequency. A voltage of the same frequency is generated across the other winding.
7. Synchro (also called selsyn) is a self synchronising device used in servomechanism as a position indicator. It has a 3-phase winding on stator and a single phase winding on rotor. The synchros can be used for maintaining synchronism between two shafts.
8. A synchro system may have a transmitter, differential and motor. It is used to permit rotation of shaft to be a function of sum or difference of the rotation of two other shafts.

## TEST POINT QUESTIONS

### A. State whether the following are true (T) or false (F):

10.1 In a reluctance motor the reluctance of air gap is a function of angular position of rotor.
10.2 A reluctance motor runs at speed less than synchronous speed.
10.3 In a hysteresis motor the rotor is a smooth cylinder of hard steel.
10.4 Hysteresis motor runs at synchronous speed.
10.5 Two phase servomotor has a wound rotor.
10.6 A two phase servomotor may rotate in one direction or the other.
10.7 Drag cup rotor has very low inertia.
10.8 Stepper motor has a single phase stator winding.
10.9 AC tachometer has a two phase stator winding.
10.10 A synchro is a miniature ac synchronous machine.

### B. Fill in the blanks with suitable words.

10.11 Reluctance torque is proportional to ......... .
10.12 Reluctance motor has a ......... stator winding.
10.13 Reluctance motor always runs at .........
10.14 In hysteresis motor the rotor does not have any .........
10.15 The two stator windings of two phase servomotor are called ......... and .........
10.16 Two phase servomotor can be analysed using .........
10.17 Two phase servomotor has a ......... rotor.
10.18 Drag cup rotor has very low .........
10.19 Stepper motor has a ......... stator winding.
10.20 DC tachometer generates a dc output voltage proportional to .........

## ANSWERS

10.1 T 10.2 F 10.3 T 10.4 T 10.5 F 10.6 T 10.7 T
10.8 F 10.9 T 10.10 T 10.11 sin $2\delta$ 10.12 split phase

10.13 synchronous speed 10.14 winding 10.15 reference winding and control winding
10.16 symmetrical components 10.17 squirrel cage 10.18 inertia
10.19 multi phase 10.20 speed of shaft

## JOB INTERVIEW QUESTIONS

10.1 What is a hysteresis motor?
10.2 Where is reluctance motor used?
10.3 What is meant by reluctance?
10.4 What is reluctance torque?
10.5 Tell some applications of hysteresis motor.
10.6 How many stator windings are there in a servo motor?
10.7 Where is a servo motor used?
10.8 How can a servomotor be analysed?
10.9 What is a drag cup rotor?
10.10 What is the use of a stepper motor?
10.11 What is the use of a dc tachometer?
10.12 What is Selsyn?

## REVIEW QUESTIONS

10.1 What is reluctance torque? Draw the torque-speed curve of a reluctance motor.
10.2 Discuss the construction of a hysteresis motor. Draw its torque-speed curve.
10.3 What is a two-phase servo motor? Where is it used? Show its connection diagram. Draw its torque-speed curve.
10.4 Derive the transfer function of a two-phase servo motor. What is the advantage of drag cup rotor?
10.5 Draw a diagram showing the construction of a stepper motor and discuss its operation?
10.6 Discuss the function of an ac tacho-meter. Explain its construction and operation.
10.7 What is a synchro? Where is it used? Draw a connection diagram of a synchro transmitter-motor system and explain its working.

## PROBLEMS

10.1 The voltages in an unbalanced two phase system are $V_a = 150 \angle 0° V$ and $V_b = 90 \angle -30° V$. Find the positive and negative sequence voltages. [$105\angle 21.79° V, 65.385\angle -36.59°$]
10.2 The positive and negative sequence voltages in a two-phase system are $188.8 \angle 11.5°$V and $66.6\angle -34.4° V$. Find the phase voltages. [$240 \angle 0° V, 150 \angle -60° V$]
10.3 The positive and negative sequence currents in a two-phase system are $0.1183\angle -31.1°$A and $0.0546 \angle -57.7° A$. Find the phase currents. [$0.169\angle -39.5° A, 0.07358\angle -101.7° A$]
10.4 The phase currents in a two-phase system are $0.3024\angle -43.63° A$ and $0.1134\angle -54.95° A$. Find the symmetrical components of currents. [$0.1716\angle -24.72° A, 0.15072\angle -65.27° A$]

**10.5 In a two-phase, 50 Hz, 2-pole servomotor the air gap powers for balanced positive sequence and balanced negative sequence operation at $V_r = V_c = 100$ V and $s = 0.5$ are 9.6 W and 6 W respectively. Find the net torque in synchronous watts and N-m if he motor is fed by voltages $V_r = 150\angle 0°$V and $V_c = 90\angle -30°$V and $s = 0.5$.**

**[8.019 synch. watts, 0.0255 N-m]**

**10.6 For the servomotor of example 10.2 plot torque-speed curves for $\frac{V_c}{V_r} = 1.0, 0.6$ and 0.3.**

**10.7 A two-pole, 50 Hz 230 V two-phase servomotor has the following parameter: $r_1 = 342\ \Omega$, $x_1 = x_2' = 72\ \Omega$, $r_2' = 1020\ \Omega$ and $X_0 = 1194\ \Omega$. When it is operating at a slip of 0.6 with $V_r = 230\ V$ and $V_c = 143.75\ \angle -60°$, find the torque and stator phase currents.**

**[0.0587 N-m, 0.208$\angle$–39.5°A, 0.0904$\angle$–101.7°A]**

**10.8 For the servomotor of problem 10.7 plot the torque-speed curves for $\frac{V_c}{V_r} = 1.0, 0.7$ and 0.5.**

Chapter **11**

# Solid State Motor Control

## 11.1 INTRODUCTION

The application of solid state electronic circuits to motor control has revolutionized the electric power industry. Almost any power control problem can be solved using these circuits. The ac to dc converters have made it possible to run dc motors from ac supply. The ability to change ac power at one frequency to another has made it possible to control the speed of induction motor. Moreover, the cost of these solid state devices has decreased and their reliability has increased over the past 10 years or so. The operation of these solid state circuits is simple and efficiency is very high.

## 11.2 DC MOTOR CONTROL USING CONVERTERS

Many industrial applications (e.g. steel mills, paper mills, textile mills, traction systems) require dc motors. Ac to dc converters using thyristors are very convenient and are extensively used. A converter is known as one quadrant converter (or semi-converter) if the polarity of output voltage cannot be reversed with a given polarity of output current. In a two-quadrant converter, the output voltage can be made either positive or negative, with a given polarity of output current. The thyristors are phase controlled to get variable output voltage. The commutation process, i.e. transfer of current from one thyristor to another in these converters is line commutation, i.e. as an incoming thyristor is turned on, it reverse biases the outgoing thyristor, thus turning it off. Another way of classifying the converters is according to the number of pulses of current, which pass through the load circuit in each cycle of the source voltage.

### 11.2.1 Single-phase Two Quadrant Bridge Converter (fully controlled)

Figure 11.1 shows the circuit configuration of a single-phase two-quadrant bridge converter feeding a separately excited dc motor. The circuit requires 4 thyristors. $R_a$ and $L_a$ are the resistance and inductance of motor armature winding. The thyristors $T_1$ and $T_2$ conduct during the positive half-cycle, while $T_3$ and $T_4$ conduct during negative half-cycle. The firing of each thyristor is delayed by an angle $\alpha$. During the positive half-cycle, current flows from the source through thyristor $T_1$ to the motor and back to the source through $T_2$. During the negative half-cycle, the current flows from the source through $T_3$ to the motor and back to the source through $T_4$. The direction of current through the motor remains the same. Thyristors $T_1$ and $T_2$ conduct during the period $\alpha$ to $(\pi + \alpha)$. At $(\pi + \alpha)$, thyristors $T_3$ and $T_4$ are triggered and supply voltage appears as reverse bias across $T_1$ and $T_2$ and turns them off. From $(\pi + \alpha)$ to $(2\pi + \alpha)$, thyristors $T_3$ and $T_4$ conduct. From $\alpha$ to $\pi$ energy flows from source to motor. However, during the interval $\pi$ to $\pi + \alpha$, output voltage and current have opposite polarity and some of the motor energy is fed back to supply.

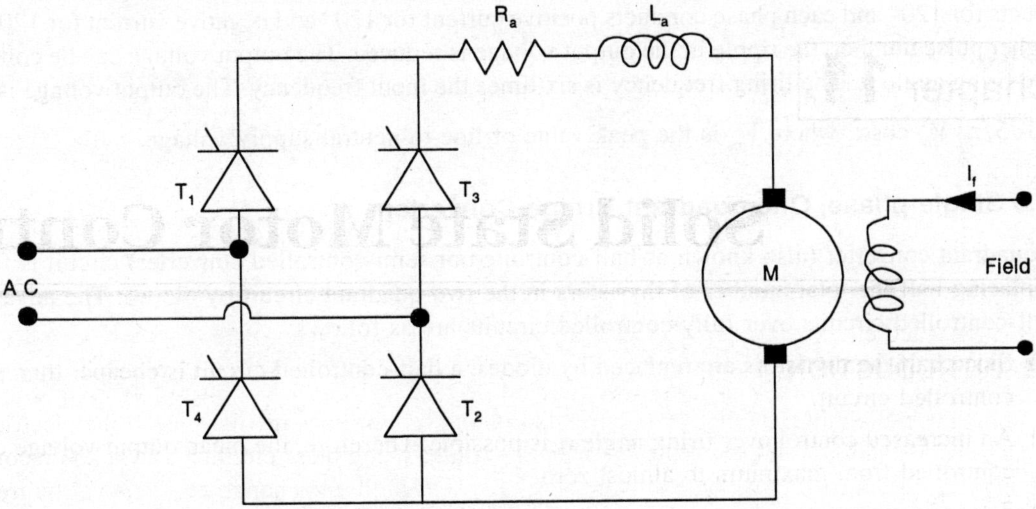


**Fig. 11.1.** Single-phase two-quadrant bridge converter

### 11.2.2 3-Phase, Six-pulse Two-quadrant Bridge Converter (fully controlled)

The single-phase converter can be used to feed small motors only. For medium and large size motors, a 3-phase converter is necessary. Figure 11.2 shows a 3-phase, six-pulse, two-quadrant bridge converter feeding a separately excited motor. Thyristors $T_1$, $T_3$, $T_5$ are called positive group, since they are turned on when the supply voltages are positive. Thyristors $T_2$, $T_4$, $T_6$, which conduct when supply voltages are negative, form the negative group. The firing angle $\alpha$ for thyristor 1 is measured from 30°, with reference to positive half-cycle of voltage of phase $R$. The firing of thyristor $T_1$ (when phase voltage of

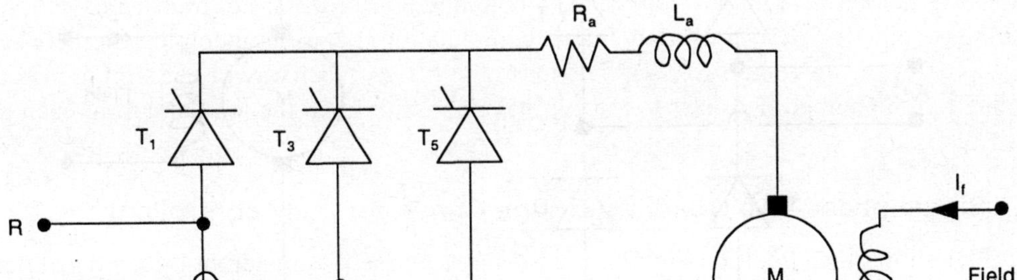


**Fig. 11.2.** Three-phase two-quadrant bridge converter

phase $R$ is more than that of phase $B$) will reverse bias thyristor $T_5$ and turn it off by line commutation. The negative group thyristors $T_2$, $T_4$, $T_6$ are turned on 180°, after the positive group thyristors $T_1$, $T_3$, $T_5$ are turned on, respectively. In each input cycle, there are 6 pulses in the output voltage. Each thyristor conducts for 120° and each phase conducts positive current for 120° and negative current for 120°. Due to higher pulse number, the ripple in the output voltage is reduced. The output voltage can be controlled by the firing angle $\alpha$. The firing frequency is six times the input frequency. The output voltage is given by $\left(3\sqrt{3}/\pi\right) V_m \cos\alpha$, where $V_m$ is the peak value of line-to-neutral supply voltage.

### 11.2.3 Single-phase, One-quadrant Bridge Converter

One quadrant converter (also known as half-controlled or semi-controlled converter) circuit is formed by replacing half the total number of thyristors in the two-quadrant circuit by diodes. The advantages of half-controlled circuits over fully controlled circuits are as follows:

(1) Since half the thyristors are replaced by diodes, a half-controlled circuit is cheaper than a fully controlled circuit.

(2) An increased control over firing angle $\alpha$ is possible. Therefore, the mean output voltage can be controlled from maximum to almost zero.

(3) The reactive power requirement of a half-controlled circuit is less than that for fully controlled one. Therefore, power factor of half-controlled circuits is better.

A fully controlled converter can be operated both as a rectifier as well as an inverter. However, a half-controlled converter can work only as a rectifier.

Figure 11.3 shows a single-phase half-controlled converter circuit. As seen this circuit has been formed by replacing thyristors $T_2$ and $T_4$ (in Fig. 11.1) by diodes $D_1$ and $D_2$.

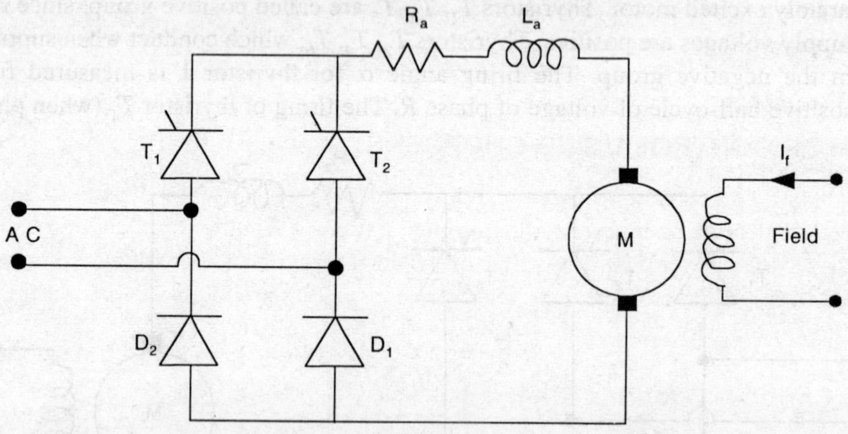


**Fig. 11.3.** Single-phase half-controlled converter

Thyristor $T_1$ is turned on at firing angle $\alpha$ in the positive half-cycle and $T_1$ and $D_1$ conduct from $\alpha$ to $\pi$. When the supply voltage reverses at $\omega t = \pi$, diode $D_2$ is forward biased and load current passes through $D_2$ and $T_1$. The supply voltage reverse biases $D_1$ and turns it off. Thus the load current free wheels during the interval $\pi$ to $\pi + \alpha$. At the instant $(\pi + \alpha)$, a trigger pulse is applied to $T_2$. Since it is forward biased, it turns on. Consequently, the supply voltage reverse biases $T_1$ and turns it off by line commutation. The load current now flows through $T_2$ and $D_2$.

### 11.2.4 3-Phase, One Quadrant Bridge Converter

Figure 11.4 shows a 3-phase, one-quadrant (half-controlled) converter. It is seen that this circuit has been formed by replacing thyristors $T_4$, $T_6$ and $T_2$ (in Fig. 11.2) by diodes $D_1$, $D_2$ and $D_3$. The operation of this circuit is in many respects similar to that of 3-phase, two-quadrant converter. However, this circuit cannot be used as inverter. The three thyristors $T_1$, $T_2$, $T_3$ have been connected to form a common cathode. The diode corresponding to the phase which ever is more negative conducts. The firing angles for $T_1$, $T_2$ and $T_3$ are measured in the same way as in two-quadrant, three-phase converter.

This circuit is very commonly used for feeding dc traction motors from ac system.

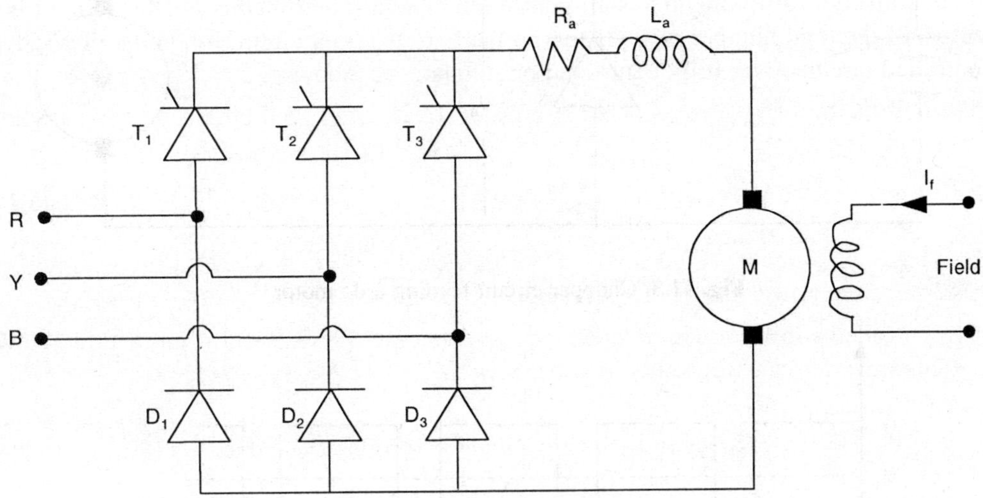


**Fig. 11.4.** Three-phase half-controlled converter

## 11.3 DC MOTOR CONTROL USING CHOPPERS

A dc chopper converts a constant voltage dc supply into a variable voltage dc supply. Such a circuit is needed for speed control of dc motors fed by storage batteries. Another application of chopper circuit is when a constant voltage supplies a number of motors, which require variable voltages to suit their individual speed-torque characteristics and requirements. Chopper circuits are more efficient than other methods of voltage control.

### 11.3.1 Basic Chopper Circuit

Figure 11.5 shows a dc motor supplied from a battery through a chopper circuit. When the thyristor is conducting, the battery voltage is applied to the motor. During the off period, when thyristor is not conducting, the applied voltage to the motor is zero. During the off state, the stored energy of the load inductance keeps the current flowing in the motor circuit through the freewheeling diode *D*. In this circuit, current can flow only from the source to the load and hence power can not be fed back from the load to the source and a regenerative braking of motor is not possible. This is known as one-quadrant chopper circuit.

Figure 11.6 (a and b) show the waveforms of voltage and current applied to the motor. Out of the total time $T$, the thyristor is conducting for time $t_1$ and is off for time $t_2$. The average value of voltage applied to the motor is

$$V_0 = (V)\left(\frac{t_1}{t_1 + t_2}\right) = V\left(\frac{t_1}{T}\right) = \alpha V \quad \text{...(11.1)}$$

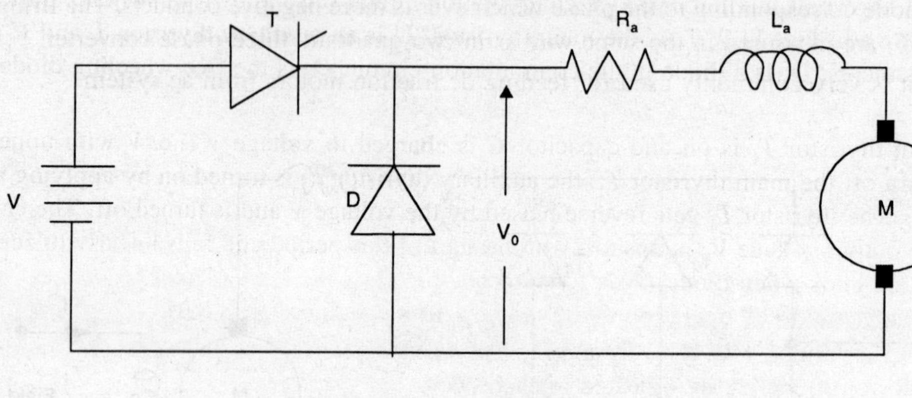


**Fig. 11.5.** Chopper circuit feeding a dc motor

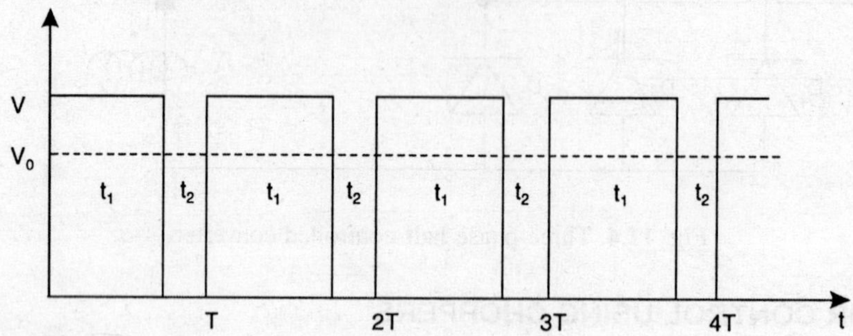


**Fig. 11.6.** Voltage and current waveforms of a dc motor fed from a chopper circuit

The ratio $t_1/T$ or $\alpha$ is known as duty cycle and controls the output voltage $V_0$ applied to the motor. By changing $\alpha$, the voltage to the motor can be changed. As shown in Fig. 11.6(b), the motor current is continuous. However, if the value of $\alpha$ is small (i.e. off period $t_2$ is large) the current may become discontinuous.

### 11.3.2 DC Chopper Commutation

The thyristor in the chopper circuit can be turned on by applying a gate signal at the appropriate instant. However, special methods are needed to turn it off, because in a chopper circuit, the source voltage does not on its own, go to zero and become negative (In a rectifier circuit the source voltage does go to zero and become negative. Therefore line commutation is possible).

One circuit, known as auxiliary voltage commutation circuit (also known as switch parallel commutation), is shown in Fig. 11.7. $T_1$ is the main thyristor, $T_2$ is an auxiliary thyristor. $L$ and $C$ are the commutating elements. $D_2$ is a diode in the commutation circuit. $D_1$ is the free wheeling diode of the chopper.

Assume that thyristor $T_1$ is on and capacitor $C$ is charged to voltage $V$ (i.e. $V$ with upper plate negative). To turn off the main thyristor $T_1$, the auxiliary thyristor $T_2$ is turned on by applying a firing pulse to its gate. The thyristor $T_1$ gets reverse biased by the voltage $V$ and is turned off. The capacitor discharges. The output voltage $V_0$ jumps to $2V$ at the start of this period and falls linearly to zero. This mode of operation ends when diode $D_1$ is forward biased. Then the load current fee wheels through diode $D_1$. The next mode of operation starts when a firing pulse is applied to the gate of $T_1$. An oscillatory circuit is formed with $L$, $C$, $T_1$ and $D_2$. The capacitor voltage at the end of this mode is $-V$. The above modes of operation are repeated in each cycle.

A number of other circuits for commutation have also be devised.

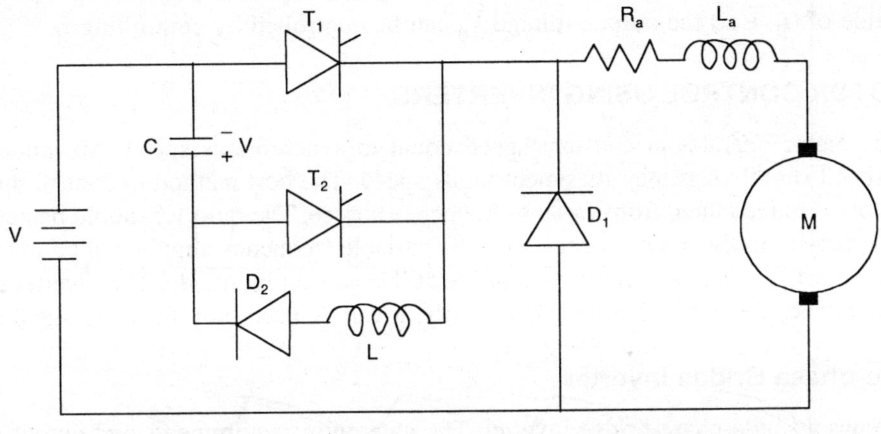


**Fig. 11.7.** Auxiliary voltage commutation circuit for chopper

### 11.3.3 Step-up Chopper

The chopper circuit of Fig. 11.7 can give an output voltage which is equal to or less than the battery voltage $V$. A circuit which can give an output voltage more than the input voltage, is known as step-up chopper and is shown in Fig. 11.8. When thyristor $T_1$ is on, the current will build up in the inductance $L$. When $T_1$ is turned off, a current $I_d$ will flow through diode $D$ and charge capacitor $C$. The capacitor will retain this charge and provide a continuous output voltage. Assume that thyristor conducts for a period $t_1$ and remains off for a period $t_2$. If current $I_d$ remains constant, the energy input to $L$ during time $t_1$ is

$$\text{Energy input} = VI_d t_1 \qquad ...(11.2)$$

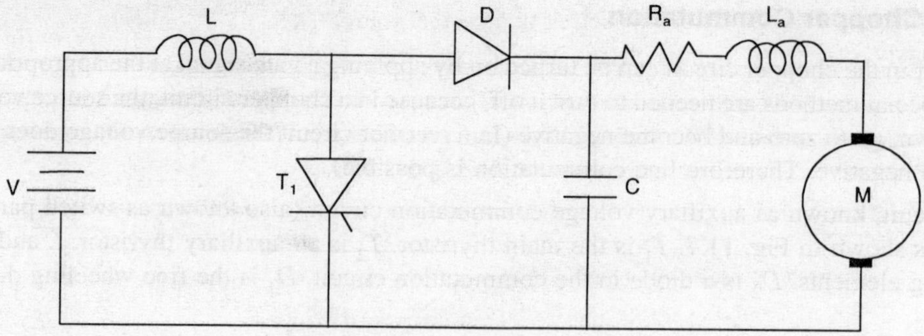


**Fig. 11.8.** Step-up chopper circuit

The energy delivered to the load is

$$\text{Energy output} = (V_0 - V)I_d t_2 \quad \text{...(11.3)}$$

Neglecting the losses, energy input and energy output must be equal

or $$VI_d t_1 = (V_0 - V)I_d t_2$$

or $$V_0 = \frac{V(t_1 + t_2)}{t_1} \quad \text{...(11.4)}$$

For a fixed value of $(t_1 + t_2)$ the output voltage $V_0$ can be controlled by controlling $t_1$.

## 11.4 AC MOTOR CONTROL USING INVERTERS

A synchronous motor operates at constant speed equal to synchronous speed. An induction motor operates at a speed slightly less than its synchronous speed. The best method to control the speeds of both these motors is to feed them from variable frequency source. The ratio *V/f* should be kept constant, so that the flux density in the motor is constant. The variable frequency supply can be obtained using a rectifier-inverter set. The rectifier converts 3-phase 50 Hz ac supply into dc. The inverter converts dc into variable frequency ac. Since the method is costly, it can be used only for large sized motors.

### 11.4.1 Single-phase Bridge Inverter

Figure 11.9 shows a single-phase bridge inverter. The gate pulse requirements and output voltage are shown in Fig. 11.10 (a and b). In the first interval *T*/2, the thyristors $T_1$ and $T_2$ are turned on and current flows from dc source through $T_1$, load, $T_2$ and returns to the dc source. In the next interval *T*/2, thyristors $T_3$ and $T_4$ are turned on and current flows from source through $T_3$, load $T_4$ and returns to the source. The direction of current in the two intervals are reverse of each other. The frequency of supply can be controlled by controlling *T*.

For a pure resistive load, current reverses at the same instant as the voltage and diodes are not needed. However for inductive and capacitive loads, the current does not reverse at the same instant as the voltage. The diodes $D_1$ and $D_2$ connected in anti-parallel with thyristors $T_1$ and $T_2$ permit load current to flow in the first-half interval. Similarly, diodes $D_3$ and $D_4$ permit the load current to flow in the second-half interval.

The thyristors require special commutation circuit to turn them off at proper instants. It is necessary to ensure that the two thyristors in series across the source (i.e. $T_1$, $T_4$ and $T_2$, $T_3$) don't conduct simultaneously, otherwise a short circuit will occur across the battery.

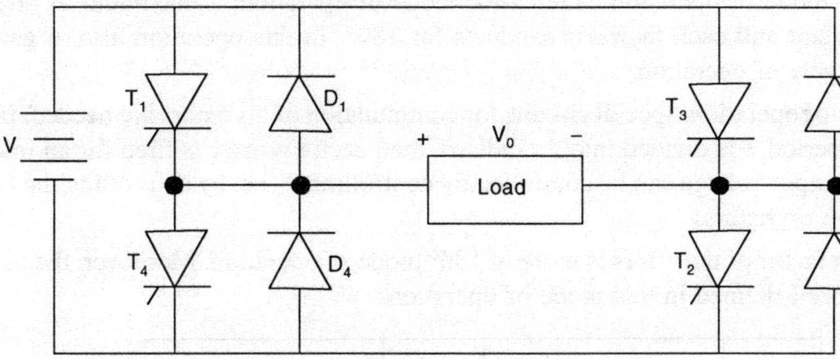


**Fig. 11.9.** Single-phase bridge inverter circuit

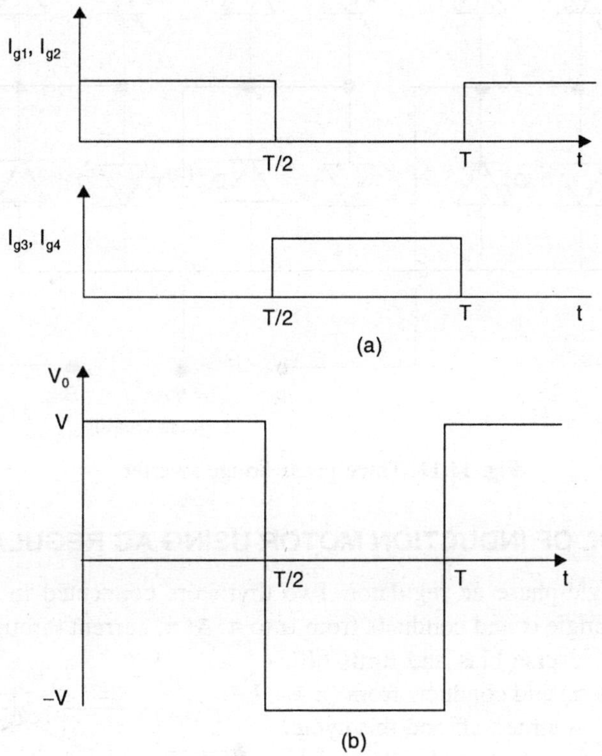


**Fig. 11.10.** Single-phase bridge inverter (a) gate pulses, (b) output-voltage

### 11.4.2 3-Phase Bridge Inverter

Figure 11.11 shows a 3-phase bridge inverter using 6 thyristors and 6 diodes.

There are two schemes of operation for this inverter. In one scheme each thyristor conducts for 120° and a maximum of two thyristors conduct at any instant. This is known as 120° mode of operation. Out of the two thyristors conducting at any instant, one is among $T_1$, $T_3$, $T_5$ and other among $T_2$, $T_4$, $T_6$. Six gate pulses are required in each cycle of operation.

In the second mode of operation called 180° mode of operation, a maximum of three thyristors conduct at any instant and each thyristor conducts for 180°. In this operation also, 6 gate pulses are required in each cycle of operation.

In both modes of operation, special circuits for commutation of thyristors are needed. If the inverter ac output voltage period $T$ is divided into $2\pi$ radians, then each thyristor is fired for an interval of $\pi/3$. The frequency of output voltage can be controlled by controlling $T$ (i.e. by controlling the time between firing of successive thyristors).

The utilisation factor of thyristors is more in 180° mode of operation. Moreover, the output voltage waveform is also well-defined in this mode of operation.

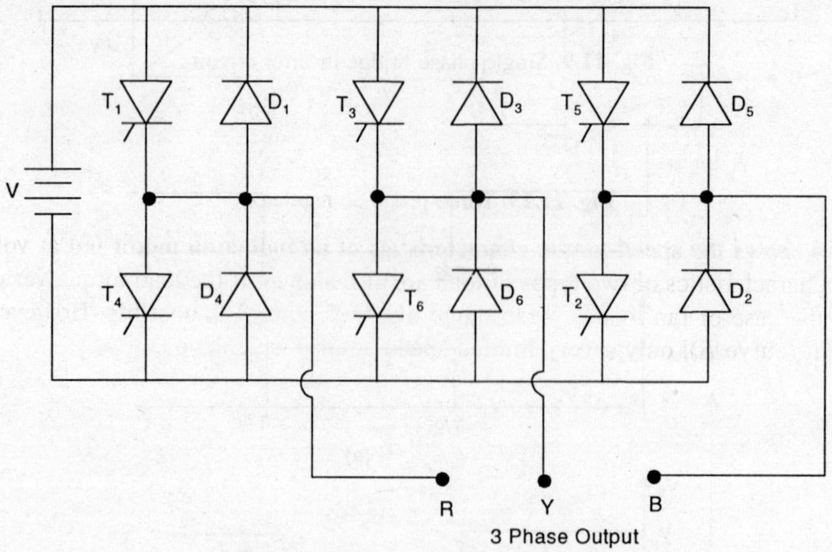


**Fig. 11.11.** Three-phase bridge inverter

## 11.5 SPEED CONTROL OF INDUCTION MOTOR USING AC REGULATOR

Figure 11.12 shows a single-phase ac regulator. Two thyristors connected in anti-parallel are used. Thyristor $T_1$ is fired at an angle $\alpha$ and conducts from $\alpha$ to $\pi$. At $\pi$, current through $T_1$ falls to zero, this thyristor is subjected to a reverse bias and turns off. Thyristor $T_2$ is fired at $(\pi + \alpha)$ and conducts from $(\pi + \alpha)$ to $2\pi$. At $2\pi$ thyristor $T_2$ is turned off and this cycle is repeated. As $\alpha$ is varied from 0 to $\pi$, the rms value of the applied voltage changes from $V$ to 0.

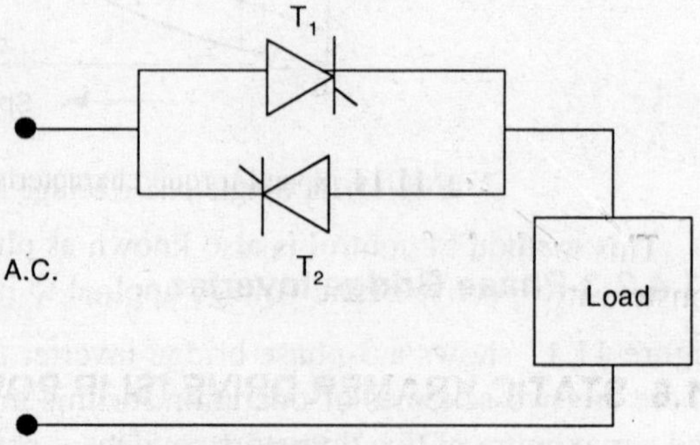


**Fig. 11.12.** Single-phase ac regulator

Figure 11.13 shows a 3-phase ac regulator feeding a 3-phase induction motor. The thyristors must be fired in proper sequence to obtain balanced currents and the firing angle must be the same for all the phases. At any time at least two thyristors conduct. The sequence for firing of thyristors is 1, 3′, 2, 1′, 3 and 2′. The interval between successive firing is 60°. The variation of firing angle controls the voltage applied to the stator of the motor.

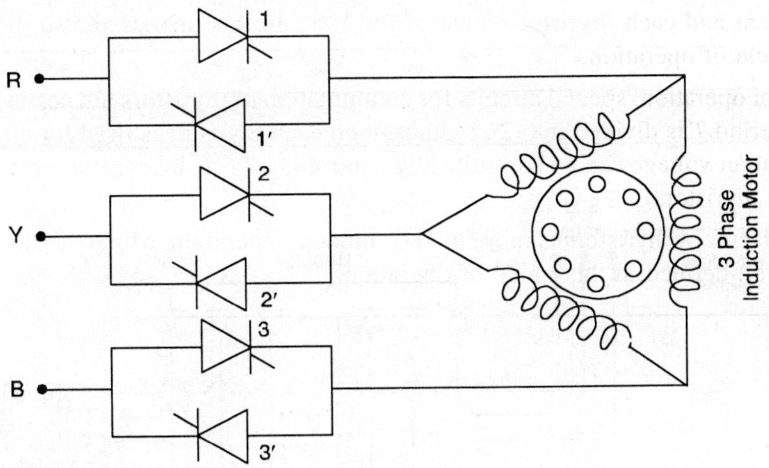


**Fig. 11.13.** Three-phase ac regulator

Figure 11.14 shows the speed-torque characteristics of an induction motor fed at voltages $V_1$, $V_2$, $V_3$, and $V_4$. The characteristics of two types of load are also shown. If the load torque varies with speed (curve $A$) as in the case of fan load, a wide range of speed control is possible. However, if the load torque if constant (curve $B$), only a very limited speed control is achieved.

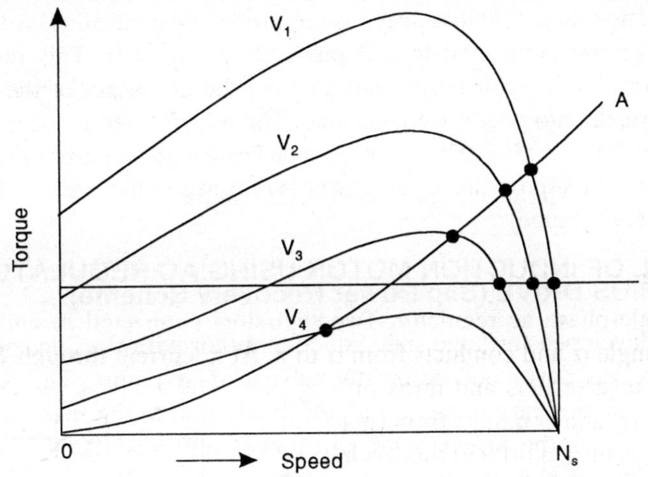


**Fig. 11.14.** Speed-torque characteristics of induction motor at different voltages

This method of control is also known as phase control. Since the speed of a synchronous motor is constant irrespective of the voltage applied to it, this method cannot be used for a synchronous motor.

## 11.6 STATIC KRAMER DRIVE (SLIP POWER RECOVERY SCHEME)

Steel rolling mills require a large size motor with flexible speed control. A drive called Kramer drive was developed for this purpose. The original drive was invented before the advent of thyristors and used a rotary converter. After the advent of thyristors the drive was modified and is called static Kramer drive. Fig. 11.15 shows the circuit.

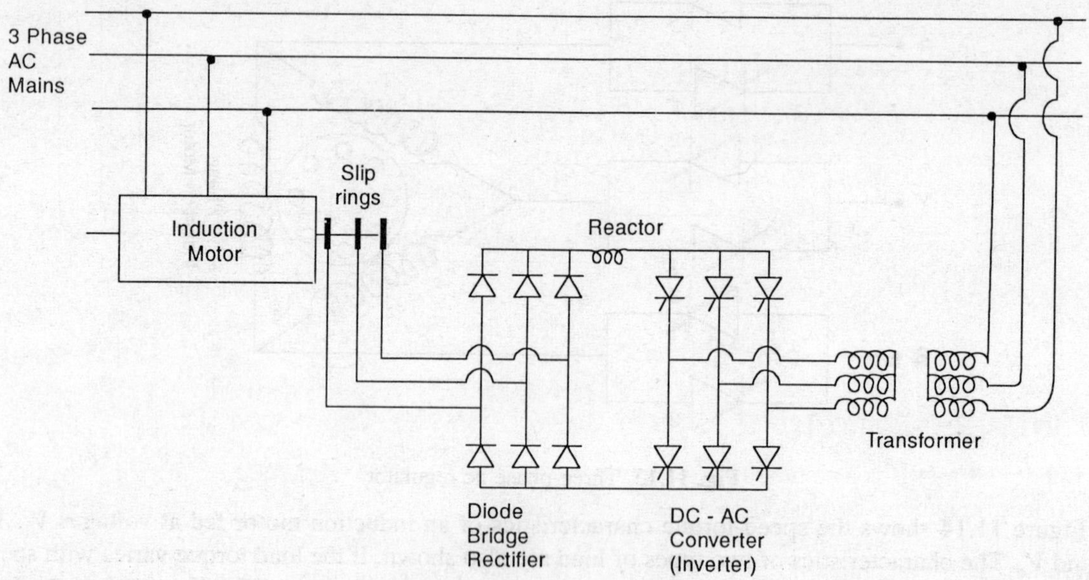


**Fig. 11.15.** Static Kramer Drive

The wound rotor induction motor (whose speed is required to be controlled) is fed from 3 phase AC mains supply. The sliprings are connected to a 3 phase diode rectifier. This rectifier converts slip frequency ac power into dc. The reactor is for smoothening the dc output of the rectifier. The dc-ac converter (inverter) converts dc into power frequency ac. The transformer changes the voltage level of ac output (from inverter) to the ac mains voltage. By proper firing angle control of inverter a wide range of speed control is possible. The slip frequency ac power is returned to the mains. Therefore the scheme is called slip power recovery scheme.

## 11.7 STATIC SCHERBIUS DRIVE (Slip Power Recovery Scheme)

Scherbius drive was also developed for large size induction motor with flexible speed control requirement. The original system was developed before the advent of thyristors. The original system used commutating machine and an auxiliary induction motor in addition to the main induction motor. The static Scherbius drive also uses the principle of slip power recovery. It uses a dual converter or cycloconverter for converting the slip frequency ac power to power frequency ac power or vice versa. Thus power can be extracted from rotor and feed back to the ac mains. In this mode the motor runs at speeds below synchronous speed. In the alternative mode of operation ac power frequency power is converted to slip frequency power and fed to slip rings. In this mode speeds above synchronous speed can also be obtained. Fig. 11.16 shows a simplified diagram.

The cycloconverter (or dual converter) converts slip frequency ac power to power frequency ac or vice versa by proper firing angle control. Thus a wide range of speed control is possible. Since the power extracted from rotor is fed back to the mains, the efficiency of the drive is also improved.

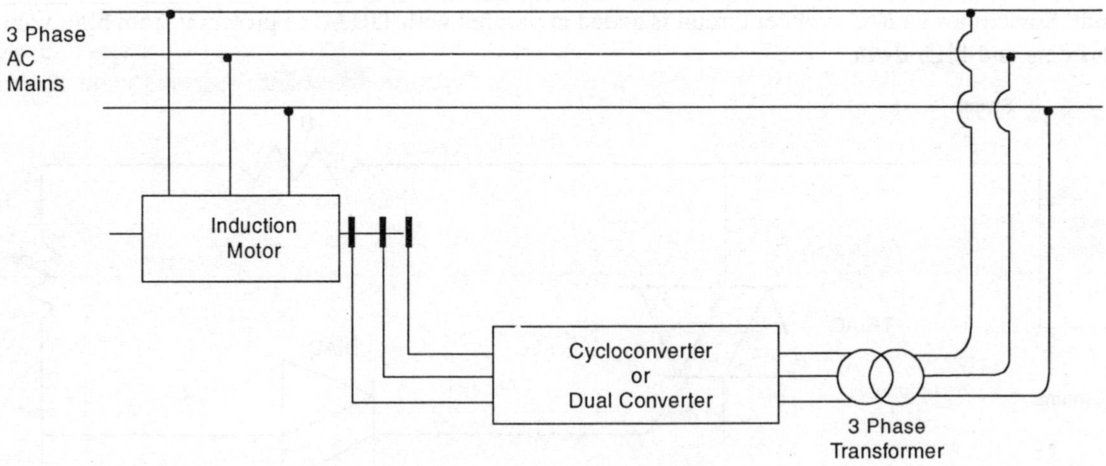


**Fig. 11.16.** Static Scherbius Drive

## 11.8 INTEGRATED MOTORS

The integrated motor is a standard ac induction motor along with an integrated frequency inverter and EMC (electromagnetic compatibility) filters. It is robust, reliable and can be operated in rough conditions. Thus it combines an induction motor with ac-ac converter (for frequency changing) integrated in the motor frame. The converter is often mounted in a box, above the motor frame, in place of terminal box. Its advantages include.

- Saving in space because a separate converter is not required.
- Reduction in installation cost because the cabling between motor and converter is eliminated.
- Elimination of motor problems due to high voltage transients.

The features like start, stop, forward, reverse, speed change are included. The power rating of this drive is limited to less than 15 kW.

## 11.9 SPEED CONTROL OF SINGLE PHASE INDUCTION MOTOR USING TRIAC (FAN REGULATOR)

The conventional speed regulator for a fan uses a resistance regulator. The regulator resistance is in series with the fan motor. When it is desired to reduce speed, the regulator resistance is increased. A higher voltage drop occurs across regulator resistance and, therefore, a lower voltage is applied to the motor and its speed decreases. The resistance regulator causes loss of energy. This loss of energy can be appreciable at low speeds.

Fig. 11.17. shows a speed control circuit using a DIAC-TRIAC pair. The DIAC is meant to trigger the TRIAC into conduction. It is preferable to use a DIAC-TRIAC pair. The R-C circuit forms the triggering network. By adjusting the resistance $R$ the voltage across capacitor $C$ can be adjusted. When voltage across $C$ exceeds the break down voltage of DIAC, it is triggered into conduction and sends a triggering signal to the gate of TRIAC. When TRIAC is turned on the motor starts. By varying $R$ the firing angle of TRIAC can be changed thus changing the voltage applied to the motor. This causes the change in speed of motor. Since the current through resistances $R_1$ and $R$ is small the energy loss is small. Sometimes an *R-C* snubber circuit is added in parallel with TRIAC to protect it from high voltage transients and high *dv/dt*.

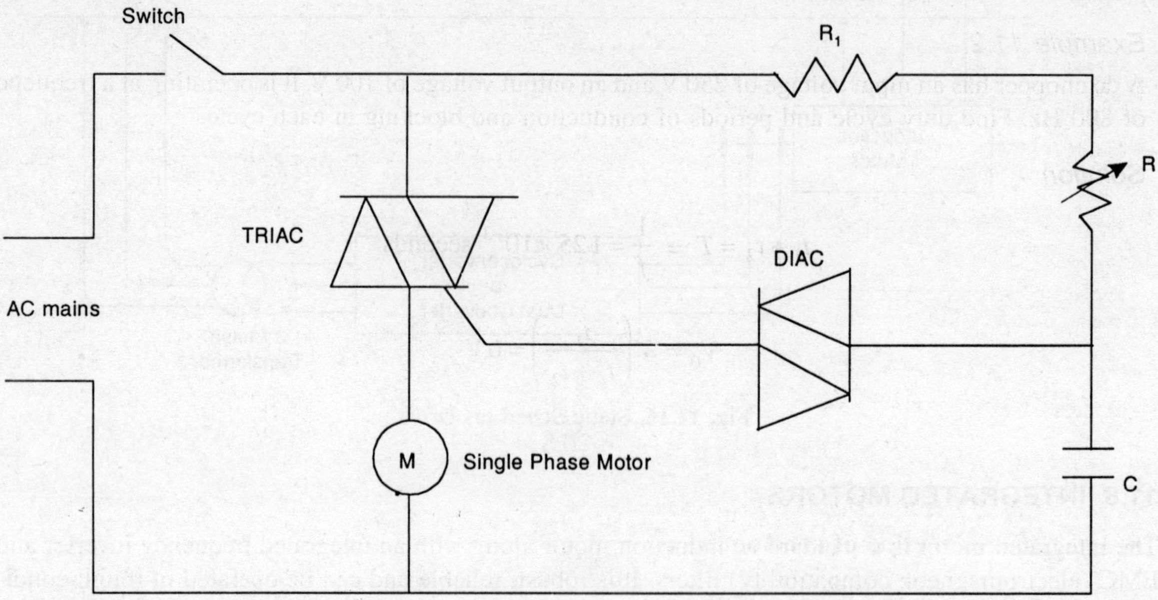


**Fig. 11.17.** Speed Control of Single Phase Induction Motor using TRIAC.

### *Example 11.1*

A single-phase 230 V, 50 Hz ac supply is rectified by a controlled thyristor and fed to a dc motor. The thyristor is continuously triggered by a dc signal. The armature resistance is 2 ohms and the back emf of the motor may be assumed to be constant at 200 V. Neglect armature inductance. Find the average current in the circuit.

### *Solution*

Thyristor will conduct when instantaneous value of source emf is more than back emf.

$$\sqrt{2} \times 230 \sin\omega t = 200$$

or

$$\omega t = 37.95° \text{ and } 142.05°$$

$$= 0.662 \text{ and } 2.48 \text{ radians}$$

$$\text{Average current} = \frac{1}{2\pi} \int_{0.662}^{2.48} \frac{\sqrt{2} \times 230 \sin \omega t - 200}{2} d(\omega t)$$

$$= \frac{1}{2\pi} \Big| -162.61 \cos \omega t - 100\omega t \Big|_{0.662}^{2.48}$$

$$= \frac{1}{2\pi} [-162.61 \cos 142.05° + 162.61 \cos 37.95° - 100(2.48 - 0.662)]$$

$$= \frac{1}{2\pi}(256.45 - 181.8) = 11.88\text{ A}.$$

*Example 11.2*

**A dc chopper has an input voltage of 250 V and an output voltage of 100 V. It is operating at a frequency of 800 Hz. Find duty cycle and periods of conduction and blocking in each cycle.**

*Solution*

$$t_1 + t_2 = T = \frac{1}{f} = 1.25 \times 10^{-3} \text{ seconds}$$

$$V_0 = V\left(\frac{t_1}{t_1 + t_2}\right) = \alpha V$$

$$\text{Duty cycle } \alpha = \frac{100}{250} = 0.4$$

$$\text{Period of condition } t_1 = \frac{V_0(t_1 + t_2)}{V}$$

$$\text{or} \quad t_1 = \frac{100(1.25 \times 10^{-3})}{250} = 0.5 \times 10^{-3} \text{ seconds}$$

$$t_2 = 1.25 \times 10^{-3} - 0.5 \times 10^{-3} = 0.75 \times 10^{-3} \text{ seconds.}$$

*Example 11.3*

**A step up chopper has an output of 300 V and an input of 250 V. The blocking period in each cycle of operation is 0.5 × $10^{-3}$ seconds. Find the period of conduction in each cycle and chopping frequency.**

*Solution*

$$V_0 = V\frac{t_1 + t_2}{t_2}$$

$$\text{or} \quad t_1 + t_2 = \frac{V_0 t_2}{V} = \frac{300 \times 0.5 \times 10^{-3}}{250} = 0.6 \times 10^{-3} \text{ seconds}$$

$$t_1 = 0.6 \times 10^{-3} - 0.5 \times 10^{-3} = 0.1 \times 10^{-3} \text{ seconds}$$

$$T = t_1 + t_2 = 0.6 \times 10^{-3} \text{ seconds}$$

$$f = \frac{1}{t} = 1.667 \times 10^{3} \text{ Hz}$$

*Example 11.4*

**A dc series motor is fed from a 250 V dc supply through a chopper operating on constant chopper**

frequency of 800 Hz. The armature and series field resistances are 0.08 ohm and 0.05 ohm respectively. The back emf of motor is 180 V. Find periods of conduction and blocking. Armature current is 50 A.

*Solution*

$$V_0 = 180 + 50(.08+.05) = 186.5\text{ V}$$

$$T = \frac{1}{800} = 1.25 \times 10^{-3} \text{ seconds}$$

$$V_0 = V\frac{t_1}{T} \quad \text{or} \quad t_1 = \frac{V_0 T}{V} = \frac{186.5 \times 1.25 \times 10^{-3}}{250} = 0.9325 \times 10^{-3} \text{ seconds}$$

$$t_2 = T - t_1 = 0.3175 \times 10^{-3} \text{ seconds.}$$

*Example 11.5*

A separately excited dc motor has the following data: rated armature current 30 A, rated voltage 250 V dc, armature resistance 0.8 ohm, rated speed 1000 rpm. It is desired to control the speed using a step down chopper. (a) Find the duty cycle so that motor may operate at 800 rpm and develop rated torque. (b) Find the duty cycle so that the motor may operate at 800 rpm and develop 60% of rated torque.

*Solution*

(a) Back emf at 1000 rpm = 250 – 30 × .8 = 226 V

$$\text{Back emf at 800 rpm} = \frac{226 \times 800}{1000} = 180.8\, V$$

$$\text{Torque} \propto \varphi\ \text{Ia}$$

A separately excited motor has a constant flux. Since torque is constant, $I_a$ is also constant, i.e. 30 A

$$V_0 = 180.8 + 30 \times 0.8 = 204.8\ V$$

$$\text{Duty cycle} = \frac{204.8}{250} = 0.8192$$

(b) Since torque is reduced to 60%, $I_a$ is also reduced to 60%

$$I_a = 0.6 \times 30 = 18\ A$$

$$V_0 = 180.8 + 18 \times 0.8 = 195.2\ V$$

$$\text{Duty cycle} = \frac{195.2}{250} = 0.7808.$$

*Example 11.6*

An RLE load is operating on a chopper circuit from a 500 V dc source. If L = 0.05 H, R = 0 and duty cycle = 0.2, find the chopping frequency to limit the amplitude of load current excursions to 10 A.

*Solution*

RLE load signifies a resistance, inductance and a motor (having back emf E)

$$V_0 = \alpha V = 0.2 \times 500 = 100\ V$$

$$V - V_0 = 500 \times 100 = 400\ V$$

This difference $V - V_0$ is responsible for setting up the flux by the inductance

$$400 = L\frac{di}{dt} = L\frac{\Delta i}{\Delta t}$$

or $$t_1 = \Delta t = \frac{L\Delta i}{400} = \frac{0.05 \times 10}{400} = 0.00125 \text{ seconds}$$

$$t_1 + t_2 = \frac{t_1}{\alpha} = \frac{0.00125}{0.2} = 0.00625$$

$$f = \frac{1}{t_1 + t_2} = 160 \text{ pulses/}s.$$

### Example 11.7

The operation of a 400 V, 50 Hz, 3-phase induction motor is controlled by a chopper operating on TRC principle. The rotor resistance is 1 ohm per phase. Due to the action of chopper, an additional resistance of 2 ohms is connected in each phase of the rotor circuit during off period of the chopper. The chopper is operating on a constant frequency of 400 Hz and slip is 4%. Find the torque developed. The stator and rotor windings are similar and have the same number of turns per phase. Neglect stator resistance and leakage reactances. Chopper off time = $1.5 \times 10^{-3}$ seconds.

### Solution

Using Eq. (7.27) and neglecting $r_1$, $x_1$ and $x_2$ we have

$$T = \frac{3V_1^2 s}{r_2'} \text{ synch. watts}$$

$$V_1 = \frac{400}{\sqrt{3}} = 230.94\ V, s = 0.04$$

Since rotor and stator windings are similar and have equal turns, rotor resistance as referred to stator is equal to rotor resistance.

$$r'_2 = 1 \text{ ohm}$$

$$t_1 + t_2 = \frac{1}{f} = \frac{1}{400} = 2.5 \times 10^{-3} \text{ seconds}$$

$$t_2 = 1.5 \times 10^{-3} \text{ seconds}$$

$$t_1 = 2.5 \times 10^{-3} - 1.5 \times 10^{-3} = 1 \times 10^{-3} \text{ seconds}$$

Effective rotor resistance as referred to stator = $R'_2$

$$R_2' = \frac{r_2' t_1 + (r_2' + R)t_2}{t_1 + t_2}$$

where $R$ is the additional resistance added in the rotor circuit during off time of the copper.

$$R_2' = \frac{1 \times 1 \times 10^{-3} + (1+2)1.5 \times 10^{-3}}{2.5 \times 10^{-3}} = 2.2\,\Omega$$

$$T = \frac{3(230.94)^2 \times 0.04}{2.2} = 2909.1 \text{ synch. watts}$$

*Example 11.8*

A six-pulse bridge converter fed from 400 V ac mains is supplying a 400 V dc motor. The voltage drop in the circuit between the converter and motor is 20 V. Find the firing angle of the converter.

*Solution*

$$\text{output voltage of converter} = 400 + 20 = 420V$$

$$V_m = \frac{400}{\sqrt{3}} \times 1.414 = 326.56\,V$$

$$\frac{3\sqrt{3}}{\pi} V_m \cos\alpha = 420$$

or $$\alpha = 38.96°$$

*Example 11.9*

A six-pulse converter is feeding a 400 V dc motor. The rated armature current is 50 A and armature resistance is 0.4 ohm. When the motor is operating at its rated speed of 800 rpm, the firing angle is 10°. It is desired to reduce the speed to 500 rpm by adjusting the firing angle. The current at 500 rpm is equal to the rated current. Find the firing angle.

*Solution*

$$\text{Back emf at 800 rpm} = 400 - 50 \times 0.4 = 380 \text{ V}$$

$$\text{Back emf at 500 rpm} = \frac{380 \times 500}{800} = 237.5\,V$$

Applied voltage to motor at 500 rpm

$$= 237.5 + 50 \times 0.4 = 257.5 \text{ V}$$

$$\text{Output voltage of converter} \propto \cos\alpha$$

$$\frac{380}{257.5} = \frac{\cos 10°}{\cos\alpha}$$

or $$\alpha = 48.14°.$$

*Example 11.10*

A 3-phase, 4-pole, 400 V, 50 Hz induction motor develops maximum torque at 10% slip. In a particular application it runs at 4% slip and drives a load whose torque varies as square of the speed. It is desired to operate the motor at 900 rpm by changing the voltage applied to its stator which is delta connected. Neglect stator impedance. Find the required stator voltage at 900 rpm speed.

*Solution*

$$r_2' = 0.1\, x_2'$$

$$N_s = 120 \times \frac{50}{4} = 1500 \text{ rpm}$$

Let $V_1$, $N_1$, $s_1$ and $T_1$ denote stator voltage, speed, slip and torque under initial conditions and $V_2$, $N_2$, $s_2$ and $T_2$ denote the new values of these quantities.

$$N_1 = 1500\,(1 - .04) = 1440 \text{ rpm}$$

$$V_1 = 400\ V$$

$$s_1 = 0.04$$

$$N_2 = 900 \text{ rpm}$$

$$s_2 = \frac{1500 - 900}{1500} = 0.4$$

$$\frac{T_2}{T_1} = \frac{900^2}{1440^2}$$

Using Eq. (7.27a)

$$T_1 = \frac{3}{2\pi n_s}\left[\frac{400^2}{\left(\frac{r_2'}{0.04}\right)^2 + x_2'^2}\right]\frac{r_2'}{0.04}$$

$$T_2 = \frac{3}{2\pi n_s}\left[\frac{V_2^2}{\left(\frac{r_2'}{0.4}\right)^2 + x_2'^2}\right]\frac{r_2'}{0.4}$$

$$\frac{T_2}{T_1} = \frac{900^2}{1440^2} = \frac{V_2^2}{400^2}\left[\frac{625 r_2'^2 + x_2'^2}{6.25 r_2'^2 + x_2'^2}\right]\left(\frac{0.04}{0.4}\right)$$

or

$$\frac{900^2}{1440^2} = \frac{V_2^2}{400^2}\left(\frac{7.25 x_2'^2}{1.0625 x_2'^2}\right)(0.1)$$

or

$$V_2 = 302.65\ V$$

## SUMMARY

1. A dc motor can be operated from ac mains by converting ac supply to dc using phase controlled thyristor rectifiers. These rectifiers may be half wave converter, semi converter, full converter or

dual converter. Three phase converters are used when the size of the motor is more than 15 kW

2. Armature voltage control is a very convenient method for speed control of dc motors. Phase controlled rectifiers and chopper circuits can provide variable dc voltage for armature voltage control.
3. Chopper drives can be operated in 'motoring mode', 'regenerative braking mode' and 'rheostatic braking mode'.
4. Speed of three phase induction motor can be controlled by stator voltage control, rotor voltage control, voltage and frequency control methods. Thyristors are used in all these methods.
5. Speed control of synchronous motor can be done by varying the frequency of ac supply. A circuit consisting of rectifier, filter and inverter is used.
6. Speed control of fan motors can be done very efficiently through DIAC-TRIAC pair.

## TEST POINT QUESTIONS

### A. State whether the following statements are true (T) or false (F).

11.1 The speed of a dc shunt motor can be varied by varying either the armature voltage or field current.

11.2 For small size motors, a 3 phase converter is preferable.

11.3 A regenerative braking method improves system efficiency.

11.4 The slip, at rated torque, of a 3 phase induction motor is about 0.4.

11.5 In a 3 phase induction motor speed variation by stator voltage control is preferable when load torque is constant.

11.6 It is advisable to operate an induction motor with constant V/f ratio.

11.7 A change in stator voltage can change the speed of synchronous motor.

11.8 The conventional resistance regulator for a fan motor is very efficient.

11.9 A cycloconverter changes frequency without a dc link.

11.10 Static Kramer drive is a slip recovery scheme.

### B. Fill in the blanks with suitable words.

11.11 The best method to vary the speed of an induction motor is to vary the input ..............

11.12 A step up chopper gives an output whose .............. is .............. than input.

11.13 A synchronous motor is generally used for .............. control.

11.14 If magnetic circuit is not saturated the torque of a dc series motor is proportional to ..............

11.15 If the field current of a dc shunt motor is reduced the speed ..............

11.16 Static scherbius drive uses a ..............

11.17 An example of slip power recovery scheme is ..............

11.18 A half controlled circuit is .............. than fully controlled circuit.

11.19 Power factor of a half controlled circuit is .............. than that of fully controlled circuit.

11.20 A half controlled converter can be operated only as ..............

## ANSWERS

11.1 T 11.2 F 11.3 T 11.4 F 11.5 F 11.6 T 11.7 F
11.8 F 11.9 T 11.10 T 11.11 frequency 11.12 voltage, more
11.13 power factor 11.14 square of armature current 11.15 increases
11.16 cycloconverter 11.17 static kramer drive 11.18 cheaper
11.19 better 11.20 rectifier.

## JOB INTERVIEW QUESTIONS

11.1 Name the two methods to change the speed of dc motors.
11.2 What is regenerative braking?
11.3 What is the difference between half controlled and fully controlled circuit?
11.4 Name the methods to control speed of induction motor.
11.5 Draw the circuit of fan regulator using a TRIAC.

## REVIEW QUESTIONS

11.1 Distinguish between one-quadrant and two-quadrant converters. Draw the connection diagrams and explain the working of single-phase one-quadrant and two-quadrant converters.
11.2 Draw connection diagrams of three-phase two-quadrant and one-quadrant converters. Explain their working.
11.3 What is a chopper? How can it be used for control of a dc motor? Why is a special commutation circuit necessary for a chopper? Draw the diagram of such a circuit.
11.4 Distinguish between step down and step up choppers. Draw diagrams and explain their working.
11.5 Explain the principle of ac motor control using converter-inverter set. For which motors is it applicable?
11.6 Draw the connection diagram and explain the working of (a) single-phase inverter, (b) three-phase inverter.
11.7 What is an ac regulator? How can it be used for speed control of an induction motor?
11.8 Explain the working of single-phase and three-phase regulators, with the help of suitable diagrams.

## PROBLEMS

11.1 A dc motor is fed from a single phase 110 V, 50 Hz ac supply through a continuously fired thyristor. The motor armature resistance is 10 ohms and its back emf may be assumed to be constant at 55.5 V. Find the average current in the circuit. [2.495 A]
11.2 A dc motor is fed from a 250 V dc supply through a chopper. The operating frequency is 1 k Hz and the average voltage across the motor is 150 V. Find duty cycle and periods of conduction and blocking in each cycle. [0.6, $0.6 \times 10^{-3}$, $0.4 \times 10^{-3}$ seconds]
11.3 A step up chopper has an output of 400 V and an input of 300 V. If the blocking period in each cycle of operation is $0.8 \times 10^{-3}$ seconds, find period of conduction in each cycle and chopper frequency. [$0.267 \times 10^{-3}$, $0.937 \times 10^{3}$Hz]

11.4 A dc series motor is fed from a 300 V dc source through a chopper operating at a frequency of 500 Hz. The average armature current is 40 A and the armature and series field resistances are 0.07 and 0.03 ohms respectively. The back emf of motor is 200 V. Find the periods of conduction and blocking in each cycle.

[$1.36 \times 10^{-3}$s, $0.64 \times 10^{-3}$ seconds]

11.5 A separately excited dc motor has the following data: rated voltage 500 V, rated armature current 40 A, armature resistance 0.5 ohm, rated speed 1200 rpm. Find the duty cycle so that (a) motor may operate at 1000 rpm and develop rated torque, (b) motor may operate at 1000 rpm and develop 70% of rated torque.

[(a) 0.84 (b) 0.828]

11.6 A dc series motor is controlled by a chopper fed from a 600 V.dc supply. The total inductance in the circuit is 0.08 H and resistance may be neglected. Find the chopping frequency to limit the amplitude of current excursions to 15 A when the duty cycle is 0.4.

[120 pulses/s]

11.7 A six-pulse bridge converter is feeding a 500 V dc motor. The voltage drop in the circuit between converter and motor is 10 V. If the converter is fed by a 440 V, 50 Hz 3-phase ac supply, find the firing angle.

[30.84°]

11.8 A six-pulse converter is feeding a dc motor having an armature resistance of 0.2 ohms and a rated armature current of 100 A. When the speed is equal to the rated speed of 1000 rpm, the voltage applied to the motor terminals is 500 V and the firing angle of the converter is zero. Find the firing angle to operate the motor at 600 rpm and rated current. [50.1°]

11.9 A chopper is used to add resistance in the rotor circuit of an induction motor. The chopper is operating at constant frequency time ratio control. The resistance of rotor winding is 0.5 ohms per phase. The chopper adds a resistance of 1.5 ohm in each rotor phase during the period when it is off. The chopper is operating at a frequency of 500 Hz. Chopper off time is $1 \times 10^{-3}$ seconds in each cycle. Find the torque developed if stator input voltage is 400 V. Assume the stator to be star connected and the stator and rotor windings to be similar with equal turns. Motor slip is 5%. [6400 synch. watts]

# Appendix A

## OBJECTIVE QUESTIONS

1. A permeable substance is one
   (a) which is a strong magnet
   (b) through which the magnetic lines of force can pass very easily
   (c) which is a good conductor
   (d) which can behave as a conductor as well as an insulator
2. Fleming's left hand rule is used to find
   (a) direction of force on a current carrying conductor in a magnetic field
   (b) direction of flux in a solenoid
   (c) direction of magnetic field due to current carrying conductor
   (d) none of above
3. If the area of hysteresis loop of a material is large, the hysteresis loss in this material will be
   (a) small (b) large
   (c) zero (d) indeterminate
4. The relative permeability of ferromagnetic material is
   (a) 1 (b) less than 1
   (c) more than 1000 (d) 0
5. An emf induced in a coil due to a change in current in a neighbouring coil is known as
   (a) self-induced emf (b) speed emf
   (c) mutually induced emf (d) none of the above
6. The induced emf is maximum if the conductor cuts the magnetic field at
   (a) 0° (b) 45°
   (c) 90° (d) an angle which is nearly 90°
7. The direction of induced emf is given by
   (a) Fleming's right hand rule (b) Fleming's left hand rule
   (c) Kirchhoff's voltage law (d) Ampere's law
8. The magnitude of emf induced in a wire does not depend on
   (a) length of wire (b) speed of wire
   (c) diameter of wire (d) flux density
9. The magnitude of statically induced emf depends on
   (a) magnitude of flux (b) rate of change of flux
   (c) resistance of coil (d) none of the above

10. An emf of 8 volts is induced in a coil of inductance 2H. The rate of change of current must be
    (a) 16 A/sec (b) 2 A/sec
    (c) 0.25 A/sec (d) 4 A/sec
11. In an iron-cored coil, the iron core is removed, so that the coil becomes an air cored coil. The inductance of coil will
    (a) increase (b) decrease
    (c) may increase or decrease (d) remain the same
12. Two coils have self-inductances of 5H and 1H, the mutual inductance being zero. If the two coils are in series, the total inductance will be
    (a) 4H (b) 5H
    (c) 6H (d) 12H
13. The yoke of a dc machine
    (a) is laminated (b) is not laminated
    (c) may or may not be laminated (d) is partially laminated
14. The pole shoes of a dc machine
    (a) are always laminated (b) are never laminated
    (c) are sometimes laminated (d) are partially laminated
15. The thickness of laminations is about
    (a) 2 mm (b) 5 mm
    (c) 0.5 mm (d) 0.1 mm
16. The ventilating ducts in most of the machines are
    (a) only radial (b) only longitudinal
    (c) both radial and longitudinal (d) neither radial nor longitudinal
17. The yoke of a dc machine
    (a) must be made of magnetic material
    (b) should preferably be of magnetic material but can be made of non-magnetic material
    (c) is always made of non-magnetic material
    (d) is partially made of magnetic material and partially of non-magnetic material
18. A commutator is made up of
    (a) iron laminations (b) copper segments
    (c) both iron laminations and copper segments (d) insulating material
19. In a dc machine, the coil span
    (a) must be equal to pole pitch (b) may or may not be equal to pole pitch
    (c) can never be equal to pole pitch (d) is equal to half pole pitch
20. The commutator pitch of a lap winding for a dc machine is
    (a) +1 (b) −1
    (c) +1 or −1 (d) none of the above
21. Equaliser rings are used in
    (a) lap wound machines (b) wave wound machines
    (c) both lap wound and wave wound machines (d) none of the above

22. Interpoles are used in
    (a) lap wound machines
    (b) wave wound machines
    (c) both lap and wave wound machines
    (d) none of the above
23. The interpole winding consists of
    (a) a large number of turns of thin wire
    (b) a small number of turns of thin wire
    (c) a large number of turns of thick wire
    (d) a small number of turns of thick wire
24. The mmf of interpole is proportional to
    (a) armature current
    (b) field current
    (c) product of armature current and field current
    (d) ratio of armature current to field current
25. The induced emf of a dc machine is
    (a) directly proportional to speed
    (b) inversely proportional to speed
    (c) proportional to square of speed
    (d) proportional to square root of speed
26. A 4-pole dc machine has a lap winding. This winding is removed and then a wave winding with the same number of turns is put. The induced emf will
    (a) increase
    (b) decrease
    (c) remain the same
    (d) may increase or decrease
27. For the production of induced emf, the field system of an electric machine
    (a) must be on stator
    (b) may be on stator or rotor
    (c) must be on rotor
    (d) should be preferably on rotor
28. The reactance voltage is
    (a) directly proportional to armature current and time of commutation
    (b) indirectly proportional to armature current and directly proportional to time of commutation
    (c) directly proportional to armature current and indirectly proportional to time of commutation
    (d) indirectly proportional to armature current and time of commutation
29. In a dc generator, the polarity of an interpole is
    (a) the polarity of the next main pole
    (b) the polarity of main pole immediately following the interpole
    (c) the polarity of main pole opposite to the interpole
    (d) none of the above
30. For providing proper commutation it is advisable to use
    (a) emf commutation
    (b) resistance commutation
    (c) both emf and resistance commutation
    (d) either of two but not both
31. In a dc generator, emf commutation can be achieved by
    (a) keeping the brushes in gnp
    (b) shifting the brushes ahead of gnp
    (c) shifting the brushes forward
    (d) shifting the brushes backward
32. The number of conductors of the compensating winding in a dc machine
    (a) is always more than the number of armature conductors per pole
    (b) is always less than the number of armature conductors/pole
    (c) may be more or less than the number of armature conductors per pole
    (d) is always 1

33. The conductors of the compensating winding are housed
    (a) in slots of armature
    (b) in the slots in the pole faces
    (c) partly in armature slots and partly in pole faces
    (d) none of the above
34. If all other quantities are the same, the number of conductors of compensating winding
    (a) is less in lap wound machine than in a wave wound machine
    (b) is more in lap wound machine than in a wave wound machine
    (c) is the same in lap wound and wave wound machines
    (d) may be more or less in any of the machines
35. A dc generator has 6 poles. A brush shift of 6° actual means a brush shift of
    (a) 6° electrical (b) 18° electrical
    (c) 36° electrical (d) 2° electrical
36. A compensating winding is used in
    (a) all dc generators (b) all dc motors
    (c) special purpose dc machines (d) all dc generators and motors
37. A 220 V dc generator is run at full speed without any excitation. The open circuit voltage will be
    (a) 0 (b) about 2 V or so
    (c) about 50 V (d) 220 V
38. In a separately excited dc generator supplying rated load, the armature reaction
    (a) is always absent (b) is always present
    (c) may be sometimes present (d) is generally present
39. If residual magnetism is present in a dc generator, the induced emf at zero speed will be
    (a) 0 (b) small
    (c) the same as rated voltage (d) high
40. A dc shunt generator has a full-load voltage regulation of 10%. If the generator is separately excited and delivers rated load, the regulation will be
    (a) 10% (b) more than 10%
    (c) less than 10% (d) 0
41. A separately excited dc generator has an open circuit voltage $E_0$ for a certain value of field current. If the field current is doubled, the open circuit voltage
    (a) will be 2 $E_0$ (b) will be less than or more than 2 $E_0$
    (c) may be 2 $E_0$ or less than 2 $E_0$ (d) may be 2 $E_0$ or more than 2 $E_0$
42. A series generator can self-excite
    (a) only if load current is zero (b) only if load current is not zero
    (c) irrespective of the value of load current (d) none of above
43. A shunt generator can self-excite
    (a) only if resistance of the field circuit is less than critical value
    (b) only if resistance of the field circuit is more than critical value
    (c) irrespective of the value of the resistance in the field circuit
    (d) only if field is open circuited

44. When a shunt generator is delivering a certain load current, its terminal voltage
   (a) will always be less than open circuit voltage
   (b) will always be more than open circuit voltage
   (c) may be more or less than open circuit voltage
   (d) will be equal to open circuit voltage
45. The terminal voltage of a series generator is 150 V when the load current is 5 A. If the load current is increased to 10 A, the terminal voltage will be
   (a) 150 V (b) less than 150 V
   (c) more than 150 V (d) any of above
46. The open circuit voltage of a compound generator is 250 V. At full-load the terminal voltage
   (a) will always be less than 250 V (b) will always be 250 V
   (c) may be greater or less than 250 V (d) may be 250 V or less or more
47. Which generators require an equaliser bar for parallel operation
   (a) series (b) shunt
   (c) compound (d) series and compound
48. Two dc shunt generators are operating in parallel. It is desired to shut down one of the generators. The proper method is
   (a) to open its main switch suddenly (b) to open its main switch slowly
   (c) to reduce its excitation to zero (d) any of above
49. The load on an over compounded dc generator is reduced to zero. Its terminal voltage
   (a) will increase (b) will decrease
   (c) may increase or decrease (d) any of above
50. A series generator is supplying its full load. A diverter is connected across the field. The diverter switch is initially open. If the diverter switch is closed, the terminal voltage
   (a) will increase (b) will decrease
   (c) will remain the same (d) any of above
51. Two dc shunt generators of same rating are operating in parallel. It is desired that they should share the load current equally. This is possible
   (a) if their open circuit emfs are equal (b) if their external characteristics are similar
   (c) if their open circuit emfs are equal and their external characteristics are similar (d) none of the above
52. The torque developed by a dc motor is proportional to
   (a) flux (b) armature current
   (c) flux and armature current (d) none of the above
53. A dc machine is required to rotate in both directions. The brushes should be
   (a) on g.n.p (b) ahead of g.n.p
   (c) shifted backward (d) any of the above
54. When the armature current of dc series motor is small, the torque is proportional to
   (a) $(I_a)^{0.5}$ (b) $I_a$
   (c) $I_a^2$ (d) $(I_a)^{1.5}$

55. The resistance of field diverter of a dc series motor is around
(a) 100 ohm (b) 10 ohm
(c) 2 ohm (d) 0.2 ohm
56. The resistance of shunt field is about
(a) 100 ohm (b) 10 ohm
(c) 2 ohm (d) 0.2 ohm
57. A dc series motor is delivering constant torque. The diverter switch is initially closed. If the diverter switch is opened, the speed
(a) will increase (b) will decrease
(c) may increase or decrease (d) may increase or decrease or remain the same
58. As the load supplied by a dc shunt motor is increased, its speed
(a) decreases slightly (b) decreases significantly
(c) remains absolutely constant (d) any of the above
59. It is desired to control the speed of a dc shunt motor such that speeds less than rated speed are obtained. The proper method of speed control is
(a) field control (b) armature resistance control
(c) shunted armature control (d) armature resistance control or shunted armature control
60. The speed control method generally used in traction motors is
(a) field diverter control (b) armature resistance control
(c) shunted armature control (d) series-parallel control
61. Modern starters used in dc shunt motors
(a) are manual (b) use time delay relays
(c) use back emf sensing relays (d) use time delay relays or back emf sensing relays
62. The electric braking which leads to conservation of energy is
(a) plugging (b) dynamic braking
(c) regenerative braking (d) none of the above
63. It is not advisable to start a series motor on no-load, otherwise
(a) excessive sparking may occur at brushes (b) line current may be very high
(c) speed may be very high (d) all of the above
64. A series motor is driving a load whose torque varies as square of the speed. When speed is 250 rpm, the line current is 15 A. If speed is 500 rpm, the line current is likely to be
(a) 15A (b) 21.2 A
(c) 30 A (d) 60 A
65. A dc shunt generator has a speed of 800 rpm when delivering 20 A at 220 V. If this machine is run as a motor and takes a line current of 20 A from 220 V supply, the speed will be
(a) 800 rpm (b) less than 800 rpm
(c) more than 800 rpm (d) 800 rpm or less
66. Swinburne test can be used only for
(a) series motor (b) shunt motor
(c) series and shunt motor (d) shunt and compound motor

67. A dc machine has an efficiency of 0.88 when delivering 20 A at 250 V. If it is used as motor taking 20 A from 250 V supply, the efficiency will be

(a) 0.88 (b) less than 0.88
(c) 0.88 or less (d) more than 0.88

68. The maximum losses in a dc machine are

(a) copper-losses (b) iron-losses
(c) mechanical losses (d) field losses

69. In Hopkinson test on two shunt machines

(a) the armature currents of both the machine are equal
(b) the armature current of generator is less than that of motor
(c) the armature current of motor is less than that of generator
(d) the armature current of motor may be more or less than that of generator

70. Two dc machines are tested by Hopkinson test. The rating of each machine is 250 kW. The power input during Hopkinson test would be around

(a) 250 kW (b) 375 kW
(c) 500 kW (d) 50 kW

71. A crossfield dc machine has

(a) only one set of brushes on $q$-axis (b) only one set of brushes on $d$-axis
(c) one set of brushes on $d$-axis and another set on $q$-axis (d) any of above

72. The number of turns in the primary winding of a single-phase transformer depends on

(a) applied voltage (b) line current
(c) both applied voltage and current (d) applied voltage, current and efficiency

73. A 50 kVA single-phase 230/2000 V transformer is fed at 230 V and is supplying 50 kVA load at 0.8 p.f. lagging. The output voltage will be

(a) 2000 V (b) 2000 V or more
(c) less than 2000 V (d) more than 2000 V

74. It is necessary that magnetic coupling of a transformer is close. This is done to ensure

(a) high efficiency (b) good regulation
(c) good regulation and high efficiency (d) good regulation and high power factor

75. The purpose of using laminated core in a transformer is

(a) to reduce copper-losses (b) to reduce all losses
(c) to reduce hysteresis losses (d) to reduce eddy current losses

76. A 10 kVA, 200/400 V, single-phase transformer is supplying its rated current of 25A. The input current is likely to be

(a) 50 A (b) less than 50 A
(c) more than 50 A (d) any one of the above

77. A 10 kVA, 200/400 V single-phase transformer is operating with secondary open circuited. The input current should be about

(a) 2 A (b) 10 A
(c) 20 A (d) 50 A

78. The power factor at primary terminals, with the secondary of a two-winding transformer open circuited, is about

(a) 0.2 (b) 0.5
(c) 0.8 (d) 1

79. A 200/400 ratio transformer has a secondary winding resistance of 0.5 ohm. The total resistance as referred to primary should be about

(a) 0.25 ohm (b) 0.5 ohm
(c) 1 ohm (d) 2 ohm

80. The secondary winding of a transformer is short circuited. It is required to circulate full-load rated current through the windings. The voltage applied to primary should be about

(a) 0.1 pu (b) 0.5 pu
(c) 0.8 pu (d) 1 pu

81. The regulation of a two-winding transformer is 5% at full-load unity p.f. At full-load 0.8 p.f lagging the regulation would be

(a) 5% (b) less than 5%
(c) more than 5% (d) 5% or more than 5%

82. The full-load efficiency of a power transformer is about

(a) 0.75 (b) 0.85
(c) 0.9 (d) 0.98

83. When the secondary winding of a transformer is short circuited, the power factor at primary terminals is about

(a) 0.2 lagging (b) 0.2 leading
(c) 0.8 lagging (d) 0.8 leading

84. The secondary winding of a transformer is open circuited. The primary is connected to a dc voltage equal to the rated primary voltage. The primary current would be about

(a) 50 pu (b) 10pu
(c) 1 pu (d) 0.1 pu

85. An auto-transformer results in saving of material if

(a) turn ratio is high (b) turn ratio is low
(c) rating of transformer is high (d) rating of transformer is low

86. A 10 kVA, 100/300 V two-winding transformer is used as an auto-transformer by connecting the two windings in series to supply a 300 V load from a 400 V ac supply. The maximum kVA which can be supplied by auto-transformer is

(a) 10 kVA (b) less than 10 kVA
(c) 40 kVA (d) 100 kVA

87. An auto-transformer is used to step down voltage $V_1$ to $V_2$. An open circuit develops in the winding. The maximum voltage across the load may become

(a) $V_1 + V_2$ (b) $V_1 - V_2$
(c) $2V_1$ (d) $V_1$

88. An auto-transformer is used to step down a voltage $V_1$ to $V_2$. The input current is $I_1$ and the output current is $I_2$. The volt-amperes transferred conductively from input to output are

(a) $V_1 I_1$ (b) $V_2 I_1$

(c) $V_2I_2 - V_2I_1$ (d) None of the above

89. An open circuit test is carried out on a 100/500 V transformer. The 500 V winding is open circuited and 100 V is applied to the other winding. The input power is 200 W. If the open circuit test is repeated with 100 V winding open circuited and 500 V applied to the other winding, the input power would be
(a) 200 W (b) 1000 W
(c) 40 W (d) 8 W

90. A two-winding transformer has iron-losses and full-load copper-losses equal to $W$ watts each. It is tested along with another similar transformer for back-to-back test. The total power input would be
(a) 4 W (b) 2 W
(c) W (d) 0.5 W

91. During a short circuit test on transformer the iron-losses are negligible because
(a) current is high (b) mutual flux is small
(c) frequency is low (d) power factor is low

92. Two transformers $A$ and $B$, having the same rating, are operated in parallel. The per unit impedance of $B$ is higher than that of $A$. Then
(a) they will share the load equally (b) load on $A$ will be more than that on $B$
(c) load on $B$ will be more than that on $A$ (d) load on $B$ may be more or less than that on $A$

93. A 100 kVA and 300 kVA transformer are operated in parallel. It is desired that they should share the load in proportion to their ratings. Then
(a) their impedances should be equal (b) their p.u impedances should be equal
(c) their copper-losses should be equal (d) their efficiencies should be equal

94. The function of tubes welded to the sides of transformers is
(a) to improve the cooling (b) to improve voltage regulation
(c) to improve efficiency (d) all the above

95. One primary and one secondary terminal of a two winding transformer are connected to an ohm-meter. The reading of ohm-meter will be
(a) 0 (b) infinity
(c) 0 or infinity (d) equal to winding resistance

96. Which of the following combinations of 3-phase transformers can operate successfully in parallel?
(a) $\Delta - Y$ and $\Delta - Y$ (b) $Y - Y$ and $\Delta - Y$
(c) $\Delta - \Delta$ and $\Delta - Y$ (d) $Y - \Delta$ and $\Delta - \Delta$

97. The current rating of secondary winding of a C T is generally
(a) 100 A (b) 50 A
(c) 10 A (d) 5 A

98. The primary and secondary windings of C T are closely coupled
(a) true (b) false

99. If the transformer window is long and narrow, the leakage flux reduces
(a) true (b) false

100. A transformer has a voltage regulation of 5% at full-load power factor 0.8 lagging. The regulation at full-load 0.8 leading p.f. would be less than 5%
(a) true (b) false

101. The all-day efficiency of a transformer depends on
(a) copper-losses (b) iron-losses
(c) load cycle (d) copper-losses, iron-losses and load cycle
102. The polarity of a transformer is important when
(a) the transformer is switched on to supply (b) the transformer is switched off
(c) the transformer is to be operated in parallel with another transformer
(d) all of the above
103. Three-phase transformer connections most commonly used are:
(a) $\Delta - \Delta$ (b) $Y - Y$
(c) $\Delta - Y$ and $Y - \Delta$ (d) $\Delta - \Delta$ and $\Delta - Y$
104. A 3-phase transformer is composed of three separate transformers connected in $\Delta - \Delta$. Its rating is 100 kVA. If one of the units gets damaged and it is operated in open delta, the rating will be
(a) 100 kVA (b) 66.6 kVA
(c) 57.7 kVA (d) 50 kVA
105. The magnitude of third harmonic in the waveshape of magnetising current of a transformer is about
(a) 50% (b) 20%
(c) 5% (d) 1%
106. The maximum value of inrush current in a transformer depends on
(a) instant of switching (b) residual flux density
(c) instant of switching and residual flux density (d) none of above
107. Transformer taps are usually provided on low voltage winding
(a) true (b) false
108. Transformer oil serves the twin functions of cooling and insulation
(a) true (b) false
109. Modern day synchronous machines use double layer winding
(a) true (b) false
110. In modern synchronous machines the phase spread is
(a) 30° (b) 60°
(c) 120° (d) 360°
111. Fractional pitch windings result in
(a) higher terminal voltage (b) better voltage waveform and saving in material
(c) higher efficiency (d) higher power factor
112. In synchronous machines, fractional slot windings are more common than integral slot windings
(a) true (b) false
113. The breadth factor for a 3-phase ac winding is about
(a) 0.96 (b) 0.85
(c) 0.7 (d) 0.5
114. The breadth factor for third harmonic is higher than that for fundamental
(a) true (b) false

115. The full-pitch coil, in a synchronous machine, has a span of 18 slots. To eliminate third harmonic, the coil span should be

(a) 9 slots (b) 12 slots

(c) 15 slots (d) 18 slots

116. The torque developed in a synchronous machine depends on

(a) stator mmf only (b) rotor mmf only

(c) both stator and rotor mmfs

(d) both stator and rotor mmf and the phase angle between these mmfs

117. If the thickness of laminations is $t$, the eddy current losses are proportional to

(a) $t$ (b) $t^2$

(c) $t^3$ (d) $\sqrt{t}$

118. An ac winding has 2 slots per pole per phase. The slot harmonics will be

(a) 5th and 7th (b) 11th and 13th

(c) 17th and 19th (d) none of the above

119. Modern high speed synchronous generators used in thermal plants have

(a) salient pole rotor (b) cylindrical rotor

(c) either salient pole or cylindrical rotor (d) none of the above

120. Salient pole machines have

(a) large number of poles and small length-to-diameter ratio

(b) small number of poles and small length-to-diameter ratio

(c) large number of poles and high length-to-diameter ratio

(d) any of above

121. Cylindrical rotor machines usually have horizontal configuration

(a) true (b) false

122. Turbo alternators use

(a) hydrogen cooling (b) air cooling

(c) nitrogen cooling (d) any of the above

123. In a hydrogen cooled synchronous machine, the pressure inside the machine

(a) is more than atmospheric pressure (b) is less than atmospheric pressure

(c) is equal to atmospheric pressure (d) is equal to or less than atmospheric pressure

124. The use of hydrogen cooling decreases the windage loss in a synchronous machine

(a) true (b) false

125. Under short circuit conditions, the power factor of a synchronous machine is

(a) 1 (b) about 0.8

(c) almost zero lagging (d) about 0.5

126. The short circuit characteristics of a synchronous machine

(a) is always linear

(b) is always non-linear

(c) is sometimes linear and sometimes non-linear

(d) depends upon the rating of the machine

127. In modern large size synchronous machines, the synchronous impedance is about
   (a) 0.2 pu (b) 0.5 pu
   (c) 1.0 pu (d) 0.05 pu
128. A 3-phase synchronous machine has an unsaturated synchronous reactance equal to 3 ohm per phase. The saturated synchronous reactance
   (a) will be more than 3 ohm (b) will be less than 3 ohm
   (c) may be more or less than 3 ohm (d) will be 3 ohm or more than 3 ohm
129. Modern day large size synchronous generators are usually operated at a power angle of about
   (a) 10° (b) 40°
   (c) 70° (d) 90°
130. If the power factor of an alternator is 0 lagging, the armature reaction is
   (a) demagnetising (b) magnetising
   (c) cross-magnetising (d) demagnetising as well as cross-magnetising
131. An alternator is delivering its rated current at rated voltage and 0.8 p.f. lagging. If it is required to deliver rated current at rated voltage and 0:8 leading p.f., the required excitation will be
   (a) more (b) less
   (c) more or less (d) the same
132. The short circuit ratio of a modern alternator is about
   (a) 0.1 (b) 0.3
   (c) 0.8 (d) 3
133. In two-reaction theory, the armature current is decomposed into two components $I_d$ and $I_p$. If $E$ and $V$ are the generated emf and terminal voltage of the machine, then
   (a) $I_d$ is perpendicular to $V$ and $I_q$ is in phase with $V$
   (b) $I_d$ is in phase with $E$ and $I_q$ is perpendicular to $E$
   (c) $I_d$ is perpendicular to $E$ and $I_q$ is in phase with $E$
   (d) $I_d$ is in phase with $V$ and $I_q$ is perpendicular to $V$
134. The reluctance power of a synchronous machine does not depend on field excitation
   (a) true (b) false
135. A salient pole synchronous machine has maximum power output when power angle δ is
   (a) 90° (b) less than 90°
   (c) either 90° or more (d) more than 90°
136. An alternator is feeding an infinite bus bar. Its prime mover is suddenly shut down. The alternator will
   (a) continue to work as alternator but the direction of rotation will reverse
   (b) come to standstill
   (c) continue to work as synchronous motor and direction of rotation will also be the same
   (d) start working as an induction motor
137. When an alternator is feeding an infinite bus bar
   (a) the excitation controls reactive power output and the governor setting controls real power output
   (b) the excitation controls the terminal voltage and the governor setting controls the real power output
   (c) the excitation controls both the real and reactive power output
   (d) the governor setting controls the terminal voltage

138. If a synchronous machine does not have damper winding, then

(a) $x_d' = x_d''$ (b) $x_d' = x_d$

(c) $x_d = x_d''$ (d) $x_d = x_d' = x_d''$

139. In a synchronous machine

(a) $x_d'' < x_d' < x_d$ (b) $x_d < x_d' < x_d''$

(c) $x_d \cong x_d'$ (d) $x_d = x_2$

140. The reactive power output of a cylindrical rotor alternator is maximum when

(a) $\delta = 90°$ (b) $\delta = 45°$

(c) $\delta = 0$ (d) $\delta$ is slightly less than 90°

141. When the real power output of a cylindrical rotor alternator is maximum, the reactive power output is

(a) $\dfrac{3V^2}{x_s}$ (b) $\dfrac{-3V^2}{x_s}$

(c) $\dfrac{3EV}{x_s}$ (d) $\dfrac{3E^2}{x_s}$

142. An under excited synchronous motor operates at

(a) lagging pf (b) unity pf

(c) leading pf (d) lagging power factor at low loads and leading power factor at high loads

143. The function of amortisseur winding in a synchronous motor is

(a) to prevent hunting (b) to provide starting torque

(c) to improve power factor (d) to prevent hunting and provide starting torque

144. When a synchronous motor is started

(a) the field winding is excited by a dc voltage (b) the field winding is open circuited initially

(c) the field winding is short circuited initially (d) any of the above

145. A synchronous motor is provided with amortisseur winding. Its field winding is short circuited and the stator is switched on to a 3-phase ac voltage. Then

(a) the machine will not start

(b) the machine will start and its speed will become equal to synchronous speed

(c) the machine will start but its speed will not be able to reach the synchronous value

(d) the machine may or may not start

146. For a fixed load on the shaft, the armature current of a synchronous motor is minimum when the power factor is

(a) less than 1 and leading (b) unity

(c) lagging (d) leading or lagging but not unity

147. A synchronous motor is supplying its rated load and is operating at unity power factor. If excitation is increased, then

(a) its load will increase (b) its power factor would remain the same

(c) its power factor would become lagging (d) its power factor would become leading

148. Synchronous motor are generally salient pole machines

(a) true (b) false

149. *V* curves of synchronous motor show the relation between
(a) armature current and terminal voltage
(b) armature current and load
(c) armature current and field current
(d) all the above

150. A synchronous motor is fed at its rated voltage and is supplying half its rated load and is operating at unity power factor. It the load is increased to full-load and the terminal voltage and excitation remain the same, then
(a) its power factor will become lagging
(b) its power factor will become leading
(c) its power factor will remain the same
(d) its power factor may remain unity or may become leading

151. As the load on a synchronous motor increases, the torque angle
(a) increases
(b) decreases
(c) remains the same
(d) may increase or decrease

152. A salient pole synchronous motor is operating at no-load. If excitation is suddenly switched off
(a) the motor will come to a stop
(b) the motor will continue to run at a speed less than synchronous speed
(c) the motor will continue to run at synchronous speed
(d) the motor may or may not continue to run

153. In a synchronous motor
(a) *E* is always less than *V*
(b) $E = V$
(c) *E* is always more than *V*
(d) *E* may be more or less than *V*

154. In a synchronous motor
(a) the speed of stator mmf is always more than that of rotor mmf
(b) rotor and stator mmf are stationary with respect to each other
(c) speed of stator mmf is always less than that of rotor mmf
(d) none of the above

155. A 50 Hz 3-phase induction motor has a full-load speed of 1440 rpm. The number of poles of the motor are
(a) 4
(b) 6
(c) 12
(d) 8

156. The usual value of slip of a 3-phase induction at no-load is about
(a) 0.01
(b) 0.05
(c) 0.1
(d) 0.3

157. The usual value of slip of a 3-phase induction motor at full-load is about
(a) 0.05
(b) 0.1
(c) 0.3
(d) 0.8

158. In a 3-phase induction motor
(a) the speed of stator field is twice the speed of rotor field
(b) the speed of stator field is very high as compared to speed of rotor field
(c) the speed of rotor field is the same as that of a stator field
(d) the speed of stator field is very low as compared to that of rotor field

159. In a 3-phase induction motor at full-load
(a) the frequency of currents in stator and rotor are the same

(b) the frequency of stator current is very low as compared to that of rotor current

(c) the frequency of rotor current is very low as compared to that of stator current

(d) the frequency of rotor current is slightly less than that of stator current

160. At full-load the speed of an induction motor

(a) is very low as compared to its synchronous speed

(b) is the same as synchronous speed

(c) is half of its synchronous speed

(d) is slightly less than its synchronous speed

161. The emf induced per phase is the rotor winding of a 3-phase induction motor is 100 V at standstill. Under full-load condition, this emf would be about

(a) 100 V (b) 50 V

(c) 4 V (d) 0.2 V

162. For a 20 kW 3-phase 400 V induction motor, the no-load current would be about

(a) 100 A (b) 40 A

(c) 10 A (d) 2 A

163. In a 3-phase induction motor, the rotor core losses are very high

(a) true (b) false

164. If the stator winding of a 3-phase induction motor is delta connected, the rotor winding

(a) should be delta connected (b) should be star connected

(c) should not be delta connected (a) may be star or delta connected

165. A 3-phase cage rotor induction motor has its stator wound for 8 poles. The rotor of this motor can be used in any other induction motor of the same size and having

(a) 4 poles (b) 8 poles

(c) 2 poles (d) any number of poles

166. In a 3-phase induction motor

(a) both stator and rotor cores are laminated (b) only stator core is laminated

(c) only rotor core is laminated (d) the stator core is laminated but the rotor core may or may not be laminated

167. Under blocked rotor condition the frequency of rotor currents in a 50 Hz 3-phase induction motor is

(a) very low (b) very high

(c) 50 (d) about 50

168. At $s = 0$, the torque of an induction motor is

(a) 0 (b) equal to full-load

(c) very high (d) nearly zero

169. Under full-load condition, the power factor of rotor circuit of a 3-phase induction motor is

(a) about 0.2 lagging (b) about 0.5 lagging

(c) almost unity (d) unity

170. An induction motor takes 40 seconds to reach its full-load if started by direct switching. If the motor is started by star-delta starter, the time taken by the motor to reach full speed will be

(a) 40 seconds (b) less than 40 seconds

(c) more than 40 seconds (d) 40 seconds or less than 40 seconds

171. An induction motor takes a line current of 60 A when started by direct switching. If an auto-transformer with 50% tapping is used, the line current would be

(a) 60 A (b) 30 A
(c) 15 A (d) 10 A

172. The speed regulation of a 3-phase induction motor from no-load to full load is about

(a) 40% (b) 20%
(c) 10% (d) 3%

173. A 3-phase induction motor has a starting torque of 150 N-m when switched on directly to supply. If a star-delta starter is used for starting, the starting torque would be

(a) 200 N-m (b) 150 N-m
(c) 100 N-m (d) 50 N-m

174. A 3-phase induction motor has a starting torque of 200 N-m when switched on directly to supply. If an auto-transformer with 50% taping is used for starting, the starting torque would be

(a) 400 N-m (b) 200 N-m
(c) 100 N-m (d) 50 N-m

175. A 3-phase induction motor takes a line current of 30 A when started by direct switching. If a star-delta starter is used, the line current would be

(a) 30 A (b) 17.32 A
(c) 10 A (d) 5 A

176. The best method of control the speed of a 3-phase induction motor over a wide range is

(a) line frequency control (b) rotor resistance control
(c) pole changing (d) none of the above

177. An induction generator does not have a field winding

(a) true (b) false

178. The power factor of an induction generator at full-load is about

(a) 0.8 lagging (b) 1
(c) 0.8 leading (d) 0.2 lagging

179. In a double cage induction motor

(a) the outer cage has higher resistance and lower inductance as compared to inner cage
(b) the outer cage has higher resistance and inductance as compared to inner age
(c) the outer cage has lower resistance and inductance as compared to inner cage
(d) the outer cage has lower resistance and higher inductance as compared to inner cage

180. A single-phase induction motor has

(a) two windings on stator (b) one winding on stator and one on rotor
(c) may have one or two windings on stator (d) two windings on rotor

181. The rotor of a single-phase induction motor

(a) may be a cage rotor or a wound rotor (b) is always a cage rotor
(c) is always a wound rotor (d) has a double cage

182. According to double revolving field theory, a pulsating field can be resolved into two revolving fields. The magnitude of each of these component fields is

(a) equal to the amplitude of pulsating field

(b) half the amplitude of pulsating field
(c) two-third the amplitude of the pulsating field
(d) none of the above

183. According to double revolving field theory a single-phase induction motor can be considered as equivalent of two hypothetical constituent motors. If the slip of one of these motors is $s$, the slip of the second is
(a) $s$ (b) $2s$
(c) $2-s$ (d) $1-s$

184. When a single-phase induction motor is operating at full-load, the torques due to forward and backward fields are equal
(a) true (b) false

185. In a single-phase induction motor, it is necessary that the auxiliary winding is disconnected after the motor has come upto rated speed
(a) true (b) false

186. In a split-phase motor
(a) *X/R* ratio of main winding is higher as compared to that of auxiliary winding
(b) *X/R* ratio of main winding is lower as compared to that of auxiliary winding
(c) *X/R* ratio of the two winding are exactly equal
(d) *X/R* ratio of the two winding are nearly equal

187. In a capacitor start motor, the capacitor is connected
(a) in series with both the windings (b) in series with auxiliary winding
(c) in series with main winding (d) in parallel with auxiliary winding

188. A capacitor start, capacitor run motor has two capacitors connected in
(a) parallel (b) series
(c) series at the time of starting and in parallel during running (d) parallel at the time of starting and in series during running

189. As compared to split phase motor, a capacitor start motor has higher starting torque
(a) true (b) false

190. A capacitor start, capacitor run motor has two capacitors
(a) both the capacitors are in the circuit during starting and running
(b) one of the capacitors is in the circuit during starting and the other is in the circuit during running
(c) both the capacitors are in the circuit during starting and one of them during running
(d) one of the capacitors is in the circuit during starting and both are in the circuit during running

191. If the centrifugal switch of a capacitor start motor fails to open,
(a) the capacitor will be destroyed
(b) the auxiliary winding will be destroyed
(c) the capacitor and auxiliary winding will be destroyed
(d) none of the above

192. A capacitor start, capacitor run motor is essentially a two-phase moor during starting as well as running
(a) true (b) false

193. A ceiling fan uses
(a) split phase motor (b) capacitor start capacitor run motor

(c) capacitor start motor
(d) universal motor

194. The motor used in electric toys is
(a) capacitor start motor
(b) split phase motor
(c) permanent split capacitor motor
(d) shaded pole motor

195. Which of the following does not use a centrifugal switch
(a) capacitor start motor
(b) capacitor start capacitor run motor
(c) permanent split capacitor motor
(d) split phase motor

196. In a capacitor start motor, the capacitor is replaced by an inductor having the same reactance as capacitor. Then
(a) the motor will start but run at a slow speed
(b) the motor will not start
(c) the motor will start and run at its rated speed
(d) the motor will be damaged

197. In a permanent split capacitor motor, the auxiliary winding is designed for a short time rating
(a) true
(b) false

198. In a capacitor start capacitor run motor, the auxiliary winding is designed for continuous rating
(a) true
(b) false

199. A schrage motor is an adjustable speed, 3-phase motor
(a) true
(b) false

200. A schrage motor does not require external resistances for starting or running
(a) true
(b) false

201. A dc series motor is switched on to ac supply
(a) the motor will not start
(b) the motor will be damaged
(c) the motor will start and run
(d) the motor will start and run but will show poor performance

202. In a single-phase series motor
(a) both stator and rotor cores are laminated
(b) only stator core is laminated
(c) only rotor core is laminated
(d) neither of the cores is laminated

203. The function of compensating winding in an ac series motor is
(a) to provide starting torque
(b) to reduce reactance of armature winding
(c) to improve the efficiency of the machine
(d) to convert it into a two-phase motor

204. An ac series motor is switched on to dc supply of rated voltage. The speed of motor will be
(a) higher than that under ac operation
(b) lower than that under ac operation
(c) equal or less than that under ac operation
(d) equal to that under ac operation

205. The electric motor generally used in house-hold food mixers is
(a) universal motor
(b) shaded pole motor
(c) capacitor start motor
(d) none of the above

206 The no-load speed of a Universal motor is limited by windage and friction
(a) true
(b) false

207. A repulsion motor uses a commutator
(a) true
(b) false

208. A repulsion motor does not have a starting torque

(a) true (b) false

209. A reluctance motor does not have a field winding

(a) true (b) false

210. A reluctance motor runs at

(a) synchronous speed (b) a speed slightly less than synchronous speed

(c) half the synchronous speed (d) above synchronous speed

211. A two-phase servo motor has

(a) wound rotor (b) cage rotor

(c) wound or cage rotor (d) a rotor similar to that in dc motors

212. A two-phase servo motor is always used

(a) in fractional h.p. range (b) in big sizes

(c) in both high rating and fractional h.p. rating (d) mostly in high ratings and sometimes in fractional h.p. rating

213. The voltages fed to the two windings of a two-phase servo motor must

(a) be in phase (b) have a phase difference of 90°

(c) have a phase difference more than 90° (d) have a phase difference of 180°

214. A two-phase servo motor develops maximum torque at

(a) synchronous speed (b) a speed less than synchronous speed

(c) a reverse speed of one-half synchronous speed

(d) a forward speed of one-half the synchronous speed

215. A stepper motor is

(a) a dc motor (b) a single-phase ac motor

(c) a two-phase motor (d) a multi-phase motor

216. The characteristic of a stepper motor is usually drawn between

(a) torque and speed (b) torque and voltage

(c) torque and current (d) torque and pulse rate

217. A synchro is used

(a) to start a stepper motor (b) to maintain synchronism between two shafts

(c) to start a schrage motor (d) to maintain constant voltage for various ac appliances

218. A synchro has

(a) a three-phase winding on rotor and a single-phase winding on stator

(b) a three-phase winding on stator and a commutator winding on rotor

(c) a three-phase winding on stator and a single-phase winding on rotor

(d) a single-phase winding on stator and a commutator winding on rotor

219. A fully controlled converter can be operated as

(a) rectifier only (b) both rectifier and inverter

(c) inverter only (d) none of the above

220. A chopper converts

(a) a constant dc voltage into a variable dc voltage

(b) a variable dc voltage into a constant dc voltage
(c) a constant ac voltage into a variable ac voltage
(d) a variable ac voltage into a constant dc voltage

## ANSWERS TO OBJECTIVE QUESTIONS

| | | | |
|---|---|---|---|
| 1 (b) | 2 (a) | 3 (b) | 4 (c) |
| 5 (c) | 6 (c) | 7 (a) | 8 (c) |
| 9 (b) | 10 (d) | 11 (b) | 12 (c) |
| 13 (b) | 14 (a) | 15 (c) | 16 (c) |
| 17 (a) | 18 (b) | 19 (b) | 20 (c) |
| 21 (a) | 22 (c) | 23 (d) | 24 (a) |
| 25 (a) | 26 (a) | 27 (b) | 28 (c) |
| 29 (a) | 30 (c) | 31 (b) | 32 (b) |
| 33 (b) | 34 (a) | 35 (b) | 36 (c) |
| 37 (b) | 38 (b) | 39 (a) | 40 (c) |
| 41 (c) | 42 (b) | 43 (a) | 44 (a) |
| 45 (c) | 46 (d) | 47 (c) | 48 (c) |
| 49 (b) | 50 (b) | 51 (c) | 52 (c) |
| 53 (a) | 54 (c) | 55 (d) | 56 (a) |
| 57 (b) | 58 (a) | 59 (d) | 60 (d) |
| 61 (d) | 62 (c) | 63 (c) | 64 (c) |
| 65 (b) | 66 (d) | 67 (c) | 68 (a) |
| 69 (b) | 70 (d) | 71 (c) | 72 (a) |
| 73 (c) | 74 (d) | 75 (d) | 76 (c) |
| 77 (a) | 78 (a) | 79 (a) | 80 (a) |
| 81 (c) | 82 (d) | 83 (a) | 84 (a) |
| 85 (b) | 86 (c) | 87 (d) | 88 (b) |
| 89 (a) | 90 (a) | 91 (b) | 92 (b) |
| 93 (b) | 94 (a) | 95 (b) | 96 (a) |
| 97 (d) | 98 (b) | 99 (a) | 100 (a) |
| 101 (d) | 102 (c) | 103 (c) | 104 (c) |
| 105 (b) | 106 (c) | 107 (b) | 108 (a) |
| 109 (a) | 110 (b) | 111 (b) | 112 (a) |
| 113 (a) | 114 (b) | 115 (b) | 116 (d) |
| 117 (b) | 118 (b) | 119 (b) | 120 (a) |
| 121 (a) | 122 (a) | 123 (a) | 124 (a) |
| 125 (c) | 126 (c) | 127 (c) | 128 (b) |
| 129 (b) | 130 (a) | 131 (b) | 132 (c) |
| 133 (c) | 134 (a) | 135 (b) | 136 (c) |
| 137 (a) | 138 (a) | 139 (a) | 140 (c) |
| 141 (b) | 142 (a) | 143 (d) | 144 (c) |
| 145 (c) | 146 (b) | 147 (d) | 148 (a) |
| 149 (c) | 150 (a) | 151 (a) | 152 (b) |
| 153 (d) | 154 (b) | 155 (a) | 156 (a) |
| 157 (a) | 158 (c) | 159 (c) | 160 (d) |
| 161 (c) | 162 (c) | 163 (b) | 164 (d) |
| 165 (d) | 166 (d) | 167 (c) | 168 (a) |

169 (c)
170 (c)
171 (c)
172 (d)
173 (d)
174 (d)
175 (c)
176 (a)
177 (a)
178 (c)
179 (a)
180 (a)
181 (b)
182 (b)
183 (c)
184 (b)
185 (b)
186 (a)
187 (b)
188 (a)
189 (a)
190 (c)
191 (c)
192 (a)
193 (c)
194 (d)
195 (c)
196 (b)
197 (b)
198 (a)
199 (a)
200 (a)
201 (d)
202 (a)
203 (b)
204 (a)
205 (a)
206 (a)
207 (a)
208 (b)
209 (a)
210 (a)
211 (b)
212 (a)
213 (b)
214 (c)
215 (d)
216 (d)
217 (b)
218 (c)
219 (b)
220 (a)

# Appendix B

## THREE PHASE CIRCUITS

### B.1 Introduction

AC systems have the advantage that it is possible to have two, three or more circuits in the same device. All these circuits operate at the same frequency but their voltage and currents are out of phase from one another.

All present day power systems are 3 phase systems. The generation, transmission and distribution systems are all 3 phase systems. Bulk of utilisation is also at 3-phase level. However single phase is used for domestic and commercial installations.

The three voltages of a 3-phase supply system can be used to feed three independent single phase circuits. This is done in domestic and commercial sector. Alternatively these three voltages may feed a 3-phase device, e.g. 3-phase induction motor. Three phase machines are cheaper than single phase machines. Three phase transmission systems and distribution systems are cheaper than single phase systems.

### B.2 Generation of 3 Phase Voltages

A 3-phase generator can be thought of as three separate single phase generators within one common frame. It has a field system on the rotor and armature windings on the stator. As discussed in Chapter 5, the field system may be cylindrical or salient pole type. The field system provides the constant magnetic flux which is necessary for the generation of emf in armature conductors.

The stator of a 3-phase generator has three separate armature windings, one for each of the three phases. The axes of these three windings are displaced by 120° in space. The relative motion between the armature winding and magnetic flux induces emfs in the armature winding as per Faraday's law. The magnetic flux linking any phase winding varies sinusoidally with time. Therefore emfs varying sinusoidally are induced in all the three phase windings. Since the three armature windings are displaced by 120° in space, the three emf waves are displaced by 120° from each other. The magnitudes of these three emfs are equal because the three windings are exactly similar and the magnetic field system is common for all the three phases.

Each phase winding has two terminals. All the six terminals, of the 3-phase winding are brought out as shown in Fig. B-1. The terminals $a$, $b$, $c$ denote the beginnings of the three windings while $a'$, $b'$, $c'$ denote the ends. If the terminals $a'$, $b'$, $c'$ are joined together to form the neutral and the terminals $a$, $b$, $c$ are brought out, we get a 3-phase star connection. Alternatively we can connect $a'b$, $b'c$ and $c'a$ to give a delta connection.

In a balanced 3-phase system, the voltage (and the resulting currents) of the 3 phases are equal in magnitude (i.e. same rms value) and have a phase difference of 120° from one another. In an unbalanced system the magnitudes may not be equal or the phase angles may not be 120°. Most of the 3 phase systems are balanced systems or assumed to be so.

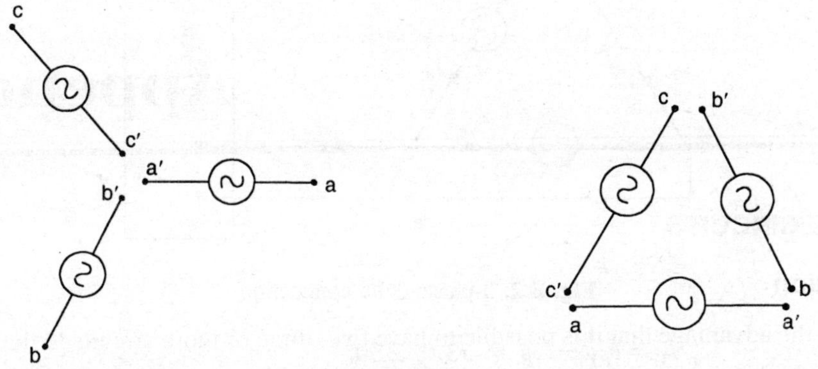


Fig. B.1

## B.3 Phase Sequence

The three terminals of 3 phase systems are denoted by abc or ABC or R Y B (red, yellow, blue). The symbols R Y B are very popular because in power stations and substations, the three leads are of red, yellow (or white) and blue colours for easy identification. The three voltages may reach their positive maximum values in the sequence R Y B or R B Y. In the sequence R Y B, the voltage of phase Y lags that of phase R by 120° and the voltage of phase B lags that of phase Y by 120° (Evidently voltage of phase B leads voltage of phase R by 240°). In the sequence R B Y, the voltage of phase B lags that of phase R by 120° and voltage of phase Y lags that of phase B by 120°. These two are referred to as phase sequence. R Y B is known as positive sequence while R B Y is known as negative sequence. The phase sequence of voltages across a three phase load is fixed by the order in which the three line terminals are connected to the load. An interchange of any two terminals reverses the phase sequence.

## B.4 Three-Phase Delta Connection

Figure B-2 shows a delta connection of 3 phase emfs. It is evident that voltage across each phase is equal to line voltage. Thus

$$V_L = V_p \quad \text{(B - 1)}$$

$I_{RY}$, $I_{YB}$ and $I_{BR}$ are the phase currents. The magnitudes of these three currents are equal but they are displaced from one another by 120°. Thus

$$I_{RY} = I_p \angle 0 \quad \text{(B - 2)}$$

$$I_{YB} = I_p \angle -120° \quad \text{(B - 3)}$$

$$I_{BR} = I_p \angle -240° = I_p \angle 120° \quad \text{(B - 4)}$$

The line currents are $I_R$, $I_Y$ and $I_B$. It is seen that

$$I_R + I_{RY} = I_{BR} \quad \text{(B - 5)}$$

or

$$I_R = I_{BR} - I_{RY} = I_P \angle -240° - I_p \angle 0$$

$$= \sqrt{3}\, I_p \angle 150° \quad \text{(B - 6)}$$

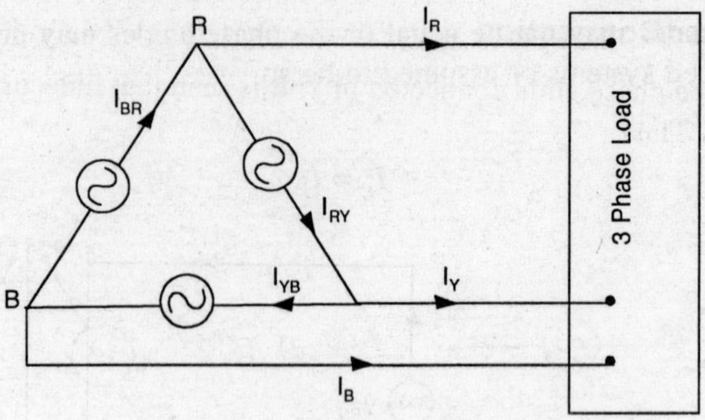


**Fig. B.2.** 3-phase delta connection

Similarly $$I_Y = I_{RY} - I_{YB} = \sqrt{3}\, I_p \angle 30° \quad \text{(B - 7)}$$

$$I_B = I_{YB} - I_{BR} = \sqrt{3}\, I_p \angle -90° \quad \text{(B - 8)}$$

It is seen that the three line currents are also balanced. The magnitude of each line current is $\sqrt{3}$ times the magnitude of phase current. Since the voltages and currents are balanced

$$V_{RY} + V_{YB} + V_{BR} = 0$$

$$I_{RY} + I_{YB} + I_{BR} = 0$$

$$I_R + I_Y + I_B = 0$$

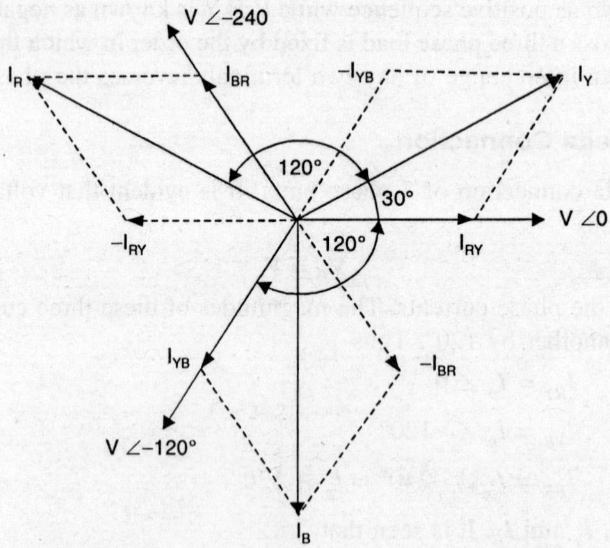


**Fig. B.3.** Phasor diagram for 3-phase delta connection

Phasor diagram is shown in Fig. B.3.

## B.5 Three Phase Star Connection

Figure B.4 shows the three phase emfs connected in Y. It is seen that line current and phase current are equal in this connection. Thus

$$I_L = I_P \tag{B - 9}$$

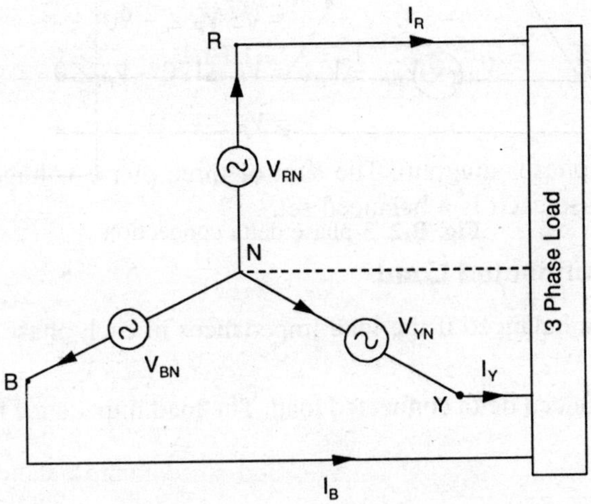


**Fig. B.4.** 3-phase star connection

Let the phase voltages be

$$V_{RN} = V_P \angle 0 \tag{B - 10}$$

$$V_{YN} = V_P \angle -120° \tag{B - 11}$$

$$V_{BN} = V_P \angle -240° = V_P \angle 120° \tag{B - 12}$$

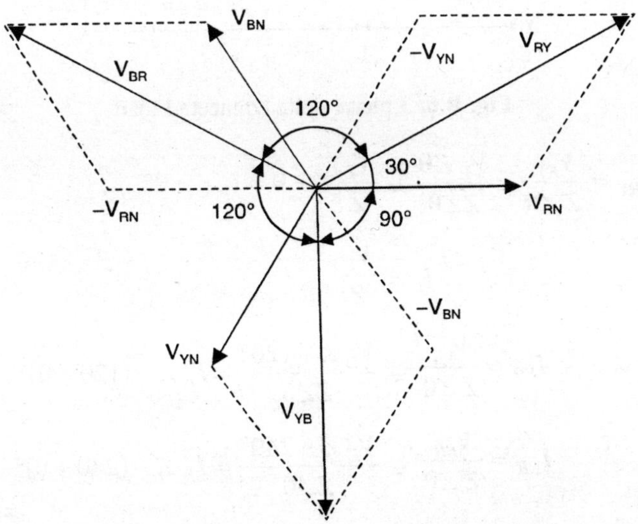


**Fig. B.5.** Phasor diagram for 3 phase star connection

$$V_{RY} = V_{RN} - V_{YN} = V_P\angle 0 - V_P\angle -120°$$
$$= \sqrt{3}\, V_P \angle 30° \qquad \text{(B - 13)}$$
$$V_{YB} = V_{YN} - V_{BN} = V_P\angle -120° - V_P\angle 120°$$
$$= \sqrt{3}\, V_P \angle -90° \qquad \text{(B - 14)}$$
$$V_{BR} = V_{BN} - V_{RN} = V_P\angle 120° - V_P\angle 0$$
$$= V_P\angle 150° \qquad \text{(B - 15)}$$

Figure B-5 shows the phasor diagram. The sum of three phase voltages, three line voltages and three currents is zero because each is a balanced set.

## B.6 Balanced Delta Connected Load

A 3 phase load is said to be balanced if the load impedances in each phase are equal in magnitude as well as phase angle.

Figure B-6 shows a balanced delta connected load. The load impedance in each phase is Z ∠ θ. The phase currents are

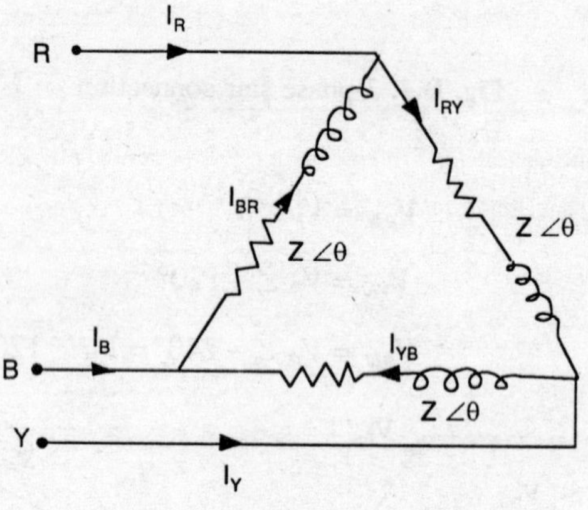


**Fig. B.6.** 3 phase delta connected load

$$I_{RY} = \frac{V_{RY}}{Z\angle\theta} = \frac{V_L\angle 0}{Z\angle\theta} = \frac{V_L}{Z}\angle -\theta = I_p\angle -\theta° \qquad \text{(B - 16)}$$

where
$$I_P = \frac{V_L}{Z}$$

$$I_{YB} = \frac{V_{YB}}{Z\angle\theta} = \frac{V_L\angle -120°}{Z\angle\theta} = I_p\angle -(120+\theta)° \qquad \text{(B - 17)}$$

$$I_{BR} = \frac{V_{BR}}{Z\angle\theta} = \frac{V_L\angle -240°}{Z\angle\theta} = I_P\angle -(240+\theta)° \qquad \text{(B - 18)}$$

$$I_R = I_{RY} - I_{BR} = \sqrt{3}\, I_P \qquad \text{(B - 19)}$$

$$I_Y = I_{YB} - I_{RY} = \sqrt{3}\, I_P \qquad \text{(B - 20)}$$

$$I_B = I_{BR} - I_{YB} = \sqrt{3}\, I_P \qquad \text{(B - 21)}$$

Figure B-7 shows the phasor diagram.

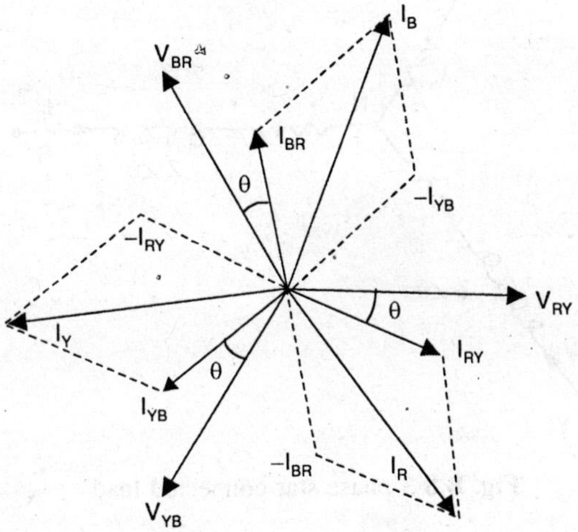


**Fig. B.7.** Phasor diagram for delta connected load

## B.7 Balanced Star Connected Load

Figure B-8 shows a 3 phase balanced star connected load having an impedance Z ∠ θ in each phase

$$I_R = \frac{V_{RN}}{Z\angle\theta} = \frac{V_P\angle\theta}{Z\angle\theta} = \frac{V_P\angle -\theta^\circ}{Z} \qquad \text{(B - 22)}$$

$$I_Y = \frac{V_{YN}}{Z\angle\theta} = \frac{V_P\angle -120^\circ}{Z\angle\theta} = \frac{V_P}{Z}\angle -(120+\theta)^\circ \qquad \text{(B - 23)}$$

$$I_B = \frac{V_{BN}}{Z\angle\theta} = \frac{V_P\angle -240^\circ}{Z\angle\theta} = \frac{V_P}{Z}\angle -(240+\theta)^\circ \qquad \text{(B - 24)}$$

The line and phase currents are equal.

Figure B-9 shows the phasor diagram.

## B.8 Power in Three Phase Circuits

Let $V_P$ and $I_P$ be the phase voltage and phase current. If phase angle between $V_P$ and $I_P$ is θ, then active, reactive and apparent power per phase are

$$P_{1\phi} = V_P I_P \cos\theta \quad \text{watts} \qquad \text{(B - 25)}$$

$$Q_{1\phi} = V_P I_P \sin\theta \quad \text{vars} \qquad \text{(B - 26)}$$

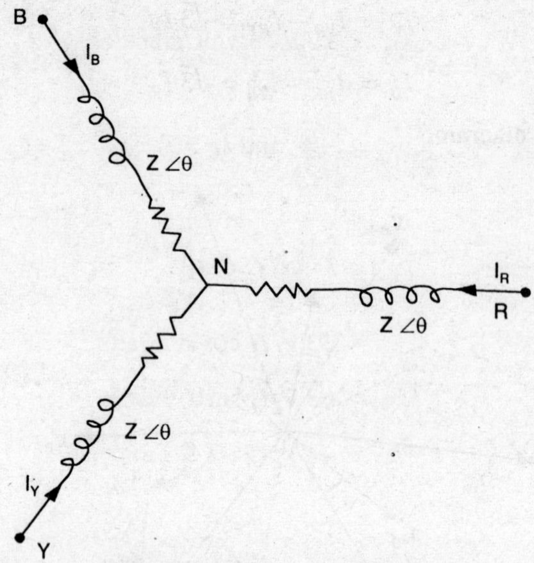


**Fig. B.8** 3 phase star connected load

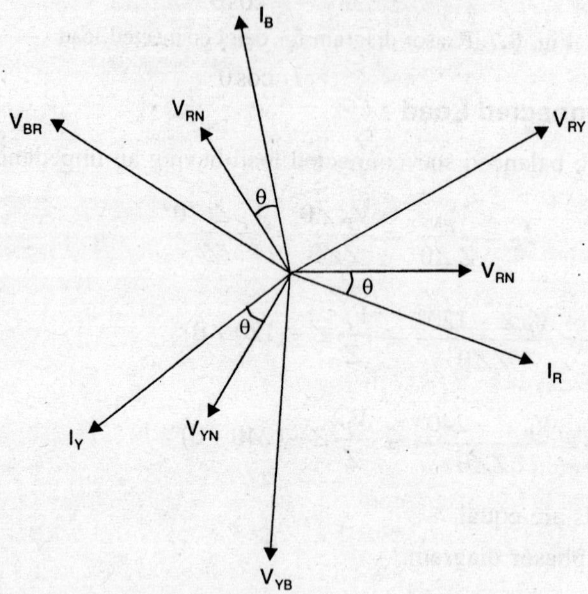


**Fig. B.9.** Phasor diagram for star connected load

$$S_{1\phi} = V_P I_P \quad \text{volt-amperes} \tag{B - 27}$$

Total 3 phase values are

$$P_{3\phi} = 3 V_P I_P \cos\theta \quad \text{watts} \tag{B - 28}$$

$$Q_{3\phi} = 3 V_P I_P \sin\theta \quad \text{vars} \tag{B - 29}$$

$$S_{3\phi} = 3V_P I_P \quad \text{volt-amperes} \tag{B - 30}$$

For star connection

$$V_P = \frac{V_L}{\sqrt{3}} \text{ and } I_P = I_L$$

Therefore

$$P_{3\phi} = 3\frac{V_L}{\sqrt{3}} \cdot I_L \cos\theta$$

$$= \sqrt{3}\,V_L I_L \cos\theta \quad \text{watts} \tag{B - 31}$$

$$Q_{3\phi} = \sqrt{3}\,V_L I_L \sin\theta \quad \text{vars} \tag{B - 32}$$

$$S_{3\phi} = \sqrt{3}\,V_L I_L \quad \text{volt-amperes} \tag{B - 33}$$

For delta connection

$$V_P = V_L \text{ and } I_P = \frac{I_L}{\sqrt{3}}$$

$$P_{3\phi} = 3V_L \cdot \frac{I_L}{\sqrt{3}} \cos\theta$$

$$= \sqrt{3}\,V_L I_L \cos\theta \quad \text{watts} \tag{B - 31}$$

$$Q_{3\phi} = \sqrt{3}\,V_L I_L \sin\theta \quad \text{vars} \tag{B - 32}$$

$$S_{3\phi} = \sqrt{3}\,V_L I_L \quad \text{volt-amperes} \tag{B - 33}$$

It is seen that total 3 phase active, reactive and apparent power are given by the same equations in both delta and star connections. The phase angle $\theta$ in all the above equations is the angle between a phase voltage and the corresponding phase current.

## B.9 Measurement of Power in Three Phase Circuits

Figure B-10 shows the connections of two wattmeter method of power measurement in 3 phase circuits. For wattmeter $W_1$, the current in its current coil is $i_R$ and the voltage across its potential coil is $(v_R - v_Y)$. For wattmeter $W_2$, the current in its current coil is $i_B$ and the voltage across its potential coil is $(v_B - v_Y)$.

Total instantaneous power $= v_R i_R + v_Y i_Y + v_B i_B$

Since

$$i_R + i_Y + i_B = 0 \tag{B - 34}$$

$$i_Y = -i_R - i_B \tag{B - 35}$$

Total instantaneous power

$$= v_R i_R + v_Y(-i_R - i_B) + v_B i_B$$

$$= i_R(v_R - v_Y) + i_B(v_B - v_Y) \tag{B - 36}$$

$i_R(v_R - v_Y)$ is the instantaneous reading of $W_1$ and $i_B(v_B - v_Y)$ is the instantaneous reading of $W_2$. The sum of two wattmeter readings gives the total 3 phase power.

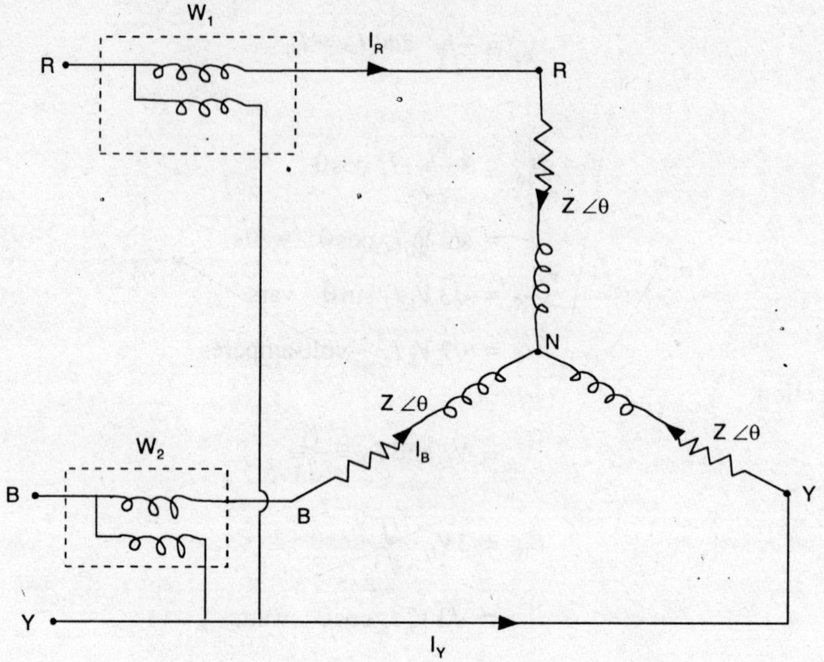


**Fig. B.10.** Connection for two wattmeter method

Figure B-11 shows the phasor diagram. For wattmeter $W_1$, the r.m.s current in its current coil is $I_R$ and the voltage across its potential coil is $V_{RY}$. For wattmeter $W_2$, the respective values are $I_B$ and $V_{BY}$. As seen in the phasor diagram, the phase angle between $I_R$ and $V_{RY}$ is $(30 + \theta)$ while that between $I_B$ and $V_{BY}$ is $(30 - \theta)$. Therefore

$$W_1 = V_{RY} I_R \cos(30 + \theta)$$

$$= \sqrt{3} V_P I_P \cos(30 + \theta) \qquad \text{(B - 37)}$$

$$W_2 = V_{BY} I_B \cos(30 - \theta)$$

$$= \sqrt{3} V_P I_P \cos(30 - \theta) \qquad \text{(B - 38)}$$

$$W_1 + W_2 = 3 V_P I_P \cos\theta \qquad \text{(B - 39)}$$

$$= \text{Total 3 phase power}$$

$$W_1 - W_2 = \sqrt{3} V_P I_P [\cos(30 + \theta) - \cos(30 - \theta)]$$

$$= -\sqrt{3} V_P I_P \sin\theta \qquad \text{(B - 40)}$$

$$\frac{W_1 - W_2}{W_1 + W_2} = \frac{-\sqrt{3} V_P I_P \sin\theta}{3 V_P I_P \cos\theta}$$

or
$$\tan\theta = \frac{\sqrt{3}\,(W_2 - W_1)}{W_1 + W_2} \qquad \text{(B - 41)}$$

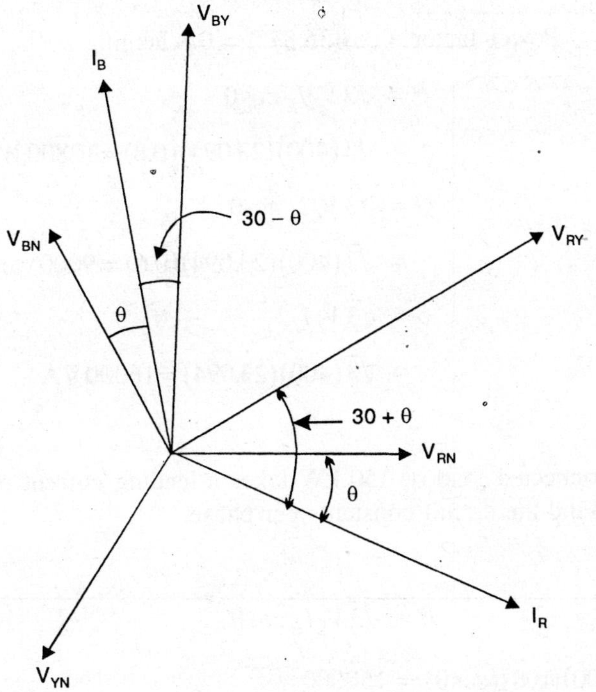


**Fig. B.11.** Phasor diagram for 2 wattmeter method

Eq. (B-41) indicates that

(a) if $\theta = 0$, $W_1 = W_2$

(b) if $\theta < 60°$, both $W_1$ and $W_2$ are positive and the total 3 phase power is the sum of wattmeter readings.

(c) if $\theta = 60°$, $W_1 = 0$

(d) if $\theta > 60°$, $W_1$ is negative. In this case it is necessary to reverse the connections of either the current coil or the potential coil of wattmeter $W_1$. Then $W_1$ is taken as negative in Eq. (B-39) and (B-41).

*Example B.1*

A balanced star connected load of $8 + j6$ ohm per phase is connected to a 3 phase 400 V system. Find line current, power factor, power, reactive power and total volt amperes.

*Solution*

$$V_L = 400 \text{ V}$$

$$V_P = \frac{400}{\sqrt{3}} = 230.94\ V$$

$$Z = 8 + j6 = 10\angle 36.87° \text{ ohms}$$

$$I_P = V_P/Z = 23.094\ A$$

$$I_L = I_P = 23.094\ A$$

$$\text{Power factor} = \cos(36.87°) = 0.8 \text{ lagging}$$

$$P = \sqrt{3}\, V_L I_L \cos\theta$$

$$= \sqrt{3}(400)(23.094)(0.8) = 12800\ W$$

$$Q = \sqrt{3}\, V_L I_L \sin\theta$$

$$= \sqrt{3}(400)(23.094)(0.6) = 9600 \text{ vars}$$

$$S = \sqrt{3}\, V_L I_L$$

$$= \sqrt{3}(400)(23.094) = 16000 \text{ VA}$$

*Example B.2*

A balanced 3 phase star connected load of 150 kW takes a leading current of 100 A when the line voltage is 1100 V, 50 Hz. Find the circuit constants per phase.

*Solution*

$$P = \sqrt{3}\, V_L I_L \cos\theta$$

or $$\sqrt{3}(1100)(100)(\cos\theta) = 150000$$

or $$\cos\theta = 0.787,\ \sin\theta = 0.617$$

$$Z = \frac{V_P}{I_P} = \frac{1100}{\sqrt{3} \times 100} = 6.35 \text{ ohm}$$

$$R = Z\cos\theta = 6.35 \times 0.787 = 5 \text{ ohm}$$

Since the power factor is leading, the other element is capacitance

$$X_c = Z\sin\theta = 6.35 \times 0.617 = 3.92 \text{ ohm}$$

$$\frac{1}{2\pi \times 50 \times C} = 3.92$$

or $$C = 8.12 \times 10^{-4} F$$

*Example B.3*

A symmetrical 3 phase 400 V system supplies a balanced delta connected load. The phase current is 20 A and the phase angle is 40° lagging. Find line current, power factor, power, reactive power, apparent power and circuit constants per phase.

*Solution*

$$I_L = \sqrt{3} I_P = \sqrt{3} \times 20 = 34.64\ A$$

$$\text{Power factor} = \cos 40° = 0.766 \text{ lagging}$$

$$P = \sqrt{3} \times 400 \times 34.64 \times 0.766 = 18383\,W$$

$$Q = \sqrt{3} \times 400 \times 34.64(\sin 40°) = 15426 \text{ vars}$$

$$S = \sqrt{3} \times 400 \times 34.64 = 23998.6\,VA$$

$$Z = 400/20 = 20\,\text{ohms}$$

$$R = Z\cos 40° = 20\cos 40° = 15.32 \text{ ohms}$$

$$X_L = Z\sin 40° = 20\sin 40° = 12.856 \text{ ohms}$$

$$2\pi \times 50 \times L = 12.856 \text{ or } L = 0.041 \text{ H}$$

*Example B.4*

A balanced delta connected load of $4 - j3$ ohm per phase is fed by a 400 V, 3 phase system. Find line current, phase current, power factor, total volt-amperes, power and reactive power.

*Solution*

$$V_P = V_L = 400\,V$$

$$I_P = \frac{400}{4 - j3} = 80A$$

$$I_L = \sqrt{3} \times 80 = 138.56\,A$$

$$\text{Power factor} = \cos(\tan^{-1}3/4) = 0.8 \text{ leading}$$

$$S = \sqrt{3}\,V_L I_L = \sqrt{3} \times 400 \times 138.56 = 95994.4\,VA$$

$$P = \sqrt{3} V_L I_L \cos\theta = 95994.4 \times 0.8 = 76795.5\,W$$

$$Q = \sqrt{3}\,V_L I_L \sin\theta = 95994.4 \times 0.6 = 57596.6 \text{ vars (capacitive)}$$

*Example B.5*

The power input to a 2000 V, 50 Hz, 3 phase induction motor is measured by two wattmeter method. The readings of the two wattmeters are 300 kW and 100 kW. The efficiency of motor is 0.8. Calculate power input, power factor, line current and power output.

*Solution*

$$W_1 = 300\,kW,\ W_2 = 100\,kW$$

$$\text{Power input} = W_1 + W_2 = 400\,kW$$

$$\tan\theta = \frac{\sqrt{3}(W_2 - W_1)}{W_2 + W_1} = \frac{\sqrt{3}(-200)}{400} = \frac{-\sqrt{3}}{2}$$

$$\text{Power factor} = \cos\theta = 0.756 \text{ lagging}$$

$$\sqrt{3}\, V_L I_L \cos\theta = P = 400 \times 1000$$

or $$\sqrt{3}(2000)(I_L)(0.756) = 400000$$

or $$I_L = 152.74 \text{ A}$$

$$\text{power output} = 400 \times 0.8 = 320 \text{ kW}$$

*Example B.6*

During no load test on a 3 phase induction motor, the readings of the two wattmeters were –0.75 kW and 3 kW. Find no load losses and power factor at no load.

*Solution*

$$\text{No load losses} = -0.75 + 3 = 2.25 \text{ kW}$$

$$\tan\theta = \frac{\sqrt{3}(3+0.75)}{3-0.75} = 2.887$$

$$\text{power factor} = \cos\theta = 0.327$$

## REVIEW QUESTIONS

1.1 Discuss the generation of 3 phase a.c. Are the windings of the three phases exactly similar?

1.2 What is understood by balanced and unbalanced 3 phase systems?

1.3 Derive the relations between phase voltages and line voltages and phase currents and line currents in (a) balanced star connection (b) balanced delta connection.

1.4 What is a balanced load? The load across the three phases is a system were $20 \angle 0$, $20 \angle 90°$ and $20 \angle -90°$ ohms. Are these loads balanced? Give reason for your answer.

1.5 Derive relation between phase and line quantities for (a) balanced star connected load (b) balanced delta connected load.

1.6 Give a diagram of connections and derive equations for total power and factor in two watt-meter method of measurement of power.

## PROBLEMS

B.1 A star connected 3 phase load has a resistance of 6 ohms and inductive reactance of 8 ohms in each phase. Line voltage is 220 V. Find line current, total power and power factor. [12.7 A, 2904 W, 0.6 lagging]

B.2 Three load impedances each $10 \angle 45°$ ohms are connected in star and fed by a 3 phase system having a line voltage of 173.2 V. Find line current and power. [10 A. 2121 W]

B.3 A balanced 3 phase star connected load is rated at 400 V, 50 Hz, 5 kVA and 0.707 lagging power factor. It is connected to a 400 V supply. Find line currents and impedance of load in each phase. [7.22 A, 22.6 + j 22.6 ohms]

B.4 A balanced 3 phase load consists of a resistance of 50 ohms in series with a 50 μF capacitor in each phase, connected in delta. It is fed by a 440 V, 3 phase system. Find phase current, line current, total power and total volt-amperes. [5.43 A, 9.40 A, 4427 W, 7164 VA]

B.5 A balanced delta connected load of 8 + j6 ohm per phase is fed by a balanced 3 phase 230 V supply. Find phase current, line current, power factor, power, reactive power and apparent power.

[23 A, 39.84 A, 0.8 lagging, 12697 W, 9522 vars, 15871 VA]

B.6 In two wattmeter method of power measurement, the readings of the two wattmeters were 6717 W and 2658 W. Find power and power factor. [9375 W, 0.8]

B.7 Find total power and power if the readings of the two wattmeters are –0.5 kW and 5 kW.

[4.5 kW, 0.427]

# Appendix C

## MAGNETIC CIRCUITS

### C.1 Magnetic Flux and Flux density

A magnetic field can be established by passing an electric current through a wire. This magnetic field can be mapped by drawing lines which indicate the direction of force on the north pole of a permanent magnet. These lines are called magnetic flux lines. The SI unit of flux is weber (Wb). It is defined as the flux which, linking a single turn coil, induces an emf of one volt when the flux is reduced to zero at a uniform rate in one second.

The flux per unit area of cross-section is the magnetic flux density (symbol B). Its SI unit is $Wb/m^2$ or tesla (T).

### C.2 Magneto motive Force and Magnetic Field Strength

Magneto motive force (i.e. MMF) of a coil is the product of number of turns in the coil (N) and the current (I). Its symbol is $\Im$ and its unit is ampere turns (AT) or simply amperes (A)

The mmf per unit length of a magnetic circuit is called magnetic field intensity (symbol H) i.e.

$$H = \frac{\Im}{l} = \frac{NI}{l} \text{ Amperes}/m \tag{C.1}$$

### C.3 Permeability (symbol μ)

It is a measure of the ease with which a the magnetic flux can be set up in a material. A material having high permeability provides an easy path for the magnetic flux. The permeability of free space is written as $\mu_0$ and equals $4\pi \times 10^{-7}$ *H/m*. For all other materials permeability is expressed as the product of $\mu_0$ and relative permeability $\mu_r$ i.e.

$$\mu = \mu_0 \, \mu_r \tag{C.2}$$

Thus relative permeability is simply a numeric which indicates the degree to which the material is a better conductor of magnetic flux than free space. Relative permeability of ferromagnetic materials is in hundreds or thousands and is not constant but varies with *H*.

### C.4 Reluctance and Permeance

Reluctance of a magnetic circuit is the ratio of mmf to flux i.e.

Reluctance = MMF/Flux

or

$$\Re = \frac{\Im}{\phi} \tag{C.3}$$

Reluctance can be calculated as

$$\Re = \frac{l}{\mu a} \tag{C.4}$$

where $l$ is the length of magnetic circuit, a is the area of cross-section and $\mu$ is the permeability.

The reciprocal of reluctance is called permeance

$$\text{Permeance} = \frac{1}{\Re} = \frac{\phi}{\Im} \tag{C.5}$$

Eqn (C.3) is similar to ohm's law of electric circuits. The quantities reluctance and resistance are analogous quantities. $\frac{1}{\mu}$ is analogous to resistivity. Permeability, therefore, is analogous to electrical conductivity.

From Eqs (C.3) and (C.4)

$$\frac{\Im}{\phi} = \frac{l}{\mu a} \quad \text{or} \quad \frac{\Im}{l} = \frac{\phi}{\mu a} \tag{C.6}$$

Since $\frac{\Im}{l}$ is magnetic field strength H and $\frac{\phi}{a}$ is flux density B, Eq (C.6) gives

$$H = \frac{1}{\mu} B$$

or

$$B = \mu H \tag{C.7}$$

## C.5 Analogy Between Electric and Magnetic circuits

Magnetic circuits and electric circuits are similar in many respects. Eq. (C.3) is, sometimes, called ohm's law for magnetic circuits. Table C.1 lists the analogous quantities in magnetic and electric circuits. However there are some important differences also between these two circuits. These are

1. Resistivity of electric materials is more or less constant. However permeability of ferromagnetic materials varies with magnetic field strength.
2. When a direct current flows in an electric circuit energy is continuously supplied to the circuit. In magnetic circuits, where a flux is established and maintained constant, such a similar situation does not prevail.
3. In ferromagnetic materials the flux density increases with H only till the state of magnetic saturation is reached. After this the increase in flux density with H is almost negligible. There is no such phenomenon in electric circuits.
4. Magnetic circuits have, invariably, an air gap to prevent saturation and also to allow an object to move in the magnetic field. However an electric circuit can not have any air gap because, in that case, the current will be zero.

**Table C.1: Analogous Quantities in electric and magnetic circuits**

| *Magnetic circuit* | *Electric circuit* |
|---|---|
| mmf $\Im$ | emf $E$ |
| flux $\phi$ | current $I$ |
| Reluctance $\Re$ | Resistance $R$ |
| Flux density $B$ | Current density |
| Permeability | Conductivity |
| Permeance | Conductance |
| Magnetic field intensity $H$ | Electric field intensity |

## C.6 Series Magnetic Circuit

A series magnetic circuit has one or more paths of ferromagnetic circuits connected in series. One or more air gaps are also present. A coil carrying current is put on a part of the length of the magnetic circuit. This current creates the necessary mmf. The different sections of ferromagnetic materials may not be uniform in cross-section.

Kirchhoff's laws are valid for magnetic circuits also. The exciting ampere turns must be equal to the sum of mmfs for the different parts of the circuit. The value of flux is the same in all the parts of the circuit. Since the cross-sections of different parts are not the same, the value of flux density will be different in different parts. From the values of flux and areas of cross-section we find the values of flux density for different parts. Then using Eq (C.7) we find the values of magnetic field strength. If the value of $\mu$ is not known, then it is necessary to use B-H curve to find H. Then

$$NI = H_1 l_1 + H_2 l_2 + H_3 l_3 + \ldots\ldots \tag{C.8}$$

where NI are the total ampere turns of the coil, $H_1$, $H_2$, $H_3$ etc. are magnetic field strengths and $l_1$, $l_2$, $l_3$... are lengths of different sections of the circuit.

## C.7 Series Parallel Magnetic Circuit

A series parallel magnetic circuit has a number of parallel magnetic paths in series with one or more magnetic paths. The flux entering a junction is equal to the flux leaving the junction. MMFs across two parallel paths are equal (just as voltages across parallel branches, in an electric circuit, are equal). The actual magnetic circuits in transformers and electric machines have series parallel configuration

### *Example C.1*

A magnetic circuit has a length of 80 cm and cross-sectional area of 4 cm². The total flux is 10 × $10^{-4}$ wb. If the coil has 100 turns find flux density, mmf for a current of 4 A and magnetic field strength.

*Solution*

$$\text{Flux density } B = \frac{\text{Flux}}{\text{area of cross-section}} = \frac{10 \times 10^{-4}}{4 \times 10^{-4}} = 2.5\ T$$

$$MMF = NI = 100 \times 4 = 400\ A$$

$$H = \frac{MMF}{\text{length}} = \frac{400}{80 \times 10^{-2}} = 500\ A/m$$

*Example C.2*

An air cored toroidal coil has 100 turns and the length is 10 cm. If the cross-sectional area is 10 cm$^2$, find the current required to set up a flux of $1 \times 10^{-6}$ wb.

*Solution*

$$B = \frac{\phi}{a} = \frac{1 \times 10^{-6}}{10 \times 10^{-4}} = 0.1 \times 10^{-2}\ T$$

$$H = \frac{B}{\mu_0} = \frac{0.1 \times 10^{-2}}{4\pi \times 10^{-7}} = 0.0795 \times 10^4\ A/m$$

$$MMF = Hl = 0.0795 \times 10^4 \times \frac{10}{100} = 79.5\ A$$

$$\text{Current} = \frac{MMF}{N} = \frac{79.5}{100} = 0.795\ A$$

*Example C.3*

The air core of the toroid in Example C.2 is replaced by iron core. Assume $\mu_r = 400$. Find the flux for the same mmf.

*Solution*

$$H = 0.0795 \times 10^4\ A/m$$

$$\mu_r = 400$$

$$B = \mu_0 \mu_r H = 4\pi \times 10^{-7} \times 400 \times 0.0795 \times 10^4$$

$$= 0.4\ T$$

$$\text{Flux} = (0.4)(10 \times 10^{-4}) = 4 \times 10^{-4}\ \text{Wb}$$

*Example C.4*

Fig. C.1 shows a rectangular magnetic circuit with square cross-section. The relative permeability of the material is 500. If it is required to set up a flux of $25 \times 10^{-4}$ Wb find the mmf and current. Assume that N = 200.

*Solution*

$$\text{Area of cross-section} = 0.1 \times 0.1 = 0.01\ \text{m}^2$$

$$B = \frac{25 \times 10^{-4}}{0.01} = 0.25\ T$$

For Iron

$$H_i = \frac{B}{\mu_0 \mu_r} = \frac{0.25}{4\pi \times 10^{-7} \times 500}$$

$$= 397.89 \text{ A/m}$$

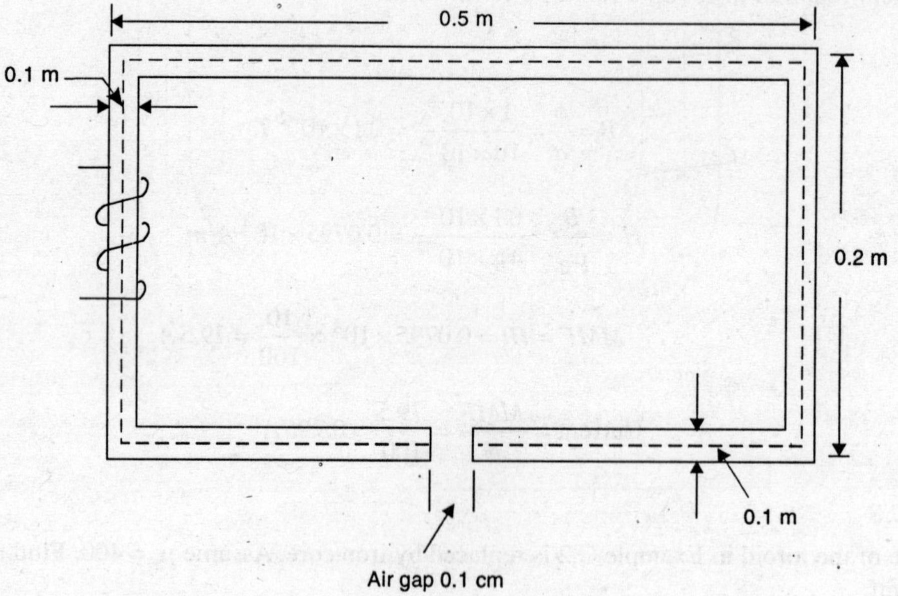


**Fig. C.1**

$$\text{Mean length of iron path} = 2(0.5 + 0.2) - 0.1 \times 10^{-2}$$

$$= 1.399 \text{ m}$$

$$\text{MMF for iron} = 397.89 \times 1.399 = 556.65 \text{ A}$$

For air gap

$$H_a = \frac{0.25}{4\pi \times 10^{-7}} = 198945\, A/m$$

$$\text{MMF for air gap} = 198945 \times 0.1 \times 10^{-2} = 198.95 \text{ A}$$

$$\text{Total mmf} = 556.65 + 198.95 = 95.6 \text{ A}$$

$$\text{Current} = \frac{MMF}{N} = \frac{755.6}{200} = 3.778\, A$$

*Example C.5*

Fig. C.2 shows a toroid consisting of two sections. The area of cross-section for both sections is 0.001 m$^2$. Relative permeability for section a is 400 and that for section $b$ is 100. Flux is $8 \times 10^{-4}$ wb. Find (a) mmf (b) current if $N = 1500$ (c) reluctance of each section.

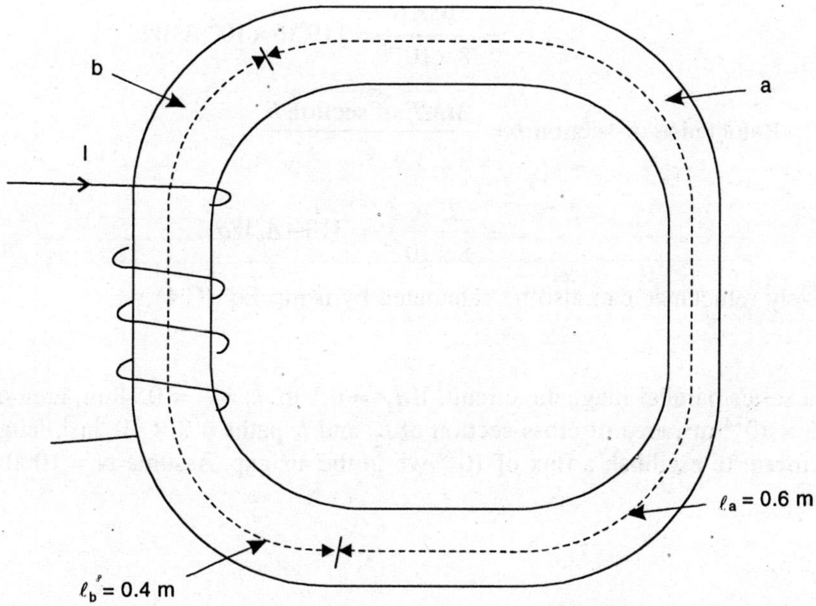


**Fig. C.2**

*Solution*

(a)

$$\phi = 8 \times 10^{-4} \text{ wb}, \; a = 0.001 \text{ m}^2$$

$$B = \frac{\phi}{a} = \frac{8 \times 10^{-4}}{0.001} = 0.8\,T$$

For section $a$

$$H_a = \frac{B}{\mu_0 \mu_r} = \frac{0.8}{4\pi \times 10^{-7} \times 400} = 1591.5\,A/m$$

$$\Im_a = H_a l_a = 1591.5 \times 0.6 = 954.9\,A$$

For section $b$

$$H_b = \frac{B}{\mu_0 \mu_r} = \frac{0.8}{4\pi \times 10^{-7} \times 100} = 6366\,A/m$$

$$\Im_b = H_b l_b = 6366 \times 0.4 = 2546.4\,A$$

$$\text{Total mmf} = 954.9 + 2546.4 = 3501.3 \text{ A}$$

(b)

$$I = \frac{MMF}{N} = \frac{3501.3}{1500} = 2.33\,A$$

(c)

$$\text{Reluctance of section } a = \frac{MMF \text{ of section } a}{\phi}$$

$$= \frac{954.9}{8 \times 10^{-4}} = 119.36 \times 10^4 \, A/Wb$$

$$\text{Reluctance of section } b = \frac{MMF \text{ of section } b}{\phi}$$

$$= \frac{2546.4}{8 \times 10^{-4}} = 318.3 \, A/Wb$$

Alternatively reluctance can also be calculated by using Eq. (C.4)

*Example C.6*

Fig. C.3 shows a series parallel magnetic circuit. If $l_1 = 0.1$ m, $l_2 = l_3 = 0.18$ m, area of cross-section of $l_1$ path = $6.25 \times 10^{-4}$ m², area of cross-section of $l_2$ and $l_3$ paths = $3 \times 10^{-4}$m², length of air gap = 2 mm find the current to establish a flux of $10^{-4}$ wb in the air gap. Assume N = 1000 and $\mu_r = 400$

*Solution*

For air gap

$$B_a = \frac{10^{-4}}{6.25 \times 10^{-4}} = 0.16 \, T$$

$$H_a = \frac{B_a}{\mu_0} = \frac{0.16}{4\pi \times 10^{-7}} = 1.27 \times 10^5 \, A/m$$

$$\Im_a = 1.27 \times 10^5 \times 2 \times 10^{-3} = 254 \text{ A}$$

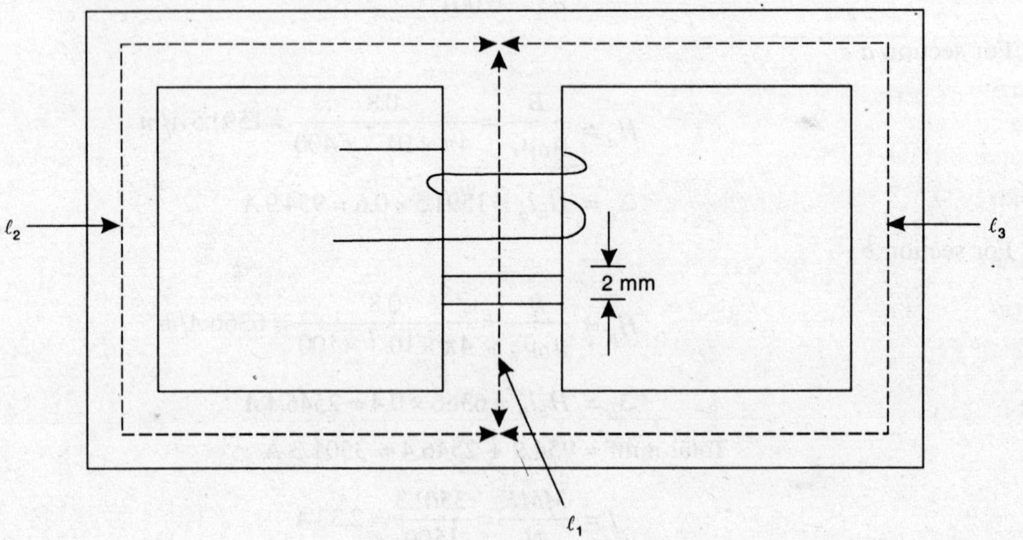


**Fig. C.3**

For path $l_1$

$$B_1 = 0.16 \text{ T}$$

$$H_1 = \frac{B_1}{\mu_0\mu_r} = \frac{0.16}{4\pi \times 10^{-7} \times 400} = 317.5\, A/m$$

$$\Im_1 = H_1 l_1 = 317.5\left(0.1 - 2\times 10^{-3}\right) = 31.12\, A$$

The paths $l_2$ and $l_3$ are similar. Therefore total flux divides equally between these two paths. It is necessary to find mmf for only one of these two parallel paths.

For path $l_2$, $\phi = 0.5 \times 10^{-4}$ Wb

$$B_2 = \frac{0.5\times 10^{-4}}{3\times 10^{-4}} = 0.167\, T$$

$$H_2 = \frac{B_2}{4\pi\times 10^{-7}\times 400} = \frac{0.167}{4\pi\times 10^{-7}\times 400} = 332.2\, A$$

$$\Im_2 = H_2 l_2 = 332.2 \times 0.18 = 59.79\, A$$

$$\text{Total mmf} = \Im_a + \Im_1 + \Im_2$$

$$= 254 + 31.12 + 59.79$$

$$= 344.91 \text{ A}$$

$$I = \frac{MMF}{N} = \frac{344.91}{1000} = 0.34491\, A$$

*Example C.7*

A magnetic core made of stalloy has the configuration as shown in Fig. C.4. The coil has 400 turns. The cross-sectional area of central limb is 1600 mm$^2$ and that of each side limb is 1000 mm$^2$. It is required to set up a flux of 1 m Wb in the air gap. Find the exciting current. The B-H curve of the material is as under:

| B(tesla) | 1 | 1.2 | 1.3 | 1.35 | 1.42 | 1.45 | 1.5 | 1.52 | 1.53 |
|---|---|---|---|---|---|---|---|---|---|
| H(A/m) | 200 | 450 | 700 | 1000 | 1500 | 2000 | 3000 | 3500 | 4000 |

*Solution*

For the air gap, flux = 1 × 10$^{-3}$ Wb

$$B_a = \frac{1\times 10^{-3}}{1000\times 10^{-6}} = 1 \text{ tesla}$$

$$H_a = \frac{B_a}{4\pi\times 10^{-7}} = \frac{1}{4\pi\times 10^{-7}} = 795.77\times 10^3\, AT/m$$

$$\Im_a = 795.77\times 10^3 \times 1.2\times 10^{-3} = 954.92\, A$$

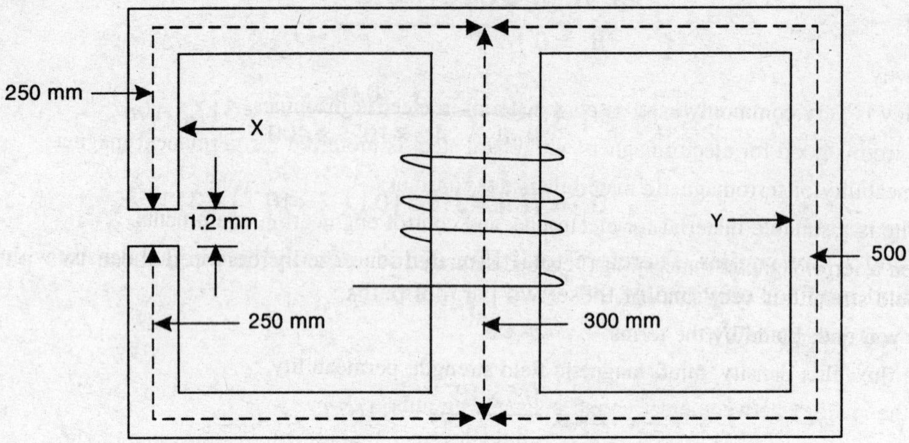


Fig. C.4

For iron portion is side limb *X*

$$B = 1 \text{ tesla}$$

$$H = 200 \text{ A/m (from the given B-H curve)}$$

$$\mathfrak{I} = 2 \times 200 \times 250 \times 10^{-3} = 100 \text{ A}$$

$$\text{Total mmf for side limb } X = 954.92 + 100 = 1054.92 \text{ A}$$

The mmf for side limb *Y* is also the same as for side limb *X* because these two side limbs are in parallel.

$$\text{H for side limb } Y = \frac{1054.92}{500 \times 10^{-3}} = 2109.84 \text{ A/m}$$

$$B = 1.455 \text{ tesla (using the given B-H curve)}$$

$$\text{Flux in side limb } Y = 1.455\ (1000 \times 10^{-6}) = 1.455 \times 10^{-3} \text{ Wb}$$

$$\text{Total flux in central limb} = 1 \times 10^{-3} + 1.455 \times 10^{-3} = 2.455 \times 10^{-3} \text{ wb}$$

$$\text{Flux density in central limb} = \frac{2.455 \times 10^{-3}}{1600 \times 10^{-6}} = 1.53 \text{ tesla}$$

$$H = 4000 \text{ A/m (using the given B-H curve)}$$

$$\text{MMF for central limb} = 4000 \times 300 \times 10^{-3} = 1200 \text{ A}$$

$$\text{Total mmf} = \text{mmf for central limb} + \text{mmf for one side limb} = 1200 + 1054.92 = 2254.92 \text{ A}$$

$$\text{Exciting current} = \frac{2254.92}{400} = 5.64\,A$$

## REVIEW QUESTIONS

C.1 Discuss why

(a) Stalloy is very commonly used as core material in electric machines.

(b) Soft iron is good for electromagnets while hard steel is preferred for permanent magnets.

(c) Permeability of ferromagnetic materials is not constant.

(d) Ferrite is a suitable material for electronics and control engineering equipments.

(e) When a ferromagnetic material has reached saturated state, the increase in flux density with increase in field strength is very small.

C.2 What do you understand by the terms:

magnetic flux, flux density, mmf, magnetic field strength, permeability.

C.3 Discuss the analogy between electric and magnetic circuits.

C.4 Discuss the procedure for finding mmf for a given magnetic circuit for a given value of flux.

## PROBLEMS

C.1 An iron ring has a cross-sectional area of $10^{-3}$ sq.m and mean circumference of 1m. The relative permeability of the material is 500 and the number of turns is 200. Find the current required to set up a flux of 0.001 Wb. [7.95 A]

C.2 The iron ring in problem C.1 has an air gap 2 mm length. Find the exciting circuit if the flux is 0.001 Wb. Neglect leakage flux. [15.9 A]

C.3 A coil has an mmf of 1000 A. The reluctance of the magnetic path is $2 \times 10^6$ A/Wb. Find the total flux. [$500 \times 10^{-6}$ Wb]

C.4 An iron ring has a mean circumference of 0.4 m and an area of $10^{-4}$ m$^2$. The number of turns is 500, flux is $6 \times 10^{-6}$ Wb and exciting current is 0.06 A. Find (a) B (b) H (c) $\mu_r$

[(a) 0.06 T (b) 75 A/m (c) 636.6]

C.5 A core has an area of 3 cm$^2$ and a length of 0.2 m. When B = 0.72 T, the permeability is $8 \times 10^{-3}$ H/m. Find reluctance of core. [83333 A/Wb]

C.6 The core in Problem C.5 has an air gap of 2 mm length in addition to 0.2 m length of iron. Find (a) reluctance of air gap (b) mmf required to keep the flux density at 0.72 T. [(a) $5.31 \times 10^6$ A/Wb (b) 1164.9 A]

C.7 An electromagnet has 100 turns. The length of core is 8 cm and current in the coil is 0.2 A. If B = 0.13 T find (a) H (b) $\mu$ (c) $\mu_r$. [(a) 250 A/m (b) $5.2 \times 10^{-4}$ H/m (c) 413.8]

C.8 A magnetic circuit has the configuration of Fig. C.3 but the dimensions are as under: air gap 2 mm $l_1$ = 20 cm, $l_2 = l_3$ = 40 cm, cross-sectional area of middle leg is 10 cm$^2$ and cross-sectional area of remaining circuit is 5 cm$^2$. If $\mu_r$ = 800, find the mmf required to set up a flux of $4 \times 10^{-4}$ wb in the air gap. Neglect leakage. [874.5 A]

C.9 A rectangular core has cross-sectional area of 20 cm$^2$. It is made of two materials cast steel 50 cm long and sheet steel 30 cm long. The relative permeability for cast steel is 1000 and that of sheet steel is 4000. The coil has 200 turns. It is required to set up a flux of 2 mWb in the core. Find mmf, exciting current and flux density. [457.57 A, 2.288 A, I T]

C.10 A core has an iron length of 0.3 m and an air gap 0.2 cm long. Flux density is 0.5 T when magnetic field strength (for iron) is 800 A/m. The cross-sectional area is 8 cm$^2$. Find mmf. [1035.77 A]

# Appendix D

## SUMMARY OF IMPORTANT FORMULAE

### Chapter 1: Basic Principles

Faraday's law of electromagnetic Induction

$$e = \frac{d\lambda}{dt} = \frac{Nd\phi}{dt}$$

Dynamically Induced EMF $\quad e = \oint E \cdot dl = \oint (v \times B) \cdot dl = Blv \sin\theta$

Statically Induced EMF $\quad e = \int_s \frac{\partial B}{\partial t} \cdot ds$

General case of Induction $\quad e = \oint (v \times B) \cdot dl - \int_s \frac{\partial B}{\partial t} \cdot ds$

Coefficient of Self Induction $\quad e = L\frac{di}{dt}$

$$L = N\frac{d\phi}{di} = \frac{N^2 \mu a}{l}$$

Coefficient of Mutual Induction $\quad e = M\frac{di}{dt}$

Force on a current carrying conductor in a Magnetic Field

$$F = B\, I\, l \sin\theta$$

Torque on a current carrying coil in a Magnetic Field

$$\text{Torque} = 2\, NBIlr$$

Hysteresis Loss $\quad P_h = k_h (\text{volume})(f)(B_m)^n = K_h f (B_m)^n$

For grained oriented silicon steel $n = 1.6$

Eddy Current Loss $\quad P_e = k_e f^2 B_m^2 t^2 \ (\text{volume}) = K_e B_m^2 f^2$

Energy Balance $$dW_e = dW_m + dW_f$$

Efficiency of Generator $$\eta = \frac{\text{Power output}}{\text{Power Input}} = \frac{\text{Power output}}{\text{Power output} + \text{Losses}}$$

Efficiency of Motor $$\eta = \frac{\text{Power output}}{\text{Power Input}} = \frac{\text{Power Input} - \text{Losses}}{\text{Power Input}}$$

Machine Rating $$kW(\text{rms value}) = \left[\frac{\Sigma(kW)^2(\text{time})}{\text{operating time} + \dfrac{\text{stand still time}}{k}}\right]^{0.5}$$

## Chapter 2: Transformers

RMS value of Induced EMF

$$E = 4.44\, f\, N\, \phi_m$$

Voltage and Current Ratios

$$\frac{E_1}{E_2} = \frac{I_2}{I_1} = \frac{N_1}{N_2}$$

No Load Current $$I_0 = \left(I_m^2 + I_W^2\right)^{0.5}$$

Secondary terminal voltage $$V_2 = E_2 - I_2(R_2 + jX_2)$$

Primary terminal voltage $$V_1 = E_1 + I_1(R_1 + jX_1)$$

Secondary resistance $R_2$ and reactance $X_2$ as referred to Primary

$$R_2' = R_2(N_1/N_2)^2$$

$$X_2' = X_2(N_1/N_2)^2$$

Primary resistance $R_1$ and reactance $X_1$ as referred to secondary

$$R_1' = R_1(N_2/N_1)^2$$

$$X_1' = X_1(N_2/N_1)^2$$

Per unit Values $$\text{Per unit value} = \frac{\text{Actual value}}{\text{Base value}}$$

$$Z_{\text{base}} = \frac{V_{\text{base}}}{I_{\text{base}}} = \frac{(V_{\text{base}})^2}{\text{Volt. Amp}_{\text{base}}}$$

Voltage Regulation for lagging power factor

$$\text{Voltage regulation} = \frac{I_2 R_{e2}}{V_2}\cos\theta_2 + \frac{I_2 X_{e2}}{V_2}\sin\theta$$

$$= R_{pu}\cos\theta_2 + X_{pu}\sin\theta_2$$

Voltage Regulation for Leading power factor

$$\text{Voltage Regulation} = R_{pu}\cos\theta_2 - X_{pu}\sin\theta_2$$

Condition for zero voltage Regulation

$$\tan\theta_2 = \frac{R_{pu}}{X_{pu}}$$

Losses and Efficiency

$$\text{Copper losses} = I_1^2 R_1 + I_2^2 R_2 = I_1^2 R_{e1} = I_2^2 R_{e2}$$

$$\text{Efficiency} = \frac{V_2 I_2 \cos\theta_2}{V_2 I_2 \cos\theta_2 + I_2^2\,\text{Re}_2 + \text{Iron losses}}$$

Condition for Maximum Efficiency

$$I_2^2 R_{e2} = \text{Iron losses}$$

$$\text{All day efficiency} = \frac{\text{Energy output in 24 hours}}{\text{Energy output in 24 hours} + \text{Energy losses in 24 hours}}$$

Open Circuit Test $$I_W = \frac{W_0}{V_0}$$

$$I_m = \left(I_0^2 - I_W^2\right)^{0.5}$$

$$R_0 = \frac{V_0}{I_W}$$

$$X_0 = \frac{V_0}{I_m}$$

Short Circuit Test $$Z_e = \frac{V_{sc}}{I_{sc}}$$

$$R_e = \frac{W_{sc}}{I_{sc}^2}$$

$$X_e = \left(Z_e^2 - R_e^2\right)^{0.5}$$

Auto Transformer

$$\frac{\text{Output } VA \text{ of auto-transformer}}{\text{Tranformed VA}} = \frac{a}{a-1} = \frac{N_1/N_2}{\frac{N_1}{N_2}-1}$$

$$\text{Ratio of weights} = 1 - \frac{1}{a} = 1 - \frac{N_2}{N_1}$$

Open Delta Connection

$$\frac{P_{\text{open } \Delta}}{P_{3\text{ phase}}} = \frac{\sqrt{3}\, V_P I_P}{3 V_P I_P} = \frac{1}{\sqrt{3}} = 0.577$$

Parallel Operation with Unequal Voltage Ratios

$$I_A = \frac{E_A Z_B + (E_A - E_B)Z}{Z_A Z_B + (Z_A + Z_B)Z}$$

$$I_B = \frac{E_B Z_A - (E_A - E_B)Z}{Z_A Z_B + (Z_A + Z_B)Z}$$

$$I = \frac{E_A Z_B + E_B Z_A}{Z_A Z_B + (Z_A + Z_B)^Z}$$

$$V = IZ = \frac{E_A Z_B + E_B Z_A}{\frac{Z_A Z_B}{Z} + Z_A + Z_B}$$

$$\text{Circulating current } I = \frac{(E_A - E_B)Z}{Z_A Z_B + (Z_A + Z_B)Z}$$

Parallel Generator Theorem

$$V\left[\frac{1}{Z} + \frac{1}{Z_A} + \frac{1}{Z_B}\right] = \frac{E_A}{Z_A} + \frac{E_B}{Z_B}$$

Three Winding Transformer

$$Z_1 = R_1 + jX_1 = 0.5(Z_{12} + Z_{13} - Z_{23})$$

$$Z_2 = R_2 + jX_2 = 0.5(Z_{12} + Z_{23} - Z_{13})$$

$$Z_3 = R_3 + jX_3 = 0.5(Z_{23} + Z_{13} - Z_{12})$$

## Chapter 3: DC Machines

Total number of armature conductors

$$Z = 2\, CN_c$$

EMF Equation $$E = \frac{Z P\phi N}{60A}$$

Reactance Voltage $$= \frac{2LI}{T_c}$$

$$\text{Demagnetising AT/pole} = \frac{I_a Z}{AP}\left(\frac{\theta}{180^\circ}\right)$$

$$\text{Cross magnetising AT/pole} = \frac{(180-2\theta)}{180}\frac{(I_a Z)}{2AP}$$

Shunt Generator $$I_a = I_L + I_f$$

Series Generator $$I_a = I_f = I_L$$

Electromagnetic Torque $$T_e = \frac{Z\phi P I_a}{2\pi A} = \frac{EI_a}{2\pi N/60}$$

Shunt Motor $$N = \frac{V - I_a R_a}{k_e \phi} = \frac{V}{k_e \phi} - \frac{T_e R_a}{k_t k_e \phi^2}$$

Series Motor $$N = \frac{V - I_a (R_a + R_{se})}{k_e \phi}$$

Shunt Motor Starter $$k^n = \frac{R_1}{R_a}$$

Brake Test $$\eta = \frac{(2\pi N)(9.81r)(W_1 - W_2)}{60\ VI_L}$$

Metadyne $$e_{ad} = k_{dq} i_q$$

$$e_{aq} = k_{qf} i_f - k_{qd} i_d$$

Amplidyne Output voltage $V_L = AV_f - [K_1(1-C) + r_d + r_c] i_d$

$$\frac{i_d}{i_f} = \frac{Ar_f}{K_1(1-C) + r_d + r_c}$$

**Chapter 4: AC Machine Fundamentals**

EMF Equation $$E = 4.44 f N \phi k_b k_p$$

Breadth factor $$k_b = \frac{\sin(q\alpha/2)}{q\sin\alpha/2}$$

Pitch factor $$k_b = \cos(\sigma/2)$$

Fundamental Component of MMF of phase a

$$F_{a_1} = \frac{4}{\pi} k_b k_p \frac{N_{ph}}{P} i_a \cos\theta$$

Maximum value of $F_{a_1}$

$$F_{\max} = \frac{4}{\pi} k_b k_p \frac{N_{ph}}{P} (\sqrt{2}I)$$

Synchronous Speed $$N_s = \frac{120f}{P}$$

$$\text{Torque } T = \left(\frac{\pi}{2}\right)\left(\frac{P}{2}\right)^2 F_2\phi_R \sin\delta$$

Maximum Efficiency condition

Constant Losses = Copper Losses

## Chapter 5: Synchronous Generators

Armature reaction voltage $$E_{ar} = -jIx_a$$

Terminal voltage Equation $$E = V + IR_a + jI(x_l + x_a)$$

$$\text{Voltage Regulation} = \frac{|E| - |V|}{|V|} \times 100$$

Synchronous Impedance

$$Z_s = \frac{\text{O.C. phase voltage for a certain excitation}}{\text{S.C. current for the same excitation}}$$

$$x_s = (Z_s^2 - R_a^2)^{0.5}$$

Power Angle Characteristics

Cylindrical Rotor $$P = \frac{3VE}{x_s}\sin\delta$$

$$Q = \frac{3VE}{x_s}\cos\delta - \frac{3V^2}{x_s}$$

Salient pole Rotor $$P = \frac{3VE}{X_d}\sin\delta + \frac{3V^2(X_d - X_q)}{2X_d X_q}\sin 2\delta$$

Synchronising Power per electrical radian

$$P_s = \frac{3VE}{x_s}\cos\delta$$

Synchronising Torque = $P_s$ synchronous watts

$$= \frac{P_s}{2\pi n_s} N\text{-}m$$

Parallel operation of Two Alternators

$$I_1 = \frac{E_1 Z_{s2} + (E_1 - E_2)Z}{Z_{s1}Z_{s2} + (Z_{s1} + Z_{s2})Z}$$

$$I_2 = \frac{E_2 Z_{s1} + (E_2 - E_1)Z}{Z_{s1}Z_{s2} + (Z_{s1} + Z_{s2})Z}$$

$$I = I_1 + I_2 = \frac{E_1 Z_{sL} + E_2 Z_{s1}}{Z_{s1}Z_{s2} + (Z_{s1} + Z_{s2})Z}$$

$$V = IZ = \frac{E_1 Z_{s2} + E_2 Z_{s1}}{\frac{Z_{s1}Z_{s2}}{Z} + Z_{s1} + Z_{s2}}$$

Circulating current $$I_s = \frac{(E_1 - E_2)Z}{Z_{s1} + Z_{s2} + (Z_{s1} + Z_{s2})Z}$$

Parallel generator theorem

$$V\left[\frac{1}{Z} + \frac{1}{Z_{s1}} + \frac{1}{Z_{s2}} + \ldots\right] = \frac{E_1}{Z_{s1}} + \frac{E_2}{Z_{s2}} + \ldots$$

RMS current after short circuit

$$I(t) = (I'' - I')e^{-t/T_d''} + (I' - I_{ss})e^{-t/T_d'} + I_{ss}$$

## Chapter 6: Synchronous Motors

Cylindrical rotor

$$V = E + IR_a + jIx_s$$

Salient Pole Rotor

$$V = E + IR_a + jI_d x_d + jI_q x_q$$

Power output (Cylindrical rotor)

$$P = \frac{3EV}{Z_s}\cos(\phi_s - \delta) - \frac{3E^2}{Z_s}\cos\phi_s$$

$$\text{Torque} = T = \frac{P}{2\pi n_s}\ \text{N-m}$$

$$Q = \frac{3V^2}{x_s} - \frac{3VE}{x_s}\cos\delta$$

Power output (Salient Pole rotor)

$$P = \frac{3EV}{X_d}\sin\delta + \frac{3V^2(X_d - X_q)}{2X_d X_q}\sin 2\delta$$

## Chapter 7: Polyphase Induction Motors

$$\text{slip } s = \frac{n_s - n}{n_s}$$

$$\text{Rotor current } I_2 = \frac{sE_2}{\left(r_2^2 + s^2 x_2^2\right)^{0.5}} = \frac{E_2}{\left[\frac{(r_2)^2}{s} e x_2^2\right]^{0.5}}$$

$$\text{Airgap Power } P_g = 3E_2 I_2 \cos\theta_2 = \frac{3I_2^2 r_2}{s}$$

$$\text{Rotor copper loss} = sP_g$$

$$\text{Torque } T = \frac{P_g}{2\pi n_s}$$

Also

$$\text{Torque } T = \left[\frac{3}{2\pi n_s}\right]\left[\frac{r_2'}{s}\right]\left[\frac{V_1^2}{\left(r_1 + \frac{r_2'}{s}\right)^2 + (x_1 + x_2')^2}\right]$$

Slip for maximum torque

$$s = \pm\frac{r_2'}{\left[r_1^2 + (x_1 + x_2')^2\right]^{0.5}}$$

Maximum torque $$T_{max} = \frac{3V_1^2}{4\pi n_s}\left[\frac{1}{r_1 + \left[r_1^2 + (x_1 + x_2')^2\right]^{0.5}}\right]$$

Condition for maximum torque at starting

$$r_2' = \left[r_1^2 + (x_1 + x_2')^2\right]^{0.5}$$

Star Delta Starter

$$\text{Starting torque} = \frac{1}{3}\ \text{(starting torque with direct starting)}$$

$$\frac{\text{Starting current per phase (star connection)}}{\text{Full load current per phase (delta connection)}} = \frac{I_s}{\sqrt{3}\, I_f}$$

Auto-transformer starter

$$\text{Starting torque} = \frac{1}{k^2}\ \text{(starting torque with direct starting)}$$

No Load Test

$$\cos\theta_0 = \frac{W_0}{3V_0 I_0}$$

$$R_0 = \frac{V_0}{I_w} = \frac{V_0}{I_0 \cos\theta_0}$$

$$X_0 = \frac{V_0}{I_m} = \frac{V_0}{I_0 \sin\theta_0}$$

Blocked Rotor Test

$$\cos\theta_{sc} = \frac{W_{sc}}{3V_{sc} I_{sc}}$$

$$r_1 + r_2' = \frac{V_{sc}}{I_{sc}} \cos\theta_{sc}$$

$$x_1 + x_2' = \frac{V_{sc}}{I_{sc}} \sin\theta_{sc}$$

Cascade Connection

Speed $$N = \frac{120 f_1}{P_1 + P_2}$$

## Chapter 8: Single Phase Induction Motor

Stator MMF

$$F = 0.5\,F_{\max}\cos(\theta - \omega t) + 0.5F_{\max}\cos(\theta + \omega t)$$

$$P_{gf} = \left(I'_{2f}\right)^2\left[\frac{r'_2}{2s}\right], W$$

$$P_{gb} = (I'_{2b})^2\left[\frac{r'_2}{2(2-s)}\right], W$$

$$T_f = \frac{P_{gf}}{2\pi n_s}\text{N-m}$$

$$T_b = \frac{P_{gb}}{2\pi n_s}\text{N-m}$$

$$\text{Net torque } T = T_f - T_b$$

$$\text{Total Rotor Copper loss } = sP_{gf} + (2-s)\,P_{gb}$$

$$\text{Mechanical power developed} = (1-s)\left(P_{gf} - P_{gb}\right)$$

No Load Test

$$\cos\theta_0 = \frac{W_0}{V_0 I_0}$$

$$V_{ab} = V_0 - [I_0 \angle -\theta_0]\left[\left(r_1 + \frac{r'_2}{4}\right) + j\left(x_1 + \frac{x'_2}{2}\right)\right]$$

$$\frac{x_0}{2} = \frac{V_{ab}}{I_0}$$

Blocked rotor test

$$R_{sc} = r_1 + r'_2 = \frac{W_{sc}}{I_{sc}^2}$$

$$Z_{sc} = \frac{V_{sc}}{I_{sc}}$$

$$x_1 + x'_2 = \left(Z_{sc}^2 - R_{sc}^2\right)^{0.5}$$

## Chapter 9: Ac Commutator Motors

AC Series Motor

Compensating Winding turns per pole $= N_c = \dfrac{Z}{2PA}$

$$\text{Speed } n = \frac{1}{C}\left[\sqrt{\left(\frac{V}{I}\right)^2 - (X' + X)^2} - R'\right]$$

Universal Motor $\quad \dfrac{n_{dc}}{n_{ac}} = \dfrac{1 - IR'/V}{\cos\theta - IR'/V}$

## Chapter 10: Special Machines

Two Phase Servomotor

$$\text{Torque } T = \frac{P_{g1} - P_{g2}}{2\pi n_s}$$

$$\text{Also Torque } T = \frac{1}{2\pi n_s}\left[\left(\frac{V_{r1}}{V_r}\right)^2 P_g - \left(\frac{V_{r2}}{V_r}\right)^2 P'_g\right]$$

$$\text{Transfer function} = \frac{K_C}{Js + (f + K_n)} = \frac{K_m}{sT_m + 1}$$

## Chapter 11: Solid State Motor Control

Step down chopper

$$V_0 = (V)\left[\frac{t_1}{t_1 + t_2}\right] = \alpha V$$

Step up Chopper

$$V_0 = \frac{V(t_1 + t_2)}{t_1}$$

## REFERENCES


1. Adkins B. et al, *'General Theory of Alternating Current Machines'* Chapman and Hall, London, 1975.
2. Alger P.L., *'Introduction to Machines'* Gordon and Breach, New York, 1970.
3. Armensky E.V. et al, *'Fractional Horse Power Electric Machines'* Mir Publishers, Moscow, 1978.
4. Brosan C.S. and Heyden J.T., *'Advanced Electrical Power and Machines'* Pitman, London, 1966.
5. Blume L.F. et al, *'Transformer Engineering'* John Wiley, New York.
6. Chapman S.J., *'Electric Machinery Fundamentals'* McGraw Hill, New York, 1987.
7. Clayton A.E., *'Performance and Design of DC Machines'* Pitman, London, 1959.
8. Concordia C., *'Synchronous Machines'* John Wiley, New York, 1958.
9. Del Toro V., *'Electric Machines and Power Systems'* Prentice Hall of India, N. Delhi, 1988.
10. Fitzgerald A.E. et al, *'Electric Machinery'* McGraw Hill International, N. Delhi, 1983.
11. Franklin A.C. and Franklin D.P., *'J and P Transformer Book'*, Butterworth, London, 11th edition 1983.
12. Gibbs W.J., *'Electric Machine Analysis using Tensors'* Pitman, London, 1967.
13. Gupta B.R. and V. Singhal, *'Fundamentals of Electric Networks'*, S. Chand, New Delhi, 2003.
14. Gupta B.R., *'Generation of Electrical Energy'* Eurasia, New Delhi, 2004.
15. Gupta B.R. *'Power System Analysis and Design'* S. Chand, New Delhi, 2003.
16. Gupta B.R., *'Principles of Electrical Engineering'*, S. Chand, New Delhi 1996.
17. Gupta B.R. and V. Singhal., *'Power Electronics'* S.K. Kataria and Sons, New Delhi, 2003.
18. Kimbark E.W., *'Power System Stability* Vol III *Synchronous Machines'*, John Wiley, New York.
19. Kosov I.L., *'Electric Machinery and Transformers'*, Prentice Hall, N.J., 1972.
20. Kusko, A., *'Solid State DC Motor Drives'*, M.I.T. Press, 1969.
21. Langsdorf, A.S., *'Theory of A.C. Machinery'* McGraw Hill, New York, 1955.
22. Lyon W.V., *'Tranient Analysis of A.C. Machines'* Chapman & Hall, London, 1954.
23. Mablekos V.E., *'Electric Machine Theory for Power Engineers'* Harper and Row, New York, 1987.
24. Mc Pherson G., *'An Introduction to Electric Machines and Transformers'*, John Wiley, New York, 1981.
25. M.I.T., *'Magnetic Circuits and Transformers'* John Wiley, New York, 1943.
26. Ramamoorthy M., *'An Introduction to Thyristors and their Applications'* East West Press, New Delhi, 1977.
27. Ram Shaw R.S., *'Power Electronics'* Chapman and Hall, London, 1973.
28. Say M.G., *'Performance and Design of AC Machines'* Pitman, London, 1983.
29. Say M.G., *'Introduction to Unified Theory of Electromagnetic Machines'* Pitman, London, 1971.
30. Siskind C.S., *'Electric Machines, Direct and Alternating Current'* McGraw Hill, New York, 1950.
31. Taylor E.O., *'Performance and Design of A.C. Commutator Motors'* Wheeler, N. Delhi, 1990.
32. Walker J.H., *'Large AC Machines'* B.H.E.L, India, 1979.
33. Westinghouse Electric Corporation, Westinghouse S.C.R. *Designer's Handbook*, Youngwood,

# Index


AC machines
  commutator 503
  fundamentals 247
  generated emf 251
  mmf waves 258
  rotating field 261
AC regulator 550
AC winding
  chorded 248
  double layer 247
  fractional slot 248
  full pitch 248
  integral slot 248
  phase spread 247
  single layer 247
Air gap line 300
Alternator (*see* synchronous generator)
Amortisseur winding 385
Amplidyne 212
Armature 143
Armature mmf
  ac machines 258
  dc machines 156
Armature reaction
  dc generator 156
  dc motor 178
  synchronous generator 289
Armature winding
  lap 148
  multiplex 146
  progressive 147
  retrogressive 147
  simplex 146
  synchronous machines 247
  wave 149
Asynchronous machine 404
Audio frequency transformer 114
Autotransformer 62

Blocked rotor test
  single phase induction motor 482
  three phase induction motor 428
Brake test
  dc motor 206
  induction motor 431
Breadth factor 253
Breakdown torque 419
Brushes 144
Brushless excitation system 286

Capacitor start motor 491
Capacitor start capacitor run motor 491
Chorded winding 248
Chopper 545
Circle diagram 432
Classification of electric machines 24
Commutating poles 145
Commutation 155
Commutator 144
Compensating winding 505
Compound generator 167
Compound motor 182
Consequent poles 443
Continuous rating 20
Control transformer 536
Core loss 268
Converter
  one quadrant 544
  single phase 543
  three phase 544
  two quadrant 543
Critical field resistance 164
Cross field machines 211
Cross field theory 489
Cumulative compounding 167
Current transformer 108
Cylindrical rotor 283

Damper winding 348
DC generator
  classification 162
  compound 167
  magnetisation characteristics 162
  open circuit characteristics 162
  parallel operation 168
  self excited 164

separately excited 163
series 166
shunt 165
DC machines
commutation 155
construction 142
crossfield 211
dynamics 215
efficiency 204
emf equation 152
losses 202
magnetic circuit 151
testing 206
DC motors
armature reaction 178
back emf 176
braking 201
commutation 178
compound 182
permanent magnet 202
shunt 179
series 181
speed control 185
starter 196
torque equation 175
Deep bar rotor 436
Differential compounding 167
Direct axis reactance 312
Double cage rotor 437
Double layer winding 247
Double revolving field theory 478
Drag cup rotor 527
Dynamic braking 201
Dynamically induced emf 2

Eddy current loss 15
Efficiency 17
Electric braking 201
Electromagnetic torque
ac machines 265
dc machines 175
Emf
dynamically induced 2
speed 2
statistically induced 4
transformer 4
Emf equation
ac machines 251
dc machines 152
transformer 29
Energy balance 16
Excitation system 286
Exciting current 33, 411

Faraday's law 1
Fleming's left hand rule 12
Fleming's right hand rule 12
Flux
ac machines 258
dc machines 161
mutual 41
Fractional hp motor 477
Free wheeling diode 546
Full pitch coil 248

Gate signal 542
Generated emf
ac machines 251
dc machines 152
Generators
ac 281
dc 162
synchronous 281
turbo 283
Grounding transformer 115

Hopkinson test 208
Hunting
alternators 347
synchronous motors 385
Hydroelectric generator 284
Hydrogen cooling 285
Hysteresis loss 15
Hysteresis motor 521

Ideal transformer 33
Inductance
leakage 35, 293
mutual 8
self 7
Induction generator 455
Induction motor, polyphase
blocked rotor test 428
break down torque 419
cage rotor 404
circle-diagram 432
cogging 440
construction 404
crawling 440
deep bar 436
double cage 437
equivalent circuit 411

exciting current 411
frame 405
leakage reactance 411
losses 413
magnetising current 411
magnetising reactance 411
no load current 411
no load test 428
phasor diagram 411
principle of operation 405
slip 406
speed control 441
starting 421, 423
torque 406, 414
torque slip curve 418
wound rotor 404
Induction motor single phase
blocked rotor test · 482
capacitor start 491
capacitor start capacitor run 491
construction 477
cross field theory 489
double revolving field theory 478
equivalent circuit 480
no load test 483
permanent split capacitor 492
shaded pole 493
starting 489
Induction regulator
single phase 103
three phase 104
Infinite bus bar 335
Interaction of magnetic fields 13
Interpoles 145
Iron loss 15
Inverter
single phase 547
three phase 548

Leakage flux 34
Leakage inductance 35
Leakage reactance 35
Lenz's law 1
Long shunt connection 168
Losses
ac machines 267
dc machines 202
induction motor 413
transformer 52
Machine ratings 21
Magnetic circuit
dc machine 151
Magnetic saturation 163
Magnetisation characteristic
ac machine 295
dc machine 162
Magnetising current
induction motor 411
transformer 33, 99
Magnetising reactance 36
Metadyne 211
MMF
concentrated coil 258
single phase winding 259
three phase winding 261
method 297
Motional emf 2
Moving coil voltage regulator 106
Mutual flux 41
Mutual inductance 8

Negative sequence reactance 344
No load current
induction motor 411
transformer 33
No load losses 58, 428
No load test 58, 428

Open circuit characteristic
ac machine 295
dc machine 162
Open circuit test 58
Open delta connection 81
Over excitation 374

Parallel generator theorem 88
Parallel operation
alternators 324
dc generators 168
transformers 87
Permanent split capacitor motor 492
Per unit value 48
Phase belt 247
Pitch factor 254
Plugging 201
Pole amplitude modulation 444
Pole changing 442
Polyphase induction motor
(*see* Induction motor-polyphase)
Potential transformer 107
Potier reactance 299
Power angle 305

Power angle characteristics
  cylindrical rotor machine 305
  salient pole machine 316
Pull out torque 419
Pulse transformer 113

Quadrature axis reactance 314

Reactance voltage 155
Rectifier circuit
  single phase 543
  three phase 544
Regenerative braking 202
Reluctance motor 520
Reluctance torque 318
Repulsion induction motor 516
Repulsion motor 515
Repulsion start induction motor 516
Rotor
  cylindrical 283
  drag up 527
  salient pole 284
  squirrel cage 404
  wound 404

Salient pole rotor 284
Saturated synchronous reactance 298
Schrage motor 504
Self excitation 164
Self inductance 7
Separately excited generator 162
Series dc generator 166
Series motor
  ac 505
  dc 181
Servo motor 522
Shaded pole motor 493
Short circuit characteristic 295
Short circuit ratio (SCR) 305
Short Circuit test
  alternator 295
  transformer 59
Short shunt connection 168
Short time rating 20
Single phase motor 478
Slip 406
Sliprings 404
Slot harmonics 269
Solid state control 542
Speed control
  dc series motor 191
  dc shunt motor 185
  single phase induction motor 497, 553
  three phase induction motor 441
  ward leonard method 195
Speed voltage 2
Spit phase motor 490
Squirrel cage rotor 404
Starter
  auto transformer 426
  dc motor 196
  induction motor 424
  star delta 425
Static excitation system 287
Stepper motor 531
Stray load loss 268
Sub-transient reactance 341, 344
Sub-transient time constant 341
Synchro 536
Synchronising 326
Synchronising power 326
Synchronising torque 326
Synchronous condenser 380
Synchronous generator
  armature reaction 289
  armature winding 247
  cylindrical rotor 283
  direct axis reactance 313
  electro magnetic torque 265
  equivalent circuit 292
  excitation system 286
  hunting 347
  losses 267
  open circuit characteristics 295
  open circuit test 295
  parallel operation 324
  power angle 305
  power angle characteristics 305, 316
  power output 305
  reactance 293
  reluctance torque 316
  salient pole 284
  short circuit characteristics 295
  short circuit ratio 305
  Synchronising 326
  Synchronous impedance 293
    time constants 346
    voltage regulation 293
Synchronous motor
  application 386
  overexcited 373
  power factor 373

starting 384
V curves 376
Synchronous reactance 293

Tachometer
ac 534
dc 534
Tertiary winding 95
Time constant
dc machine 215
synchronous machine 346
Torque
ac machine 265
dc machine 175
reluctance 317
Transformer
all day efficiency 55
amorphous metal core 131
approximate equivalent circuit 38
audio frequency 114
auto 62
back to back test 61, 79
breather 110
conservator 110
construction 32
cooling 108
core 32
core loss 33
coupled circuit 41
current 108
delta-delta 70
delta-star 71
dry type 131
efficiency 53
emf equation 29
energy efficiency 55
equivalent circuit 35
equivalent reactance 38
equivalent resistance 38
exciting current 33
frequency response 112
grounding 115
ideal 33
inrush current 100
leakage flux 34
leakage reactance 35
losses 52
magnetising current 33, 99
magnetising reactance 36
mutual flux 41
noise 110
no load test 58
oil 109
open circuit test 58
open delta connection 81
parallel operation 87
phasor diagram 35
polarity 57
potential 107
principle 29
pulse 113
ratio of transformation 30
regenerative test 61
scott connection 82
shell type 33
short circuit test 59
star delta 70
star star 69
sumpner's test 61
taps 102
three phase 68
three winding 95
types 32
V connection 81
voltage regulation 49
wide band 111
Tooth harmonics 269
Transient reactance 341, 343
Two reaction theory 312

Universal motor 508

V connection 81
V curves 376
Voltage regulation
synchronous machines 292
transformer 49

Ward leonard control 195
Wide band transformer 111
Winding
ac machines 247
chorded 248
dc machine 148
fractional pitch 248
fractional slot 248
full pitch 248
integral slot 248
lap 148
wave 149
Winding factor 253
Wound rotor 404

Yoke 143

Zero sequence reactance 293